Systems Engineering in Context

Stephen Adams • Peter A. Beling
James H. Lambert • William T. Scherer
Cody H. Fleming
Editors

Systems Engineering in Context

Proceedings of the 16th Annual Conference on Systems Engineering Research

Springer

Editors
Stephen Adams
Department of Systems and Information Engineering
University of Virginia
Charlottesville, VA, USA

Peter A. Beling
Department of Systems and Information Engineering
University of Virginia
Charlottesville, VA, USA

James H. Lambert
Department of Systems and Information Engineering
University of Virginia
Charlottesville, VA, USA

William T. Scherer
Department of Systems and Information Engineering
University of Virginia
Charlottesville, VA, USA

Cody H. Fleming
Mechanical and Aerospace Engineering
University of Virginia
Charlottesville, VA, USA

ISBN 978-3-030-00113-1 ISBN 978-3-030-00114-8 (eBook)
https://doi.org/10.1007/978-3-030-00114-8

Library of Congress Control Number: 2019930841

This Springer imprint is published by the registered company Springer Nature Switzerland AG
The registered company address is: Gewerbestrasse 11, 6330 Cham, Switzerland

Preface

The papers that comprise this volume were presented at the 16th Annual Conference on Systems Engineering Research (CSER 2018). The conference was held on May 8–9, 2018, at the University of Virginia (UVA) in Charlottesville, Virginia, USA. The co-chairs of the conference were Peter A. Beling (UVA) and William T. Scherer (UVA). The honorary chair of the conference was Yacov Y. Haimes (UVA). The program chair was Stephen Adams (UVA), and the panel chair was Cody Fleming (UVA). The publication chairs were James H. Lambert (UVA) and Brian Park (UVA).

The CSER 2018 Program Committee had participation of a wide range of universities and companies, including Carl Elks (Virginia Commonwealth University), Cody Fleming (UVA), Yacov Y. Haimes (UVA), James H. Lambert (UVA), Brian Park (UVA), Michael Farnsworth (UVA), David Long (Vitech Corp.), Tom McDermott (Georgia Tech), Chris Paredis (Clemson University), Alejandro Salado (Virginia Tech), Lu Xiao (Stevens Institute of Technology), Ye Yang (Stevens Institute of Technology), Steve Y. Yang (Stevens Institute of Technology), Roger Georges Ghanem (University of Southern California), and N. Peter Whitehead (The MITRE Corporation).

The 2018 theme "Systems in Context" is inclusive of topics across systems engineering, industrial engineering, systems acquisition, engineering management, operations research, engineering systems, and related fields. Few systems operate independently of other systems and stakeholders. Understanding the interactions with other entities and the context in which a system will be used is paramount to the design process. The conference featured papers from a wide range of domains and topics including:

- Systems in context and applications of systems engineering concepts
- Theoretical foundations of systems engineering
- Systems-of-systems design
- Integration of systems
- Complex and large-scale systems

- Human factors, human-machine interaction, and cyber-human systems
- Simulation integration into systems engineering
- Optimization and multi-criteria decision-making
- Risk and resilience
- Distributed decision-making and control
- Model-based systems, stochastic modeling, and state estimation
- Data-driven decision-making in large-scale systems
- Data collection, storing, and handling for large-scale systems
- Sensor and data fusion in large-scale systems
- Autonomous systems
- Safety and security systems
- Transportation systems
- Cyber-physical systems and the Internet of Things

We would like to thank all of the authors and presenters for participating in CSER 2018 and contributing to a positive learning experience in Charlottesville. We would also like to thank all that contributed to organizing the conference.

Charlottesville, VA, USA
May 2018

Stephen Adams
Peter A. Beling
James H. Lambert
William T. Scherer
Cody H. Fleming

Contents

About the Editors

Stephen Adams PhD is a senior research scientist in the Systems and Information Engineering Department at the University of Virginia. His research interests include feature selection, machine learning with cost, sequential decision-making, reinforcement learning, and probabilistic modeling of systems. He is currently part of the Adaptive Decision Systems Lab at UVA, and his research is applied to several domains including activity recognition, prognostics and health management for manufacturing systems, psychology, cybersecurity, data trustworthiness, natural language processing, and predictive modeling of destination given user geo-information data.

Peter A. Beling is a professor in the Engineering Systems and Environment (ESE) Department at the University of Virginia (UVA). Dr. Beling's research interests are in the area of decision-making in complex systems, with emphasis on artificial intelligence systems and on model-based approaches to system-of-systems design and assessment. His research has found application in a variety of domains, including mission-focused cybersecurity, reconnaissance and surveillance, prognostic and diagnostic systems, and financial decision-making. He directs the UVA site of the Center for Visual and Decision Informatics, a National Science Foundation Industry/University Cooperative Research Center, and the Adaptive Decision Systems Laboratory, which focuses on data analytics and decision support in cyber-physical systems. He is a member of UVA's Data Science Institute and its Link Lab for Cyber-Physical Systems. Dr. Beling has served in numerous editorial roles and has been a member of five National Academies panels. Dr. Beling received his Ph.D. in operations research from the University of California at Berkeley.

James H. Lambert FIEEE, FASCE, FSRA, DWRE, PE, MINCOSE, MAAAS is a professor of systems and information engineering at the University of Virginia, USA. He is the chair of the SRA Fifth World Congress on Risk in Cape Town, South Africa, in 2019. He is a past president (2015–2016) of the Society for Risk Analysis and a past chair (2015) of the SRA Annual Meeting. He is a founding editor-in-chief of the Springer journal *Environment Systems and Decisions* and area editor of the Wiley journal *Risk Analysis* and serves on the Editorial Board of the ASCE/ASME *Journal of Risk and Uncertainty in Engineering Systems*. He is a founding co-chair of the IEEE Technical Committee on Social and Economic Security. He has PhD and MS from the University of Virginia (1994, 1990) and BSE (1988) from Princeton University. He serves on the Technical Advisory Committee of the Commonwealth Center for Advanced Logistics Systems.

William T. Scherer SEBS, ME, PhD is an expert in systems engineering and stochastic control. Professor Scherer has served on the University of Virginia Department of Systems and Information Engineering Faculty since 1986. He has authored and co-authored numerous publications on intelligent decision support systems, combinatorial optimization, and stochastic control. His current research focuses on systems engineering methodology, financial engineering, and intelligent transportation systems. His co-authored book, *How To Do Systems Analysis*, was published by Wiley in 2007, and his follow-on book, *How To Do Systems Analysis: Primer and Casebook*, was also published by Wiley in 2016. He was also a visiting professor at the Darden Graduate School of Business in 2001–2002 and the president of the IEEE Intelligent Transportation Systems (ITS) Society 2007–2008.

Cody H. Fleming is an assistant professor of systems engineering and aerospace engineering at the University of Virginia. He received his PhD in aeronautics and astronautics at the Massachusetts Institute of Technology. Broadly, Dr. Fleming is interested in system modeling and analysis during all phases of development of complex systems, particularly those with high levels of automation. He has investigated several next-generation air traffic management initiatives during the early concept formation stages as well as safety assurance and algorithm development for driverless and connected vehicles. Related research interests lie in modern feedback control, dynamics, and modeling as well as model-based systems engineering and system assurance. In addition, Cody worked for several years in the aerospace industry, developing large satellite systems and laser systems in all phases of systems engineering including design, analysis, testing, integration, and validation and verification.

Chapter 1
Toward Achieving Architecture Alignment of Business and IT: A Portfolio Decision Analysis Approach

Mengmeng Zhang, Honghui Chen, and Kalle Lyytinen

1.1 Introduction

The fit between the business domain and IT domain of a firm contributes to "maximizing the return value of IT investments," [1, 2] "helping to improve IT usage," [3, 4] and "improving overall company performance" [5, 6]. However, there are currently different ways of aligning the business domain and IT domain; for example, strategic alignment [7–9], architecture alignment [10, 11], and business alignment [12, 13] are some of the many types of alignment problems discussed and investigated in the literature. While most research and practice have focused on strategic alignment to synchronize IT and business plans, our research suggests that architecture alignment is equally important. Architecture alignment brings strategic alignment to the practical level and creates value from architecture design and management.

Several alignment dimensions, such as social [14, 15], cultural [14, 16], and structural [17, 18], were often considered in the various types of alignment problems. This study focuses on the structural dimension of architecture alignment without regard for the other dimensions. In this paper, architecture alignment will refer to a structural fit between business elements (e.g., business processes, activities, and business structures) in business architecture and IT elements (e.g., applications, services, and software components) in IT architecture.

Two research questions (RQ) will be addressed in this paper:

M. Zhang (✉) · H. Chen
Science and Technology on Information Systems Engineering Laboratory, National University of Defense Technology, Changsha, People's Republic of China
e-mail: mxz391@case.edu

K. Lyytinen
Weatherhead School of Management, Case Western Reserve University, Cleveland, OH, USA

S. Adams et al. (eds.), *Systems Engineering in Context*,
https://doi.org/10.1007/978-3-030-00114-8_1

RQ1: What research methods are described in the literature, and what are the advantages and disadvantages of each method?
RQ2: What approach can act as a complementary method to deal with the disadvantages of existing research methods?

These two questions will be discussed throughout the study. In the following section, we will analyze the status quo of architecture alignment, which will answer RQ1. In Sect. 1.3, we will introduce a portfolio decision analysis (PDA) method, which extends existing methods and helps to answer RQ2. In Sect. 1.4, we will discuss the ways the method meets the alignment criteria. Section 1.5 draws conclusions and presents directions for future research.

1.2 Theoretical Background

The effective alignment of business architecture and IT architecture is essential for the daily operations of a company. When effective, this alignment helps to achieve the strategic alignment and to improve the company's overall performance. A sustainable architecture alignment will have a positive impact on the other alignment issues. This section explores the existing research on architecture alignment and analyzes its implications.

After collecting and screening the literature on architecture alignment, the volume of literature turned out to be small. A series of representative papers are specifically discussed in this paper. They have been chosen either because they have been widely cited, or because they provide a particular perspective on architecture alignment. Table 1.1 summarizes the papers discussed below, their research questions, key techniques, and research methods.

According to Table 1.1, several techniques, such as ontology, metamodel, and SOA, were usually used to achieve the architecture alignment. Ontology describes a domain model by associating meaning with its terms and relations [29, 30]. To increase the consistency of separate concerns, ontologies were used to represent and integrate multiple architecture description languages. The metamodel is a high-level abstraction of architecture contents; it owns the ability to describe relationships for both sides (business and IT). SOA is a coarse-grained, loosely coupled system architecture which supports dynamic enterprise application integration. It is beneficial for mutual understanding between business departments and IT departments. These articles proposed significant implications for the research of architecture alignment.

In a rapidly digitalized world, a company is facing severe competitive challenges due to the dynamic market environment [32, 33], pervasive digital technologies [34], and multiple interdependent relationships [35]. The research of architecture alignment should embrace the dynamic changes and other unforeseeable situations. Five effective alignment criteria (*consistency, traceability, complexity, agility, performance*) are introduced in this paper, which are also existed in the above articles. First, a consistent description between business domain and IT domain should be

Table 1.1 Representative research on architecture alignment

Authors	Research questions	Key techniques	Research methods
1. EA frameworks (e.g., TOGAF [19], DoDAF [20], MoDAF [21])	How can we provide a unified method and tool for developing an enterprise architecture by any organization?	Metamodel; ontology	The relationships of different viewpoints are bound by various metamodels and models
2. Bodhuin et al. [22]	Faced with unexpectable changes, how are the business processes and software systems coevolved?	Impact analysis	The misalignment of business processes and systems are detected, and the impact of modifications is discussed
3. Shishkov et al. [23]	How can we align the application software and business processes during initial software design phase?	Service-oriented architecture (SOA)	A service-oriented software implementation process with an initial business model is proposed
4. Strnadl [24]	How can we address the complexity and competition of today's IT architecture and propose a high-level architecture model?	Metamodel	A four-layer process-driven architecture model containing process, information, services, and technology is proposed
5. Chen [11]	How can the BITA be achieved with the help of SOA?	SOA	A three-layer, multidisciplinary BITAM-SOA service engineering schematic is developed
6. Aier and Winter [10]	How can we propose mapping artifacts to enable linkages of business and IT artifacts?	Domain clustering algorithm	The mapping mechanisms of business architecture and IT architecture are decoupled by a clustering algorithm
7. Franke et al. [17]	How are EA frameworks extended for dependency analysis into the realm of quantitative method?	Fault tree; Bayesian network	The fault tree and Bayesian networks are combined to extend the dependency of DoDAF artifacts
8. Elvesater et al. [25]	How can we ensure a good connection and mapping between business models and IT models?	SOA	The method to map the SoaML with BPMN is described
9. Kang et al. [26]	How are the relationships among business strategies, business processes, and enterprise resources recognized?	Metamodel; ontology	A business enterprise architecture is proposed which included the business architecture and resources architecture

(continued)

Table 1.1 (continued)

Authors	Research questions	Key techniques	Research methods
10. Delgado et al. [27]	How do different participants from different organizations collaborate to reach a common goal?	SOA	An approach for the automatic generation of SoaML in BPMN is proposed
11. Iacob et al. [28]	Does ArchiMate language support the modeling of business strategy concepts and IT portfolio valuation?	Metamodel; programming algorithm	The ArchiMate is extended from value-related concepts and optimizes the application portfolio with a binary integer algorithm
12. Antunes et al. [29, 30]	How can we integrate different enterprise architecture domains and analyze the resulting models?	Ontology	A core domain independent language which is extended by several domain-specific languages is presented
13. Hinkelmann et al. [31]	How can the company react to continuous and unexpected changes for BITA?	Metamodel; ontology	Both human interpretable graphical enterprise architecture and machine interpretable enterprise ontologies are considered

involved. Then to deal with the changes, complexity, and unexpected events in a company, the architecture alignment should be adapted effectively and realigned quickly. Meanwhile, the alignment should be discussed to ensure a better company performance. Each of the criteria contributes to an effective alignment between business and IT. Here are the definitions of the criteria:

- *Consistency* [23, 29, 30]: the desired relationships between models that address separate concerns, for example, business viewpoint and IT viewpoint
- *Traceability* [23, 29, 30]: an appropriate reflection on how changes in the business environment impact the IT and vice versa
- *Complexity* [10, 24]: an effective description for increasing applications, mounting information, and multiform relationships
- *Agility* [11, 31]: the reactions to the unexpected events and the return to alignment and the ability to deal with a fair degree of uncertainty
- *Performance* [5, 6, 28]: the achievement of the best returns on investment

All of the five criteria reflect the alignment requirements to address the dynamic changes and to achieve a sustainable competitive advantage. We argue that any approach should involve the above five criteria. Table 1.2 lists the criteria discussed in each of the articles. Their relationships were divided into three types. In Table 1.2, all of the papers discussed *consistency*. Some discussed the issue of *consistency* in a

Table 1.2 Comparative analysis of papers and alignment criteria

	1	2	3	4	5	6	7	8	9	10	11	12	13
Consistency	●	○	●	●	○	○	○	○	●	○	●	●	○
Traceability	●	○	●	○	●			●	●	●	○	●	○
Complexity	–	●	–	●		●	–	–	–	–	●	○	○
Agility	–	–	○	○	●	–	–	–	○	–	–	–	●
Performance	–	–	–	–	–	–	●	–	–	–	●	–	–

– = It is not discussed in the research
○ = It is not discussed in a detailed form
● = It is discussed in a detailed form

detailed form, while others only referred to it briefly. The majority of the research related to *traceability*. The articles discussed how to react to changes and to adjust the business, and IT contents were discussed in these research. However, six studies discussed *complexity*. For example, Bodhuin [22] adopted an impact analysis algorithm to define the complete set of actions to be performed on all the related components of a system. Aier and Winter [10] proposed a domain clustering algorithm to decouple the many-to-many relationships between business architecture and IT architecture. Similarly, five studies talked about *agility*. Chen [11] discussed multiple examples, which showed how different modules are activated in response to different change events either from top-down or bottom-up. Hinkelmann [31] pointed out that combining metamodel with ontology is beneficial for dealing with the four dimensions of agility: dynamism, adaptability, flexibility, and awareness. With regard to *performance*, only two studies proposed specific methods to improve it. Franke [17] provided a quantitative probability-based method to calculate the "dependency" for realizing business goals. Iacob [28] optimized the application portfolio to minimize systems maintenance costs and implementation costs with a binary integer algorithm.

In summary, none of the above studies referred to all of the five criteria. And the *performance* criterion received the least attention. The results show that complementary methods should be introduced to extend existing research, especially with regard to *performance*. A portfolio decision analysis (PDA) method, which is popular in the field of projects management, is introduced in this paper. The PDA method will be combined with EA to meet all of the five criteria.

1.3 Architecture Alignment with PDA

This section aims to adopt a PDA approach to address the limitations of existing research. In the last section, we saw that architecture alignment should be evaluated in each separate period when facing dynamic changes. We deem that PDA can act as a dynamic mechanism to align business architecture and IT architecture. In this section, we will focus on how to select an effective IT portfolio that aligns with business processes dynamically.

1.3.1 PDA

PDA exists widely in the field of projects planning and management. It refers to a body of theory, methods, and practices which seeks to help decision-makers make informed multiple selections from a discrete set of alternatives through mathematical modeling that accounts for relevant constraints, preferences, and uncertainties [36, 37]. In the research stream of PDA, Systems Analysis Laboratory at Aalto University School of Science has done considerable research. Salo and his coauthors issued a book to explain PDA's definition, source, quality framework, and some relevant techniques [38]. Liesio presented the robust portfolio modeling [36] and scenario-based portfolio selection [39] to deal with incomplete information. In order to validate the feasibility of methods, Salo and Liesio applied the PDA in a Scandinavian Research Program [40]. They also studied the strategy to produce a portfolio in a telecommunication company [41]. Kangaspunta, Salo, and Liesio studied the weapon system portfolio with a cost-efficient analysis [42]. Based on these studies, we can conclude that the maturity and feasibility of PDA increase along with the growing attention on this topic.

There are some similarities between the objects discussed in this paper (business architecture and IT architecture) and objects discussed in prior research (projects, products, and weapons). Business elements like processes and activities can be regarded as decision goals, and IT elements like services and applications can be regarded as discrete alternatives. Then, according to the constraints, budgets, and costs among them, PDA can be used to select IT portfolios to support business processes in a cost-efficient way. This will improve the *performance* which was one of the criteria we discussed in the last section.

1.3.2 Applying PDA to EA

Generally speaking, the EA frameworks, such as TOGAF [19], DoDAF [20], and MoDAF [21] (in Table 1.1), describe and design EA on the basis of a holistic metamodel, multiple architecture viewpoints, and dozens of architecture models. They can help achieve *consistency* by designing dependency models (e.g., OV5, SV5a in DoDAF) and *traceability* by architecture governance and migration planning (e.g., ADM in TOGAF). However, their descriptions for alignment (e.g., matrix, mapping tables) are quite simple, which represent several deficiencies. First, it is hard to describe the increasing complexity of business elements and IT elements. Second, it can't react to changes quickly, especially when faced with uncertain information. Third, it didn't explain the performance of architecture alignment in the periods of architecture design and evolution.

PDA is applied to EA to address the above gaps. The framework is sketchily drawn in Fig. 1.1 with the objective to select an IT portfolio to support business activities dynamically. It contains three layers. From the business viewpoint, the business processes are composed of business activities. The business processes are

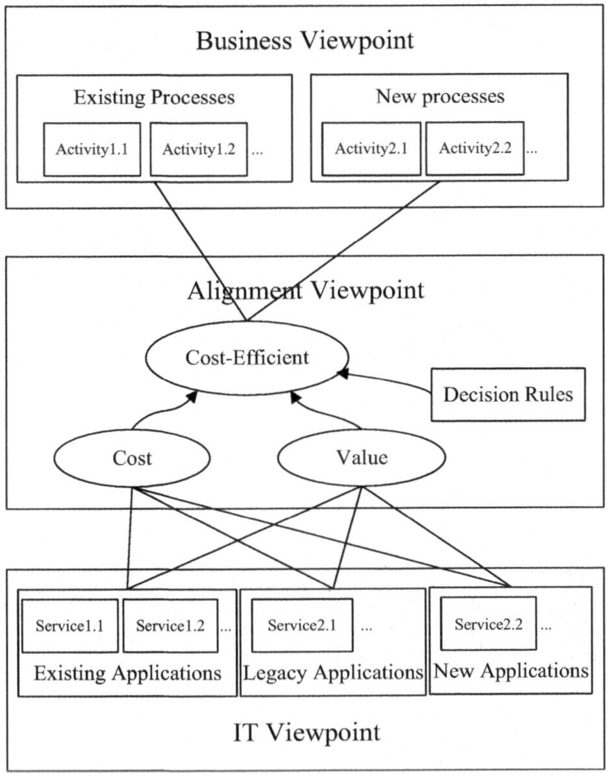

Fig. 1.1 Applying PDA to EA

divided into two types: existing processes (*EP*) represent the activities for current business requirements and new processes (*NP*) represent the reaction to business changes. From the IT viewpoint, the applications are composed of services. The applications are divided into three types: existing applications (*EA*) represent the services for current IT functions; legacy applications (*LA*) represent the applications that have become obsolete but can also be used in the future; and new applications (*NA*) are introduced when IT changes or new business requirements occur. The alignment viewpoint represents how to select an efficient service portfolio to support business processes with PDA. The combination of EA and PDA can be further explained in the following four steps.

1. Gathering the elements in business viewpoint and IT viewpoint

   In the process of EA evolution, the elements in the business viewpoint or IT viewpoint are constantly changing. To determine an optimal architecture alignment, all of the relative elements need to be extracted and listed. For the business viewpoint, activities involving the existing processes and new processes are displayed. For the IT viewpoint, the services composing the existing applications, legacy applications, and new applications are displayed. The objective is to select a cost-efficient service portfolio for supporting both the existing and new business activities.

2. Collecting data

   This step involves collecting the data required for PDA. There are three primary kinds of data. The first is the costs (c) for maintaining, updating, or creating the applications. The second is the values (ν) of services that support business activities. And the third is the weights (ω) for all of the activities. The data can be gathered from EA documentations or other interview-related methods. The collected data can be presented as an exact point or interval. The formula applying PDA to architecture alignment is presented as follows:

$$\max \sum_{a=1}^{m+n} \omega_a \nu \{X(EA) \cup X(LA) \cup X(NA)\}_a y_a$$
$$e.g., \sum_{i=1}^{p} c(EA)_i x_i + \sum_{j=1}^{q} c(LA)_j x_j + \sum_{k=1}^{s} c(NA)_k x_k \leq \text{Budget}$$
$$x_i, x_j, x_k, y_a = 1, 0$$
$$x_{EA_i} + x_{LA_j} + x_{NA_k} \leq 1, \text{if } EA_i, LA_j, NA_k \text{align} \{EP \cup NP\}_a$$

   Here, m is the number of EP, n is the number of NP, p is the number of EA, q is the number of LA, s is the number of NA, Budget represents the total investment, and align represents the supporting relationships between applications and processes.

3. Getting a cost-efficient IT portfolio

   The above formula should be solved by a mathematical programming algorithm. The research on PDA has provided abundant references. If the data were collected as points, a preference programming method [42] should be used to acquire the IT portfolio. If the data were collected as interval values, a robust portfolio modeling method [36, 39] has been proposed by Liesio. Furthermore, if there is more than one suitable IT portfolio, additional decision rules, such as maximin or minimax regret decision rules [36], should be used to acquire the most cost-efficient IT portfolio.

4. Adjusting the EA models

   After acquiring one cost-efficient IT portfolio, the new business elements and IT elements in this portfolio should be reflected in the EA models. The corresponding EA models will be updated (e.g., OV5, SV5a in DoDAF). The redundant applications should be put in the legacy list (e.g., legacy system list in DoDAF), which may be used for addressing the next changes. Therefore, the business architecture and IT architecture are better aligned with each other when reacting to changes.

Based on the above four steps, PDA is applied to EA to get a better architecture alignment in the dynamic environment. The research findings of PDA are considered in this process. In this paper, PDA is embedded in the EA evolution process to sustain the architecture alignment continuously.

1.4 Discussion

In this section, we will mainly discuss whether our proposed method meets the five criteria in Sect. 1.2. Because EA frameworks were used to develop the business architecture and IT architecture, the *consistency* and *traceability* are satisfied. We will mainly discuss the other three criteria.

Currently, the *complexity* is increasing with the more and more IT elements and dynamic business and IT environment. In this paper, all of the business and IT elements, including the legacy and new, were listed to be optimized. The outcome was displayed in the EA models, and other elements were put on the legacy list. The complexity is mitigated during this process. On the other hand, the complexity of the algorithm is also low according to the research findings on PDA [36, 38–40]. Furthermore, to deal with the dynamic environment, PDA was used to achieve the optimized IT portfolio. The sustainable alignment will be automatic if we input the needed data. On the other hand, the interval data of costs, values, and weights are also allowed in the PDA research, which increased the ability to address uncertain information. Therefore, the *agility* of the architecture alignment is improved with the proposed method. Meanwhile, the *performance* is also considered in this paper. With the selected IT portfolio, the supporting values for business processes are rather high, while the overall costs are rather low. In summary, applying PDA to EA extends the existing research on architecture alignment. It can realize the architecture alignment through the above five criteria.

1.5 Conclusion

In response to the various alignment types and alignment dimensions discussed in the literature, this paper focuses on architecture alignment from the structural dimension. Following the research on architecture alignment, this paper can help practitioners and researchers make sense of and achieve the architecture alignment. This paper contributes in several ways. We collected the existing research on architecture alignment and analyzed their implications. We extracted five criteria to meet the architecture alignment and check the existing literature with them. Furthermore, we introduced and applied the PDA approach, which helped extend the current research through the above five criteria.

The proposed method does have its limitations, however. On one hand, this paper mainly discussed the combination of EA and PDA, but has not explained each of them in detail. On the other hand, this paper evaluated the proposed method with the five criteria. The practical cases should be conducted in order to verify the criteria and identify any other effective criteria. The above limitations provide a starting point for future work. The detailed combination process of EA and PDA should be explained in further research. A suitable case study should be introduced for the proposed method. Furthermore, the techniques for collecting different kinds of data need also to be determined.

References

1. Avison, D., Jones, J., Powell, P., & Wilson, D. (2004). Using and validating the strategic alignment model. *Journal of Strategic Information Systems, 13*, 223–246.
2. Kashanchi, R., & Toland, J. (2006). Can ITIL contribute to IT/business alignment? An initial investigation. *Wirtschaftsinformatik, 48*(5), 340–348.
3. Beimborn, D., Franke, J., Wagner, H., & Weitzel, T. (2007). *The influence of alignment on the post-implementation success of a core banking information system: An embedded case study*. In Proceedings of the 40th Annual Hawaii International Conference on System Sciences (HICSS'07), Waikoloa, Hawaii.
4. Wagner, H., Beimborn, D., Franke, J., &Weitzel, T. (2006). *IT business alignment and IT usage in operational processes: A retail banking case*. In Proceedings of the 39th Annual Hawaii International Conference on System Sciences (HICSS'06), Koloa, Kauai.
5. Gerow, J. E., Grover, V., Thatcher, J. B., & Roth, P. L. (2014). Looking toward the future of IT-business strategic alignment through the past: A meta-analysis. *MIS Quarterly, 38*(4), 1059–1085.
6. Luftman, J., Lyytinen, K., & ben Zvi, T. (2017). Enhancing the measurement of information technology (IT) business alignment and its influence on company performance. *Journal of Information Technology, 32*(1), 26–46.
7. Baker, J., Jones, D. R., Cao, Q., & Song, J. (2011). Conceptualizing the dynamic strategic alignment competency. *Journal of the Association for Information Systems, 12*(4), 299.
8. Henderson, J. C., & Venkatraman, H. (1993). Strategic alignment: Leveraging information technology for transforming organizations. *IBM Systems Journal, 32*(1), 472–484.
9. Luftman, J., & Kempaiah, R. (2007). An update on business-IT alignment: "A line" has been drawn. *MIS Quarterly Executive, 6*(3), 165.
10. Aier, S., & Winter, R. (2009). Virtual decoupling for IT/business alignment–conceptual foundations, architecture design and implementation example. *Business & Information Systems Engineering, 1*(2), 150–163.
11. Chen, H. M. (2008, January 7). Towards service engineering: Service orientation and business-IT alignment. In *Proceedings of the 41st Annual Hawaii International Conference on System Sciences* (p. 114). IEEE.
12. Fritscher, B., & Pigneur, Y. (2011, June 20). Business IT alignment from business model to enterprise architecture. In *International Conference on Advanced Information Systems Engineering* (pp. 4–15). Berlin: Springer.
13. Hinkelmann, K., & Pasquini, A. (2014, August 2). Supporting business and IT alignment by modeling business and IT strategy and its relations to enterprise architecture. In *Enterprise Systems Conference (ES)* (pp. 149–154). IEEE.
14. Schlosser, F., Wagner, H. T., & Coltman, T. (2012, January 4). Reconsidering the dimensions of business-IT alignment. In *2012 45th Hawaii International Conference on System Science (HICSS)* (pp. 5053–5061). IEEE.
15. Wagner, H. T., & Weitzel, T. (2012). How to achieve operational business-IT alignment: Insights from a global aerospace firm. *MIS Quarterly Executive, 11*(1), 25.
16. Chan, Y. E., & Reich, B. H. (2007). IT alignment: What have we learned? *Journal of Information Technology, 22*(4), 297–315.
17. Franke, U., Flores, W. R., & Johnson, P. (2009, March 22). Enterprise architecture dependency analysis using fault trees and Bayesian networks. In *Proceedings of the 2009 Spring Simulation Multiconference* (p. 55). Society for Computer Simulation International.

18. Hugoson, M. Å., & Pessi, K. (2011, June 15). Operational and structural business IT alignment. In *International Conference on Business Information Systems* (pp. 196–207). Berlin: Springer.
19. Haren, V. (2011, December 7). TOGAF Version 9.1. Van Haren Publishing.
20. DoD AS. (2009). *DoD architecture framework version 2.0* (DoDAF V2.0). Washington, DC: Department of Defense.
21. Biggs, B. (2005). Ministry of defence architectural framework (MODAF).
22. Bodhuin, T., Esposito, R., Pacelli, C., & Tortorella, M. (2004, June). Impact analysis for supporting the co-evolution of business processes and supporting software systems. In *CAiSE Workshops* (Vol. 2, pp. 146–150).
23. Shishkov, B., Van Sinderen, M., & Quartel, D. (2006, October). SOA-driven business-software alignment. In *2006 IEEE International Conference on e-Business Engineering (ICEBE'06).* (pp. 86–94). IEEE.
24. Strnadl, C. F. (2006). Aligning business and it: The process-driven architecture model. *Information Systems Management, 23*(4), 67–77.
25. Elvesater, B., Panfilenko, D., Jacobi, S., & Hahn, C. (2010, October 3). Aligning business and IT models in service-oriented architectures using BPMN and SoaML. In *Proceedings of the First International Workshop on Model-Driven Interoperability* (pp. 61–68). ACM.
26. Kang, D., Lee, J., & Kim, K. (2010). Alignment of business enterprise architectures using fact-based ontologies. *Expert Systems with Applications, 37*(4), 3274–3283.
27. Delgado, A., Ruiz, F., de Guzmán, I. G., & Piattini, M. (2012, March 26). Model transformations for business-IT alignment: From collaborative business process to SoaML service model. In *Proceedings of the 27th Annual ACM Symposium on Applied Computing* (pp. 1720–1722). ACM.
28. Iacob, M. E., Quartel, D., & Jonkers, H. (2012, September 10). Capturing business strategy and value in enterprise architecture to support portfolio valuation. In *2012 IEEE 16th International Enterprise Distributed Object Computing Conference (EDOC)* (pp. 11–20). IEEE.
29. Antunes, G., Caetano, A., Bakhshandeh, M., Mayer, R., & Borbinha, J. (2013, June 19). *Using ontologies for enterprise architecture model alignment.* In Proceedings of the 4th Workshop on Business and IT Alignment (BITA 2013), Poznan, Poland.
30. Antunes, G., Caetano, A., Bakhshandeh, M., Mayer, R., & Borbinha, J. (2013, June 19). Using ontologies to integrate multiple enterprise architecture domains. In *International Conference on Business Information Systems* (pp. 61–72). Berlin: Springer.
31. Hinkelmann, K., Gerber, A., Karagiannis, D., Thoenssen, B., Van der Merwe, A., & Woitsch, R. (2016). A new paradigm for the continuous alignment of business and IT: Combining enterprise architecture modelling and enterprise ontology. *Computers in Industry, 79*, 77–86.
32. Tanriverdi, H., & Lim, S. Y. (2017). How to survive and thrive in complex, hypercompetitive, and disruptive ecosystems? The roles of IS-enabled capabilities. In *Proceedings of the 38th International Conference on Information Systems*, Seoul, South Korea (pp. 1–21).
33. Tanriverdi, H., Rai, A., & Venkatraman, N. (2010). Research commentary—reframing the dominant quests of information systems strategy research for complex adaptive business systems. *Information Systems Research, 21*(4), 822–834.
34. Yoo, Y., Boland Jr., R. J., Lyytinen, K., & Majchrzak, A. (2012). Organizing for innovation in the digitized world. *Organization Science, 23*(5), 1398–1408.
35. Allen, P. M., & Varga, L. (2006). A co-evolutionary complex systems perspective on information systems. *Journal of Information Technology, 21*(4), 229–238.
36. Liesiö, J., Mild, P., & Salo, A. (2007). Preference programming for robust portfolio modeling and project selection. *European Journal of Operational Research, 181*(3), 1488–1505.
37. Salo, A., Keisler, J., & Morton, A. (2011). *An invitation to portfolio decision analysis* (pp. 3–27). New York: Springer.
38. Salo, A., Keisler, J., & Morton, A. (Eds.). (2011). *Portfolio decision analysis: Improved methods for resource allocation.* New York: Springer Science & Business Media.

39. Liesiö, J., & Salo, A. (2012). Scenario-based portfolio selection of investment projects with incomplete probability and utility information. *European Journal of Operational Research, 217* (1), 162–172.
40. Salo, A., & Liesiö, J. (2006). A case study in participatory priority setting for a Scandinavian research program. *International Journal of Information Technology & Decision Making, 5*(1), 65–88.
41. Lindstedt, M., Liesio, J., & Salo, A. (2008). Participatory development of a strategic product portfolio in a telecommunication company. *International Journal of Technology Management, 42*(3), 250–266.
42. Kangaspunta, J., Liesiö, J., Mild, P., & Salo, A. (2012). Cost-efficiency analysis of weapon system portfolios. *European Journal of Operational Research, 223*, 264–275.

Chapter 2
Emergent Network of Systems Implication for System Design

James R. Enos and Roshanak Nilchiani

2.1 Introduction

In systems engineering, nonfunctional attributes, often referred to as the ilities, are an important component of the system design process and increase the system's value to its stakeholders. Additionally, systems are becoming more interconnected and dependent on other systems for operation and often form a network of systems that delivers additional, emergent value to its stakeholders. Within the DoD, this problem becomes amplified as systems have been developed and deployed over the past several decades by independent services without the consideration of how they all operate together on the battlefield. Since its inception in the 1950s, systems engineering and systems architecture have developed tools and methods to manage the complexity of modern systems. However, traditional systems engineering and architecture tools and methods are not equipped to handle the hundreds of systems with thousands of connections in these emergent networks of systems. Both of these observations present an opportunity to improve the design process for systems to incorporate these attributes and environmental considerations into the systems design process.

Although the systems engineering ilities are important to the design of systems, it is difficult to determine which of the ilities to use. The literature presents dozens of ilities, but it is difficult to determine which ones are important to a system's stakeholders and will result in a successful system. Within Department of Defense (DoD) systems, previous work identified several important ilities that are present in successful systems to include extensibility, flexibility, interoperability, robustness, and versatility [1]. The next challenge becomes how to incorporate these attributes into a systems design, a very difficult task given that these system attributes often manifest after a system is deployed and operational. Another work examined the

---

J. R. Enos (✉) · R. Nilchiani
School of Systems and Enterprises, Stevens Institute of Technology, Hoboken, NJ, USA
e-mail: jenos@stevens.eud

S. Adams et al. (eds.), *Systems Engineering in Context*,
https://doi.org/10.1007/978-3-030-00114-8_2

application of social network analysis tools and methods to network of systems that connect through information flows, physical connections, and shared resources. This work applied social network analysis to a network of DoD systems to better understand the interoperability and versatility, two systems engineering ilities, of individual systems within the networks [2]. It is important to accept that future systems will operate in a network of diverse systems and understand how the addition of a new system would affect the overall network of systems. Social network analysis tools enable the evaluation and analysis of both the overall network of systems and individual systems within the network.

This paper examines how the insights gained from this previous work can apply to the design of systems to incorporate ilities early in the development process of a system. It consists of four additional sections that examine the relevant literature, the incorporation of ilities into design, effects on the network of systems, and proposed future work. First, it proposes methods for incorporating ilities into the design of a system to increase the value robustness of the system. It also examines how system design can incorporate the findings of the application of social network analysis to networks of systems. This paper proposes that several network level considerations should be included in the design process; however, it does not go as far as to model the potential impacts of these types of designs on networks of systems. Future work could expand this research by examining historical design process that took similar approach to incorporate the ilities, or nonfunctional attributes, into consideration when designing a system. Additionally, future work could model the effect on a network of systems of the introduction of a new system to determine which designs improve performance, increase reliability, increase complexity, or introduce vulnerabilities into the network.

2.2 Literature Review

This section presents a review of the relevant literature to include a discussion on systems engineering, grounded theory, and social network analysis. The systems engineering portion of the literature review discusses the methods systems engineers use to manage complexity of systems to include the ilities and system architecture. The ilities provide a means to capture the nonfunctional, qualitative nature of systems engineer and are essential to the design of successful systems. The grounded theory section discusses how researchers develop hypotheses about a set of data by first examining and coding the data to generate an emergent theory. Finally, the portion on social network analysis discusses social network analysis methods and how the method applies to networks of systems.

2.2.1 Systems Engineering

As technology progresses and systems are more interdependent, systems engineers face increasing complexity of systems and their environment. Systems engineering differs from traditional engineering disciplines in that system engineers must consider the system in its entirety as well as its operational environment [3]. The International Council on Systems Engineering (INCOSE) [4] defines systems engineering as "an interdisciplinary approach and means to enable the realization of successful systems. It focuses on defining customer needs and required functionality early in the development cycle, documenting requirements, and then proceeding with design synthesis and system validation while considering the complete problem." Within the field of systems engineering, one area that receives substantial attention is the modeling of complex, networked systems [5]. These systems often demonstrate emergent behavior and operate outside of the intended design parameters to meet the changing needs of their stakeholders.

The ilities are one tool that systems engineers have developed to manage the complexity and qualitative nature of systems engineering as a construct for assessing nonfunctional attributes of a system. These nonfunctional attributes of system have become critical to the design of systems, and systems engineers have incorporated them into the design process [6]. One problem with using the ilities in the design process is that these properties often manifest themselves after the system is operational [7]. The literature describes dozens of ilities; however, a few appear to be critical to the design of systems to include extensibility, flexibility, interoperability, quality, robustness, and versatility. A challenge to the application of the ilities in systems design is that these attributes often have a range of definitions in the literature. de Weck, Ross, and Rhodes [7] identified relationships between the ilities to demonstrate that these attributes do not exist in isolation and the incorporation of these attributes in design lead to additional desirable attributes of systems. Table 2.1 presents definitions for a subset of the systems engineering ilities for this paper.

Another systems engineering tool to manage complexity is system architecture that enables system architects to understand, design, and manage complex [9]. As such, system architects attempt to understand and document the internal connections

Table 2.1 Subset of systems engineering ility definitions

Ility	Definition
Extensibility	"The ability to accommodate new features after design" [7]
Flexibility	"The ability of a system to be changed by a system-external change agent" [6]
Interoperability	"The ability to effectively interact with other systems" [7]
Quality	"Ability to deliver requirements at a "high" level, as perceived by people relative to other alternatives that deliver the same requirements" [8]
Robustness	"The ability of a system to maintain its level and set of specification parameters in the context of changing system external and internal forces" [6]
Versatility	"The ability of a system to satisfy diverse expectations on the system without the need for changing form" [6]

of a system as well as how the system interacts with external systems. Maier and Rechtin [10] describe a system's architecture as "whatever aspects of physical structure, behavior, cost, human organization, or other elements are needed to clarify the client's priorities." A system's architecture consists of three major components: function, what the system does; physical, what the system is; and operational, how the functions consume resources [11]. Similar to systems engineering, system architects begin with the system of interest and decompose the architecture into levels of details to understand the inner workings of the system. A system's architecture documents interactions with elements outside the system boundary, but generally only one order away from the system of interest. At the enterprise level, some efforts have attempted to combine independent system architectures to understand how systems provide value in a network of systems [12]. These efforts are often limited because of the complexity of analyzing a network of systems, a lack of data for individual systems, and a lack of commonality between the individual architectures.

Architecture frameworks provide a means to ensure commonality between individual system architectures and serve as a communication tool for managing complexity by presenting a manageable amount from a common set of data [13]. These frameworks standardize multiple architectures through specific taxonomies, artifacts, and terminologies [14]. The Department of Defense Architecture Framework (DoDAF) is one of many frameworks that architects use to capture multiple perspectives on a system's architecture. DoDAF specifies eight viewpoints to capture data relevant to capability development, integration of systems, military operations, and the program management aspects of a system. These viewpoints include all data and information, standards, capability, operations, services, systems, and project viewpoints [15]. The DoD designed DoDAF to meet the needs of a diverse set of stakeholders involved in the development, validation, delivery, and sustainment of warfighting capabilities. Like other frameworks, the intent of the DoDAF is to assist decision-makers by abstracting essential pieces of information and presenting them in manageable pieces depending on the stakeholder's perspective [15].

2.2.2 *Grounded Theory*

Ground theory is a research method for developing a hypothesis from systematically obtained and analyzed data through induction and abduction [16]. One of the main benefits of grounded theory is the ability to analyze textual data, classify that data, and develop a theory from the categorized data. The theory uses an iterative process of simultaneously collecting and analyzing data so that the researcher analyzes the data as they collect the data set [17]. The process begins with induction as researchers study several cases to develop conceptual categories based on patterns contained within the data [18]. In abduction, researchers conduct a few observations to develop an initial hypothesis and then strengthen the hypothesis with empirical evidence [19]. Grounded theory emerged from the social sciences; however, it has expanded to other fields to include healthcare, information systems, and even engineering.

A fundamental aspect of grounded theory is coding the data set in a meaningful manner, especially since grounded theory often focuses on qualitative, textual data that a researcher must interpret prior to analyzing [20]. Coding is the process for naming, categorizing, summarizing, and accounting for segments of data identified throughout the research [18]. An initial hypothesis emerges from the data as researchers code and combine these individual segments of data into clusters and groups [21]. Researchers use constant comparison, a fundamental process in grounded theory, to organize new data against the developed categories [22]. They constantly compare new data until they reach a point of theoretical saturation as a theory emerges from the coded and categorized data [16]. For this paper, a case study of five DoD systems provide the basis for developing the grounded theory to include the B-52 bomber, A-10 attack aircraft, M-1A2 Abrams tank, UH-60 Blackhawk helicopter, and the E-2E Hawkeye. For each system, a review of 20 sources provides the data to support the theory and a subset of the system engineering ilities provides the codes for textual data contained in these articles.

2.2.3 Social Network Analysis

Social network analysis focuses on networks that are nodes of individuals, groups, or organizations that share a variety of relationships. In social networks, nodes may connect through one or more types of interdependencies that include shared interests, social contacts, membership in organizations, participation in events, family ties, or financial ties [23]. Social network analysis tools calculate valuable metrics about the network itself (centralization, density, reach, and cohesion) and individual nodes (centrality, degree, closeness, bridge, and equivalence) [24]. An important aspect of social network analysis is that multiple relationship types combine to determine the underlying structure of the network that may not be evident in any individual network [25]. Both social network analysis and networks of systems face similar challenges as they are dynamic and often have fuzzy boundaries and it is likely that the data on the network is not complete [26]. Social network analysis has overcome these challenges through advanced algorithms, improved computer processes, and visualization to incorporate hundreds of nodes and relationships.

With the domain of systems engineering, various works have identified the need to analyze networks of systems that share various resources and connections. Ring et al. [12] identified a need to merge individual system architectures to better understand the network of DoD systems that operate together in order to provide value to a range of stakeholders. This notion is similar to the notion of systems of systems engineering, which describes how systems can operate together, as designed or otherwise, to deliver emergent behavior and value to the system of system's stakeholders [27]. Other work expanded on this notion to discuss how social network analysis methods and visualization tools could apply to these networks of systems to identify important systems within the network and potentially identify vulnerabilities to the network [28]. Social network analysis potentially provides a means to quantify, and thus compare, the interoperability and versatility of systems [2, 29]. The methods from social network analysis also provide a means to understand interoperability within systems of systems to identify important component systems in DoD systems of systems [30]. To expand beyond the DoD, another work in the area applied a similar method to an infrastructure network where various types of systems interoperate to provide value to the customers [31].

2.3 Implications for System Design

The field of systems engineering has accepted that the nonfunctional attributes are important considerations in the design of systems alongside the traditional functional requirements that drive design. The ilities are an important construct in defining and understanding nonfunctional attributes of systems, and engineers are beginning to incorporate them into the systems design process [7]. However, the difficulty becomes which of the dozens of ilities to incorporate into the design of systems. Previous work identified several important ilities to DoD systems that effect the longevity of the systems and result in systems providing robust value over time for the DoD [1]. These ilities include changeability (extensibility and flexibility), interoperability, robustness, and versatility. Given that the future environment will change over time, a system must be changeable to accommodate new capabilities or upgrade existing capabilities to successfully operate in the changing environment. The new system must also be able to operate with other systems in the network of systems along several lines to include information flows, physical connections, and shared resources. It is likely that systems will face new threats in the future that affect the performance of the system, so new systems must be robust to mitigate the effects of these emergent threats. Finally, the versatility of a system becomes an important factor in the decision to extend the life of a system, so systems designed for a single purpose may not be the best option to meet certain requirements.

The ability of a system to change to meet new requirements or threats is essential to ensuring that a system provides value to stakeholders over time. Changeability generally comes in three means: extensibility, the ability of a system to add new capability; flexibility, the ability of a system to change via external change agents; or adaptability, the ability of a system to change from internal agents [32]. To accept new capabilities onto a legacy system, that system must have the ability to accept new technology based on the size, weight, power, and cooling (SWaP-C) constraints of an existing system. So, systems engineers should consider leaving a buffer for the SWaP-C attributes when designing systems to allow the system to accept new technologies and not use all of the SWaP-C during the design process. To enable flexibility and adaptability, systems engineers should consider advances in modular architectures to ensure that sub-systems and components can easily be changed out as technological advances improve the performance of components. In the literature describing the ilities, modularity demonstrates a strong link with the flexibility of a system [7]. Although the attributes of changeability often manifest themselves after the deployment of a system, system engineers can incorporate these attributes into the design of systems. An example of this from DoD systems is the UH-60 Blackhawk helicopter. Although the initial design of the helicopter remained the same over time, other services like the Navy and Air Force have modified the aircraft to incorporate new technology and improve on existing capability in order to meet their requirements [33].

Another important attribute to consider during the system design process is the interoperability of a system, which extends beyond just information flows to include physical connections and shared resources. Often, systems engineers focus solely on the information flow and data sharing aspects of interoperability as evident in the various definitions of interoperability. However, the physical connections and shared resources are equally important and often account for a larger percentage of the connections between systems [2]. Additionally, it is likely that newly developed systems will interact with legacy systems, which will require new systems to comply to older data standards or upgrades of legacy systems to ensure communication between these groups of systems. In the worst case the development of a new system would require the development of a second system in order to interoperate with legacy systems. By considering the physical connections between systems as a component of interoperability, systems engineers should account for size and weight constraints of legacy systems in addition to the information flows between systems. Improvements to the M-1 Abrams over time demonstrate the importance of maintaining interoperability as the Army incorporated GPS technology, battlefield awareness capability, and improved information sharing systems into the design of the latest Abrams tanks [33].

The remaining two attributes, robustness and versatility, are difficult to incorporate into the design process as they manifest themselves after the DoD deploys a system. Robustness focuses on the ability of a system to adapt to a changing environment and provide the same level of performance. This description is slightly different than value robustness as it focuses on performance parameters, whereas, a system that demonstrates value robustness may deliver value to stakeholders in a different manner [7]. In order for a system to be robust, it must consider and anticipate emergent threats the system could face in possible future scenarios. It is possible for systems engineers to design in robustness by embedding a margin of safety when setting design parameters to ensure that threat systems cannot outperform newly developed systems, which may take decades to produce and deploy. Another means to combat a changing environment is designing in versatility. The DoD often designs single purpose systems that are specifically built to meet an operational need; however, these systems display versatility by expanding their operational activities to accomplish a range of roles. A good example of this is the B-52, designed as a heavy, strategic bomber, but the Air Force has expanded its role to include reconnaissance, close air support, and even maritime surveillance [1]. So, as systems engineers consider the roles a system is designed to perform, it is likely that the user will expand these roles to accomplish other operational activities.

2.4 Designing Networks of Systems

From a network perspective, systems engineers should take several considerations into account when designing systems that will operate within a network of systems. The first consideration should be understanding how a new system increases or, in some cases, degrades the overall performance of the network of systems. Next, systems engineers could consider a new system's effect on the reliability of the overall network. Likewise, systems engineers could examine the impact a new system will have on the complexity of a network and determine if that is a benefit to its stakeholders. Additionally, systems engineers could determine if the introduction of a new system into the network introduces additional vulnerabilities to the network that an advisory could exploit. Systems engineers should consider these types of attributes as they design systems that will fit within a network of systems to provide value to its stakeholders.

At the network level, a new system has the potential to improve, or in some cases degrade, the overall performance of a system of systems or a network of systems. In system of systems, emergent behavior describes the additional value that systems of systems provide to their stakeholders that becomes more than just the sum of its parts [27]. This concept could hold true for networks of systems as the individual nodes, systems in this case, operate together to provided additional value for the DoD. Although each system itself provides some value to the DoD, it is not until they operate together on the battlefield that the true value becomes evident. As the DoD introduces new systems into the network, they may have a positive or adverse effect on the overall operation of the network of systems. A new system may contribute additional capability to the DoD in a positive manner that increases the overall performance of the DoD. However, these systems may also place a large burden on the network of systems to a point that the introduction of a new system requires other systems to operate just to sustain the new system and the overall performance decreases.

Another important aspect of a network of systems is understanding the reliability of a network of systems and how systems could impact the overall reliability. In a network of systems, especially through an operational activity lens, systems engineers could identify the overall reliability of a network and identify where some systems could be substituted for other systems. Systems engineers should consider if a new system increase or decrease the reliability of the overall network. Potentially a new system could replace other systems or serve as a backup for systems in the network. In the event of a failure of one of the systems in the network, this alternative system could pick up the performance requirements for the failed system to ensure that the network continues to operate.

Similar to the reliability of a network, the addition of a new system into the network could impact the complexity of the network of systems. Inherently, the more connections in a network, the more complex that network becomes as systems rely on many other systems and connections to operate. If systems engineers are not diligent in how they add new systems to a network of systems, they could introduce additional complexity in to the network of systems that will make it more difficult to understand and operate. They could also increase the complexity of the network by developing new systems that require a second new system to bridge the gap between the new system and legacy systems. However, they consider the complexity of a network when designing new systems; it may be possible to reduce the complexity of a network by introducing new systems that decrease the connections between systems. Given that systems operate in a network environment, it is important for systems engineers to consider how a new system will integrate into the existing network of legacy systems and the impact of the new systems on the complexity of the network.

The final consideration that systems engineers should consider during the design and development of new systems is if the new system introduces vulnerabilities into the network of legacy systems. A major application of traditional social network analysis is in the assessment of terrorist and criminal networks to identify vulnerabilities that law enforcement can take advantage of [26]. Just like criminal networks, networks of systems could contain vulnerabilities that affect the performance of the overall network, and systems engineers should take caution not to introduce unnecessary vulnerabilities into a network. As system engineers are developing new systems that will operate in an existing network of legacy systems, they should consider if that system introduces new vulnerabilities into the network. Likewise, they could also assess the network of legacy systems to identify where vulnerabilities exist and determine if it is necessary to develop a new system to account for and mitigate those vulnerabilities.

2.5 Conclusion and Future Work

This paper builds on previous work that identified important nonfunctional attributes, or ilities, for DoD systems that have been extended beyond their planned retirement date. First, it provided a review of the literature to include a focus on the systems engineering ilities that provide a means for systems engineers to understand and manage the complexity of modern systems. Additionally, the paper examines the application of social network analysis tools and methods to networks of systems to better understand and potentially quantify the interoperability and versatility of a system. From this perspective, it provides considerations for systems engineers as they design and develop new systems that must operate within this networked environment of new and legacy system. It considers how systems engineers can design in these important attributes to systems to ensure the successful delivery of value to their stakeholders over time. Also, it presents considerations for systems that operate in an emergent network of systems to understand how a new system may affect the overall network of systems.

Literature on the ilities describes the importance of these nonfunctional attributes and their impact on the design process. This paper focuses on the changeability, interoperability, robustness, and versatility of a system. It is likely that the environment and the requirements for a system will change over time and an important aspect of a system is its ability to change in order to meet these changing factors. Systems change through flexibility, adaptability, and extensibility, but these factors often occur after a system is in operation. However, system engineer can account for these attributes and design them into a system through the use of modular architectures and ensuring a buffer of SWaP-C during the design process. Likewise, systems engineers should consider all elements of interoperability of a system during the design process to include information flows, physical connections, and shared resources between systems. Finally, systems engineers should consider possible future threats against a system to ensure its robustness and potentially its versatility to accomplish other operational activities outside of its intended design.

Additionally, there is a lot of previous work that focuses on understanding how systems operate in a network of systems and the type of attributes that lead to successful systems over time. Given that systems will operate in a networked environment, systems engineers should take the effect a new system has on the legacy network of systems into account when developing and designing new systems. They could consider the impact that a new system has on the performance, complexity, and reliability of a network as systems as well as the potential to introduce vulnerabilities into the network. Social network analysis metrics provide a means to quantify these effects and could be used by systems engineers in the design and development of new systems. Likewise, systems engineers could use these tools and methods to identify vulnerabilities or redundancies within the legacy network of systems to drive the development of new systems.

This paper provides several opportunities for future work in the area of incorporating ilities and the impact of a system on a network of legacy systems. First the work is focused on DoD systems; however, this could be expanded to examine other networked systems to include cyber networks, infrastructure networks, or other groups of systems that operate together to provide value to stakeholders. Additionally, the work could be further refined to determine when and where to inject these considerations into the various systems engineering design processes. For the network analysis portion of the work, a more detailed modeling effort could be undertaken to determine the effect new systems have on the overall network of systems. This could potentially validate the use of these metrics to measure the performance, complexity, and reliability of a network of system. It appears that these considerations are important to the design of modern systems and further work could expand on this paper to further the research in this area.

References

1. Enos, J. R., & Nilchiani, R. R. (2017). *A tale of two aircraft: How non-functional attributes impact a system's lifecycle*. In 2017 International Annual Conference on American Society for Engineering Management, Huntsville, AL.

2. Enos, J. R., & Nilchiani, R. (2017). *Using social network analysis to quantify interoperability and versatility in the U.S. Air Force Network of Systems*. SAE Aerospace Technical Papers.
3. Kossiakoff, A., Sweet, W., Seymour, S., & Biemer, S. (2011). *Systems engineering principles and practice* (2nd ed.). Hoboken, NJ: Wiley.
4. INCOSE. (2007). *Systems engineering handbook v3.1*. International Council on Systems Engineering.
5. Rouse, W. (2007). Complex engineered, organizational and natural systems. *Systems Engineering, 10*, 260.
6. McManus, H., Richards, M., Ross, A., & Hastings, D. (2009). *A framework for incorporating "ilities" in tradespace studies*. In American Institute of Aeronautics and Astronautics.
7. de Weck, O. L., Ross, A. M., & Rhodes, D. H. (2012). *Investigating relationships and semantic sets amongst system lifecycle properties (ilities)*. In Third International Engineering Systems Symposium, Delft, NL.
8. ESD Symposium Committee. (2002). *ESD terms and definitions (version 12)*. Cambridge, MA: Massachusetts Institute of Technology.
9. Crawley, E., de Weck, O., Eppinger, S., Magee, C., Moses, J., Seering, W., et al. (2004). *The influence of architecture in engineering systems*. Cambridge, MA: MIT Press.
10. Maier, M. W., & Rechtin, E. (2002). *The art of systems architecting* (2nd ed.). New York: CRC Press.
11. Buede, D. (2000). *The engineering design of systems: Models and methods*. New York: Wiley.
12. Ring, S., Nicholson, D., Thilenius, J., & Harris, S. (2004). *An activity-based methodology for development and analysis of integrated DoD architectures - "the art of architecture"*. Bedford, MA: MITRE.
13. Richards, M., Shah, N., Hasting, D., & Rhodes, D. (2006). *Managing complexity with the Department of Defense Architecture Framework: Development of a dynamic system architecture model*. In Conference on Systems Engineering Research, Los Angeles, CA.
14. Friedenthal, S., Moore, A., & Steiner, R. (2012). *A practical guide to SysML: The systems modeling language*. New York: Morgan Kaufmann OMG Press.
15. DoD Chief Information Officer. (2010, August). *DoD Architecture Framework version 2.02*. Retrieved March, 2014, from http://www.dodcio.defense.gov/dodaf20.aspx
16. Glaser, B., & Strauss, A. (1967). *The discovery of grounded theory: Strategies for qualitative research*. London, England: Aldine Transaction.
17. Cho, J. Y., & Lee, E.-H. (2014). Reducing confusion about grounded theory and qualitative content analysis: Similarities and differences. *The Qualitative Report, 19*, 1–20.
18. Thornburg, R., & Charmaz, K. (2011). Grounded theory and theoretical coding. In *The SAGE handbook of qualitative data analysis* (pp. 153–160). Los Angeles, CA: Sage.
19. Timmermans, S., & Tavory, I. (2012). Theory construction in qualitative research: From grounded theory to abductive analysis. *Sociological Theory, 30*(3), 167–186.
20. Birks, M., & Mills, J. (2015). *Grounded theory: A practical guide* (2nd ed.). Los Angeles, CA: Sage.
21. Fendt, J., & Sachs, W. (2008). Grounded theory method in management research: Users' perspectives. *Organizational Research Methods, 11*(3), 430–455.
22. Urquhart, C., & Fernandez, W. (2013). Using grounded theory method in information systems: The researcher as blank slate and other myths. *Journal of Information Technology, 28*, 224–236.
23. Serrat, O. (2010). *Social network analysis*. Washington, DC: Asian Development Bank.
24. Freeman, L. (2004). *The development of social network analysis*. Vancouver, BC: Empirical Press.
25. White, H., Boorman, S., & Breiger, R. (1976). Social structure from multiple networks. *American Journal of Sociology, 81*, 730–780.
26. Everton, S. (2012). *Disrupting dark networks*. New York: Cambridge University Press.
27. Boardman, J., & Sauser, B. (2006). *System of systems – the meaning of of*. In 2006 IEEE/SMC International Conference on System of Systems Engineering, Los Angeles, CA.

28. Enos, J. R. (2016). *Merging system architecture and social network analysis to better understand emergent networks of systems*. In 2016 International Annual Conference of the American Society for Engineering Management, Charlotte, NC.
29. Enos, J., & Nilchiani, R. (2017). *Understanding how social network analysis can provide insights into emergent networks of systems*. In Conference on Systems Engineering Research 2017, Redondo Beach, CA.
30. Enos, J. R., & Nilchiani, R. R. (2017). *Using social network analysis to quantify interoperability in a large system of systems*. In 12th Annual System of Systems Engineering conference, Waikoloa, HI.
31. Enos, J. R., Mansouri, M., & Nilchiani, R. (2017). *Applying social network analysis to systems of systems: The case of the 2016 Puerto Rico power outage*. In 12th Annual System of Systems Engineering conference, Waikoloa, HI.
32. Beesemyer, J. C., Ross, A. M., & Rhodes, D. H. (2012). *An empirical investigation of system changes to frame links between design decisions and ilities*. In Conference on Systems Engineering Research (CSER), St. Louis, MO.
33. U.S. Army. (2016). *Weapon system handbook 2016*. Washington, DC: Department of Defense.

Chapter 3
Using Systems Engineering Ilities to Better Understand Resiliency

James R. Enos

Nomenclature

t_d Time of disruption
$t_{d'}$ Time of degraded performance
t_r Time of recovery
t_f Time of future performance

3.1 Introduction

Within the Department of Defense (DoD), the concept of resiliency in engineered systems is becoming increasingly important as a means to develop systems that not only meet current performance objectives but can also meet future unanticipated needs and threats. The DoD uses the term engineered resilient systems to describe systems that are resilient in nature and have been designed with resiliency as a major consideration [1]. In 2014, the Secretary of Defense published guidance that identified engineered resilient systems as a priority research area to focus on agile, cost-effective designs that the DoD can easily modify to meet future threats [2]. Given the current fiscal environment in the DoD, systems must be able to meet a diverse range of stakeholder needs and provide value to the DoD in a changing operational and threat environment. Thus, the need for engineered resilient systems is greater now than ever.

The concept of resiliency emerges from the social sciences and often describes natural systems; however, it is equally applicable to engineered systems. This paper

J. R. Enos (✉)
Department of Systems Engineering, U.S. Military Academy, West Point, NY, USA
e-mail: jenos@stevens.eud

S. Adams et al. (eds.), *Systems Engineering in Context*,
https://doi.org/10.1007/978-3-030-00114-8_3

examines resiliency through a lens of the systems engineering ilities to determine how these nonfunctional attributes of systems can help to understand resiliency in engineered systems. As systems become more complex, systems engineers face increasing complexity and have developed the ilities, or nonfunctional attributes, of a system to understand and account for the qualitative aspects of systems engineering. The application of the ilities to engineered resilient systems assists systems engineers in understanding both the effects of resiliency and the means by which systems achieve resiliency. The ilities of quality, robustness, and agility link to the resiliency performance curve to understand how these attributes describe resilient systems. Additionally, ilities such as repairability, extensibility, adaptability, flexibility, and versatility describe how systems react after a disruption to achieve resiliency. Through an understanding of these nonfunctional attributes of systems, systems engineers may be able to improve design processes to achieve engineered resilient systems.

This paper contains three additional sections to include a literature review, a discussion on the ilities that comprise resiliency, and a conclusion to propose future work in this area. The literature review focuses on resilient engineered systems and provides an introduction to systems engineering and the ilities. The next section goes into more detail on the ilities that describe the effects and means for achieving an engineered resilient system. Finally, the last section discusses the initial findings from this paper and proposes future work to examine example systems from inside and outside of the DoD that demonstrate these nonfunctional attributes and achieve resiliency over time.

3.2 Literature Review

This section presents a review of the relevant literature to include a discussion on the resiliency of engineered systems and systems engineering with a focus on the ilities. The portion on resiliency focuses on the definition of resiliency in engineered systems. Although resiliency applies to a range of disciplines from systems engineering to psychology, this review focuses on engineered resilient systems. It examines a few examples of methods and applications with a concentration on systems within the DoD. The second portion of the literature review examines systems engineering and the methods that engineers use to confront and manage complexity and uncertainty. One of the main tools systems engineers use for this is the construct of the ilities, or nonfunctional attributes. Several of the ilities contribute to a better understanding of resiliency in engineered systems.

3.2.1 Resiliency

For engineered systems, resiliency is "the ability of a system to adapt affordably and perform effectively across a wide range of operational contexts, where context is defined by mission, environment, threat, and force disposition" [3]. Moving beyond resiliency, engineered resiliency systems are specifically designed systems to account for the changing environment systems and must operate within in order to provide value to stakeholders after disruption to system operations [1]. The literature identifies critical links between engineered resiliency systems and several of the systems engineering ilities, specifically flexibility, robustness, and versatility [4]. A significant amount of literature, outside of the systems engineering domain, discusses human resiliency and the resiliency of natural systems [5]. Francis and Bekera [6] proposed a framework for assessing the resiliency of system based on the system performance.

From a DoD perspective, resiliency describes a system's ability to "anticipate, resist, absorb, respond to, adapt to, and recover from a disturbance from either natural or man-made events" [1]. This perspective includes two main portions: first the ability to react to a disturbance and then the ability to return to the previous level of functionality after the disturbance. Some other works in the DoD focused on incorporating the attributes of resiliency into the early design of systems during the analysis of alternatives phase of requirements generation. This work found that incorporating various nonfunctional attributes into the design of systems increases the resiliency of the system and that engineers should move beyond the evaluation of a few point solutions to a larger tradespace exploration [7]. Tradespace exploration tools provide a means to evaluate thousands of design alternatives to understand the resiliency of a system's design under uncertainty and provide systems engineers the ability to engineer resiliency into a system [8]. Systems engineers can also leverage model-based systems engineering to explore the tradespace and determine how the ilities influence a system's resiliency [5].

The general concept behind resiliency focuses on a system's performance over time and the reaction of the system to a disruption to the system's performance. After a disruption to the system, engineers and managers react to the disruption in order to restore the system to an acceptable level of operation. Figure 3.1 presents an adaptation of Francis and Bekera's work that examines the system performance as a reaction to a disruption at t_d, a time of degraded performance until $t_{d'}$ until a recovery at t_r, and finally a future state equilibrium at t_f [6].

The literature also provides several examples of engineering resiliency into a range of systems to demonstrate the importance of this system attribute. Klingensmith and Madni [9] describe architecting cyber systems for security and resiliency with a focus on various system attributes leading to a cyber-secure system. Leonard [10] also focused on DoD cyber systems and found that the DoD needs to move beyond developing survivable cyber systems to incorporate resiliency in their cyber systems. Other works focused on developing a resilient autonomous vehicle network through a model-based approach to account for uncertainty [11]. In the DoD, this application of resiliency to automated vehicles is incredibly important as these systems will join the ranks of human-driven vehicles and must be able to react and respond to disruptions without a failure of the mission; additionally, these systems must learn from their environment to deliver full value to the stakeholders [12]. These examples focus on mission resiliency for systems; however, platform

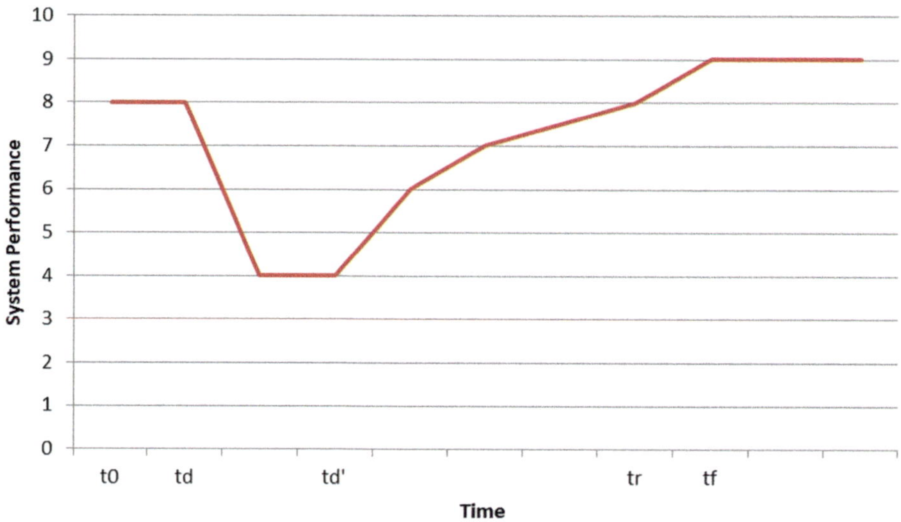

Fig. 3.1 System performance after a disruption (Adapted from [6])

resiliency is also an important attribute for systems to react to disruptions over a greater duration of time [5]. Despite a focus on DoD systems, resiliency also applies to systems outside of the DoD, and efforts to model resiliency also apply to other systems such as infrastructure networks [13].

3.2.2 Systems Engineering

The International Council on Systems Engineering (INCOSE) [14] defines systems engineering as "an interdisciplinary approach and means to enable the realization of successful systems. It focuses on defining customer needs and required functionality early in the development cycle, documenting requirements, and then proceeding with design synthesis and system validation while considering the complete problem." During the design phase, systems engineers translate stakeholder needs and wants into system requirements, develop the functional architecture of a system, and allocate these functions to physical aspects of the system [15]. To accomplish this, systems engineers decompose a system's function and form, design these decomposed elements, and then combine them to achieve the overall function of the system. However, systems engineers face increasing complexity of systems as technology progresses and systems become more and more interconnected.

Systems engineers have developed the ilities, a construct for assessing the nonfunctional attributes of a systems, as a means to manage the complexity and the qualitative nature of systems engineering. They recognized the criticality of these nontraditional design criteria and now include these attributes with the tradition considerations of form and function [16]. However, one of the major problems with

Table 3.1 Ility definitions

Ility	Definition
Adaptability	"The ability of a system to be changed by a system-internal change agent" [16]
Agility	"The ability to change in a timely fashion" [17]
Changeability	"The ability of a system to alter its form—and consequently possibly its function—at an acceptable level of resource expenditure" [16]
Extensibility	"The ability to accommodate new features after design" [17]
Flexibility	"The ability of a system to be changed by a system-external change agent" [16]
Quality	"The ability to deliver requirements at a "high" level, as perceived by people relative to other alternatives that deliver the same requirements" [19]
Repairability	"The ability to be returned to the original state of function when some function is lost" [19]
Robustness	"The ability of a system to maintain its level and set of specification parameters in the context of changing system external and internal forces" [16]
Versatility	"The ability of a system to satisfy diverse expectations on the system without the need for changing form" [16]

using the ilities in the design of a system is that these attributes often arise after the system is in operation [17]. Further study of the ilities examines how system-level ilities begin to emerge from the subsystem level, where systems engineers can design in these nonfunctional attributes [18]. Table 3.1 presents a subset of the systems engineering ilities along with a definition for each ility that provides the foundation for this paper.

3.3 Ilities of Resiliency

This section of the paper discusses the roles that ilities serve in understanding and achieving resiliency in engineered systems. First, it applies several ilities to the resiliency performance curve to understand the effects of resiliency in engineered systems. These ilities assist systems engineers in understanding how a system's nonfunctional attributes can contribute to resiliency. Next, the section discusses the means by which systems achieve resiliency through changeability or versatility. These ilities allow systems to continue to meet stakeholder's expectations and in some cases exceed the previous performance level to demonstrate resiliency.

3.3.1 Ility Effects in Resiliency

This section describes the ilities that represent the effects of resiliency of engineered systems to include quality, agility, and robustness. Quality describes the ability of a system to meet the expectations of its stakeholders and focuses on the performance of a system. Agility is an attribute that describes how quickly a system can respond

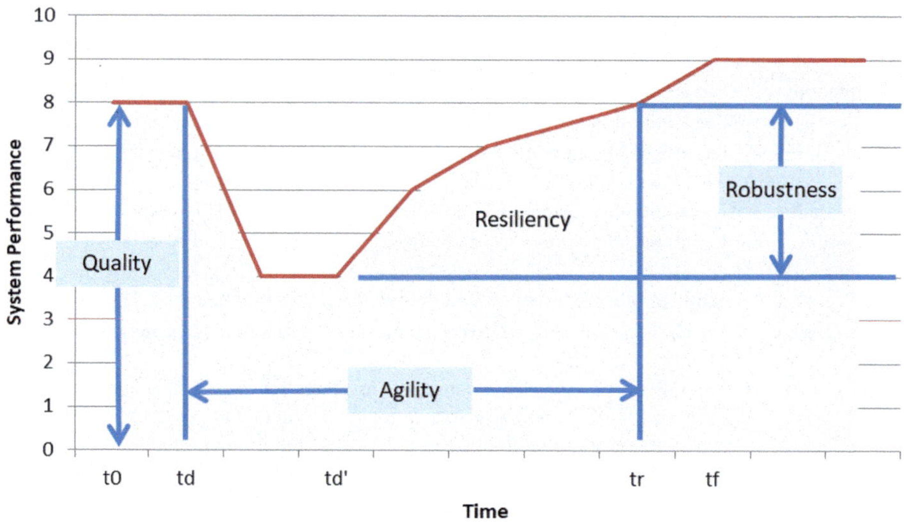

Fig. 3.2 Ilities of resiliency

to a disruption and change to return to the previous level of performance. Finally, robustness describes how much a disruption degrades the performance of a system. Figure 3.2 presents an overlay of the ilities on the resiliency performance curve to demonstrate how the ilities help to better understand resiliency. The initial performance of a system is consistent with the definition of quality and the ability to provide value to a system's stakeholders. After a disruption, the degradation in performance represents the system's robustness. The agility of a system allows it to change after a disruption, and the time aspect of this ility demonstrates the amount of time required to return to the original level of performance. Finally, the area under the resiliency curve represents the systems resiliency as a result of the other ilities.

The first ility to consider when examining resiliency through this lens is quality, which is the "ability to deliver requirements at a "high" level, as perceived by people relative to other alternatives that deliver the same requirements" [19]. Although there are several definitions of quality, from a systems engineering perspective, it is truly the ability to meet stakeholder requirements at a high level and serves as the starting point for the resiliency performance curve. The initial quality of a system provides the benchmark for which the system should return to after a disruption. In some cases, the system returns to an even higher level of performance, due to added capability or versatility, after a disruption, which in turn increases the quality of the system from a stakeholder perspective.

The next ility that helps to understand the resiliency of an engineered system is agility that is "the ability to change in a timely fashion" [17]. Although the literature does not have a definitive definition of agility, it is one of the few ilities that considers time when describing the nonfunctional attributes of a system. The literature often links agility to the flexibility of a system but with the additional descriptor of being a rapid or timely change to a system [20]. Ryan et al. [21] closely link these two ilities with the distinction being that agility is flexibility that is quick rather than easy. One of the means to achieve agility is the modular design of systems so that engineers can quickly change system components to improve the performance of the system [22]. It is also important to note that a large part of the literature focuses on agile systems engineering, which is separate from this description of an agile system [23]. Agile systems are often confused with agile systems engineering which traces its roots to agile software engineering and is based on a desire to increase the speed at which systems are engineered [22].

McManus et al. [16] define robustness as the ability of a system to maintain its level and set of specification parameters in the context of changing system external and internal forces. Like other ilities, the definition of robustness is currently up for debate and is similar to the literature on resiliency. Additionally, the literature often describes robustness in the context of the flexibility or adaptability of a system; however, it does not require a change to the system as in the other ilities [21]. The literature also differentiates a robust system, one that can continue to provide value to unforeseen changes, and a robust design, a system that operates through a range of known conditions [23]. De Weck, Ross, and Rhodes [17] note that robustness correlates with flexibility, adaptability, and extensibility and is related to the quality of a system. The quantification of robustness is difficult given that it relies heavily on stakeholder value, which could include several different attributes that define a system's utility. Ross, Rhodes, and Hastings [24] describe the method of using Pareto Trace to identify designs that provide passive value robustness as they lie on the Pareto front in tradespaces across various epochs. This method enables systems engineers to identify designs that provide the best value for the cost given changing context in each epoch. Fitzgerald and Ross [25] expand on this work to measure robustness based on an effective normalized Pareto Trace and Fuzzy Pareto shift to acknowledge changeability and account for passive robustness of a system. The literature on robustness is consistent in the view that robustness is a desirable attribute of systems that enables them to continue to provide value to stakeholders under changing requirements and environmental conditions.

After considering how system ilities assist systems engineers in understanding the resiliency of engineering systems, it is possible to quantify resiliency and potentially compare the resiliency of system designs. Effectively, the area under the performance curve could represent the resiliency of a system. By using the area under the curve, systems engineers could compare various designs of systems to engineer and field a resilient system. Systems engineers often evaluate the performance of systems with a normalized value or utility function, which would provide a consistent comparison along the vertical axis of this graph. Likewise, if engineers compared design options and their reaction to a disturbance along the same time line, the horizontal axis remains constant. Under these conditions, systems engineers could compare the area under the resiliency curve of the design options to determine which design is more resilient. These designs could increase resiliency through increased agility that decreases the time to return to normal operation or through a more robust system that minimizes the effect of the disturbance on the system's performance. Through an evaluation of the resiliency of a system, systems engineers could introduce this concept into the design process and leverage commonly accepted system attributes to achieve engineered resilient systems.

3.3.2 Ility Means for Resiliency

This section describes the ilities that represent the means by which systems achieve resiliency to include repairability, extensibility, adaptability, flexibility, and versatility. Given that resiliency occurs after some disruption to a system's normal operation, these ilities enable systems to return to their previous level of functionality to meet stakeholder expectations. Repairability describes how system operators make repairs to a system after a disruption and is generally a short-term, immediate fix to the problem. Extensibility is more in-depth than just making repairs after a disruption and focuses on the ability of a system to accept new capabilities to return a system to its previous performance level or in some cases even higher performance. Adaptability and flexibility both require changes to the system to meet stakeholder requirements and can return a system to its previous level of performance. Finally, versatility describes a system's ability to meet a range of expectations, so one aspect of performance may decrease after a disruption, but overall value may remain constant as stakeholders can use a versatile system in another role.

The literature defines repairability as "the ability to be returned to the original state of function when some function is lost" [19]. Although the systems engineering literature defines this as an independent system attribute, repairability is closely linked to the reliability and availability of a system. When linked to reliability, the repairability of a system determines if engineers can return a system to full functionality and the amount of time that a system is undergoing repairs [26]. In the reliability literature, a repairable system is one in which operators can return a system to full functionality without the need to replace the entire system, and the quantification of this attribute considers both the probability of needing a repair and the time required to repair one or multiple components of a system [27]. Additional work that identified the repairability of components improves the overall reliability of a system as these components can quickly be returned to full operating capability, thus making a system more reliable overall [28]. Within the DoD, repairability is considered in the Sustainment Key Performance Parameter as part of the maintainability of a system as a desirable attribute to ensure that DoD systems are available for operations [29].

Extensibility is the "ability to accommodate new features after design" and focuses on adding new capabilities to a system [17]. Some authors consider extensibility as part of flexibility as the addition of new capability since external actors instigate the change to the system [21]. The systems engineering literature on extensibility is not very robust; however, outside of systems engineering, the medical and computer science fields often mention extensibility. In the medical field, extensibility aligns with the flexibility of muscles as it describes the ability of a joint to move and gain a larger range of motion [30]. In the computer science field, extensibility refers to the ability of a software system to export services and resources to other applications [31].

Adaptability is another factor of change that focuses on changes to a system's form or function caused by internal change agents to the system [21]. The major difference between adaptability and flexibility is the source of the change to the system [32]. Adaptability allows systems to continue to provide value to stakeholders through the system life cycle by altering the form or function of the system [33]. Additionally, systems engineers can design adaptability into systems by developing systems that operators can change during operation to meet environmental conditions. In an evaluation of architectures for intelligence, surveillance, and reconnaissance systems, Pape et al. [34] noted that a system may be adaptable as it allows users to select sensors based on conditions. Likewise, the variable wing sweep design of the F-14 enables the pilot to adapt the system to achieve desired speeds and alter the performance of the aircraft based on conditions [33].

Flexibility is similar to adaptability in that it accounts for a system's ability to change form or function; however, in the case of flexibility, the change agent is external to the system. Ryan, Jacques, and Colombi [21] examined a range of literature on flexibility and found that nearly all of the literature attributes the change to a system to external sources when discussing flexibility. Given that the systems engineering literature has well-defined flexibility, additional works focused on quantifying the effect of a flexible system. Nilchiani and Hastings [35] presented a six-element framework for measuring flexibility and applied this framework to space systems. In the DoD realm, the literature examined quantifying flexibility in system of systems through the use of architecture products [36].

McManus et al. [16] define versatility as the "ability of a system to satisfy diverse expectations on the system without the need for changing form." A versatile system meets a wide range of needs and requirements but is distinct from flexibility in that it does not require a fundamental change to the system [21]. Versatility is present in the initial design and fielding of a system, even though it may not be directly related to design requirements and specifications [4]. Additionally, the literature identifies various links between versatility and other ilities, especially the attributes of latent value and the robustness of a system [17]. As systems increase in complexity and cost, versatility is becoming a more desirable attribute of systems that face a rapidly changing environment. Systems perform multiple roles to keep pace with changing requirements without the need to expend resources to change a system's design [37].

3.4 Conclusion and Future Work

Systems engineers have adopted the construct of the ilities as a means to address the nonfunctional, qualitative nature of systems engineering and manage complexity. These ilities provide systems engineers with an ability to better understand resiliency in terms of the effects of resiliency as well as the means to achieve resiliency. Within the DoD, the need to develop engineered resilient systems, those system with resiliency intentionally designed into the system, is increasing to manage limited budgets and unknown future threats. Through an assessment of the ilities and the application of several ilities to the resiliency performance curve, systems engineers can improve their understanding of resiliency and incorporate these attributes into the design of systems.

The resiliency performance curve depicts how a system responds to a disturbance and the time required to return the system to its original performance level. The ilities of quality, agility, and robustness assist systems engineers in understanding how a system reacts to a disturbance and provides the potential to quantify and compare the resiliency of alternative system designs. First, the quality of a system describes the initial performance of a system and the ability of that system to meet the stakeholder requirements to a high level. After a disturbance, the attribute of agility, with its focus on time, describes the time required to make changes to the system to return it to full functionality. Finally, the robustness of a system describes how much the disturbance affects the system's performance. The area under the resiliency curve could represent the resiliency of alternative system designs, assuming a consistent performance function and time frame. This aspect of resiliency enables systems engineers to determine the resiliency of design alternatives in order to engineer resilient systems during the design phase.

The paper also provides links to several ilities to describe the means by which a system can achieve resiliency and return to, or in some cases exceeds, the original system performance. The first attribute is the repairability of a system, which characterizes the ability of a system to be returned to its normal operations after a disruption. Systems that are changeable, through extensibility, adaptability, or flexibility, are able to react to disruptions and return to full functionality at an appropriate level of resource expenditures. In some cases, the ility of versatility captures a system's ability to recover after a disruption as it describes the ability of a system to meet a range of stakeholder requirements. Versatile systems can continue to provide value to stakeholders for unaffected performance criteria, or stakeholders could use the system for another function.

Future work should expand on this application of the ilities to better understand the resiliency of an engineered system. There are several examples of resilient system design within the DoD to include the B-52 bomber and the M-1 Abrams main battle tank. Both of these systems continue to deliver value to the DoD despite changing threat and operational environments. Likewise, systems like the F-117 stealth fighter provide an opportunity to demonstrate how systems that cannot change in response to disruptions will be retired as they no longer meet stakeholder needs. In addition to future work focused on DoD systems, this work could expand to engineered systems outside of the DoD to determine if these observations are generalizable.

References

1. Goerger, S. R., Madni, A. M., & Eslinger, O. J. (2014). *Engineered resilient systems: A DoD perspective*. In 2014 Conference on Systems Engineering Research, Redondo Beach, CA.
2. Department of Defense. (2014). *DoD research and engineering enterprise* (p. 9). Washington D.C.: Author.
3. Neches, R., & Madni, A. M. (2013). Towards affordably adaptable and effective systems. *Systems Engineering, 16*(2), 224–234.
4. Sitterle, V. B., Freeman, D. F., Goerger, S. R., & Ender, T. R. (2015). *Systems engineering resiliency: Guiding tradespace exploration within an engineered resilient systems context*. In 2015 Conference on Systems Engineering Research, Hoboken, NJ.
5. Buchanan, R. K., Goerger, S. S., Rinaudo, C. H., Parnell, G., Ross, A., & Sitterle, V. (2015). Resilience in engineered resilient systems. *Journal of Defense Modeling and Simulation, 12*(1), 57–66.
6. Francis, R., & Bekera, B. (2014). A metric and frameworks for resiliency analysis of engineered and infrastructure systems. *Reliability Engineering and System Safety, 121*, 90–103.
7. Small, C., Parnell, G., Pohl, E., Goerger, S., Cottam, B., Specking, E., et al. (2017). *Engineering resilience for complex systems*. In 15th Annual Conference on Systems Engineering Research, Redondo Beach, CA.
8. Spero, E., Avera, M. P., Valdez, P. E., & Goerger, S. R. (2014). *Tradespace exploration for the engineering of resilient systems*. In Conference on Systems Engineering Research (CSER 2014), Redondo Beach, CA.
9. Klingensmith, K., & Madni, A. M. (2017). *Architecting cyber-secure, resilient system-of-systems*. In 15th Annual Conference on Systems Engineering Research, Redondo Beach, CA.
10. Leonard, W. (2017). *Resilient cyber secure systems and system of systems: Implications for the department of defense*. In 15th Annual Conference on Systems Engineering Research, Redondo Beach, CA.
11. Madni, A. M., Sievers, M. W., Humann, J., Ordoukhanian, E., D'Ambrosio, J., & Sundaram, P. (2017). *Model-based approach for engineering resilient system-of-systems: Application to autonomous vehicle networks*. In 15th Annual Conference on Systems Engineering Research, Redondo Beach, CA.
12. Klingensmith, K., & Madni, A. M. (2017). *Resilience concepts for architecting an autonomous military vehicle system-of-systems*. In 15th Annual Conference on Systems Engineering Research, Redondo Beach, CA.
13. Ouyang, M., & Wang, Z. (2015). Resilience assessment of interdependent infrastructure systems: With a focus on joint restoration modeling and analysis. *Reliability Engineering and System Safety, 141*, 74–82.
14. INCOSE. (2007). *Systems engineering handbook v3.1*. San Diego, CA: Author.
15. Buede, D. (2000). *The engineering design of systems: Models and methods*. New York: Wiley.
16. McManus, H., Richards, M., Ross, A., & Hastings, D. (2009). *A framework for incorporating ilities in tradespace studies*, In American Institute of Aeronautics and Astronautics.
17. de Weck, O. L., Ross, A. M., & Rhodes, D. H. (2012). *Investigating relationships and semantic sets amongst system lifecycle properties (ilities)*. In Third International Engineering Systems Symposium, Delft, NL.
18. Lee, J. Y., & Collins, G. J. (2017). On using ilities of non-functional properties for subsystems and components. *Systems, 5*, 47.
19. ESD Symposium Committee. (2002). *ESD terms and definitions (version 12)*. Cambridge, MA: Massachusetts Institute of Technology.
20. Bauer, W., Elezi, F., Roth, M., & Maurer, M.. (2015). *Determination of the required product platform flexibility from a change perspective*. In 9th Annual IEEE International Systems Conference.
21. Ryan, E. T., Jacques, D. R., & Colombi, J. M. (2012). An ontological framework for clarifying flexibility-related terminology via literature survey. *Systems Engineering, 16*, 99–110.

22. Schapiro, S. B., & Henry, M. H. (2012). *Engineering agile systems through architectural modularity*. In 2012 International Systems Conference, Vancouver, BC.
23. Turner, A. J., Monahana, W., & Cottera, M. (2017). *Quantifying the ilities: A literature review of robustness, interoperability, and agility*. In 15th Annual Conference on Systems Engineering Research, Redondo Beach, CA.
24. Ross, A. M., Rhodes, D. H., & Hastings, D. E. (2009). *Using Pareto Trace to determine system passive value robustness*. In 3rd Annual IEEE International Systems Conference, Vancouver, B.C.
25. Fitzgerald, M. E., & Ross, A. M. (2012). *Mitigating contextual uncertainties with valuable changeability analysis in the multi-epoch domain*. In SysCon2012—IEEE International Systems Conference, Vancouver, BC.
26. Maciejewski, H., & Caban, D. (2008). Estimation of repairable system availability within fixed time horizon. *Reliability Engineering and System Safety, 93*, 100–106.
27. Garmabaki, A., Ahmadi, A., Block, J., Pham, H., & Kumar, U. (2016). A reliability decision framework for multiple repairable units. *Reliability Engineering and System Safety, 150*, 78–88.
28. Levitin, G., Jai, H., Ding, Y., Song, Y., & Dai, Y. (2017). Reliability of multi-state systems with free access to repairable standby elements. *Reliability Engineering and System Safety, 167*, 192–197.
29. Joint Chiefs of Staff. (2015). *Manual for the operation of the joint capability integration and development system (JCIDS)*. Washington, DC: Department of Defense.
30. Ballantyne, F., Fryer, G., & McLaughlin, P. (2003). The effect of muscle energy technique on hamstring extensibility: The mechanism of altered flexibility. *Journal of Osteopathic Medicine, 6*(2), 59–63.
31. Bershad, B. N., Savage, S., Pardyak, P., Sirer, E. G., Fiuczynski, M. E., Becker, D., et al. (1995). *Extensibility, safety and performance in the SPIN operating system*. In Fifteenth ACM Symposium on Operating Systems Principles, Copper Mountain, CO.
32. Ross, A., Rhodes, D., & Hastings, D. (2008). Defining changeability: Reconciling flexibility, adaptability, scalability, modifiability, and robustness for maintaining system lifecycle value. *Systems Engineering, 11*(3), 246–262.
33. Beesemyer, J. C., Ross, A. M., & Rhodes, D. H. (2012). *An empirical investigation of system changes to frame links between design decisions and ilities*. In Conference on Systems Engineering Research (CSER), St. Louis, MO.
34. Pape, L., Giammarco, K., Colombi, J., Dagli, C., Kilicay-Ergin, N., & Rebovich, G. (2013). *A fuzzy evaluation method for system of systems meta-architectures*. In 2013 Conference on Systems Engineering Research, Atlanta, GA.
35. Nilchiani, R., & Hastings, D. (2007). Measuring the value of flexibility in space systems: A six-element framework. *Systems Engineering, 10*(1), 26–44.
36. Agarwal, S., Pape, L. E., Dagli, C. H., Ergin, N. K., Enke, D., Gosavi, A., et al. (2015). *Flexible and intelligent learning architectures for SoS (FILA-SoS): Architectural evolution in systems-of-systems*. In 2015 Conference on Systems Engineering Research, Hoboken, NJ.
37. Calvano, C. N., & John, P. (2004). Systems engineering in an age of complexity. *Systems Engineering, 7*(1), 25–34.

Chapter 4
Military Enterprise Wireless Communications Architecture Power Analysis

Howen Fernando, Ronald Giachetti, and Anthony Pollman

Nomenclature

D	Duty cycle [dimensionless]
EWC	Enterprise wireless communications
MOSA	Modular open-source architecture
MWC	Military wireless communications
P_{con}	Power consumption [W·s]
$P_{contotal}$	Total power consumption system [W·s]
P_{total}	Total instantaneous power [W]
P_{xxx}	Instantaneous power for component *xxx* [W]
RHM	Radio head module
RF	Radio frequency
SCA	Software communications architecture
SDR	Software-defined radio

4.1 Introduction

Increasing demand for wireless communications to support deployed, military operations results in increased energy requirements for these missions. The appetite for greater connectivity on the modern battlefield, to support "net-centric" operations, only adds to energy demands for communications. In most cases,

H. Fernando
Space and Naval Warfare Systems Command, San Diego, CA, USA

R. Giachetti · A. Pollman (✉)
Naval Postgraduate School, Monterey, CA, USA
e-mail: regiache@nps.edu

S. Adams et al. (eds.), *Systems Engineering in Context*,
https://doi.org/10.1007/978-3-030-00114-8_4

expeditionary command and control nodes cannot rely on existing infrastructure and must provide their own power. Diesel generators generally provide on-site power. Vulnerable, costly, ground-based logistics convoys supply the fuel.

Currently, military wireless communications (MWC) employ dedicated, stand-alone radio networks with each terminal having its own power supply and processors. In contrast, the industry has adopted an enterprise wireless communications (*EWC*) architecture for cellular base stations [1, 2]. The US Army and Navy have established research and development projects to investigate the *EWC* architecture for expeditionary networks. However, the extant military research focuses on providing connectivity while failing to address the potential power savings of EWC architecture [3]. This paper demonstrates replacement of the current expeditionary MWC architecture with an *EWC* architecture of software-defined radios (SDRs) has the potential to reduce overall power consumption by reducing powered-equipment redundancies such as processors and power supplies. This paper contributes to the literature on *EWC* architecture by conducting analysis to estimate the potential power reduction resulting from adoption of an *EWC* architecture for expeditionary, military operations.

4.1.1 Current Military Wireless Communications (MWC) Architecture

Under the current MWC architecture, an expeditionary node employs numerous, stand-alone, half-duplex radios. Each radio (or SDR) corresponds to a command and control network and serves a dedicated role or function. High-frequency radio networks are employed for over-the-horizon communications, for example, ship-to-shore communications. Very-high-frequency radio networks perform medium-range communications, for example, ground-tactical communications or convoy control. And, ultrahigh-frequency radio networks perform line-of-site communications, for example, air-to-ground communications. Figure 4.1 is a schematic representation of the current MWC architecture highlighting the hardware redundancies inherent in the architecture. The schematic also shows major components, along with notation, used for power modeling and simulation.

To simplify power modeling and simulation, four general radio system types are defined based on their ability to perform power management. In general, and by design, current short-range radio systems use older technology and cannot perform power management. Medium-range tactical systems can perform limited power management. Long-range tactical systems employ newer technology, can perform power management, and use a medium complexity waveform. Finally, long-range tactical anti-jam systems employ newer technology, can perform power management, and use a complex anti-jam waveform.

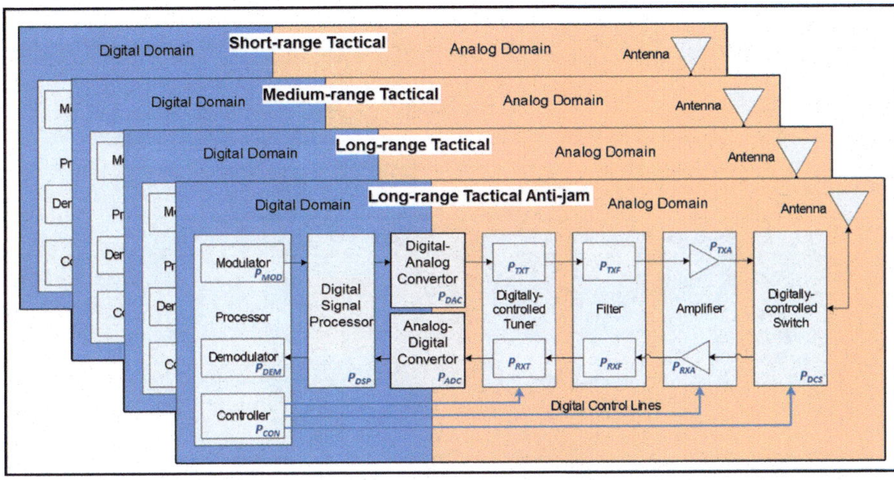

Fig. 4.1 Schematic of software-defined radios (SDRs) employed using the current military wireless communications (MWC) architecture. Adapted from Fernando [4]

4.1.2 Technological Advancements Enabling Adoption of an Enterprise Wireless Communications Architecture

Technologies have matured such that the realization of an EWC architecture for expeditionary applications is possible. Key enabling technologies include software-defined radios (SDRs), enterprise processing systems, modular open systems architectures (MOSA), and high-speed fiber optic serial interfaces.

The introduction of analog-to-digital converters and digital-to-analog converters and performance improvements in general-purpose processing brought about SDRs [5, 6]. Early digital radios allowed the use of simple waveforms in specialized digital subsystems to implement modulators and demodulators. Digital radios modulate digital information into an analog signal for transmission and digitize demodulated incoming analog signals. In the 1990s, general-purpose processor technology improved enough to allow a wide variety of waveforms to run. This innovation gave rise to the SDR [6]. Today's SDRs can implement complex modulation and demodulation algorithms in software to increase the amount of digital information transmitted and received at any given frequency and time.

Advancements in processing performance and high-speed processing node connectivity have brought about an enormous increase in processing capability [7]. Processing nodes now have multiple processors with multiple processing cores that can process more data, faster than ever before. Moreover, processing capability scales up with the addition of processing nodes that communicate with other processing nodes at very high speeds. Interconnected processing nodes form a high-performance computing environment called a cluster. A large cluster is the

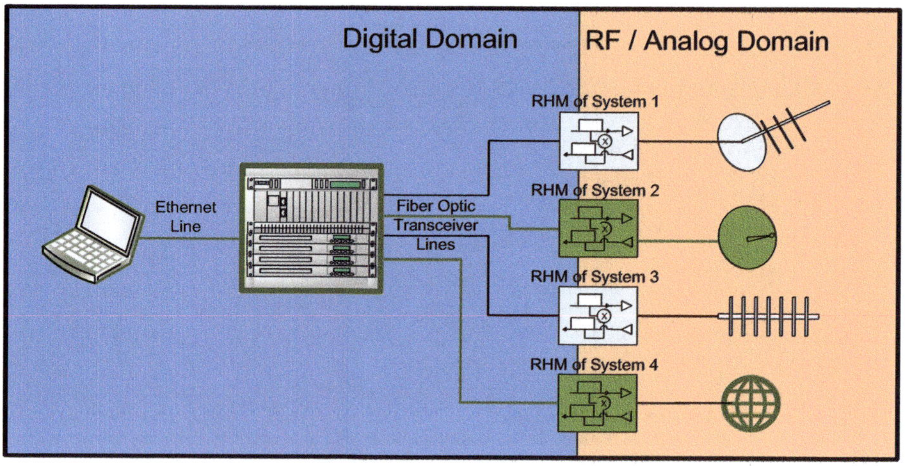

Fig. 4.2 Schematic of multiple software-defined radios (SDRs) employed using an enterprise wireless communications (EWC) architecture. Adapted from Fernando [4]

technology behind cloud computing. A cluster, or enterprise processing system, can process multiple waveforms simultaneously [7].

MOSA provides the mechanism to use interoperable waveform software modules in an enterprise environment. It can also provide an abstraction between the hardware and software to allow plug and play capability for any waveform to run within an enterprise. The Department of Defense's Software Communications Architecture (SCA) is a specific example of MOSA that creates a highly modularized architecture that detaches the waveform application from the underlying hardware platform [8]. SCA simplifies the implementation of multiple waveforms in SDRs.

Finally, the enterprise server, implementing the SCA, needs to transmit and receive digital signals at very high speeds. The digital signals, which require more than 10 Gbps of data throughput, would be too expensive to implement via parallel interfaces. The latest serial ports known as field programmable gate arrays are implemented using fiber optics and have this threshold. Furthermore, industry leaders in this technology have announced bandwidths of 1 Tbps over dual-mode serial ports [9].

Figure 4.2 is a schematic of SDRs employed using an EWC architecture. The major components are radio head modules (RHM), an enterprise server, and terminals. The RHM translates radio frequency (RF) signals to digital signals and vice versa. The enterprise server implements the software functions of a SDR. The enterprise server is connected to each RHM by high-speed fiber optics. The server modulates information for transmission, and the RHM synthesizes the modulated data into signals for the antenna to radiate. Likewise, the antenna receives the signals, the RHM digitizes the signals, and the enterprise server demodulates the digitized radio frequencies. The terminals execute applications to perform higher-level functions like user interfaces, integrated displays, and maintenance functions.

In the EWC architecture, RHMs (a single RHM can replace several radios) replace the numerous, stand-alone, dedicated half-duplex radios of the current MWC architecture. In the figure, four RHMs are shown to depict the four general radio system types.

The EWC architecture has all the advantages of an enterprise system [6], such as improved reliability, maintainability, and affordability. An enterprise architecture allows for easier and inexpensive upgrade of the enterprise server when performance and higher-efficiency processors become available. Moreover, the software is portable from the older enterprise system to the next-generation enterprise system [6]. Currently, expeditionary nodes employ a collection of disparate, dedicated communication systems, which have redundant power and processing systems. An enterprise system incorporates redundant processing subsystems that remain powered off when not required and could thereby reduce the power requirement of the current architecture while providing the same command and control capabilities.

4.2 Modelling and Simulation

4.2.1 Power Consumption Modeling

Power consumption is the amount of energy used for a given time interval, and total system power consumption is the fundamental comparison for this work. Power consumption is a function of instantaneous power, power modes, duty cycle, and radio system type.

Instantaneous power consumption is the power used by a system component at an instant in time. Equation (4.1) expresses the total instantaneous power of each component of an SDR (with the subscripts correspond to each component, left to right and top to bottom in Fig. 4.1: modulator, demodulator, controller, digital signal processor, digital-analog converter, analog-digital converter, transmit tuner, receive tuner, transmit filter, receive filter, transmit amplifier, receive amplifier, and digitally controlled switch, respectively).

$$\begin{aligned} P_{\text{total}}[W] = {} & P_{\text{MOD}} + P_{\text{DEM}} + P_{\text{CON}} + P_{\text{DSP}} + P_{\text{DAC}} + P_{\text{ADC}} \\ & + P_{\text{TXT}} + P_{\text{RXT}} + P_{\text{TXF}} + P_{\text{RXF}} + P_{\text{TXA}} + P_{\text{TXR}} + P_{\text{DCS}} \end{aligned} \tag{4.1}$$

For ease of simulation, we chose a differential approach, to obtain the following expression for power consumption as a function of instantaneous power:

$$P_{\text{con}}[W \cdot s] = P_{\text{total}} \Delta t \tag{4.2}$$

Since radios are half-duplex, transmit components are not powered during receive (and vice versa). More modern radios can also do power management. We refer to these as power modes and define four: sleep, standby, transmit, and receive. The amount of time that the HRM is transmitting or receiving is the duty cycle. To

Table 4.1 Duty cycles used for this work

Mode	Duty cycle		
	Light usage (%)	Medium usage (%)	Heavy usage (%)
Sleep	10	0	0
Standby	30	20	0
Transmit	30	40	50
Receive	30	40	50
Total	100	100	100

Validated by Uhm and Fountain [10]

simplify modeling, we defined three duty cycles: light, medium, and heavy usage. Finally, power requirements vary depending on which of the four general radio types are being used (due to power management capabilities). To implement these considerations mathematically, a constant D (for duty cycle), is used to modify Eq. (4.2). Equation (4.3), coupled with tabulated power values, can be used to calculate power consumption for a system of SDRs.

$$P_{\text{totalcon}}[W \cdot s] = P_{\text{con}} D \tag{4.3}$$

4.2.2 *Consumption Simulation*

Power consumption simulation requires instantaneous power and duty cycle values with which to evaluate Eq. (4.3). Table 4.1 summarizes duty cycle profiles. Tables 4.2 and 4.3 summarize the instantaneous power values used for this work. The component differences in Tables 4.2 and 4.3 are architecture related, namely, power is shared in EWC but not under the current MWC architecture. The component power values and power supply efficiencies were found in the literature and verified against publicly available technical specification [1, 10–13]. Subject matter experts then validated both the component power requirements and duty cycle values [10]. Simulation was conducted with via spreadsheet. Interested readers are directed to Fernando [4] for a more nuanced discussion of these considerations and implementation.

4.3 Results

The intent of modeling and simulations was to determine the potential power savings that could be realized from replacing the current MWC architecture with an EWC architecture. We used data from Table 4.1 to evaluate the current MWC architecture with Eq. (4.3) for four SDRs operating in parallel, one for each of the four general radio types, operating for a week at light, medium, and heavy usage, respectively.

Table 4.2 Mode-based instantaneous component power requirements for SDRs operating under the current MWC architecture [1, 10–13]

Mode	MOD (W)	DEM (W)	CON (W)	DSP (W)	DAC (W)	ADC (W)	TXT (W)	RXT (W)	TXF (W)	RXF (W)	TXA (W)	RXA (W)	DCS (W)	Power Supply Efficiency	Total Instant. Power (W)
Short-range Tactical															
Sleep	2.0	2.0	0.4	1.0	1.5	1.5	0.4	0.4	0.5	0.5	11.0	2.8	1.1	50%	50.1
Standby	2.0	2.0	0.4	1.0	1.5	1.5	0.4	0.4	0.5	0.5	11.0	2.8	1.0	50%	49.9
Transmit	15.0	2.0	0.5	1.0	1.5	1.5	0.4	0.4	0.5	0.5	11.0	2.8	1.0	50%	76.1
Receive	2.0	15.0	0.5	1.0	1.5	1.5	0.4	0.4	0.5	0.5	11.0	2.8	1.0	50%	76.1
Medium-range Tactical															
Sleep	0.2	0.2	0.2	0.3	0.2	0.2	0.1	0.1	0.3	0.3	44.0	11.0	0.5	50%	115.2
Standby	1.5	1.5	0.4	0.6	1.0	1.0	0.3	0.3	0.5	0.5	44.0	11.0	0.5	50%	126.2
Transmit	12.0	1.5	0.6	1.2	1.0	1.0	0.3	0.3	0.5	0.5	44.0	11.0	1.0	50%	149.8
Receive	1.5	12.0	0.6	1.2	1.0	1.0	0.3	0.3	0.5	0.5	44.0	11.0	1.0	50%	149.8
Long-range Tactical															
Sleep	0.1	0.1	0.1	0.2	0.1	0.1	0.1	0.1	0.4	0.4	80.0	20.0	0.1	50%	203.6
Standby	0.5	0.5	0.3	1.0	0.8	0.8	0.2	0.2	0.4	0.4	80.0	20.0	0.5	50%	211.2
Transmit	30.0	0.5	0.5	2.0	0.8	0.8	0.2	0.2	0.4	0.4	80.0	20.0	1.0	50%	273.6
Receive	0.5	30.0	0.5	2.0	0.8	0.8	0.2	0.2	0.4	0.4	80.0	20.0	1.0	50%	273.6
Long-range Tactical Anti-jam															
Sleep	0.1	0.1	0.1	0.2	0.1	0.1	0.1	0.1	0.4	0.4	80.0	20.0	0.1	50%	203.6
Standby	0.5	0.5	0.3	1.0	0.8	0.8	0.2	0.2	0.4	0.4	80.0	20.0	0.5	50%	211.2
Transmit	30.0	0.5	0.5	2.0	0.8	0.8	0.2	0.2	0.4	0.4	80.0	20.0	1.0	50%	273.6
Receive	0.5	60.0	0.5	4.0	0.8	0.8	0.2	0.2	0.4	0.4	80.0	20.0	1.0	50%	337.6

Figure 4.1 shows this architecture. This time scale is long enough for duty cycles, like those defined, to emerge from real-world operations. We used data in Table 4.2 to evaluate the EWC architecture with Eq. (4.3) for a single SDR with four HRMs, one HRM for each of the four general radio types, operating for a week at light, medium, and heavy usage, respectively. Figure 4.2 shows this architecture. Table 4.4 summarizes simulation results.

4.4 Architecture Power Consumption Discussion

The simulation results in Table 4.4 and comparison of MWC and EWC power usage values reveals reduced total energy usage for an EWC architecture regardless of scenario. As one might expect, the degree of power-energy reduction varies depending on radio system type and duty cycle.

Percent reduction is the energy reduction that could be realized by replacing the current MWC architecture with a EWC architecture. These values were calculated using the equation shown in Table 4.4. These values yield a lower and upper power reduction bound of 6% and 31%, depending on the radio type. Average power reduction is the average of the four *percent reduction* values.

Table 4.3 Mode-based instantaneous component power requirements for SDRs operating under an EWC architecture [1, 10–13]

	Enterprise				*Radio Head Module*														
Mode	MOD (W)	DEM (W)	Power Supply Efficiency	Ent. Sub-total Power (W)	High-speed Serial (W)	CON (W)	DSP (W)	DAC (W)	ADC (W)	TXT (W)	RXT (W)	TXF (W)	RXF (W)	TXA (W)	RXA (W)	DCS (W)	Power Supply Efficiency	RHM Sub-total Power (W)	Total Instant. Power (W)
Short-range Tactical																			
Sleep	0.1	0.1	78%	0.3	0.1	0.1	0.1	0.1	0.1	0.1	0.1	0.4	0.4	11.0	2.5	0.1	50%	28.2	28.5
Standby	0.5	0.5	78%	1.3	0.2	0.3	0.5	0.8	0.8	0.2	0.2	0.4	0.4	11.0	2.8	0.5	50%	36.1	37.4
Transmit	10.0	0.5	78%	13.5	0.4	0.5	1.0	0.8	0.8	0.2	0.2	0.4	0.4	11.0	2.8	1.0	50%	38.9	52.4
Receive	0.5	10.0	78%	13.5	0.4	0.5	1.0	0.8	0.8	0.2	0.2	0.4	0.4	11.0	2.8	1.0	50%	38.9	52.4
Medium-range Tactical																			
Sleep	0.1	0.1	78%	0.3	0.1	0.1	0.1	0.1	0.1	0.1	0.1	0.4	0.4	44.0	10.0	0.1	50%	103.0	103.3
Standby	0.5	0.5	78%	1.3	0.2	0.3	0.5	0.8	0.8	0.2	0.2	0.4	0.4	44.0	10.0	0.5	50%	109.0	110.3
Transmit	10.0	0.5	78%	13.6	0.4	0.5	1.0	0.8	0.8	0.2	0.2	0.4	0.4	44.0	10.0	1.0	50%	111.0	124.6
Receive	0.5	10.0	78%	13.6	0.4	0.5	1.0	0.8	0.8	0.2	0.2	0.4	0.4	44.0	10.0	1.0	50%	111.0	124.6
Long-range Tactical																			
Sleep	0.1	0.1	78%	0.3	0.1	0.1	0.2	0.1	0.1	0.1	0.1	0.4	0.4	80.0	20.0	0.1	50%	203.0	203.3
Standby	0.5	0.5	78%	1.3	0.2	0.3	1.0	0.8	0.8	0.2	0.2	0.4	0.4	80.0	20.0	0.5	50%	210.0	211.3
Transmit	30.0	0.5	78%	39.4	0.4	0.5	2.0	0.8	0.8	0.2	0.2	0.4	0.4	80.0	20.0	1.0	50%	213.0	252.4
Receive	0.5	30.0	78%	39.4	0.4	0.5	2.0	0.8	0.8	0.2	0.2	0.4	0.4	80.0	20.0	1.0	50%	213.0	252.4
Long-range Tactical Anti-jam																			
Sleep	0.1	0.1	78%	0.3	0.1	0.1	0.2	0.1	0.1	0.1	0.1	0.4	0.4	80.0	20.0	0.1	50%	203.0	203.3
Standby	0.5	0.5	78%	1.3	0.2	0.3	1.0	0.8	0.8	0.2	0.2	0.4	0.4	80.0	20.0	0.5	50%	210.0	211.3
Transmit	30.0	0.5	78%	39.4	0.4	0.5	2.0	0.8	0.8	0.2	0.2	0.4	0.4	80.0	20.0	1.0	50%	213.0	252.4
Receive	0.5	60.0	78%	78.1	0.4	0.5	4.0	0.8	0.8	0.2	0.2	0.4	0.4	80.0	20.0	1.0	50%	217.0	295.1

Table 4.4 Summary of weekly power consumption simulations for MWC and EWC architectures for expeditionary command and control

	Current MWC Architecture			EWC Architecture				
Mode	Light Usage	Medium Usage	Heavy Usage	Light Usage	Medium Usage	Heavy Usage		
Short-range Tactical								
Sleep	842 W·Hr	0 W·Hr	0 W·Hr	478 W·Hr	0 W·Hr	0 W·Hr		
Standby	2515 W·Hr	1677 W·Hr	0 W·Hr	1884 W·Hr	1256 W·Hr	0 W·Hr	Percent Reduction	
Transmit	3835 W·Hr	5114 W·Hr	6392 W·Hr	2643 W·Hr	3525 W·Hr	4406 W·Hr	(1 - ∑EWC Totals /	
Receive	3835 W·Hr	5114 W·Hr	6392 W·Hr	2643 W·Hr	3525 W·Hr	4406 W·Hr	∑MWC Totals) = 0.31	
Total	11028 W·Hr	11904 W·Hr	12785 W·Hr	7649 W·Hr	8305 W·Hr	8811 W·Hr		
Medium-range Tactical								
Sleep	1935 W·Hr	0 W·Hr	0 W·Hr	1738 W·Hr	0 W·Hr	0 W·Hr		
Standby	6360 W·Hr	4240 W·Hr	0 W·Hr	5538 W·Hr	3692 W·Hr	0 W·Hr	Percent Reduction	
Transmit	7550 W·Hr	10067 W·Hr	12583 W·Hr	6297 W·Hr	8397 W·Hr	10496 W·Hr	(1 - ∑EWC Totals /	
Receive	7550 W·Hr	10067 W·Hr	12583 W·Hr	6297 W·Hr	8397 W·Hr	10496 W·Hr	∑MWC Totals) = 0.16	Average Power
Total	23396 W·Hr	24373 W·Hr	25166 W·Hr	19871 W·Hr	20485 W·Hr	20991 W·Hr		Reduction
Long-range Tactical								Percentage = 0.15
Sleep	3420 W·Hr	0 W·Hr	0 W·Hr	3421 W·Hr	0 W·Hr	0 W·Hr		(15%)
Standby	10644 W·Hr	7096 W·Hr	0 W·Hr	10629 W·Hr	7086 W·Hr	0 W·Hr	Percent Reduction	
Transmit	13789 W·Hr	18386 W·Hr	22982 W·Hr	12739 W·Hr	16985 W·Hr	21231 W·Hr	(1 - ∑EWC Totals /	
Receive	13789 W·Hr	18386 W·Hr	22982 W·Hr	12739 W·Hr	16985 W·Hr	21231 W·Hr	∑MWC Totals) = 0.06	
Total	41644 W·Hr	43868 W·Hr	45965 W·Hr	39528 W·Hr	41056 W·Hr	42463 W·Hr		
Long-range Tactical Anti-jam								
Sleep	3420 W·Hr	0 W·Hr	0 W·Hr	3421 W·Hr	0 W·Hr	0 W·Hr		
Standby	10644 W·Hr	7096 W·Hr	0 W·Hr	10629 W·Hr	7085 W·Hr	0 W·Hr	Percent Reduction	
Transmit	13789 W·Hr	18386 W·Hr	22982 W·Hr	12739 W·Hr	16985 W·Hr	21231 W·Hr	(1 - ∑EWC Totals /	
Receive	17015 W·Hr	22687 W·Hr	28358 W·Hr	14891 W·Hr	19855 W·Hr	24819 W·Hr	∑MWC Totals) = 0.09	
Total	44869 W·Hr	48169 W·Hr	51341 W·Hr	41681 W·Hr	43926 W·Hr	46050 W·Hr		

The simulation estimates an average energy reduction of 15% across the entire simulation matrix. Restricting calculations to the medium usage numbers alone, with the assumption that this is the most probable operational scenario, an 11% power reduction is expected. These results suggest adoption of an expeditionary EWC architecture for SDRs would result in a meaningful reduction in energy requirements for expeditionary nodes without compromising requisite operational command and control capabilities.

4.5 Conclusions and Future Work

This paper presents a straightforward modeling, simulation, and power analysis demonstrating that an expeditionary EWC architecture would reduce operational energy requirements when compared to the current MWC architecture. This gain is, in part, due to the sharing and full utilization of power supplies and processing. In general, the underutilization of processing reduces efficiency [12]. However, modeling processor utilization is complicated due to the randomness of waveform processing occurrence. Additionally, waveform-specific processing specifications, in instructions per second (IPS), were unobtainable and would vary for different types of processors, making it necessary to test actual hardware and implement actual waveforms in software. Nonetheless, the simple approach used for this work likely provides a good first approximation for potential power reduction.

A follow-on proof-of-concept effort is necessary to prove these results with actual hardware. Actual hardware implementation would provide absolute data and determine any added power savings from enterprise processing. Real systems would also demonstrate possible emergent capabilities in network management, dynamic spectrum allocation, coalition interoperability, and electronic warfare. Modifying the current model and performing sensitivity analysis could potentially bind the impact of waveform randomness or IPS differences and inform the hardware test regimen.

The EWC architecture concept is already a reality in the commercial world. The cellular industry is driving future efficiency enhancements of RHMs, which include efficient technologies in power supplies and amplifiers. The push for efficiency in the enterprise computing industry is already in crescendo. As this paper demonstrates, adoption by the military can result in noticeable power savings, which is a valuable benefit to expeditionary operations where fuel must often be transported to support operations.

Acknowledgments This work was sponsored by Naval Postgraduate School mission funds under the joint technical supervision of the Dean of Research and the Space and Naval Warfare Systems Command.

References

1. Arnold, O., Richter, F., Fettweis, G., & Blume, O. (2010). *Power consumption modelling of different base station types in heterogeneous cellular networks*. In IEEE Proceedings of the Future Network and Mobile Summit, Florence, Italy.
2. Conte, A. (2012). *Power consumption of base stations*. In Towards Real Energy-efficient Network Design (TREND) Plenary Meeting, Ghent, Belgium.
3. Hasik, J. (2018). The whole network needs to mesh in wartime. *Atlantic Council*. Retrieved January 5, 2018, from www.atlanticcouncil.org/blogs/defense-industrialist/the-whole-network-needs-to-mesh-in-wartime
4. Fernando, H. Q. (2017). *Power analysis of an enterprise wireless communications architecture*. School (thesis), Naval Postgrad, Monterey, CA.
5. Brannon, B. (2018). Software defined radio. *Analog Devices*. Retrieved January 5, 2018, from https://pdfs.semanticscholar.org/9145/5e9f73b2fcdc1e1c6dc01ef09ea9e1182da4.pdf
6. Reed, J. H. (2002). *Software radio: A modern approach to radio engineering*. Upper Saddle River, NJ: Prentice Hall.
7. Goldworm, B., & Skamarock, A. (2007). *Blade servers and virtualization: Transforming enterprise computing while cutting costs*. Indianapolis, IN: Wiley.
8. DOD Waveform Standards Directorate - Joint Tactical Networking Center. (2016). *Software communications architecture specifications: User's guide*. Ver. 4.1. San Diego: DOD Waveform Standards Directorate - Joint Tactical Networking Center.
9. Xilinx, Inc. (2018). *Xilinx announces integration of 56G PAM4 tranciever technology into Virtex UltraScale + FPGAs*. Retrieved January 5, 2018, from https://www.xilinx.com/news/press/2017/xilinx-announces-integration-of-56g-pam4-transceiver-technology-into-its-virtex-ultrascale-fpgas.html
10. Uhm, M., & Fountain, T. (2017, November 15). Personal interview.
11. Acopian. (2018). Linear power supply vs. switching power supply vs. unregulated power supply [specification sheet]. *Acopian*. Retrieved January 5, 2018, from https://www.acopian.com/linear- power-supply-vs-switching-power-supply-vs-unregulated-power-supply.html
12. Baliga, J., Ayre, R., Hinton, K., & Tucker, R. S. (2011). Energy consumption in wired and wireless access networks. *IEEE Communications Magazine, 49*(6), 70–77.
13. Thomas, D., McPherson, R., & Irvine, R. (2016). *Power analysis of local transmission technologies*. In IEEE Prime, Lisbon, Portugal.

Chapter 5
Grand Challenges in Systems Engineering Education

Steven Hoffenson, Peggy Brouse, Don S. Gelosh, Michael Pafford, Larry D. Strawser, Jon Wade, and Ariela Sofer

5.1 Introduction

This paper describes the methodology and outcomes from a series of three workshops organized by the International Council of Systems Engineering (INCOSE) Academic Council. Distinguished systems engineering (SE) researchers and professionals from across the United States were invited to identify research gaps and provide recommendations on closing those gaps in the area of SE education. The paper begins by describing ongoing education efforts by the INCOSE, the concept of global Grand Challenges, and the resultant workshop series. The paper goes on to describe how Workshop I defined the problem, Workshop II defined the research gaps, and then Workshop III developed the necessary research directions. The paper concludes by describing the recommended research threads to meet the grand challenges in systems engineering education.

5.1.1 Grand Challenges in Systems Engineering

The concept of global Grand Challenges can be used to focus technical, educational, and humanitarian efforts [1–6]. In 2014, the INCOSE developed a Systems

S. Hoffenson (✉) · J. Wade
Stevens Institute of Technology, Hoboken, NJ, USA
e-mail: shoffens@stevens.edu

P. Brouse · A. Sofer
George Mason University, Fairfax, VA, USA

D. S. Gelosh
Worcester Polytechnic Institute, Worcester, MA, USA

M. Pafford · L. D. Strawser
Johns Hopkins University, Baltimore, MD, USA

S. Adams et al. (eds.), *Systems Engineering in Context*,
https://doi.org/10.1007/978-3-030-00114-8_5

Engineering Vision 2025 [7], which coupled societal needs to systems challenges and then to gaps in SE capabilities. Subsequently, the "INCOSE Systems Engineering Research Future" project [8] was initiated by the INCOSE Academic Council, which used the framework of the SE Vision 2025. This project aimed to achieve the following objectives:

- Build communities among academia, industry, and government from numerous domains, dedicated to tackling the major global systems challenges confronting humankind for societal good.
- Excite, inspire, and guide systems engineering research in these communities.
- Achieve consensus among these communities to establish the priorities of systems engineering research.
- Provide the means by which to create synergy in these systems engineering research efforts such that progress can be measured against these objectives.

5.1.2 INCOSE Grand Challenge Workshop Series

This project produced a series of three workshops (October 2016, March 2017, and October 2017) with participants comprised of SE researchers and professionals from academia, industry, and government. The participants in the first workshop used the list of societal needs in the INCOSE SE Vision 2025 to select the following four Grand Challenges to be explored by the project: (1) access to education, (2) access to healthcare, (3) access to clean water, and (4) security and safety. The first workshop also adopted the five desired characteristics of societal needs Grand Challenges first proposed by the National Council of Teachers of Mathematics and added a sixth as follows:

1. Represent complex and extremely difficult questions that are solvable (potentially within 10–20 years).
2. Improve quality of life through positive educational, social, and economic outcomes potentially affecting millions of people.
3. Involve multiple research projects across many subdisciplines to be satisfactorily addressed.
4. Require measurable outcomes so that progress and completion can be identified.
5. Compel popular support by encouraging the public to relate to, understand, and appreciate the outcomes of the effort.
6. Be systemic in nature.

5.1.3 Systems Engineering Education

The INCOSE Academic Council held another, earlier series of workshops in 2015 [9] and 2016 [10], with the aim to enhance the ability of engineering education to meet future societal needs by infusing SE knowledge and skills into general engineering education. These workshops were attended by participants from academia, government, and industry, and the following questions were addressed:

- Why: Describe the value of SE to all engineers in a language which other disciplines can understand and accept.
- What: Identify the key areas of SE and systems thinking that is needed at every level of undergraduate engineering education, to identify existing gaps in that knowledge and how to fill them.
- How: Identify useful approaches to infuse the knowledge into engineering education and even possibly begin to create practical materials which other faculty could use.

The progress made in these workshops resulted in well-attended pilot workshops at the 2017 meeting of the American Society for Engineering Education (ASEE). These focused on integrating SE foundations in first-year engineering undergraduate courses as well as the capstone experience [11].

5.2 Problem Definition: Workshop I

The first workshop in the Grand Challenges series set the stage for the four focus areas by selecting Grand Challenges from the SE Vision 2025 based on participant interest and identifying obstacles and relevant research questions. Access to education was one of the four challenges selected, along with food and clean water, access to healthcare, and security and safety. The results of this workshop are well documented [8], and the education challenge is summarized here.

The education Grand Challenge working group was established with the purpose to "reform education systems to address gaps in systems skills in individuals." Here, systems skills refer to a set of knowledge, skills, and abilities (KSAs) that align with the INCOSE competency framework. A road map for addressing this Grand Challenge was defined, illustrated in Fig. 5.1, which begins with the identification of the

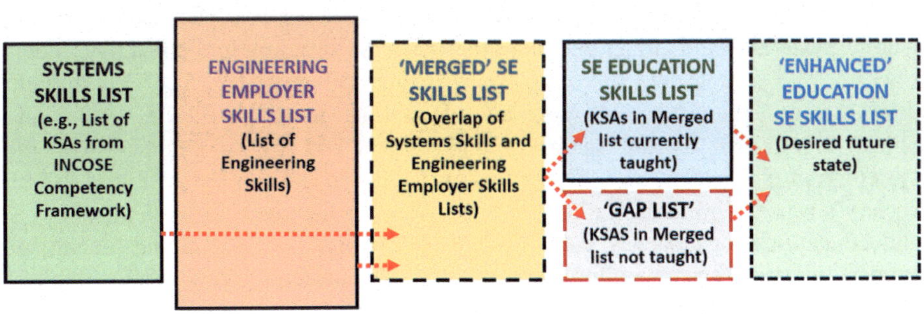

Fig. 5.1 Illustration of process to identify KSA gaps and reach an improved future state of systems engineering (SE) education

needed KSAs from industry, understanding the gaps between those KSAs and those of graduates and professionals, and developing a plan for what needs to be done to close those gaps. The endpoint, or desired future state of SE and education, involves graduates and professionals with improved systems skills that add collaborative, long-term value to society.

After defining the problem and road map, a number of problems with the current state were identified, considering the origins of the current educational systems, changing technology and industrial landscapes, the increasing complexity of society as a system of systems, and resistance to change at all levels. One of the key obstacles to improving SE education is that society as a whole does not share a common understanding of systems and therefore may not recognize the value of improving SE skills. A number of research questions were then created that could address these problems and obstacles.

5.3 Research Gaps: Workshop II

The second workshop began by distilling the list of research questions from the previous workshop down to five that were perceived to be the most critical. A corresponding set of necessary SE capabilities was then defined, and the current state of the art was discussed. Finally, the research needs, or SE gaps, in these five areas were discussed. A summary of the discussions and conclusions is provided in Table 5.1.

5.4 Research Directions: Workshop III

5.4.1 Synthesis of Research Needs

To organize these research gaps and identify key stakeholders and their relationships to the gaps, a systemigram was developed during the third workshop of this series, shown in Fig. 5.2. The purpose of systemigrams, which is a portmanteau of the words "systemic" and "diagram," is to help provide structure to a collective understanding of a system and to communicate strategic intent [12]. The systemigram development began by identifying an overarching narrative of the system and objectives, known as the mainstay and represented by the elements of the systemigram shaded in gray. The mainstay begins in the top left of Fig. 5.2 with society's value of systems thinking, which motivates the needs for systems thinking and educational approaches that teach systems thinking. Educational techniques support improved systems thinking in education, resulting in more systems thinkers in the world, engineering profession, and SE domain. This in turn supports higher levels of systems thinking in society, which is expected to lead to better problem understanding and decision-making across a broad range of societal challenges.

Table 5.1 Key research questions and gaps defined during Workshop II for the education grand challenge

Research questions	Necessary SE capabilities	Current state of the art	SE gaps
What are the necessary KSAs for the future? How are these affected by culture, economic trends, and technological change?	Forecasting demands on the future workforce; KSAs to support these demands in the face of culture, economic, and technical change	A nascent set of modeling and simulation capabilities being developed in socio-technical areas	Abilities to model long-term socio-technical systems and to address culture and value-based decisions; an ontology for use by SE for cultural issues
What are the most effective curriculum and pedagogy to support these objectives?	Assessing educational outcomes and determining how curriculum and pedagogy affect these	Systems thinking (ST) is in some curricula, but limited work has analyzed education using ST	Assessing human outcomes, in education and otherwise; correlating human actions with developmental outcomes
How do you create a regenerative/evolving system in the face of the change-resistant status quo?	Abilities to create self-adaptive, evolving, self-sustaining systems that can cope with a range of unforeseen challenges	Only preliminary work in the area of the modeling of adaptive systems and the use of influence to guide emergence in systems	Developing self-adaptive, evolving, self-sustaining systems; monitoring system behaviors and providing the necessary modifications
How do we introduce and support creativity in SE and engineering?	Taking the human element into account in SE and engineering; cultivating creativity throughout the life cycle	SE is beginning to embrace human aspects, both the creators of the system and those impacted by it; design thinking addresses these issues	Understanding human elements both in the system development and the deployed system; balancing creativity and rigor
How do we cultivate critical interpersonal, social, and collaborative skills?	Cultivating the interpersonal capabilities needed for collaborative work	Little has been done formally in SE; a number of books have been written, particularly on teamwork in engineering	Abilities to determine how to effectively support interpersonal capabilities in collaborative work

To the lower left of the mainstay are the identified challenges related to systems thinking and systems modeling capabilities. These include the ability to cope with dynamic and unpredictable systems, including in particular those that involve humans, and they also include interpersonal and interdisciplinary skills that systems engineers need to properly bring together diverse groups of people that are working on the same problem. These capabilities can support more accurate and meaningful socio-technical system models, which can enable improved assessments of the value of systems thinking and therefore provide evidence on the value of systems thinking to society, which feeds back into the beginning of the mainstay. The other side of the systemigram, in the upper right of Fig. 5.2, illustrates how different stakeholders

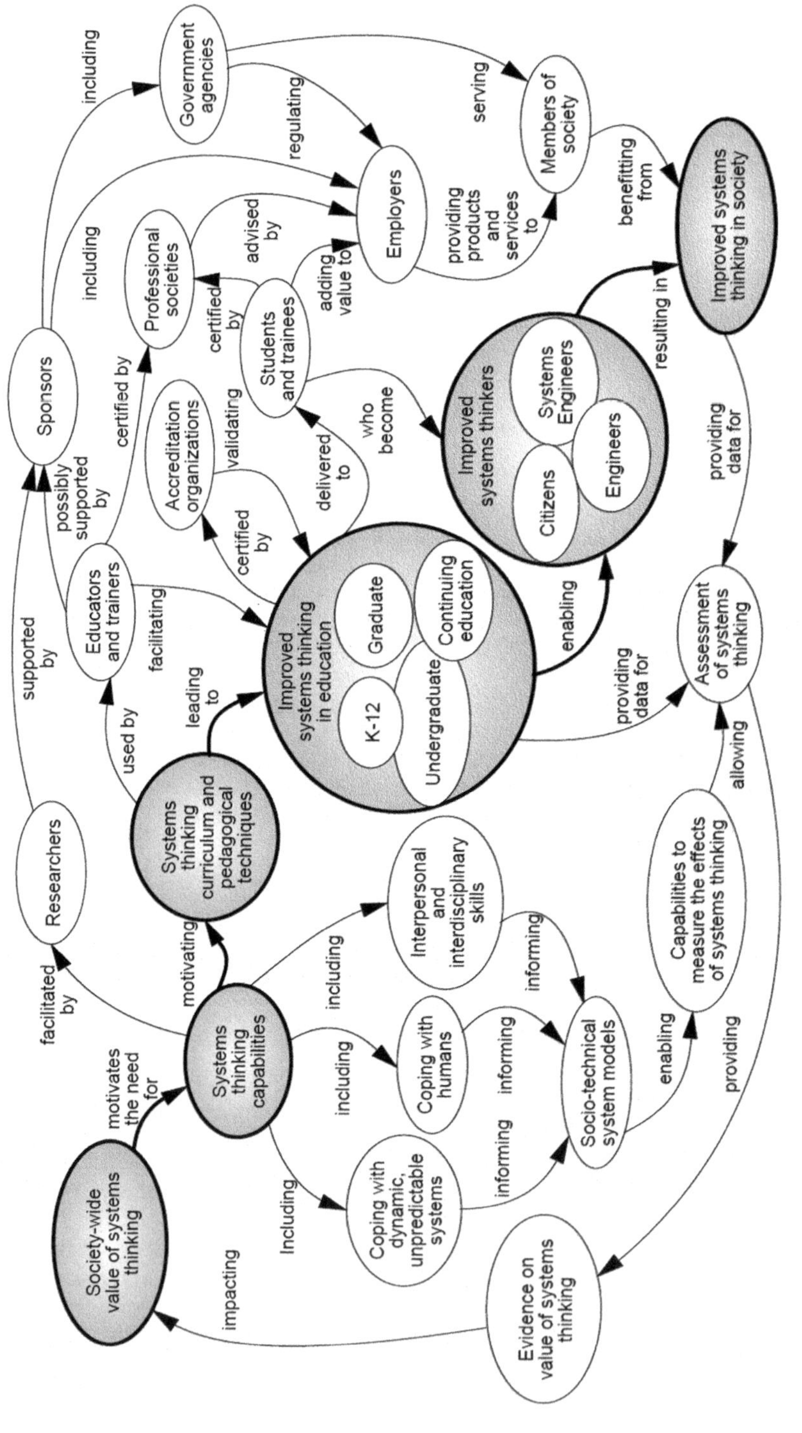

Fig. 5.2 Systemigram illustrating the education Grand Challenges and key stakeholders

interact with one another to support the mainstay. These include practitioners, researchers, and organizations, detailed further in the next subsection.

5.4.2 Stakeholders and External Organizations

The external organizations that are key stakeholders and/or potential collaborators are shown on the upper right of Fig. 5.2. When developing the systemigram, the node of educators and trainers was a natural first response to the question: "Who are the key stakeholders?" This group of stakeholders facilitates the exchange of knowledge from those who have the knowledge to those who need it. This led to considering students and trainees—those who need the knowledge—as another group of stakeholders. Then came the organizations who employ the students and trainees: commercial industries such as defense, communications, power, and information technology. Additionally, organizations that accredit engineering programs, such as the Engineering Accreditation Commission (EAC) of ABET, play a key role in shaping educational curricula.

Professional societies are also key stakeholders due to their impact on advancing the state of the art in engineering and engineering education. Examples of these organizations include the American Society for Engineering Education (ASEE), International Council on Systems Engineering (INCOSE), and Institute of Electrical and Electronics Engineers (IEEE) representing engineering and the American Management Association (AMA) and Project Management Institute (PMI) representing nonengineering disciplines.

Another important group of key stakeholders include government agencies who develop and publish policy and guidance and support innovation in engineering and education. Examples of these stakeholders include the US Department of Education, the National Institute of Standards and Technology, and the National Science Foundation. Research centers are also key stakeholders; some examples are University Affiliated Research Centers (UARCs) such as the Johns Hopkins University Applied Physics Laboratory and the Systems Engineering Research Center hosted by the Stevens Institute of Technology and involving more than 20 universities. Federally funded research and development centers (FFRDCs) such as the MITRE Corporation and the Rand Corporation, as well as national laboratories, are also key research stakeholders.

The last two groups of key stakeholders are those who sponsor and fund the initiatives and research being conducted and members of society who benefit in general from the advancement of the state of the art in engineering education.

5.4.3 Preliminary Research Threads

Reviewing the research needs and stakeholders resulted in the identification of several research threads and corresponding topics. Some research has been completed in several of these areas, which will briefly be reviewed. The focus of this section is on researching ways to efficiently and effectively collect, process, analyze, share, and report actionable data on skill sets for engineering educators, students, trainers, and practitioners. This data collection research need has been discussed several times, including at the Academic Forum at Worcester Polytechnic Institute in May 2015 [9]. The following subsections describe the six identified research threads related to SE education.

5.4.3.1 Study the Intersection of SE and Human Factors

Academic programs currently exist that combine SE, design, and human factors. For example, a BS in Human Systems Engineering is offered at Arizona State University that "focuses on a field that combines engineering and psychology to design systems consistent with human capabilities and limitations. In other words, making technology that works for humans" [13].

There is, however, very little published in this area. Notable efforts by Gabriele [14] address this topic at K-12 levels using experimental activities. These findings include "busy classroom teachers really need sets of student-ready materials, which they can cover together with their students. . .papers and other workshop results need to be converted by text-book publishers to ideas and activities suitable not only across disciplines, but at multiple hierarchical levels grades 1–12. Theoreticians, publishers, and teachers all have different purposes and needs."

To advance this area, several research gaps were identified in the workshops including the ability to understand the human element both in developing the system and the deployed system, providing the proper balance between creativity and rigor, exploration and exploitation in system conception, design, and sustainment. Little has been done formally in SE in this area. There are a number of books that have been written on the subject, particularly in the area of teamwork in engineering. Another research area is the ability to determine how to effectively support interpersonal capabilities necessary for successful collaborative work, considering cultural and personal differences, and translate this into both professional and educational contexts.

5.4.3.2 Study the Role of SE in Adaptive Systems

Complex and adaptive systems are expected to see substantial growth in the near future, and SE is expected to be a critical component. A number of research thrusts along this thread are underway, including the Complex Adaptive Systems

Engineering (CASE) methodology developed by MITRE [15] and exploring complexity in Complex Adaptive Systems of Systems [16]. Recently, the topic of "Complex Adaptive Systems Science meets Systems Engineering" was discussed at the World Congress on Industrial Automation [17].

In addition, other research has been done by the NASA Systems Engineering Research Consortium [18]. This consortium was established at the end of 2010 "with a broad research question to enable several subordinate research projects to address different aspects of systems engineering. The primary research question can be stated as: "What are the fundamental engineering and organizational relationships of systems engineering?"

Only preliminary work in the area of the modeling of adaptive systems and the use of influence to guide emergence in systems has been pursued. A gap identified in the workshop is the ability to create a self-adaptive, evolving, self-sustaining system to cope with a range of unforeseen challenges. Also needed is the ability to monitor the behavior of these systems and provide the necessary modifications to direct them to successful behaviors.

5.4.3.3 Map KSAs with Accreditation and Learning Outcomes

This area is of particular need to educators. There are distinct sets of SE knowledge, skills, and abilities or KSAs (such as those from the INCOSE, the UK Ministry of Defence, and the Systems Engineering Professional (SEP) Certification) that aim to meet industry needs, which may or may not be well-aligned with accreditation organizations and curriculum development guidelines (such as the EAC of ABET, Systems Engineering Body of Knowledge (SEBoK), and Graduate Reference Curriculum for Systems Engineering (GRCSE) [19]). There is a need to map the set of KSAs needed to effectively enable, encourage, and assess the initial education of engineering concepts, ideas, theories, definitions, specifications, standards, or rules, with the set of KSAs needed to effectively enable, encourage, and assess continuous professional training of engineering principles which are, for the most part, learned in context through engineering practice and experience.

Little work has been undertaken in mapping the KSAs. In the workshops, a need was identified to research the feasibility of mapping SE practitioner KSA sets (mainly based on the INCOSE/UK Ministry of Defence Competency List, INCOSE SEP Certification Criteria, and Open Source Industry SE-related Job Qualification Lists) with engineering student curricula (mainly based on EAC of ABET and SEBoK GRCSE criteria). To facilitate this mapping, several gaps were identified, including the need to determine the KSAs most desired by SE employers. This could be done by surveying INCOSE member groups or by mining job applications.

5.4.3.4 Study the Value of Systems Thinking in Other Fields

There was considerable interest in all three workshops as to the role of systems thinking in systems development. Several research areas have been pursued along this thread, including Kasser's work on holistic thinking [20], a review of literature addressing systems thinking by Monat and Gannon [21], and Wise's book section on "Systems Engineers as the Engineers of Tomorrow" [22].

To facilitate this thread, several gaps were identified including ongoing research in introducing systems thinking to students, educators, and trainers by the INCOSE Systems Science Working Group, as well as the benefit of systems thinking in other disciplines. For example, a Systems Engineering Principles course is being taught in the second year of George Mason University's BS in Cyber Security Engineering program [23].

5.4.3.5 Study the Role of Nature Versus Nurture Regarding Systems Thinking

As a continuation of the previous thread, the question arose regarding whether systems thinking is an innate characteristic (nature) or whether it can be learned (nurture). David Long has addressed this in "Nature or Nurture? Expanding the Systems Engineering Population" [24]. Systems thinking has been added to educational curricula [25], but limited work has been done in the analysis of education with systems thinking approaches. A gap identified in this thread is the capability in measurement and assessment of human outcomes, in education and otherwise, or the ability to correlate human actions with developmental outcomes. Also, capabilities in the areas of cultural assessments and understanding are not well-understood.

5.4.3.6 Study the Role of SE in Adaptive Systems

There is a nascent set of modeling and simulation capabilities being developed in socio-technical areas. Richey et al. addressed a complex socio-technical systems approach to educational policies for the future workforce [26]. The gaps identified are the abilities to model long-term socio-technical systems, to address culture and value-based decisions, and to develop an ontology for use by SE in the areas of culture and related issues.

5.5 Discussion

5.5.1 *Key Ideas and Directions Forward*

This paper pursued the direction given in a series of three workshops organized by the INCOSE Academic Council to support the INCOSE concept of global Grand Challenges. Research questions and gaps were first identified, then a systemigram was created to structure the problem and present a road map for the Grand Challenge, and finally key research threads were recommended along with a road map for moving forward. These key research threads were then placed in the context of the existing literature to more specifically define the current gaps and suggested topics for research.

5.5.2 *Common Directions Across Grand Challenge Areas*

As was discussed in Sect. 5.1.2, in addition to the education Grand Challenge, three other Grand Challenges were explored by the Academic Council: clean water, healthcare, and security and safety. In similar fashion to the education Grand Challenge, the key problems and research threads for each of these were identified, followed by identification of current gaps in SE research and suggested topics for future SE research. Although each of these Grand Challenges identified specific topics for SE research relevant to their specific challenge, there were two major SE research needs common to all four Grand Challenges:

- SE research is needed to support the development of self-adaptive, evolving, and self-sustaining systems.
- SE-driven research of human factors is needed to better understand long-term socio-technical systems and address culture- and value-based decisions.

Acknowledgments The authors would like to acknowledge the support from the INCOSE Academic Council for the three Grand Challenges in engineering workshops, as well as the contributions from the participants of the workshops.

References

1. Stephan, M. L., Fish, M. C., Chval, K. B., Herbel-Eisenmann, B., Wanko, J. J., Konold, C., et al. (2015). Grand challenges and opportunities in mathematics education research. *Journal for Research in Mathematics Education, 46*(2), 134–146.
2. Varmus, H., Klausner, R., Zerhouni, E., Acharya, T., Daar, A. S., & Singer, P. A. (2003). Grand challenges in global health. *Science, 302*(5644), 398–399.
3. National Academy of Engineering. (2008). *Grand challenges for engineering* [Internet]. [Cited 18 Jan 2018]. Available from http://www.engineeringchallenges.org

4. Uehara, E., Flynn, M., Fong, R., Brekke, J., Barth, R. P., Coulton, C., et al. (2013). Grand challenges for social work. *Journal of the Society for Social Work and Research, 4*(3), 165–170.
5. U.S. Agency for International Development. (2011). *Grand challenges for development.* http://www.usaid.gov
6. The Institution of Engineering and Technology. (2013). *Global grand challenges.* http://www.raeng.org.uk/publications/other/ggcs-report
7. Friedenthal, S., Beihoff, B., Nichols, D., Oster, C., Paredis, C., Stoewer, H., et al. (2014). *A world in motion: Systems engineering vision 2025.* San Diego, CA: International Council on Systems Engineering.
8. Wade, J., Adcock, R., McDermott, T., & Strawser, L. (2018). Future systems engineering research directions. In A. M. Madni, B. Boehm, et al. (Eds.), *Disciplinary convergence in systems engineering research* (pp. 1165–1179). New York: Springer.
9. International Council on Systems Engineering. (2016). *INCOSE Spring Academic Forum* [Internet]. [Cited 18 Jan 2018]. Available from https://web.wpi.edu/research/seli/incose61.html
10. International Council on Systems Engineering. (n.d.). [Internet]. [Cited 18 Jan 2018]. Available from http://washingtondc.eventful.com/events/incose-academic-forum-2016-/E0-001-091901388-6@2016050207
11. Squires, A. F., Looft, F. J., Fortney, W. B., Simoni, M., & Robinson-Bryant, F. (2017). *Wednesday workshop: Part 1—support for faculty on how to integrate an introduction to systems engineering foundations for first year engineering students; part 2—best practices for integrating systems engineering into the capstone experience.* In ASEE Annual Conference & Exposition. Available from https://www.asee.org/public/conferences/78/registration/view_session?session_id=8014
12. Boardman, J., & Sauser, B. (2008). *Systems thinking: Coping with 21st century problems.* Boca Raton, FL: CRC Press.
13. Arizona State University. (n.d.). *Human systems engineering* [Internet]. [Cited 18 Jan 2018]. Available from https://explore.engineering.asu.edu/undergraduate-degrees/human-systems
14. Gabriele, S. F. (2018). Agency and causal factors in social system behavior: Advancing human systems engineering with general system theory. In A. M. Madni, B. Boehm, et al. (Eds.), *Disciplinary convergence in systems engineering research* (pp. 713–726). New York: Springer.
15. White, B. E. (2016). *A complex adaptive systems engineering (CASE) methodology—the ten-year update.* In: Annual IEEE Systems Conference (SysCon).
16. Johnson, B., & Hernandez, A. (2016). Exploring engineered complex adaptive systems of systems. *Procedia Computer Science, 95*, 58–65.
17. Doyle, M. J., Watz, E., & Portrey, A. M. (2015). Merging worlds: Complex adaptive systems science meets systems engineering: a foundation for complex adaptive agent-based modeling architectures. In *Proceedings of the 48th Annual Simulation Symposium* (pp. 86–93). Society for Computer Simulation International.
18. Watson, M. D., & Farrington, P. A. (2016). *NASA systems engineering research consortium: Defining the path to elegance in systems.* In: Conference on Systems Engineering Research.
19. Pyster, A., Olwell, D. H., Ferris, T. L. J., Hutchison, N., Enck, S., Anthony, J., Henry, D., & Squires, A. (Eds.). (2015). *Graduate Reference Curriculum for Systems Engineering (GRCSE™) v1.1.* Trustees of the Stevens Institute of Technology. Available from www.bkcase.org/grcse
20. Kasser, J. E. (2010). *Holistic thinking and how it can produce innovative solutions to difficult problems.* In Proceedings of the 7th Bi-annual European Systems Engineering Conference (EuSEC). Stockholm, Sweden.
21. Monat, J. P., & Gannon, T. F. (2015). What is systems thinking? A review of selected literature plus recommendations. *American Journal of Systems Science, 4*(1), 11–26.
22. Wise, C. K. (2010). Engineers of tomorrow: Holistic-thinking system engineers. In *Holistic Engineering Education* (pp. 227–241). New York: Springer.

23. George Mason University. (n.d.). *Cyber Security Engineering, BS* [Internet]. [Cited 19 Jan 2018]. Available from https://catalog.gmu.edu/colleges-schools/engineering/cyber-security-engineering-bs/
24. Long, D. (2015). *Nature or nurture? Expanding the systems engineering population* [Internet]. [Cited 18 Jan 2018]. Available from http://community.vitechcorp.com/index.php/nature-or-nurture-expanding-the-systems-engineering-population.aspx
25. Lannon, C. (n.d.). *Revitalizing the schools: A systems thinking approach* [Internet]. [Cited 18 Jan 2018]. Available from https://thesystemsthinker.com/revitalizing-the-schools-a-systems-thinking-approach/
26. Richey, M., Nance, M., Hanneman, L., Hubbard, W., Madni, A. M., & Spraragen, M. (2014). A complex sociotechnical systems approach to provisioning educational policies for future workforce. *Procedia Computer Science, 28*, 857–864.

Chapter 6
The Joint Optimization of Critical Interdependent Infrastructure of an Electricity-Water-Gas System

Jie Cheng, Qishuai Liu, Qing Hui, and Fred Choobineh

Nomenclature

Index and Constant

a_n,b_n	Parameters for pricewise linearization in kW and \$, respectively
c_1,c_2,c_3	Cost coefficients for quadratic electric power curve in \$/kW^2, \$/kW, and \$, respectively
h_g	Power coefficient for gas load flow in kW/(m^3/h)
h_w	Power coefficient for water load flow in kW/(m^3/h)
N^s	Number of points for piecewise linearization
N^T	Number of time steps
p_p^{ref}	Nominal pipe pressure reference in Pa
r_e'	Pseudo electricity rate in \$/kWh
r_e	Finalized electricity rate in \$/kWh
r_g	O&M cost coefficient for gas storage system in \$/m^3
r_p	O&M cost coefficient for pipe system in \$/Pa
r_s	Coefficient of per unit cost of gas supply in \$/unit
r_w	O&M cost coefficient for water system in \$/m^3
S_p^{ref}	Nominal pipe storage reference in m^3

J. Cheng
University of Nebraska-Lincoln, Lincoln, NE, USA

CalEnergy Operating Corporation, Calipatria, CA, USA

Q. Liu · Q. Hui (✉) · F. Choobineh
University of Nebraska-Lincoln, Lincoln, NE, USA
e-mail: qing.hui@unl.edu

S. Adams et al. (eds.), *Systems Engineering in Context*,
https://doi.org/10.1007/978-3-030-00114-8_6

t	Time index
T	Time step constant in hour
V_g	Volume of one unit gas in gas transportation in m^3/unit

Variables

$\lambda_n(t)$	Ancillary variable for pricewise linearization
$L_r(t)$	Residential electricity load at time t in kW
$L_g(t)$	Gas load at time t in m^3/h
$L_w(t)$	Water load at time t in m^3/h
$m(t)$	Gas transportation decision variable at time t in per unit
$p_p(t)$	Pipe pressure status at time t in Pa
$P_e(t)$	Residential electric load (non-infrastructure electric load) at time t in kW
$P_g(t)$	Gas system electric load at time t in kW
$P_w(t)$	Water system electric load at time t in kW
$Q_g(t)$	Gas flow rate at time t in m^3/h
$Q_w(t)$	Water flow rate at time t in m^3/h
$S_g(t)$	Gas tank storage status at time t in m^3
$S_p(t)$	Gas pipe storage status at time t in m^3
$S_w(t)$	Water storage status at time t in m^3

6.1 Introduction

The national critical infrastructures are those that provide the essential services and serve as the backbone of the nation's economy, security, and health. The proper functioning and coordination of these infrastructures is essential for the nation's economic development and security since these infrastructures cater to the basic needs of the population.

The electricity, water, and gas (EWG) systems are among the most critical infrastructures of a community, and traditionally each has been planned, designed, and operated isolated from each other. Most studies and optimization of these systems have been done individually and within the scope of a homogeneous system.

The homogeneous system optimization can be found in the following literatures. The optimal power flow is optimized by the distributed algorithms based on the multi-agent system [1]. The models and methods of optimal control of water distribution systems is summarized in [2]. The optimal operation of water distribution is studied in [3], considering the water shortage and the quality constraints. A stochastic optimal control frame for natural gas network operation is proposed in [4]. Genetic algorithm and model predictive control are employed to solve the optimal operation of pipelines in [5, 6], respectively.

However, the EWG systems are functionally and economically linked. For example, water and gas systems use electrical energy for its operation, and electrical system may rely on water for its cooling and gas as a source of fuel. Moreover, the EWG systems not only share some common features, but they also share some complementary characteristics. The power system is a real-time balanced system, where at any given time, the generation is equated with the load and the system does not have the benefit of a storage in the transient form. The water system is an asynchronous system, where the water storage that provides water pressure by height naturally provides a storage buffer for the gap between the water production and the water load. The natural gas is compressible in the pipelines. It forms a big inertia system to absorb supply fluctuation within a certain range. Thus, it is insufficient to only consider one system within the scope of optimization, since these infrastructures are interdependent.

Some bilateral interdependent system studies can be found in [7–9]. However, trilateral interdependent systems are rarely studied.

This paper emphasizes on the energy linkage and load balance among/between the three systems. The power, water, gas system will be modeled and optimized separately first as a benchmark. Then, the joint economic operation of an EWG system as a whole system is studied. Finally, the economic performance is compared.

6.2 Critical Infrastructure

The homeland security defines the critical infrastructure as the infrastructure that provides the essential services underpinning the American society and serves as the backbone of our nation's economy, security, and health [10]. The electricity, gas, oil, water, transportation, and communication systems are the common critical infrastructures that are associated with and have fundamentally shaped our daily lives. Other less tangible critical infrastructures include the public health, food safety, financial services, and security services (police, military).

We selected the EWG system because they share some common features, such as the natural monopoly in one area, non-sufficient competition, major energy consumers, critical element to the residents, continuous supply in physical networks, simultaneous load balance, and no storage in the end users. The power, water, and gas distribution systems are introduced as follows, respectively.

6.2.1 Power System

The power system includes the power generation, transmission and distribution, and the load. The power is generated by various generation resources, and transmitted through the transmission and distribution networks, and reaches to the end users. The electricity is a real-time commodity, where the generation and consumption happen simultaneously.

In this paper, the aggregated model is used to represent generation and load. The generation from different sources and the load from different end users are aggregated as one bulk generator and on bulk load. Thus, a single generator and a single load can represent the behavior of the power supply and power consumption pattern in the represented area.

6.2.2 Water System

The water system, as shown in Fig. 6.1a, includes the water sources/reservoir, treatment plant, pump station, storage tank/tower, distribution network, and commercial and residential loads. The water level in the water tower has to be maintained in a certain range in order for the outlet water pressure be maintained in the required

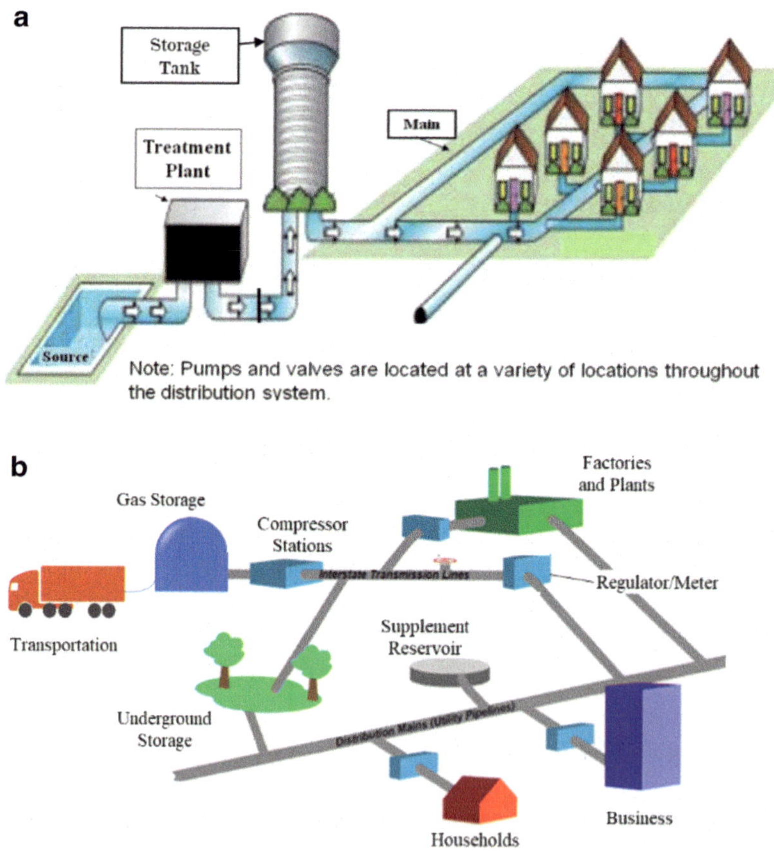

Fig. 6.1 Layout of (**a**) water system network and (**b**) gas system network [11, 12]

pressure range for the customers. The single provider and one aggregated load model are used in this system.

6.2.3 Gas System

The gas system, as shown in Fig. 6.1b, includes the gas transportation, tank storage, compression station, distribution pipelines, and commercial and residential end users.

Because the gas is compressible, the pipelines contain a certain amount of natural gas that can be considered as another type of storage. Gas pressure has to be maintained within the allowable range for the satisfaction of end users. The single provider and one aggregated load model are used in this system.

6.3 Separate and Joint Optimizations

In order to analyze the performance of different operation strategies of these systems, two cases are considered for simulation.

Case 1. Independent System Operation This case assumes that the three systems are operated independently. The system models are constructed and optimized, separately. The objective of the optimization is to minimize the respective system cost consisting of electricity-related energy cost and operation and maintenance (O&M) costs. The scheme is described in Fig. 6.2a.

In this case, a pseudo electricity rate is assumed to facilitate the calculation of water and gas system electric load since the power system cannot be optimized before the optimal water and gas loads are obtained. Then, water and gas system electric load (infrastructure electric load) and the residential electric load (non-infrastructure electric load) are aggregated in each hour and form a load curve. The total electric cost (infrastructure + non-infrastructure) is calculated through a quadratic polynomial cost curve. The total electric cost is divided by the total amount of the electricity to obtain a finalized flat power rate. Finally, a flat electricity rate is finalized and used to substitute the pseudo electricity rate to recalculate the real electricity cost for three systems. The order of optimization is organized as the water, gas, and power systems as described in Sects. 6.3.1, 6.3.2, and 6.3.3, respectively.

Case 2. Interdependent Optimization This case inherits the same balance functions and capacity constraints constructed in Case 1. The only difference is that the objective function will not be minimized until all the constraints are formulated and listed. The scheme is described in Fig. 6.2b.

Then, the problem is optimized with the programming method, and the final results give the breakdowns of the optimal cost values and the corresponding

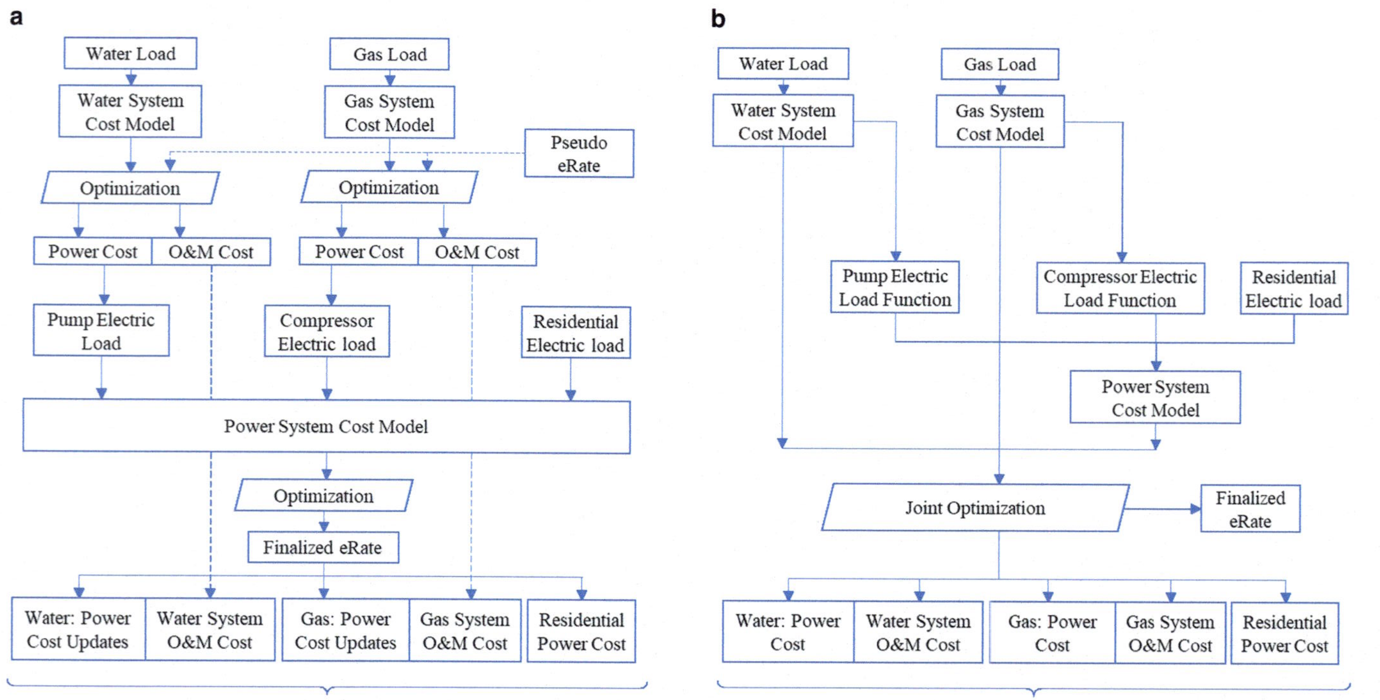

Fig. 6.2 Total system cost of Electricity-Water-Gas system. Flowcharts of (**a**) independent system operation and (**b**) interdependent optimization

operation strategies. The finalized electricity rate will be calculated as a by-product. This case is explained in Sect. 6.3.4.

6.3.1 Water Balance Optimization

The water storage balance is described in (6.1), where the storage status for the next time step is the existing water balance, plus flow-in, minus flow-out in the same period. The electric power for the corresponding flow rate is depicted in (6.2). In order to simulate the sustainable cycle of water balance, the final state of the storage should be equal to the initial state of the storage. The storage capacity and the flow capacity should meet their respective limits. In the water system optimization, the known variable is the water load for N^T time steps, and the decision variable is the water flow into the storage at time t. The objective function consists of the electric cost and the O&M cost as shown in (6.4). The problem can be solved by the linear programming method.

$$S_w(t) = S_w(t-1) + Q_w(t) \cdot T - L_w(t) \cdot T \tag{6.1}$$

$$P_w(t) = h_w \cdot Q_w(t) \tag{6.2}$$

$$t \in \{0, 1, 2, 3, \ldots\ldots, N^T\} \tag{6.3}$$

$$\text{Min } Z_w = \sum_{i=1}^{N^T} r_e' \cdot h_w \cdot Q_w(t) + \sum_{i=1}^{N^T} r_w \cdot S_w(t) \tag{6.4}$$

6.3.2 Gas Balance Optimization

The gas storage balance and pipe storage balance are described in (6.5) and (6.6), respectively. The electric power for the corresponding gas flow rate is depicted in (6.7). The relationship between the pipe gas pressure and pipe storage status is depicted in (6.8). It is noted that the O&M cost for the pipe is proportional to the gas pressure in the pipe, which is an indirect relationship to the status of pipe storage.

In the gas system optimization, the known variable is the gas load vector. The decision variables are the transportation decision variable and the gas flow vector (time series). It is noted that only the integer number of gas transportation is allowed in this study.

The objective function consists of the power cost, the gas transportation cost, and the O&M costs for both storage and the pipe, as shown in (6.10). The problem can be solved by the mixed integer linear programming method.

$$S_g(t) = S_g(t-1) + m(t) \cdot V_g \cdot T - Q_g(t) \cdot T \tag{6.5}$$

$$S_p(t) = S_p(t-1) + Q_g(t) \cdot T - L_g(t) \cdot T \tag{6.6}$$

$$P_g(t) = h_g \cdot Q_g(t) \tag{6.7}$$

$$p_p(t) = S_p(t) \frac{p_p^{\text{ref}}}{S_p^{\text{ref}}} \tag{6.8}$$

$$m(t) \in \text{non-negative integer} \tag{6.9}$$

$$\text{Min } Z_g = \sum_{i=1}^{N^T} r_e' \cdot h_g \cdot Q_g(t) + \sum_{i=1}^{N^T} r_s \cdot m(t) + \sum_{i=1}^{N^T} r_g \cdot S_g(t) + \sum_{i=1}^{N^T} r_p \cdot p_p(t) \tag{6.10}$$

6.3.3 Power Balance

The aggregated electric demand is depicted in (6.11), consisting of water electric load, gas electric load, and the residential electric load (non-infrastructure electric load). The generation capacity constraint is described in (6.12). The single representative generation model is used, and the coefficients of the quadratic cost function are given as the known parameters c_1, c_2, and c_3. Total electric load is the summation of infrastructure power load and non-infrastructure power load. The total electric cost is described in (6.13). The finalized electricity rate is calculated in (6.14).

$$P_e(t) = h_w \cdot Q_w(t) + h_g \cdot Q_g(t) + L_r(t) \tag{6.11}$$

$$0 \leq P_e(t) \leq P_e^{\text{UP}} \tag{6.12}$$

$$\text{Min } Z_e = \sum_{i=1}^{N^T} c_1 \cdot P_e^2(t) + c_2 \cdot P_e(t) + c_3 \tag{6.13}$$

$$r_e = \frac{Z_e}{\sum_{i=1}^{N^T} P_e(t)} \tag{6.14}$$

It is noted that because the simultaneous balance of the power system with non-storage is considered, the decision variable of hourly power generation is determined by the residential electric load and the optimization results of the water flow and gas flow vectors.

Since the finalized electricity rate may be different from the pseudo electricity rate, the finalized electricity rate will be used to substitute the pseudo electricity rate and update the optimal values of the objective functions in the water and gas system independent optimizations. No optimization method is involved in this step. Only arithmetic calculation is needed.

6.3.4 Joint Optimization

The joint optimization inherits all previous balance equations and the constraints but not the objective functions. The pseudo electricity rate will not be used because the finalized electricity rate will be calculated as a by-product of the programming. The water and gas systems will not execute the optimization separately because the total electric cost will be optimized in the aggregated objective function as shown in (6.15).

$$\text{Min } Z_T = \sum_{i=1}^{N^T} c_1 \cdot P_e^2(t) + c_2 \cdot P_e(t) + c_3 + \sum_{i=1}^{N^T} r_w \cdot S_w(t) + \sum_{i=1}^{N^T} r_s \cdot m(t) + \sum_{i=1}^{N^T} r_g \cdot S_g(t) + \sum_{i=1}^{N^T} r_p \cdot p_p(t) \tag{6.15}$$

The joint optimization involves the nonlinear term in its objective function, which may introduce difficulty in problem-solving. A piecewise linearization technique is used to linearize the objective function.

Assume vector a_n is the index setting points in the quadratic cost curve and vector b_n is the value points on the curve. By introducing the ancillary variable $\lambda_n(t)$, the objective function can be expressed by (6.20). Additional constraints [13] are described in (6.16)–(6.19). Then the formulation only includes the linear constraints and objective function. The problem can be solved by the mixed integer linear programming method.

$$0 \le a_n \le P_e^{\text{UP}} \tag{6.16}$$

$$b_n = c_1 \cdot a_n^2 + c_2 \cdot a_n + c_3 \tag{6.17}$$

$$n \in \{0, 1, 2, 3, \ldots\ldots, N^s\} \tag{6.18}$$

$$\sum_{n=1}^{N^s} \lambda_n(t) = 1 \text{ for all } t \tag{6.19}$$

$$\text{Min } Z_e = \sum_{i=1}^{N^T} \sum_{n=1}^{N^s} b_n \cdot \lambda_n(t) + \sum_{i=1}^{N^T} r_w \cdot S_w(t) + \sum_{i=1}^{N^T} r_s \cdot m(t) + \sum_{i=1}^{N^T} r_g \cdot S_g(t) + \sum_{i=1}^{N^T} r_p \cdot p_p(t) \tag{6.20}$$

6.4 Numerical Simulation Study

6.4.1 Conditions and Parameters

The water load, gas load, and residential electric load are obtained from the prior research works in the area of EWG systems [14–16] and normalized, respectively, by scaling down to the range to match a city load like the size of Lincoln, Nebraska. The curve is shown in Fig. 6.3.

The simulation horizon is 24 h, where 1-h time steps are considered. Pseudo electricity rate is assumed to be 0.124$/kWh. The per unit volume of gas transportation is 500 m^3 if converted into the normal supply status in the pipeline.

6.4.2 Result and Analysis

The power generation curves in two cases are compared in Fig. 6.4a, where the joint optimization significantly reduced the power fluctuation by shifting a portion of the electric load from peak hours to off-peak hours. Specifically, the non-infrastructure electric load is shifted to avoid the peak load hours.

The water power load and gas power load in two cases are compared in Fig. 6.4b, c. It is observed that since hour 16–20 are peak hours, the water and gas load shifted a portion of the loads from peak hours to off-peak hours in joint optimization. This means the joint optimization allows the water and gas systems to perceive the power cost signal and adjust its operation strategy accordingly.

The states of the water storage, the gas tank storage, and the gas pipe storage are compared in Fig. 6.4d–f. It can be seen that the average storage in the water tank and gas pipe are increased, compared to the independent optimization. This illustrates that the power system utilizes the storage facilities in the water and gas systems to redistribute its load through the time axis.

Because the water storage state is proportional to the water pressure, the simulation result also indicates the residence will have improved water pressure as a

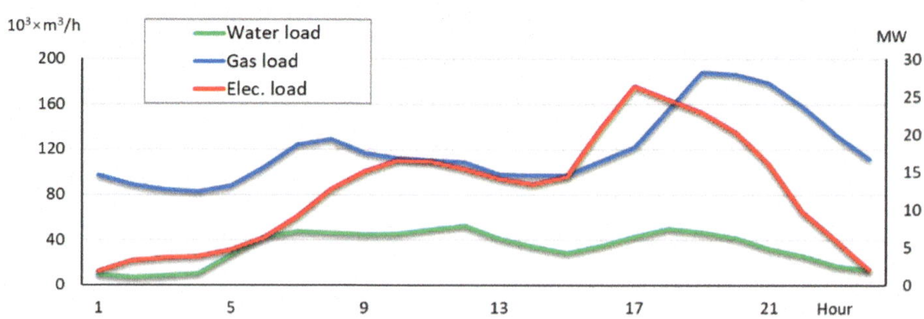

Fig. 6.3 Water load, gas load, and residential electricity load

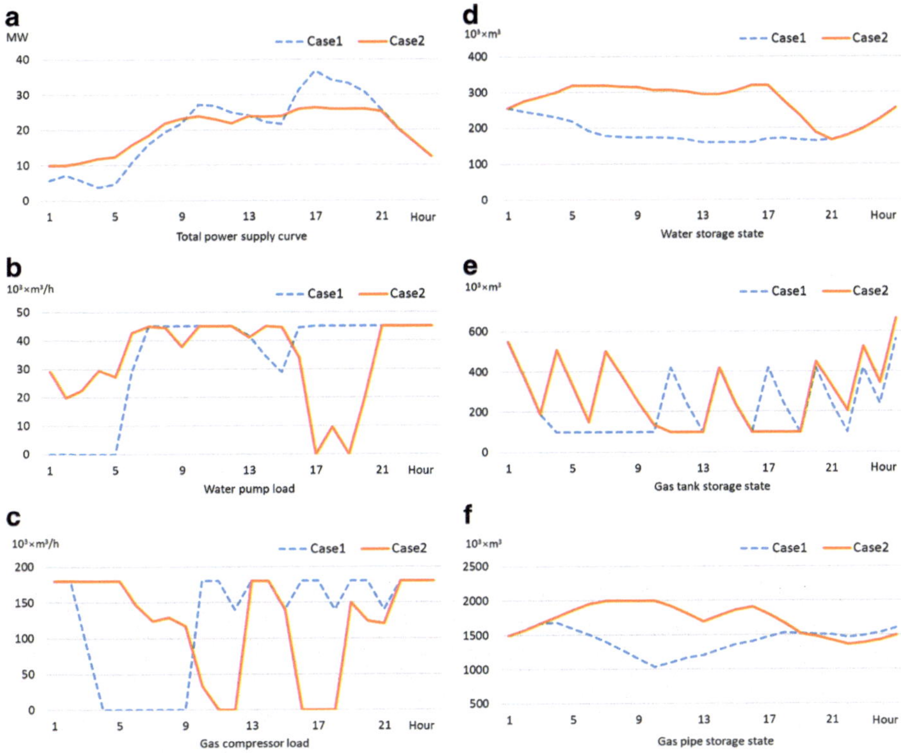

Fig. 6.4 Comparison of (**a**) total power supply curve, (**b**) water electric load, (**c**) gas electric load, (**d**) water storage, (**e**) gas tank storage, and (**f**) gas pipe storage

Table 6.1 Total cost breakdown and comparison

	Case1	Case2	Rate of change (%)
Water O&M cost (A1)	$810	$1193	32.1
Gas O&M cost (A2)	$7107	$7986	11.0
Water electric cost (B1)	$18,453	$16,124	−14.4
Gas electric cost (B2)	$9640	$8129	−18.6
Residential electric cost (B3)	$45,983	$40,181	−14.4
Total EWG cost	$81,994	$73,613	−11.4
Final electric rate	$0.154	$0.134	−14.7

by-product of the joint optimization. The situation is similar to the gas system, where an improved gas pressure is achieved by the co-optimization.

The explicit comparison of the components of the entire system cost is given in Table 6.1. It is found that the O&M costs for the water and gas systems are increased by 32.1% and 11.0%, respectively. However, the electric costs are reduced by 14.4% and 18.6%, respectively. And, the residential electric cost is reduced by 14.4%.

The total EWG cost is calculated as the summation of all the O&M cost and electric cost above, i.e., A1 + A2 + B1 + B2 + B3 in Table 6.1. The joint operation leads to a reduction on the total EWG cost by 11.4%, and the finalized electricity rate is therefore reduced from 0.154$/kWh to 0.134$/kWh.

6.5 Conclusion

This paper proposed the schemes for the separate operation and the joint optimization of the electricity-water-gas interdependent system. Both models share the same constraints but have different structures of the objective function and workflow. The separate operation model optimized the system before the final electric rate is calculated. The joint model did not optimize the objective function until all the constraints and the objective functions are formalized and merged.

The numerical study showed the reduction on both the finalized electricity rate and the total EWG system cost. This is because the water and gas systems intrinsically included the storages in their networks. Those storages could serve as the load shifting capacities or regulation resources for the power system. The joint optimization utilizes these resources to shift and redistribute the load through the time axis. The utilization of spare capacity of storage helped to reduce the power generation cost at the expense of a slight increase in O&M cost.

References

1. Liu, Y., Qu, Z., Xin, H., & Gan, D. (2017). Distributed real-time optimal power flow control in smart grid. *IEEE Transactions on Power Systems, 32*(5), 3403–3414.
2. Ormsbee, L. E., & Lansey, K. E. (1994). Optimal control of water supply pumping systems. *Journal of Water Resources Planning and Management, 120*(2), 237–252.
3. Solgi, M., Bozorg-Haddad, O., Seifollahi-Aghmiuni, S., Ghasemi-Abiazani, P., & Lóaiciga, H. A. (2016). Optimal operation of water distribution networks under water shortage considering water quality. *Journal of Pipeline Systems Engineering and Practice, 7*(3), 04016005.
4. Zavala, V. M. (2014). Stochastic optimal control model for natural gas networks. *Computers and Chemical Engineering, 64*, 103–113.
5. Afshar, M. H. & Rohani, M. (2009) Optimal operation of pipeline systems using genetic algorithm. In *Evolutionary computation, 2009. CEC'09. IEEE congress on* (pp. 1399–1405). IEEE.
6. Gopalakrishnan, A., & Biegler, L. T. (2013). Economic nonlinear model predictive control for periodic optimal operation of gas pipeline networks. *Computers and Chemical Engineering, 52*, 90–99.
7. Bagchi, A., Sprintson, A., Guikema, S., Bristow, E., & Brumbelow, K. (2010). Modeling performance of interdependent power and water networks during urban fire events. In *Communication, control, and computing (Allerton), 2010 48th annual allerton conference on* (pp. 1637–1644). IEEE.
8. Santhosh, A., Farid, A., Adegbege, A., & Youcef-Toumi, K. (2012). *Simultaneous co-optimization for the economic dispatch of power and water networks*. In Advances in

power system control, operation and management (APSCOM 2012), 9th IET international conference on, Hong Kong, China. IET.
9. He, C., Wu, L., Liu, T., & Bie, Z. (2017). Robust co-optimization planning of interdependent electricity and natural gas systems with a joint n-1 and probabilistic reliability criterion. *IEEE Transactions on Power Systems, 99*, 1–1.
10. Homeland Security. (2013). Critical infrastructure security and resilience.
11. Water Distribution Systems. (2017). https://www.epa.gov/dwsixyearreview/drinking-water-distribution-systems
12. Natural Gas Distribution. (2014). https://www.eversource.com/Content/ema-g/residential/safety/gas-safety-tips/gas-pipeline-safety%0A
13. Bradley, H. (1977). Nonlinear programming. In *Applied mathematical programming* (pp. 419–464).
14. Dutt, G. & Tanides, C. (1999). *Hourly demand curves for residential end uses in Argentina and potential for load management*. In Proceedings of l5eme. Congres International des Reseaux Electriques de Distribution.
15. Gurung, T. R., Stewart, R. A., Beal, C. D., & Sharma, A. K. (2015). Smart meter enabled water end-use demand data: Platform for the enhanced infrastructure planning of contemporary urban water supply networks. *Journal of Cleaner Production, 87*, 642–654.
16. Zhang, X., Che, L., Shahidehpour, M., Alabdulwahab, A., & Abusorrah, A. (2016). Electricity-natural gas operation planning with hourly demand response for deployment of flexible ramp. *IEEE Transactions on Sustainable Energy, 7*(3), 996–1004.

Chapter 7
Architecture Models for Coordination of Unmanned Air and Ground Vehicles Conducting Humanitarian Assistance and Disaster Relief

Wyatt T. Middleton, Gregory Miller, and Anthony Pollman

7.1 Introduction

As the usability and accessibility of unmanned vehicles increases, it is highly likely the Department of Defense (DOD) will use unmanned vehicles to be "first on the beach and first through the door on tomorrow's battlefield" [1]. In recent studies, the exploration of a team of heterogeneous unmanned systems conducting collaborative autonomy in order to execute specific military operations has increased. This technological scenario is trending upward among various nations and will continue to redefine the future of unmanned systems missions. However, to date, few articles discuss the methods necessary to explore what is technologically feasible or discuss the operational utility of the collaborative unmanned systems. In addition, few studies quantifying changes in performance specifications for heterogeneous unmanned vehicle swarms with their impact on operations. The original motivation for this research comes directly from the Commandant of the Marine Corps, who stated the US military "ha[s] to focus on the capabilities required for near-peer competitors, including cyber, information warfare, electronic warfare, unmanned air and ground systems and robots" [2]. A key step to creating a collaborative autonomous system is the development of the system architecture for the cross-domain system.

This paper contributes a comprehensive architecture for cross-domain UGV and UAV conducting expeditionary operations, which uses the relationship of system architecture products and model-based systems engineering analysis to analyze system feasibility and system performance. For the purpose of this study, a cross-domain system is defined as an unmanned system that can operate autonomously or remotely, alone or in a swarm, in at least two of the six defined domains: land, sea, air, space, cyberspace, and electromagnetic spectrum. However, this research will focus specifically on the operation of unmanned vehicles in the land and air domain.

W. T. Middleton (✉) · G. Miller · A. Pollman
Naval Postgraduate School, Monterey, CA, USA

S. Adams et al. (eds.), *Systems Engineering in Context*,
https://doi.org/10.1007/978-3-030-00114-8_7

This report will process the system's operational requirements into preferred system configurations using a detailed experiment. The experiments will inform operational effectiveness metrics for a pairing of heterogeneous unmanned systems. The MBSE MEASA methodology also highlights the feasibility of the utilization of a UAV-UGV team collaboratively conducting structured, rudimentary tasks that may be present in a given mission scenario.

7.2 Model-Based Systems Engineering Methodology for Employing Architecture in Systems Analysis

The Model-Based Systems Engineering Methodology for Employing Architecture in System Analysis (MBSE MEASA) is a baseline analysis method following the development of SysML products. SysML is defined as "a general-purpose graphical modeling language that supports the analysis, specification, design, verification, and validation for complex systems" [3]. Use of the MBSE MEASA approach is only appropriate when the stakeholder analysis has been conducted and the system's operational requirements have been defined. The MBSE MEASA has the capability to conduct a system analysis, determining the system's technical feasibility and examining the operational effectiveness of the system. The results of the analysis can then be implemented back into the system architecture based on the general system's engineering iteration process, leading to the optimal system solution. This method "defines the use of architecture to support analysis (and vice versa) to ensure that behaviors represented in the models and simulation created in the System Analysis Domain can be traced to functions prescribed in the System Architecture Domain" [4]. By using this analysis, the system's comprehensive framework is traceable to the physical and functional components of the system. The MBSE MEASA yields a comprehensive framework for the development of system architecture products, external simulation models, and the cycle of the systems engineering process past the abilities of any current system engineering approach [4].

Figure 7.1 is a graphical representation of the stages of the MBSE MEASA. The MBSE MEASA is divided into five stages: Requirements Analysis, Functional Architecture, Physical Architecture, Model Definition (Operational)-Model Definition (Synthesis), and Model Analysis. Figure 7.1 also includes some of the SysML products of the system under research that will be generated. The first stage of MBSE MEASA is to conduct a requirements analysis, a very important stage of the iterative process that must be done efficiently. INCOSE (2010) states that the "system requirements are the foundation of the system definition and form the basis for the architectural design, integration, and verification" [6]. This paper provides the MEASA SysML product generation for the collaborative autonomy of the unmanned vehicle systems conducting search and rescue efforts which decompose amphibious operations. Following the requirements analysis and the functional architecture for the system, the system design is developed. During this phase, the

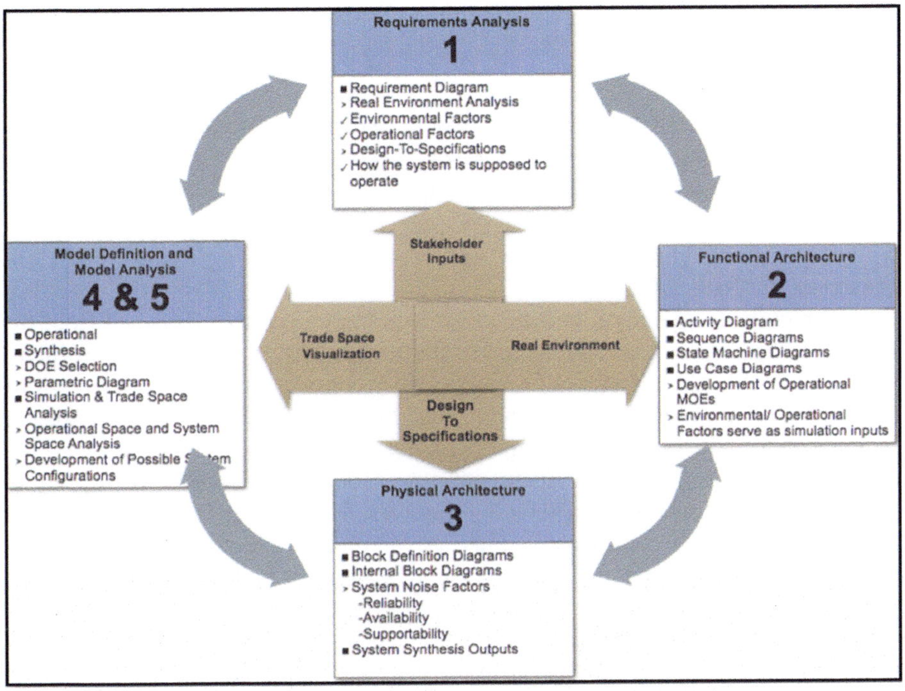

Fig. 7.1 MBSE MEASA methodology. Adapted from Beery [4]

system is decomposed and partitioned into various critical functions that meet the system requirements. These critical functions in the system's decomposition will capture its complexity [6]. Once the system's functional architecture is generated, the physical components of the system can be applied to the architecture, resulting in the physical architecture products. Finally, a system model can be developed. A unique function of the MBSE MEASA is that the analysis "utilizes the combined functional and physical architecture products as a basis for the development of external models and simulations" [4]. This process allows for the model to be generated based on the system's architecture. This enables the model to focus on the system's operational effectiveness by evaluating physical and functional components of the system. As stated earlier, this paper focuses on the development of the system's architecture for the unmanned vehicle systems and the iterative process of implementing SysML products in order to help determine the system's technical feasibility and operational effectiveness.

7.3 Introduction to Humanitarian Assistance and Disaster Relief Operations

This research used humanitarian assistance and disaster relief (HA/DR) as the context for exploring the utilization of MBSE MEASA. The MBSE MEASA allows for the analysis of various levels of collaborative autonomy while conducting HA/DR amphibious operations. Figure 7.2 illustrates expeditionary operations with emphasis on amphibious operations. It also identifies amphibious operations of interest to MBSE MEASA.

An amphibious operation is defined as "a military operation launched from the sea by an amphibious force, embarked in ships or craft with the primary purpose of introducing a landing force (LF) ashore to accomplish the assigned mission" [5]. Conducting amphibious operations allows the US military to strategically mobilize their military forces, giving the military the capacity to strike at a position of choice and assert a strong projection of power for ground forces. The majority of amphibious operations are carried out by Navy and Marine Corp forces. Amphibious vessels such as amphibious assault ships, amphibious transport docks, and amphibious command ships were specifically designed to support the Marine Corps doctrine to operate from sea to land. These ships may support assaults, raids, withdrawals, demonstrations and humanitarian assistance and disaster relief operations. As natural disasters strike, the US military has proven to be a key member of the relief support system. Within the scope of the recent natural disasters, for which the US military has provided support, there has been a focused effort to avoid placing service members in life-threatening environments while conducting humanitarian assistance and disaster relief operations. Operationally, unmanned air vehicles and unmanned ground vehicles should keep service members out of harm's way by performing some or all HA/DR missions.

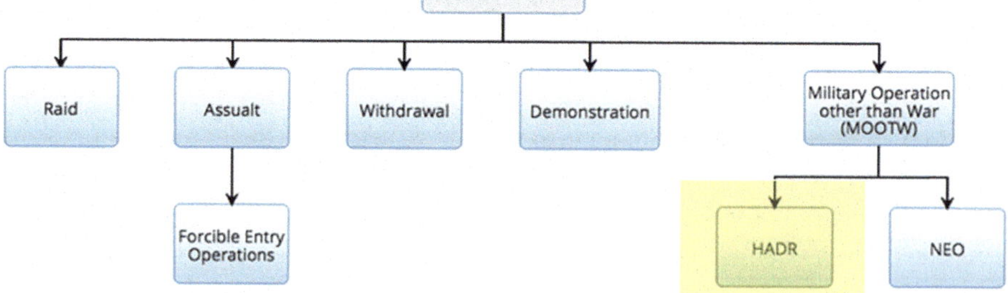

Fig. 7.2 Amphibious operations activities. Adapted from: Office of the Chairman of the Joint Chiefs of Staff (OCJCS) [5]

7.4 MEASA SysML Product Generation for Cross-Domain Unmanned Vehicles Conducting HA/DR

This paper conducts a requirements analysis (Stage one), develops the functional hierarchy (Stage two), and develops the physical hierarchy (Stage three) for a cross-domain unmanned vehicle system conducting HA/DR operations and implementing the following stages into the system analysis. Stages one to three develop the comprehensive framework for the system which can then be related to an external model, simulations, or field experiment to determine operational effectiveness and technical feasibility. Stages four and five represent the model, simulations, and or field experiment used to complete the MEASA. This paper does not present the results of Stages four and five of the MBSE MEASA. It highlights only the first three stages of MBSE MEASA by identifying system requirements and developing the system's architectures. Future work will fully implement the entire methodology of MBSE MEASA.

7.4.1 *Requirements Analysis*

The purpose of the requirements analysis is to take the inputs of the stakeholders and transform that stakeholder-centric view into a technical view that can provide a system that fulfills those primary activities. The foreign and domestic policies such as the Chief of Naval Operations Instruction (OPNAV) 3500.38B guide US military forces on how to effieicently conduct HA/DR without violating humanitarian or operational laws. The instrunctions included in OPNAV 3500.38B define the system's operational requirements. The requirement analysis for HA/DR is represented by the governing policies currently in place:

- Expeditionary Operations.
  - Amphibious Operations.

    Amphibious Raid *(Marine Corps Tactical Task (MCT) 1.3.2.2).*
    Amphibious Assault *(MCT 1.3.2.3).*
    Amphibious Withdrawal *(MCT 1.3.2.7).*
    Amphibious Demonstrations *(MCT 1.3.2.1).*
    Military Operations other than War (MOOTW) *(MCT 1.6.6).*

    Provide Humanitarian Support [Assistance] (*Navy Tactical task (NTA) 4.7.8).*
    Provide Disaster Relief Support (*NTA 4.7.9).*
    Conduct Non-combatant Evacuation Operations (NEO)-*(MCT 1.6.6.6).*
    Conduct Anti-Terrorism Operations-*(MCT 1.6.6.1).*
    Conduct Anti-Terrorism Enabling/Support Operations-*(MCT 1.6.6.2).*
    Conduct Counter-Terrorist (CT) Operations-*(MCT 1.6.6.3).*
    Implement Anti-Terrorism Measures-*(MCT 1.6.6.4).*

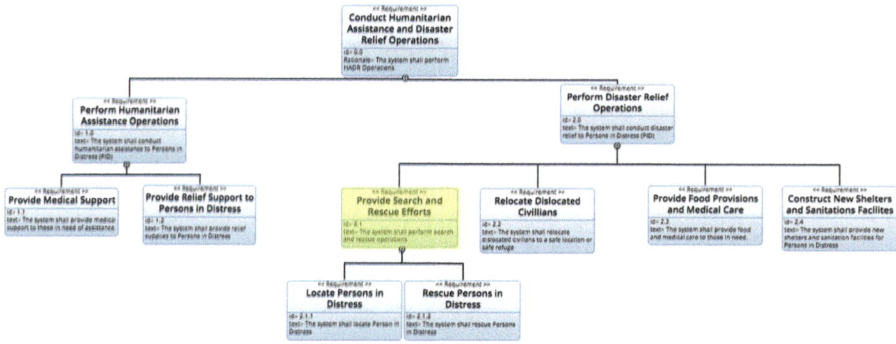

Fig. 7.3 Requirements diagram

Support Anti-Terrorism Measures-*(MCT 1.6.6.5).*
Conduct Peace Operations-*(MCT 1.6.6.8).*
Conduct Security and Stability Operations (SASO)-*(MCT 1.6.6.9).*

Another way to graphically represent what the system must accomplish is to generate a SysML requirement diagram. The requirement diagram highlights the functionality of the system as well as its capabilites, operational requirements, and performance conditions as shown in Fig. 7.3. The collaborative autonomy of the unmanned vehicle system will take place while providing search and rescue efforts.

7.4.2 Functional Architecture Products

After the problem is clearly stated, the system must be designed. System design is composed of functional architecture, physical architecture, design configurations, and modelling and simulation. Functional architecture is defined as "an arrangement of functions and their sub-functions and interfaces (internal and external) that defines the execution sequencing, conditions for control or data flow, and the performance requirements to satisfy the requirements baseline" [7]. The functional architecture describes the operations or functions that will be carried out in order to meet mission objectives. It also breaks down the capabilities of the system to ensure unnecessary operational activities are reduced. Reducing unnecessary activities results in fluent system updates and a cost-effective system. Figure 7.4 depicts the functional decomposition of a team of unmanned vehicles conducting search and rescue operations autonomously.

The UGV and UAV collaborate with one another in order to locate the persons in distress and provide assistance. The high-level functions for the UGV and UAV include: sense, navigate, communicate, and move and control. Each vehicle will be equipped with similar electro-optic capabilities. Ideally, the UAV will locate the

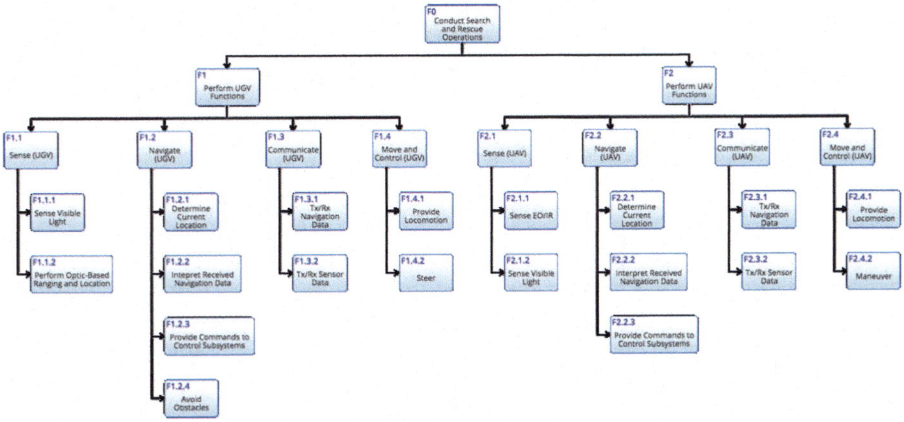

Fig. 7.4 Functional architecture for HA/DR operations

Persons in Distress (PID) and then command and control (C2) will launch the UGV to rendezvous with the PID to provide relief support and rescue assistance. In order to keep the human interface in the communication loop, all data gathered from the UAV and UGV will be transmitted back to C2, where the Joint Task Force (JTF) commander will make the operational decisions.

The MBSE MEASA functional architecture also consists of developing any of the following diagrams: activity diagrams, sequence diagrams, state machine diagrams, and use case diagrams. The use case diagram depicts all personnel involved in the operation and can be used as a tool to help identify issues in system control or system implementation. All of the components in the SysML use case diagram illustrate what functional components of the system must be completed in order to meet the primary activity.

Figure 7.5 identifies all US military forces involved in HA/DR. The combination of these two diagrams depict what functions the system must execute and the key members of the system. The products generated in this phase capture how the system will satisfy the system requirements outlined earlier. The SysML diagrams generated in this section help portray traceability which highlights the expected utilization of each diagram. Both diagrams presented will help ensure the technical feasibility of the design.

7.4.3 Physical Architecture Products

Once the system has been divided into functions and sub-functions, the physical architecture can be developed. The physical architecture provides the baseline for all the required component resources. These architectures are often graphically represented by block definition diagrams. Block definition diagrams provide a

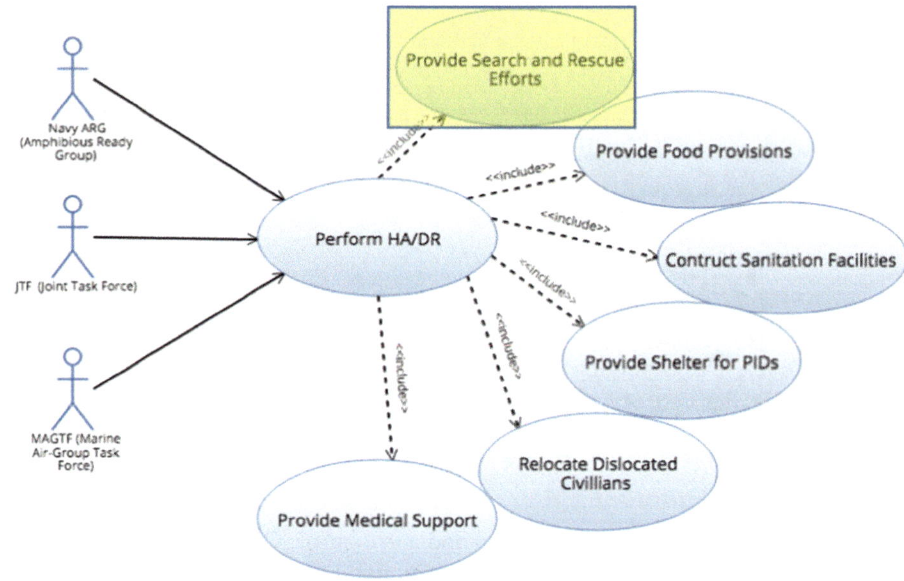

Fig. 7.5 Use case diagram for HA/DR

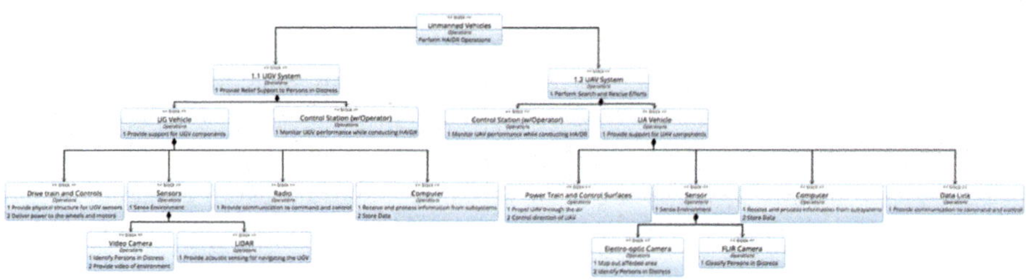

Fig. 7.6 Block definition diagram for unmanned vehicle systems

graphical representation of the system's components. Once one has defined the high-level components, then a decomposition of those components can be conducted which will examine the physical makeup of the system more in depth. There are several benefits of using the SysML block definition diagrams. One benefit is how the relationship of each component is shown in the diagram. This allows the architect to know the decompositional relationship of all the components in the system (Fig. 7.6).

The UGV system physical components are the UG vehicle and the control station with an operator. The UG vehicle is broken down into the following components: drive train and controls, sensors (LiDAR and video camera), radio, and computer. The LiDAR's optical sensing ability enables the UGV with the capability of obstacle avoidance. The UAV system is composed of the UA vehicle and the control station with an operator. The UA vehicle is broken down into the following components: power train and control surfaces, sensors (electro-optic camera and FLIR camera), data link, and computer. The FLIR camera allows the mission to be conducted during low visibility and also help to the identify and classify a PID. Both operators are included in the physical loop to monitor the vehicle's performance and to help identify PIDs.

7.5 Conclusion

The MEASA methodology develops an iterative and systematic approach to developing a solution for the collaborative system. By focusing on Stages one, two, and three, the operational requirements were clarified and the functional and physical architectures were developed. This allows those inquiring about collaborative autonomous systems conducting HA/DR missions to have the critical functions and physical components of the system identified, organized, and analyzed. Multiple designs for potential cross-domain unmanned vehicle systems maybe generated and analyzed using this methodology. Future work will consist of developing a field experiment using the SysML products generated in this paper to determine the operations effectiveness of the system based on the architecture and provide a cost-effective option of constructing a collaborative autonomous system that has military utility and technical feasibility. This paper presented the functional and physical architecture required to execute amphibious missions such as HA/DR, which will catalyze further development for the use of robotics within expeditionary operations.

References

1. Cantelli, L., Mangiameli, M., Melita, C. D., & Muscato, G. UAV/UGV cooperation for surveying operations in humanitarian demining. In *Paper presented at the 2013 IEEE international symposium on safety, security, and rescue robotics (SSRR)*, Linkoping, Sweden (pp. 1–6). http://ieeexplore.ieee.org/stamp/stamp.jsp?arnumber=6719363. Accessed 18 May 2017.
2. Kreisher, O. (2017). *Neller: 'We need a fifth-generation Marine Corps'*. Seapower. http://seapowermagazine.org/stories/20170112-neller.html
3. Friedenthal, S., Moore, A., & Steiner, R. (2009). *A practical guide to SysML. The systems modeling language*. San Francisco, CA: Morgan Kaufmann.
4. Beery, P. T. (2016). *A model-based systems engineering methodology for employing architecture in system analysis: developing simulation models using systems modeling language products to link architecture and analysis*. PhD dissertation, Naval Postgraduate School. http://hdl.handle.net/10945/49363. Accessed 9 Sept 2017.
5. Office of the Chairman of the Joint Chiefs of Staff (OCJCS). (2009). *Amphibious Operations (Joint publication [JP] 3-02)*. Washington, DC: Department of Defense.
6. Walden, D. D., Roedler, G. J., Forsberg, K. J., Douglas Hamelin, R., & Shortell, T. M. (2015). *INCOSE systems engineering handbook: A guide for system life cycle processes and activities* (4th ed.). Hoboken, NJ: Wiley.
7. IEEE. (2005). IEEE Std 1220-2005. *Standard for the application and management of the systems engineering process*.

Chapter 8
A Dynamic Model of Military Suicide for Improved Understanding of Effects and Integration of Interventions

Ronald F. A. Woodaman

8.1 Introduction

While suicide has always been a pernicious and tragic burden on society, recent military conflicts have seen a near doubling of the rate of suicide among US military personnel [1]. Despite an increased investment in suicide prevention [2] and an end to the war in Iraq in 2011, rates reached an all-time high in 2012 and remain elevated [3, 4]. Suicide prevention interventions come in many varieties but what works and what does not is difficult to ascertain [5]. There is both a lack of coherency among the Department's many efforts and a lack of measures to determine whether these efforts individually or collectively are having an effect [6].

To help address these challenges, and drawing from insights gained while providing program evaluation and analytical support for the Defense Suicide Prevention Office, we have developed the Stress-Injury Pathway to Military Suicide (SIPMS) systems model. From an organizational standpoint, it serves as a common framework for understanding and defining the role of the diverse suicide prevention functions. In contrast with many of the other models of suicide found in the literature, SIPMS offers a computational approach to model suicide on a population level. Thus, SIPMS can yield insights into the macro dynamics in suicide: why rates in the military have climbed so high and have remained high, despite the drawdown of US involvement in combat and the Department of Defense's heavy investment in preventions. Looking forward, it provides a structural basis for forecasting suicide rates, a quantitative means to evaluate the likely impact of proposed interventions, and a baseline from which to validate implemented interventions.

SIPMS integrates the combat and operational stress continuum [7], the rudiments of Joiner's interpersonal-psychological theory of suicide [8], and the mathematics of epidemiology [9]. The resulting model is strategic versus tactical, working at a

R. F. A. Woodaman (✉)
Innovative Decisions, Inc., Vienna, VA, USA
e-mail: rwoodaman@innovativedecisions.com

S. Adams et al. (eds.), *Systems Engineering in Context*,
https://doi.org/10.1007/978-3-030-00114-8_8

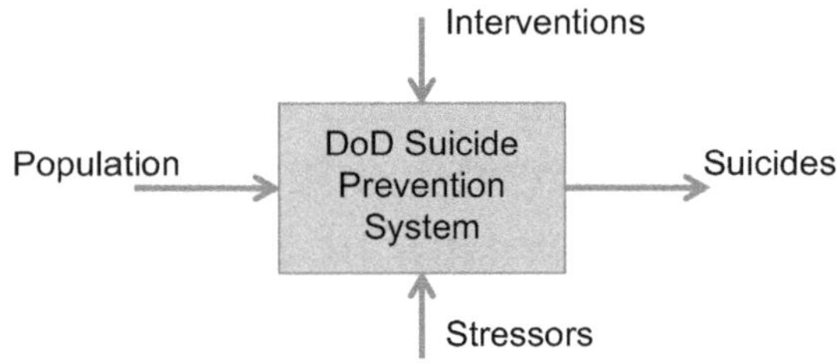

Fig. 8.1 IDEF0 representation of DoD suicide prevention system (http://www.idef.com/idef0.htm)

population level versus an individual one. Consequently, it enables decision-makers in the public health arena to holistically consider the range of interventions, their interactions, and plausible impacts.

Ordinarily, it is difficult to measure the effectiveness of a particular prevention intervention, due to confounding from other prevention interventions. Even were one to try to estimate the effectiveness of the overall portfolio, one must deal with the counterfactual; there is no firm basis for determining whether or not suicide rates are expected to rise or fall. We illustrate this conundrum via a simplistic model of the DoD suicide prevention system (Fig. 8.1).

Although abstract, this IDEF0 representation conveys powerful concepts. It is natural for many observers to infer that a significant increase in suicides must be the result of a failure of the prevention system. The IDEF0 model says that suicides are a function of three macro variables: the population, the interventions, and the environmental stressors. All three macro variables must be accounted for when attempting to attribute the driver behind a change in the output. We are not aware of any extant means of doing this for the Department of Defense. SIPMS seeks to fill this role. As we will see, it provides a means to represent the composition of the population, the vector of interventions, and the stressor signal while quantitatively measuring their effects on the output: suicides. But first, we must briefly discuss mathematical approaches to modeling health problems like suicide.

8.2 Mathematical Models for Modeling Public Health Problems

Modern epidemic models got their start with the work of McKendrick and Kermack [10]. Epidemic models are simple means of describing the transmission of communicable diseases among individuals in a population. They commonly divide the population into *compartments*—what in other literature might be called a *state space*. The most basic is the so-called SIR model which partitions the population by the susceptible (S), the infected (I), and the removed (R). The total population N equals $S(t) + I(t) + R(t)$, so the model is closed. The population flows across the states: $S \rightarrow I \rightarrow R$. This system is described via a system of differential equations, where the infection rate is proportional to the contact rate between the susceptible population and the infected population. Individuals depart the infected state at a rate of the inverse of the average infectious period. Departures result in those that are cured—and now immune—and those that have died. McKendrick and Kermack showed that their models were able to predict the progression of several documented epidemics. Epidemic models have found utility beyond disease modeling, such as in the epidemic theory of ghettos [11].

Another mathematical concept for modeling dynamic systems is the Markov chain [12]. Markov chains model random processes where the system transitions from one state to another. Markov chains have been used to model a great variety of human health problems with the most direct association with SIPMS being stochastic population processes, where the system is defined by the proportion of the population distributed across the state space [13–15]. Yip et al. used a Markov chain illness and death model to depict suicide dynamics in an abstract population [16]. This simple model had only four states while allowing for replacement of deaths by births. Using this model, they conducted an experiment where they illustrated that moderately effective public health interventions on the public at large had a greater effect on the suicide rate than very effective interventions targeted on high-risk individuals. SIPMS was developed independent of Yip et al., but both approaches share many of the same concepts. Qualitatively, SIPMS is focused on the military population, relies on COSC and IPT for its states, and is richer in its outputs and dynamics.

Markov chains share much with system dynamics, an approach that features stocks, flows, feedback, and time delays to model complex nonlinear systems. Developed originally by MIT's Jay Forrester in the 1950s to model cycles in General Electric's labor force, system dynamics has been applied to a variety of domains, to include public health [17]. Atkinson et al. argue for a broader implementation of system dynamics in modeling complex public health problems, to include suicide prevention [18].

Atkinson, Guetz, and Wein developed a stochastic model to forecast the number of PTSD cases that the Department of Veterans Affairs (VA) could expect over time due to ongoing combat operations [19]. They sought to estimate the number of troops exposed to varying levels of traumatic stress. Employing a *strength failure model* common in manufacturing, they allow individuals to accumulate stress while deployed and recover between deployments. Individuals whose cumulative traumatic stress exceeded a threshold are *broken* and, after a random period of lag, become symptomatic. The lag, while random, was affected by how soon after their breaking point the individuals departed the service. While there are similarities in that both models employ a stress signal as input, the mechanisms within the models and the outputs are quite different. SIPMS models the effects of interventions and their effect on suicide rates, while the PTSD model forecasts the number of PTSD cases, diagnosed and undiagnosed, over time.

8.3 SIPMS State Space: Modeling Stressor Effects

During Operation *Iraqi Freedom*, leaders and medical professionals in the US Navy and Marine Corps collaborated in developing a new public health protection tool to respond to the volume of post-traumatic stress disorder (PTSD) cases. A fundamental shift was severely operational, and combat stress injuries should be viewed in the same light as physical injuries and illnesses. This result is the combat and operational stress continuum (COSC) [7]. COSC can be used to partition the population into four stress-exposure states: Ready, Reacting, Injured, and Ill (Fig. 8.2).

The states provide the basis for service members to understand the effects of stress both upon themselves and their fellows. Note that the difference between the Injured state and the Ill state is one of persistence: Ill individuals are those whose Injured state has become a chronic condition which can be diagnosed by a medical professional. Since from a leadership perspective, both states require intervention, for our modeling purposes, we combine these states into one: Injured/Ill.

An individual in the Ready state when exposed to a strong stressor signal will naturally react, thus transitioning to the Reacting state. Most people are resilient: like a person with a strong immune system when exposed to a potential infection , the resilient service member will experience some symptoms but ultimately return to Ready.

To depict this logic graphically, we use a node and arc chart with nodes representing COSC states and arcs representing transition between the states. Computationally, we assign to each arc the transition rate from one state to the next. The arrow from Ready to Reacting is the *Stressor rate*—the rate at which unstressed individuals experience significant, stressful events. The arrow from Reacting back to Ready is the *Resilience rate*—the rate at which Reacting individuals recover without needing medical intervention.

Some individuals do not recover to Ready. These individuals transition from Reacting to Injured/Ill. We call this the *Morbidity rate*. To return to Ready,

READY	REACTING	INJURED	ILL
DEFINITION • Adaptive coping • Effective functioning • Well-being FEATURES • In control • Calm and steady • Getting the job done • Playing • Sense of humor • Sleeping enough • Ethical and moral behavior	DEFINITION • Mild and transient distress or loss of function FEATURES • Anxious • Irritable, angry • Worrying • Cutting corners • Poor sleep • Poor mental focus • Social isolation • Too loud and hyperactive	DEFINITION • More severe and persistent distress or loss of function TYPES • Trauma • Fatigue • Grief • Moral injury FEATURES • Loss of control • Can't sleep • Panic or rage • Apathy • Shame or guilt	DEFINITION • Clinical mental disorders • Unhealed stress injuries TYPES • PTSD • Depression • Anxiety • Substance abuse FEATURES • Symptoms persist > 60 days after return from deployment

Fig. 8.2 Combat and operational stress continuum [7]

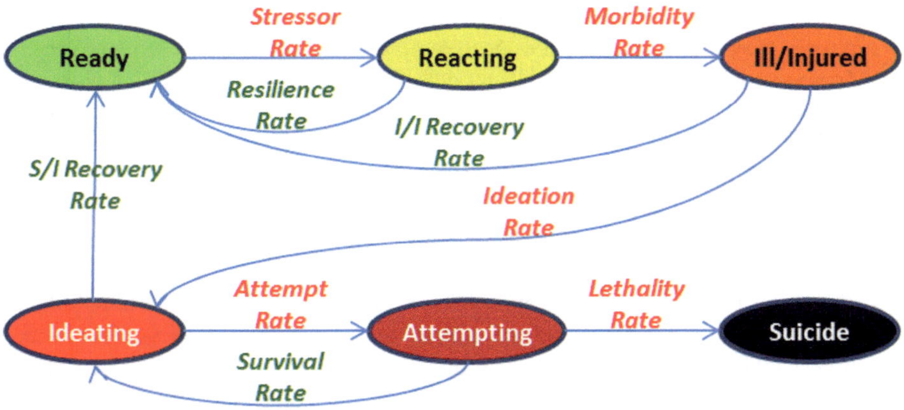

Fig. 8.3 Basic SIPMS model

individuals will require treatment. We make the simplistic assumption that recovery proceeds from Injured/Ill straight back to Ready; the Reacting state is exclusively for those Ready individuals recently exposed to the stressor signal. The rate from Injured/Ill back to Ready is the *Recovery rate* and captures all aspects involved in recovery, to include stigma reduction, help-seeking, and treatment.

Next, we incorporate states related to suicide. Joiner's interpersonal-psychological theory (IPT) of suicide states that individuals kill themselves because (a) they no longer want to live and (b) they have acquired the capability to end their lives. Acquired capability comes about by a variety of means, to include combat exposure which may function to lower the individuals' fear of dying by their own hand while increasing their competence in affecting the means of their own death. The desire to die, on the other hand, arises because of two simultaneously persistent factors: thwarted belongingness and perceived burdensomeness. Thwarted belongingness may be manifest from social withdrawal, loss from death, or loss of social support. Perceived burdensomeness can come about from distress from physical illness, low self-esteem, agitation, shame, and self-blame. Interestingly, these two factors—isolation and lack of self-worth—are common themes in the progression of stress injury. IPT then contributes three states: Ideating, the state of seriously contemplating suicide; Attempting, actively seeking one's death; and Suicide. Combining these with the previous three states yields the following graphical model with transitions (Fig. 8.3).

From within the Injured/Ill population, those at risk of suicide would first transition to the Ideating stage as a function of the *Ideation rate*. Transitions from Ideating to Attempting occur at the *Attempt rate*. Conversely, those ideating may still seek help, get treatment, and recover. Those Attempting would die as a function of the *Lethality rate*: the proportion of suicide attempts that result in Suicide. The *Survival rate* is the strict complement of the *Lethality rate*; there are no other possible transitions from this state. Attempt survivors return to the Ideating state. From there they may attempt again, or they may seek help and go onto recovery.

Mapping back to our basic IDEF model, this basic SIPMS model captures the effect of the environment (the stressor signal), but we do not capture explicitly the effect of interventions. This forces us to first consider the subpopulation of service members with preexisting risk factors, which we call the Vulnerable state [20]. To address these, the military services employ a variety of educational programs aimed at building service members' resilience to stressors [6]. We incorporate the higher-risk Vulnerable state and the concept of resilience programs by providing two possible transitions: *Resilience Building rate* for transitions from Vulnerable to Ready and the stressor rate which transitions individuals to Reacting. The former captures a fundamental hypothesis of military resilience programs: that the vulnerability of those with preexisting conditions can be mitigated. The stressor rate is the same rate as before since COSC argues that all people when exposed to a pervasive stressor will enter the reacting mode. However, since we expect the vulnerable population to have a lower resilience rate, we partition Reacting into two states—one for the previously Ready and one for the previously Vulnerable. Similarly, we need two different morbidity rates. To reduce complexity, we keep Injured/Ill as a single state, assuming that all individuals reaching this state are equally morbid and would have the same recovery and ideation rates.

DoD resilience programs exist that focus on helping those in treatment to complete their treatments and rebuild resilience. This yields the notion that there is a transitory state where individuals do not go straight back to the Ready but, like the Vulnerable, are more susceptible to stressors. We call this the Recovering. While there are similarities between the Vulnerable group and the Recovering group, the key difference is that the Recovering group is one that the Department can clearly identify.

To better address recovery, we introduce explicit states for addressing treatment but make separate whether treatment is for Injured/Ill or for Ideating. This explicitly forces the system to consider at what rate those who need treatment are actually getting it and the rate at which that treatment fails (*Recidivism*). Hoge and Castro stated that the "most important challenges in suicide prevention are stigma surrounding mental illness, negative perceptions of treatment, and other barriers. . .that result in the majority of service members and veterans not accessing care when needed or dropping out prematurely" [21]. The accompanying figure depicts an advanced version of the model that builds further on these ideas. It highlights, for example, that there are two kinds of suicide attempts: observed attempts that usually result in treatment and hidden attempts where we presume the individual returns to Ideating (Fig. 8.4).

With these systems of nodes and arcs, we can numerically model SIPMS as a Markov chain. In contrast to standard Markov chain analysis, our system is semi-closed as we allow personnel to enter (enlistments or commissions) and depart (discharges). Importantly, we allow the stressor rate to vary over time, as the force encounters changes to rates of deployment and involvement in combat. Other rates, such as the rate at which Ill/Injured seeks care, could be varied, but in this brief paper, we only consider variation in the stressor rate.

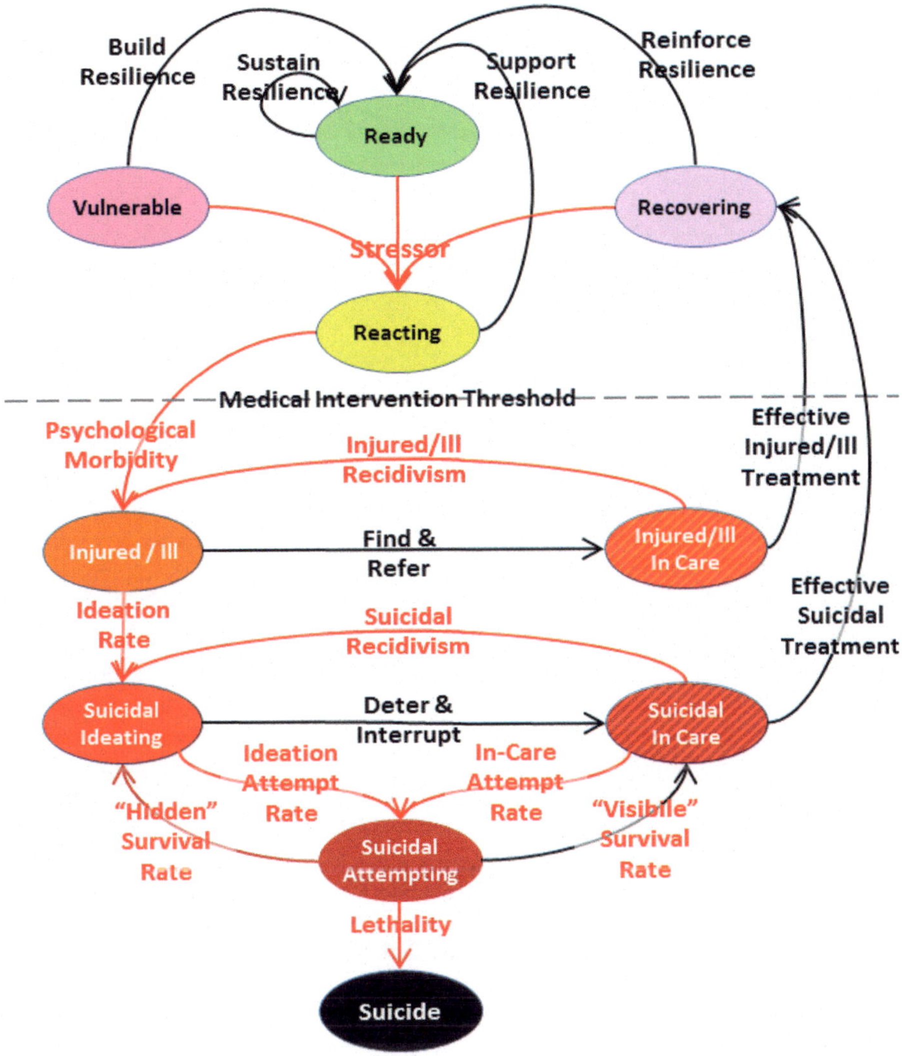

Fig. 8.4 Advanced SIPMS model

8.4 Evidence of Model Validity

We employ a speculative simulation using SIPMS to show how it may explain the persistence of high rates of suicide in the active military and link these results to other surveillance data. Actual numerical values on the state populations and transition rates cannot be found in any one place, though some rates are found in studies focused on narrowly defined populations. The only variable allowed to vary with

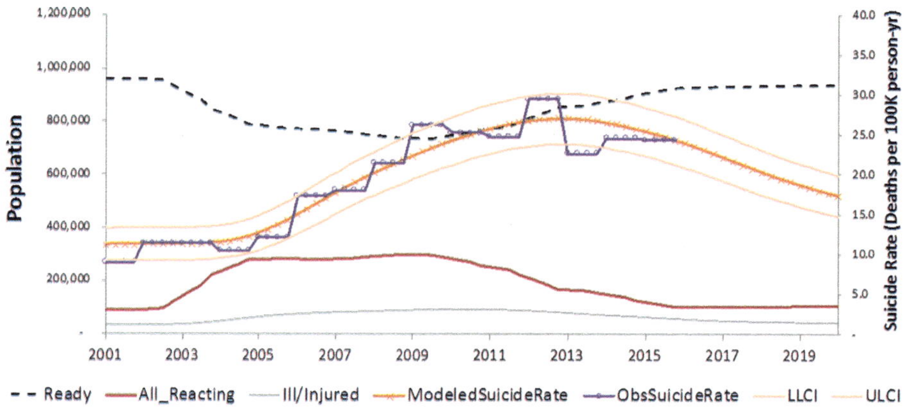

Fig. 8.5 Speculative simulation with advanced SIPMS model

time was the stressor signal, for which we chose the proportion of the Army's active duty force deployed by year (deployment rates were kept constant after 2015 due to lack of data) [22]. The other transition rates were kept invariant. These and the initial populations were based on expert-informed conjecture but purposefully adjusted to show a plausible fit between modeled suicides and the actual suicide rate observed within the US Army over the recent decade and a half. The time step is year quarters (Fig. 8.5).

The simulation tells us that the persistently higher suicide rate is due to a large number of untreated Ill/Injured (lowest line on the graph) who remain in the service long after exposure and only slowly exit. This is echoed in the latest reports which show an increase in suicide rates in the military *since* 2012 among the 30–39-year-old demographic [4]. News coverage of the past decade has repeatedly hit at the high suicide rate in the military, with a recent article quoting experts who say that this may simply be the "new normal" [23]. SIPMS argues that we are seeing a long-lagged effect with many high-risk individuals persisting in the service well after their stressor exposure. The speculative simulation predicts that the rate only returns to pre-war levels sometime after 2020. While this is just a kind of thought experiment, if the lagged effect is true, we should see evidence for it in other surveillance data.

In the figure below, we plot the annual suicide rate (DoDSuRate) versus the annual number of new mental health diagnoses [24] and compare to a similar contrast from SIPMS (ModelSuRate). SIPMS's logic predicts that the relationship should be stronger between lagged years. An increase in stressor exposure in year $t-1$ should lead to increased diagnoses in year t and a resulting increase in suicides year $t+1$. The top panel is from the model data. The bottom panel is from surveillance data. The left charts show a no lag relationship; the right has a 1-year lag. The model shows a stronger correlation, as expected. Looking in the lower panel, we see that as predicted, the lagged relation was stronger: an increase in mental health diagnoses predicts a rise the following year in suicides (Fig. 8.6).

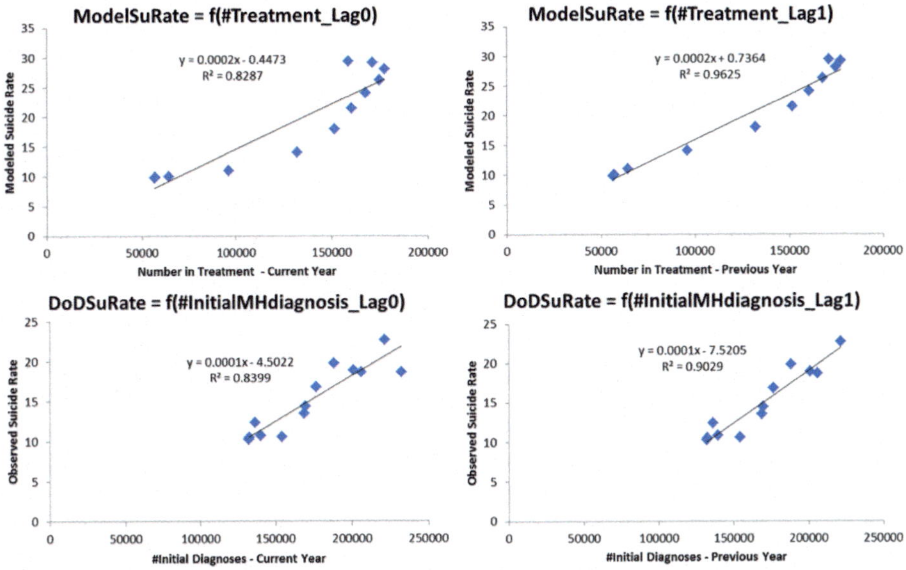

Fig. 8.6 Comparing modeled lagged relationship (ModelSuRate vs Treatment, Lag0 and Lag1) to observed surveillance data (DoDSuRate vs InitialMHdiagnosis, Lag0 and Lag1) [24]

8.5 Conclusion

This paper seeks to introduce a variety of new ideas to broaden the holistic understanding and evaluation of suicide prevention within the Department of Defense. First, we integrate a truncated version of the combat and operational stress continuum with the rudiments of the interpersonal-psychological theory of suicide to create the stress-injury pathway to military suicide model. This is a diagnosis-agnostic model of suicide, one focused on the effect of stress on the force and its symptoms as they affect suicide. Our goal is to understand how interventions might influence this pathway to reduce the negative flows. We expand on this model to incorporate resilience and explicitly represent the effect of treatment. Treatment is decomposed between the treatments of Injured/Ill from those who are experiencing suicidal ideations. The utility of this approach, we maintain, lies not in its ability to aid in identifying and treating specific individuals—which may frustrate those in the clinical community—but in modeling and understanding the effectiveness of the collective suicide prevention system in order to implement far-reaching systemic changes. The model presented is not complete. Indeed, we note that there is no one-size-fits-all model. Rather, we are advocating for the model's methodology. For some modeling questions, we may need to decompose a particular state or incorporate more arcs to address a particular issue in question.

We advocate three key uses of this model: integration of suicide prevention activities, to include research; cost-benefit analyses of proposed programs; and improved surveillance of suicide within the military. Integration of suicide prevention activities is difficult because it is not clear how all the different efforts work together. Again, the model may be a valuable complement to these efforts since it provides an integrative environment within which to test hypotheses and understand, at least, the potential impact of an intervention prior to expending research funds or investing in new programs. Furthermore, the flexibility of the methodology allows researchers to "zoom in" by decomposing states and transitions into finer detail to examine specific issues, perhaps in combination with examining select demographic groups.

Closely related to integration is cost-benefit analysis. As in other applications of system dynamics in healthcare [25], we believe this approach lends itself to cost-benefit analysis at the service and department level within the military, since the military deals with a semi-closed system and a pre-screened population. The model's simulation approach allows for the easy implementation of parametric analyses and design of experiment studies to identify cost-effective mixes of interventions.

Surveillance systems should lend themselves to understanding what may happen in the future and where to focus preventive measures. From a behavioral health perspective, SIPMS can provide the Department with a new paradigm for surveillance. Via the tracking of a set of state-transition rates and subpopulations across key defined states, the Department will be able to proactively surveil mental health and suicide, determining where the bottlenecks are and identifying opportunities for improvement.

As shown in some of the examples in this paper, SIPMS is amendable to being fit to historical suicide rates. As the model is validating against previous years' data, it is a natural step to use the model to forecast future rates. These can provide the basis for better informing leadership and the American public about the dynamics of military suicide. They can also provide the basis for validating effectiveness claims by implemented interventions.

Recent research has shown that recently separated veterans are at 41–61% elevated risk of suicide [26]. Outside of the Department, SIPMS may be adapted to serve in a similar role for the VA, helping to predict the number of high-risk individuals, by category, that are departing the military in a given year. For a VA version of SIPMS, the entries (departures from the DoD model) would presumably maintain their SIPMS state. New states might be introduced to address other aspects of the veteran population, such as whether they use the VA health system or not. Outside of the military and veteran domain, the methodology used for SIPMS may be useful for other institutional suicide prevention systems, such as local school systems, prison systems, university student bodies, and managed healthcare systems.

8.6 Future Research

We would welcome the opportunity to conduct a study of how to implement SIPMS within a service, the Department of Defense, or the VA. Clearly, it will require various forms of estimation since many of these states or only indirectly observable.

As further evidence of SIPMS potential utility, we have developed a regression model based on SIPMS concepts, wherein we employ casualty rates by year (versus deployment rates) and economic data as various proxies for stressor signals. Preliminary results show that this model—using observation data versus speculative data as in the simulation described here—can accurately predict suicide rates by enlistment year cohort. While self-funded, we hope to complete the research and seek publication in the near future.

Acknowledgments We would like to acknowledge the following colleagues at the Defense Suicide Prevention Office who contributed ideas and encouragement: William Crowley, Matthew Sampson, Edwin Middlebrook, Katerina Brancato, David Krizov, Charles Twardy, Marie Rowland, and Michelle Cornette.

References

1. Ramchand, R., Acosta, J., Burns, R., Jaycox, L., & Pernin, C. (2011). *The war within: Preventing suicide in the U.S. military*. Santa Monica, CA: RAND Corporation.
2. U. S. Department of Defense. (2013). Message from Secretary Hagel on Suicide Prevention Month [Press release]. Retrieved April 9, 2018, from http://archive.defense.gov/Speeches/Speech.aspx?SpeechID=1801
3. U. S. Department of Defense, Defense Centers of Excellence for Psychological Health & Traumatic Brain Injury. (2014). *Department of Defense Suicide Event Report, Calendar Year 2013 Annual Report*. Retrieved April 9, 2018, from http://t2health.dcoe.mil/programs/dodser
4. U. S. Department of Defense, Defense Suicide Prevention Office. (2017). *Department of Defense Quarterly Suicide Report, Calendar Year 2017 Fourth Quarter*. Retrieved April 9, 2018, from http://www.dspo.mil/Prevention/Data-Surveillance/Quarterly-Reports/
5. Mann, J., Apter, A., Bertolote, J., Beautrais, A., Currier, D., Haas, A., Hegerl, U., Lonngvist, J., Malone, K., Marusic, A., Mehlum, L., Patton, G., Phillips, M., Rutz, W., Rihmer, Z., Schmidtke, A., Shaffer, D., Silverman, M., Takahashi, Y., Varnik, A., Wasserman, D., Yip, P., & Hendin, H. (2005). Suicide prevention strategies: A systematic review. *JAMA, 294*(16), 2064–2074.
6. Institute of Medicine. (2014). *Preventing psychological disorders in service members and their families: An assessment of programs*. Washington, DC: The National Academies Press.
7. Nash, W. (2011). Marine Corps and Navy combat and operational stress continuum model: A tool for leaders. In E. C. Ritchie (Ed.), *Combat and operational behavioral health* (pp. 107–119). Washington, DC: Borden Institute Textbook of Military Psychiatry.
8. Joiner, T. (2005). *Why people die by suicide*. Cambridge, MA: Harvard University Press.
9. Brauer, F., & Castillo-Chávez, C. (2001). *Mathematical models in population biology and epidemiology*. New York: Springer.
10. Kermack, W., & McKendrick, A. (1927). A contribution to the mathematical theory of epidemics. *Proceedings of the Royal Society A: Mathematical, Physical and Engineering Sciences, 115*(772), 700–721.

11. Crane, J. (1991). The epidemic theory of Ghettos and the neighborhood effects on dropping out and teenage childbearing. *The American Journal of Sociology, 96*(5), 1226–1259.
12. Markov, A. (1971). Extension of the limit theorems of probability theory to a sum of variables connected in a chain. Reprinted in Appendix B of: Howard R. *Dynamic probabilistic systems, volume 1: Markov chains*. New York: Wiley.
13. Pegels, C., & Jelmert, A. (1970). An evaluation of blood-inventory policies: A Markov chain application. *Operations Research, 18*, 1087–1098.
14. Meredith, J. (1974). Program evaluation in a hospital for mentally retarded persons. *American Journal of Mental Deficiency, 78*, 471–481.
15. Renshaw, A. (2011). *Stochastic population processes: Analysis, approximations, simulations*. Oxford: Oxford University Press.
16. Yip, P., So, B., Kawachi, I., & Zhang, Y. (2014). A Markov chain model for studying suicide dynamics: an illustration of the Rose theorem. *BMC Public Health, 14*, 625.
17. Homer, J., & Hirsch, G. (2006). System dynamics modeling for public health: Background and opportunities. *Am J Public Health, 96*(3), 452–458.
18. Atkinson, J., Page, A., Wells, R., Milat, A., & Wilson, A. (2015). A modeling tool for policy analysis to support the design of efficient and effective policy responses for complex public health problems. *Implementation Science, 10*, 26.
19. Atkinson, M., Guetz, A., & Wein, L. (2009). A dynamic model for posttraumatic stress disorder among U.S. troops in operation Iraqi Freedom. *Management Science, 55*(9), 1454–1468.
20. Ingram, R., & Luxton, D. (2005). Vulnerability-stress models. In B. L. Hankin & J. R. Z. Abela (Eds.), *Development of psychopathology: A vulnerability stress perspective* (pp. 32–46). Thousand Oaks, CA: Sage Publications Inc.
21. Hoge, C., & Castro, C. (2012). Preventing suicides in US service members and veterans: Concerns after a decade of war. *Journal of the American Medical Association, 308*(7), 671–672.
22. Belasco, A. (2009). *Troop levels in the Afghan and Iraq wars, FY2001-FY2012: Cost and other potential issues*. Congressional Research Service, 7-5700, R40682.
23. Zoroya, G. (2016). Experts worry high military suicide rates are 'new normal'. *USA Today*. http://www.usatoday.com/. April 10, 2018.
24. Armed Forces Health Surveillance Center. (2014). Numbers and proportions of U.S. military members in treatment for mental disorders over time, active component, January 2000-September 2013. *Medical Surveillance Monthly Report, 21*(5), 2–7.
25. Brailsford, S., Lattimer, V., Tarnaras, P., & Turnbull, J. (2004). Emergency and on-demand health care: Modeling a large complex system. *The Journal of the Operational Research Society, 55*, 34–42.
26. Kang, H., Bullman, T., Smolenski, D., Skopp, N., Gahm, G., & Reger, M. (2015). Suicide risk among 1.3 million veterans who were on active duty during the Iraq and Afghanistan wars. *Annals of Epidemiology, 25*(2), 96–100.

Chapter 9
Foundational and Precautionary Considerations for Value-Driven Tradespace Exploration and Analysis

Adam M. Ross, Matthew E. Fitzgerald, and Randy K. Buchanan

9.1 Introduction

A key challenge in modern acquisition is balancing the appropriate commitment of significant resources against the uncertain development of complex systems with promised capabilities. On top of this challenge is the uncertain value of these systems in the face of uncertain and changing threats and opportunities. Once made, the consequences of an acquisition decision will echo throughout the system's life cycle, both enabling and limiting the ultimate value that it can deliver. Resilient decisions and systems would increase the ability of the DoD to deal with both anticipated and emergent threats and opportunities. A first step is considering a broad array of possible decisions and understanding the consequences of choosing one over another. Cost capability analysis (CCA) is one such approach, broadening the acquisition question into a generalized trade-off between costs and benefits of alternatives, allowing for situational awareness of what is possible and what is affordable.

Value-driven tradespace exploration and analysis (TSE&A) is a generalization of CCA, encoded as a set of techniques that enable the data-driven discovery of superior and resilient systems that are more affordable and more capable. Harnessing modern computing technologies, and leveraging the state of the art in systems engineering, decision analysis, and visual analytics, formalized TSE&A is poised to become a game-changing capability for guiding and supporting complex and impactful acquisition decisions, both in the DoD and beyond. Providing a solid foundation for TSE&A is essential for ensuring not only its appropriate role in supporting data-driven decision-making, but also its effective application across domains and decision types.

A. M. Ross (✉) · M. E. Fitzgerald
The Perduco Group, Beavercreek, OH, USA
e-mail: adam.ross@theperducogroup.com

R. K. Buchanan
Institute for Systems Engineering Research, US Army Engineer Research and Development Center, Vicksburg, MS, USA

S. Adams et al. (eds.), *Systems Engineering in Context*,
https://doi.org/10.1007/978-3-030-00114-8_9

9.2 Foundations

Tradespace exploration and analysis is an approach whereby users delay making selection decisions about what alternative "best" satisfies a particular problem (or need) at hand. Instead, a large set of possible alternatives are simultaneously considered, with evidence gathered as to each of their abilities to "solve the problem." Exploration involves attempting to better characterize some aspect of this flow (e.g., what alternatives should I consider? What data best characterizes compelling evidence for each alternative? What are my most important criteria for selecting better alternatives? How do I know when I have an answer?). Analysis involves an algorithmic and usually quantitative generation and reduction of data associated with the alternatives in order to provide further evidence for choosing particular alternatives as a solution. Value-driven TSE&A explicitly uses the ends (what makes a solution good?) to orient and drive the creation of the alternatives, thereby increasing the likelihood of finding "better" solutions. In this way "values" are judgments that are layered on top of the evaluations of the alternatives, which represent a key activity in decision-making that always occurs. Explicitly considering values in TSE&A enables their impact to be explored (e.g., how do changes in my range "requirement" impact the attractiveness of alternatives?). Core mechanics, key strengths, and misconceptions of value-driven TSE&A will now be discussed.

9.2.1 Core Mechanics

Based on a synthesis of the literature, interviews, and the authors' research experience based on 15 years of conducting TSE&A studies, Fig. 9.1 describes a general three-layer approach for conducting a TSE&A study [1–7]. The first layer—*define*—encompasses the identification and definition of the decision to be made (addressing either a problem or opportunity). This would entail determining the key stakeholders, project scope, constraints on possible solutions, as well as concepts for

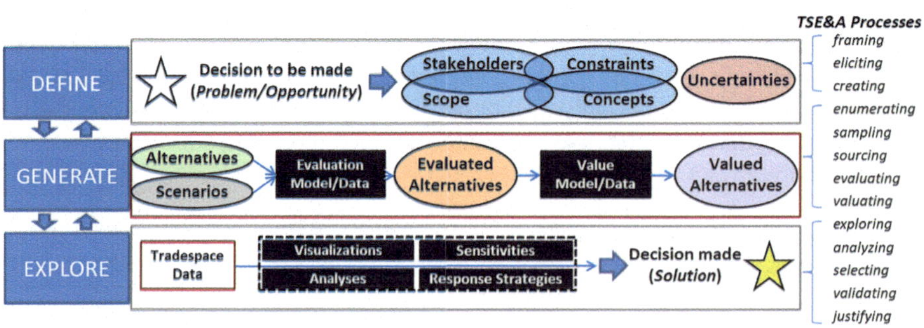

Fig. 9.1 The three key layers in general tradespace exploration and analysis (TSE&A) along with core concepts and associated processes

Table 9.1 Tradespace exploration and analysis (TSE&A) processes descriptions

Process	Description
Framing	The definition of the driving problem to be solved, including relevant stakeholders who specify how alternatives should be judged, scope extent of what to consider in the study, as well as study constraints
Eliciting	The definition of specific judgment criteria and limitations on possible solutions as specified by stakeholders, policy, and other data (e.g., MOEs, MOPs, TPMs, inherited technologies and systems, budget, schedule, etc.), as well as key uncertainties
Creating	The definition of the alternative (e.g., design) space from which particular alternatives can be specified (e.g., spanned by a design variable set), as well as the potential scenario space (e.g., contexts and needs, both short run and long run)
Enumerating	The definition of particular alternatives from the alternative space and particular scenarios from the scenario space
Sampling	The definition of a particular subset of alternatives from the enumerated set intended for evaluation (e.g., through design of experiments or expert rules), including definition of particular subset of scenarios from the enumerated set
Sourcing	The determination of how alternatives will be evaluated within and across scenarios, usually consisting of the identification and procurement of appropriate model(s) and/or dataset(s)
Evaluating	The evaluation of sampled alternatives via evaluative model(s)/data in terms of desired metrics (e.g., performance and resource-related metrics) often done through modeling and simulation
Valuating	The valuation of evaluated alternatives via value model(s) in terms of aggregate perceived benefits and costs as specified by stakeholders (e.g., utility and life cycle cost)
Exploring and analyzing	The intentional investigation of relationships and patterns between the input space and outputs space, resulting in knowledge and insights for the analyst/decision maker
Selecting	The decision on one or more alternatives as "answering the question" posed by the study, for example, providing the "best" benefit at a given cost across considered use contexts
Validating	The act of confirming the selected alternative(s) actually address(es) the driving problem (i.e., satisfy stakeholder needs)
Justifying	The intentional act of compiling the evidence needed to support the solution selection recommendation(s)

potential alternatives and the uncertainties that might impact the success of a solution. The second layer—*generate*—encompasses the data-generation layer of TSE&A, where alternatives and scenarios are specified, evaluated, and valued to provide the basis for decision-making in the next layer. The third layer—*explore*—encompasses the primary data exploration and analysis-supported decision-making activities, usually involving some combination of visualizations, analyses, sensitivity explorations, and consideration of uncertainty response strategies. Each of the layers both feed-forward and feed-back, as users seek to gain insights and confidence in the results. The processes associated with each layer shown in the figure are described in Table 9.1.

9.2.2 Key Strengths

The strengths of TSE&A as an approach for conceptual design and acquisition decision support are generally well-known, as they have been communicated extensively throughout the history of the field. Some recent examples follow. A recent paper addressing the use of TSE&A broadly identified four key positive outcomes of using the approach [8]:

- Knowledge development—learning the underlying cause-and-effect relationships with value
- Refining preferences—"design by shopping" paradigm allows the stakeholder to gradually reveal their own preferences rather than requiring an exact elicitation [9]
- Choosing a concept—the central act of making a decision
- Advancing the state of the art—incorporates modern concepts such as value-focused thinking and an expanded scope far beyond traditional point-design studies

Another perspective, following a survey of TSE&A tools, identified six main functions enabled by TSE&A that support the decision-making process [7]:

- Capturing value—using value-focused methods to customize for user perception
- Multidisciplinary optimization—allowing for maximization of chosen objective functions
- Statistical data analysis—identifying patterns and developing forecasts
- Visualization—displaying information with intuitive graphics
- Decision analysis—evaluating sets of alternatives against elicited preferences
- Project/portfolio management—scoping to include resource allocation across a portfolio of assets

These strengths only scratch the surface, but most stem from delaying premature decision-making, revisiting and questioning assumptions, enabling knowledge gain through developing intuition about potentially complex relationships between what stakeholders ask for and what is technically possible, fostering open dialogue around problem framing and solution approaches, and identifying and making explicit the characteristics of a good solution.

9.2.3 Misconceptions

TSE&A, as a growing field, has experienced substantial change over the past 10–15 years. The sophistication of the TSE&A research community (and practitioners) has increased dramatically, yet some misconceptions have persisted from the early days of TSE&A. These misconceptions are mostly perceived weaknesses of "classic" or "one-shot" TSE&A as performed by inexperienced designers and engineers, when current best practices have addressed these by various means. The following subsections will highlight a handful of these concerns and how experienced TSE&A practitioners can address them.

9.2.3.1 "TSE Will Stifle Innovation"

When viewing TSE&A as a tool for design, some designers feel that a rigid modeling and analysis structure such as TSE&A has a negative effect on innovation—limiting designer creativity and expertise by forcing the solution to come from the set of enumerated and evaluated designs. From the perspective of the TSE&A community, this is a straw man argument: the exploration of TSE&A has always been first and foremost a technique to be used as a *decision aid* between potential identified solutions, not a tool for assisting or replacing the innovation/identification of new solution concepts. That said, the stages of TSE&A prior to the evaluation of alternatives still rely heavily on designer creativity through the innovation and enumeration of system concepts and their associated controllable design variables and feasible levels. If new or innovative solutions to a problem are desired, then innovation must be applied at this point in the design process—not during the exploration and analysis stage. However, lessons learned during exploration *can* be utilized to drive innovation when defining the next set of concepts, if TSE&A is iterated.

It is also possible to utilize TSE&A as a means of guiding targeted innovation. Because the exploration and analysis of TSE&A is (often) performed on simulated data, it is possible to compare not only currently actionable/feasible solutions but also *hypothetical* solutions. For example, when designing a new aircraft, currently impossible levels of engine fuel efficiency and/or thrust could be included in the tradespace evaluations. These solutions will nominally be superior to known-feasible solutions and can be compared against them on value-oriented dimensions to capture the possible value of technological innovation. The value added of different innovations can also be compared to each other to direct innovation research resources into the most promising areas of design.

9.2.3.2 "A Tradespace Covers Only a Fraction of Possible Solutions"

This argument suggests that the "best" answer from a tradespace is not truly the "best." It is primarily targeted at two aspects of the tradespace: [1] the large number of controllable design variables that must be omitted for the sake of managing computational effort, and [2] the discretization of continuous variables. The small nugget of truth in this argument (likely based on observations of inexperienced TSE&A practitioners) is that some people do ascribe too much confidence in the "bestness" of the answer that is chosen via TSE&A. However, proper TSE&A practice addresses these concerns in two important ways. First, there are techniques available for TSE&A practitioners to quickly and accurately identify the most important value-driving design variables, such as the design value mapping

(DVM) [10]. It is relatively uncontroversial to suggest that a small handful of decisions often determines the vast majority of value (the Pareto principle), and therefore the most impactful value trade-offs can be captured as long as this subset of drivers can be incorporated into the tradespace (e.g., wingspan is more important to model than the head type of the screws on an aircraft). Second, most modern TSE&A is designed to be iterative—exploring and locking in "big" decisions before iterating with successively "smaller" decisions before finalizing a concept and entering detailed design. Any deviations from "bestness" can be removed in these iterations as scope tightens and detail improves, including those deviations derived from discretizing continuous variables—either by increasing the fineness of the discretization or eventually moving into a low-level optimization.

In general, the constraints of procedural generation/enumeration and evaluation of system alternatives—a finite set of design variables, limited to the domain for which the model is valid and used to assess only tractably modeled attributes—do limit the scope of TSE&A from considering the infinite space of conceptual design. However, these limitations are not unique to TSE&A but are rather a fundamental aspect of all model-based engineering efforts, including multivariable optimization. Such a limitation in scope is the price that must be paid to allow computers to evaluate potential system designs faster and cheaper than via human expertise or prototyping.

9.2.3.3 "Trade-offs Can Mislead or Be Actively Manipulated"

This argument is a challenge to the fundamental value of performing trade-off analysis, suggesting that such analysis is unreliable either by chance or by deliberate manipulation. For many years, practitioners were taught that the "knee" of the Pareto front in the standard benefit-cost tradespace was the first place to look for the "best" solution, and to this day the "knee" has a particular perceptual attraction to many TSE&A practitioners—it appears visually to be the best "bang for your buck" solution. However, when either dimension of the benefit-cost space uses a non-ratio scale (including most non-monetized value models), the "knee" is not meaningfully superior to other Pareto efficient alternatives. Accordingly, recent TSE&A research has deemphasized the "knee" in favor of more holistic analysis of the Pareto front in order to avoid the potential for misleading analysis based on purely visual circumstances [4, 5]. The penetration of this idea into practice has been slowly following.

Because the "knee" is generally identified visually, it is also subject to change depending on the chosen axis limits and scale. This does open the door for potential deliberate manipulation, to place a favored alternative on the "knee" to give it additional attention. However, such behavior should not be considered a fault of TSE&A but rather the result of unscrupulous, advocacy-based analysis. In fact, such advocacy bias is substantially more difficult to implement in the many-alternative framework of TSE&A than in either point-design studies (e.g., the common "not good enough," "too expensive," and "just right" three-solution study) or multivariable optimizations (where the parameters can be tweaked until the preferred solution is returned).

9.2.3.4 "Utility Theory Measures Total Preference and Should Not Separate Cost and Benefit"

This final argument is a mathematical challenge to the principle of using utility theory to model subcategories of value, but is unfounded. The original von Neumann-Morgenstern utility theory [11] very specifically has only two postulates: [1] the subject can always determine which of two packages they prefer or if they are indifferent between them (*complete ordering*) and [2] that this evaluation can include probabilistic events (*lotteries*). They go on to describe the creation of a vector basis of utility based on the weakening of the complete ordering postulate. This would be the case if the subject is unable to say which of two benefit-cost packages they prefer (e.g., low-cost, low-benefit or high-cost, high-benefit) or displays intransitive preferences in those dimensions. In such a case, separate utilities for each dimension—with complete ordering *within* each—can form a utility vector with no adverse effects beyond the accompanying increase in complexity. This type of incomplete ordering between benefit and cost has been observed in the field of complex system design and is justification for the conceptual positioning of TSE&A as a means of trading benefit and cost, either or both of which can be modeled with utility theory if they are comprised of multiple attributes. Incomplete ordering is also the main barrier to the reduction of value to a meaningful one-dimensional function, with the goal of layering "isovalue" curves on top of the benefit-cost tradespace. In practice, such isovalue curves can be estimated by asking stakeholders to identify alternatives to which they are indifferent.

9.3 Precautions

Despite the strengths of TSE&A, practitioners must continue to take appropriate precautions. Certain problem features present challenges that require careful management of both human and computing effort—failure to do so can result in faulty insights and/or inferior decision-making that fails to align with stakeholder preferences. The following subsection will discuss a few of these challenges and how they can be managed with effective practice.

9.3.1 Visualization Scalability

One challenge faced when applying TSE&A with many alternatives is the effective visualization of each alternative. In particular, as the number of alternatives

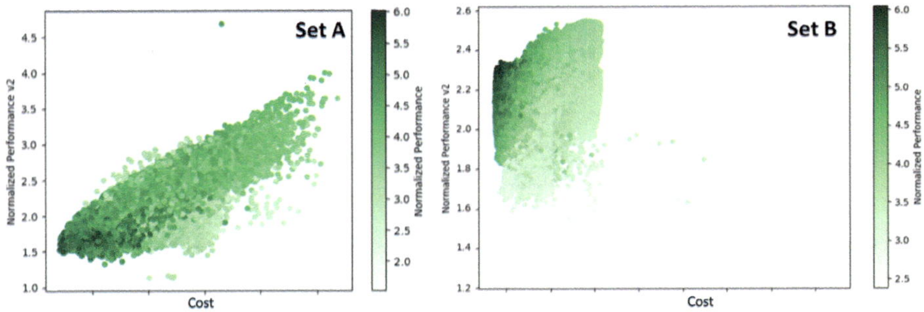

Fig. 9.2 Example of Normalized Performance v2 vs Cost tradespace scatterplots (colored by Normalized Performance v1)

increases, many visualizations that rely on one-to-one visual representation demonstrate significant *occlusion*, in which some data is covered by overlapping data. This is a common problem in TSE&A when looking at large design spaces. Consider an example with two different samplings of the design space for the same study (Set A = design of experiment (DOE)-sampled and Set B = genetic algorithm (GA)-sampled). The occlusion becomes noticeably worse when moving from the $O(10^4)$ designs of Set A to the $O(10^5)$ of Set B, as fewer points are individually identifiable. In this case, occlusion limits our ability to see how densely populated different areas of the tradespace are, because the entire tradespace appears as a solid block. Occlusion can also create misleading visualizations, commonly when looking for patterns in features such as color, as in Fig. 9.2.

In these plots, Normalized Performance v1 (NP) is indicated by color, and Normalized Performance v2 (NP2) has taken the y-axis. First, we can see that the tradespace takes a common shape, with more NP2 benefit becoming achievable as more cost is spent. However, based on the color, we might infer very different insights from these two plots about the relationship between NP and NP2. In the Set A plot, it appears that NP and NP2 are roughly inversely proportional, with the highest NP scores (darkest colors) appearing in the lower left corner. In the Set B plot, it appears that the same designs are best in both NP and NP2, at least in the low-cost regime. However, neither of these conclusions is reliable given the large amount of occlusion: the colors that can be seen are merely the points that were plotted “on top” and may not reflect the points underneath. Importantly, this is not a problem that can be addressed simply by increasing computational power. One approach for addressing occlusion in large datasets is the use of a binned tradespace as in Fig. 9.3.

These two plots show the same data but presented in a hexagonal grid. Each hexagon summarizes all of the points contained in its bounds. The size of the colored area (how “filled” the hex is) corresponds to the *number* of contained points and enables a more accurate reading of the density of points in different regions. The *color* represents a summary statistic for the color-value of all the contained points: in this case, the average NP. Now we can see with confidence that the surface-level

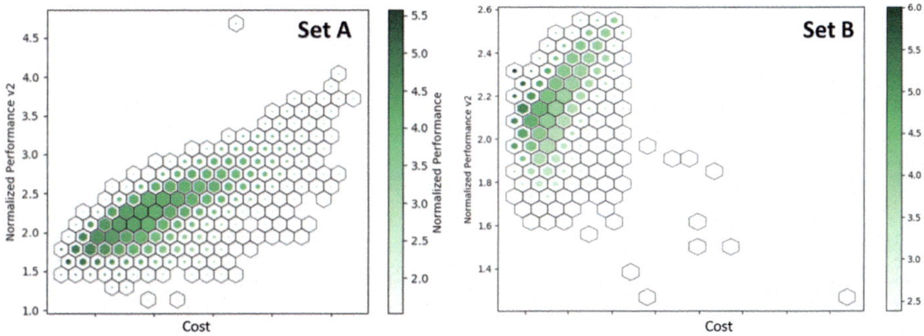

Fig. 9.3 Example of Normalized Performance v2 vs. Cost binned tradespaces (colored by mean Normalized Performance v1)

insight we drew from the colored scatterplot of Set A was not very strong: the lower-cost designs do have a higher average NP than the high-cost designs but not by very much. On the other hand, the Set B insight appears to have been correct, with the highest-NP hexes located in the upper-left corner where NP2 is also high. What could explain the difference between the two? The upper-left portion of the Set B tradespace is outside the domain of the Set A tradespace; the GA found strictly superior designs to the DOE in that region, and they are very good in both benefit metrics. The portion of the Set B tradespace that does overlap with Set A also has roughly uniform NP.

We can also see that the Pareto front is more densely populated in Set B than Set A (larger filled hex sizes on the edge of the tradespace), as expected from an intelligently searching GA compared to a DOE, which is likely to hit many more medium-attractiveness designs. However, the bulk of the Set B points are still located in the center of the tradespace. It is not reasonable to expect the Pareto front to be the *densest* part of the tradespace—it is, after all, the region that pushes the boundaries of design, and the random mutations occurring in a genetic algorithm are individually unlikely to be optimal—but in this case, it appears that none of the hexes on the Pareto front have more than ~10% of the designs that the eight or so "full" hexes in the center do. This may merit further investigation to check if the GA is enumerating new alternatives effectively: for example, this pattern could be caused by an excessively high mutation rate or a selection method that insufficiently weights individual fitness.

9.3.2 Organizational Buy-In and Effort for Value Modeling

A significant part of TSE&A's value as an approach comes from its alignment with the principles of value-focused thinking and value-driven design. If the true value proposition of the system as defined by system stakeholders and decision-makers is not captured, the insights derived from TSE&A will be strictly mechanical—how design variables impact attributes—and lack the connection to what is desired from the system. As such, maximizing the benefit of TSE&A requires participation from stakeholders. Unlike other analysis techniques that may be able to be completed with full effectiveness by engineers and analysts alone, practitioners seeking to fully realize the benefits of TSE&A (and justify the effort needed) must gather support from beyond the analyst, to include key stakeholders.

Unfortunately, capturing value propositions for complex systems can be effort intensive, particularly when considering large stakeholder networks or "big" problems such as systems-of-systems or portfolios. Because stakeholders and decision-makers are often busy, they frequently balk at the prospect of eliciting and revising value statements for incorporation into a value model. As such, perhaps the most useful leverage to practitioners of TSE&A is organizational buy-in: a culture of value-focused thinking that supports the effort to make decisions that align with perceived value instead of just technical parameters alone.

9.3.3 Stakeholder Turnover for Long-Lived Programs

Building upon the prior challenge, even if stakeholders do participate in the value modeling aspects of TSE&A, an additional challenge remains: development timelines for complex systems are frequently longer than the timescale of leadership roles (particularly in the government). Though some programs are completed with a single program manager from start to finish, rarely do the key stakeholders that those managers must satisfy remain the same. Stakeholder turnover can result in dramatic and potentially unpredictable changes in the value proposition for a system. Organizational buy-in can help here again, if it ensures that the new stakeholder is at least willing to put in the effort to define and model their own preferences (needed for TSE&A to continue). However, because some design decisions are locked in gradually, the threat of stakeholder turnover also requires an increased focus on resilience concepts such as changeability or versatility. If TSE&A is used only to optimize performance for a single scenario, the resulting system will likely be brittle to any changes in stakeholder preference. Practitioners should incorporate uncertainty frameworks and analyses (e.g., Epoch-Era Analysis, sensitivity analysis, etc.) into their TSE&A studies, or they risk being unable to remain relevant and insightful.

Alternatively, this problem could be addressed by changing the way such programs are conducted: either through longer-term leadership commitments or fixed program directions or requirements that new leadership is not allowed to alter. Of course, the barriers to this type of organizational change are incredibly high—especially for the government—so TSE&A practitioners should concentrate on managing stakeholder uncertainty with resilience.

9.3.4 Incorporating Preexisting Modeling/Simulation/ Analysis

Because modeling and simulation (or data sourcing more generally) is typically the most time-consuming part of a TSE&A study, practitioners often must incorporate preexisting models into the evaluation of the tradespace. This type of model reuse can save considerable time and effort but, if not treated with care, can be performed incorrectly and lead to faulty analysis. The central task is to connect the outputs of these evaluative models with the corresponding attributes of the value models defined by the stakeholders. Often, the evaluative models were not created with value-focused thinking in mind, causing their outputs to be more technical (or less applied) than the value-oriented terms a stakeholder might request. For example, a technical model of a car might return "handling" as a measure of side force that can be applied while making a turn without slipping, while a stakeholder might think of "handling" as a shorthand for smoothness of ride. A TSE&A practitioner must ensure that these types of mismatches are addressed via conversion models (i.e., a translation from engineering-speak to value-speak). When time permits models to be developed specifically for a given TSE&A study, such a conversion model may not be required. However, this can cause practitioners to possibly forget the importance of such a conversion when performing future studies with repurposed models.

Similarly, using preexisting models will typically involve creating a composite model of multiple different models—for example, recycling both an engine model and a weight/sizing model to evaluate a car. However, combining preexisting models can be a trap for unwary TSE&A practitioners, as they can differ in ways that prevent a straightforward combination. Some models may rely on different assumptions and may be unreconcilable. Others may have different timescales of simulation, necessitating either interpolation or forecasting (based on their dependencies in the composite model). Others may have varying degrees of uncertainty, which then requires careful statistical accounting of uncertainty as it propagates through the model. TSE&A practitioners must be aware of these threats, any of which could compromise the effectiveness of the study. This is a contributing factor to why a TSE&A practitioner should have a working understanding of modeling and simulation: analyzing the data properly requires understanding (and vetting) the source of the data and cannot be fully separated from the evaluation task [12].

9.3.5 Training: Moving Beyond Tool Development

The expertise necessary to run a believable and insightful TSE&A study should not be underestimated. The TSE&A practitioner needs to be familiar with evaluative modeling/simulation, value modeling, visualization, and statistics to be able to integrate their contributions to TSE&A. It is easy to think of TSE&A as a mechanical in/out task akin to solving a math problem for the "best" alternative, but this ignores the difficulty of setting up a tradespace. There are a minimum of 12 distinct tasks necessary to create and analyze a tradespace listed in Table 9.1, many of which require or benefit from human intuition and expertise. The previous subsection also identified some of the practical challenges TSE&A practitioners face with respect to collecting data through multiple models, such that the data remains valid for tradespace analysis. In fact, "soft factors" such as "trust" and "confidence" have been recognized as increasingly important in order to leverage model-centric techniques to support consequential decision-making [13].

As such, there is a definite risk with assigning TSE&A tasks to new, untrained analysts. Though the value of sophisticated and intuitive TSE&A tools and software is appreciated by many people [7, 14], TSE&A training is equally important given the significant impact that the human-in-the-loop has on the result of a study. Organizations that do not acknowledge the importance of training staff to both rigorously assemble models/datasets for TSE&A and deeply understand/perform the exploration and analysis tasks will struggle to realize the full potential of TSE&A.

9.4 Conclusion

Sitting at the nexus of advancements in computation, decision analysis, visual analytics, model-centric engineering, and even behavioral economics, TSE&A is poised to offer compelling and powerful insights to the design and acquisition communities, helping to curtail and manage the complexity of real decision-making in dynamic and uncertain high-dimensional problem spaces. But the real power of TSE&A stems not from its ability to manipulate large quantities of data in a structured manner, but rather from its ability to encourage explicit consideration and communication of how we both frame and seek solutions to our problems and how it broadens our perspective around what a "good" solution might look like over time. Fostering open dialogue in the community around foundations, best practices, challenges, and research opportunities can help move TSE&A toward becoming an essential part of any critical decision-making endeavor.

Acknowledgments This material is based upon work supported, in whole or in part, by the US Department of Defense through the Engineered Resilient Systems (ERS) Program. Any opinions, findings, and conclusions or recommendations expressed in this material are those of the author(s) and do not necessarily reflect the views of the US Department of Defense.

References

1. Ross, A. M. & Hastings, D. E. (2005, July). *The tradespace exploration paradigm*. In INCOSE Int'l Symp 2005. Rochester, NY.
2. Ross, A. M., McManus, H. L., Rhodes, D. H., & Hastings, D. E. (2010). *Revisting the tradespace exploration paradigm: structuring the exploration process*. In AIAA Space 2010. Anaheim, CA.

3. Ricci, N., Schaffner, M. A., Ross, A. M., Rhodes, D.H., & Fitzgerald, M. E. (2014). *Exploring stakeholder value models via interactive visualization*. In 12th Conference on System Engineering Research (CSER14). Redondo Beach, CA.
4. Ross, A. M., Rhodes, D. H., & Fitzgerald, M. E. (2015). *Interactive value model trading for resilient systems decisions*. In 13th Conference on System Engineering Research (CSER15). Hoboken, NJ.
5. Ross, A. M., Fitzgerald, M. E., & Rhodes, D. H. (2016). *Interactive evaluative model trading for resilient systems decisions*. In 14th Conference on System Engineering Research (CSER16). Huntsville, AL.
6. Ender, T., Goerger, S., McGroarty, C., & Spero, E. (2014). *Engineered resilient systems: tradespace enabled decision making*. In 17th Annual NDIA Systems Engineering Conference, Springfield, VA.
7. Spero, E., Avera, M. P., Valdez, P. E., & Goerger, S. R. (2014). *Tradespace exploration for the engineering of resilient systems*. In 12th Conference on System Engineering Research (CSER14). Redondo Beach, CA.
8. Collopy, P. D. (2017). *Tradespace exploration: promise and limits*. In 15th Conference on System Engineering Research (CSER17). Redondo Beach, CA.
9. Balling, R. (1999). *Design by shopping: a new paradigm?* In Proceedings of 3rd World Congress of Structural and Multidisciplinary Optimization (WCSMO-3). Buffalo, NY.
10. Ross, A. M. (2006). *Managing unarticulated value: changeability in multi-attribute tradespace exploration*. PhD dissertation, MIT, Cambridge.
11. Von Neumann, J., & Morgenstern, O. (1953). *Theory of games and economic behavior* (3rd ed.). Princeton, NJ: Princeton University Press.
12. Reymondet, L., Rhodes, D. H., & Ross, A. M. (2016). *Considerations for model curation in model-centric systems engineering*. In IEEE SysCon 2016. Orlando, FL.
13. Rhodes, D. H., & Ross, A. M. (2016, July). *A vision for human-model interaction in model-centric systems engineering*. In INCOSE Int'l Symp 2016. Edinburgh, UK.
14. Spero, E., Bloebaum, C. L., German, B. J., Pyster, A., & Ross, A. M. (2014). *A research agenda for tradespace exploration and analysis of engineered resilient systems*. In 12th Conference on System Engineering Research (CSER14). Redondo Beach, CA.

Chapter 10
Designing Engineered Resilient Systems Using Set-Based Design

Zephan Wade, Gregory S. Parnell, Simon R. Goerger, Ed Pohl, and Eric Specking

10.1 Introduction

10.1.1 Engineered Resilient Systems

The Department of Defense (DoD) and the engineered resilient system (ERS) program seek to leverage the capabilities of a model-based engineering (MBE) integrated framework early in the design process to improve decision-making in the analysis of alternatives (AoA). As a military acquisition policy requirement, the DoD uses AoAs to ensure multiple design alternatives have been analyzed prior to making major investment decisions [1]. Continued improvements in computing capabilities have made it possible for the increased use of MBE and model-based systems engineering tools to generate numerous system design variants, simulate their performance, and analyze the results in a shorter time frame [2]. However, as the number of system design variants analyzed continues to grow, the subsequent analysis of the resulting large volumes of data can become computationally expensive and time-consuming.

10.1.2 Set-Based Design

Tradespace exploration (TSE) techniques support the design and development of more resilient systems by providing researchers, analysts, and decision-makers with a more complete understanding of capabilities, gaps, and potential trade-offs

Z. Wade (✉) · G. S. Parnell · E. Pohl · E. Specking
University of Arkansas, Fayetteville, AR, USA
e-mail: zwwade@uark.edu

S. R. Goerger
Engineer Research and Development Center, Vicksburg, MS, USA

S. Adams et al. (eds.), *Systems Engineering in Context*,
https://doi.org/10.1007/978-3-030-00114-8_10

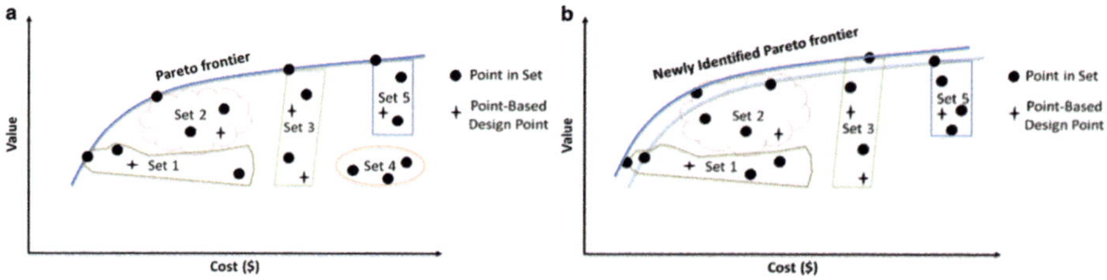

Fig. 10.1 (**a**) Set-based design with sets of alternatives defined, compared to point-based design. (**b**) Set-based design iteration, the tradespace no longer exploring dominated Set 4 and generating more alternatives in the efficient sets

between proposed system alternatives. Furthermore, decisions may be made throughout a system's life cycle; thus, redefining the systems required capabilities as well as its producibility, performance, sustainability, and cost [3]. A goal of TSE during the DoD acquisition process is to provide decision-makers with a more complete understanding of candidate system component choices and their impacts on various missions across the joint war-fighting environments [4]. To perform effective TSE requires a large set of alternatives which in turn lends itself to the use of set-based design techniques.

Compared to point-based design techniques, tradespace exploration with set-based design considers a considerably larger number of possible solutions and enables converging on a final solution or solution subset. Set-based design is a technique of explicitly considering sets of design choices. A design set is a group of design alternatives classified by sharing one or more, but not all, specified design choice(s). Considering a larger number of solutions from the start increases the likelihood of identifying the viable solutions along the efficient or Pareto frontier. An illustration of the motivation for set-based design may be seen in Fig. 10.1a. The investment to more fully define and explore the tradespace in the beginning provides the opportunity to converge more quickly toward discerning a solution on or near the Pareto frontier that may have been missed in a traditional point-based design process. For DoD and ERS, set-based design is useful for projects with a large number of design variables, tight coupling among design variables, conflicting requirements, flexibility in requirements allowing for trade-offs, or technologies and design problems that are not well understood [5].

Design spaces are multidimensional abstractions often containing design decisions of discrete and continuous variables. By definition of continuous variables, the design space is often infinite, and while set-based design greatly expands the Pareto frontier exploration beyond point-based design, set-based design still only explores a finite number of alternatives and may not enumerate an alternative defining the "absolute" Pareto frontier. A potential advancement to address this issue is a technique used in this paper termed iterative set-based design. Iterative set-based design is the process of eliminating inefficient sets and refining efficient sets. Inefficient sets are sets which are clearly dominated through statistical mean comparison in the tradespace, i.e., having lesser or equal value and higher cost. The dominated sets correspond to inefficient design choice(s) and are removed from the design space. Re-initializing set-based design with thousands of alternatives increases the distribution of exploration in efficient sets, potentially exploring more of the design space. Figure 10.1b illustrates iterative set-based design using Fig. 10.1a, where Set 4 has been identified as an inefficient set and the design choice defining Set 4 has been removed. Removing Set 4 resulted in more points identified in efficient Sets 1, 2, 3, and 5 and an expanded Pareto frontier.

Using TSE with set-based design to gain a more complete understanding of the capabilities and limitations of multiple variants, analysts can create a tradespace with quantitative and qualitative characteristics and performance variables or ilities. In systems engineering, "ilities" are desired system properties. "Ilities" describe the system (nonfunctional) rather than specific system behaviors" [6]. Ilities are measured in terms of value added and may include factors such as availability, reliability, survivability, and resilience. Many of these factors can be considered in the aggregate or subdivided to add additional fidelity based on time. In traditional point-based design, each alternative is measured holistically in each performance category. This must also be true with set-based design but with the additional requirement that performance tradeoffs are understood. Set-based design must have a continually improving sense of the trade-offs within the design space in terms of system performance and value by extension. This may be accomplished by defining performance functions through modeling and simulation. Though this requirement does add effort to model building, this requirement generates flexibility to change powerful insights for AoAs. Cost may also be considered in the aggregate of total life cycle cost or disaggregated into smaller time frames and amounts based on system life cycle phases such as research and development, acquisition, operation and sustainment, and decommissioning. Each cost factor is likely to have a varying degree of confidence based on how far in the future the cost will be incurred or, in the case of operations and sustainment costs, the frequency of use and environment of employment. This paper will use multiple objective decision analysis (MODA) to define the value of a given alternative in the tradespace and to understand the importance and value impact of each factor [7].

10.1.3 Resilience in an Engineered System

As an ility, resilience is a factor of interest to the DoD that may be disaggregated to provide additional insights into resilient system alternatives. In the ERS program, an engineered resilient system can be defined as "a system that is able to successfully complete its planned mission(s) in the face of a disruption (environmental or adversarial) and has capabilities allowing it to successfully complete future missions with evolving threats" [8]. Two quantifiable types of resilience are proposed in this definition: mission resilience and platform resilience which contribute to the short- and long-term capability of the system. The illustration in the paper will incorporate mission resilience; defined as the expected performance maintained and recovered by the system in the face of potential performance loss, the system suffers under a

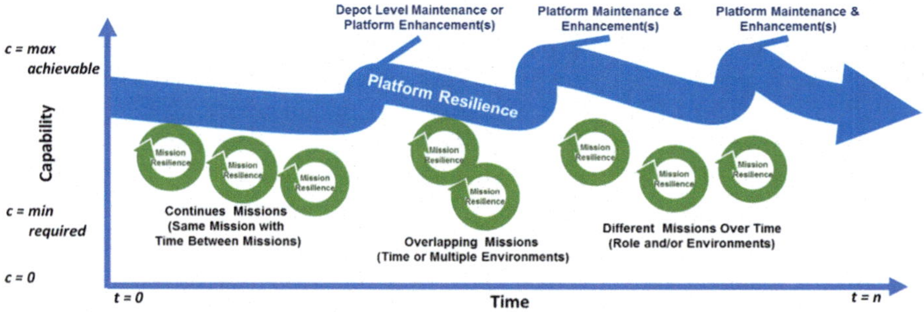

Fig. 10.2 Interaction of mission and platform resilience contributing to capability over time

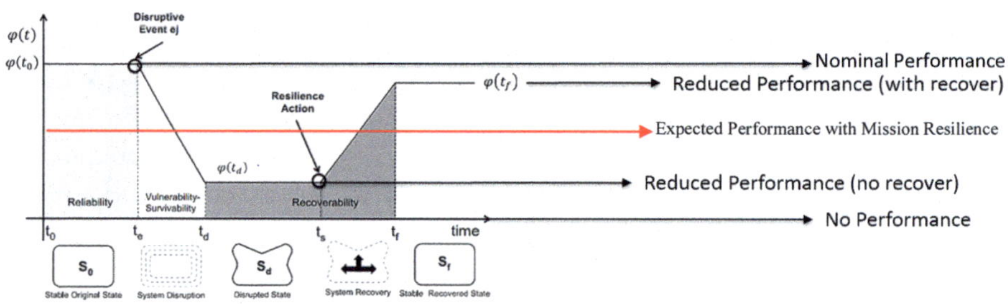

Fig. 10.3 System performance at sequential states of time given a threat [9]

threat. The contribution and interaction of mission and platform resilience within the resilience ility may be seen in Fig. 10.2.

The focus of this paper is primarily set-based design with integrated mission resilience; platform resilience is reserved for future work. Figure 10.3 illustrates system performance in a single objective at various system states with respect to mission resilience. The calculation of expected performance given a threat may be seen in Eq. (10.1) as the formulaic addition of resilience into the mission chain and is a modification of Specking et al. [8]. In a time period which allows for a recovery action, the expected performance with resilience is within the value measures of the MODA value model.

$$E[P] = \frac{1}{n}\sum_{t=1}^{n} ART_t\big(R(W_t)p + R(S_t)\big(R(E_t)p((1-d_t)+(d_t r_t)) \\ +((1-R(E_t))p(1-d_t)) \tag{10.1}$$

Equation (10.1) calculates the expected value of performance (P) with mission resilience shown in Fig. 10.3 for a single alternative in a single performance measure. The expected mission resilient performance is the average performance value given threats (t) 1 to n weighted by probability of threat occurrence (T). Variables include availability (A) and reliability (R) calculated using the standard methods. Equation (10.1) includes probabilities of withstand (W), survive (S), and restoration (E). Withstand is the ability of a system to absorb a threat without affecting performance. Survive is the ability of a system to maintain some level of performance in the face of a threat. Restoration is the ability of a system to restore some previously lost level of performance after a threat has occurred. Equation (10.1) also defines three additional elements, nominal performance (p), degradation caused by a threat (d), and recoverability from a threat (r). Nominal performance is the performance calculated by a performance function before system properties and threats are considered. Degradation is the immediate loss in system performance given a threat has been encountered, expressed in a percentage of nominal performance. Finally, recoverability is the percentage of performance lost in degradation from a threat able to be recovered within the mission time frame [8]. All of these parameters are measurable through modeling and simulation.

This paper will illustrate ERS set-based design by adapting and updating a squad enhancement study for ERS by MacCalman [10]. The squad enhancement study consists of four independent systems: recon unmanned aerial vehicles (UAVs), rifle advancements, body armor advancements, and robots. The following section will describe the modeling methodology for the systems, resilience calculations, the use of set-based design, results, and insights from the illustration. Finally, conclusions and future work will be summarized.

10.2 Squad Enhancement Illustration

10.2.1 Squad Enhancement Design Space and Modeling

The squad enhancement illustration is an example of complex system modeling. In total, the design space is composed of 30 design choices across 4 independent systems including 14 discrete design characteristics and 16 continuous design variables. By definition, the design space is a 30-dimensional abstraction with infinite possibilities of alternatives. To relate the design space to the tradespace, a value model and cost model were defined. The value model created is based on the functions outlined by MacCalman and expanded with objectives and value measures in the MODA model. A cost model was built from the combination of the independent cost models for each system based on cost estimations from MacCalman's study. In the MODA implementation, nine measurable objectives were defined from the four functions, and swing weights were assigned using a swing weight matrix [6]. As mentioned in the introduction, performance functions were enumerated to enable effective set-based design. These relations were obtained from both physics equations (e.g., squad speed as a function of enhancement weight) and agent-based simulation data gathered by MacCalman with regression analysis performed (e.g., beyond line of sight awareness as a function of UAV abilities, robot abilities, and rifle scope abilities) [10].

10.2.2 Mission Resilience Integration

The first step in integrating mission resilience in the value model is identifying which performance measures could be directly affected by an environmental or adversarial threat. Based on MacCalman's work, it was determined that the natural performance measures were susceptible to threat: beyond line of sight capability, detection distance, and maneuverability (three of nine measures). To calculate mission resilience according to Eq. (10.1), performance functions for the ilities necessary in Eq. (10.1) were defined through simulation data found in MacCalman's work and diagramed as the probability tree seen in Fig. 10.4. Other probability trees similar to Fig. 10.4 were created for each mission resilience performance measure. The probability tree is similar to those commonly found in the DoD for system performance with the added branch of partial survivability (probability of survive in Eq. (10.1)) with an associated chance of a recovery state given sufficient time.

To calculate mission resilience in Eq. (10.1), this particular model assumes the worst-case threat is certain to occur. This assumption is one that may also be made outside of this illustration, but may not be necessary due to dedicated threat assessments and mission simulations, particularly common in the area of national defense. By using the worst-case assumption, the demonstration allows for obvious analysis of mission resilience across the alternatives. In Fig. 10.4, the expected performance with mission resilience value is calculated by solving the mission chain as a probability tree. Without mission resilience, beyond line of sight performance calculated from Fig. 10.4 in the mission chain is 0.27 (nominal performance times full survival times reliability times availability). With mission resilience, solving Eq. (10.1) yields beyond line of sight performance as 0.34 for Fig. 10.4. The value model interprets these quantities without resilience as a 2.78 value score and with resilience a 6.11 value score in the beyond line of sight performance measure. If any performance with or without mission resilience is lower than the minimum acceptable performance in a single value measure, as defined by the value model, the entire alternative is infeasible. In this illustration, no alternatives were deemed infeasible due to this condition.

10.2.3 Set-Based Design and Iterative Set-Based Design

Set-based design was implemented to better understand the complex design space with the increased complexity from calculating mission resilience. The Fig. 10.5 demonstration uses probability management as a Monte Carlo simulation software to generate 10,000 alternatives in the design space [11]. Probability management does this by creating 10,000 repeatable pseudorandom numbers uniformly distributed in each design choice in the design space. An alternative is generated by the combination of the continuous or discrete design choices selected from the random number in each of the 30 design dimensions. Each alternative has an associated value calculated

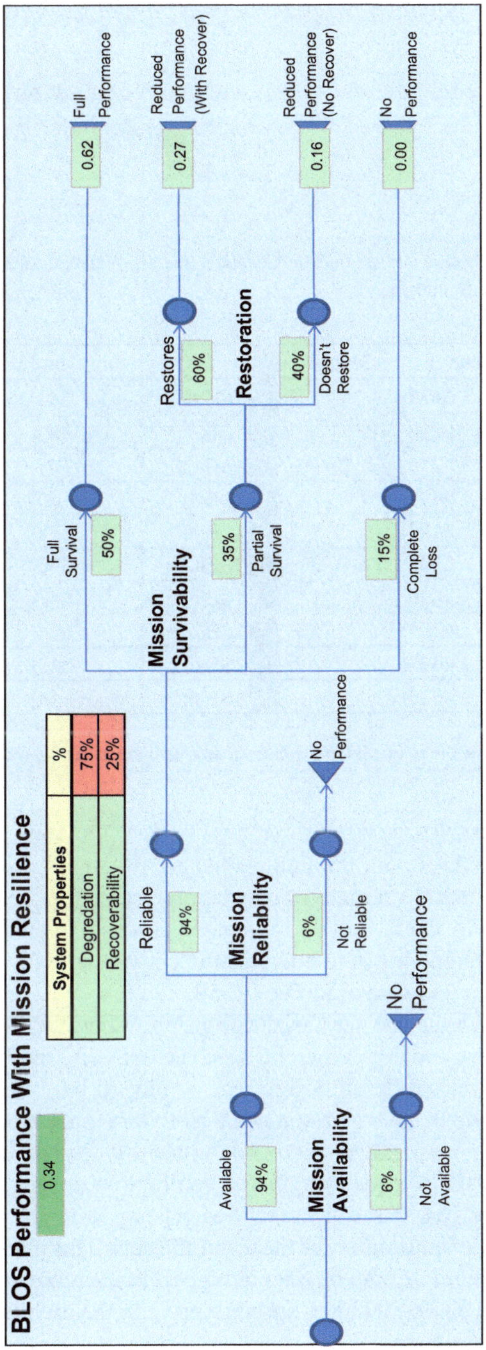

Fig. 10.4 Mission resilience calculation implicit in the performance calculation for the beyond line of sight value measure

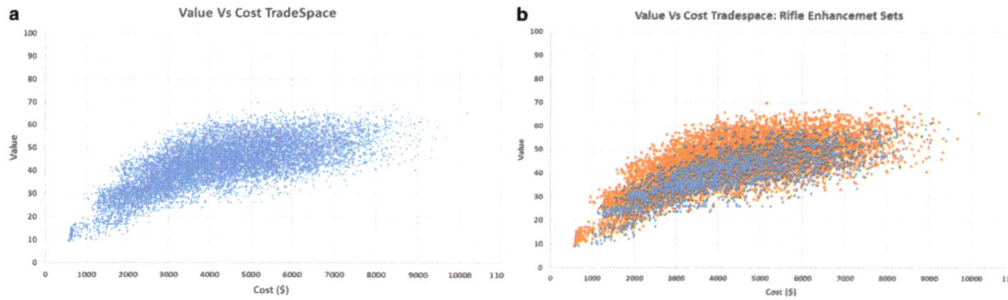

Fig. 10.5 (**a**) 10,000 alternatives in the tradespace. (**b**) Same alternatives colored as defined sets of the rifle enhancement design choice

Set	Num Obs	Mean ($)	Groups
1SBin:10	683	5472	A
1SBin:9	690	5398	A B
1SBin:8	641	5215	B C
1SBin:7	680	5190	B C
1SBin:6	652	5127	C D
1SBin:5	698	5007	C D
1SBin:4	647	4890	D E
1SBin:3	674	4701	F
1SBin:2	617	4639	F
1SBin:1	714	4635	

Set	Num Obs	Mean (Val)	Groups
1SBin:10	683	45.8	A
1SBin:9	690	45.9	A
1SBin:8	641	45.4	A
1SBin:7	680	46.3	A
1SBin:6	652	46.4	A
1SBin:5	698	45.9	A
1SBin:4	647	46.2	A
1SBin:3	674	45.7	A
1SBin:2	617	46.1	A
1SBin:1	714	46.4	A

Fig. 10.6 Set definition, number of observations, means, and statistical grouping. Inefficient sets are highlighted

from performance functions with implicit resilience and an associated cost from the cost model. Thus, all 10,000 alternatives may be represented in the value versus cost tradespace as seen in Fig. 10.5a and disaggregated as sets. An example of identified sets grouped by the design choice to enhance the combat rifle (orange points) or to not enhance (blue points) is shown in Fig. 10.5b.

In this illustration, the initial set-based design run identified 40 unique efficient points lying within presumably efficient sets. To classify inefficient and refine efficient sets, an iterative set-based design was employed by examining the probability management outputs using a statistical software package, Minitab. Every pseudorandom number was grouped into sets either by discrete design choices or partitioned into intervals of equal length for continuous design variables. Of the 30 design dimensions, 184 sets were identified. Using Minitab, Tukey's pairwise comparison grouped the value and cost means of the sets. This allowed for detection of inefficient sets; an example may be seen in Fig. 10.6, the recognition of inefficient sets sharing the design choice within a speed interval in the architectures of the first UAV.

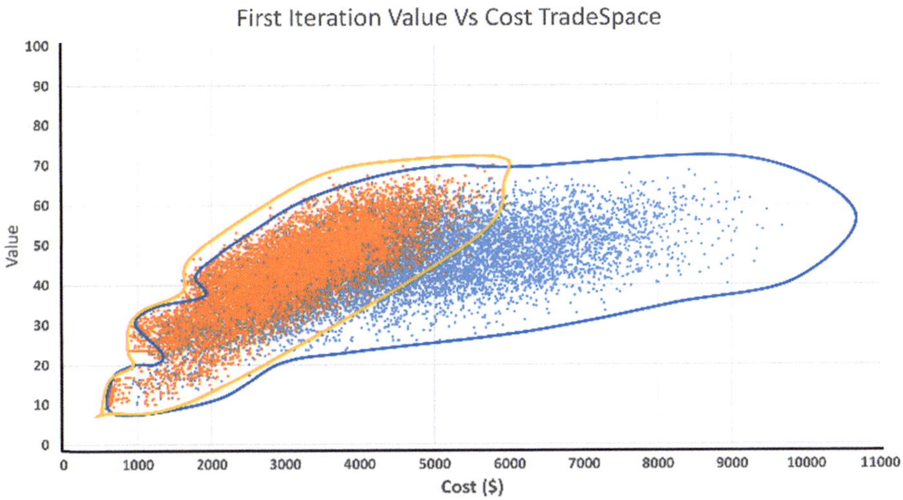

Fig. 10.7 First run of set-based design alternatives and area are in blue; after iteration set-based design alternatives and area are in orange

The inefficient sets including those in Fig. 10.6, totaling 40 of the 184 sets, were eliminated as possible design choices. By re-initializing set-based design, the remaining efficient sets are refined through more investigation. Using Fig. 10.6 as an example, the efficient sets in the initial set-based design run totaled 2,005 alternatives. In the second run, the total alternatives sharing one of those design choices equaled 6,696 alternatives. This additional investigation revealed 44 previously unidentified efficient points, and 23 of the 40 efficient points identified in the first run are no longer on the Pareto frontier. A visual graphing of the change in investigated tradespace may be seen in Fig. 10.7, where the alternatives and area of the tradespace colored in blue are the results of the initial set-based design run and the tradespace colored in orange are the results of the first iteration in iterative set-based design.

In this illustration, iterative set-based design is performed once. However, the statistical methods to detect inefficient sets may again be applied to the resulting data, and perhaps additional inefficient sets may be eliminated. Further research is needed to identify how many iterations are required for a desired confidence interval.

10.2.4 Illustration Insights

The first insight provided by the squad enhancement illustration is the ability of the model to handle the complexity of the illustration tradespace. The model first generates 10,000 alternatives out of an infinite design space and then calculates 10,000 value and 10,000 cost measures while performing 120,000 performance estimates and 40,000 cost estimates. After an update to the design parameters, it was able to recalculate all new values and costs in near real time. The architecture of the models used in the illustration was able to handle such complexity in the processing platform of Microsoft Excel using SIP math. It also appears to meet the expectation of being adaptable to requirements changes from the increasing knowledge learned from simulation and test data.

The second insight gleaned from the illustration is the merit of a mission resilience calculation. In terms of the example listed in the previous section (Fig. 10.4), incorporating resilience into the design alternative measured 0.06 additional performance which is not found in traditional mission chains (a 21% increase). Of additional importance is the measurement of this change in a value model which enables understanding of the additional resilience performance in terms of value. In Fig. 10.4, mission resilience measured 3.33 additional value score (a 119% increase). This distinction enables decision-makers in the design process to understand mission resilience options through the lens of increasing system value and affordability.

The third insight gathered from the illustration is the potential for set-based design to better identify the Pareto frontier compared to point-based design. The initial run of set-based design generated 40 unique efficient points. In the event that no further analysis was necessary and the adaptability component of set-based design was not required, it is highly improbable point-based trade-off analysis would reveal 40 efficient solutions to the decision-maker. In the DoD context, it is rare for such a study to even consider a total of 40 alternatives. Thus, from this illustration, it appears the increased investment in creating performance functions for set-based design analysis may yield more valuable, more affordable alternatives for the decision-maker to consider.

The fourth and final insight provided by the squad enhancement illustration is the ability of iterative set-based design to explore efficient sets in greater depth and identify even more efficient solutions than set-based design. The statistical methods used in iterative set-based design benefit from the large amount of observations in each set for comparison. Iterative set-based design appears to follow its mathematical principles in this illustration, clearly recreating the tradespace away from inefficient sets with a greater focus on the efficient sets. In identifying 44 additional efficient solutions, with the potential for more given iterations, iterative set-based design could at some point guide decision-making toward the "absolute" Pareto frontier of alternatives.

10.3 Conclusions and Future Work

This illustration has demonstrated the potential of set-based design for designing more resilient systems, specifically incorporating mission resilience. Set-based design was able to handle the increased complexity and requirements in designing a model with mission resilience. The illustration generated 10,000 alternatives in the design space which were able to be associated and evaluated in the value versus cost tradespace. The mission resilience method implicitly in value proved successful with the assumption of the worst-case occurrence. Thus, the TSE performed in the illustration successfully explored concepts and identified designs with the potential to be resilient systems. Set-based design expanded the tradespace much further than traditional point-based design. In addition, the 40 efficient points identified in this illustration is greater than will typically be found in most point-based design studies and opens the potential for superior designs to be considered and selected by a decision-maker. The illustration demonstrated iterative set-based design as a tool to eliminate inefficient sets and refine efficient sets. Iterative set-based design displayed the potential to further explore the efficient frontier in a system including a system with resilience, perhaps approaching the "absolute" efficient frontier with repeated iterations. Finally, the illustration demonstrated the usability of set-based design in creating an adaptive model which provides valuable insights into changes, mission resilience factors, and efficient solutions in near real time.

Future work for designing engineered resilient systems with set-based design includes three main areas of interest. The first is the removal of the worst-case occurrence assumption. One way this may be accomplished is through threat assessments and simulations in an active demonstration model. The second is the continued iteration of iterative set-based design. It is theoretically possible that repeated iterations would yield a narrow tradespace where the differences across all sets (as narrowly defined as they may be) are not statistically significant. It is conceivable that the model could be considered to have found the "absolute" efficient frontier and certainly the most efficient sets possible. Finally, future work proposals include the definition and evaluation of explicit, platform resilience. With the incorporation of platform resilience, the full disaggregation of resilience may then be considered within set-based design and be able to fully satisfy the goals and purpose of the engineered resilient systems program.

References

1. U.S. Office of Management and Budget. (2008). *Circular no. A–11, preparation, submission and execution of the budget*. Washington, DC: Executive Office of the President.
2. Rinaudo, C., Buchanan, R. K., & Barnett, S. K. (2016). Considerations for analyzing resiliency in systems engineering. *2016 Industrial & Systems Engineering Research Sessions (ISERC)*. Anaheim, CA: Institute of Industrial Engineering (IIE).
3. Kelley, D., Goerger, S. R., & Buchanan, R. K. (2016). *Developing requirements for tradespace exploration and analysis tools*. Institute of Industrial Engineers (IIE) Annual Conference, 2016 Industrial and Systems Engineering Research Sessions (ISERC), *Anaheim, CA*.
4. Spero, E., Avera, M., Valdez, P., & Goerger, S. (2014). *Tradespace exploration for the engineering of resilient systems*. Conference on Systems Engineering Research (CSER).
5. GovEvents. (2017). *Design sciences series: Set-based design*. Retrieved from GovEvents. https://www.govevents.com/details/24509/design-sciences-series-set-based-design/
6. Long, A. (2012). *Proposed unified 'ility' definition framework*. Presentation at the 15th Annual Systems Engineering Conference, San Diego, CA.

7. Parnell, G. S. (2016). *Trade-off analytics: Creating and exploring the system tradespace*. Hoboken, NJ: Wiley.
8. Specking, E., Cilli, M., Parnell, G., Wade, Z., Cottam, C., & Small, C. (2017). E. P. Tech Report: Graphical Representation of Resilient Engineered Systems.
9. Henry, D., & Ramirez-Marquez, J. E. (2012). Generic metrics and quantitative approaches for system resilience as a function of time. *Reliability Engineering & System Safety, 99*, 114–122.
10. Macalman, A., Kwak, H., McDonald, M., Upton, S., Grider, C., Hill, R., et al. (2015). *Illuminating tradespace decisions using efficient experimental space-filling designs for the engineered resilient system architecture*. West Point, NY: Operations Research Center.
11. Probability Management SIPmath. (2018). Retrieved from Probability Management. http://probabilitymanagement.org/sip-math.html

Chapter 11
Candidate Architecture for an Active Orbital Space Debris Removal System

Rastislav Hronco and Wilson N. Felder

11.1 Introduction

A very common approach to perceived system-level problems is to apply a convenient or self-evident solution without considering the complete system within its context. This results in the almost comical repetition of press conferences in which system managers across a wide variety of industries and communities can be observed promising that "we have taken steps to ensure that this (insert unfortunate result here) will never happen again," as they report on some one-time fix that fails to deal with the underlying system context. To illustrate the application of a systems approach to this type of challenge, we have selected a topical problem—the proliferation of orbital debris—to illustrate a different approach. Our intent in this paper is not to provide a solution to the specific problem of space debris, but to demonstrate that the application of structured systems engineering techniques leads to potential approaches that have not been previously considered. We will begin with a brief review of the space debris challenge in general. We will then provide an analysis of this challenge through the use of the novelty/complexity/technology/pace (NCTP) framework. This high-level analysis suggests a nontraditional approach to debris mitigation involving on-orbit capture of debris along with on-orbit additive manufacture of replacement parts for use in orbital maintenance operations.

R. Hronco (✉) · W. N. Felder
Stevens Institute of Technology, Hoboken, NJ, USA
e-mail: rhronco@stevens.edu

S. Adams et al. (eds.), *Systems Engineering in Context*,
https://doi.org/10.1007/978-3-030-00114-8_11

11.2 Background

Ever since Sputnik in 1957, humanity has been putting objects into space with every rocket launch. The majority of these objects are destined for the Earth's orbit, while others are passing through the Earth's orbit and moving on to deeper space. Some of these objects such as satellites and manned capsules serve a purpose; others such as inactive spacecraft, fairings, rocket bodies, payload covers, material released during any stage of ascent or deployment of spacecraft, object collisions and explosions, and many others do not. In general, this is the accepted definition of orbital debris: something that is in the Earth's orbit, but which no longer has a useful purpose.

Orbital debris poses a threat to the continued use of local Earth space. This threat is very broad in nature, affecting everything from ground-based science missions using telescopes, to every form of science mission and also all government and commercial endeavors in space. Examples of science missions would be the Earth Observations Satellites (EOS), which provide all of the weather data and any activities on or around the International Space Station (ISS). Examples of government endeavors would be of anything related to military activities and in the realm of national security. Examples of commercial endeavors are increasing by the day, from science missions to money-making endeavors in communications (radio, TV, telephone, etc.), transportation, and exploration, forming the beginning of the commercialization of space. As we commercialize space and move out into the solar system, we need to lay a foundation for design, use, and disposition of all we take into space. This foundation needs to start now with the immediate need to clean up the near-Earth orbits which we have neglected for so long. This approach reflects growing appreciation for the need to exercise life cycle governance of complex aerospace systems [1].

In general, for the first 25 years of spaceflight, the useful orbits have been thought of as a vast open ocean when you can launch an endless number of ships. Although the space community did realize that the orbital resource was finite and that there would be issues in the future, it was believed that with the slow increase in numbers of rockets and satellites, we had centuries before debris would become an issue. Over the years, however, we have found that vehicles returning from the orbit showed damage by debris at a greater than expected frequency. The focus turned from understanding meteorite and large object collisions to that of small manufactured objects. Once the focus changed, the list of possible debris began to grow to include nonoperational spacecraft, launch vehicle stages, mission-related debris, fragmentation debris of spacecraft, and rocket bodies. As we learned more about the debris situation from ground radars and telescopes and the US Space Surveillance Network, additional items were added to the list, such as coolant residue, paint flecks, and slag from solid rockets. Then in 2007 the Chinese shot down one of their own satellites (Fengyun-1C). This single event increased the amount of debris in the LEO orbit by over 3000 pieces, as shown in Table 11.1 [2].

Then in 2009 an accidental collision between an operational US satellite (Iridium 33) and an inactive Russian satellite (Cosmos 2251) added an additional 2000 pieces of debris to the LEO orbit. At this point some orbital models start predicting that the

Table 11.1 Top 10 breakups as of 2016

Rank	International designator		Common name	Year of breakup	Altitude of breakup	Cataloged debris	Debris in orbit	Assessed cause of breakup
1	1999	25	Fengyun-1C	2007	850	3428	2880	Intentional collision
2	1993	36	Cosmos 2251	2009	790	1668	1141	Accidental collision
3	1994	29	STEP-2 rocket body	1996	625	754	84	Accidental explosion
4	1997	51	Iridium 33	2009	790	628	364	Accidental collision
5	2006	26	Cosmos 2421	2008	410	509	0	Unknown
6	1986	19	SPOT-1 rocket body	1986	805	498	32	Accidental explosion
7	1965	82	OV2–1/LCS 2 rocket body	1965	740	473	33	Accidental explosion
8	1999	57	CBERS 1/SACI 1 rocket body	2000	740	431	210	Accidental explosion
9	1970	25	Nimbus 4 rocket body	1970	1075	376	235	Accidental explosion
10	2001	49	TES rocket body	2001	670	372	80	Accidental explosion
Total						**9137**	**5059**	

*As of 04 January 2016

Kessler syndrome has begun. The Kessler Syndrome predicts the cascade of events that will, over time, lead to the LEO, MEO, GTO, and GEO orbits unusable for spacecraft and will make it difficult if not impossible for man to safely leave the Earth. Approximately 95% of the objects in this illustration are orbital debris, i.e., not functional satellites. The dots represent the current location of each item. The orbital debris dots are scaled according to the image size of the graphic to optimize their visibility and are not scaled to Earth. This image was generated from a distant oblique vantage point to provide a good view of the object population in the geosynchronous region (around 35,785 km altitude). Note that the larger population of objects over the northern hemisphere is due mostly to Russian objects in high-inclination, high-eccentricity orbits [3]. With these breakup events, there are many scientists that now believe that passive debris removal is no longer a viable option. Even if we no longer launch anything into orbit, the debris clouds will continue to grow until there is a uniformly distributed debris field at all the orbits. The international community consensus is that some kind of active debris removal (ADR) is the only way to clean up the orbits so that we can continue with all of our endeavors in space.

11.3 Problem and Approach

Cleaning up orbital debris is straightforward; you simply have to find a way to track an object, capture it, and remove it. Yet no proven solutions exist to the problem of debris removal. Using the novelty/complexity/technology/pace (NCTP) framework [4–6], we have categorized the previously suggested options and show that these all have similar framework challenges. We next present a high-level concept of operations (CONOPS) depicting a possible end-to-end solution, based on recycling and reusing the existing debris in each of the Earth's orbits. The concept uses captured debris as the source material in the additive manufacturing process. An additional feature is the use of localized vision systems to perform local navigation and capture of the debris. An expanded NCTP framework analysis including additional cost-related axes demonstrates that this concept is superior to approaches previously considered. We conclude by offering a notional top-level functional architecture for the proposed system.

As space becomes more approachable to the commercial and educational sectors, concepts for ADR also have been increasing. In the past space was an arena in which only the government's scientists and militaries were able to be effective, thus limiting the feasibility of the approaches taken to debris. Past practice has depended on guidelines that depend on passivation as the technique to meet the objective of not generating additional debris. Included in these guidelines are items such as burning residual propellants to depletion, venting propellant lines and tanks, venting pressurized systems, preventing recharging of batteries or other energy storage systems, and de-energizing control moment gyroscopes [7]. Several active debris removal (ADR) methods have been presented to the industry that range from the self-evident robot capture approach to more creative and exotic ones such as using ground-based or space-based lasers.

There are several tether-type solutions, one of which is based on electrodynamic properties. When a conductive wire moves through the Earth's magnetic field and generates electric potential, the current generated causes the object to drop in altitude eventually burning up in the atmosphere. The other tether solution is one based on changing an object's momentum, using a tether, and swinging it into a lower orbit for disposal. Both the tether options are novel, complex to implement, and costly to use [8].

The laser removal methods are based on destroying a small fraction of a piece of debris which creates a thrust and causes an orbit alteration. The alteration would be enough to shorten the time in orbit. This approach is suited for smaller debris which would be easier to partially vaporize. This approach is very novel, but as with the tethers, it is extremely complex, and lasers come with a large price tag [8].

There are also concepts where additional means of propulsion are attached to the debris item in order to either lower or raise the object out of harm's way. This is also costly and complex to implement. Other concepts involve adding solar sails to increase the drag on the object and decrease its orbit (this approach applies to LEO only) and again costly and complex to implement.

Then there are the "space tug" or on-orbit services (OOS) types of systems, using highly maneuverable spacecraft to capture and re-orbit larger objects to disposal altitudes, either by capturing and dragging them to a new orbit or servicing the debris to a point where it can take itself to the disposal orbit. Even though this approach has additional incentive, it again would be very challenging and very costly to implement [8].

All of these approaches increase the potential risk of generating additional debris. De-orbiting approaches and the 25-year rule for new spacecraft increase the reentry risks. The highest of these risks are human death if a large enough piece makes it through the atmosphere. NASA has found that there has been, on average, one piece of debris from the LEO orbit entering the atmosphere every day for the past 50 years [8].

11.4 Analysis Using NCTP

With a good breadth of the proposed ADR solutions, they are placed in the NCTP framework diagram, and we can review their complexity, technology, novelty, and pace for ADR. Figure 11.1 shows the NCTP framework with all the ADR solutions overlaid on each other and with each ADR solution in its own framework.

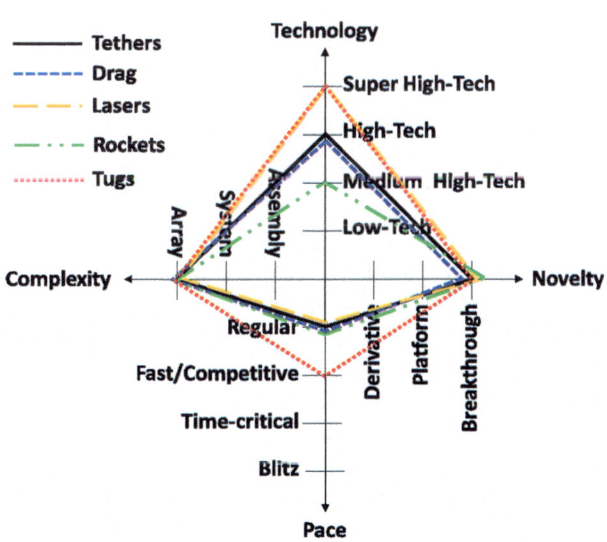

Fig. 11.1 NCTP classification for potential solutions

11.4.1 Complexity

In general, the solutions are all very complex (array) systems of systems that require their own support structure as with any space mission. These systems of systems would work together to accomplish the overall goal, including everything involved with design, Preliminary Design Review (PDR), Critical Design Review (CDR), and Technology Readiness Level (TRL) and also every aspect of the launch flow, launch vehicle selection, testing, integration, and launch. Then you have the commanding and telemetry aspects of these concepts, which need ground stations, maintenance plans, and retirement plans. The above complexity analysis results in almost the same levels for all of these approaches.

11.4.2 Technology

We see a slight difference in the technology area among a few of these concepts; most are categorized as high-tech, with the laser and tug options being classified as super high-tech. The laser approach has some unique technical challenges; for a space-based laser, it has some extreme challenges that have never been attempted in space. For the ground-based approach, the challenge is only slightly less since some of the concepts have already been proven in other military applications, but the power and targeting required to hit orbital debris would be entirely new. There is also commonality again in that they are technically challenging because you are sending something into space, but this has been happening since the late 1950s. The most challenging part of tether, drag, rocket, and tug approaches is the need to track, maneuver, and capture a piece of debris, all of which are unproven.

11.4.3 Novelty

To categorize satellites in the area of novelty in the NCTP vernacular is a bit of a perspective conversation one has to have. The definition of “breakthrough” is an introduction of a new concept or a new idea, or a new use of a product, which customers have never seen before. The definition of “platform” [4] is a new generation of an existing product or family of products. So, the classification can be either. One can look at the options from the perspective of a new satellite with a new payload on it, which can also be interpreted as just a new generation of the existing satellite product. On the other hand, if you think of each satellite being a unique specific application that has to work as a whole and has never been done before, then it’s definitely a breakthrough. Breakthrough is the way these approaches should be categorized, they may have ties back to the technology that normal satellites use, but there is a breakthrough in the concept of operations for these ADR system concepts.

11.4.4 Pace

The NCTP category of pace is an interesting one as it relates to ADR, unlike the Apollo 13 crisis categorization where time is critical. These options for the most part have not had a pace, since ADR has never been seen as a viable option to the debris problem. But now with more attention on the problem of debris and the commercialization of space, all these options are now on the map as regular pace. But since there is some urgency and due to the potential for some commercial competition to remove debris to potentially make a profit, by charging the various governments to protect their high-value assets in the orbit, the category for the space tug should be moved to fast-competitive.

11.5 Evaluation of Proposed Concept Using Extended NCTP Framework

Since these proposed solutions to the ADR problem are similar in their NCTP categorizations, the recommendation is to look at this problem with some project management perspective, thus requiring the addition of legs to the NCTP framework, focused on overall costs. The addition of legs to the NCTP framework was an idea that grew out of discussions with colleagues as a way to help further differentiate similar NCTP frameworks. The first is the "development cost"; this will be defined as the total cost associated with the research and development and deployment of the ADR solution. The second is the "operations and maintenance" leg, which is defined as the cost associated with post-deployment of the system. With these additional legs, only one of the options stands out as being feasible, the ground-based laser. See Fig. 11.2 below for the new NCTP framework. We can see this since all the other options are all unproven concepts that will all be costly (high) to develop, as with any satellite, and we already know that the O&M of most satellite systems is medium to high. The ground-based laser has already been developed for military uses. Since the military developed the laser technology, they also developed the O&M infrastructure. Also, the estimated cost per shot of this previously developed chemical laser was only $1000 USD [9], primarily the cost of the chemicals. Even though the tactical high-energy laser was canceled for its military application, the technology was proven and successful [10]. With innovations in laser technologies, such as solid-state lasers (SSL), the cost per shot will go down.

Despite the obvious advantages, lasers are still perceived of as a weapon, which you could use to shoot down satellites and aircraft. This view is shared by all the major countries in the world and it would be extremely difficult to change. International collaboration, agreements, and treaties would all have to be worked out before this can become a viable solution. There is another drawback to success of

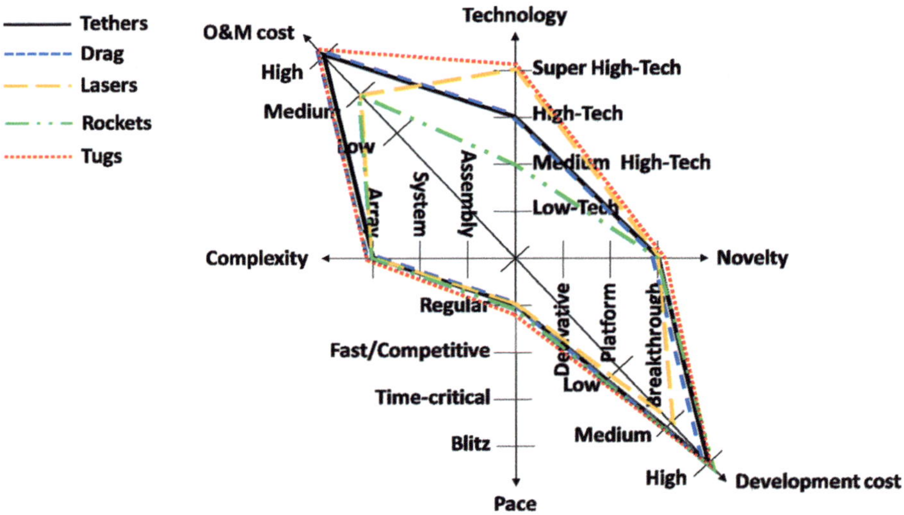

Fig. 11.2 NCTP Classification for the proposed ADR solutions with additional legs

commercial travel to space; the private sector will find a way to capitalize on the removal of orbital debris.

11.6 Alternative Concept (ADRSS) Development

The common goal for all ADRs is to remove debris from the orbit; the approach that the above ADR systems have taken is to destroy the debris by some means. There is another alternative to destroying the debris; it is reusing it. Our proposed architecture takes advantage of the reuse of materials captured in space. The following sections describe this architecture, its benefits, and challenges.

Compared to the concepts analyzed earlier, the proposed Active Debris Removal and Recycle Spacecraft (ADRRS) shifts the end point for a piece of debris. The focus of the other concepts is to deorbit the debris and have it burn up into the atmosphere. The ADRRS concept is to take the captured debris and process it into usable raw materials for additive manufacturing (3D printing) in space, saving millions of dollars in raw materials for new construction in space and additional millions of dollars in design, build, and launch of new spacecraft. The majority of the materials used on spacecraft and launch vehicles have been metallic in nature, primarily aluminum or an aluminum alloy [11]. There are industry-level 3D printers already printing space quality parts with aluminum among other alloys and metals, which makes the ADRRS end-to-end CONOPS an attractive approach to orbital debris removal.

Elements from existing concepts of operations (CONOPS) for the active debris removal (ADR) options described earlier are summarized below. The ADRRS

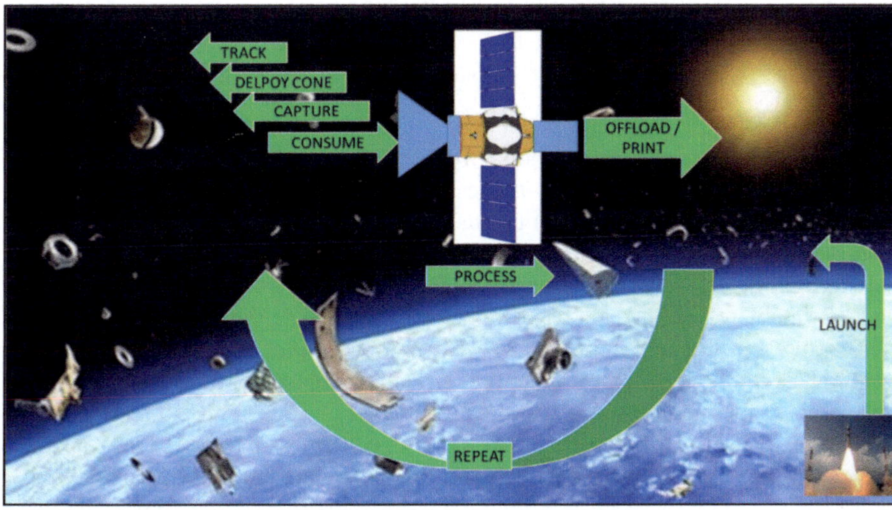

Fig. 11.3 High-level OV-1 diagram of the Active Debris Removal and Recycle Spacecraft (ADRRS)

CONOPS differences are pointed out and elaborated on and also depicted in Fig. 11.3, the ADRSS OV-1 diagram.

In general, all the space-based ADR concepts share the same CONOPS which are as follows:

1. Launch the ADR system (unless it's a ground-based laser).
2. Macroscale. Based on the available catalog of debris, a target is identified.
3. Microscale. Vehicle will find possible debris using its own array of sensors to track and rendezvous with the debris.
4. Capture the debris.
5. Perform deorbit solution. This is the unique task for the type of ADR system, which will deorbit the debris and in most cases the ADR system.
6. Repeat steps 2–5. This step applies to the tug option or a service mission option.

The next steps are the ones that are added to the overall CONOPS for the proposed ADRRS which will begin with the first four steps above and continue with the ones below.

7. Take in or consume the captured debris.
8. Process the debris into raw materials.
   (a) Break down the debris mechanically
   (b) Break down the debris thermally
   (c) Process material into three categories: Metals, plastics, and composites.
9. Report back to user of the amount of raw material on board.

10. Commanded to either off-load the raw material or to locally print something useful.
11. Repeat steps 2–10 (omit steps 5 and 6).

The ADRRS concept shares challenges with other options that require the object to be captured by another spacecraft. Many of these common challenges can be solved by evolving existing technologies. For example, when the ADRRS is close enough to the debris target, it can use its local rendezvous hardware and software. The system will use several technologies such as single-beam laser rangefinder (LRF), light detection and ranging (LiDAR), and onboard active radar to track the target debris in real time to match its trajectory [12]. The deployment of a device that can capture the targeted debris is also a common challenge; for the ADRSS we envision a hypothetical flexible cone-type structure. This cone is flexible in nature so that debris shapes and their attitudes are not an issue during capture. It has the ability to close on its target by being electrically energized. It is then retracted into the main body of the spacecraft. Internal to the spacecraft, there are areas that break down the debris. This is envisioned to be similar in nature to the way electronic waste is processed on Earth where the materials are first mechanically broken apart into smaller physical pieces. Then a thermal breakdown happens with the various metals and plastics liquefied and separated, perhaps using a super high efficient induction heating system. Once this separation is complete, the liquid materials are then cooled into the small micron-sized spheres required for 3D printers.

A top-level functional decomposition of the ADRRS system is shown in Fig. 11.4, which shows the breakdown of the ADRRS system into its main components. Satellites can be broken down into two main parts, the "bus" and the "payload." The "bus" is the satellite itself; it will have the power, communication, and propulsion subsystems. The "payload" is the mission of the satellite and will contain the capture, the processing, and the recycling subsystems. This decomposition makes it possible to swap out a piece of the satellite in favor of another piece or removing a piece that isn't needed. For example, the ADRRS system can be modified by removing the 3D printer. This now forces the raw material to be

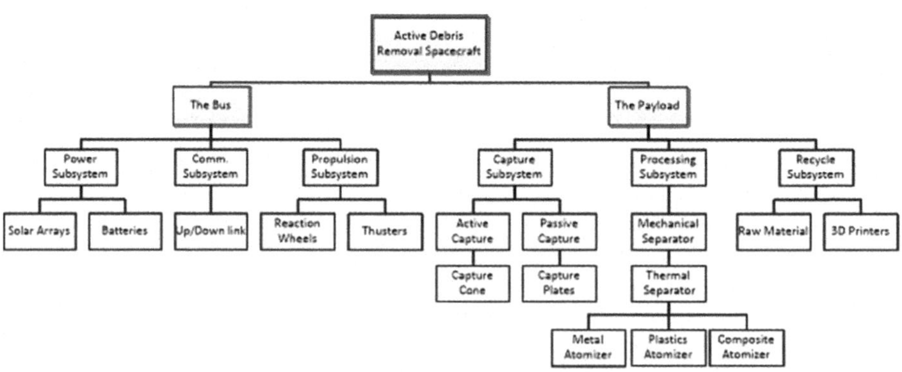

Fig. 11.4 Functional decomposition of the ADR spacecraft

off-loaded instead of being used to directly print parts. The off-load would happen by docking with the ISS or some other orbiting station which can utilize the raw materials to print needed various components.

The ability to dock with the ISS opens up other options for the extension of the ADRRS system, such as the ability to be refueled to continue its mission. Also, once NASA turns the ISS over to the private sector, then the addition of new items to its structure will be needed. The ADRRS would provide abundant 3D raw material being delivered to the ISS at a much better price than sending up material from the surface. The benefits from combining ADR with recycling are twofold, in that it not only cleans up orbits but also provides cost-effective building materials for future space development. Even though the NTCP framework for the ADRRS will look very similar to the other ADR concepts, with high O&M costs, high development, high complexity, high novelty, super-high technology, and a competitive pace, the ability to turn debris into raw materials is the main reason why the ADRRS concept has the potential to rise above the rest of the ADR concepts. The commercial world needs proper motivation to set foot into a new arena; money and cost-effectiveness is this motivation.

11.7 A Note on Debris

The shuttle stopped flying in 2011. Since then there has been very little new information collected on Micrometeoroid and Orbital Debris (MMOD) strikes on spacecraft. The shuttle and the items it brought back from orbit were the primary sources for MMOD information from LEO. Without real-world examples of actual orbital debris strikes, we must rely on modeling, simulation, and high-speed ballistic impact testing, none of which are perfect, but which produce useful data. The new sensor Space Debris Sensor on the ISS improves our situational awareness of the MMOD environment by accurately measuring the submillimeter region in order to provide reliable MMOD impact risk assessments for critical US space assets. This sensor will be one of the few sources of quantitative debris strike data, which will lead to a better understanding of the debris environment, leading to effective spacecraft shielding.

11.8 Conclusions and Recommendations

"It is neither technically nor economically feasible to clean up space, action must be taken to prevent the creation of new debris," is quoted from the European Conference on Space Debris on April 5–7, 1993 (p. 107). The quote encapsulates what has been the mindset of the international space community with regard to orbital debris. This paper has provided a concept and pointed to recent research that shows the above statement to be incorrect, or at least shifting into the realm of possibility. As the privatization and commercialization of space become more prevalent, we will see that options like the ADRRS become reality. The one thing that is clear and has been stated many times is that "corrective action must begin now to forestall the development of a serious problem in the future" [13]. As a spacefaring civilization, we must take care not to contaminate our environment, wherever we may "boldly go."

In developing the ADRRS concept, we noted several shortfalls in technology readiness that would contribute to making the ADRRS technically viable; thus the following areas are recommended for further research. 3D printing or additive manufacturing in space has come a long way and there is a printer onboard the ISS, but it is limited to nonmetallic parts. Something that would be useful in space is 4D printing where you can print something in a small print space that over time will transform into its final desired shape. The most popular method for metal 3D printing is laser sintering; this method is primarily used for the creation of durable metal parts for aerospace industry. Inconel is the metal alloy that is preferred as it is resistant to high heat and stresses. It is also extremely difficult to work with conventional machinery and thus lends itself to 3D printing, additive technology. Further research in this area and its applications in space are also recommended.

Once you have printed a component, you will also need to perform some type of assembly in orbit. Further research needs to be done in the area of assembly techniques in space. This research could yield effective ways to assemble larger objects such as satellites, human habitats, and deep-space vehicles.

In addition to NASA's existing best practices and recommendations as it pertains to the area of shielding, the recommendation is to continue enhancing the Whipple shields that are used and make them more prevalent on all spacecraft, with the stuffed Whipple shield, which have filler in the gaps, with specific focus on more absorbent and lighter-weight materials. These materials should also have the capability to capture and hold the debris. This will help with orbital cleanup of smaller particles.

References

1. Felder, W. N. (2013). A framework for verification and validation of complex aerospace systems. In *2013 aviation technology, integration, and operations conference*. Los Angeles, CA: American Institute of Aeronautics and Astronautics.
2. Anz-Meador, P. (2016, April). Top Ten Satellite Breakups Reevaluated. In: *Orbital Debris Quarterly News* (Vol. 20, 1&2, pp. 5–6).
3. NASA. (2010). [Online]. http://unifiedpoptheory.com/nasa-releases-space-junk-ibook-about-orbital-debris-management-and-risk-mitigation/orbital_debris_1960_2010/
4. Shenhar, A. J. (2001). One size does not fit all projects: Exploring classical contingency domains. *Management Science, 47*(3), 394–414.
5. Sauser, B. J. (2006). A return to the moon: A system engineering management framework and the success of lunar prospector. In *Systems research forum* (Vol. 1, pp. 27–33). Singapore: World Scientific.

6. Sauser, B. J., Reilly, R. R., & Shenhar, A. J. (2009). Why projects fail? How contingency theory can provide new insights–a comparative analysis of NASA's Mars climate orbiter loss. *International Journal of Project Management, 27*(7), 665–679.
7. NASA. (2012). NASA-STD 8719.14 revision A with change 1. NASA-STD-8719.14A.
8. Johnson, N. L., Hoffman, E., & Forsgren, R. (2012). *Orbital Debris Management & Risk Mitigation*. Washington, DC: NASA Academy of Program/Project & Engineering Leadership (APPEL).
9. IMDA. IMDA [Online]. http://imda.org.il/English/Defenses/defense_System.asp?missileId=30
10. Broad, W. J. (2006, July 30). *U.S. and Israel shelved laser as a defense*. New York: The New Your Times.
11. Dunn, B. D. (2016). *Materials and processes for spacecraft and high reliability applications* (p. 667). Cham: Springer International Publishing.
12. Nayak, M., Beck, J., & Udrea, B. (2013). Design of relative motion and attitude profiles for three-dimensional resident space object imaging with a laser rangefinder. In *2013 IEEE Aerospace Conference* (pp. 1–6). Big Sky, MT: IEEE.
13. Anonymous. (1981, July). *Space Debris: An AIAA Position Paper*. AIAA Technical Committee on Space Systems.

Chapter 12
How Do Knowledge Domains Come Together in Systems?

Swaminathan Natarajan, Anand Kumar, Subhrojyoti Roy Chaudhuri, Niruj Mohan Ramanujam, Rajiv Banerjee, Amar Banerjee, and Puneet Patwari

12.1 Motivation and Overview

Engineering is a knowledge-centric activity. The engineering of large systems, such as radio telescopes, involves dozens of knowledge domains e.g. interferometry, signal processing, networking, power, civil construction, control systems, software, availability, safety, security, cost, schedule etc. Each domain has its own models and vocabulary of concepts. A single engineering decision may impact multiple domains. As engineers, we manage this complexity operationally using our (partly tacit) understanding, by creating overall system models, multiple domain-specific models and views, and maintaining and managing consistency among all of them. But what are the underlying principles that govern how all these knowledge domains relate to each other?

Our experience with the Square Kilometer Array (SKA) radio telescope project made it clear to us that there was a fundamental theoretical gap in this area. The SKA project is currently towards the end of a 4-year design phase involving around 500 engineers and scientists in a dozen countries, divided into ~10 subsystem teams. SysML models are created for the overall system and each subsystem. In addition, each team builds domain-specific models using MATLAB, CAD/CAM tools, networking tools, requirements tools, project management tools etc. Each tool has an underlying informational model, yet we are unable to integrate all of them to create a unified model/view that covers all aspects of the system and the associated project. Of course, the situation is not unique to this project, all systems engineering efforts face the same challenge, that at best there are a few vendor-specific integrations among particular tools, but we do not have the fundamental understanding

S. Natarajan (✉) · A. Kumar · S. R. Chaudhuri · A. Banerjee · P. Patwari
Tata Consultancy Services Research, Pune, India
e-mail: swami.n@tcs.com

N. M. Ramanujam · R. Banerjee
National Centre for Radio Astrophysics-TIFR, Pune, India

S. Adams et al. (eds.), *Systems Engineering in Context*,
https://doi.org/10.1007/978-3-030-00114-8_12

needed to reconcile all the disparate system viewpoints into a comprehensive set of system models/views with mutual consistency relationships, and associations with applicable knowledge in each domain. The specific challenges and state-of-the-art are discussed in the subsequent sections.

There are other fundamental gaps as well that the systems research community would like to close. Systems engineering is widely viewed as an empirical discipline, without a foundational theory that provides the basis for practice. It is intuitively clear that systems science is that foundation, yet it has proven challenging to establish the linkage. Along the same lines, systems modelling languages such as SysML are viewed as operationally defined, without well-defined deep semantics that map the model to a comprehensive reasoning structure.

12.1.1 Overview of the Conceptual Model

Our inquiry into the nature of knowledge domains and how they arise led us to the following conceptual structure:

- A distinctive characteristic of systems engineering (and software engineering) compared to other engineering disciplines is that creating solutions in other disciplines such as electrical engineering depends primarily on knowledge in that particular discipline, whereas creating a good systems or software engineering solution requires an understanding of the various other disciplines involved and how engineered artefacts in these disciplines interact. This indicates why we have had difficulty in building theoretical foundations for these fields: a theory of how multiple knowledge domains come together in systems is critical to creating foundations for systems and software engineering.
- Knowledge of interest to engineering (in science, engineering and human activity domains such as banking, sociology and education) can be viewed as *systems knowledge*, generated based on the *systems axiom*: *behavior arises from processes operating over structures, in particular contexts.*
- Systems knowledge is generated by observing or synthesizing systems and looking for patterns of relationships between structures, processes and the resulting outcomes (behavior, associated characteristics and consequences) across groups of observations.
- There appear to be two major approaches to grouping observations for the purposes of knowledge generation: *generalization* and *abstraction*. Generalization groups observations based on commonality of the entities involved, thereby building knowledge about types of wholes (e.g. dogs, banks, telescopes, network routers). Abstraction focuses on particular aspects of interest (e.g. optics, security, network routing) and builds knowledge, abstracting away all information not considered relevant to that aspect.
- Knowledge about wholes includes typical characteristics and concerns associated with the whole, and the aspects that address those concerns i.e. knowledge about wholes includes how its concepts map to relevant aspects. Wholes knowledge

also includes the collection of *context roles* that relate to this whole e.g. knowledge about dish antennas includes roles such as power supply, network link, sky, reference signal, scientist, operator etc., and *role profiles* associated with each role that capture how that role relates to the dish antenna, in terms of structural relationships, interactions, processes, assumptions about characteristics etc. This key insight that *context relationships need to be bundled with entities to produce wholes* results in a composition-friendly formulation: each of these context roles need to be bound to other wholes, whose characteristics must satisfy the role profile. Since, in principle, each whole can belong to a different knowledge domain, this drives how multiple knowledge domains in the context of a system. Role profile bindings define *view mappings* that create bridge across knowledge domain ontologies, based on their relative roles in the particular system of interest.

- We can visualize the relationships involved in the creation and use of knowledge in terms of four worlds: the real world of experience, a system models world (and associated general systems knowledge domains), a world of type knowledge domains, and a world of aspect knowledge domains. Systems engineering involves creating solutions in the real world, using elements available in the real world, based on knowledge in aspect knowledge domains and type knowledge domains. This leads to a conceptual model for systems engineering based on first principles, which maps well to current empirical practice and can be used to reason about practice issues.

It should be emphasized that individually none of the above insights are novel, all of them are pretty much common knowledge in engineering and systems science. What is novel is the articulation and synthesis into a conceptual model that seems promising in its ability to address some of the significant challenges in systems engineering. It should also be mentioned that currently the scope of the work is limited to mechanistic systems. This research is still work-in-progress, and considerable effort is still needed to refine these seed ideas, including improving conceptual rigour, and synthesizing it with existing knowledge and research in systems science and systems engineering. We have initiated a project in the INCOSE systems science working group to explore and refine these ideas. The purpose of this paper is to share the initial thinking with the systems engineering research community to invite feedback, and participation in improving and evolving this preliminary thinking.

The rest of this paper explains the conceptual structure in more detail. Section 12.2 discusses the current challenges in bridging across knowledge domains, and reviews the state-of-the-art. Section 12.3 presents the four worlds model of systems engineering. Section 12.4 indicates some applications of the conceptual structure: creating view mappings to bridge across domain ontologies, creating a conceptual model of systems engineering, and working out a semantic basis for system modelling. Section 12.5 concludes the paper with a discussion of plans for refining the model.

12.2 Challenges in Bridging Domain Ontological Models, the State of the Art

There are several challenges in bridging domain-specific models and integrating them with systems models:

- *Vocabulary*: Each domain has its own vocabulary and concepts. An antenna may be described as a receiver in signal processing, as a dish in the physical viewpoint, and as a load in the power viewpoint. The mappings between these concepts comes from knowledge of the specific system and context, so it is not just a matter of establishing translations across vocabularies directly at the level of domains.
- *Semantic flattening*: The SysML model, which has a view of the entire system, could potentially bridge the gap, since concepts in every other domain and viewpoint map to the corresponding specific system elements. But the SysML model is itself captured in terms of generic systems concepts, such as entities and attributes, losing the semantics of specific domains: both antennas and routers are entities, latency and availability are both attributes, and so on. So the semantic richness needed for bridging is missing.
- *Loss of information*: There may not be a one-to-one relationship between elements across models. Multiple elements in a view may be aggregated in another, or partitioned differently, complicating mappings.
- *Semantic mismatch*: Concepts in one space may not precisely match concepts in another space, even if they have the same name. For example, providers and consumers may have different definitions for when a service is considered to be available. This gap is NOT addressed in this work, it needs to be handled separately.
- *Viewpoints*: Systems can be described from many viewpoints, and knowledge about any particular domain can also be captured and expressed from different viewpoints. Proliferation of viewpoints makes it difficult to create definitive ontologies for domains.

12.2.1 The State of the Art

The area of model-centric engineering [1] is actively working on this issue. Sowa [2] provides a foundational discussion of the nature of conceptual structures and knowledge representation. Shani et al. [3] discusses semantic mediation, an approach for establishing bridges between domain ontologies. Blackburn and Denno [4] recommends the creation of a problem-formulation ontology for a domain of discourse. This involves identifying all the knowledge domains relevant to addressing a particular class of problems (e.g. radio telescope engineering) and combining them to develop a unified ontology for the problem in which their mutual interrelationships are worked out. This is powerful, but creates a combinatorial explosion, and challenges in exchange of information across these discourse-specific ontologies. Ideally, we would like to obtain insights that facilitate a more modular approach. Ideally, such a modular approach should include both systematic derivation of desired views from core ontologies for a domain, and ways to bridge across knowledge domains. The work on aspect weaving [5] addresses reconciling and synthesizing solutions that originate in multiple aspects Aspect weaving is challenging, since aspect solutions can interact in complex ways. We focus on a much simpler problem: starting with knowledge about types of wholes, and discussing how the whole relates to its various aspects.

Hall and Fagen [6], and Hall [7] discuss basic concepts and definitions in systems engineering. Mobus and Kalton [8] provide a synthesis of the extensive knowledge that has been generated in the area of systems science. Hybertson's Model-Oriented Systems Engineering Science (MOSES) framework [9] discusses how systems science principles and modeling can be applied to systems engineering across a wide range of problem classes. Our work complements the existing body of knowledge with a first principles inquiry into the nature of knowledge domains, their role in systems engineering, and how systems science provides the conceptual framework within which we solve systems engineering problems.

12.3 The Four Worlds Model of the Relationship Between Systems, Knowledge and Engineering

This section presents a model of how knowledge domains arise, and how this knowledge is used in systems engineering. We use the term systems knowledge to refer to any knowledge that is derived from observation of systems (including trying to design systems and observing results), with the aim of constructing linear or non-linear (i.e. network/cyclic) causal explanations of behavior in terms of the structures, processes and context interactions that produce it. Implicit in this is the notion of boundaries, that we are observing a whole that has internal structure, and internal and external processes (i.e. interactions with context) which produces outcomes of interest. Thus a system (together with relevant abstractions of its environment) is viewed as a whole that is composed of interacting wholes, and which may be a part of one or more larger wholes. Our starting point assertion is that knowledge of interest to systems engineering, including science, engineering and people systems (such as insurance, economics) knowledge, is systems knowledge.

We generate knowledge by identifying commonalities across groups of observations of systems. As shown in Fig. 12.1, we can distinguish between two major approaches to grouping of observations: focusing on entities and generalizing based on entity commonality, versus focusing on particular aspects of interest and abstracting out all information not considered relevant to the aspect. The first approach, for instance, generates knowledge about entities such as candles (their structure, properties, usages and so on), while the second, given a candle on a table

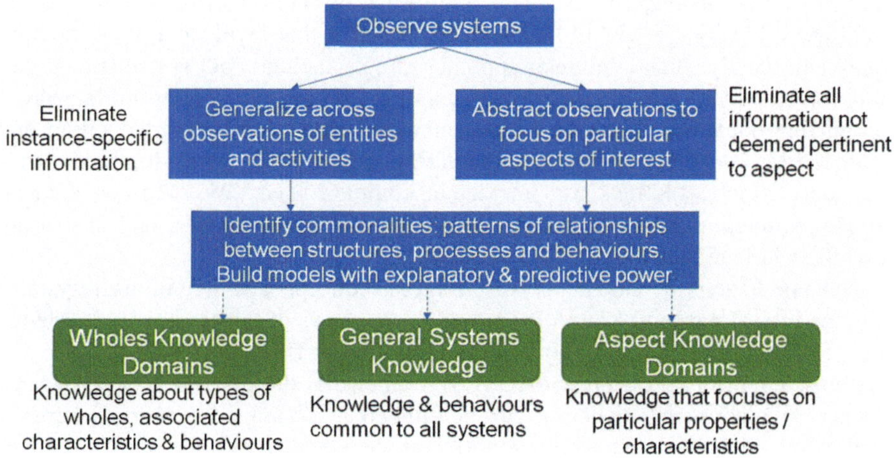

Fig. 12.1 How knowledge domains arise

whose light reflects in a mirror to form an image, might abstract it to a light source interacting with a reflecting surface to produce an image. The first approach produces knowledge domains about types of wholes (e.g. network routers), while the second produces theoretical knowledge about aspects of interest (e.g. network routing).

As we shall see, we regard wholes as an assemblage or collection of parts, together with environment interactions, that is regarded as a single entity i.e. it asserts a collective identity. [Note: Herbert Simon's principle of near-decomposability indicates that this collective identity involves some degree of approximation/abstraction]. An aspect is an abstraction that focuses on a set of related properties, characteristics or activities of interest.

12.3.1 The Four Worlds Model

This leads us to conceptualize the relationship between systems and knowledge in terms of four worlds, as shown in Fig. 12.2. There are phenomena in Real World of experience that we observe and describe, thereby generating a Model World. There are gaps between the real world and model world, partly because of the observation filter (focus of interest), and partly because of limitations of observation and description. From these observations, we generate Wholes Knowledge Domains (knowledge about types of wholes) based on generalization, and Aspect Knowledge Domains (knowledge about aspects) based on abstraction. Wholes knowledge builds on aspects knowledge e.g. when we describe a telescope, we describe its construction, its signal processing behavior, its power usage characteristics, each in terms of concepts from the corresponding aspect domain i.e. the nature of Wholes Domain

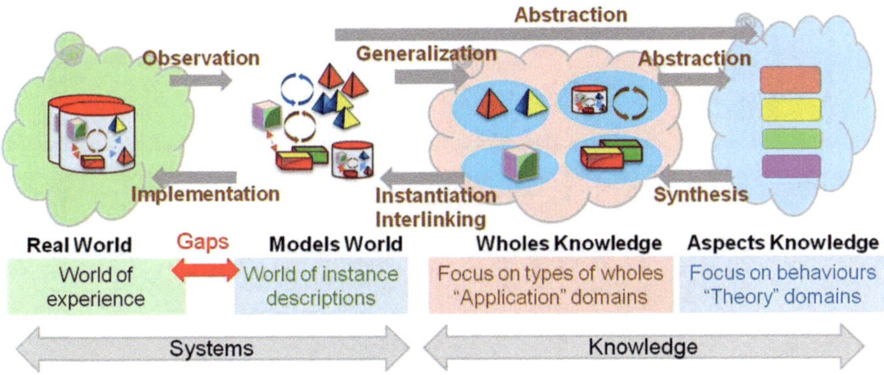

Fig. 12.2 The relationship between systems, knowledge and engineering

Knowledge is that it builds on various relevant Aspect Domains, synthesizing them into a consistent whole (e.g. the same properties may show up in multiple aspects). Entities in the Model world are instances of the types in Wholes knowledge domains, and engineering involves implementing these models in the real world. Thus the left to right arrows indicate how knowledge is generated, and the right to left arrows indicate how knowledge is used in engineering. To successfully achieve desired real world behaviors, engineering must include practices to close gaps between the model world and real world.

12.3.2 *The Nature of Wholes Knowledge*

The key to the value of the conceptual model is the nature of knowledge about types of wholes. As shown in Fig. 12.3, we understand a whole in terms of its internal attributes, state and operations, constituent parts (such as Feeds and Motor) and relationships among them (not shown to reduce clutter), relevant aspects (signal reception, signal processing etc), internal and external processes (shown as brown loops), properties arising from these processes (attributes are also properties), and most importantly, its assumptions about its environment in terms of a collection of roles, with associated role profiles (that we sometimes refer to as lenses, to signify that they define the abstracted view of another whole as seen from the whole of interest).

While the other elements are pretty much standard in any block model (e.g. in SysML), we point out that abstractions of environmental dependencies are in fact part of our knowledge about types of wholes (such as dogs, dish antennas and network routers), and propose that they should be included when capturing Wholes knowledge, and in systems modelling. Knowledge typically frames these relationships in terms of particular roles played by the context relative to the whole. In the dish antenna example of Fig. 12.3, Sky is the source of signal inputs & noise, Signal

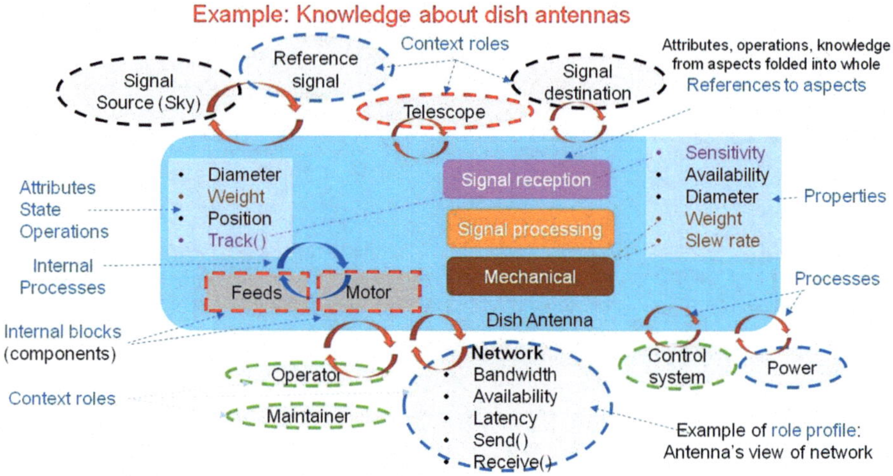

Fig. 12.3 The nature of knowledge about types of wholes

destination may be a beamformer or correlator. Telescope is the hierarchical parent of the dish antenna, Feed and Motor are children (actually they are also roles that need to be bound to concrete instances, and also have assumptions associated with them). Reference Signal, Network and Power Supply are dependencies to be satisfied. Control system may be viewed either as a dependency or as an operational management role, and Operator and Maintainer are operational management and life cycle management roles respectively.

It is noteworthy that the roles reflect a multi-dimensional view of the environment of the whole, as indicated by colour codes: Sky and Signal destination are part of functional behavior, Reference signal, Power Supply and Network are dependencies, Telescope, Feed and Motor capture structural hierarchy (important to operational management), while Control System, Operator and Maintainer are related to the life cycle and the management of operations. At the level of Telescope, we may also have Value dimensions in the form of a Scientist role. Thus role relationships enable modelling to capture the multi-dimensional complexity of real world systems (current practice typically does not separate out environment dimensions).

Each role has a role profile: a collection of assumptions made by the whole (Dish Antenna) about that role, including characteristics, structural relationships, operations provided and processes that span the whole and the role. Each role profile is an abstract collection of assumptions that need to be satisfied by any concrete whole that is bound to that role. These constitute a lens through which the whole views the other role. When the abstract role is bound to a concrete whole e.g. the Power Supply role is bound to a Transformer or Generator that provides the power, the characteristics of the bound object must match the abstract role profile. This is a bidirectional obligation i.e. Dish Antenna must also match their Power Load role profile.

Our assertion is that this is the nature of knowledge about wholes, that a textbook on dish antennas, for example, would discuss not only the internals of dish antennas, but also these other roles and how they affect the dish antenna. We suggest that in systems modelling, it would be useful to model each whole this way, so that it is complete and self-contained with respect to behavioral reasoning: such a model includes all the information needed to reason about the properties of the whole arise from the internal and external processes operating over the structures, including the fact that the reasoning may need to draw upon knowledge in various aspects.

12.4 Applications of the Conceptual Model

These foundational ideas—the distinction between knowledge domains that relate to types of wholes, the four worlds model of how knowledge flows into engineering, and the above model that include context roles and associated role profiles as part of the knowledge and model of wholes—suggest solutions to several vexing challenges in systems engineering. This section discusses two of them: bridging across domain ontologies in the context of particular systems, and creating a conceptual model of systems engineering. Since these foundational ideas are systems science ideas, it also promises to close another gap: how systems science provides the basis for systems engineering.

12.4.1 View Mappings: Bridging Across Domain Ontological Models in the Context of Particular Systems

Systems may be viewed as consisting of interacting wholes. Relationships between wholes includes peer relationships, hierarchical whole-part relationships, and intent-object relationships (one whole operating on another as input or output). A fourth kind of relationship is abstract whole—concrete whole, the relationship between an abstract entity and its technological realization. For example, a design may include an abstract block that represents an adder (that adds numbers), but in order for it to be realizable in the real world, we must map this abstract block to a concrete technology e.g. mechanical adder or electronic adder (note that there may be multiple levels of abstraction/technology binding). A fifth type of role relationship that we have identified is Intent-Object, the relationship between a block and the inputs/outputs on which it operates. For example, we can apply a variety of Intents to the Dish Antenna, such as construction, verification, transportation, fault diagnosis, calibration and so on. In general, the number of Intents that can be applied to a Whole is extremely large: blocks such verification system and construction system may operate on the Whole. For all the types of role relationships, each whole relates to the other whole in terms of a role they play and associated role profile. Table 12.1 shows the types of role relationships that arise in systems, and their implications for view mappings. It is not clear whether this list of relative role types is complete.

Table 12.1 Typical relative roles in systems among entities/aspects, and their implications for view mappings

Relationship type	Relative roles	Approach to view mapping
Whole-part	System—component	Vocabularies overlap
Whole-aspect	Block—concern	Whole aware of concerns and associated mappings
Aspect-aspect	Concern—concern	Derive by composing view mappings
Whole-whole	Provider-consumer collaboration	Bidirectional view mappings (lens → concrete) instantiated based on role bindings
Abstraction-realization	Functional block—technology block	Interpretation mappings between technological elements and elements in abstraction space
Intent-object	Block—input Block—output Activity—block	Object often unaware of intent. Bind intent lens (e.g. sales) to object based on taxonomical type of object (e.g. sales of product vs. sales of services)

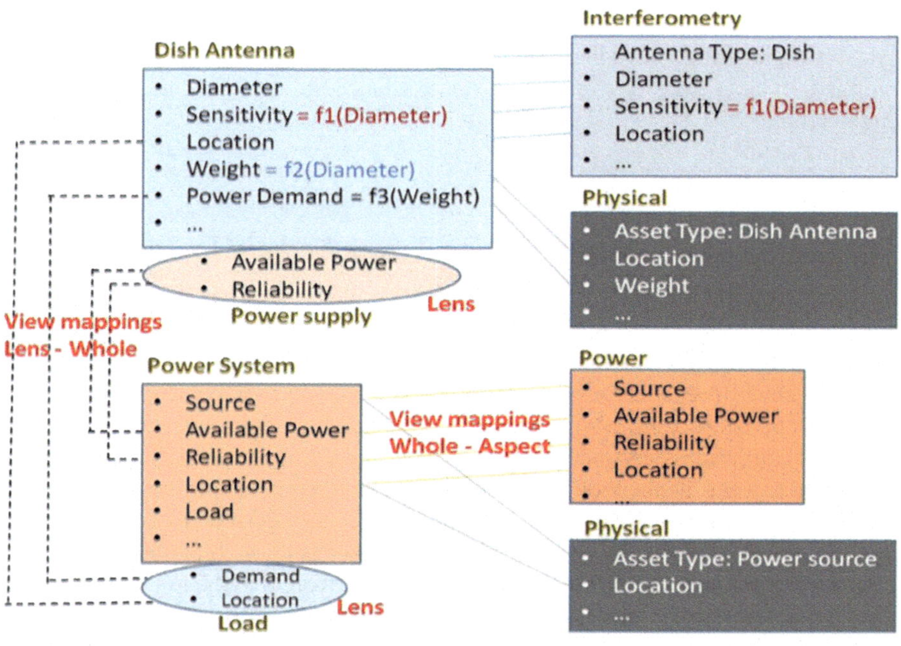

Fig. 12.4 An example of view mappings: binding lenses to concrete model elements

A view mapping binds the elements of a role profile (characteristics, operations, processes, structural relationships e.g. interfaces, connectors) to concrete elements in the whole that plays that role. Figure 12.4 shows an example, in which the Power Supply role of the Dish Antenna is bound to the Power System, and the Power Load role of the Power System is bound to the Dish Antenna. It also shows examples of view mappings that connect each whole to relevant aspects. Such bridging enables the integrated reasoning needed for engineering. For example, the Power System needs to know the power demand, which is known as part of dish antenna knowledge to depend on antenna weight, which is known in the physical aspect to depend on the diameter, which the interferometry aspect relates to sensitivity. Thus we infer a relationship between desired antenna sensitivity and required power system capacity based on knowledge.

It should be noted that the relative roles of wholes arise from the particular system design. The underlying principle is that in general, *bridges between knowledge domain ontologies need to be instantiated based on their relative roles in particular systems*. Of course, it may be possible to establish some types of view mappings (whole-part, whole-aspect) without reference to particular systems. An implication of viewing causal knowledge in all domains as systems knowledge is that we can use systems concepts as an upper ontology when developing domain ontologies for engineering-related purposes. This would simplify the job of bridging between domain ontologies.

In systems modeling, we can associate a *semantic type* with each block that links it to the ontological model of the corresponding Whole e.g. instead of simply viewing a dish antenna as an entity, the systems model would link it to a Dish Antenna ontological in a Wholes knowledge domain such as SignalReceivers. Semantic type linkages can be extended to each element of the block: attributes, operations, relationships etc. Combined with the above approach that establishes bridges between the ontologies, it enables an integrated systems modeling approach that spans multiple domains. It is relatively straightforward to integrate such an integrated model with domain-specific tools, since several of the challenges identified in Section 12.2 are addressed (except semantic mismatch, which is a problem of social agreement + concept scope definition). This integrated model includes view mappings to Aspect knowledge domains (the mapping from Wholes to Aspects can be captured directly as part of knowledge), so that desired views can be generated automatically. Thus the above approach holds promise for enabling the type of integrated system modeling that was the primary motivator for this work.

12.4.2 A Conceptual Model of Systems Engineering

The problem of systems engineering is to generate solutions that improve situations in real world in accordance with desired stakeholder value, using elements available in real world, based on knowledge in Wholes domains and Aspect domains. This leads to the conceptual view of systems engineering shown in Fig. 12.5. This is a declarative model of information items (and real world instances) linked by mutual consistency relationships—it does not discuss the problem of how to establish these consistency relationships.

Engineering problem formulation includes an assertion that the characteristics of the modified situation resulting from introduction of the proposed engineering solution provides acceptable stakeholder value. The information items in knowledge

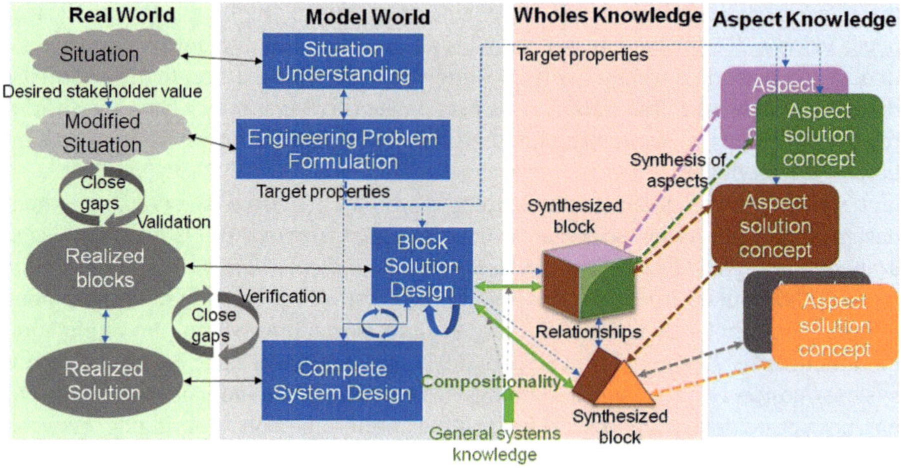

Fig. 12.5 A conceptual model of systems engineering in terms of informational consistency relationships

space must be consistent with the applicable knowledge in those spaces. The synthesized block concept must be consistent with each of the aspect solution concepts, and with other related blocks. Block designs must satisfy compositionality constraints in addition to being consistent with the synthesized block concept, as discussed in the next paragraph, followed by a discussion of the consistency relationships between system design and block design.

If we satisfy all these consistency relationships, we have a solution in the model world that we deem correct based on applicable knowledge (including informal and tacit knowledge, which participate in human reasoning), but it may still have gaps with actual behavior in real world, because of gaps in problem understanding, gaps in knowledge (incompleteness, approximations, limits to applicability etc) and gaps in realization. These gaps must be closed by engineering practices such as verification, validation, and continuous monitoring of gaps between system behavior and stakeholder value.

A synthesized block concept weaves together structures and processes from various aspects to propose a block that will deliver all of the desired characteristics, based on patterns (and analysis/synthesis) knowledge in individual aspects and the Wholes domain of the block. However, there are a variety of systems phenomena that may affect the desired behavior, including the effects of concurrency and interplay, variety, short-term and long-term dynamics. Knowledge is based on the results of a single process (set), but a typical block may have multiple threads of activity that can potentially interact, both due to concurrency, and because of multiple concern dimensions: functional behavior, life cycle activities, operational management, business operations etc.

In addition to the expected input and output spaces, there may be spontaneous processes (e.g. rusting), undesired inputs (e.g. security threats), undesired outputs (e.g. network hardware faults may result in spurious traffic) and undesired outcomes (e.g. safety threats). Block synthesis and binding to other blocks creates a combinatorial explosion of variety (cases/scenarios). The network of internal processes in the block and external processes involving context dynamics produces short-term dynamics behavior, including possibly emergent behaviors. Contexts are potentially open and may evolve, leading to complex long-term dynamics. Compositionality of block design involves using systems science knowledge to account for these potential modifiers on block behavior.

Evolving a complete system design from an initial block involves the repeated application of three activities: decomposition, dependency closure and realization (implementation to technology binding), so that every role slot of every whole is bound either to another whole or to the solution context, and the blocks resulting from the design are available, acquirable or implementable in the real world. Binding one whole to another involves ensuring that the role profiles assumptions of each block are consistent with the design of the other block.

This is a conceptual model of systems engineering derived from first principles that provides a basis for explaining and reasoning about empirical practice. It defines the problem of systems engineering in terms of a collection of informational elements with mutual consistency relationships. We can take the collection of activities identified in any given model of systems engineering e.g. the ISO/IEC 15288 standard, and map the activities and deliverables to this model. Preliminary attempts at such mappings have been successful.

12.5 Conclusion

We have described a strawman conceptual structure that explicates the relationship between systems, knowledge and engineering. The key idea is to distinguish between Wholes knowledge domains and Aspect knowledge domains. By augmenting entities with roles and role profiles that capture all their assumptions about their environment, we can frame them as Wholes that are complete with respect to reasoning. By binding wholes to each other based on matching the characteristics of each whole to the role profile desired by the other whole, we create the basis for bridging across the knowledge domains of each whole. This is a modular, composable model of system blocks. A system consists of a multi-dimensional network of such blocks which is closed with respect to decomposition, dependencies and realization. The four worlds model brings together all these concepts to produce a first principles conceptual model of systems engineering.

This work holds out the promise of modular integrated system models linked to modular knowledge models. We have successfully built a few small PoCs based on these ideas, and also validated them against some large projects in our organization that developed integrated knowledge models spanning multiple domains. Given the vast scope, these are still really seed ideas that need to be refined, validated and matured. We have initiated a project aimed at this within the INCOSE systems science working group. We are also interested in developing a mathematical formulation of this conceptual model, but as of now we only have a few starter ideas towards such a model. We would appreciate validation, inputs, suggestions and feedback from the SE research community. People interested in contributing to the project are requested to contact the authors.

Acknowledgements Many, many people have contributed to these ideas over the years, at the TCS Systems Research Lab, Business Systems and Cybernetics Centre, NCRA-TIFR, IIIT Hyderabad, as well as interactions at ISO SC7 and INCOSE SSWG.

References

1. Blackburn, M., & Verma, D. (2017). *Transforming systems engineering through model centric engineering*. Hoboken, NJ: Stevens Institute of Technology.
2. Sowa, J. F. (2000). *Knowledge representation: Logical, philosophical, and computational foundations* (Vol. 13). Pacific Grove, CA: Brooks/Cole.
3. Shani, U., Jacobs, S., Wengrowicz, N., & Dori, D. (2016). *Engaging ontologies to break MBSE tools boundaries through semantic mediation*. In 2016 conference on systems engineering research.
4. Blackburn, M. R., & Denno, P. O. (2015). Using semantic web technologies for integrating domain specific modeling and analytical tools. *Procedia Computer Science, 61*, 141–146.
5. Wand, M., Kiczales, G., & Dutchyn, C. (2004). A semantics for advice and dynamic join points in aspect-oriented programming. *ACM Transactions on Programming Languages and Systems (TOPLAS), 26*(5), 890–910.
6. Hall, A. D., & Fagen, R. E. (1956). Definition of system. *General Systems, 1*(1), 18–28.
7. Hall, A. D. (1989). *Metasystems methodology: A new synthesis and unification*. Oxford: Pergamon Press.
8. Mobus, G., & Kalton, M. (2014). *Principles of systems science*. Berlin: Springer.
9. Hybertson, D. (2016). *Model-oriented systems engineering science: A unifying framework for traditional and complex systems*. Boca Raton, FL: CRC Press.

Chapter 13
Review and Constructive Definitions for Mathematically Engineered Systems as Categorical Interpretation

Chris Klesges

13.1 Introduction

Systems theory, particularly in connection with engineering, is used in a variety of ways, yet it yields solutions by providing naturalized (as in naturally numbered) for semantics and syntax. Subsequent "specification" are central to model-based systems engineering [1, 2]. Mathematical foundations, following from formal proof, provided consistency method for specification. These foundations were originally underlain by mathematical primitives obtained in set and (functional) relationship theories, which led to so-called mathematical systems engineering, model-based systems engineering, and simulations. This chapter surrounds interest in formalizing expression of system design, engineering, and aspects using categories to capture mathematical context. Here this will mean a conceptual system and extending itself in type theoretic ways, effectively "context topdown."

The motivation for this research begins by considering an alternative formalization language and subsequent deductive language. General consistency with (engineered) system primitives are desirable, yet alternative expressions, constructions, and interpretations must be possible. The theory of categories resulted from abstractions developed in algebraic methods, primarily expanding the primitives of "equivalence" [3] and generalizing "group" notions [4]. Further development through logic [5] and type systems [6] are of concern here as these methods provide a descriptive domain and (set) proposition analogue respectfully. Central system aspects are provided that make it possible to construct categorical definitions used in theory.

This work was supported both from an internship at MITRE Corporation and from research assistance through the Systems Engineering Research Center (SERC) under contract HQ0034-13-D-0004.

C. Klesges (✉)
Socio Technical Systems, Stevens Institute of Technology, Hoboken, NJ, USA
e-mail: cklesges@stevens.edu

S. Adams et al. (eds.), *Systems Engineering in Context*,
https://doi.org/10.1007/978-3-030-00114-8_13

Shown here is a consistent interpretation using categories against fundamentals of model-based systems, as described by Wymore. Specifically, this is interpreted in Wymore [7] through the deductive system of Awodey [8] and labelled as constructions. These constructions then show "algebraic" abstractions can be contextual to engineered systems, and category theory would provide useful linguistic. Further categorical types allow potential specification of engineered systems, and as system science has shown application [9], this represents opportunities for theory transfer over the language. This type methodology is thought an "extensive" or "extensional" alternative for mathematical systems theory given it is derived from abstract algebra, homology, and extended graphs.

13.2 Procedural Presumptions

13.2.1 Elementary Consideration for Categories

Some conceptualization behind a category's fundamental notions to aid applied identification is presumed. The intuition here is a "preordered setting for structure." Systems theory as a philosophical stance often invokes some structured environment via space as "field," order through "index," and "pre-" via domain or "perspective." These are consistent primitives, so categories maintain the referent idea(s). This then hypothesizes a classification means where set/element operations present limitations, e.g., networks, topology, "complex" fields.

One then considers kinds of mappings that "preserve structure" or "morphisms" in some sense. These usually present as "natural pairings": e.g., homeomorph (openness preservation) in topology, homomorphic (operation preservation) in groups, and several more [10, 11]. These are "self-recursive" in the sense of the structure such that it extends in a denumerable way, i.e., by counting said morphisms. Then for systems that involve topological notions means for specifying the "openness" context to "closed" set specification become (theoretically) available. For engineering, this notion can be seen in fluid dynamics, where "laminar" and "nonlaminar" present different structures and equations. Yet one can extend usage of a desired class using "unitless" numbers (i.e., measuring the Reynold's number), that says which model(ing) upon context. Similar mechanics pursue categorically different methods depending on "static" mechanics or "dynamic" mechanics. Even enterprises require certain "social structuring/ordering" to be maintained at least on project instances. Then for fluid-mechanic-enterprise, these contextual parallels are presumably minimally necessary.

13.2.2 Comparison Method

The works mentioned are used as independent, deductive chains in the proof-theoretic sense. This comparison method provides a "parallel consistency" as the conceptual steps needed to satisfy each converge on similar construct. Set and category seemingly describe different aspects [12, 13], yet here the "parallel" appear as constructively modellable so represents (system) engineering method. The pragmatic stance is that, while each captures different concepts, engineering types realize themselves similarly. Differing constructions in the needed category are noted. It is assumed that each is complete ("up to isomorphism") own definitions, and specific constructions are identified using its primitives.

The choice deductive language is due to the presentation of categories in a way analogous to types and functionals. This makes it easier to notice similar representations, yet often a category is discussed around homological sets and (uni-)valent axiomatics. One note is there are dialects depending on purposes. For instance, there are types discussed as type of category ("categorical type"), yet categories can discuss types in the elementary sense, often "type of types" [14]. Observations are that homotopy type theory [15], with notions from order theory, offer synthesis hypothesis. But depending on "algebraic orientation", this is similar to different representations for complex analysis.

The two works are presented as "sides" and in order of theorem usage. The proof chains are presented, labeled atop each other, and "construction" is used for definitions here. Conceptual checks are provided throughout to explain a "system" interpretation. Attention is given to categorical descriptions as the model-based system is thought to be more familiar, and the "constructive analysis" searches for overlapping application.

13.3 Theorem Presentation

13.3.1 Category Primitives

The initial construction requires a formal definition of category, given below from Awoday:

A-Definition 1 (Category) A "collection" of objects and arrows with the following conditions (Awoday Def. 1.1, p. 4):

1. For each object A, there is an "identity (arrow),"

$$id : A \rightarrow A.$$

2. For each arrow $f, f: A \rightarrow B$ is indicated by domain $\text{dom}(f) = A$ and codomain cod $(f) = B$.[1]

[1] Conditions 1 and 2 make for "functional-like" relations where the "collection" is (injectively) graphical, with each object as the vertex and the arrow an edge.

3. Arrows are "compositional," that $f : A \rightarrow B$ and $g : B \rightarrow C$ there is $gf : A \rightarrow C$ called 'composite' (e.g., the ability to functionally decompose)
4. Arrows are "associative," (as in associative property from arithmetic but for maps) for all $f : A \rightarrow B$, $g : B \rightarrow C$, $h : C \rightarrow D$

$$h(gf) = (hg)f.$$

5. Arrows are "unitary," or can collapse identity from (co)domain, for all $f : A \rightarrow B$, there exists id_A and id_B

$$fid_A = id_B f = f.$$

The "collection" of the category is intended to replicate the abstract algebra of a discourse domain. For instance, the sets and relations have a regular structure to discourse, so one abstracts the "grammar" or 'structure', in fact "sets and relations" is an example category. However varieties of structure can be communicated in an environment, and one claims types and classes of categories (by their axiomatics) as one does for types and classes of sets. The highlights are that the category captures the structure over the arrows, so declares morphisms instead of pure (objective) sets. So in context, one can claim associations without (necessarily) particular setting, and to (re)gain an objective (set) description the "(full) structure type," "category of (blank)," is usually declared.

This allows abstracted description where set models by themselves might be insufficient or inefficient for a given environment. Yet set-analogous descriptions are useful to describe such systematized collections. The restrictions on categories allow for a foundational description [3] and for a declaration of an abstracted, preordered structure. This has a syntax through the "collection" into the "set of objects" with "homo(logical)-set of arrows" (Awoday Def. 1.11, p. 24). Used here are basic descriptions for categories and structural aspects.

A-Definition 2 (Homological Designation of Categories) The following conditions with labels:

- "Small" if both "collection of object" C_0 and "collection of arrows" C_1 are (finite) sets
- "Large" if C_0 or C_f is otherwise not (finitely) settable
- "Locally small" if for each object pair their collection of arrows is a finite "hom-set"

A-Definition 3 (Epimorphism) For category $C, f : A \rightarrow B' \in C_1$ is "epimorphic" if any $i,j : B \rightarrow C$, $if = jfi = j^2$.

A-Definition 4 (Functor) A functor between categories $F : S \rightarrow Z$ is a mapping of "objects to objects" and "arrows to arrows" in such a way that:

[2] This can be thought of as "surjective on structure."

- $F(f: A \to B) = F(f) : F(A) \to F(B)$ preserves the (co)domain
- $F(id_A) = id_{F(A)}$ preserves identity arrows
- $F(gf) = F(g)F(f)$ preserves the composition

This captures some semantic intuition from systems by sharing definita. The object and arrow designation abstract the descriptive aspects to functional system languages whether operations or mappings. The environments are designated as described ("small" system) and as actual ("larger" system), and realization validity question is finding an interpretation [4, 8, 15] ("local small" homology-setting).[3] The "aleatory basis or space" from engineering shares the small category's finite specification, particularly to be modeled. Sometimes what systems engineering colloquially means by "modeled" in generous sense is finding this small categorization. Then the collection needs (but does not necessarily have) normal properties for algebraic closure: "composite," associative," and "unit" aspects.

Making these mappings to binary associations restricts them to graphical statements, which allows both presentation and useful proofing. Note the similarity to "many-to-one," "functional" relations, but here with relations-of-relations to unitary foundation. And binary (structural) permutations are normal presumptions or end-goal for systems engineering. The epimorphism mirrors monotonic representations, but here on the resultant structure. So n-ary structures still reduce to binary upon obtaining a result, not (necessarily) on input or usage. This focuses the discussion on "epistemic" boundaries and ordering resultant descriptions within engineering philosophy.

Lastly, the notion of functionals in (real, continuous) system environments has a categorical representation in "functors." The concern of mapping objects to objects in functions, then functions to functions, has a similar notion, categories just ask that the identity (i.e., "preorder") be maintained. "Model-based" discusses the same in functional set relational form explicitly using the "set of reals," notably "large" compared to rational sets. Both (set, relation) and (reals, continuous functions) form categories with corresponding transformations, so one can generalize with the (categorical) algebra. Often discussion uses "intensive/intension" to refer to functional, model-based perspective while "extensive/extension" to a functorial, categorical perspective.

13.3.2 System Definition

The desire then is forming a constructed definition for "engineering system," "system" here throughout. The "epimorphic" results are of central concern, as a system should result in enumerable permutation with clear labeling, "determinable" results. The belief is that, because system science concerns "choice" mappings,

[3]Category development stems from areas that deal with continua, namely, in logical typings and algebraic topology.

system engineering has a "backward" concern to construct these. This "reversing" is called "duality" in categorical logic, and how many categorical proofs if done in one construction (monomorphic) can be reproduced in its dual (epimorphic). Used here are the dual of categorical statements, hypothesizing base transfer between system science and engineering.

The epimorphism presents, then, a "right cancellation" in theory for an engineering "structure" based on (desired) output descriptors. While general experience highlights the various descriptions a system can present, the system activity finds those designations that allow greater permutation through more operations, connections, or modules, for example. Systems architecture is the disciplinary term for this activity in design, and system verification and validation seeks this upon construction. The additional plus is that epimorphic conditions are necessary for categorical proofing, showing a shared-ness at least among technical specifications.

Then let us compare Wymore's definition of a system below with its categorical construction, and consider the "system as category" definition[4]:

W-Definition 1 (System) A system is a set Z with elements and conditions:

- S, set of "state(ment)s" of the system
- P, set of "input state(ment)s" of the system
- $F : R \rightarrow P$, admissible set of input functions[5]
- $M : S \rightarrow S$, set of transition functions
- T, "time scale" index
- $\sigma : FXT \rightarrow M$, set of set transition functions

1. identity, $id_m \in M$ such that

$$\forall f \in F, \sigma(f, 0) = id_m$$

2. if $f \in F$; and $s,t,s + t \in T$, then

$$\sigma(f \rightarrow s, t)\sigma(f, s) = \sigma(f, s + t)$$

3. if $f,g \in F$, $s \in T$, and $f(t) = g(t)$, then

$$\sigma(f, s) = \sigma(g, s) \text{for all} \, t \in R(s)$$

***Sys*-Construction 1 ("System" or *Sys* as collection in Category)** Is a category with:

[4]His term for "assemblage" is simply the categorical "collection" without further conditions.

[5]R stands for set of real numbers.

- Objects: S is a large category, Z a small category
- Arrows: $\sigma : S \rightarrow Z$ such that Sys_{σ} is locally small
- "Identity," "Composition," "Associativity," and "Unit" are proscribed by definition
  - σ is "epimorphic"
  - σ is "functorial"

As seen, this "system in category" describes relational effects in similar fashion but in algebraic terms. The conceptual expansion is the focus on mapping rather than objective declaration. This allows for entertaining environments for systems irrespective of (declared) objective structure, as one encapsulates the "permutation profile," via "patterns," independent "paradigms." As one compares, say, the Cauchy and Dedekind definitions of "real numbers," the specified objects are defined differently, yet as far as structural properties are concerned, they are the "same field."[6] There is then a convergent "holism" in the sense that structural properties consume otherwise independently defined models, so concerns about "algebraic sameness" hinges on their study for this "convergence," or "homology."

Wymore defines sets with "type of finite functions" using "morphisms" separately for comparisons. The large(-to-small) category there is the indexed reals (R, T) whose structure shows the same field as functionally defined sets, so it matches intent. But because categories have to do with independent structures, this homology can be defined without regard to structure; e.g., networks are large compared with sets, yet the "graph" construct allows a shared topic, "homotopy." Category usage usually rather identifies domains and then homotopy considering their unique classification. A comparative study of analyses just considers the reals as a field for objective results, and similarly, other determinants might be contextual where categories can express these class convergent properties. This epistemology provides observations for engineering of seemingly independent model objects offering a "systematic" means for engineering.

The benefit to systems is that "top-down" constraints then have semantics and allows "structural" properties expressed in those morphisms, which a set system must fulfill. This would require in practice presumably a "pretheorized situation" to claim structural invariance, but this is not uncommon in engineering. Yet this could relate otherwise disparate spaces across interdisciplinary domains by showing the same topics yielding classifiable specifications, much like categories do for mathematics. While this may be trivial or unnecessary for certain environments (e.g., mechanical environments), some (e.g., social systems) do not present such "nice" set vectors. Yet it is still common to describe (social) group actions through accessible permutations. This representation might take context, qualified by structure and quantified through categories, into presentation by morphism. Then independently these can be researched as to how domains express these (or can they), or a domain

[6]Similarly, different intervals yet the same "space," different coordinate integers yet the same "shape" [12].

known to have such structure available can have constructive procedure. For example circuits express various structuring, yet these are usually brought to diagrams as to which connections result into which overall effects.

Collected, then, are several observations on touch points for system theory. These have some common sensing that is at least analogous to spatial interpretations:

- *S* is discussed as the "semantic space," "large system," or "natural system" effectively with an interpretive (universal) structure.
- (Z,σ) is then the "(constructed) aleatory space" or "formalized σ space" as experienced by engineers.[7]
- For σ this "constructor" allows for handling "structure" by associations presenting (trans)finite closure:
  - As a set equivalence relates objects, an epimorphism relates "(co)equivalent structure" [3] as one cancels any resultant "σ words" in the algebraic sense.
  - The functorial description ensures this epimorphism is maintained over any functional as well as objective results, hence "resultant structure."
- "System identification" which labels ("black box," "white box," and "model") mirrors ("epistemic," "aleatory," and "construct").
- One major difference is the indexing becomes an agreement on dimensioning. The "family index" is time in most system *T*., and is implicitly assumed. A real index can be expressed as real numbers have categorization, yet there seems to be a "(minimal) dimensional (agreement) problem."

This generally supports the idea that there is a (dual) space to "conceptual context" that provides a universal(ized) description for model(ing) structure. There might not be a unique way to discuss networks, but given that graphs are a shared system language, it presents a universal algebraic means for the aforementioned contextual space. This language of universals then provides a principled method for engineered systems, and categories are a unique basis for this discussion.

13.3.3 Universal Properties for Sys *from Category Theory*

A universal here merges the logical and linguistic usages as it is asserted over a class in which observes a grammatical rules set. So "uni-verse" or "universal property" might be read as "singular linguistic" that holds over a category and thus its (co) domains. In model-based engineering, ideally research observe a grammar that covers a domain, and then systematic domains can be collapsed and classified to such "model grammar." The universal in Wymore's system extendeds the grammar of groupings and finite states, i.e., "finite state machines." grammars, which he

---

[7]The linguistic usage of categories treats these as "semantics," "interpretation," and "syntax" labels, respectfully.

observes as becoming a language for specification. "Algebraic closure," usually over continuous functions, forms a basis for programming aligned with the parsimonious "model basis." The engineering usage is that these aspects guide similar procedures in ever "higher order": programming, specification, configuration, etc.

The approach in categories aligns universal aspect first at least in a particular context: design, architecture, terminology [2]. To consider aspects up to a specific context, this (singular) algebra can be extended creating a greater universe. This extended linguistic has its own "systematic grammar" by the general homology, likely what general systems tries to observe?. The "nice" condition is that each approach shares the same type hierarchy, "univalence," but this usually represents stronger conditions between the internality and externality to systems. Yet this gives an ideal in some sense so concepts qua mathematics require their requirements qua conditions.

Considered is how categories provide the language to satisfy "extensional" specifications, while allowing consistency with a given model-based typing. In system practice, one could rely on an ongoing testing procedure to ensure external matches internal in system praxis, yet method inspiration can be seen in usage from the sciences [9, 16, 17]. So if methodology is similar enough, system science observations might immediately transfer. Here "coequalizer" relations (a kind of "structural equivalence") are proposed as this implies epimorphic preservation (Awodey Def. 3.18 and Prop 3.19, pp. 66, 67), yet begs a methodological question.[8]

Here this gives *Sys* as a category an easier presentation, much like equivalence in sets. Then the type classification of space based on relation in pursued, here as an equivalence *qua* categories. Propositions explain space in terms of universals through "higher-order" relations, inevitably leveraging the association between (system) spaces. As universal algebras do for model-based systems, here grouping universals serve the same function for the categorical *Sys*. For an example of usage, model interoperability closure was encountered for industrial systems that contextualize industrial processes, then leverages these (co)equalizing relations. The associated "industrial space" then offers unique categorization with minimal expressions needed for database activities and optimality [18].

The approach is then progressing to "contextualize space" to objects (processes). Comparing again in parallel, below is an associating transition or grouping of transition functions.

W-Theorem 1 (Homomorph and Relation Between Sets of Transition Functions) *Let Z_1, Z_2, be assemblages related by homomorph $h : Z_2 \to Z_1$. Then there exists an onto mapping $\psi : M_2 \to M_1$ with the following composition property: For $(g,t),(f,s) \in F_2XT_2$ and $\sigma_2(g,t)\sigma_2(f,s) \in M_2$, $\psi(\sigma_2(g,t)\sigma_2(f,s)) = (\psi(\sigma_2(g,t)))(\psi(\sigma_2(f,s)))$*

[8]Should one identify epimorphisms directly or through coequalizers? Possibly different perspectives align with each of these or others in concepts.

A-Definition 5 (Groups in a Category) A homomorphism $h : G \to H$ of groups in *Sys* consists of an arrow in *Sys* that:

- Preserves associative products *m*, $m(m(x,y),z) = m(x,m(y,z))$
- Preserves unity for the products *u*, $m(x,u) = x = m(u,x)$
- Preserves the inverse with respect to associative products *i*, $m(x,ix) = u = m(ix,x)$[9]

A-Proposition 1 *Abelian-ness of Category of Groups, Groups in the Category of Groups are exactly the abelian groups.*

Here product composition under homomorphism is mirrored. The product is defined in each (Wymore Def. 1.4, p. 17) although given the construction for *Sys* coproducts (the dual of products) are used (Awodey 2.1–2.2, pp. 49, 50). However, the universalizing property is the same against the generalized epimorphic, resultant definidum. One does not necessarily need the full group structure as Wymore does, but rather one needs to identify the subgroup aspects in a categorized manner. This implies that there is not (necessarily) uniqueness to grouping and association, so uniqueness might need be declared or researched. Then the associative "abelian" property must be extended over Wymore's system and is a general typing of a category.[10]

The result is that state transitions can be associated over the permutation of products, so the state-language "(re)universalizes" over their usage. These determinations establishes their own structure conditions which are further required for *Sys* implementation. Thus, one can independently state the behavior of (co)products, thereby adding context(s) depending on the composed system environment. These could then transfer as rule sets upon design or natural system testing: for each (x,y) product type, this G-grouping and P-properties are/should be uniquely determined." As coproducts are spatial dimension properties, categories maintain the descriptive generalization.

One begins a "type of type" concern in the underlying logic here. Categories generalize the structure, giving it a "kind" then allowing associative typing of that structure: "structure," "group," then product type. The structure type indicates a constructed category, while the other their associative and spatial extensions, "spatial behavior" to be bold. As long as the types are ordered, this maintains a good description in the language. However, just like the difference between formal "universal" [14] and stronger "univalence" [15], weak and strong propositions, respectfully, can be inserted. There is a question as how to encode these system linguistics into a mathematical logic, yet one needed condition is respecting the "chain of types" for any (Z,σ) theory.

This type-of-types declaration is often expressed as "qualitative requirements." "Technical functions" are a type (of) specification, and a type (of) interface to said functions (e.g., APIs) is then a "type of type" or "quality" aspect. Categories allow in

---

[9]As Wymore converges over structure, a homomorphism is defined similarly, but in models as set functions equipped with the operand property.

[10]Note, the reals are grouped under "basic" operations in Wymore.

this way a spatial interpretation and may help to better distinguish formal references in system praxis. Hence the qualitative aspects to the systematic space are "contextual" in nature yet require a kind of "type matching" where intuitive type theory might be of use [14]. Just as we mentally construct points into lines into shapes, the design requirements need shared, constructive aspects to align specification to documentation to module, etc.[11]

For category theory, this "universal" property appears to hold with respect to a category.[12] Here by being able to use the product ubiquitously across a structure, "for all of category" is a unique product association, so it respects the "universal" of the composition. This then gives a classifiable property for the category and becomes useful for simulations across domains. In wanting a set (index) to track over, say, combined vector modeled environments, one needs to assume compositional factors for said simulation. The existence of a common algebraic tracing over the tensor products (i.e., matrix operations) keeps a "single simulative algebra," or a "monoid (al)" system, then allowing a "free(ly leverageable) algebra" to the space.

Categorical interpretation helps designate the typing of the simulation space, as opposed to mechanical systems, which capture over vector algebra, thermodynamical over statistical, and biological over relational analytics. The composition of these models is then an open question particularly with respect to combined simulations. Within each model, the composition is fairly free, yet combined information theoretic limits and hierarchy complicate the picture. However, given a particular situation, one could conceivably match these limitations, a "shared, situational limit," "colimit." Further examples on this "compositional typing" can be found in [19, 20].

Further categorizing for model and simulation development may be found in Ref. [21]. Proceeding with "strictly monoidal" and "closed Cartesian," descriptions are justified here as they imply traditional computational spaces matching Wymore"s finite-state machine determinants. Yet for categorical language, generality is not necessarily lost.

Then one wishes to relate system designs by some comparison of universes by the "like output structure" given by simulative colimits. An equivalence of output over systems then gives a classification mean via homomorphic comparisons. In categories this is known as a "congruence." Here "output" is generalized to "terminal" in categories (Awoday Def. 2.9, p. 33), and Wymore's "output analysis" has replication under "congruence."[13]

W-Definition 2 (Output Equivalence by Homomorphism) The assemblage Z_1 is an "output alternative" (p, Q, ξ_1, ξ_2) to assemblage Z_2 if:

- $p \in R^{+}$, and $T_1 = pXT_2$

---

[11] Not to mention the validity process about the actual artifacts, processes, and space.

[12] A "for all unique (of a class)" quantifier condition.

[13] Note: Congruence is usually presented using a monomorphism in (system) scientific classifications.

- Q is a nonempty set
- $\xi_1 \in F(S_1, Q)$
- $\xi_2 \in F(S_2, Q)$
- For every $g \in F_2, y \in S_2$, there exists $f \in F_1, x \in S_1$, such that

  $\xi_1(\sigma_1(f, pXt)(x)) = \xi_2(\sigma_2(g, t)(y))$ for every $t \in T_2$

W-Theorem 2 ([Duality] Relationship Between Homomorphism and Alternative) *If the assemblage Z_1 is a homomorph of the assemblage Z_2, then Z_2 is an "output alternative" to Z_1. Further, if Z_1 is isomorphic to Z_2, then Z_1 is equivalent to Z_2.*

A-Definition 6 (Congruence on *Sys* [Terminal]) A "congruence" on a category is an equivalence relation f~g on arrows such that

- $f \sim g$ implies $\text{dom}(f) = \text{dom}(g)$ and $\text{cod}(f) = \text{cod}(g)$
- $f \sim g$ implies bfa bga for all arrow $a : A \to X$ and $b : Y \to B$, where dom $(f) = X = \text{dom}(g)$ and $\text{cod}(f) = Y = \text{cod}(g)$

A-Proposition 2 *Factorization of Functor in Category Every functor $F : C \to D$ factors as $F \cong F^* \pi$,*

[Sorry. Ignored `\begin{tikzcd} ... \end{tikzcd}`*]*where π is bijective on objects and surjective on Hom-sets, and F_B : Hom$(A, B) \to (FA, FB)$ for all A, $B \in C/$ ker (F), where ker(F) is a "kernel category."

The two notions of equivalence establish shared resultant functions/arrows over a system. Much like an equivalence in a set, one obtains an equivalence of output in systems. This then appears as classification by transformations that then provide a "factorization" on the category, or "quotient." Wymore then uses these shared relations to establish an extended duality, immediately gaining an (isomorphic) equivalence class. The class by "kernel quotient" is analogous here and can be thought of as "engineering interfaces" to operation within a spatial structure, or at least seen upon the results obtained.

This can be seen as a "goal-seeking" analog to the intent on finding congruent outputs by this "chaining analysis." The set-based representation approaches this transformation from the set statements where categories treat the output transformations as given, as is declared to be fundamental in the collection. The *categorical* "classification by system terminal" then establishes a "universal factoring" here in the scheme for given *Sys*, so a "type of system" is one that acts congruently "in the end" with "coequalized terminal." Thus specifying output equivalence classes (plural) is propositional rather than set theorem: matching limitations to colimit, equalized structure "coequalizers," and stated in the "kernel category."

The algebraic intuition could then establish an ideal "contextual engineering interface" presenting maximals/minimals to relational statements. Again much like design seeks model reduction for component specification, there is an implied simulation reduction for factorization. This puts pressure on matching simulations with objective system factors; otherwise, results-independent model combinations arise, i.e., "emergent properties." In the end, a system is not satisfactory if system extensions are continually disparate from intensive schemes.

This allows a specification for these factorizations by underlying relations. Just as groups can be finitely presented in algebra, the congruences allow for a "finite presentation of categories." Twofold usage then becomes possible as system identification might be described: formal congruence by means of injective graphical presentation and categorical factoring often expressed in soft system methods. These paired functional and qualitative requirements respect this, and both fundamentally factor for *Sys* via underlying universals, similarly to functional equivalence over model-based types.

This establishes a "graphical indexing" under a "diagram" in categorical parlance. Establishing a diagram or diagrammatic type is the analog to the family index for sets. This is a descriptive aspect of the ordering, compositional patterning previously described, and "diagrams of (some) type" then become an organizing principle for aligning system functionals. This is shared behind many "systems thinking" methodologies, often offering (graphical) diagrams encapsulating composition and dimensional factors.

This organization by colimitation and congruence theory for systems theorems is generalized over "push-outs," extending under homomorphisms, and aligning (co) products:

W-Theorem 3 (Systems Implication by Homomorph) *Let assemblages be related by the homomorph* $h : Z_2 \rightarrow Z_1$. *If* Z_2 *is a system, then so is* Z_1.

A-Proposition 3 (Category Push-Out Lemma) *If a category has binary products and coequalizers, then it has push-outs.*

A-Proposition 4 (Push-Out Corollary) *The push-out of a commutative triangle is a commutative triangle.*[14]

"Push-out" describes the same effect intended by Wymore where one can extend this system identification. One can "reuse" the constructed aleatory either establishing extended description or refining the specification. The categorical requirement is that this must maintain the same semantic interpretation, hence what needs to be "pushed-out." This effectively classifies the universal epistemic of referent system (design). The prospect is that these could be subsumed by homomorphic interaction to other (ideal) system space(s), often the intent in "-ilities" research. For example, the maintenance aspects are usually in reference to systems and might be guided by their associations, units, and compositions "up to the referent system/structure."

One aspect in system inference that is used but not explicitly stated in Awoday is ordering. An order inference is needed in Wymore, yet as orderings are precaptured in categories, this becomes difficult to explicate. These might be subsumed or inferred by categorical declarations, yet a separate investigation is conjectured in

[14]Pull-back statements are used, yet this generalizes to push-outs using coequalizers.

applying categories. The general notion of "order adjoint" accomplishes this by declaration, yet this is not (necessarily) implied without further context.[15] However, order can easily be its own proposition by homomorphic relations as both deductive languages show (Wymore Thm 4.4, p. 139). This is likely a necessary construction in discussing systems categorically.

***Sys*-Construction 2 (Order Through Homomorphism)** *[Sorry. Ignored* `\begin {prooftree} ... \end{prooftree}`*]*where order preservation is inferred for each operation maintained over the declared system.

Decision procedures are conjectured to handle these potential problems, if only in situated environments. Yet this represents a potential problem with theory as such, which various uncertainty and possibility theorems show. The interpretation here is that orderings are not (necessarily) unique or certain depending on the domain, yet there is then means for arriving at principal ordering, common to tradespace and choice procedures. Similar in design, one could refer to "system boundary" between both internal (e.g., components, modules, domains) and external (e.g., linkages, dynamics, transitions) aspects. Then how does one separate fully whether an extension is further internal representation or further external interaction? This appears to be a self-reference to the language, which makes completion suspect.

Yet the boundary is easily established using duality (if internal not external and vice versa), yet these conceptual dualities need justification and coordination. It is believed that this leads to many uncertainty and validity questions, which might be resolved by this "order, boundary duality." The problem seems to be that internal and external types identify separately which can be seen how set and category handle independent concepts, and "complex systems" theory seems to observe this difficulty with intensional and extensional identification [12]. For instance, cognitive systems, equipped with memory and communication, blend internal and external representations. Internal representation can be covered further in topos theory [10], automata theory [22], and general decomposition [23].

13.3.4 Completion by Distinguishing System States

However, additional concern focuses on purely extending identification for system specification. As a classification was obtained, accompanying complete specification is needed. To accomplish this, extending the system definition fully requires a "completed spatial description." This finite "fully stating of a system" allows several aspects, namely formal validation. Thus, completion and "adjoining" procedures are compared for each. There are conceivably multiple means for this, and Wymore essentially "repeats" the output classification with a choice procedure (Wymore Def.

[15]There is also a notion of antifoundation, situated order [8], as are Galois and Grothendieck underpinnings.

4.6–Thm 4.8). Matching in categories are monomorphism -and epimorphism imply isomorphism (Awoday Prop 2.6, p. 31) generalizing "definitive" sets (Wymore, p. 152–166).

The goal is for the "well-described" condition to establish an isomorphism/ bijective functor between the "epistemic" and "aleatory." The implied structure-to-structure description offers finite functional/functorial objects or "distinguishable states." For example ideally system architecture gives an identification on the epistemic system and conversely allows a complete labelling to system effects, "adjoining" the system description.[16]

The "distinguished states" condition then describes this "duality closure" as a "*reductio ad* bijectivity."

W-Theorem 4 (General [Discrete] Reduction Theorem) *Let Z be a system such that $T = R^+$ (or $T = I^+$ for discrete) and F has an order-preserving input base G without periodic elements. Let Q nonempty and $\psi \in F(S, Q)$ exist. Then there exists a system Z^{**} and $\psi^{*}* \in F(S^{*}*, Q)$ with the following aspects:*

- *Z^{**} is homomorphic to Z for some functional grouping and some associative ordering*
- *Z^{**} is output-equivalent to Z*
- *Z^{**} is distinguished by the input scheme $(Q, \psi^{*}*)$*[17]

***Sys*-Construction 3 ([Discrete] Category for *Sys*)** Let $\sigma : S \to Z$, where *S* is the poset category (or set for discrete).[18]

Then the closed "algebraic description" for the system requires a basic specification triple for *Sys*:

***Sys*-Construction 4** Specification tuple for *Sys* specification

- "Diagram type," associative composition graph
- "Functorial requirements," by σ homology
- "Group factors," by finite setting

The general reduction against the discrete category determines an adjoining space, as does poset for analytic environments. The bijectivity between specification and system conceptually allows the descriptive language a sense of completion, either finite or transfinite, respectively. Requirements documentation in model-based is a finite set linguistic "adjoining" objective elements of a system, and the algebraic version is similar for *Sys* between a spatial description and system functionals. Just as system V&V tests the results over objective configurations, there can be similar spatial conditions yielding contextual structuring.

[16]Such extensions do appear to be nonunique, for example, "Kan," "strong," and "weak" extensions have all been observed in the categorical literature.

[17]Wymore centrally establishes duality in Corollary 4.4 (p. 170).

[18]Additionally having σ be from the groups gives a "linear representation" [8].

A categorical construction can be specified by associative, group type, and functorial requirements. The idea is that a "holistic engineered system" is an adjunction condition about the (denumerable) model and simulation, via basis and factors, respectively. This allows all to complete in a "similar fashion," hence a "cocompletion" procedure; e.g., the complete realizable descriptions of the system match the programmed completion in the model and simulation. Such a presumption seems common (and frankly necessary) to philosophical stances in system engineering activities, and conceivably further reduction can be achieved through more "powerful" descriptions: "intensive determination" with "extensive monoids."

13.3.5 System Coupling

Finally, the extensive aspect is considered in realizing further system specification by "connecting" completed systems in *Sys*, or "systems of systems." Intuitively relating these is common for combined systems in some sense by "coupling" "existing" systems. This can be thought of in the direct sense as describing further connections to physical systems or an analysis by extending semantics across or within identified systems.[19] Traditional systems consider "coupling" input/initial to output/terminal for simplicity, yet this need not be the case in either representation.

Below are constructions defined by functor mapping between systems as objects of *Sys*.

***Sys*-Construction 5 (Feedback or 1-*Sys* Couple)** Let $\omega_\sigma : Sys \rightarrow Sys$ be a functorial mapping [on terminals of (*S*,*Z*)] self-permutation [onto initials of (*S*,*Z*)].[20]

***Sys*-Construction 6 (Couple or n-System Functor)** Let $\Omega_{i,j} : Sys_i \rightarrow Sys_j$ be functors on/onto disjoint systems $Sys_i \in SYS$. Term the collection objects $n{-}Sys$ and the collection of couples $\sum : Sys \rightarrow Sys$.[21]

***Sys*-Construction 7 (Free System of System Representation)** Let *SYS* be a category whose SYS_0 are the $n{-}Sys$ objects and whose arrows SYS_Σ are the $\sum : Sys \rightarrow Sys$ arrows. The representational functor [or "forgetful functor"] is defined by $UMP : SYS \rightarrow Graph$, where *Graph* is the category of graphs.

This "universal mapping" maintains the representation of principles, categorized through *Graphs*. Given the same categorized interpretation the universals can be maintained without sacrificing the categorical sense of completion. Then observe that one can continue an injective "mapping" to graphical semantics, with couple to edge and system to vertex. This "coupling" functor is a "system as object" to "system as object" mapping of permutations, and "feedback" is this but for the

[19]The latter is a shared interest in categorical logic and algebraic geometry.

[20]Wymore Def. 5.1, pp. 202–3.

[21]Wymore Def. 5.2, p. 211.

same system. Thus one gains a "higher-order" collection of *Sys* which is a "category of category" definition.

Taking each *Sys* as vertex and couple/feedback to the edge, one obtains a collection as originally defined, thus recursing the *Sys* description. The remaining definida then are conjectured (necessary) to hold presuming the following underlying adjoined system descriptions:

1. Each system has an identity (distinguished by state) and feedback couples expressed an "identifiable" mapping (i.e., self-permutation).
2. The functor $\sum : Sys \rightarrow Sys$, expressible in Cat and UMP, maintains a finite representation and is thus "locally small."
3. The resultant functor Σ produces from permutations of σ_i "s is itself an epimorphism" [4] "right adjoint."
4. Finite, small categories are still a "small" category.

By this "coupling" functor, a "*Sys* of *Sys*" appears to be expressible through algebraic conditioning, matching the original system construction. This generalizes to a "free category," as in freely permuted over the couplings. This "free system category" then is maintained by organized combinations over the epimorphic permutations, each with adjoining semantic interpretations. This has an analogous intuition on later model-based system engineering work [1], as the proofs were all up to isomorphism to a discrete system. Thus, again conceptually, there seems to be a matching reciprocal supplemental proceduring. The sense is that maintaining an algebraic tracing through system coupling allows for a "systematic basis":

- Reuse of shared diagrammatic typing
- Tracing of shared homological sets by functors
- Finite objects over finite identified systems

The generalized requirement is for this ongoing cocompletion, given that well-defined connections appear as such [24, 25].

Inevitably the preservation allows for the expression of a fundamental theorem for categories that guarantees (co)completion by a "Yoneda embedding."[22]

***Sys*-Theorem 5 (Yoneda Lemma/Theorem for *Sys* Couple)** *The Yoneda embedding is the functor $y : SYS \rightarrow \text{Sets}^{SYS*}$ taking $Sys \in SYS$ to the "contravariant representable" functor:*

$$y(Sys) = \text{Hom}_{SYS}(-, Sys) : SYS^* \rightarrow \text{Sets}$$

and taking $f : Sys_1 \rightarrow Sys_2$ to the transformation:[23]

[22]Note that the normal presentation of the embedding is "flipped" given the focus on the epimorphism/right adjoint.

[23]The corresponding lemma (Awoday 8.2, p. 188) and theorem (Awoday 8.5, p. 193) show this is a full, faithful, injective "embedding."

$$y(f) = \mathrm{Hom}_{Sys}(-,f) : \mathrm{Hom}_{Sys}(-, Sys_1) \rightarrow \mathrm{Hom}_{Sys}(-, Sys_2).$$

This "contravariant" functor generalizes the computational and vector use of the term as one finds a representation that maintains the ordering of the type theory used, then determines a "cocompletion" for the permutations on *Sys* by this functor in categories taking objects and functions used in the generalized "cohomology." This describes a "completing diagram" often expressed in the intent of "linkage maps," "system grams," or other graphical tracking for system permutations. This Set^{Sys*} representation then matches the end goal of "systems of systems" engineering by "diagrams that are algebraic" in Set^{Sys}, and, as presumed, the duality represents the (co)completion. The implication is that this represents necessary formal conditioning and pragmatics for designing environment(s).

The formalization then defends a natural representation for a system of systems in the category of categories using its core theorem. This might be further explored through other categorical constructions, yet one loses the "nicely viewed" underpinnings to (discrete) model-based systems explicated here. The benefit is that a "system (out) of systems" identification can still be embedded in a structural representation using category theory. This might lead to further exploration maintaining this base category if nothing else as index. The expressed "free category" is involves invoking the several categorical presumptions and typings, yet conceptually covers specifying objects and objective transformations. Interesting applications can be seen in vectored couples for matrices [11] and "string diagramming" for physical systems [26], which might align with engineered system descriptions here.

Second, it guarantees conditioning for structural requirements as expected within an engineered systems sense. Several classifications were used for "clearer" categories, yet the same basic principles are still required for extended results. There are a wealth of questions on how system descriptions appear over underlying typings and ordering of said types. However, this still supports the general result that, given adjoining specifications with epimorphic functors, one still has means for an engineering context (economics and practically aside).

Yet mindful of algorithmic complexity theory, one wonders at a system's own algebraic complexity. Where are these language problems in systems? How is a measurable or classifiable sense to be obtained? Can categories or underlying algebra guide representations, at least under resources or typing? This investigation might yield interesting "engineering theory" under operational variety, analogous to complexity classes for automata. The spatial changes observed across "complex," particularly "sociotechnical," environments need judge internal and external orderings independently. Then how difficult are systematically declared situations, and do they have an algebraic character [20]?

13.4 Discussion

13.4.1 System qua *Categories*

Given the theoretic expressions, further investigation using categorical primitives would seem warranted. Definitional aspects of system environments ordered through categories might at least provide further linguistic analysis to engineering theory. Systems thinking with respect to elicitation might be one avenue in the discipline of universal codifying. Universals appear satisfactory for constraint problems [27] is pragmatically in usage. Diagrammatic methods and the general semantics of graphs in engineering are then one possible path, requiring matching with (types, classes) or (association, homology). There are notions of situated language [1] particularly bicategories [28] at least providing ontological representations [29], so elicitation methods in various respects [30–32] could incorporate this primitive. Either way the centrality of adjoining descriptions to a model simulation following similar adjoining semantics would necessitate "higher-level" guidance-ing [33]: control and observation, aleatory and epistemic, system of interest and environment.

As previously mentioned, the aspects of system duality and boundary appear central. This requires a "placement," possibly similar to choice in sets, yet would be "foundational" in the sense of finding formal language. However, this begs the complexity question here with algebraic complexity. Taking the complex numericals, one might imagine similar complex universals, each with adjoining universals yet conjoined with another universe. There is an interesting set incompleteness explanation via categories [34, 35] further complicating the linguistic picture. Yet "sociotechnical" systems, which are order and language-generating systems, would appear to minimally require this formalization. A "social space" having its conjoined language semantics and "technical space" its domain might each have algebraic solutions, yet not categorically in each other's universal interpretations.

13.4.2 Pragmatics and Application Questions

Pragmatics in applied environments yield several interesting representation questions. A conceptual take between objects and trajectories seems pertinent as to whether to specify a trajectory's *minor* points or object's *minor* continuum. Categories and sets respectively seem good terminologies language for each, yet pragmatically there is a sense for each. For system environments needing high degrees of organization, i.e., system of systems [36, 37], an ongoing question is an analytic/synthetic take on whichever is of interest. Additionally there are differences in senses and concepts particularly between engineers, management, and usage, that require coordinates/co-contextuals. The desire for frames and frameworks in (enterprise) system engineering can be appreciated in this light. This may integrate well with systems thinking studies as to how such needed primitives can be incorporated into a language or specification process.

Likewise, the initial intuition plays a spatial role that requires a context. The identification on varieties of systems requires abstraction on otherwise intuitive "patterns" [38]. Primitives of space have received interesting encapsulation in a categorical continuum, yet these at the moment seem to remain *in mathematica* with open questions as to their (ideal) application. If information systems theory is a guide, one conjecture is that categories allow at least "encoding" space to an underlying geometry. An approach to research might begin with existing intuitive theory and find sensible bases for formalized categorical mapping. For instance, informational economics has remained only a semiformal approach in engineering despite being theoretically well described. One conjecture is that the underlying Harsanyi type space might require better "encoding" into "mental models" using categorical correspondence.

13.5 Conclusions

There is a strong stance for categories for system engineering particularly for the abstract algebraic notions. Category theory is used here because it shares primitives with model-based systems theory and represents a natural supplement to that type of engineering. Definable constructions and intuition is shared among the works provided and present cooperatively up to certain aspects. Several constructions appear, hinging on epimorphic designation, bijective descriptions, and system coupling according to functorial procedures. Through duality, homo-/isomorphism, and functors both systems and system of system representations can be obtained with theoretic embedding for specification. This allows a basic triple for specification and development through diagrams, homology, and finite groupings, and observes Yoneda's lemma/theorem over extensive universals. This raises interesting system questions across pragmatic environments through algebraic methods expressed in categorical semantics. Particular attention should be paid to orderings, internal vs external, and sub-/superhierarchy, possibly via algebraic complex theory, particularly within "sociotechnical" systems.

Acknowledgment Special thanks to my advisor Michael Pennock for always maintaining his critical realism and for being constantly available. Thank you to my fellow students for their social and academic support. I acknowledge MITRE, where the initial intuition was developed, for its sponsorship, SERC for investigating its semantic use, and Spencer Breiner and Eswaran Subrahmanian of the NIST working group for helpful discussions and references relevant to applying categories in systems. Finally, as in all things, I express my eternal gratitude to friends and family for their unconditional love and support.

References

1. Wymore, W. (1993). *Model-based systems engineering*. Boca Raton, FL: CRC Press.
2. Rashevsky, N. (1961). Biological epimorphism, adequate design, and the problem of regeneration. *Bulletin of Mathematical Biology, 23*(2), 109–113.
3. MacLane, S. (1978). *Categories for the working mathematician*. New York: Springer.
4. MacLane, S., & Birkhoff, G. (1999). *Algebra*. Providence, RI: AMS Chelsea.
5. Univalent Foundations Program. (2013). *Homotopy type theory*. Institute for Advanced Study. Retrieved from https://homotopytypetheory.org/book
6. Barwise, J., et al. (1991). *Situation theory and its applications* (Vol. 2). Stanford, CA: CLSI Publications.
7. Wymore, W. (1967). *A mathematical theory of systems engineering - The elements*. New York: Wiley.
8. Awoday, S. (2010). *Category theory, oxford logic guides*. New York: Oxford Press.
9. Spivak, D. (2014). *Category theory for the sciences*. Cambridge, MA: MIT.
10. Goldblatt, R. (2006). *Topoi: Categorical analysis of logic*. Mineola, NY: Dover.
11. Baez, J., & Stay, M. (2009). Physics, topology, logic and computation: A rosetta stone. ArXiv.org, arXiv:0903.0340v3[quant-ph].
12. Cloutier, R., & Verma, D. (2007). Applying the concept of patterns to systems architecture. *Systems Engineering, 10*(2), 138–154.
13. Lawvere, F. (2006). Diagonal arguments and Cartesian closed categories. *Reprints in Theory and Applications of Categories, 15*, 1–13.
14. Martin-Lof, P. (1984). *Intuitionistic type theory*. Napoli: Bibliopolis.
15. Aczel, P. (1988). *Non well founded sets*. CSLI Lecture Notes, Standford University. Retrieved from irafs.org/courses/materials/aczel_set_theory.pdf
16. Rosen, R. (1978). *Fundamentals of measurement and representation of natural systems*. New York: North-Holland.
17. Breiner, S., Jones, A., Spivak, D., Subrahmanian, E., & Wisnesky, R. (2017). Using category theory to facilitate multiple manufacturing service database integration. *Journal of Computing and Information Science in Engineering, 17*(2), 11.
18. Spivak, D., & Kent, R. (2011). Ologs: A categorical framework for knowledge representation. ArXiv.org, arXiv:1102.1889v2[cs.LO].
19. Arbib, M., & Manes, E. (1974). Foundations of system theory: Decomposable systems. *Automatica, 10*(3), 285–302.
20. Rouse, W. (2005). Enterprises as systems: Essential challenges and approaches to transformation. *Systems Engineering, 8*(2), 138–150.
21. Awoday, S., & Warren, M. (1990). Homotopy theoretic models of identity type. *Mathematical Proceedings of the Cambridge Philosophical Society, 146*, 45.
22. Breiner, S., Subrahmanian, E., & Jones, A. (2018). Categorical foundations for systems engineering. In *Disciplinary convergence in systems engineering research*. Berlin: Springer.
23. Airbib, M., & Manes, E. (1975). *Arrows, structures, and functors*. New York: Academic.
24. Baez, J., Foley, J., Moeller, J., & Pollard, B. (2018). Network models. ArXiv.org, arXiv:1711.00037v2 [math.CT].
25. Perez, M., & Spivak, D. (2015). Toward formalizing ologs: Linguistic structures, instantiations, and mappings. ArXiv.org, arXiv:1503.08326v2[math.CT].
26. Lawvere, F. (1969). Adjointness in foundations. *Dialectica, 23*, 281–296.
27. Yanofsky, S. (2003). A universal approach to self-referential paradoxes, incompleteness and fixed points. ArXiv.org, arXiv: arXiv:math/0305282.
28. Spivak, D. (2016). The steady states of coupled dynamical systems compose according to matrix arithmetic. ArXiv.org, arXiv:1512.00802v2[math.DS].
29. Salado, A., Nilchiani, R., & Verma, D. (2017). A contribution to the scientific foundations of systems engineering: Solution spaces and requirements. *Journal of System Science and Systems Engineering, 26*(5), 549–589.

30. Zinovy, D., & Maibaum, T. (2012). Category theory and model-driven engineering: From formal semantics to design patterns and beyond. ArXiv.org, arXiv:1209.1433v1[cs.SE].
31. Pennock, M. (2017). An analysis of the requirements for the composition of engineering models. Pre-print.
32. Pennock, M., & Rouse, W. (2016). Epistemology of enterprises. *Systems Engineering, 19*(1), 24–43.
33. Hovmand, P. (2014). *Community based system dynamics*. New York: Springer.
34. Lane, J., Dahmann, J., Rebovich, G., & Lowry, R. (2010). Key system of systems engineering artifacts to guide engineering activities. In *NDIA systems engineering conference*.
35. Dahmann, J., Rebovich, G., Lane, J., Lowry, R., & Baldwin, K. (2011). *An implementers' view of systems engineering for systems of systems*. Systems Conference (SysCon) International.
36. Pennock, M., Bodner, D., Rouse, W., Gaffney, C., Hinkle, J., Klesges, C., et al. (2016). *Enterprise systems analysis TR-103*. Hoboken, NJ SERC-2016-TR-103: Systems Engineering Research Center.
37. Pennock, M., Bodner, D., Rouse, W., Cardoso, J., & Klesges, C. (2017). *Enterprise systems analysis TR-106*. Hoboken, NJ: Systems Engineering Research Center. SERC-2017-TR-106.
38. Maier, M. (1998). Architecting principles for systems of systems. *Systems Engineering, 1*(4), 267–284.
39. Phillips, S., & Wilson, W. (2010). Categorical compositionality: A category theory explanation for the systematicity of human cognition. *PLoS Computational Biology, 6*(7), e1000858. https://doi.org/10.1371/journal.pcbi.1000858
40. Pennock, M., Bodner, D., & Rouse, W. (2017). Lessons learned from evaluating an enterprise modeling methodology. *IEEE Systems Journal, 12*(2), 1219–1229.
41. Landry, E., & Marquis, J. (2005). Categories in context: Historical, foundational, and philosophical. *Philosophia Mathematica, 13*(1), 1–43.
42. Awodey, S. (2013). Structuralism, invariance, and univalence. *Philosophia Mathematica, 22* (1), 1–11.
43. Jeavons, P., Cohen, D., & Pearson, J. (1998). Constraints and universal algebra. *Annals of Mathematical and Artificial Intelligence, 24*, 51–67.

Chapter 14
Systems Engineering Grand Challenges in Security and Safety

Tom McDermott, Larry D. Strawser, Daryl Farber, Mike Yokell, and Mark Walker

14.1 Introduction

This paper describes the methodology and outcomes from a series of three workshops organized by the International Council of Systems Engineering (INCOSE) Academic Council. Distinguished systems engineering (SE) researchers and professionals from across the United States were invited to explore the global Grand Challenges from the SE perspective as related to security and safety, identify research gaps, and provide recommendations on closing those gaps. The paper begins by describing the concept of global Grand Challenges and the resultant workshop series. The paper goes on to describe how Workshop I defined the problem, how Workshop II defined the research gaps, and then how Workshop III developed the necessary research directions. The paper concludes by describing the recommended research threads to meet the Grand Challenges in systems engineering research relating to security and safety.

T. McDermott (✉)
Georgia Institute of Technology, Atlanta, GA, USA
e-mail: mcdermott@gatech.edu

L. D. Strawser
Johns Hopkins University, Baltimore, MD, USA

D. Farber
Pennsylvania State University, State College, PA, USA

M. Yokell
Lockheed Martin Corporation, Fort Worth, TX, USA

M. Walker
BCT LLC, Annapolis Junction, MD, USA

S. Adams et al. (eds.), *Systems Engineering in Context*,
https://doi.org/10.1007/978-3-030-00114-8_14

14.1.1 Grand Challenges in Systems Engineering

The concept of global Grand Challenges can be used to focus technical, educational, and humanitarian efforts [1–6]. In 2014, INCOSE developed a systems engineering vision for 2025 [7], which coupled societal needs to systems challenges and then to gaps in SE capabilities. Subsequently, the "INCOSE Systems Engineering Research Future" project [8] was initiated by the Academic Council of INCOSE, which used the framework of the SE Vision 2025. This project aimed to achieve the following objectives:

- Build communities among academia, industry, and government from numerous domains, dedicated to tackling the major global systems challenges confronting humankind for societal good.
- Excite, inspire, and guide systems engineering research in these communities.
- Achieve consensus among these communities to establish the priorities of systems engineering research.
- Provide the means by which to create synergy in these systems engineering research efforts such that progress can be measured against these objectives.

14.1.2 INCOSE Grand Challenge Workshop Series

This project produced a series of three workshops (October 2016, March 2017, and October 2017) with participants comprised of SE researchers and professionals from academia, industry, and government. The participants in the first workshop used the list of societal needs in the INCOSE SE Vision 2025 to select the following four Grand Challenges to be explored by the project: (1) access to education, (2) access to healthcare, (3) access to clean water, and (4) security and safety. The first workshop also adopted the five desired characteristics of societal needs Grand Challenges first proposed by the National Council of Teachers of Mathematics and added a sixth as follows:

1. Represent complex and extremely difficult questions that are solvable (potentially within 10–20 years).
2. Improve quality of life through positive educational, social, and economic outcomes potentially affecting millions of people.
3. Involve multiple research projects across many subdisciplines to be satisfactorily addressed.
4. Require measurable outcomes so that progress and completion can be identified.
5. Compel popular support by encouraging the public to relate to, understand, and appreciate the outcomes of the effort.
6. Is systemic in nature.

14.2 Problem Definition: Workshop I

The first workshop in the series set the stage for the four focus areas by selecting Grand Challenges from the SE Vision 2025 based on participant interest and identifying obstacles and relevant research questions. Security and safety was one of the four challenges selected, along with access to education, food, and clean water and access to healthcare. The full results of this workshop were previously documented [8], and the security and safety challenge is summarized here.

The security and safety group identified two Grand Challenges, which were both derived from the National Association of Engineering (NAE) Grand Challenges "Prevent Nuclear Terror" and "Secure Cyberspace." These were generalized as two themes: "resilience to catastrophic events" and "systemic security." Resilience to catastrophic events represents the need for SE methods, processes, and tools that adequately capture both the system and its external context, in order to address systemic effects. Because such events are necessarily addressed by combined policy, economic, and technical strategies, future systems engineering tools must capture both technical and nontechnical solution sets. For systemic security, there is recognition that the traditional methods of hierarchical decomposition of structure and associated perimeter defense strategies must be supplemented by new/future perspectives, process- and scenario-driven approaches, and more emphasis on failure modes (cyber warfare, catastrophic events, unintended events, etc.). Future systems engineering tools and design techniques need to emphasize the implementation to consider the heterarchical nature of cyber systems and their context.

The workshop participants developed a set of short vision statements reflecting a future state of systems engineering in which we capture system security and safety value and risk metrics into all types of systems and all levels and phases of decision analysis:

- Future systems engineering will engineer system solutions at the speed of the threats we see today.
- Systems will be designed to adapt to internal and external threats instead of reacting to them.
- System models will underlie all design practices, capture all system behaviors, be adaptive, and be able to evaluate systemic effects both within the system and out to external use contexts.
- Continuous experimentation with system vulnerabilities and threat scenarios (red and blue teams) will become a standard part of systems engineering at all phases of design, integration, test, and deployment.
- Model-based systems knowledge of both the system and current context will always reflect the current states of the system and its history.
- In order to make effective decisions, future systems engineering methods, processes, and tools will span the engineering, economics, human, and policy domains.

- Best practices from systems engineering (technology and data driven) and engineering design (user experience driven) will be merged so that all systems balance solutions and external context drivers.

  The following research questions were captured as an output of Workshop I:

1. How will SE categorize security and safety-related challenges if structural decomposition methods no longer work?
2. How will SE characterize capabilities and gaps in this domain?
3. What methods and tools will SE need to address simultaneously the human, physical, and informational aspects of the system?
4. Can SE address uncertainty in the system analysis process, particularly uncertainty of external system drivers?
5. How does one model real-time, realistic operational environments of changing systems and changing contexts?
6. How does one design governance methods for ensuring system security when the individual components are not secure, effectively emphasizing heterarchical system structures?
7. How does one "reverse engineer" the architecture of an existing system in order to learn its interfaces and potential vulnerabilities? Particularly brownfield systems (power grid, etc.)?
8. How does one measure what you don't know and plan to evolve as the system scales?
9. How does a system expose the necessary information to monitor system security in a private way?
10. How does one design a system to self-analyze failures and heal itself?
11. How can SE education engage the different domains at this level of design?

14.3 Research Gaps: Workshop II

At the second workshop, the group derived a number of gaps in current SE processes, which need to be addressed to meet the envisioned future state:

1. Current SE processes often assume static life cycle models and do not address the rapid emergence cycles now being seen in both the cyber threat environment and in commercial system development processes. As a result safety and security analyses are not regularly updated, particularly in response to emergent threats.
2. The "VEE" model paradigm does not represent the use of or the need for continuous engineering and test strategies. There is not a current SE process that balances continuous development, integration, and test with incremental safety and security analyses.
3. Most systems don't have or consider self-healing capabilities that correct externally driven failure modes. There is not currently an effective process to value such capabilities as a trade study factor in system architecture decisions.

4. System abstractions as one moves up and down component tiers tend to become ingrained into static structural boundaries in the system design process. As a result dynamic threat exploitation across these boundaries is not considered after the design is frozen. Current SE processes do not support "dynamic" abstractions such as new system boundaries or behaviors that emerge as the result of unplanned external system exploitation.
5. SE methods and tools lack language commonality and support for combining multidisciplinary expertise when considering internal system boundaries and the changing external context as a common system definition.
6. SE does not explicitly consider methods and tools to address rapid emergence and scaling in information domains. System behaviors are increasingly determined by information that is provided or collected external to the system. This data is also subject to rapid changes and shifts in context.
7. Current SE safety and security methods do not represent "bad actors" within the system boundary as a general source of system failure. As protection or resilience to such failures could rapidly become unaffordable, holistic approaches to value protection or resilience functions based on system level criticality must be developed.
8. There are many ontological barriers that limit design team interaction and design methods in the physical system versus cybersecurity domain, and there are limited methods and tools to manage their interaction. The effects of cyberattacks have had minimal consideration to date in the design of cyber physical systems, and the value or utility of functions that provide resilience to cyberattacks is not well understood.
9. SE does not explicitly consider methods and tools to encode ethical decision-making into the system operational capabilities. Systems these days are often faced with ethical choices, and the design community does not have the methods to support system trades associated with these attributes (often resulting in agnostic response to ethical dilemmas).
10. There is a need for rapid synthesis of data/information from both physical components and all data (IT) sources to information needed for decision-makers at all levels of the organization(s). This included data from the lowest level sensors on up and all data/IT sources. Rapid information development and dissemination are critical.
11. System design must include physical protections for catastrophic events that can cause catastrophic failures over wide areas of the earth. These include solar, nuclear, and cyber.

14.4 Research Directions: Workshop III

Workshops I and II produced a consistent view of SE challenges and needs in a world where the external context of safety and security is subject to much more rapid changes, and a much larger set of threats than previous SE processes, methods, and tools were designed to support. As with many other SE attributes, there is a need to shift from more efficient to more flexible, resilient practices, designs, and decisions, as well as to more fully consider the sociotechnical implications of systems and their context. Workshop III used the identified research questions and gaps to discuss potential research directions. These are not exhaustive but intended to serve future discussions. The following six research threads were identified as important to security and safety of future systems.

1. *Increasingly dynamic context.* Methods and processes that bring more agility to development and certification of critical systems should be a focus of future SE research. In the systems development process, agile "build-measure-learn" processes must have an accompanying agile "build-audit-control" model that goes beyond user stories and considers safety and security oversight of the development and modification processes. MIT's System-Theoretic Accident Model and Processes (STAMP) is an example of such research [9]. In addition, the emphasis of initial safety and security scenarios development prior to design is essential to work with customers and establish the basis for system design initiatives.

   SE life cycle models need to be adaptive to emerging threats from outside the system, such as cyberattacks, climate changes, economic shifts, political shifts, etc. One mental model for future systems is the "pipeline to platform approach"—systems that are built to endure (platforms)/built to change (apps) [10]. The community needs to address more generalized architectures and specialized functions, systems that are designed to be invulnerable but can be rapidly fixed. This includes methods and tools in the system-of-systems domain that address platform-to-platform interactions.

   Future SE processes need to support "dynamic abstractions" that can adequately deal with changing system boundaries. As today's systems mature, boundary relationships shift, but the abstractions used to design and measure system performance remain static. We need system models and simulations that reflect emergence between boundaries, changes in data flows, performance, and system languages/mathematical constructs that combine disciplines and methodologies.

2. *Human-machine interaction, bad actors, and trust.* Methods are needed to formally address human-machine interaction, not as static use cases but as "human-machine conversations." SE focuses on the human as the purposeful user of the machine, and ignores in the development process all other human interactions—including threats. Formal models of trust need to be developed within both intended and non-intended interactions, which require an integration of human activity models with system models.

   SE assurance methods must consider malicious actors within the system boundary—the insider threat. SE must consider countering modes and effects that emerge from these actors. This means that systems must be aware of their human actors (both good and bad) and monitor combined human and machine behaviors as part of their assurance design [11]. These are interactions between physical and data/informational processes. As these will be complex relationships, graph theoretic methods might be explored in addition to continuous monitoring of these interfaces for threat-related activities/changes.

3. *Modeling societal and organizational context.* Modeling and simulation must encode the societal context and impacts of solutions as tradespace alternatives. Architectures that can simulate post-conditions of a deployed system and feed that back into the model are needed. This will change how we instrument systems, shifting collection of data toward measures that are critical constructs of the system model (often called the digital twin). Models and simulations that interact with larger models and simulations of social context are also needed. Future SE tools that model "noise" from signals and erroneous information versus encoded truth may become useful when modeled at the system and system-of-systems level. All of these needs imply a shift from modeling "the system" toward modeling the system and the external context it resides in, in both environment and timeframe [12, 13].

   The need for organizations and management to fund and support the continuous evolution of system models throughout their life cycle is essential for these changes to be captured and understood. Rapid and agile system evolution requires the equivalent rapid updating of the system models. This is needed in a SoS with many operational and in development systems, otherwise chaos results along with wasted time and funds. Note that the threats to systems will continue to evolve rapidly; the models, design changes, and implementations must be evolving in an equal or more rapid manner.

   In the future it will be useful to integrate system models with narrative scenarios as a bridge between the system and its sociotechnical context. Future tools should extract meaning from text and formal representations of situations or concepts of operations (CONOPs)—using key constructs and concepts—that capture both the structure and meaning that result. This will require advances in machine learning methods and tools [14]. Future SE tools need to partially or fully automate model building by observing system behaviors and combining these observations with linked knowledge.

   CONOPs and their related scenarios, requirements, system models, designs, processes, etc. need to be continuously updated and available to developers, maintainers, management, etc. to assure that all personnel that manage, develop, and support the system(s) have access to current information for their decision needs. CONOPs (Operational Concept Documents) should be computer based to allow rapid updating of the scenarios, etc. and changes that reflect the need for design updates, etc.; this is essential for cyber security/software in addition to other failure mode impacts, process changes, physical changes, and other future safety and security challenges.

4. *Heterarchy, adaptive learning, and emergence.* System-of-systems research must advance, and research is needed particularly in the domain of human and autonomous systems, where machines will soon join the ranks with humans as drivers of complex adaptive behaviors. These systems must assure security and safety, as well as other attributes like ethics and social responsibility. There is a general

need for the SE community to embrace heterarchical models of complex systems and create methods and tools that allow heterarchical processes to be addressed in system requirements, architectures, design, and process evolution. New modeling and simulation frameworks should be designed to evaluate effective design principles that address trust and scalability in both the system-system and human-system dimensions, thereby increasing their operational utility. Most existing SE methods/processes assume hierarchical decomposition strategies, which are not incorrect but are insufficient for future systems. Even systems-of-systems research lacks a framework to address systems that learn and adapt themselves.

SE methods that explicitly consider adaptive learning and emergence must be developed to manage the complexity of future cyber-enabled systems and automated human-machine systems. Both human-machine and independent machine-machine adaptation are important. Current approaches to mitigate these challenges have been isolated within traditional simulation formulations and environments and associated siloed scientific disciplines; most typically pure engineering fields are insufficient to address the nature of the problem space. The community needs to develop the foundations for an integrated "social" systems science. This will lead to a more relevant fundamental study and characterization of the "social behavior of complex and autonomous systems" and unintended consequences of these behaviors (whether constructive or in conflict). This needs to lead to the development of new scenarios development concepts and content to drive the future systems designs.

5. *Self-healing systems and condition-based adaptation*. Research is needed on methods that automate the healing process of a compromised system [15]. Breakthroughs may come from biological-inspired processes for sensing, classifying, and recovering health, related research on "effectors" of these recoveries, methods to encode these change effectors into system components, methods and means to create system self-learning by immersing the system in different environments, multi-stage models, and simulation methods for such systems.

   Future systems should be able to define and then sense preconditions of abnormal operations at any level. Research is needed on how to model this information, particularly for nonphysical processes. Research is also needed on how to capture the information on external conditions and use it to train adaptive models. These preconditions need to be updated throughout the system's life cycle and also be reflected in the CONOP scenarios, use cases, models, data flows, etc. automatically along with identifying system design/development needed changes.

6. *Libraries, patterns, and model sharing*. Like software, system models are abstractions of physical phenomena. SE methods and tools should capture these abstractions as basic functions supporting standard system models that are created from shared libraries that reflect normally used functions and communications [16, 17]. This would help increase both the efficiency of the design activities and the resilience of new systems, drawing from the history of other successful

systems. Due to the complexity of systems today and with even greater complexity expected in the future, the SE community must develop and share both functional and formal design patterns that can be used to compose new system models and designs. Functional patterns are necessary early in the design process when architectural decisions are being made, to reduce errors in requirements. Formal methods are needed to assure the as-designed system is free from errors of construction. The SE community should lead the development of complex system modeling federations for sharing data and associated research and design processes [13]. Metamodeling needs to become a natural SE research and design skill.

14.5 Common Directions Across Grand Challenge Areas

As was discussed in Sect. 14.1, in addition to the security and safety Grand Challenge, three other Grand Challenges were explored by the Academic Council: clean water, healthcare, and education. In similar fashion to the security and safety Grand Challenge, the key problems and research threads for each of these were identified, followed by identification of current gaps in SE research and suggested topics for future SE research. Although each of these Grand Challenges identified specific topics for SE research relevant to their specific challenge, there were three major necessary SE research areas common to all four Grand Challenges:

- SE research is needed to understand the human element in developing the system and the deployed system.
- With the increasing dynamic context, research is needed on the role of SE in developing self-adaptive, evolving, and self-sustaining systems.
- Study is needed to determine the role of SE in human factors by researching models of long-term sociotechnical systems to address culture- and value-based decisions.

Each of these research areas should be addressed from the perspective of engineering systems, which according to Kossiakoff et al. [18] is:

> A broader and robust perspective to systems approaches to solve very extensive complex engineering problems by integrating engineering, management, and social science approaches using advanced modeling methodologies .... The intent is to tackle some of society's grandest challenges with significant global impact by investigating ways in which engineering systems behave and interact with one another including social, economic, and environmental factors. This approach encompasses engineering, social science, and management processes without the implied rigidity of systems engineering.

Acknowledgments The authors would like to acknowledge the contributions and support from the INCOSE Academic Forum and from the other participants in the three Grand Challenges in systems engineering workshops.

References

1. Stephan, M. L., Fish, M. C., Chval, K. B., Herbel-Eisenmann, B., Wanko, J. J., Konold, C., et al. (2015). Grand challenges and opportunities in mathematics education research. *Journal for Research in Mathematics Education, 46*(2), 134–146.
2. Varmus, H., Klausner, R., Zerhouni, E., Acharya, T., Daar, A. S., & Singer, P. A. (2003). Grand challenges in global health. *Science, 302*(5644), 398–399.
3. National Academy of Engineering. (2008). *Grand challenges for engineering* [Internet]. Available from, http://www.engineeringchallenges.org [cited 18 Jan 2018].
4. Uehara, E., Flynn, M., Fong, R., Brekke, J., Barth, R. P., Coulton, C., et al. (2013). Grand challenges for social work. *Journal of the Society for Social Work and Research, 4*(3), 165–170.
5. U.S. Agency for International Development. (2011). Grand challenges for development. Available from http://www.usaid.gov
6. The Institution of Engineering and Technology. (2013). Global grand challenges. Available from http://www.raeng.org.uk/publications/other/ggcs-report
7. Friedenthal, S., Beihoff, B., Nichols, D., Oster, C., Paredis, C., Stoewer, H., et al. (2014). *A world in motion: Systems engineering vision 2025*. San Diego, CA: International Council on Systems Engineering.
8. Wade, J., Adcock, R., McDermott, T., & Strawser, L. (2018). Future systems engineering research directions. In A. M. Madni, B. Boehm, R. Ghanem, D. Erwin, & M. J. Wheaton (Eds.), *Disciplinary convergence in systems engineering research* (pp. 1165–1179). Berlin, Germany: Springer.
9. MIT Partnership for a Systems Approach to Safety. Available from https://psas.scripts.mit.edu/home/
10. van Alstyne, M., Parker, G., & Choudery, S. (2016, April). Pipelines, platforms, and the new rules of strategy. *Harvard Business Review*, 54–60.
11. Horowitz, B., & Lucero, D. (2016). *System-aware cyber security: A systems engineering approach for enhancing cyber security*. INCOSE INSIGHT.
12. Curry, M., & Ross, A. M. (2015). *Considerations for an extended framework for interactive Epoch-Era analysis*. 13th Conference on Systems Engineering Research, Hoboken, NJ.
13. Folds, D., Hutto, C., & McDermott, T. (2017). Toward next generation social analytics: A platform for analysis of quantitative, qualitative, geospatial, and temporal factors of community resilience. *International Journal on Advances in Internet Technology, 10*(1–2), 70–86 ISSN 1942–2652.
14. Riedl, M. (2017). Computational narrative intelligence: Past, present, and future. Medium, October 24. Available from https://medium.com/@mark_riedl/computational-narrative-intelligence-past-present-and-future-99e58cf25ffa/
15. Ghosh, D., Sharman, R., Rao, H. R., & Upadhyaya, S. (2007). Self-healing systems—Survey and synthesis. *Decision Support Systems, 42*, 2164–2185.
16. Schuchardt, K., Didier, B. T., Elsethagen, T., Sun, L., Gurumoorthi, V., Chase, J., et al. (2007). Basis set exchange: A community database for computational sciences. *Journal of Chemical Information and Modeling, 47*(3), 1045–1052. https://doi.org/10.1021/ci600510j
17. Hirtz, J., Stone, R., McAdams, D., Szykman, S., & Wood, K. (2002). A functional basis for engineering design: Reconciling and evolving previous efforts. *Research in Engineering Design, 13*(2), 65–82.
18. Kossiakoff, A., Sweet, W. N., Seymour, S. J., Biemer, S. M. (2011). *Systems engineering principles and practice* (2nd ed., pp. 33–34). Hoboken, NJ: Wiley.

Chapter 15
Application of the Rayleigh Model to Predict Information Technology Program Cost and Schedule Performance

Rebekah Yang, Justin Coe, Benjamin Dryer, Matthew Cumpian, Maeve Budi, Matt Prebble, Mary (Jocelyn) Cox, and Shaelynn Hales

Nomenclature

α	Time parameter
$C(t)$	Cumulative cost
D	Cost parameter
e	Napier's constant
t	Time

15.1 Introduction

Cost and schedule growth is a major challenge for government acquisition programs, and being able to predict this growth early will allow government officials to take corrective actions to minimize the impact of technical or programmatic challenges. The United States (US) government performs systems engineering and program management on a diverse portfolio of information technology (IT) development and modernization efforts; however, many initiatives often encounter both technical and programmatic challenges [1]. IT projects are often highly visible because of their large size, importance, and proportional funding. In the 2017 High Risk Report sent to Congress, the Government Accountability Office (GAO) highlighted the need for improving the monitoring of IT investments in order to inform how government chief information officers assess investment risk and report incremental development status [2].

In addition to GAO findings, the Office of the Inspector General (OIG) regularly performs audits assessing IT program management. OIG reports have underscored

R. Yang · J. Coe · B. Dryer · M. Cumpian · M. Budi · M. Prebble · M. Cox · S. Hales (✉)
CNA, Arlington, VA, USA
e-mail: haless@cna.org

S. Adams et al. (eds.), *Systems Engineering in Context*,
https://doi.org/10.1007/978-3-030-00114-8_15

the importance of understanding software development to keep IT programs on budget and on schedule throughout their life cycles. Across the various GAO reports and OIG audits, there are a number of common themes highlighting the inadequate management and oversight processes and the unavailability or non-standardized data used to support program management decisions [3–5]. Research conducted by CNA in 2006 on the US Navy's management system for acquisitions identified several problems with the program review process, including a lack of a proactive approach to identifying programs at risk for future problems, such as schedule slips and/or cost growth [6]. A key recommendation from the report was that the Navy should be more proactive in its program monitoring and evaluation using industry standard approaches, such as monthly program reporting and the earned value management system (EVMS).

15.1.1 Earned Value Management System

EVMS attempts to integrate cost, schedule, and scope to analyze project performance and progress. It requires that metrics are reported at fixed intervals, allowing for analysis of both lagging and leading metrics (Fig. 15.1).

Three measurements form the basis for EVMS metrics and analysis: Budgeted Cost of Work Scheduled (BCWS), Actual Cost of Work Performed (ACWP), and Budgeted Cost of Work Performed (BCWP). While BCWS and ACWP are universally used to estimate cost and schedule performance, BCWP integrates scope measurements to measure and predict project performance more accurately. This can be analyzed against the BCWS and ACWP to generate lagging metrics for project performance: schedule variance (SV) and cost variance (CV). While SV and CV can be useful indicators for current project performance, they depend heavily on the accuracy of the EVMS data—specifically, the budgeted cost for work elements.

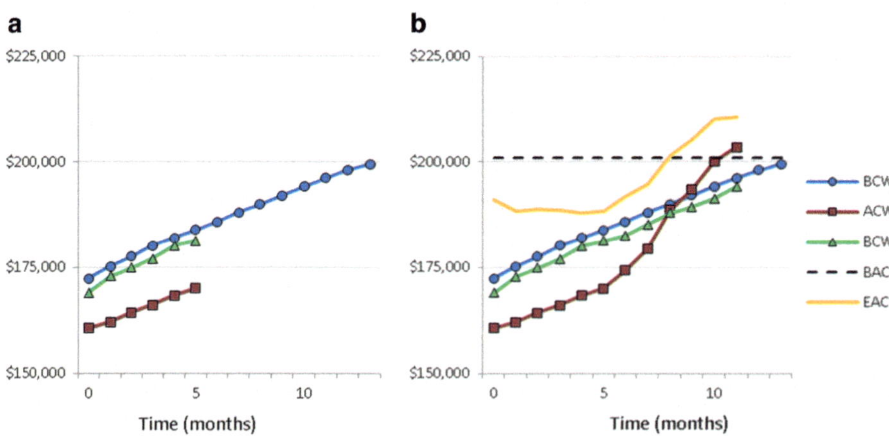

Fig. 15.1 (**a**) EVMS lagging metrics. (**b**) EVMS leading metrics

EVMS data can also be used to generate leading metrics for project performance, predicting cost and schedule overruns. The Estimate at Completion (EAC) is a prediction of the final project cost, using the historical performance thus far and the Budget at Completion (BAC). The EAC can be predicted using a number of methods; a commonly used formulation is shown in Eq. (15.1):

$$\text{EAC} = \frac{\text{BAC}}{\text{BCWP}/\text{ACWP}} \tag{15.1}$$

The Variance at Completion (VAC) is the difference between the budgeted and estimated final cost, projecting the total cost overrun. The equation for VAC is shown in Eq. (15.2):

$$\text{VAC} = \text{BAC} - \text{EAC} \tag{15.2}$$

EAC, along with other EVMS leading metrics, can be used to forecast potential cost and schedule overruns; however, they have several issues. The biggest problem with EVMS leading metrics is that they use past behavior to estimate future performance, regardless of the differences in work as a project develops. In addition, as with EVMS lagging metrics, EVMS leading metrics are highly dependent on the accuracy of the budgeted cost for work elements.

The Federal Acquisition Regulation (FAR) requires an EVMS for major development acquisitions and requires contractors to submit EVMS monthly reports for those contracts [7]. Federal agencies can also have their own EVMS reporting policy which may expand upon the FAR EVMS requirements. The availability and widespread use of EVMS base data makes it an ideal data source for analyzing project cost and schedule, while analysis techniques based on statistical distributions can be used to supplement EVMS metrics. Past research conducted by CNA identified the Rayleigh model as a way of improving the forecasting of contract costs and schedules in research and development acquisition programs, leveraging the readily available EVMS data.

15.1.2 Rayleigh Model

The basis of the model used in this study is the Rayleigh distribution. The Rayleigh distribution is a continuous probability distribution used in various domains of science, including medical imaging (e.g., magnetic resonance imaging) and nutrition [8, 9]. The Rayleigh probability density function is asymmetric and single-peaked, while its cumulative distribution function (CDF) is an s-shaped curve. The shape of the Rayleigh CDF mimics traditional project spending profiles: a slow ramp-up at the project start, peak spending as complexity increases, and a gradual tail-off as the project nears completion [10]. For this reason, the Rayleigh CDF has been adapted to

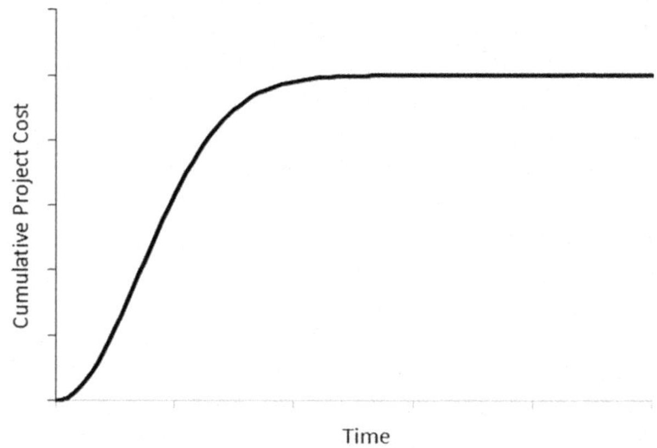

Fig. 15.2 Sample spending profile represented by the Rayleigh model

create the Rayleigh model by applying cost and time scaling parameters to the CDF. While the Rayleigh model is not a probability distribution, it utilizes the shape of the Rayleigh CDF to model project spending over time. Previous applications of the Rayleigh model include modeling the spending profiles of software development projects as well as Department of Defense (DOD) research and development (R&D) programs [11–14]. Figure 15.2 shows a typical s-shaped curve associated with a sample Rayleigh model.

Project behavior can be forecasted by fitting actual cost and schedule data to the Rayleigh model and evaluating against the estimated final budget and completion data. The equation for the Rayleigh model is shown in Eq. (15.3):

$$C(t) = D\left(1 - e^{-\alpha t^2}\right) \tag{15.3}$$

The cumulative cost and time in Eq. (15.3) can be derived from traditionally reported metrics, such as the EVMS, ACWP, and reporting dates. The cost and time parameters need to be determined empirically. The cost parameter is based on the total cost of the project. As a project progresses in time, Eq. (15.3) converges to the value of the cost parameter, representing the cumulative final cost. The time parameter is determined by the duration of the project. With larger time parameters, the exponential term in Eq. (15.3) will decrease more rapidly, resulting in the model more quickly reaching the final cumulative cost. To make certain that a Rayleigh model accurately captures project behavior, project managers must ensure that cost and time parameters are determined without bias.

Unlike EVMS metrics, the Rayleigh model does not rely on the budgeted estimates for work or the measured performance against the predicted behavior. Using the Rayleigh model is another method of generating predictive metrics based on historically observed trends in program performance. Program managers can use the Rayleigh model to evaluate past and projected project performance. One of the key benefits of the Rayleigh model is that it is an early indicator of cost and schedule overruns: with only three or more data points, a Rayleigh model can be used to generate results for comparison with the planned schedule and budget. Furthermore, use of the Rayleigh model does not require the full set of EVMS data; this expands its applicability to all IT programs/projects within a portfolio.

In the past, CNA investigated the use of the Rayleigh model as a program management tool within the DOD by evaluating its implementation on 107 Acquisition Category I (ACAT I) R&D contracts [10]. In the study, the predictive accuracy of the Rayleigh model in determining the final cost and duration of each contract was compared with EVMS metrics or contractor and project management predictions. It was found that the Rayleigh model more accurately reflects the cost-schedule profiles of large-scale programs, providing critical insight about potential cost and schedule overruns. Overall, the researchers concluded that "the Rayleigh model is a powerful tool for analyzing contract execution and plan validity" and it is a "very useful addition to the decision-makers toolkit that provides him with very early warning of potential plan or contract execution issues [10]."

15.1.3 Objectives

Based on the persistent issues with cost and schedule growth in many government IT programs [2–5], the purpose of this analysis was to investigate the ability of the Rayleigh model to predict future cost or schedule growth in civil agency IT programs. A majority of existing IT project management requirements and industry best practices used to test and monitor project performance are lagging metrics, which are easy to measure and report but tend to be output oriented. Adding leading metrics to performance reporting will help predict trends and outcomes. Leading metrics are often harder to measure and interpret, increasing the need to identify simple yet meaningful metrics that help answer the questions raised by project managers and executives. This capability would assist government decision-makers in identifying programs that require closer monitoring or course corrections to ensure completion of the program on schedule and within budget. This study uses the Rayleigh model to leverage data that are already being collected to reduce the overall reporting burden to the programs while providing value in the form of leading indicators to help provide early warning signs on program health. Building on the previous research on R&D programs, the usefulness of the Rayleigh model for managing government IT programs is evaluated. Using case studies, CNA evaluated the Rayleigh model's ability to provide meaningful leading metrics beyond those calculated through traditional EVMS metrics.

15.2 Methods

Many federal agencies maintain large-scale IT programs that are tracked at the project level and may be ideal candidates for fitting to the Rayleigh model. To assess the performance of the Rayleigh model in IT projects, CNA selected two federal programs and conducted a conceptual case study analysis using methods identified in past CNA research [10] of R&D programs. Due to the sensitive nature of cost overruns and schedule slips, the names of the programs chosen were anonymized to Program A and Program B. The first case study performed on Program A involved a large-scale, completed IT development that had a 5-year period of execution. The second case study, Program B, involved an ongoing IT development that has encountered multiple program performance challenges.

The cost and schedule data for these programs were obtained from the Office of Management and Budget (OMB) IT Dashboard. The dashboard maintains an open-source repository of information on federal IT investments from 2013 to the present. Using data from the dashboard, ACWP, BCWP, and BCWS were calculated for Program A and Program B, along with program start and end dates and total estimated budgets. The IT Dashboard provides an annual snapshot of the program's cost and schedule data broken down by individual projects and associated tasks. The individual spending profiles for the projects were extrapolated and superimposed to form a harmonized dataset of data points representing an overall project spending profile within the overall program that could be used in fitting the Rayleigh model. Both Program A and Program B consist of a single IT project; however larger programs may contain multiple projects, which would need to be modeled at the individual project level for best results. Furthermore, because the Rayleigh model is most applicable to R&D work, the project cost and schedule data were filtered and offset as needed, so only tasks related to software development were included as part of the project (e.g., tasks related to conceptualization and planning activities were not included).

CNA previously developed the Executive's Cost and Schedule Analysis (XCASA) tool that uses the Rayleigh model to fit EVMS data, for DOD ACAT I programs, using contract reporting data [10]. XCASA has since been updated to the Indicator for Schedule Analysis and Cost (ISAAC) tool, which will include a wider range of input cost and schedule data to support predictive forecasting for major IT acquisitions. ISAAC was used to fit a Rayleigh model to each project's cost and schedule data over time. The tool generates a Rayleigh model of the project spending profile using nonlinear least-squares analysis to estimate model parameters based on given cost and schedule data. For projects that are not yet complete, ISAAC is thus able to extrapolate and predict the estimated remaining spending profile of the project with the generated model.

For the two programs evaluated, the predictive use of the Rayleigh model was evaluated at standard intervals over the course of the project. A new Rayleigh model was fitted in ISAAC with each additional data point (approximately in monthly intervals), and the corresponding estimated final cost was predicted with a 95% confidence interval. This procedure simulated the real-time evaluations that would have been made during project execution. Cost metrics generated from the Rayleigh models were compared against the same metrics generated from traditional EVMS calculations for each new data point. The cost metric was defined in this analysis as the Cost Variance at Completion (CVAC), which is expressed as a percentage and shown in Eq. (15.4):

$$\text{CVAC} = \frac{\text{BAC} - \text{EAC}}{\text{BAC}} = \frac{\text{VAC}}{\text{BAC}} \tag{15.4}$$

Using this definition, a negative CVAC indicates a final program cost that is over budget, while a positive CVAC indicates a final program cost that is under budget. A percentage-based metric such as CVAC is used in this analysis because it can be more useful than an absolute cost-based VAC metric when evaluating multiple programs across a federal IT portfolio.

15.3 Findings and Discussion

According to the IT Dashboard, the first case study, Program A, was completed successfully and resulted in a positive final project evaluation from the chief information officer. The IT Dashboard provides overall lagging program metrics, including percentage cost variance as the percentage difference in ACWP from BCWP and percentage schedule variance as the percentage difference in budgeted days of work scheduled and actual days of scheduled. Programs with variances less than 10% are considered low risk, while programs with variances between 10% and 30% are medium risk, and those with variances over 30% are considered high risk. In Program A, the final IT Dashboard percentage cost variance was −3.8%, and the final percentage schedule variance was 0%. To determine whether the Rayleigh model could accurately predict the cost-schedule profile and final estimated cost of Program A, a Rayleigh model was fitted to the IT Dashboard data with the addition of each new data point throughout the course of the project. Figure 15.3 shows examples of the fitted Rayleigh model at various points throughout the project. In Fig. 15.3d, the model fully fitted to 230 data points is shown to have an r-squared value of 0.99. The final fitted Rayleigh model for Program A shows a final extrapolated cost that was slightly over the actual cost ($96.9 million vs. $93.4 million), while the final extrapolated schedule exceeded the actual duration by 1 month (December vs. November). A Kolmogorov-Smirnov test was used to assess whether the project data follow the Rayleigh model. The test failed to reject the null hypothesis ($p = 0.39$) that the distributions are approximately the same, indicating that the Rayleigh model is appropriate for the data.

Figure 15.4 shows the real-time CVAC values of the project as predicted by the Rayleigh model with corresponding 95% confidence interval as well as the CVAC values from three traditional methods [8] of calculating EAC (e.g., EAC1 is defined in Eq. (15.1)). In this case, the Rayleigh model was able to predict a cost overrun at the beginning of development as the spending accelerated more quickly than the trend fitted by the Rayleigh model. The estimated CVAC is extremely high, which demonstrates how sensitive the Rayleigh model is to initial project data. Nonetheless, the Rayleigh model is able to predict a budget overrun earlier than the traditional EVMS metrics, and the Rayleigh model is also able to predict the budget overrun consistently, unlike EAC1 and EAC2. In the following years, the project

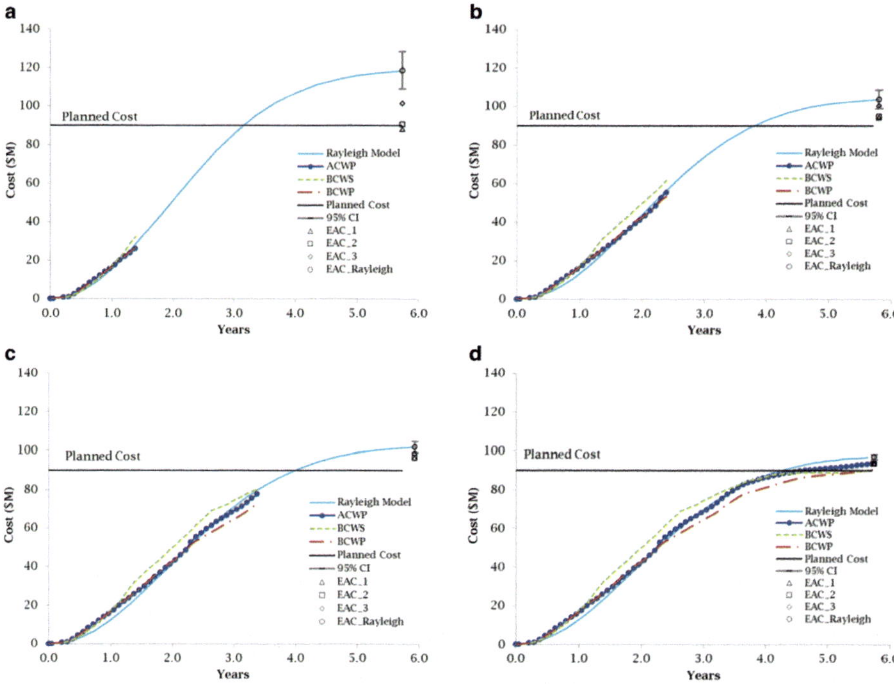

Fig. 15.3 Rayleigh model fitted to the final Program A after approximately (**a**) 1.5 years, (**b**) 2.5 years, (**c**) 3.5 years, and (**d**) completion

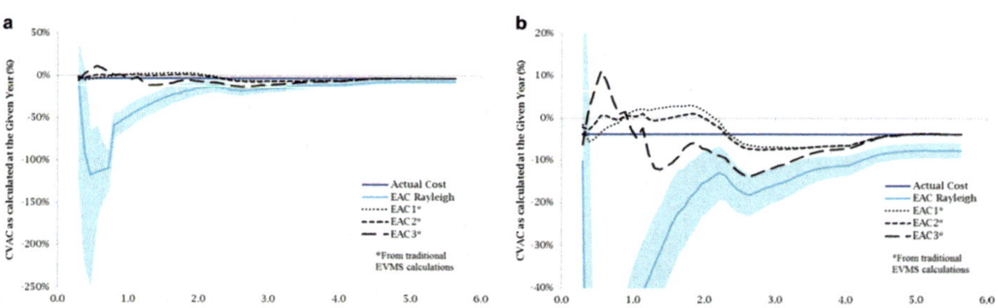

Fig. 15.4 CVAC as would be calculated throughout the course of Program A, showing (**a**) full and (**b**) close-up view

spending continued more linearly and eventually plateaued as expected. As the program progressed, this behavior gradually conformed to the curve fitted by the Rayleigh model, and the CVAC predicted from the Rayleigh model converges toward the actual CVAC.

The second case study, Program B, involves a continuing program that has encountered several project performance issues throughout its IT development phase. The current percentage cost variance and percentage schedule variance as

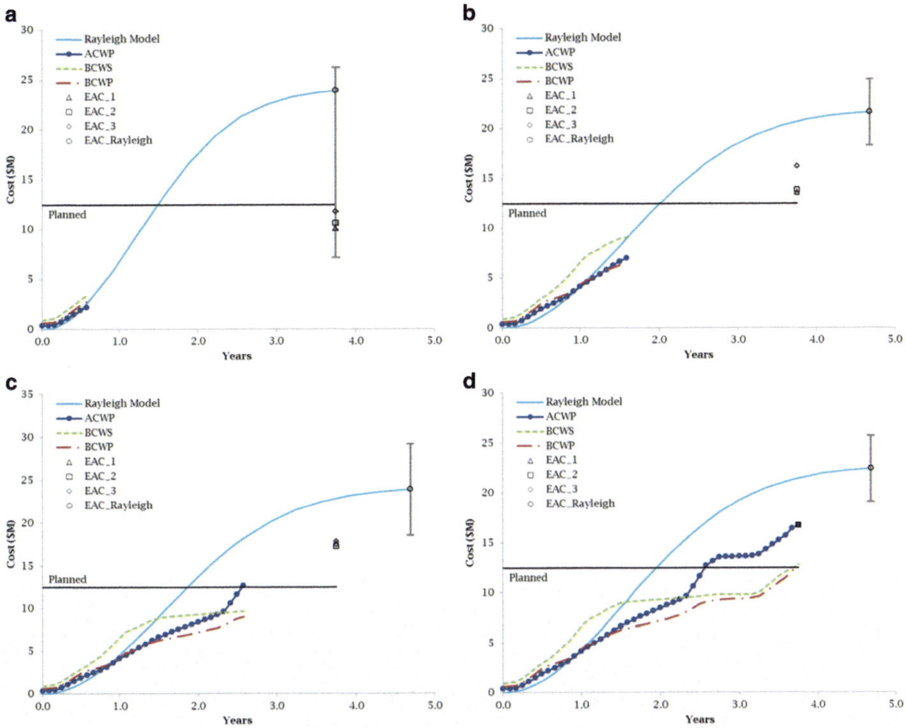

Fig. 15.5 Rayleigh model fitted to the final Program B after approximately (**a**) 0.5 years, (**b**) 1.5 years, (**c**) 2.5 years, and (**d**) current status

of July 2017 reported to the IT Dashboard are significant (−39.6% and −15.6%, respectively), indicating that the contract is high risk. As of July 2017, Fig. 15.5 shows that the Rayleigh models fitted to Program B's data consistently predicts both an overrun of both final cost and completion date throughout the course of the project. Note that this program has changed in scope and has been rebaselined, so the use of a single Rayleigh model is limited in that it does not easily fit the nonstandard cost-schedule profile, and, because the program is still ongoing, the final cost and schedule are still unknown. While the contract seems to follow the Rayleigh model at the beginning of the program, the work completed does not increase as expected at the ramp-up stage. The reduced work completed results in additional spending and consequential program scope changes and revised baselines. At the time of evaluation, the Rayleigh model is no longer able to predict the trajectory of the program because of the program's irregular cost-schedule relationship.

In Fig. 15.6, the comparisons of CVAC over time are also quickly predicted to be high risk by the Rayleigh model at the beginning of the program, while the cost risk is not apparent using the other traditionally calculated EAC methods until 2 years into the development phase. Thus, again, the Rayleigh model is able to show warnings of cost and schedule slippage earlier than traditional EVMS metrics.

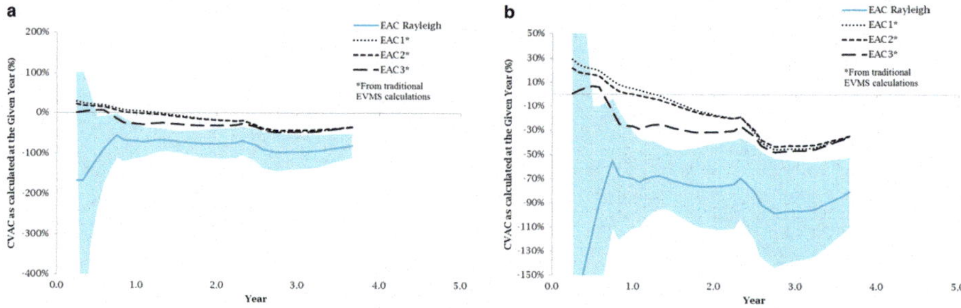

Fig. 15.6 CVAC as would be calculated throughout the course of Program B, showing (**a**) full and (**b**) close-up view

15.4 Conclusion

The Rayleigh model historically has been demonstrated to accurately model both large-scale R&D programs and software development projects. It works particularly well for forecasting software development projects because it can account for the slow ramp-up, peak spending as complexity increases, and a gradual tail-off toward completion. In addition to projecting final cost and schedule, variations in the shape of the curves can identify risk areas for additional monitoring. The Rayleigh model relies on accurate and consistent cost and schedule data to fit the distribution, but, because the model is used to fit data already collected through traditional program management techniques, there is relatively low additional effort required to realize benefits. Application of this approach across IT portfolio management can provide key insight to help program managers make informed decisions and allow decision-makers to prioritize resources across a portfolio based on forecasted execution profiles.

Based on the initial case studies evaluated in this research, using the Rayleigh model to evaluate civil agency IT projects to improve IT program management has potential. The Rayleigh model mimicked the cost and schedule profile of a successful IT project and provided early and consistent, though initially exaggerated, indications of cost and schedule slippage. Due to its unique benefits, the Rayleigh model can be used by IT portfolio managers to supplement existing EVMS metrics in order to better measure project cost and schedule. Performing this analysis on the data that the government already collects, identifying predictive metrics, and applying the Rayleigh model to forecast future performance can help identify programs or projects that are in need of course correction and can hopefully reduce or eliminate the cost and/or schedule growth. The two programs examined in this paper adhere to the traditional waterfall software development methodology. Future research could evaluate the application of the Rayleigh model to IT projects conducted using the agile software development approach, a methodology that has been gaining significant interest in government agencies.

Acknowledgment The authors would like to thank Julianne Nelson, Ph.D., for providing background information on the use and implementation of the XCASA tool to generate Rayleigh models for the analysis as well as Maryann Shane, Ph.D., for providing additional review of the paper.

References

1. Patanakul, P., Kwak, Y. H., Zwikael, O., & Liu, M. (2016). What impacts the performance of large-scale government projects? *International Journal of Project Management, 34*(3), 452–466. https://doi.org/10.1016/j.ijproman.2015.12.001
2. U.S. Government Accountability Office. (2017). *High-Risk Series: Progress on Many High-Risk Areas, While Substantial Efforts Needed on Others*. GAO-17-317. Report to Congressional Committees. 15 Feb 2017. Washington, D.C.: U.S. Government Publishing Office.
3. U.S. Office of Inspector General (2014). FAA lacks the metrics and data needed to accurately measure the outcomes of its controller productivity initiatives. AV-2014-062. 9 July 2014. Washington D.C.: U.S. Government Publishing Office.
4. U.S. Office of Inspector General (2012). The success of FAA's long-term plan for air traffic facility realignments and consolidations depends on addressing key technical, financial, and workforce challenges. AV-2012-151. 17 July 2012. Washington D.C.: U.S. Government Publishing Office.
5. U.S. Government Accountability Office. (2012). Air traffic control modernization: Management challenges associated with program costs and schedules could hinder NextGen implementation. GAO-12-223. Report to Congressional Committees. 16 Feb 2012. Washington, D.C.: U.S. Government Publishing Office.
6. Christle, G. E., & Birchler, D. A. (2006). *Strategic management system for navy acquisition* (p. 50). Alexandria, VA: CNA Corporation.
7. Earned Value Management System, 48 C.F.R. § 34.2 2018.
8. den Dekker, A. J., & Sijbers, J. (2014). Data distributions in magnetic resonance images: A review. *Physica Medica, 30*(7), 725–741. https://doi.org/10.1016/j.ejmp.2014.05.002
9. Ahmadi, H. (2017). A mathematical function for the description of nutrient-response curve. *PLoS One, 12*(11), e0187292. https://doi.org/10.1371/journal.pone.0187292
10. Davis, D., Christle, G., & Abba, W. (2009). *Using the Rayleigh Model to assess future acquisition contract performance and overall contract risk*. Alexandria, VA: CNA Corporation.
11. Putnam, L. H. (1978). A general empirical solution to the macro software sizing and estimating problem. *IEEE Transactions on Software Engineering, SE-4*(4), 345–361.
12. Watkins III, H. (1982). An application of Rayleigh curve theory to contract cost estimation and control. Naval Postgraduate School. Monterey, California.
13. Abernethy, T. S. (1984). An application of the Rayleigh distribution to contract cost data. Naval Postgraduate School. Monterey, California.
14. Lee, D., Hogue, M., & Hoffman, D. (1993). Time histories of expenditures for defense acquisition programs in the development phase. In *Annual Meeting of the International Society for Parametric Analysis*.

Chapter 16
Integrating Set-Based Design into Cost Analysis

Randy K. Buchanan, James E. Richards, Christina H. Rinaudo, and Simon R. Goerger

16.1 Introduction

The Department of Defense (DoD) science and technology communities are incorporating model-based engineering early into the design process to improve decision-making in the analysis of alternatives (AoA). Analysis of alternatives is a DoD requirement of military acquisition policy to ensure that multiple design alternatives have been analyzed prior to making costly investment decisions [1]. AoA through tradespace exploration (TSE) supports engineering design and development by providing analysts and decision-makers with an understanding of capabilities, gaps, and potential compromises required to facilitate the realization of system objectives. It informs decision-makers regarding contending system options and the implications of decisions on multiple missions across joint war fighting environments [2].

Traditionally, point-based design quickly converges to a solution that is singular in the solution space, necessitating modification of the chosen solution later to meet the design objectives. Leveraging tradespace exploration of set-based designs (SBD) allows for consideration of all solutions within given parameters and enables filtering of possibilities to converge at a final solution. Design decisions that occur before Milestone A can have a significant impact on the cost and affordability of a program, and cost is a key component within an AoA tradespace. Set-based design is useful for the DoD and projects with a large number of design variables, tight coupling among design variables, conflicting requirements, flexibility in requirements allowing for trades, or technologies and design problems that are not well understood [3]. The integration of life cycle cost (LCC) with model-based engineering (MBE) and SBD requires a strategic assessment and selection of engineering

R. K. Buchanan (✉) · J. E. Richards · C. H. Rinaudo · S. R. Goerger
U.S. Army Engineer Research and Development Center, Vicksburg, MS, USA
e-mail: randy.k.buchanan@erdc.dren.mil

S. Adams et al. (eds.), *Systems Engineering in Context*,
https://doi.org/10.1007/978-3-030-00114-8_16

parameters in conjunction with associated cost elements. Approaches to enabling cost domain generation for SBD range from creating a cost model specifically for the analysis at hand to simply repurposing an existing cost model.

16.2 Set-Based Design

Point-based design processes have demonstrated that later iterations to refine that solution can be time-consuming and costly and lead to a suboptimal design [4, 5]. Alternatively, if the tradespace is fully defined and explored in the beginning of design analysis, the design moves quickly toward convergence, and the identification of an ultimate solution that may have been overlooked in a traditional point-based design process becomes apparent. Exploring the tradespace of a set-based design allows for consideration of all solutions within given parameters and enables filtering of possibilities to converge at a final solution. SBD includes a broad set of design values and multiple sets of design solutions. Toyota introduced set-based concurrent engineering decades ago and instigated a substantial cultural shift [6]. Today, the DoD has particular interest in using set-based design to improve the AoA process. SBD has been applied to several programs within the Navy, including the Ship-to-Shore connector, amphibious combat vehicle, and the small surface combatant task force [3]. The goal of SBD is to develop sets of alternatives that span the design space rather than a limited number of specific designs. As the design process progresses, some design decisions can be delayed until adequate information is available to eliminate inferior sets of solutions.

Applying a set-based design concept to cost analysis requires identifying a manageable set of design parameters that have been determined to be principal factors in achieving maximum design value. Other requirements include consideration of factors that impart importance in differentiating designs and utilizing an analysis that identifies total value [7]. Bernstein states that the principle concepts to implement SBD are the following: (1) consider a large number of design alternatives by understanding the design space, (2) allow specialists to consider a design from their own perspective, and (3) use the intersection between individual sets to optimize a design and establish feasibility before commitment [8]. In consideration of the many direct and indirect mechanisms that affect cost, it was important for this research to establish a working baseline in which some targeted input parameters to cost could be transformed from point-based to set-based and establish the feasibility of the transformation process. A notional military vehicle development, the Light Reconnaissance Vehicle (LRV), was selected to implement the concept and provide further insight into directions the work should continue.

16.3 Cost Models

A cost model can be a mathematical model or set of parametric equations used for estimating the costs of a given project. Parametric estimating models provide a mathematical representation of cost relationships for identifying a logical and predictable correlation between the physical or functional characteristics of a project (plant, process system, etc.) and its resultant cost [9]. A life cycle cost estimate is a detailed estimate of acquisition and total ownership costs (TOC) for an acquisition program. Cost models accept input parameters that describe attributes as well as physical resource requirements. A cost model produces an output of resource requirements in terms of cost and time. Traditional cost models operate with singular input values, point-based design parameters, or perhaps a very small number of point-based "excursion" cost model evaluations. Numerous cost models exist throughout the DoD acquisition communities. Popular tools include Microsoft Excel, Automated Cost Estimating Integrated Tools (ACEIT), and reference manuals for cost estimating relationships (CERs). However, there is no standard format for a cost model and no central repository for completed models, though data and model sharing activities continue to increase. The result is that completed models cannot be easily shared and reused among different modeling groups and disparate modeling frameworks.

Integrating set-based design into life cycle cost involves treating selected inputs as a range of values and can be approached from different perspectives. While a variety of tools exist to collaboratively link cost analysts and engineers, the integration of these often complex model sets is still difficult. To address this issue, prior DoD Engineered Resilient Systems (ERS) research developed a software module to link design tools with a life cycle cost model. This linkage provides the ability to perform early costing of design solutions for initial or early requirements to inform early design and programmatic decision-making [10]. Coalescing life cycle cost and SBD requires a strategic assessment and selection of engineering parameters in conjunction with associated cost elements. Approaches to enabling cost domain generation for SBD range from creating a cost model specifically for the analysis at hand to simply repurposing an existing cost model. The methodology used for guiding this research is shown in Fig. 16.1. The stakeholder needs drive the requirements generation process for LCC early tradespace generation. Concurrent efforts then proceed via module development, various cost approaches, and LCC demonstrations. The ERS LCC development exists within TradeBuilder and the ERS integrated tradespace framework. Finally, cost flows to other tradespace framework implementation in DoD acquisition environments, including high-performance computing (HPC) and related tools such as Galaxy [10].

Incorporating life cycle cost analysis early within the SBD process informs requirement prioritization, design decisions, early design investment, and budget and affordability analysis. Figure 16.2 illustrates the cost estimating process as outlined by the Government Accounting Office (GAO) [11]. Both data and existing cost models are leveraged in the "Obtain the data" step of the process. The MBE occurs in the "Define the program" step and is where TradeBuilder links the costs

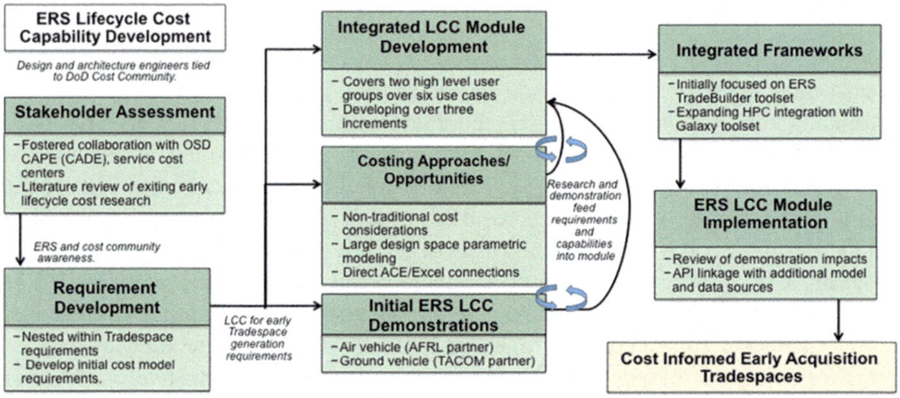

Fig. 16.1 ERS life cycle cost research methodology [10]

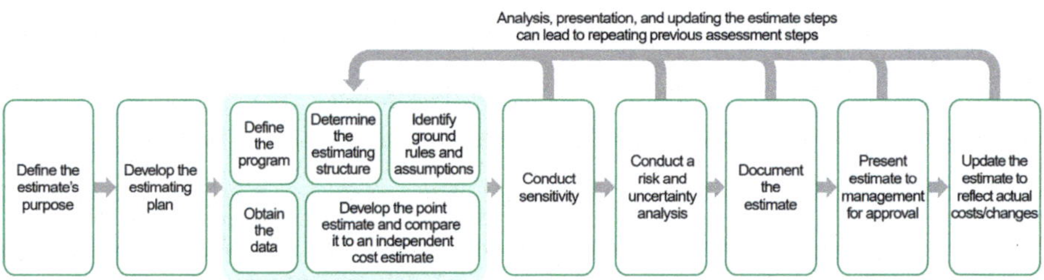

Fig. 16.2 The cost estimating process [11]

with physical models. The goal is for early cost analysis using parametric relationships to focus on design variables which will be available within engineering models during early concept development. Cost analysis for SBD relies on the broader tradespace generation functionality to address uncertainty quantification and sensitivity analysis coupled with the other physics-based models, which is a large benefit to using an integrated framework.

16.4 Light Reconnaissance Vehicle (LRV) Example

The LRV is a notional ground vehicle undergoing model-based engineering analysis by systems engineers at the US Army Tank Automotive Research, Development and Engineering Center (TARDEC) located in Warren, Michigan [12]. This research team collaborated with the US Army Tank-Automotive and Armaments Command (TACOM) Cost and Systems Analysis Division to integrate a cost model with the TARDEC engineering model resulting in an integrated cost and model-based engineering tradespace [10].

The collaboration with the TACOM cost analysis group leveraged a Technomics cost research contract to support the effort. The cost methodology (Fig. 16.3)

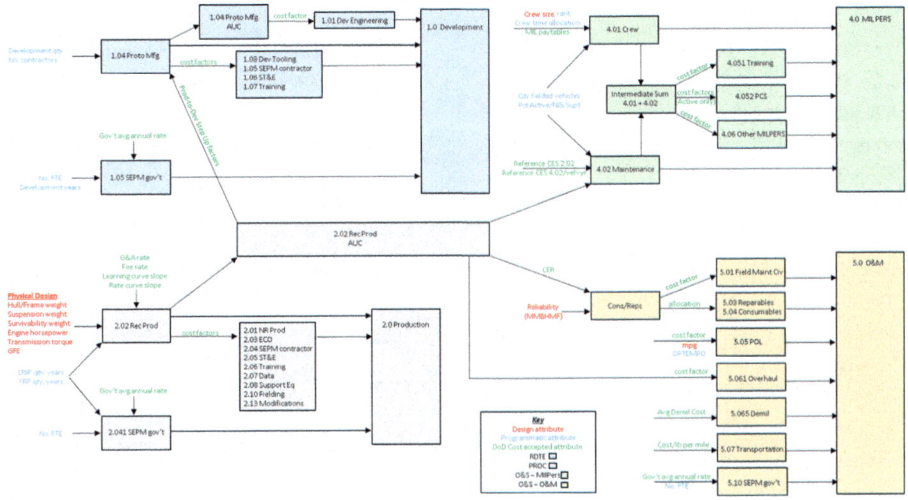

Fig. 16.3 Cost methodology [13]

illustrates the flow from physical design parameters (red) combined with programmatic cost drivers (blue) and the cost factors generated from historic data (green) prior to integration [13]. One key element of the integration effort was to ensure that the cost model was sensitive to the TARDEC and requirements community engineering parameters of highest interest. The integration approach provided default programmatic attributes (such as number of prototypes purchased for testing) which can be adjusted in future analysis but did not inhibit early tradespace generation.

A notional value model was developed to exemplify how cost versus value may be considered in an LRV AoA. The Infantry Brigade Combat Team (IBCT) program reports requirements for the LRV served as an initial starting point to determine stakeholder prioritized requirements [14]. Priorities established for previous vehicle development (such as the MRAP and the Joint Light Tactical Vehicle (JLTV)) were leveraged to develop notional stakeholder priorities for use within the LRV value model. During the development of the MRAP, stakeholders valued survivability as the highest priority [15]. In the JLTV development, the stakeholder requirements initially included payload, variant/configuration, protection, weight range, reliability, and cost [16]. Using these requirement priorities as guides for value model analysis, overall utility value scores were calculated for the generated design alternatives.

16.5 Integration of SBD into Cost Model

Cost analysts create inputs (cost drivers) and outputs (cost estimating relationships) that work together to form a functional cost model. Design cost drivers are inputs that act as physical design parameters for a given cost model. Inputs used for establishing a set-based approach for this application are shown in Table 16.1. The

Table 16.1 Set-based inputs

Set-based inputs for design	Additional set-based inputs as cost drivers
Max weight	MMBOMF
Vehicle height	OPTEMPO
Vehicle length	Miles per gallon
Vehicle width	Peacetime idle
Center of gravity (CG) height	Idle burn rate
CG fore-aft location	Crew
Wheelbase	Number of prototypes
Track width	Buy quantity
Armor weight	Useful life
Tire type	
Suspension stiffness	
Suspension damping	
Mounted weapon type	
Surveillance equipment	
Engine	

variables available from the design model are shown in the first column, while additional variables also available for cost analysis are shown in the second column. The set-based inputs for design were leveraged from a previous notional LRV tradespace analysis [12].

Transitioning to set-based value inputs requires setting upper and lower boundaries for each chosen input. For example, when considering an input such as mean miles between operational mission failure (MMBOMF) or Operational Tempo (OPTEMPO) in set-based design analysis, instead of selecting a single value, a lower and upper limit would be considered, as well as a randomly distributed set-based value.

One framework which allows for integrated model execution is TradeBuilder, a DoD research software tradespace exploration toolkit. To enter input information, the user must access a TradeBuilder user interface that is linked to inputs and outputs of integrated models within the project. The upper, lower, and default values for each parameter are required. Once entered, TradeBuilder executes a tradespace based on one of several available designs of experiment methodologies, such as the Latin hypercube used in this example.

16.6 Results and Analysis

Comparing aggregated value against total life cycle cost illustrates how early design decisions have potential impacts on value and cost. An example LRV tradespace (Fig. 16.4) displays how five general vehicle configurations drive the groupings of total cost. This approach follows concept exploration outlined by Parnell et al. [17],

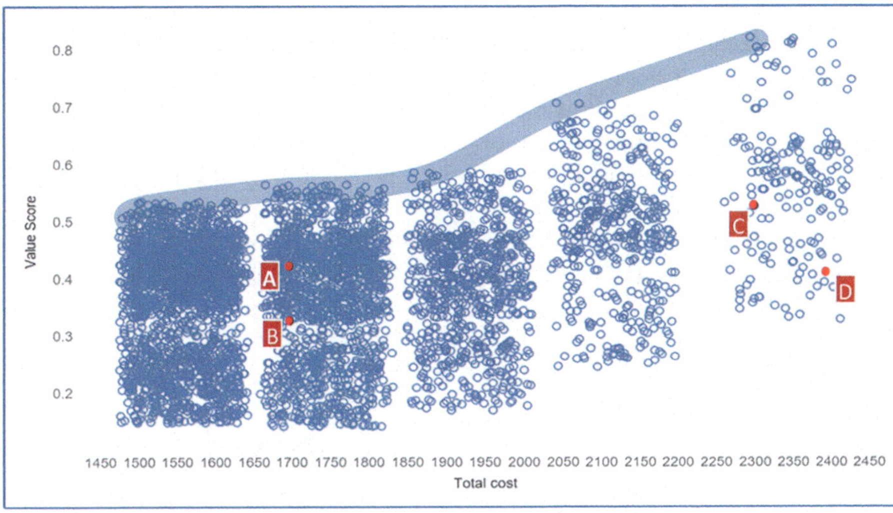

Fig. 16.4 LRV tradespace visualization. (**a**) Roll on/roll off, low weight, low survivability, medium lethality package, large cargo size. (**b**) Roll on/roll off, low weight, low survivability, low lethality, medium cargo size. (**c**) Sling load with extra armor, high weight, medium survivability, high lethality package, low cargo size. (**d**) Sling load with extra armor, high weight, medium survivability, low lethality package, low cargo size

in which the value functions included notional swing weight and value function parameters. The efficient frontier (top shaded area) defines Pareto efficient designs differentiated from dominated designs. This type of visualization may be used to inform decision-makers and requirements generators.

Relaxing some requirements up front allows for greater range of capability but at higher cost. Examples of point-based designs are indicated with red dots. Examples include a CH-47 Chinook helicopter internally transportable (roll-on/roll-off) LRV and a sling load transportable LRV. In a point-based approach, if the cost constraint is relaxed so that greater value and yet more expensive system is chosen, the design is locked into that region. If an affordability constraint is later applied, it may be too late to achieve the original value associated with roll-on/roll-off functionality as it was before the requirement was relaxed. In this case, one can easily envision a final design that neither rolls on/off nor achieves the other desired benefits that originally pushed for the extra armor requirement change in the case of the LRV.

In the case that the cost model analyst is unaware of the physical design attributes and models of interest to the engineers, sensitivity and other attribute disparities may become an issue. In addition to technical vocabulary spanning cost and engineering communities, synchronization is also important when implementing the stakeholder value model. Previous research integrated, analyzed, and generated utility value scores for vehicles such as the Mine-Resistant Ambush Protected (MRAP), Stryker, and armored personnel carrier (M113A3) but identified the need to also integrate life cycle cost [18]. This demonstrates the importance of cost-informed SBD.

To reflect programmatic life cycle costs, the cost model should be applicable for all elements of the life cycle and include development, production, operations and support, as well as disposal costs. These costs should be distributed by year within the data set, even if cost and production schedules are gross estimates. A basic understanding of the impacts of time-phased program costs is critical to affordability analysis. In order to determine if a program can fit into the budget, yearly outlays must be estimated. While time phasing can become an analytically tedious element of cost analysis, the need is clearly outlined in DoDI 5000.02 (Jan 7, 2015) which identifies the Milestone A requirement for an affordability analysis in addition to a cost analysis.

16.7 Conclusion

Informed decisions can be made early in the acquisition process by integrating cost models with model-based engineering and performance models. The majority of cost models currently available focus on a point-based design cost analysis. As a result, many of the cost models use a mixture of parametric relationships as well as "pass through" values, which, taken together, limit the general use of these cost models with SBD methodologies. Integrating the cost model with the SBD framework provides a rich tradespace from which early design phase decision-makers can maintain regions of the design space that may have been neglected in a point-based approach.

Implementing the set-based design into the LRV cost model provided an insight into the potential for improving AoA. Finding parametric consistency between engineering models and cost models, when created separately, will continue to be a hurdle to overcome. Facilitating a methodology to enable communication between these two modeling communities is critical to successful tradespace analysis. When trying to integrate a cost model into an integrated modeling framework, there are several informing outcomes which this research highlighted. There is often a technical vocabulary mismatch between the cost and engineering modeling communities. For example, the gross vehicle weight in one model may not include the same components as gross vehicle weight in another model. Executing the model may reveal restrictions on design variables in terms of their ability and degree to which they may be varied. For instance, a cost model uses a cost driver which the engineering model is holding constant, or the engineering model may vary a parameter to which the cost model is insensitive. Synchronizing sensitivity and parameter bounds requires diligence. Finally, some model components may require expert knowledge input which is difficult to automate. For example, calibration factors in a cost model or a stability constraint in a vehicle performance model do not lend themselves to a hands-off computation design sweep.

The ability to display aggregated value against cost is very powerful to decision-makers, particularly when grouped by concept sets, which show how early design decisions, often made arbitrarily before an AoA is commissioned, have potentially significant impacts on value and cost. Within the LRV demonstration, one counter-intuitive conclusion is that it may not be helpful to relax the "roll-on/roll-off" constraint as it shifts potential value albeit at a greater cost which may result in an ultimately unaffordable solution. Essentially, by relaxing some requirements, it will allow for greater capability at higher cost. If the cost constraint is reapplied later, it may be too late to achieve the original requirement that was relaxed. Future research should evaluate the intersection of a top-down life cycle affordability analysis and SBD design methodology using a time-phased MBE-linked parametric cost model.

Acknowledgment The authors would like to acknowledge the Department of Defense Engineered Resilient Systems program for its support of this research. Additionally, this paper is based on collaborative research with the US Army Tank-Automotive and Armaments Command (TACOM) Cost and Systems Analysis Division, Technomics Inc., and the US Army Tank Automotive Research, Development and Engineering Center (TARDEC) Analytics—Computational Methods and System Behavior (CMSB) Team.

References

1. U.S. Office of Management and Budget. (2008). *Circular no. A–11, preparation, submission and execution of the budget*. Washington, DC: Executive Office of the President.
2. Spero, E., Avera, M., Valdez, P., & Goerger, S. (2014). Tradespace exploration for the engineering of resilient systems. *2014 Conference on Systems Engineering Research Procedia Computer Science, 28*, 591–600.
3. GovEvents. *Design sciences series: Set-based design*. Retrieved November 7, 2017, from https://www.govevents.com/details/24509/design-sciences-series-set-based-design/
4. Iansiti, M. (1995). Shooting the rapids: Managing product development in turbulent environments. *California Management Review, 38*, 37–58.
5. Kalyanaram, G., & Krishnan, V. (1997). Deliberate product definition: Customizing the product definition process. *Journal of Marketing Research, 34*(2), 276–285.
6. Sobek, D. K., Ward, A. C., & Liker, J. K. (1999). Toyota's principles of set-based concurrent engineering. *Sloan Management Review, 40*(2), 67–83.
7. Singer, D. J., Doerry, N., & Buckley, M. E. (2009). What is set-based design? *Naval Engineers Journal, 121*(4), 31–43.
8. Berstein, J. I. (1998). *Design methods in the aerospace industry: Looking for evidence of set-based practices*. Massachusetts Institute of Technology: Master of Science thesis.
9. NASA. (2015). *NASA cost estimating handbook*. Washington, DC: NASA.
10. Richards, J., Kelley, D., Hardin, D., & Church, H. (2017). *Generating the cost domain of the tradespace for lifecycle cost analysis*. Vicksburg, MS: Internal ERDC.
11. Government Accounting Office. (2009). *GAO cost estimating and assessment guide: Best practices for developing and managing capital program costs*. United States Government Accountability Office, Applied Research and Methods.
12. Castanier, M., Pokoyoway, A., & Bronstetter, G. (2016). *Using ERS tools for trade space exploration of military ground vehicles*. NDIA Systems Engineering Conference. Springfield, VA.
13. Cherwonik, J. (2017). *Engineered resilient systems (ERS) lifecycle cost analysis for trade-space generation*. Vicksburg, MS: Internal ERDC.
14. Feickert, A. (2017). *Infantry Brigade Combat Team (IBCT) mobility, reconnaissance, and firepower programs*. Washington, DC: Congressional Research Service.

15. Blakeman, S., Gibbs, A., & Jeynathan, J. (2008). *Study of the mine resistant ambush protected (MRAP) vehicle program as a model for rapid defense acquisitions (MBA Professional Report)*. Monterey, CA: Naval Post Graduate School.
16. Canaley, W. P. (2013). *Joint light tactical vehicle: A case study*. Carlisle, PA: U.S. Army War College.
17. Parnell, G. S. (2016). *Trade-off analytics: Creating and exploring the system tradespace* (p. 367). Hoboken, NJ: Wiley.
18. Rinaudo, C. H., Buchanan, R. K., & Barnett, S. K. Applying the robustness (broad utility) workflow to assess resiliency for engineered resilient systems. *2016 Conference on Systems Engineering Research*.

Chapter 17
A Proposed Application of Systems Engineering Principles to Healthcare Decision-Making: Evaluating Patient Decision Regret in the Treatment of Oropharyngeal Cancer

Robert T. Aarhus Jr. and Edward Huang

Nomenclature

CES	Complex engineered systems
DA	Decision aid
DR	Decision regret
HNC	Head and neck cancer
HPV	Human papillomavirus
MDM	Medical decision-making
MDO	Multidisciplinary design optimization
OPC	Oropharyngeal cancer
QoL	Quality of life
SDM	Shared decision-making

17.1 Background

Medical decision-making (MDM), particularly in complex and chronic disease processes, is often characterized by an incomplete clinical picture and inherent uncertainty as to which treatments will produce the desired results for both physicians and patients [1]. A representative framework for MDM might be envisioned as the engineering of a series of interventions, planned and executed over time, aimed at maintaining or improving the health of an individual. This process is constrained by the standard of care [2], a medicolegal term referring to the actions of a reasonable, board-certified specialist consisting of acceptable practices established by evidence-

R. T. Aarhus Jr. (✉) · E. Huang
George Mason University, Fairfax, VA, USA
e-mail: raarhus@masonlive.gmu.edu

S. Adams et al. (eds.), *Systems Engineering in Context*,
https://doi.org/10.1007/978-3-030-00114-8_17

based medicine [3]. Decisions may be further modified or limited due to the policies and activities of entities outside of the patient-physician relationship; the willingness of insurance companies and government programs to fund procedures or lengthy hospital stays, the legality and availability of certain drugs, and the presence or absence of allied health and community-based services may influence the selection of treatments. The acceptability of proposed interventions to the patient and their caregivers or family may also play a role in the MDM process [4].

17.2 Model

Multidisciplinary design optimization (MDO) frameworks have been successfully applied to complex engineered systems (CES) in disciplines such as aerospace and automotive industries [5]. Although MDO has been discussed in the context of medical device development [6], a search of the literature found in PubMed (https://www.ncbi.nlm.nih.gov/pubmed/) and Medline (https://health.ebsco.com/products/medline) databases reveals scant mention of its use in other healthcare fields. Nevertheless, MDO may be applicable to MDM in the context of decision processes between multiple treatment modalities for various diseases.

A variety of definitions exist for CES. Braha et al. [7] describe CES as self-organized systems that "arise from evolutionary processes of selection in the context of actual tasks". Sosa et al. [8] associate CES with networks of interdependent components. One online resource [9] lists the characteristics of CES as a collection of autonomous, composite, emergent, nonlinear, open, and interactive entities.

Taking the infrastructure of a transportation network for a major metropolitan city as an example [10], CES can be seen to arise from an initially diverse set of technologies—pedestrian, animal-based transport, early automobiles, locomotives, and airplanes—which over time increased in their sophistication and interconnectedness. Modern transportation systems are highly regulated entities requiring the skilled participation of both designers and operators to optimize cost and maintain reliable scheduling; interactions between systems may be substantially structured themselves (e.g., restrictions on taxi and limousine service to airports or rationing of urban parking facilities where public transportation is available).

MDM can likewise be imagined as a CES. Initially, the structures of medical practice ranged across a host of independent practitioners, some of dubious skill and quality, and facilities whose procedures would today be considered nothing short of barbaric. Over time, both governmental bodies (such as the Food and Drug Administration (FDA) and the Department of Health and Human Services) and oversight organizations (such as the Joint Commission, an independent, nonprofit entity performing accreditation services for a wide variety of healthcare institutions), as well as the certification bodies of various medical specialties for physicians and other healthcare providers, arose to evaluate and monitor the practice and performance of both individuals and facilities [11]. At the same time, certain allied health and

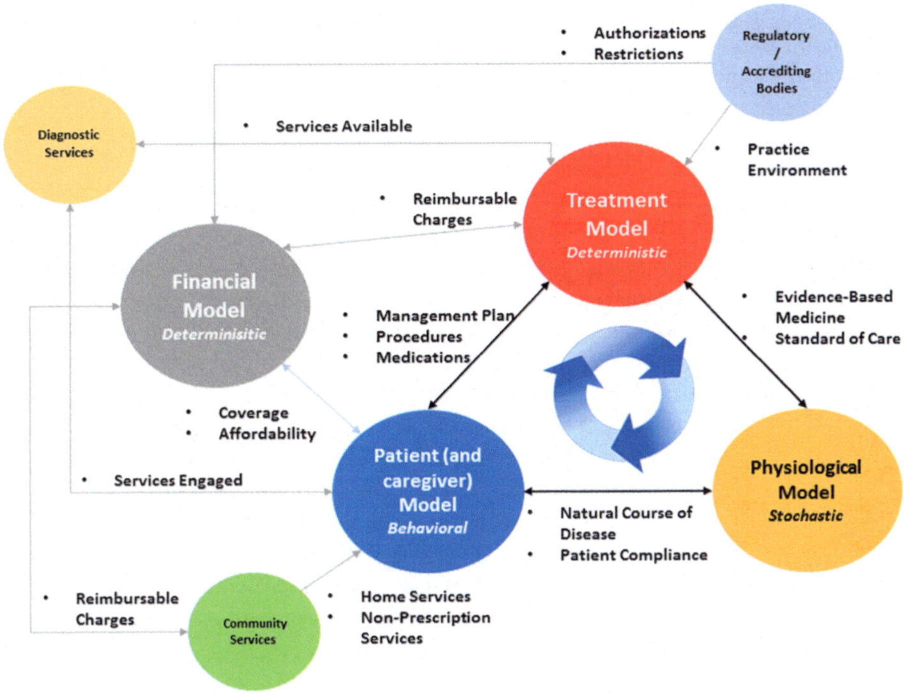

Fig. 17.1 Simplified model of entities influencing complex MDM [12]

community services evolved in response to the need for reliable, though less highly regulated, operations.

Figure 17.1 illustrates a simplified model of many of the key entities required for the diagnosis, monitoring, and treatment of disease, some of which require the participation of multiple components in a complex environment; those involving interventions and therapies for cancer and other serious chronic illnesses often do. As such, these entities can be seen to compose a CES, with each component acting independently according to its objectives but also in concert with others to provide quality healthcare for individuals.

The core of the MDM process rests with the interaction of the physician/ healthcare treatment system, the patient (and, sometimes, the patient's family), and the patient's physiology. It is important to separate the patient from his or her physiology, as the patient's body may not react to interventions in a way desired by the patient or the physician—even if the patient and physician are in agreement concerning their wishes. The three models are formulated differently. The healthcare system works on a deterministic basis, guided by evidence-based medicine and the standard of care, and implements its strategy through a series of plans, procedures, and medications meant to guide the patient's health to a desired state. The patient, for their part, is to cooperate with this plan—but sometimes, particularly in chronic disease, patients may not be compliant with medication or lifestyle modification regimens [13]. As stated, the patient's body may conform to expectations or may fail to respond to a course of treatment. Finally, the availability of caregivers or the interaction of the patient's family may have to be taken into consideration when developing or executing the patient's action plan [14].

Influencing the decision space surrounding these elements are a number of external stakeholder actors; these entities may expand or limit action space through the services they provide or the restrictions they enforce. These are represented by rule-based models that, in future implementations, will have an aggregate feedback component which may change rules according to optimization of stakeholder objectives. For example, insurance companies may discover that certain de-intensification strategies, currently not authorized for payment, may reduce long-term patient costs and therefore be favored over more drastic interventions. Government (e.g.,, the FDA) and nongovernmental agencies (such as the Joint Commission) may place restrictions on a variety of potentially available actions, thereby reducing physician and patient choice. Advanced diagnostic and treatment services, particularly those of high complexity and limited availability, may cause additional expense or inconvenience to the patient, if they are accessible at all; some geographic locations may not be able to provide these services, reducing options for healthcare providers. Community-based and privately funded entities such as skilled nursing facilities or organizations who offer support to allow patients to remain in their homes can influence the impact of postsurgical or long-term care plans.

This concept provides the basis for an MDO framework in representing cancer treatment. In order to reduce complexity to a manageable level, the approach focuses on a specific subset of cancer—that of oropharyngeal cancer (OPC) and, in particular, OPC which arises from infection with human papillomavirus (HPV).

17.3 OPC and Decision Regret

Head and neck cancers (HNC), of which OPC is a subset, represent approximately 4% of all cancers worldwide [15]. These cancers are often seen in patients with an extensive history of tobacco and alcohol use, although their frequency is on the decline in the United States as patients turn to healthier behaviors. However, the incidence of OPC has been increasing as a result of infection with HPV, with the number of HPV-positive OPC cases expected to exceed that of HPV-related cervical cancer by 2020 [16]. Figure 17.2 illustrates the location where OPC occurs; unlike cancers of the oral cavity (mouth, front of tongue, hard palette), OPC can be more difficult to identify, delaying treatment and increasing the likelihood that it will not be found until the disease has substantially progressed.

One of the hallmarks of HNC and OPC treatment is a high level of decision regret (DR) following interventions [18]. DR has been defined as "remorse or distress over a decision" [19] and "the difference between the utility of an action taken and utility of an alternative action" [20]. Various treatments for OPC, which can include combinations of radiation, chemotherapy, and surgery, often leave patients with

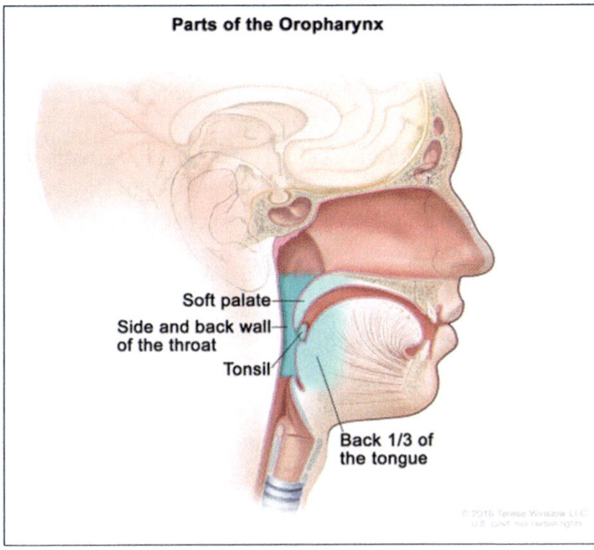

Fig. 17.2 Parts of the oropharynx [17]

the challenges of disfigurement, difficulty swallowing, degradation of speech, or poorer outcomes with respect to physical and functional health than treatments for other disease processes [21]. Patients do not always understand the trade-offs involved in the selection of therapies, and while no therapy can guarantee absence of sequelae, certain treatments (including the choice of no treatment) may be better aligned to individual patient preferences.

The characteristics of HPV-positive and HPV-negative OPC are sufficiently different to consider them individual disease processes [22]. HPV-positive cases are more frequently seen in younger patients and have substantially higher 5-year survival rates than HPV-negative OPC. As a result, certain "de-intensification" therapies are being investigated to reduce treatment effect severity and improve quality of life (QoL) among HPV-positive OPC patients [23].

Controlling DR in HPV-positive OPC may be possible through shared decision-making (SDM) processes between physicians and patients. Researchers have found that SDM activities, particularly those employing the use of decision aids (DA), can reduce the likelihood of future DR in cancer patients [24, 25]. As patients with HPV-positive OPC will often live with the treatment effects for many more years than the typical HNC patient, ensuring SDM is conducted with the best possible longitudinal data is essential to maintaining the best quality decisions consistent with patient preferences.

17.4 Representing MDM in HPV-Positive OPC

The particular challenge for the treatment of cancer is the stochastic nature of the physiological model. Physicians must confront the inherent uncertainty as to which regimens will produce desired results, as patients will not respond to therapies identically [26]. As a result, the longitudinal trajectory of these treatments often involves monitoring and reassessment of progress, periods of patient remission and relapse, salvage therapies, and palliative measures undertaken should the disease prove refractory to further efforts. The collection of possible pathways and states over time, and the determination of which policies should be pursued, can be computationally intractable.

As such, one approach to modeling the MDM process is to construct a network of pathways representing patient physiological states, connected by probabilistic arcs based on the likelihood of transition between states given available treatment actions. The action space is constrained by elements such as the standard of care, regulatory practices, patient demographics and geographic location which influence healthcare access and availability of procedures, and the affordability of these procedures to the individual patient. The calculation of the physical and psychosocial value of these states is based on three weighted functions, to be collected using interview instruments with individual patients:

- Longevity
- Quality of life (QoL)
- Patient expectations versus experience (hereinafter called regret risk).

Many cancer patients initially identify longevity as the most important factor in the selection of their care [27]. Simply put, longevity is concerned only with length of life, regardless of the current physical condition. Often, this priority changes as treatments are initiated or continued, and QoL begins to take on increased significance to the patient [28]. This transformation can be seen as one source of DR, as patients begin to appreciate the physical, financial, and emotional strains of therapeutic protocols and ponder whether a different pathway would have been more compatible with their treatment goals.

QoL is a standardized measure based primarily on physical symptoms (psychosocial experiences may sometimes be included), derived from population input, and represents patient experience in the disease process. Scaled from 1 ("perfect health") to 0 (an arbitrary cutoff point reflecting death), where values less than 0 are possible, it can be multiplied by years, resulting in the quality-adjusted life year (QALY). QALY values are typically used to compare outcomes of different treatments. Generic instruments such as the EQ-5D [29] can be used to calculate QALY values, while other tools specific to disease domains are also available [30].

A core hypothesis is that regret risk can function as a proxy measure for DR (as DR can only be ascertained post-intervention) and that patients who are risk averse will be more likely to choose treatments with smaller variances in outcome (and, in particular, less dramatic swings between expectation and experience).

Finding optimal policies requires the evaluation of the state space across time; early models used stochastic, finite-horizon dynamic programming, but as the complexity of the network increases, approximate dynamic programming techniques may have to be employed.

For every state, its contribution can be calculated as a series of weighted functions, represented by the equation

$$h(S_t, a) = -b_t(a) + w_L g_L + w_{\mathrm{QoL}}\left(g_{\mathrm{QoL}}(S_t)\right) - w_{\mathrm{RR}}\left(g_{\mathrm{RR}}\left(S_t' - S_t\right)\right)$$

where S_t is the vector of physiological and psychosocial values of the current state, b_t is a QoL penalty associated with the chosen treatment a, and w represents the weights assigned to factors of longevity (L), QoL, and regret risk (RR), where

$$w_L + w_{\mathrm{QoL}} + w_{\mathrm{RR}} = 1.$$

Functions for longevity, QoL, and RR take into consideration potentially nonlinear relationships of the elements in vectors S_t. The vector S_t' represents the expectations of the patient for the particular future health state; these values can be influenced through SDM and the use of a DA, bringing a more realistic assessment of future conditions.

The value of a given state in this finite-horizon stochastic optimization problem is determined by backward recursion from end states (usually representing a specifically determined cutoff point in time or the death of the patient) through the cumulative function

$$f(S_t) = \max_a \left(h(S_t, a) + \sum_{S_{t+1}} p^a f(S_{t+1}) \right)$$

where p^a is the probability of reaching the future state S_{t+1} given the action a.

The optimal solution to this algorithm will identify the policy, that is, series of actions chosen at each point in time, which should be targeted for implementation, starting with the first action (usually which choice of treatment options to be employed, such as surgery or radiotherapy).

In practice, as part of the initial consultation process with a physician (e.g.,, a surgeon or radiation oncologist), the patient would be offered the opportunity to complete a preference statement relating to the importance of physical and psychosocial functioning, longevity, availability of family and caregivers, and other questions to help determine those factors most important to the individual. Applying the algorithm (which takes into account the influence of external stakeholders and physiological aspects beyond the patient's control) to the patient's known diagnosis, including OPC staging, the preferences would help shape a "user-friendly sensitivity analysis" of alternatives for consideration and in particular the interventions best tolerated by the patient. This is presented to the consulting physician as a series of pre-decisional, counterfactual narratives that identify which of the three major factors influenced the optimal pathway. Using this, the physician can discuss with the patient those elements which are of concern and, as part of the SDM process, help guide the patient to the series of interventions least likely to produce post-decisional regret.

17.5 Implementation Example

A problem [31] was developed to explore dynamics of the patient preference realm using test data. The target patient was a 52-year-old female diagnosed with advanced non-HPV OPC; 3-year survival for this patient was estimated at approximately 50% [32]. The patient was afforded two choices—surgical intervention (immediate or at a time of the patient's choosing) and nonsurgical/palliative measures. Patient QoL was linked to the survival curve (an exponential function); a variable QoL intervention penalty for surgery was imposed to reflect the effects of the intervention if it was chosen. Two different sets of patient expectations—one informed to outcomes that were approximately equal to historical patterns and the other an uninformed and optimistic for chances of recovery and long-term survival—were evaluated.

The survival curves gave the patient an expected lifespan of approximately 3.158 years (SD 0.489) for the choice of immediate surgery and 1.24 years (SD 0.193) for the palliative option. Results of our simulation for the weighting of longevity, QoL, and regret risk showed (see Fig. 17.3):

- The model was most sensitive to the surgery penalty function, with greater penalties resulting in the election of nonsurgical/palliative care
- Regret risk was prominent in influencing the nonsurgical option when neither QoL nor longevity was a strongly weighted decision factor

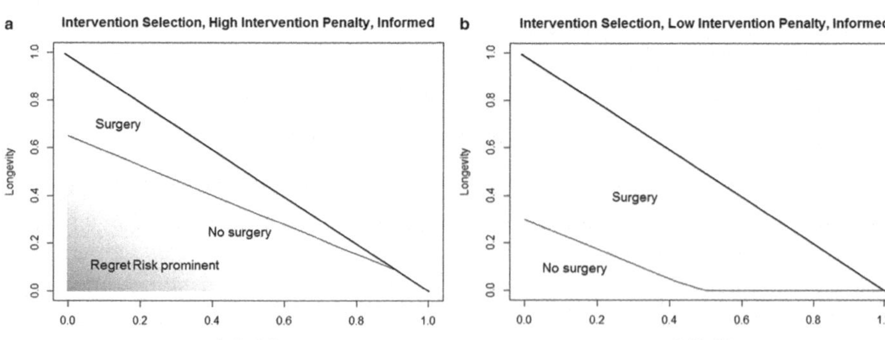

Fig. 17.3 Results of intervention selection when (**a**) a high intervention penalty and (**b**) a low intervention penalty were assessed, based on indicated preferences for QoL and longevity weighting

- Optimistic expectations changed decisions slightly in favor of surgery, but only near the center of the curves where neither longevity nor QoL weights predominated.

The simulation as constructed confirmed our overall expectation that the treatment burden imposed by intervention was acceptable if the patient favored longevity.

17.6 Conclusion and Future Work

Patients must confront uncertainty in assessing which treatment option for cancer is consistent with their preferences for functionality and longevity. In OPC, this decision sometimes is looked upon retrospectively with regret when outcomes do not conform to expectations. In developing a prospective tool to anticipate this regret, the actions of numerous stakeholders, in addition to the patient-physician decision dynamic, may influence the result of the intervention; the interaction of these factors, which are likely too complex for the patient to fully comprehend at the time of treatment selection, can be assessed using algorithms to help guide the physician in the SDM process.

We first intend to gain fidelity on the various alternative treatment processes through collection of outcome data from standard interventions (from patient and practice health records) and de-intensification strategies (largely from clinical trials) to use in our model. Our focus is on finding policies that reduce the potential for DR given a range of patient preferences for longevity, QoL, and regret risk.

As we confirm our results with the physician and patient community, we will explore the activities of other entities for optimization within their preference structures, leading to a better implementation of the MDO concept. Public policy implications, potential savings for insurance and government programs, better coordination for community support activities, and improved guidelines for caregivers and family members could be derived from findings.

References

1. Sox, H. C., Higgins, M. C., & Owens, D. K. (1988). *Medical decision making*. Boston: Butterworths.
2. Berlin, L. (1998). Malpractice issues in radiology. *AJR, 170*(6), 1417–1422.
3. Sackett, D. L. (1997). Evidence-based medicine. *Seminars in Perinatology, 21*(1), 3–5.
4. Yun, Y. H., Lee, M. K., Park, S., Lee, J. L., Park, J., Choi, Y. S., et al. (2011). Use of a decision aid to help caregivers discuss terminal disease status with a family member with cancer: A randomized controlled trial. *Journal of Clinical Oncology, 29*(36), 4811–4819.
5. Simpson, T. W., & Martins, J. R. R. A. (2011). Multidisciplinary design optimization for complex engineered systems: Report from a National Science Foundation Workshop. *Journal of Mechanical Design, 133*(10), 101002.

6. Hede, S., Nunes, M. J. L., Ferreira, P. F. V., & Rocha, L. A. (2013). Incorporating sustainability in decision-making for medical device development. *Technology in Society, 35*(4), 276–293.
7. Braha, D., Minai, A. A., & Bar-yam, Y. (2006). *Complex engineered systems*. Springer: New York.
8. Sosa, M., Mihm, J., & Browning, T. (2011). Degree distribution and quality in complex engineered systems. *Journal of Mechanical Design, 133*(10), 101008.
9. Complexity Labs. (n.d.). *Complex engineered systems*. Retreived November 9, 2017, from http://complexitylabs.io/complex-engineered-system/
10. Complexity Labs. (n.d.). Complex engineered systems. Retreived November 9, 2017, from http://complexitylabs.io/complex-engineered-system/
11. Rakich, J. S., Darr, K., & Longest, B. B. (1985). *Managing health services organizations*. Philadelphia: W.B. Saunders.
12. Aarhus, R., & Huang, E. (2017). *Use of a multidisciplinary design optimization approach to model treatment decisions in oropharyngeal cancer* (submitted) (p. 7). Washington, DC: INCOSE International Symposium.
13. Kimmel, S., & Troxel, A. (2012). Novel incentive-based approaches to adherence. *Clinical Trials, 9*(6), 689–694.
14. Gill, S. S., Frew, J., Fry, A., Adam, J., Paleri, V., Dobrowsky, W., et al. (2011). Priorities for the head and neck cancer patient, their companion and members of the multidisciplinary team and decision regret. *Clinical Oncology, 23*(8), 518–524.
15. Ojo, B., Genden, E., Teng, M., Milbury, K., Misiukiewicz, K., & Badr, H. (2012). A systematic review of head and neck cancer quality of life assessment instruments. *Oral Oncology, 48*(10), 923–937.
16. Oskam, I. M., Verdonck-de Leeuw, I. M., Aaronson, N. K., Witte, B. I., de Bree, R., Doornaert, P., ... Leemans, C. R. (2013). Prospective evaluation of health-related quality of life in long-term oral and oropharyngeal cancer survivors and the perceived need for supportive care. *Oral Oncology, 49*(5), 443–448.
17. National Cancer Institute. (n.d.). *Oropharyngeal cancer treatment*. Retrieved November 20, 2017, from https://www.cancer.gov/types/head-and-neck/patient/oropharyngeal-treatment-pdq
18. Sarkar, S., Datta, S. S., Mallick, I., Chatterjee, S. S., Shrimali, R. K., Achari, R., et al. (2015). OC-0286: Patient's choice of treatment options with CTRT in head neck cancer: a study of patient preference and decision regret. *Radiotherapy and Oncology, 115*, S145.
19. Brehaut, J. C., O'Connor, A. M., Wood, T. J., Hack, T. F., Siminoff, L., Gordon, E., et al. (2003). Validation of a decision regret scale. *Medical Decision Making, 23*(4), 281–292.
20. Hernandez, J. M., Tsalatsanis, A., Humphries, L. A., Miladinovic, B., Djulbegovic, B., & Velanovich, V. (2014). Defining optimum treatment of patients with pancreatic adenocarcinoma using regret-based decision curve analysis. *Annals of Surgery, 259*(6), 1208–1214.
21. Oskam, I. M., Verdonck-de Leeuw, I. M., Aaronson, N. K., Witte, B. I., de Bree, R., Doornaert, P., ... Leemans, C.R. (2013). Prospective evaluation of health-related quality of life in long-term oral and oropharyngeal cancer survivors and the perceived need for supportive care. *Oral Oncology, 49*(5), 443–448.
22. Bhatia, A., & Burtness, B. (2015). Human papillomavirus-associated oropharyngeal cancer: Defining risk groups and clinical trials. *Journal of Clinical Oncology, 33*(29), 3243–3250.
23. Chera, B. S., & Amdur, R. J. (2018). Current status and future directions of treatment deintensification in human papilloma virus-associated oropharyngeal squamous cell carcinoma. *Seminars in Radiation Oncology, 28*(1), 27–34.
24. Feldman-Stewart, D., Tong, C., Siemens, R., Alibhai, S., Pickles, T., Robinson, J., et al. (2012). The impact of explicit values clarification exercises in a patient decision aid emerges after the decision is actually made: Evidence from a randomized controlled trial. *Medical Decision Making, 32*(4), 616–626.

25. Chiou, C. P., & Chung, Y. C. (2012). Effectiveness of multimedia interactive patient education on knowledge, uncertainty and decision-making in patients with end-stage renal disease. *Journal of Clinical Nursing, 21*(9–10), 1223–1231.
26. Sox, H. C., Higgins, M. C., & Owens, D. K. (1988). Medical decision making. Boston: Butterworths.
27. Tschiesner, U., Sabariego, C., Linseisen, E., Becker, S., Stier-Jarmer, M., Cieza, A., et al. (2013). Priorities of head and neck cancer patients: a patient survey based on the brief ICF core set for HNC. *European Archives of Oto-Rhino-Laryngology, 270*(12), 3133–3142.
28. Davies, L., Rhodes, L. A., Grossman, D. C., Rosenberg, M. C., & Stevens, D. P. (2010). Decision making in head and neck cancer care. *The Laryngoscope, 120*(12), 2434–2445.
29. Mortimer, D., & Segal, L. (2008). Comparing the incomparable? A systematic review of competing techniques for converting descriptive measures of health status into QALY-weights. *Medical Decision Making, 28*(1), 66–89.
30. Ojo, B., Genden, E., Teng, M., Milbury, K., Misiukiewicz, K., & Badr, H. (2012). A systematic review of head and neck cancer quality of life assessment instruments. *Oral Oncology, 48*(10), 923–937.
31. Aarhus, R., & Huang, E. (2017). Use of a multidisciplinary design optimization approach to model treatment decisions in oropharyngeal cancer (submitted) (p. 7). Washington, DC: INCOSE International Symposium.
32. Cohen, E., LaMonte, S. J., Erb, N. L., Beckman, K. L., Sadeghi, N., Hutcheson, K. A., et al. (2016). American Cancer Society Head and Neck Cancer Survivorship Care Guideline. *CA: A Cancer Journal for Clinicians, 66*(3), 204–239.

Chapter 18
How Frustrations About Oversight Can Lead to Inefficiencies in Contractor Work

Samantha Brainard and Zoe Szajnfarber

18.1 Introduction

Oversight monitoring activities require that contractor engineers provide information to the government about the work they perform [1]. These activities mitigate any information asymmetries that could exist between the government and its contractors, enabling the government to evaluate the contractor's performance [2]. These oversight activities include, but are not limited to, audits, budgets, meetings, technical verification and validation activities, and other reporting activities per the contract between the government and its contractors. Oversight can also require that contractors use specific reporting systems and military grade hardware components to comply with the government's Federal Acquisition Regulations (FAR).

Stakeholders agree that oversight is a necessary part of the acquisitions process, but it does incur additional program costs [3]. The extent of these added costs is hotly debated, and many acquisitions reform efforts have been made in recent decades with the intent of reducing any unnecessary burden [4]. Few studies have been conducted to empirically measure oversight's burden, and they have been focused on its objective impact—specifically measuring the time spent performing specific oversight-related tasks [5]. Scholars, however, contend that compliance with regulations incurs a subjective burden too; and the feelings that people have about regulations can impact how they do their work [6].

While scholars have studied how individuals feel when they comply with regulations or work in bureaucratic organizations, they have not shown how these feelings connect with added time on the work they perform [7]. Technical professionals can feel constrained when complying with rules because they limit their

S. Brainard (✉) · Z. Szajnfarber
Department of Engineering Management and Systems Engineering, The George Washington University, Washington, DC, USA
e-mail: samm@gwu.edu

S. Adams et al. (eds.), *Systems Engineering in Context*,
https://doi.org/10.1007/978-3-030-00114-8_18

autonomy and can make them feel alienated from their work [8]. While scholars have posited that these feelings result in inefficient work practices, they have not identified how these inefficiencies manifest in work practices [9].

To that end, we present the findings from an in-depth, qualitative study of a major US aerospace company to understand how frustrations about complying with oversight regulations can lead to extra time spent on contractor engineering work. These findings are important for systems engineers because many modern, large-scale complex systems are developed for a US government customer. Understanding how contractors respond to frustration by adding time to the work they perform is also important for systems engineers who work to identify and mitigate inefficient work processes.

18.2 Literature Review

This section reviews three classes of prior work that are relevant to this research: [1] scholarly studies about the burden of oversight; [2] studies about the way that technical professional feels when complying with rules or working in bureaucratic organizations; and [3] studies conducted to study the burden of regulations.

There exist a few, publicly available studies about the burden of oversight on defense acquisitions [10]. Work to measure the objective impact of oversight has measured burden as the time spent performing a set of oversight-related work tasks [11]. These studies' findings, however, present a limited scope of oversight's burden because they are only focused on some, but not all, of the oversight-related work tasks performed by these organizations [12]. Moreover, they have not captured the costs associated with the direct costs of reporting systems or specific hardware. As a result, their measurements of burden could be understated. Recent work has investigated how oversight-related work can require engineers spend more time performing certain activities and can require additional tasks performed in response to requests from the government [13]. These inefficiencies add to oversight's burden as well. This work, however, looked at inefficiencies that stem from being asked to do something by the government rather than time that could be added by the contractor engineers in response to frustrations.

Another body of literature has studied how technical professionals feel when complying with rules or working in bureaucracies. Technical professionals are highly trained individuals who value professional autonomy, discretion in their time allocation, and reputation as upholding the professional standards of their field [14]. When working in any organization, these technical professionals face a professional-bureaucratic dilemma. This occurs because the organization's needs and goals are embodied in its rules, procedures, and structures; but those needs and goals can conflict with the professional's desires for autonomy and discretion in their time allocation [15]. This loss of autonomy and control can influence the professional's level of job satisfaction and impact their job performance [16]. While the majority of the work describing the professional-bureaucratic dilemma took place in the mid-twentieth century, recent work surveying engineers working for a US military depot echoed their findings. Their survey findings showed that engineers working in a bureaucratic environment who felt a strong commitment to their goals as engineers rather than the goals of their employer had low job satisfaction and low job performance as compared to other workers at the depot. As a result, these engineers felt that their day-to-day work was extremely frustrating and performed their work inefficiently [17].

Literature on public administration has also studied how the perception of burdensome rules and regulations can lead to lower job satisfaction, performance, and inefficiency. The scholarly study of red tape suggests that rules and regulations have both a subjective and objective component. Objectively, rules make individuals spend time performing certain tasks, while, subjectively, people develop their own opinions about whether these rules are effective or ineffective [18]. Red tape is a subject-dependent concept—while one group of people could view a rule as burdensome, another group might find the rule extremely important and useful [19]. Scholarly red tape studies have focused on understanding how a group (or groups) perceives burdensome rules [20]. The factors they have identified as contributing to this perception include, but are not limited to, whether the individual works in the public or private sector, the number of employees working for an organization, the amount of written rules in an organization, the levels of review and authority between managers and decision-makers, and other variables that shape the context in which individual performs his or her work [21].

Some red tape scholars have focused on the subjective reasons why individuals perceive rules as burdensome. Explanatory models of red tape assert that managers' perceptions of red tape are related to their feelings of administrative delay and work alienation [22]. These feelings include loss of autonomy and/or feelings that an individual's work has no meaning [23]. Administrative delay refers to a feeling that processes are inefficient and take too much time to complete [24]. Other scholars have also noted that when people feel alienated from their work, they will shirk on their responsibilities and will find ways to work around the rules they find constraining [25]. However, these studies have not unpacked how added time can arise in work as a result of frustrations about rules.

In summary, previous work has shown that compliance with rules can make engineers feel alienated in their work; and, in response to their feelings, they can do their job inefficiently. Those studies, however, have not elaborated upon how these feelings make work inefficient. To that end, this work specifically looks at the ways that frustrations about oversight result in contractors spending additional time on their engineering work.

18.3 Research Approach

We used qualitative research methods to study how feelings about oversight can lead to inefficiencies in contractor engineering work [26]. Since previous work has not connected feelings about oversight with impacts on oversight's burden, there is a need to inductively develop the mechanisms through which extra time can be added to engineering work [27]. In this section, we describe our data and analysis.

Our data for this research was semi-structured interviews with contractors working for a major US aerospace company. This firm is representative of firms working for the US government. We did a deep dive into this single-case setting in order to thoroughly understand it [28]. We interviewed 15 participants as part of this work. We used a purposive sampling technique to ensure that we interviewed participants working across multiple managerial levels and in different roles within the organization [29]. Interviews were set up with the help from an informant who identified people who would be willing to voluntarily participate in our research. All possible participants who were invited to participate in our study chose to do so. The interviews we conducted with each participant aimed to understand what role the participant played in the oversight process [30]. Interviews asked questions on topics that ranged from the activities each participant performs as part of his or her work responsibilities, the interactions with stakeholders involved in space systems acquisitions, and the ways that oversight impacts the work he or she performs. Interviews lasted between 60 and 90 min. Interviews were recorded and transcribed; in all 20 h of interviews were transcribed, resulting in over 150 pages of transcriptions. Each participant was assigned an identifier to protect his or her anonymity in the data. In the following sections, we will use these identifiers when recalling a participant's quote. This demonstrates variety in our quoting sources, transparency and consistency in our findings, and maintains traceability to quotes from different participants [31].

In addition to these interviews, we spent over 6 weeks on site and were briefed about the company's processes. The activities this company performs adhere to well-documented acquisition regulations and processes; and we used these documents to supplement our understanding of our research context [32].

We analyzed our data in multiple stages using open coding techniques [33]. We began by coding our interview transcripts to identify instances when participants were frustrated. Their frustration was marked by instances where they said they were frustrated, sounded frustrated when listening to the interview recording, or used words that described a state of frustration including, but not limited to, frustration, annoyance, disagreement, anger, unease, or unhappiness. We then coded those instances when their frustrations led to behaviors that were burdensome (added extra time). These burdens included spending additional time on an activity or performing new, time-consuming tasks. We grouped these codes into categories and iterated upon them multiple times, comparing and refining them across participants to improve the interpretation of the emerging mechanisms [34]. The table below provides example quotations from our transcripts to show our coding process (Table 18.1).

18.4 Findings

We found that frustration can lead to objective inefficiencies in terms of added process time (not just feelings of alienation) because contractors often spend additional work time on activities in response to frustrations about oversight. While

Table 18.1 Process to code instances of frustration-related inefficiencies

Quotation from transcript	Frustration	Extra time spent on activity	Mechanism
"The government guys are coming in saying you have an issue because you didn't do this analysis right and it's always smoke and people like me have to go deal with it. You can lose 2 days arguing and I don't have time to do it" [8]	Disagree about analysis	Argue about the results from the analysis	Argue
"The government wants more extensive testing for this part. It takes more time to do it their way, plus we're unhappy about it. I have to spend time fighting with them" [10]	Disagree about test requirements	Argue about how to perform tests	Argue

previous studies have shown that frustrations lead to inefficiencies in work, we describe the mechanisms through which added time can occur at a more granular level. We identify four mechanisms through which this additional time occurs: evade, drag, argue, and work-around. Examples of quotations describing these mechanisms can be seen in Table 18.2.

When contractors are frustrated by oversight, they can try to avoid interactions with certain members of the oversight team by evading them. This can make activities take longer by spending time waiting to work with other individuals. For instance, an engineer might not answer the phone to avoid speaking with a certain member of the oversight team if they consider that person annoying or overly inquisitive. One interviewee described wanting to avoid having meetings with a certain member of the government team because of that person's attitude. He talked about this scenario stating, "There's always at least one person who throws fits at a review . . .so when we hear reviews we sigh and don't want to do them because of this one guy. We'll try not to have [the meeting]" [13]. Another engineer discussed actively spending time to avoid working with certain government team members so he could interact with another member of the government team. He said, "We have some government people that request incredible amounts of oversight-related work. They want extremely detailed data and oversight reports where others are not as detailed. We work to avoid interactions with those people" [9].

Another type of inefficiency when avoiding work is the drag mechanism; this occurs when engineers take longer to do certain tasks or push them off until another time. This can occur when an engineer does not like doing a certain activity or if they do not like doing an activity for a specific government team member. An engineer talked about how he often pushes off certain writing reports because he does not like doing them. He said, "I don't like doing them, so I keep putting them off" [9]. Another engineer described avoiding making PowerPoint presentations for meetings because he would have rather been doing other work instead. He said "The biggest problem is the PowerPoint presentations the engineers have to make for the government. . .we could actually be working on something else so I wait to do them" [14].

Table 18.2 Mechanisms identified as frustration-driven inefficiencies in contractor work

Frustration-driven inefficiency	How time is added	Quotation example
Evade	Spend time avoiding interactions with USG team member	"There is a person on the government team that has made their way fairly high up but likes to micromanage us. . . It's very unnerving and I don't like working with them, I try to avoid them and I'll wait to work with their colleagues instead" [3]
		"A lot of the time you'll see younger people come in arrogantly with this attitude and that burns bridges. We avoid them and try not to interact" [15]
Drag	Spend more time to complete activity or to avoid doing a specific task	"We used to get the email questions from a certain guy in the at the Air Force and that's where you'd get 10–15 deep . . . But after you get 15 you're like wow that's a lot. So we took our time answering them because he was so annoying" [12]
		"They question everything and say I'm wrong. It's bad. I don't trust any of them so I'll take longer to do what they want me to do" [14]
Argue	Spend time arguing about how work should be done	"The requirements are too high industry wide, and if you talk to any company about the requirements they'll say that they're horrible. . . we're fighting with the FFRDC to change the requirements all the time" [2]
		"I'm going to a meeting about a requirement to argue with the FFRDC about what the requirement should be as opposed to what they want us to test to. A lot of times I'm trying to mediate between what the requirement says and what actually should happen" [10]
Work-around	Spend time finding alternate ways to do activity	"Anytime you ask the government something, it'll take 100 hours to answer my 5-minute question and it'll delay when I can start the work I need to do. We have to look for a way to work around their process because it would take months to get the information, but it takes a while to find another way to do things" [3]
		"If we're trying to make a change to meet a deadline, it will take 6–7 months to get that change on contract; and then I have to intervene. It takes me time to find ways to get around that by the book one to move the process so we can get back to work" [8]

When contractors disagree with what the government wants them to do, they will sometimes spend extra time arguing with the government team. Examples of issues where these disagreements arise include the extent of iterations of tests required, the testing specifications (e.g., time or temperature) for a particular part, or the methodology required for an analysis. An engineer described how the difference between his company's interpretation of test requirements and the government team's interpretation led to an argument in his interview. He said, "The government looked at these test procedures and said 'You can't do this work this way.' And we came back saying that we need to do the tests like this—so we've been arguing, and now it's a stand-off and we can't do our work either way" [4]. In this scenario, the interviewee described both the added time spent arguing and the added time waiting to do work while a decision was made about how to proceed. The arguments between the contractors and the government team can make meetings or conversations take longer than they otherwise would. They can also catalyze new meetings, questions, or conversations with the government that add time to a contractor's workload. Another engineer described how these arguments add extra time to their workload because they can require time spent in additional meetings to discuss technical differences. She said, "We're insulted by this review... the government's reviews rarely show that we've made an error. It's annoying to our folks to have another meeting just to argue about good data" [6].

Contractors will also spend extra time to find ways to work around certain processes. While their intent might be to decrease the amount of time spent doing some activities (such as getting information from the government or fast-tracking an approval process), the process of finding a way around the rules could add to their workload. In fact, they might find that they are not able to find a solution to their problem and they have to follow a process anyway. An engineer described the time he spent trying to find ways around rigid processes in his interview. He said, "There's a lot of red tape in what we do here...I have to find a way to get to the top government guys to make a decision, but it takes a while to figure out how to get there" [15].

18.5 Discussion

In this work, we identify the mechanisms through which contractors add time to work processes when frustrated by oversight. Past studies about the subjective burden of complying with government rules contend that frustrations about work lead to people doing their work inefficiently. We expand upon this idea to identify the mechanisms through which time is added. Those mechanisms include evade, drag, argue, and work-around. We show how these mechanisms arise in response to frustrations about what contractors are required to do or as a result of negative interactions with members of the government team. Understanding how contractors' frustrations about oversight can add time to the work they perform is important when studying the burden of government oversight on engineering work.

These findings are important for systems engineers because they show how responses to oversight add time to engineering work processes. Systems engineers often look at objective reasons for added time on processes (e.g., more iterations of tests required, duplicate processes). We show that what engineers are asked to do—or how they are asked to do it—has an impact on how they spend their time on their work. This is important for systems engineers because it shows the need for future research to understand the link between subjective and objective burdens of rules and their impact on engineering processes.

Understanding how contractors add time to the work they perform in response to oversight frustrations is an important part of studying the burden of oversight. Previous studies have focused on measuring the time spent performing activities requested by the government and any inefficiencies that arise from working with the government. Studies have also noted that companies have their own layers of internal oversight to check work before it is submitted to the government. This work shows that there is an additional layer of burden that can occur when contractors are frustrated by oversight. In fact, there could be another layer of frustration-driven inefficiencies due to internal oversight too. Measuring the impact of all of these types of added time is necessary to determine the burden of government oversight on contractor engineering work.

Stakeholders can use these findings to inform strategies to mitigate the amount of frustration-driven inefficiencies that arise in contractor work. Targeting why negative feelings about oversight exist could help reduce the amount of time engineers spend evading, dragging, arguing, or working around. One way this could be done is through explaining why the government is requiring the engineers perform certain tasks. Research has shown that when people understand why they have to comply with rules, they are happier to comply with what they are being asked to do [35]. Another strategy would also be to encourage better relationships between the contractor and the government teams. More frequent or more cordial interaction between the two parties could facilitate a friendlier relationship. We contend that when engineers feel that they are being trusted by the government or they enjoy working with their government counterparts, they will spend less time avoiding interactions or adding extra time to their work in other ways.

The frustration-driven inefficiencies described in this research are only one part of the burden of oversight. We do not contend that these types of inefficiencies are a major source of additional time, but they do contribute to the additional time spent on government oversight-related work. We identify them in this paper because they can contribute to the labor costs of a contract, which is the least well-understood part of the burden of oversight [36]. As previously stated, the burden of oversight includes much more than just added time due to frustrations (e.g., additional work tasks, preparations for meetings with the government); but we add to this body of work by showing that feelings about oversight contribute to actions that further the impact of oversight's burden on engineering work. Future work should be conducted to measure the impact of this extra time. We believe our observations and results are representative of contractors working in the US aerospace industry. We chose our setting for its representativeness and would expect similar findings in across other contractors. We also believe that these frustration-driven inefficiencies could also manifest in the work performed across other stakeholders in the acquisitions value stream. Future work should be done to investigate how feelings about oversight impact the time spent on oversight-related tasks across these groups.

18.6 Conclusion

This paper presented a study to understand how contractors spend additional time on engineering work in response to frustrations with government oversight. Previous studies contend that compliance with rules can make engineers perform their work less efficiently because they feel alienated from their work. We expand upon their work to show how these inefficiencies arise using qualitative methods to inductively develop the mechanisms through which added time occurs. We revealed four frustration-driven inefficiencies that add time to contractor work: evade, drag, argue, and work-around; and we discuss their implications on the burden of oversight. The results from this study provide rich insights into how feelings about oversight can result in actions that add burden. Scholars can use these findings to inform future studies to measure the burden of government oversight on engineering work.

Acknowledgment This work was funded by the NASA Systems Engineering Consortium.

References

1. Wilson, J. Q. (1991). *Bureaucracy: What government agencies do and why they do it.* New York: Basic Books.
2. Eisenhardt, K. M. (1989, January). Agency theory: An assessment and review. *The Academy of Management Review. 14*(1): 57. https://doi.org/10.2307/258191. Wilson, J. M. Bureaucracy: What government agencies do and why they do it.
3. Wilson, J. Q. (2001). Bureaucracy: What government agencies do and why they do it; Drezner, J. A., Blickstein, I., Raman, R., McKernan, M., Hertzman, M., Bradley, M. A., et al. (2007). Measuring the statutory and regulatory constraints on department of defense acquisition: An empirical analysis. (Santa Monica, CA: RAND Corporation.
4. Drezner, J. A., Blickstein, I., Raman, R., McKernan, M., Hertzman, M., Bradley, M. A., et al. (2007). Measuring the statutory and regulatory constraints on department of defense acquisition; Gavrieli, D. (2008). Regulatory burden: A systems analysis of weapons acquisition in the U.S. Department of Defense. Los Angeles, CA: University of Southern California.
5. Coopers and Lybrand and TASC. (1994, December). The DoD regulatory cost premium: A quantitative assessment. Author; Drezner, J. A., Blickstein, I., Raman, R., McKernan, M., Hertzman, M., Bradley, M. A., et al. (2007). *Measuring the statutory and regulatory constraints on department of defense acquisition*. Santa Monica, CA: RAND Corporation; Brainard, S. M., & Szajnfarber, Z. (2017). Understanding the burden of government oversight on engineering work: Adding empirical data to the debate, *Space Policy, 42*, 70–80. https://doi.org/10.1016/j.spacepol.2017.07.001.

6. Bozeman, B., & Feeney, M. K. (2011). *Rules and red tape: A prism for public administration theory and research*. Armonk, NY: M.E. Sharpe Pandey, S. K., & Welch, E. W. (2005, November 1). Beyond stereotypes: A multistage model of managerial perceptions of red tape. *Administration & Society, 37*(5), 542–75. https://doi.org/10.1177/0095399705278594.
7. Bozeman and Feeney. *Rules and Red Tape*. Frey, B. S. (1993). Shirking or work morale?: The impact of regulating. *European Economic Review, 37*(8), 1523–1532.
8. Kornhauser, W. (1962). *Scientists in Industry: Conflict and Accommodation*. Berkeley, CA: University of Califorinia Press; Miller, G. A. (1967, October). Professionals in bureaucracy: Alienation among industrial scientists and engineers. *American Sociological Review, 32*(5), 755, https://doi.org/10.2307/2092023; Baugh, S.G., & Roberts, R.M. (1994, May). Professional and organizational commitment among engineers: Conflicting or complementing? *IEEE Transactions on Engineering Management, 41*(2), 108–14. https://doi.org/10.1109/17.293377.
9. Frey, B. S. (1993). Shirking or Work Morale? *European Economic Review, 37*(8), 1523–1532; Pandey, S. K., & Welch, E. W. (2005). Beyond stereotypes: A multistage model of managerial perceptions of red tape. *Administration & Society, 37*(5), 542–75; Bozeman, B., & Feeney, M. K. *Rules and Red Tape*.
10. Coopers, & Lybrand. TASC. (1994). *The DoD Regulatory cost premium*; Author. Drezner, J. A., Blickstein, I., Raman, R., McKernan, M., Hertzman, M., Bradley, M. A., et al. (2006). *Measuring the statutory and regulatory constraints on department of defense acquisition*. Santa Monica, CA: RAND Corporation; Gavrieli, D. (2007). *Regulatory burden: A systems analysis of weapons acquisition in the U.S. Department of Defense*. Los Angeles, CA: University of Southern California; Brainard, S. M., & Szajnfarber, Z. (2017). Understanding the Burden of Government Oversight on Engineering Work. *Space Policy, 42*, 70–80; Brainard, S. M., & Szajnfarber, Z. (2018). *How government oversight adds time to contractor engineering work*. Hoboken, NJ: Wiley.
11. Coopers, & Lybrand. TASC. (1994). *The DoD Regulatory Cost Premium*. Author; Drezner, J. A., Blickstein, I., Raman, R., McKernan, M., Hertzman, M., Bradley, M. A., et al. (2006). *Measuring the Statutory and Regulatory Constraints on Department of Defense Acquisition*. Santa Monica, CA: RAND Corporation; Brainard, S. M., & Szajnfarber, Z. (2017). Understanding the burden of government oversight on engineering work. *Space policy, 42*, 70–80.
12. Gavrieli, D. (2007). *Regulatory burden: A systems analysis of weapons acquisition in the U.S. Department of Defense*. Los Angeles, CA: University of Southern California; Brainard, S. M., & Szajnfarber, Z. (2018). *How government oversight adds time to contractor engineering work*. Hoboken, NJ: Wiley.
13. Brainard, S. M., & Szajnfarber, Z. (2018). *How government oversight adds time to contractor engineering work*. Hoboken, NJ: Wiley.
14. Kornhauser, W. (1982). *Scientists in industry: Conflict and accommodation*. Westport, CT: Greenwood Press. Miller. *Professionals in bureaucracy*; Bignon, I., & Szajnfarber, Z. (2015). Technical professionals' identities in the R&D context: Beyond the scientist versus engineer dichotomy. *IEEE Transactions on Engineering Management, 62*(4), 517–28. https://doi.org/10.1109/TEM.2015.2455056.
15. Blau, P. M., & Richard Scott, W. (1962). *Formal organizations: A comparative approach, stanford business classics*. Stanford, CA: Stanford Business Books Hall, R. H. (1968). Professionalization and bureaucratization. *American Sociological Review, 33*(1), 92. https://doi.org/10.2307/2092242; Sorensen, J. E., & Sorensen, T. L. (1974). The conflict of professionals in bureaucratic organizations. *Administrative Science Quarterly, 19*(1), 98. https://doi.org/10.2307/2391790.
16. Miller, G. A. (1967). Professionals in bureaucracy: Alienation among industrial scientists and engineers. *American Sociological Review*, 755–768.
17. Baugh, S. G., & Roberts, R. M. (1994). Professional and organizational commitment among engineers: Conflicting or complementing? *IEEE Transactions on Engineering Management, 41* (2), 108–114.

18. Gouldner, A. (1963). Red Tape as a Social Problem. In R. Merton, et al (Ed.), *Reader in bureaucracy*. New York: The Free Press of Glencoe; Bozeman, B., & Feeney, M. K. (2011). Why red tape is not so black and white In *Rules and Red Tape: A Prism for Public Administration Theory and Research* (pp. 19–50). Armonk, NY: M.E. Sharpe.
19. Kaufman, H. (1977). *Red tape, its origins, uses, and abuses, the brookings classics*. Washington, D.C: Brookings Institution Press.
20. Bozeman, B., & Feeney, M. K. *Why red tape is not so black and white*; Bozeman, B. (2012). multidimensional red tape: A theory coda. *International Public Management Journal, 15*(3), 245–65. https://doi.org/10.1080/10967494.2012.725283.
21. Pandey, S. K., & Scott, P. G. (2002). Red tape: A review and assessment of concepts and measures. *Journal of Public Administration Research & Theory, 12*(4), 553.
22. Pandey, S., & Welch, E. (2005). Beyond stereotypes. *Administration & Society, 37*(5), 542–575; Welch, E. W., & Pandey, S. K. (2007). E-government and bureaucracy: Toward a better understanding of intranet implementation and its effect on red tape. *Journal of Public Administration Research and Theory, 17*(3), 379–404.
23. DeHart-Davis, L., & Pandey, S. K. (2005). Red tape and public employees: Does perceived rule dysfunction alienate managers? *Journal of Public Administration Research and Theory, 15*(1), 133–148 Seeman, M. (1959). On the meaning of alienation. *American Sociological Review, 24* (6), 783. https://doi.org/10.2307/2088565. Pandey, S., & Welch, E. *Beyond stereotypes*.
24. Pandey, S., & Welch, E. (2005). Beyond stereotypes. *Administration & Society, 37*(5), 542–575.
25. Frey, B. S. (1993). Shirking or work morale? *European Economic Review, 37*(8), 1523–1532.
26. Szajnfarber, Z., & Gralla, E. (2017). Qualitative methods for engineering systems: Why we need them and how to use them. *Systems Engineering, 20*(6), 497–511. https://doi.org/10.1002/sys.21412
27. Eisenhardt, K. M., & Graebner, M. E. (2007). Theory building from cases: Opportunities and challenges. *Academy of Management Journal, 50*(1), 25–32 Eisenhardt, K. M. (1989). Building theories from case study. *Research Academy of Management Review, 14*(4), 532–550; Corbin, J. M. (2008). *Basics of qualitative research: Techniques and procedures for developing grounded theory* (3rd ed.). Los Angeles, CA: Sage Publications, Inc.
28. Yin, R. K. (2003). Applications of case study research. In *Applied social research methods series* (2nd ed.). Thousand Oaks, CA: Sage Publications.
29. Babbie, E. R. (2010). *The practice of social research* (12th ed.). Belmont, CA: Wadsworth Cengage.
30. Rubin, H. J., & Rubin, I. (2012). *Qualitative interviewing: The art of hearing data* (3rd ed.). Thousand Oaks, CA: SAGE.
31. Yin, R. K. (2012). *Applications of case study research*. Los Angeles, CA: SAGE.
32. The Aerospace Corporation. (2012). *Mission assurance guide*. El Segundo, CA: Author; U.S. Government. (2018). *48 C.F.R. chapter 1 - federal acquisition regulations*. https://www.acquisition.gov/browsefar; US Government. (2018). 48 C.F.R. *Chapter 2 - Defense acquisition regulations system, department of defense*. https://www.acq.osd.mil/dpap/dars/dfarspgi/current/index.html
33. Corbin, J., Strauss, A., & Strauss, A. L. (1984). Basics of qualitative research. In M. B. Miles & A. M. Huberman (Eds.), *Qualitative data analysis: A sourcebook of new methods*. Beverly Hills, CA: Sage Publications.
34. Miles, M. B., & Huberman, A. M. (1994). *Qualitative data analysis*. Los Angeles, CA: Sage Publications.
35. DeHart-Davis, L. (2009). Green tape: A theory of effective organizational rules. *Journal of Public Administration Research and Theory, 19*(2), 361–384. https://doi.org/10.1093/jopart/mun004
36. Brainard, S., & Szajnfarber, Z. (2017). Understanding the burden of government oversight on engineering work: Adding empirical data to the debate. *Space Policy, 42*, 70–80.

Chapter 19
A Method to Use the SoS Explorer Application with Fuzzy-Genetic Algorithms to Support Military Veterans within Higher Education

Garrett P. Coffey and Cihan Dagli

19.1 Introduction

Military-affiliated students are entering higher education in ever-increasing numbers; and if we consider all of the armed conflicts that have occurred—and are still occurring—these numbers aren't likely to let up anytime soon. Since 2009, we've had over 600,000 veterans enter our nation's universities [1]. Service-connected incentives like the *Post-9/11 GI Bill* and the newly passed *Forever GI Bill* provide our service members (SMs) and veterans with the means to enter into a program of study, and they are doing so in droves. There is a great push to get them into and through college; for example, our government has already spent over $65 billion on the college educations of those who have fought in our two most recent conflicts in Iraq and Afghanistan [2].

However, there are some areas of concern, both for students and higher education professionals. The needs of our student veterans are different than the needs of our "traditional students" (young adults, who enter college straight out of high school), and there are gaps in how institutions of higher learning are addressing them. Currently, there are differences in rates of retention between traditional students and student veterans. According to the Department of Veterans Affairs, the national average retention rate for all students is 68.3%. At Missouri S&T, the rate for traditional students is 87.1% and for student veterans is 69.2% [3]. We are doing better than nationally overall, but there is a discrepancy between these two groups of students.

While student veterans are eligible for all standard services the university offers, their needs as students with a military background are different. As our service members move from the highly disciplined and structured world of the military to the more open-ended and organic world of a college campus, many of them go

G. P. Coffey (✉) · C. Dagli
Missouri University of Science and Technology, Rolla, MO, USA
e-mail: coffeyg@mst.edu

S. Adams et al. (eds.), *Systems Engineering in Context*,
https://doi.org/10.1007/978-3-030-00114-8_19

through a type of culture shock. This shock is consistently called *transition*. Transition is the process through which our student veterans must readapt themselves to civilian society and develop the new skills they will need to find gainful employment. One of the most common ways they seek these new skills is through higher education and earning a college degree. Navigating this *transition* is one of the most difficult things a service member will have to do after leaving the service, and many new student veterans are finding that the service of our education institutions doesn't explicitly meet their needs [4].

There are many issues new student veterans may have to cope with. These can include difficulty obtaining VA benefits; potential injuries or disabilities that may require treatment; having been away from institutions for so long that they need to "retrain" their "student mind-set;" difficulty finding old transcripts and records; being commuter students; having family, work, or other obligations; difficulty connecting with their new (often 18-year-old) peers; many similar issues to first-generation students; being unaware of other campus facilities or benefits; and, if they are Guard or Reserve soldiers, having to complete their military duties and training concurrently [5]. These difficulties at the least result in increased and unneeded stress and at the worst can lead to lower retention rates for the group as a whole.

Many institutions, recognizing not only the strengths and benefits our veterans and service members can bring to their institution, are also making a good faith effort to become more welcoming and inclusive to those who have risked much to defend their country. There are many good ideas and plans publically available, such as the American Council on Education's *Toolkit for Veteran-Friendly Institutions*, the US Department of Education's *8 Keys to Veterans Success*, and an executive order signed by President Obama which created the *Principles of Excellence*, as well as additional programs that give a good starting point for institutions in developing into a more "veteran-friendly" institution. Furthermore, in order to directly assist these new students and to guide them on the path of their educations, many institutions have implemented programs and established offices to support our student veterans and service members. However, with legislators slashing budgets and alumni greatly limiting how much they are donating, institutions must be more careful in how they appropriate resources to both effectively serve our student veterans and ensure that their resources are used wisely.

In order to manage the goals of supporting our students within the regulatory and financial constraints of our university, the Missouri University of Science and Technology, the authors needed a new methodology to manage the complexity inherent in such a large bureaucracy. They decided to seek a new way to apply systems engineering principles in the support of their goal of supporting our military-connected students in their institution in a way that both works within the existing structure and also achieves their goals at minimal cost and effort—things not always achieved within a bureaucracy. To do so, they utilized a tool called the "SoS Explorer Application," developed by the Systems Engineering Department at S&T. The SoS (System of Systems) Explorer utilizes a fuzzy intelligent learning architecture to explore ways in which to optimize the *dynamic and adaptive system*, that is, the bureaucracy within a higher education institution—to meet a specific goal. This goal being to develop, improve and realign the current institutional infrastructure so that it can meet the needs of our incoming student veterans to support them in achieving their educational goals.

The "SoS Explorer Application" is used to generate alternative meta-architectures, and a fuzzy-genetic algorithm is utilized to evaluate and assess the architectures chosen. The goal within the application (and for the authors) is to maximize several Key Performance Attributes/Measures (KPAs/KPMs) in order to evaluate our meta-architectures. These KPAs are Accuracy [of data gathered], Efficacy [are the programs effective], Speed [of the system in performing its capabilities], Affordability, and Validity [that we are doing what our students need]. Those KPAs are used to evaluate the effectiveness of the architecture in meeting its goal. The "little" capabilities that support our main one are gathering data, sending data, offering [academic] support, offering [staff or advising] guidance, and managing data. The characteristics used for the systems managing this data are *collection speed*, *affordability*, *performance*, *scalability*, *send rate*, *awareness* [of new student needs], and if our student veterans' *known needs* are *met*.

The systems within our meta-architecture are based around the current structure of the university, with several offices performing essential functions and other secondary offices that can be brought in as needed. The capabilities of these offices are then decomposed based upon our five capabilities. Some offices have more capabilities than others, but that won't have any effect on how the meta-architecture is generated. Only the characteristics inherent to those capabilities will. The SoS Explorer will then evaluate which of these offices and capabilities should be brought in to support our goal and how they are connected in support of our tasks. There are a total of 21 systems, each with one capability; while there may be several systems that make up a campus office, they all may, or may not, be accepted into the meta-architecture.

19.2 Background

The SoS Explorer Application is a versatile and powerful tool; however, it requires a thorough understanding of the underlying mathematics and algorithms to be most effectively employed, concomitant with a thorough understanding of the real-world system it is being used to model. The authors based their methodology on work performed by David Curry and Dr. Dagli in their article about the SoS Explorer [6]. The application allows the authors to define their Key Performance Measures (Table 19.1); establish characteristics, capabilities, and feasible interfaces; evaluate them all; and determine an optimized solution via a meta-architecture [6]. In so doing this, we must work within the structure of a university—specifically Missouri S&T—its offices, policies, procedures, and abilities, and these will affect our constraints. The System of Systems we design will also be bound by the specific processes of the university as they matriculate new students and allow the students to sign up for classes and benefits. For example, the admissions office will be required

Table 19.1 List of KPAs/KPMs and their descriptions

Key performance attributes/measures	Description
Accuracy	A measure describing the data we gather and models we build and if they reflect the needs of our reality. A higher score is better
Affordability	A measure of how cost-effective our plan is. A higher score is best
Efficacy	A measure of how effective our programs and policies are at meeting our goals and the needs of a student. A higher score is better
Speed	A measure of how quickly we are able to achieve our objectives and complete tasks, since some student support is time-sensitive. A higher score is better
Validity	A measure of our understanding of the needs of our students and if we are capable of meeting them. A higher score is better

to "gather" new students—but that also makes it the ideal place to gather information. The registrar's office is responsible for certifying a student's VA benefits and determining class schedules, so they would be ideal to manage the databases of several classes of data. Most offices are capable of handling several capabilities at once; however, we are seeking to determine the best structure to meet the capabilities we desire—namely, supporting student veterans—so some offices may be utilized multiple times, whereas others may be used only once or not at all.

The purpose of the SoS Explorer Application and the fuzzy-genetic algorithm we use to evaluate the meta-architectures is to generate alternatives for how the different offices will talk to each other, exchange information, and coordinate resources to meet our overall goal. A complete overhaul of the university system is beyond the scope of this paper and would be infeasible; however a "realignment" of what the offices already accomplish is more feasible and cost-effective.

As stated, the authors utilized the SoS Explorer Application. It handles the writing of the attributes for each individual system, their evaluation, and finally the System of System's end optimized architecture. First, each system, which represents the portion of a single campus office responsible for a specific capability, is input into the App. The characteristics of this system and its feasible interfaces are then input next. This is repeated for each office of import, until we have a general representation of the university.

The fuzzy logic and genetic algorithm comes into play after the initial setup is complete. The application then auto-generates the base code that will be used to ascertain the scores for the Key Performance Attributes and find the most effective interfaces, all the while taking into consideration any necessary constraints. The generated code also has modules and functions designed to interface with MATLAB's Fuzzy Logic Designer, which assists in evaluating each KPA and its overall objective final score. The authors strongly referenced work performed by Curry, Dagli, and Lesinski in their work "to assess meta-architectures for non-line of sight fires battlefield capability" [7]. Their article uses genetic algorithms to encode various meta-architectures which are then evaluated by a Fuzzy Inference System to determine the ideal meta-architecture. The basics of the genetic algorithm are included in the auto-code, and it is used to generate various chromosomes representing the different meta-architectures that are then evaluated via fuzzy logic. For the final, overall objective, the authors used fuzzy membership functions of *very unacceptable*, *unacceptable*, *mediocre*, *tolerable*, *good*, *desirable*, and *very desirable*. We used a Single-Objective Genetic Algorithm (SOGA) methodology here—with the single objective being the overall objective score—so it will eventually converge upon one ideal meta-architecture [7].

The scoring of each KPA is handled by a custom matlab code and algorithm the authors input into the SoS Explorer files which manage the changes to each KPA when interfaces are created or destroyed; the purpose of this code is to represent the strengths or weaknesses of these interfaces. There are additional lines of code that include the mathematical programming equations (below) and the overall algorithm to allow the App to determine the scores of the KPAs after a chromosome is created. After the chromosome is scored, the data are taken through the Fuzzy Inference System (FIS) to determine fitness within the App to evaluate the final scores. The genetic algorithm runs again and creates offspring or mutations based upon its internal functions. Eventually, the system converges upon an ideal "overall" score.

19.3 Methodology

Here we are presenting a fuzzy-genetic algorithm approach to determine the best SoS meta-architecture to meet our needs. The SoS Explorer is the primary engine we are using to determine this meta-architecture. It will take the raw materials of our inputs, put them together utilizing genetic algorithms, evaluate them using our algorithms and fuzzy logic, and reiterate until a potential solution is converged upon. The App will need our systems' characteristic scores, their capabilities, and feasible interfaces between other systems. Which system accomplishes which capability can be seen in Fig. 19.3.

The systems' characteristic scores were not based upon raw categorical data but for simplicity were determined by internal comparison against other offices of the university, valued between 1 and 10. For example, for the capability "offering guidance," the *affordability* characteristic would be markedly higher for an office that uses personnel who are paid less than for one that uses a PhD for the same task. Granted, the PhD may have higher scores in other characteristics so it may be the better option overall. Another example would be the "data-gathering" capability. The admissions office would have the best scores in this since they have the raw man power and systems needed to gather a large amount of data quickly, via their task of matriculating over 1000 students a year.

The systems' capabilities were touched on above, but basically our system offices were divided up based upon their different capabilities as they related to their overall missions. Certain offices may be able to support our student veterans in many areas, whereas others may only be slightly involved. Finally, the 21 systems have the potential to interface—bidirectionally—with any other, especially considering that an interface may realistically be a phone call away. However, the feasible interfaces for this SoS were determined based upon already established information paths along with others that would be simple to implement or scale up. The goal is to prioritize improvements to already existing infrastructure and to minimize having to build from scratch if at all possible.

The genetic algorithm is utilized by encoding the SoS architecture as a binary string, with a length of 231. The first 21 bits of information indicate whether that particular system is activated, and the remaining 210 bits indicate which interfaces are active and where. There is a constraint written that no system without at least one interface can be active. When the optimization is being performed, the App will generate the chromosome and assign to it which system(s) and interface(s) are active.

Next, we'll need to allow the App to determine the scores of the KPAs based upon which systems and interfaces are active. With this, the Fuzzy Inference System can evaluate that specific architecture. In order to determine the KPA scores, an equation was created based upon mathematical modelling principles. The equation is below.

Equations

$$\text{For } Sys_1 \rightarrow Sys_n$$

$$X_{1\rightarrow n}\{Ch_1 + \ldots Ch_n\} \tag{19.1}$$

If $X_2 > 0$, *for KPA Validity*:

$$\{[Ch_1 + (\delta \times X_2)] + \ldots [Ch_n + (\delta \times X_2)]\} \tag{19.2}$$

If $X_2 > 0$, *for KPAs Accuracy, Affordability, Efficacy, & Speed:*

$$\left\{\left[Ch_1 + \left(\sqrt{\delta} \times X_2\right)\right] + \ldots \left[Ch_n + \left(\sqrt{\delta} \times X_2\right)\right]\right\} \tag{19.3}$$

Here, the X_1 indicates whether the system is activated or not and would be a Boolean value. "Ch_n" indicates a particular characteristic score; X_2 is an integer and indicates how many interfaces have been added to that particular system, which will change the delta score accordingly; and δ indicates how that particular score is changed based upon adding an interface (can be positive or negative). For all KPAs except Validity, we take the square root of the *delta* to describe how too many interfaces may not greatly increase the efficacy of that particular capability—it would only provide diminishing returns. The characteristic scores are added up individually and placed into the appropriate KPA based upon which ones relate to it. Without new interfaces, the original scores are what are used to calculate the KPA score (see Table 19.2).

Then finally, the chromosome of the genetic algorithm has a few constraints added into it. The first being a general function to remove infeasible interfaces from the architecture. The second ensures that every capability is referenced at least once, and the final ensures that no system operates entirely in isolation.

Table 19.2 List of KPAs and the capabilities and characteristics associated with them

KPAs	Capabilities	Characteristics
Accuracy	Gather/store data Disseminate data Manage data	Collection speed Performance Scalability Awareness
Affordability	Gather/store data Offer support Offer guidance	Affordability Performance Scalability
Efficacy	Gather/store data Offer support Offer guidance	Performance Awareness Known needs met
Speed	Gather/store data Disseminate data Offer guidance	Collection speed Performance Scalability Send rate
Validity	Gather/store data Disseminate data Offer guidance	Collection speed Performance Scalability Send rate

After a chromosome is created and determined to be viable, it undergoes a fuzzy evaluation; we use a Mamdani Fuzzy Inference System (FIS) to evaluate each architecture and determine its fitness. It takes the scores of each KPA and places them within a triangular membership function with five possibilities, low, mediocre, average, good, and high, ranked from worst to best. Each KPA is evaluated, and an output function is determined which is based upon seven membership functions: very unacceptable, acceptable, mediocre, tolerable, good, desirable, and very desirable. The function is "defuzzified" using the centroid method, and a crisp answer is determined. Figure 19.1 displays the Fuzzy Logic Designer used from MATLAB, and it shows the five KPAs, the Mamdani function for the overall objective, and the output. Figure 19.2 shows the surfaces used within the Fuzzy Logic Designer in order to assist the evaluation of the chromosome. Finally, the App optimizes all of this using a simple SOGA (Single-Objective Genetic Algorithm) method, with up to 10,000 evaluations, a population and offspring of 40, a 0.025% probability of mutation, and a 100% probability of crossover.

19.4 Results

The system was put together and the evaluation was run; within 4 min, an optimization was found. The KPAs decided upon by the meta-architecture had values of (with a max of 300) Accuracy, 252.65; Efficacy, 264.21; Speed, 276.03; Validity, 241; Affordability, 302.65, with an overall score of 238.64. We had all 21 systems participating, most with at least one interface and several with more than four. Finally, we had a total of 36 interfaces working in concert to support our mission.

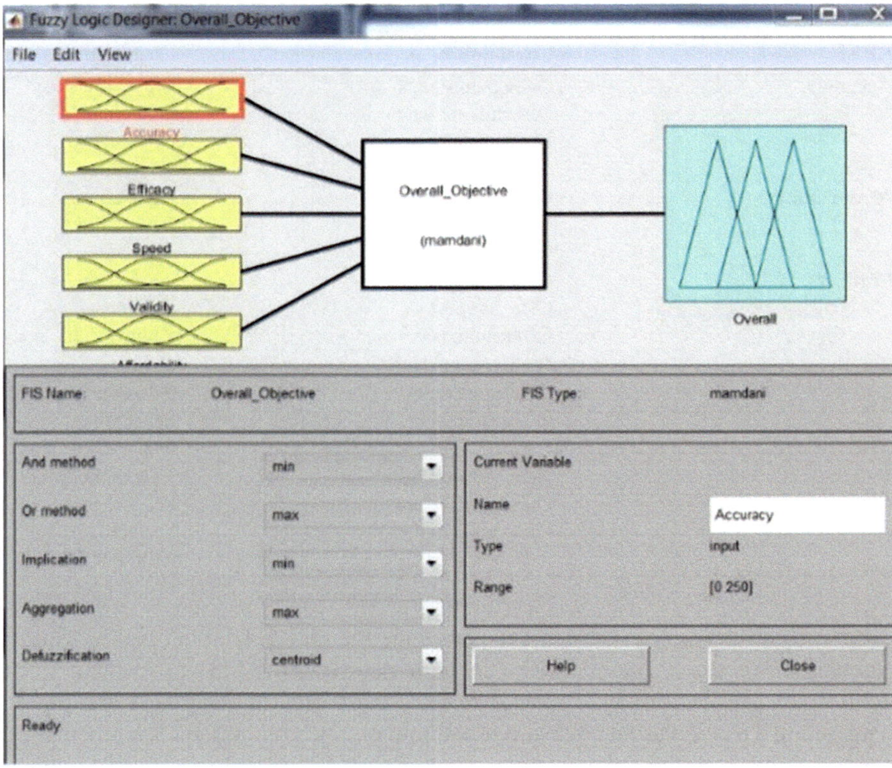

Fig. 19.1 View of the Fuzzy Logic Designer with KPAs and settings chosen

19.5 Conclusion

The paper offers an example of using a fuzzy-genetic algorithm to evaluate the fairly common, but still complex, architecture of a university campus toward the pursuit of a specific goal. The goal in this case is to build an effective data-gathering and support infrastructure to benefit our student veterans and service members as they persist toward their academic goals. We utilize the SoS Explorer Application, along with an understanding of fuzzy logic, genetic algorithms, and mathematical programming to generate specific architectures, evaluate them, and seek an optimized result. Our architectures are evaluated based upon several Key Performance Attributes, namely, Accuracy, Affordability, Efficacy, Validity, and Speed. The capabilities and characteristics of 21 synthetic systems are used to determine these KPAs, based upon whether a specific system is placed within the meta-architecture and how many interfaces are connected to it. This architecture is encoded into a genetic algorithms chromosome, and it is evaluated by a Fuzzy Inference System. Eventually a solution is converged upon; the selected architecture can be seen in Figs. 19.3 and 19.4.

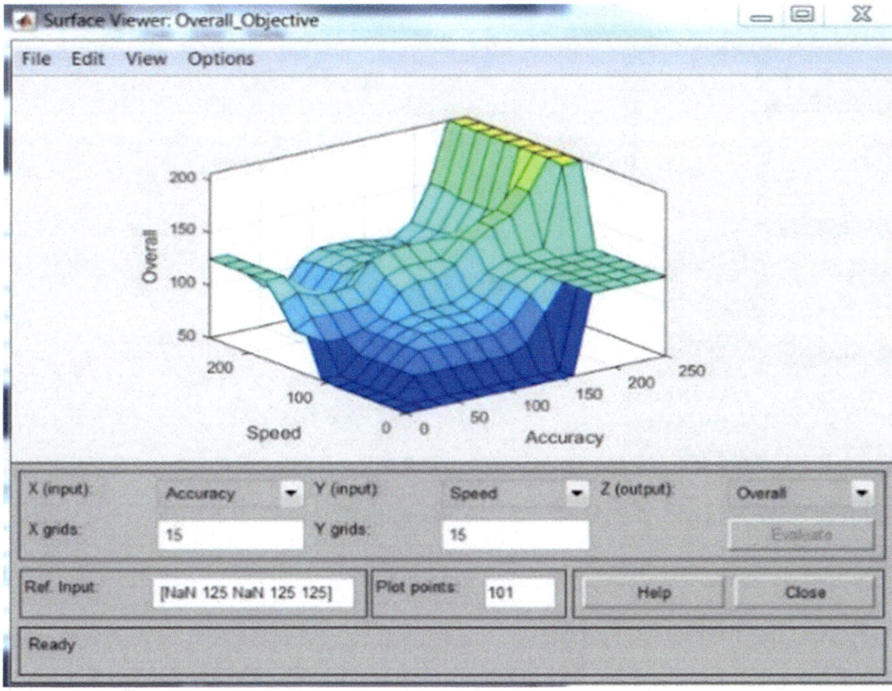

Fig. 19.2 View of a surface used in the Fuzzy Logic Designer based upon the rules established

	gather and store data		disseminate data			offer support		offer staff guidance			manage data	
System	Has	im:os	Has	Time	Cost	Has	TimeCost	Has	Time	Cost	Has	TimeCost
AD - GD	TRUE		FALSE			FALSE		FALSE			FALSE	
AD - D	FALSE		TRUE			FALSE		FALSE			FALSE	
Reg - GD	TRUE		FALSE			FALSE		FALSE			FALSE	
Reg - SD	FALSE		TRUE			FALSE		FALSE			FALSE	
Reg - MD	FALSE		FALSE			FALSE		FALSE			TRUE	
AS - OS	FALSE		FALSE			TRUE		FALSE			FALSE	
AS - OG	FALSE		FALSE			FALSE		TRUE			FALSE	
Orientation - GD	TRUE		FALSE			FALSE		FALSE			FALSE	
Orientation - OG	FALSE		FALSE			FALSE		TRUE			FALSE	
SVRC - GD	TRUE		FALSE			FALSE		FALSE			FALSE	
SVRC - OS	FALSE		FALSE			TRUE		FALSE			FALSE	
Advisor - GD	TRUE		FALSE			FALSE		FALSE			FALSE	
Advisor - SD	FALSE		TRUE			FALSE		FALSE			FALSE	
Advisor - OG	FALSE		FALSE			FALSE		TRUE			FALSE	
Advisor - MD	FALSE		FALSE			FALSE		FALSE			TRUE	
CDSW - OS	FALSE		FALSE			TRUE		FALSE			FALSE	
Faculty - OS	FALSE		FALSE			TRUE		FALSE			FALSE	
COER - OS	FALSE		FALSE			TRUE		FALSE			FALSE	
COER - OG	FALSE		FALSE			FALSE		TRUE			FALSE	
COER - DD	FALSE		TRUE			FALSE		FALSE			FALSE	
COER - GD	TRUE		FALSE			FALSE		FALSE			FALSE	

AD = Admissions, Reg = Registrars, AS = Academic Support, SVRC = Student Veterans Resource Center
Advisor = SVRC's Admin, CDSW = Counseling Office, COER = Career Office
GD = Gather Data, DD = Disseminate Data, OS = Offer Support, OG = Offer Guidance, MD = Manage Data

Fig. 19.3 Pic of the listing of systems and which capabilities they accomplish

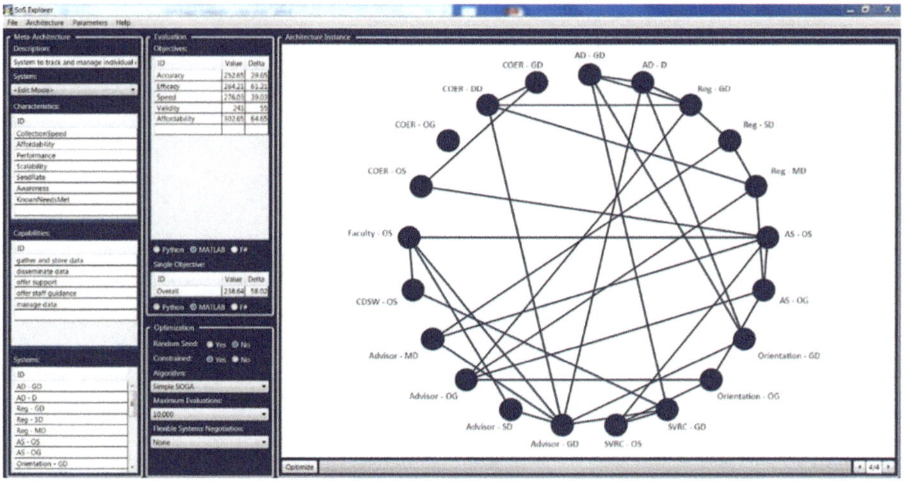

Fig. 19.4 Screenshot of SoS Explorer interface and selected architecture diagram

Our solution offers several interesting insights based upon how it is established within the application itself. The first of which is probably the most important and proves the validity of our model—it accurately reflects reality. Several of the interfaces, and the "busier" connected system nodes (with several connecting to one system), are reflective of the current university's architecture as its administrators optimized it to matriculate many students. For example, we notice that the systems, AD-GD, AD-DD, Reg-GD, and Orientation-GD [shorthand explained below in figures], are all connected within one degree of separation. This is indicative of the university's current architecture—determined solely by characteristics and the algorithm. There are other examples, but we need not go into them all here.

The results of this data do offer immense possibilities for further study, for example, by developing the mathematical modelling equations to best fit a bureaucracy. We could also use direct data—such as dollar amounts in costs—instead of comparative scores, in order to most accurately reflect the structure of a university. Finally, every model must be evaluated as toward its results. We must be willing to develop both quantitative and qualitative scores to determine how well our system performs in the real world and find ways to compare this over time and adjust and improve when needed. Initial results look good, but only time will tell.

Acknowledgments The corresponding author would like to thank Dr. Dagli for his consistent support and patience as the author developed this evaluation and David Curry for his support and in helping the author understand the nuances of the SoS Explorer and the mathematical programming equations to effectively evaluate the meta-architecture.

References

1. Gilbert, C. K., & Griffin, K. A. (2015). Better transitions for troops: An application of Schlossberg's transition framework to analyses of barriers and institutional support structures for student veterans. *The Journal of Higher Education, 86*(1), 71–97. The Ohio State University.
2. DiRamio, D. (2017). What's next for student veterans – Moving from transition to academic success. In W. A. Lang & T. O'Donnell (Eds.), *Completing the mission II: A study of veteran students' Progress toward degree attainment in the Post-9/11 Era*. Chicago, IL: Pat Tillman Foundation. Chapter 7.
3. GI Bill® Comparison Tool. (2018). *Department of Veterans Affairs*, searched retention statistics on Missouri S&T. Retrieved January, 2018, from https://www.vets.gov/gi-bill-comparison-tool/profile/11930125
4. Harmening, D. S., & Semer, C. (2015). Exploring significant factors that impact the academic success of student veterans in higher education. *Journal of Higher Education Theory and Practice, 15*(7), 31–43.
5. Cole, J. S., & Kim, Y. M. (2013, December). *Student Veterans/Service Members' Engagement in College in University Life and Education*. Report by the American Council on Education, Center of Policy Research and Strategy and National Survey of Student Engagement.
6. Curry, D. M., & Dagli, C. H. (2018). SoS explorer: A tool for system-of-systems architecting. In A. Madni, B. Boehm, R. Ghanem, D. Erwin, & M. Wheaton (Eds.), *Disciplinary convergence in systems engineering research* (pp. 187–196). Cham: Springer.
7. Gene, L., Steven, C., & Cihan, D. (2016). A fuzzy genetic algorithm approach to generate and assess meta-architectures for non-line of site fires battlefield capability. In *2016 IEEE congress on, evolutionary computation (CEC)* (pp. 2396–2401). Piscataway, NJ: IEEE.

Chapter 20
A Systems Engineering Approach to the Development of a Domain-Specific Language for Functional Reference Architectures

Richard Wise and Erika Brimhall

20.1 Introduction

As model-based systems engineering becomes more prevalent, challenges arise associated with variations across models. Different models developed using the same generic graphical modeling language (such as the Unified Modeling Language (UML) [1] or the Systems Modeling Language (SysML) [2]) use significantly different syntax and semantics to convey common concepts, even if the models use the same architectural framework (such as DoDAF [3] or TOGAF [4]), apply to the same domain, and are created by the same organization; this disparity leads to challenges in model interpretation and difficulties in comparing and analyzing models. The reverse is also true and can be even more problematic: The same terminology found in different models can refer to vastly different concepts, leading to the ambiguity and difficulty of distinguishing meaningful differences between aspects of different models. In addition, modeling language extensions and their associated rules are often developed from scratch each time a new model is needed, reducing the efficiency of new model development.

Standardized processes exist for the development of functional reference architectures [5], but without a consistent domain-specific language, such models are unlikely to be compatible and reusable unless they are first translated to match each other's concepts, relationships, and constraints—and this conversion could result in interpretation problems and the loss of important information. For new models, a more efficient and less risky approach is to simply develop the different models using the same domain-specific language (DSL).

A standardized DSL cannot meet every project's needs but can be very beneficial when developing multiple reference architectures within a given domain. The US Department of Defense (DoD) defines reference architecture as "an authoritative

R. Wise (✉) · E. Brimhall
Georgia Tech Research Institute, Atlanta, GA, USA
e-mail: richard.wise@gtri.gatech.edu

S. Adams et al. (eds.), *Systems Engineering in Context*,
https://doi.org/10.1007/978-3-030-00114-8_20

source of information about a specific subject area that guides and constrains the instantiations of multiple architectures and solutions" [6]; a reference architecture can then be used to realize a solution architecture based on specific stakeholder requirements.

The DSL described in this paper uses a Meta Object Facility (MOF)-based metamodel [12] implemented as a custom UML profile with well-formedness rules written in Object Constraint Language (OCL) [7] to enable the creation of reference architectures. The example provided supports the functional reference architectures generated using the process described in [5], including relationships to the external libraries needed to fully describe those functions. Although these reference architectures are used for the realization of aircraft software product lines, the systems engineering approach presented herein can be tailored as needed and applied to define domain-specific languages for any domain.

20.2 A Systems Engineering Approach

The development of a domain-specific language, just like the development of any product, is not a trivial endeavor. There is a risk that the product might not meet the needs of the stakeholders over the expected life cycle of the product. Potential stakeholders of the domain-specific language include program managers, software and system architecture modelers, and model analysts. The needs of the stakeholders range from those who simply need to review and understand the model to those who need a consistent structure from which to write software algorithms to interrogate the model for insight. An approach is needed to ensure that the product meets not only the required technical aspects but also the "ilities" in order to provide value and effectiveness for all users from application to deprecation of the DSL.

A systems engineering approach was taken to aid in the development of the DSL to ensure that the resulting product met the needs of the stakeholders. The approach follows the systems engineering "Vee" model [8] including the technical and technical management processes: stakeholder needs and requirements definition, architecture definition, decision management, implementation, and verification; the standard model was, of course, tailored for this work. The tailored approach, discussed in the following sections, begins with defining goals for the DSL, followed by describing the DSL architecture, and then performing decision analysis, detailed design, and implementation of the DSL and finally comes full circle with verification and validation of the DSL against the goals.

The DSL described in this paper was developed to support the functional analysis process formalized in [5]. It can be viewed as a case study in the novel application of the systems engineering process to the design and development of a DSL. The systems engineering approach discussed herein could be taken to guide the development of any DSL regardless of the domain, so that the end result effectively meets the needs of the stakeholders across the particular domain.

20.2.1 Goals of a Functional Reference Architecture DSL

The first step of the systems engineering approach begins with the elicitation and refinement of goals for a functional reference architecture (FRA) domain-specific language from key stakeholders including program managers, system and software modelers, and analysis algorithm and tool developers. The overarching goal of the DSL is to promote interoperability between users of a functional reference architecture. This means that the FRA should be easily and unambiguously understood by humans as well as interpretable by various software applications needed to interface with the FRA model.

The first specific goal in achieving interoperability is to provide the concrete syntax and semantics that precisely convey the concepts necessary in the creation of FRAs in accordance with the process defined in [5]. The syntax and semantics should unambiguously describe each of the definition components as part of the standard function definition also defined in [5] and should facilitate a data-driven functional description defined in [9]. The DSL syntax and semantics should also provide a means for capturing supplementary information, including accessing information like source document references in model libraries in a read-only manner, without affecting the state of those library elements.

The concrete syntax is insufficient by itself to ensure conformance of proper usage of the syntax for a given FRA. Therefore, the second goal is to provide an accompanying set of well-formedness rules describing the constraints on the application and usage of the concrete syntax. The well-formedness rules should also be expressed in a manner that is interpretable by software performing validation on the FRA model.

Another goal is that the DSL should be able to model structural as well as behavioral aspects of a function, such as functional decomposition hierarchy output, from following the process described in [5] and data flow through the functions described in [9].

To maximize acceptance, the syntax, semantics, and well-formedness rules should be architecture modeling tool agnostic, and the DSL should be implemented in a language familiar to the systems engineering and software engineering communities. This implies that the DSL should be UML-based as that is the de facto language in modeling system and software architectures.

In addition, the DSL should contain the minimum number of concepts needed to describe a FRA and minimal deviation from the standard definition and traditional usage of the UML metaclasses from which the DSL elements are extended. The original UML metaclasses should be used as intended following sound modeling practice and patterns; the DSL simply constrains the subset of UML metaclasses needed and provides additional semantics on top of the generic UML metaclass.

Lastly, to enable long-term maintenance, the DSL should be extensible meaning that it provides the facility to add new concepts as the architecture development process evolves. For example, the process described in [5] defines functions from a conceptual point of view. Future development may require that the FRA be described in multiple abstraction layers, for example, conceptual, logical, and physical [10]. The DSL should be able to accommodate these new concepts without affecting previous models using an older version of the DSL.

20.2.2 Functional Reference Architecture Domain Metamodel

The second phase of the approach is defining an architecture intended to satisfy the goals stated in the previous phase. This architecture is expressed as the abstract syntax of a FRA domain metamodel. A metamodel is simply a model of a model and defined as "a description of the abstract syntax of a language, capturing its concepts and relationships, using modelling infrastructure." [11] The metamodel describes the concepts important to the architectural description of the domain. The relationships between those concepts "defines how concepts can be combined to produce meaningful expressions in the language." [11] Metamodels are often complimented with textual constraints, often expressed using OCL, to specify well-formedness rules for the application of the metamodel concepts in the domain. Establishment of such a FRA domain metamodel, hereafter referred to as the domain metamodel, ensures that all reference architectures created using the DSL, regardless of mission context or model developer, are interoperable across model users, human or machine, by way of common syntax, semantics, and form.

The domain metamodel was created using MOF syntax and notation [12] and codifies the concepts identified in [5]. The following abstract syntax diagrams depict those concepts, their relationships, and the rules that constrain them. For brevity, some of the concepts, namely, abstract, base concepts, are not discussed as they are sufficiently described on the diagrams. The well-formedness rules are presented in a subsequent section.

The first diagram, Fig. 20.1, depicts the basic constituents of a ReferenceArchitecture: Library, Operand, Function, ViewTypeTrace, and AuxiliaryElementTrace as well as their base concepts.

A ReferenceArchitecture is a NamedElement and is defined as "an authoritative source of information about a specific subject area that guides and constrains the instantiations of multiple architectures and solutions" [6]. It is a top-level repository of information needed to describe a FRA. A ReferenceArchitecture contains one to many Libraries. A Library is also a NamedElement and is an organizational and grouping construct [13], a namespace for Element subtypes [1]. It contains many Operand subtypes, Functions, ViewTypeTraces, and AuxiliaryElementTrace subtypes as well as zero to many Libraries allowing for a recursive organizational structure. Although not enforced in well-formedness rules, good modeling practice should ensure that contents of a Library are homogeneous. For example, a Library containing Functions should not also contain Operands.

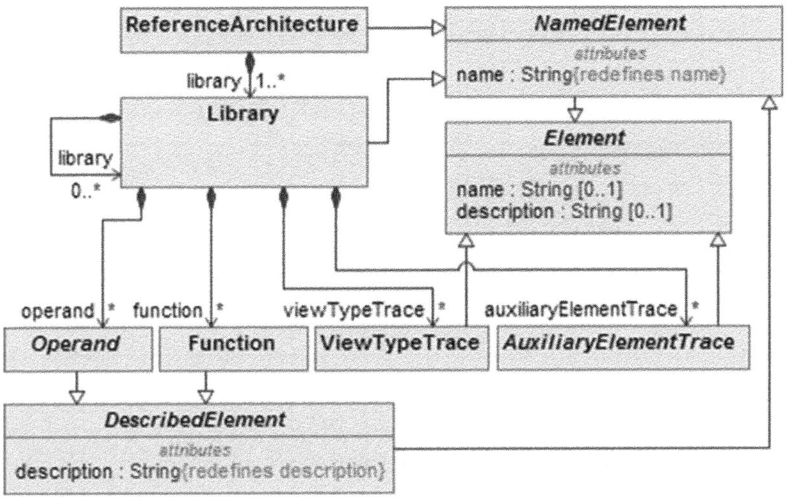

Fig. 20.1 Reference architecture constituents

Functions are the first class citizens of a FRA. They are the building blocks of a functional decomposition and describe what a system does logically [14]. A Function is a DescribedElement and is defined as an action that transforms inputs and generates outputs such as materials, energies, or information (or a combination of them) [15]. An Operand is a key concept in the description of a Function. An Operand is also a DescribedElement and is an object that may be created, modified, or consumed by the process (transformation) part of a function [16]; Operands are function inputs and outputs. A ViewTypeTrace is an Element relating Data (a structured form of Information) to a ViewType element in a data model. Last, an AuxiliaryElementTrace is an Element relating a Function to an AuxiliaryElement subtype, providing a means to supplement the description of a Function through the reference to the concept at the target end of the relationship.

The next two diagrams in Fig. 20.2 present the Operand and its subtypes and the mechanism to map a Data element in the ReferenceArchitecture to an element in a data model.

The subtypes of Operand are Material, Energy, and Information. Material is an Operand and described as the substance of all physical objects [17]. Energy is also an Operand and energizes the operation and movement of all active system components [17]. Information is an Operand representing the content of all knowledge and communication [17] and is divided into two parts: Signals and Data. Signal represents the raw, unprocessed form of Information. It is defined as an electrical impulse or radio wave transmitted or received [18].

Data is the other form of information and is one of the more noteworthy Operand subtypes. Data, in general, is structured (has attributes) and relational (participates in associations), and it is consumable and operated upon by software and software-like

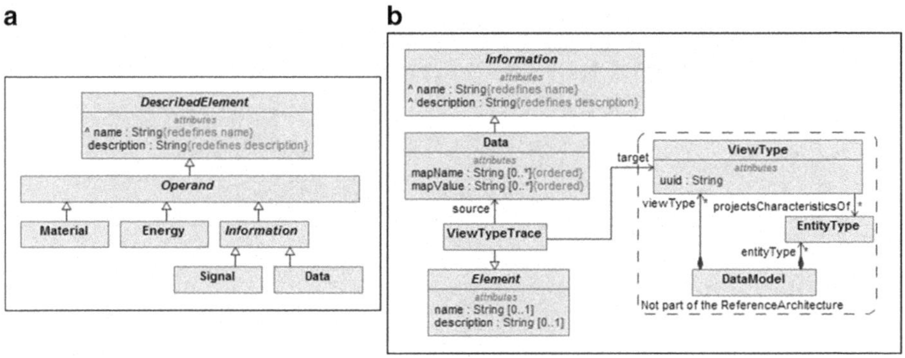

Fig. 20.2 (**a**) Operand subtypes; (**b**) data-to-data model mapping

processes. Data begins to establish the data interface requirements for a Function and is tied to a data model for proper semantic description [9]. The metamodel does not address how to model data nor is the reference architecture intended to contain a data model; the metamodel addresses how to connect a functional model to a data model as shown in diagram b of Fig. 20.2.

In keeping with the implementation-agnostic approach described in [5] as well as the data-driven functional architecture goal of the DSL, a ViewTypeTrace, shown in b of Fig. 20.2, is established as the concept by which the source, Data, relates to the target, a ViewType contained within a UML-based data model. A ViewType is loosely defined as an element that projects and encapsulates characteristics or attributes of many EntityTypes also contained within a data model for a specific purpose. An EntityType is the fundamental building block for describing the structure of data in an Entity Data Model [19]. Data may be the source of many ViewTypeTraces with the targets of the ViewTypeTraces being the possible implementation ViewTypes contained within different data models in each of a different type of data architecture, such as Future Airborne Capability Environment (FACE™) [20]. In the event that the data model is not UML-based, Data contains two attributes, mapName and mapValue, which facilitate mapping the Data element to a ViewType element in an arbitrary data structure by way of universally unique identifier name/value pairs. Both of these mechanisms afford the ability to delay data model implementation of Data until later in the design process.

The third abstract syntax diagram, Fig. 20.3, depicts the constituent elements that define a Function. A Function, as part of a FRA, is a member of a functional decomposition which is the primary output of the process described in [5].

Enabling the transformation of inputs to outputs, a Function has one to many OperandEntryPorts and OperandExitPorts. The OperandEntryPort and OperandExitPort are both kinds of OperandPort, itself a TypedElement, and represent the interaction point on the boundary of the Function where an Operand subtype can enter and exit a Function. An OperandPort redefines the type of relationship constraining the type of the OperandPort to be an Operand subtype. The OperandPort has a direction attribute typed by the enumeration datatype, OperandPortDirectionKind, with

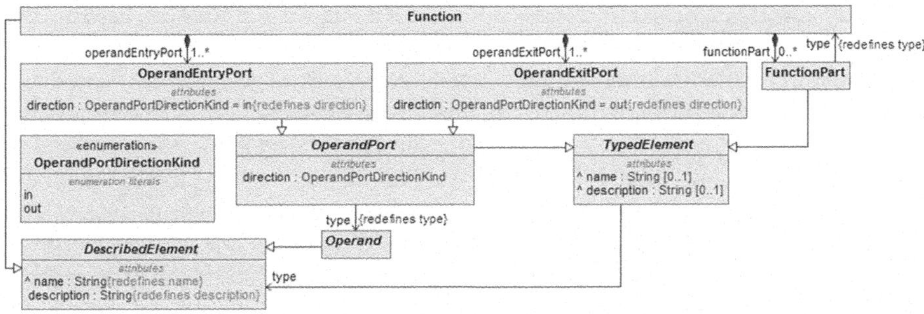

Fig. 20.3 Function constituent elements

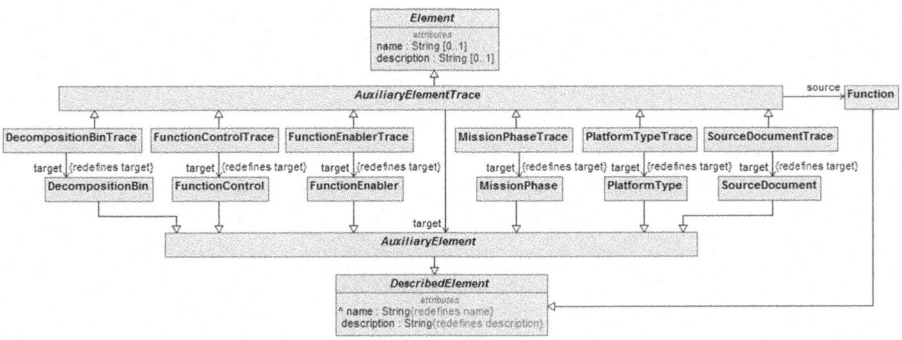

Fig. 20.4 AuxiliaryElementTrace definition and subtypes

enumeration literals, "in" and "out." The OperandEntryPort redefines the direction attribute and constrains its value to be "in." Likewise, the OperandExitPort constrains the value of the direction attribute to be "out."

A Function has zero to many FunctionParts, a kind of TypedElement, providing the mechanism for a Function to be a member of a functional decomposition. A FunctionPart is used to represent the hierarchical composition of Functions allowing for a hierarchical structure of any depth and represents the usage, role, or invocation of Functions participating in the transformation of a composite Function's inputs into outputs [21]. The FunctionPart redefines the type relationship of its supertype and constrains the type to be a Function. The lower end of the multiplicity of a FunctionPart, zero, applies when a Function is not decomposed further.

An AuxiliaryElementTrace subtype (Fig. 20.4) grants the ability to supplement the definition of a Function by reference to a respective AuxiliaryElement subtype. These AuxiliaryElements—DecompositionBin, FunctionControl, FunctionEnabler, MissionPhase, PlatformType, and SourceDocument—although components of the standard function definition presented in [5], are not part of the FRA itself. This is an intentional design decision because these elements can exist in other architecture models or libraries previously developed or being developed concurrently. For example, a SourceDocument, representing a reference to information used in the description of the function, might exist as part of an existing model library such as Unified Joint Task List libraries made available as part of MagicDraw® products [22]. The enumeration of the different types of AuxiliaryElementTrace concepts rather than a single, common AuxiliaryElementTrace concept grants the flexibility to use multiple exiting model libraries as long as the target element of the AuxiliaryElementTrace subtype fits the intended role of the concept target. The definitions of the AuxiliaryElementTrace subtypes are the same as the respective targets. For example, a MissionPhaseTrace relates to a MissionPhase representing the element that best characterizes the states or significant conditions of the system applicable to the Function [21]. Similarly, the remaining definitions are located in the standard function definition presented in [5].

20.2.3 Domain-Specific Language Selection

As with any architecture, there are multiple ways to implement the architecture, each one trading off requirements flowed down from stakeholders. Decision analysis is needed to select the optimal implementation of the architecture, thereby maximizing stakeholder benefit. The Domain metamodel presented in the previous section is no different. In theory, a domain-specific language could be developed directly from the abstract syntax of the metamodel through the use of metacase tools [23]. However, UML with extensions is one of the most widely used graphical modeling languages in software and system architecture modeling. Therefore, the functional reference architecture DSL will be an extension of UML.

The traditional representation of the three-tier modeling hierarchy [23] is extended in Fig. 20.5 to show the mapping of domain concepts to UML metaclasses and extension, as needed, via UML stereotypes.

As shown in Fig. 20.5, each layer of the three-tier modeling hierarchy is an instance of the layer above with the meta-metamodel layer, M3, being an instance of itself [24]. The metamodel layer, M2, is divided into three sublayers, the Domain

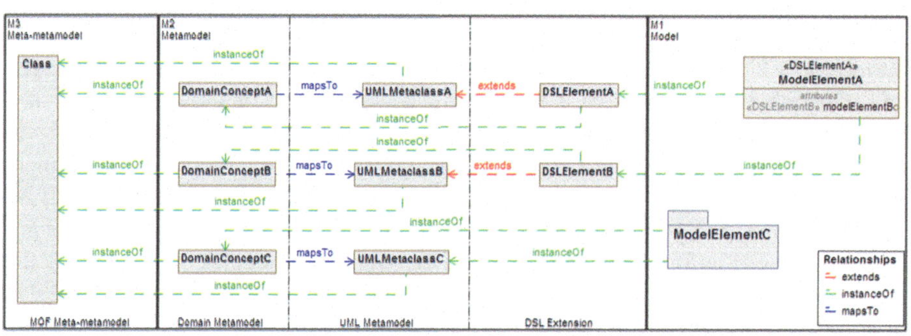

Fig. 20.5 Three-tier modeling hierarchy

Metamodel, the UML Metamodel, and the DSL Extension. Both the Domain Metamodel and the UML Metamodel are instances of the M3 MOF meta-metamodel. The concepts defined in the Domain Metamodel are mapped to the UML metaclasses that most appropriately implement the syntax, semantics, and notation of the domain concept. UML, by its very nature, is generic and on its own can often be too ambiguous for a specific domain. This has led to the creation of multiple domain-specific languages such as SysML, the Unified Profile for DoDAF and MoDAF (UPDM) [25], Service-oriented Architecture Modeling Language (SoaML) [26], and many others. DSL extensions, via UML's profiling mechanism, extend the basic UML metaclass by adding additional properties and constraints specific to the FRA domain. The elements in the Model layer, M1, are an instance of, in the meta-sense, and conform to the metamodel described in the layer above. As illustrated in Fig. 20.5, not every model element is an instance of a DSL extension. In some cases, the UML metaclass sufficiently implements the semantics and notation of the Domain concept.

As listed above, many standardized UML-based DSLs already exist each one with the goal of extending UML for application in a specific domain. It is quite possible that one of the existing, standardized DSLs contains the syntax, semantics, and well-formedness rules that would ensure conformance of a FRA model described using the DSL to the Domain Metamodel. Therefore, it was prudent to conduct a trade study, following the concept exploration methodology described in [27], to select the optimal UML-based implementation of the Domain metamodel based on the goals previously stated. Not all of the steps of the trade study are discussed in detail in this paper, but a few are worth elaborating. Notable steps of the trade study include employing the Technique for Ordered Preference by Similarity to the Ideal Solution (TOPSIS) [28] to select the optimal DSL from among the following candidate DSLs: SysML, UPDM, UML Profile-based Integrated Architecture (UPIA) [29], UML, EAST-ADL [30], and the custom DSL, UML for functional reference architecture or UML4FRA. The goals listed in Sect. 2.1 above were turned into evaluation criteria. The study concluded with analyzing the results and selecting a DSL. The analysis is subjective due to stakeholder weightings introduced in requirement priorities. The requirement priorities were varied to test the sensitivity of the results to those changes. The custom DSL, UML4FRA, was the most resilient to change in requirement priorities and therefore selected as the optimal candidate DSL.

20.2.4 Metamodel Implementation

The next phase of the approach is Implementation. With the optimal DSL selected, it was possible to begin full implementation of the metamodel; this included creating a UML profile containing domain-specific stereotypes and writing the Object Constraint Language (OCL) constraints providing the well-formedness rules constraining the use of the DSL stereotypes in a FRA. A stereotype was created

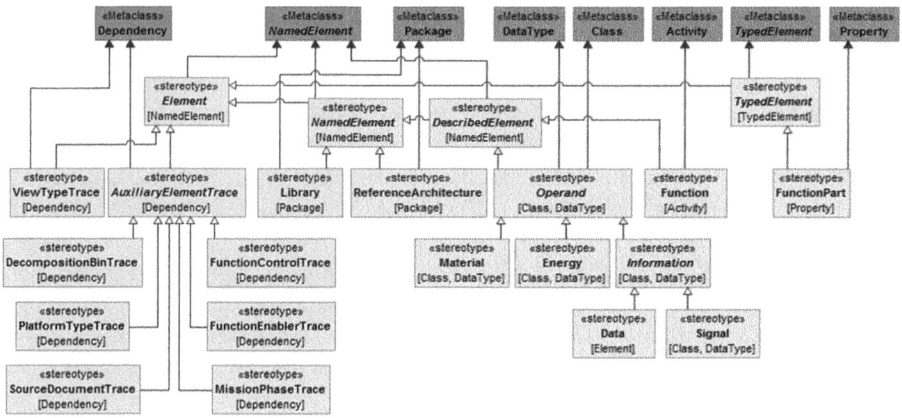

Fig. 20.6 UML4FRA stereotypes

for each of the concepts presented in the metamodel diagrams except for the OperandPort subtypes. Stereotypes were not created for the OperandEntryPort and OperandExitPort domain concepts because the standard UML Parameter when owned by a Function, or UML Activity, precisely and unambiguously implements those Domain metamodel concepts.

The stereotypes extend the most appropriate UML metaclass or in some cases multiple metaclasses. For example, both the ReferenceArchitecture and Library stereotypes are organizational and namespace constructs and consequently extend the UML Package metaclass. The Activity UML metaclass is the standard representation of behavior. It can be depicted on Class diagrams with a rectangular box notation, similar to a Class; can have structural features, such as properties; and can be the member end of Associations depicting the Function as a member in a functional decomposition tree. Therefore, it is the most appropriate element to which a Function stereotype should be extended. Subsequently, a FunctionPart, owned by a Function, extends the UML Property metaclass. OperandPort subtypes implemented as UML Parameters can be typed by either UML DataTypes or UML Classes driving the decision to extend the Operand from both the UML DataType and UML Class metaclasses. The extensions from UML DataType and Class are still fitting, though, for the implementation of the Operand concept. Both the ViewTypeTrace and the AuxiliaryElementTrace stereotypes extend the UML Dependency metaclass. Figure 20.6 presents the taxonomy of UML4FRA stereotypes.

The constraints provided by the stereotype metatypes and extended metaclasses are insufficient to describe the well-formedness rules necessary to ensure conformance of a FRA developed using the UML4FRA profile to the Domain metamodel. All of the metamodel abstract syntax diagrams were analyzed, and well-formedness rules were written in natural language describing how particular concept elements are constrained. The natural language rules, still too ambiguous to be useful, were refined into OCL constraints, the de facto constraint language for UML models,

Table 20.1 Example UML4FRA OCL constraints

Diagram	Constrained element	^^Natural language rule	OCL constraint
...	...	...	...
Fig. 20.3	OperandPort	The type of an OperandPort subtype must be an operand subtype	`self.type.oclIsKindOf(Operand)`
Fig. 20.4	AuxiliaryElementTrace	The source of an AuxiliaryElementTrace subtype must be a function	`self.client->forAll(a|a.oclIsTypeOf(Function))`
...	...	...	...

specifying the invariant on the metamodel concepts and subsequent stereotypes. An example of some of the rules and their OCL constraint equivalents are shown in Table 20.1.

The OCL constraints listed in Table 20.1 can be included as part of an .ocl script file or as the specification of UML constraints. The latter was chosen for this work. Specific implementation of the UML constraints varies by modeling tool; however the majority will flag a non-conformant element and give some indication of the constraint failure, such as indicating that a Function does not have at least one OperandPort.

20.2.5 DSL Verification

Verification of the implementation, the UML4FRA profile, and accompanying OCL constraints, or FRA DSL, of the architecture, the Domain metamodel, is needed to ensure that the implementation meets the DSL goals reflected in the abstract syntax of the Domain metamodel. A verification methodology, shown in Fig. 20.7, was developed to ensure conformance of an FRA described using the FRA DSL to the Domain metamodel.

As previously described, well-formedness rules were written to describe the constraints imposed on the usage of the metamodel concepts. These natural language rules were captured as SysML requirements and were traced to the metamodel diagrams for bookkeeping. UML4FRAConstraints contained as part of a suite of WellFormednessRulesForUML4FRA were written in OCL syntax to refine those natural language rules in a form usable by a modeling tool's model validation engine. The UML4FRAConstraints were traced to the natural language requirements, again for bookkeeping. The UML4FRAConstraints constrain the usage of UML4FRAStereotypes contained within the UML4FRAProfile. These stereotypes are then applied to constituent ReferenceArchitectureElements within a FunctionalReferenceArchitecture. A TestScenario was developed for each natural language well-formedness rule extracted from the abstract syntax diagrams. The

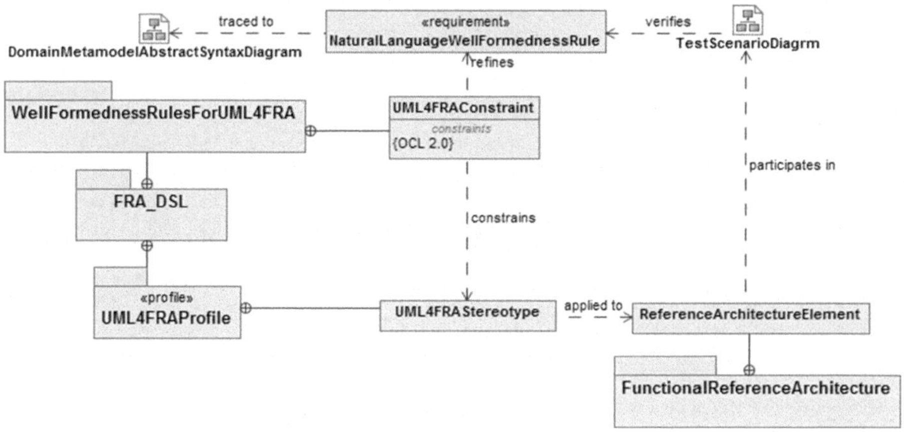

Fig. 20.7 DSL verification methodology

TestScenarios consist of participating ReferenceArchitectureElements in a usage pattern representative of the well-formedness rule.

20.3 Conclusion and Future Work

The approach described above was created based on input from multiple stakeholders, and it was refined and matured based on feedback from subject matter experts in systems engineering and other engineering disciplines. This approach can be applied to support open architectures across an entire industry, or it can simply be applied to MBSE within an organization.

As demonstrated above, this approach meets the needs of FRAs. It provides concrete syntax and semantics which unambiguously support the functions, operands, libraries, and other concepts referenced in the specified literature. The OCL constraints specify the well-formedness rules needed for validating the syntax of a model generated with a DSL. This approach can be implemented using a variety of UML modeling tools, and the extensions it makes to traditional UML simply support the desired concepts. The DSL is also extensible like traditional UML, so it can grow to support future needs.

The custom profile, OCL constraints, and design patterns in the described FRA DSL were used to create three functional reference architectures for aircraft, including one implementation-agnostic functional architecture, a mission-specific architecture, and an architecture supporting multiple missions for a particular aircraft. The generated architectures were able to support each project's requirements, which included interoperability with functional analysis software, thus providing validation of the DSL approach for these projects within the aircraft domain. The use of the DSL helped save time creating the metamodels and models to support these projects, made it easier for different groups to understand the models, and simplified the analysis of the model output.

The described metamodel approach has successfully addressed the needs where it has been applied, and further development of this approach would provide additional benefits and expand opportunities for its implementation. A broader range of metamodel applications and additional stakeholder involvement would provide further validation and could possibly yield opportunities for metamodel extensions. Similarly, the conceptual-level metamodel approach described in this paper could be extended to also address logical and physical/platform-level concerns. In addition, the efficiency of metamodel updates could be improved by automating the workflow, which currently involves manually translating the metamodel into a profile and OCL constraints.

References

1. Object Management Group. (2015). OMG unified modeling language TM (OMG UML), version 2.5. https://doi.org/10.1007/s002870050092.
2. Object Management Group. (2015). OMG systems modeling language (OMG SysML™), version 1.4.
3. DoD Deputy Chief Information Officer. DoD architecture framework version 2.02. http://dodcio.defense.gov/Library/DoD-Architecture-Framework/
4. The Open Group. TOGAF 9.1. http://pubs.opengroup.org/architecture/togaf9-doc/arch/. Published 2011.
5. Brimhall, E., Simko, R., Wise, R., Huggins, J., & Matteson, W. (2016). A systematic process for functional decomposition in the absence of formal requirements. *26th Annual INCOSE International Symposium, 1*(26), 1204–1218. https://doi.org/10.1002/j.2334-5837.2016.00221.x
6. US Dept. of Defence/Office of the DoD CIO. (2010). Reference architecture description. http://dodcio.defense.gov/Portals/0/Documents/DIEA/Ref_Archi_Description_Final_v1_18Jun10.pdf.
7. Object Management Group. (2014). Object constraint language. https://doi.org/10.1167/7.9.852.
8. INCOSE. (2015). *Systems engineering handbook 4E*. Hoboken, NJ, USA: Wiley.
9. Battigaglia, A., Brimhall, E., & Ogle, T. (2017). Conceptual data modeling for the functional decomposition of mission capabilities. *INCOSE International Symposium, 27*(1), 935–949. https://doi.org/10.1002/j.2334-5837.2017.00404.x
10. Truyen F. (2006). The fast guide to model driven architecture the basics of model driven architecture. *Practice*, 12. http://www.omg.org/mda/mda_files/Cephas_MDA_Fast_Guide.pdf
11. Paige, R. F., Kolovos, D. S., & Polack, F. A. C. (2014). A tutorial on metamodelling for grammar researchers. *Science of Computer Programming, 96*(P4), 396–416. https://doi.org/10.1016/j.scico.2014.05.007
12. Object Management Group. (2006). Meta-object facility specification (2.0) (pp. 1–2).
13. Fowler M. (2004). UML distilled: A brief guide to the standard object modeling language. *Pearson Paravia Bruno Mondad*, 175. https://doi.org/10.1109/MS.2005.81.
14. Department of Defense SMC. (2001). *Systems engineering fundamentals*. https://doi.org/10.1016/j.cmpb.2010.05.002
15. Faisandier, A. (2013). Engineering and architecting multi-disciplinary systems. In *Systems architecture and design* (Vol. 3). Belberaud, France: Sinergy'Com.

16. Crawley, E., Cameron, B., & Selva, D. (2015). *System architecture: Strategy and product development for complex systems* (1st ed.). London: Pearson.
17. Kossiakoff, A., Sweet, W. N., Seymour, S. J., & Biemer, S. M. (2011). *Systems engineering principles and practice* (Vol. 102). Hoboken, NJ, USA: Wiley. https://doi.org/10.1016/j.cmpb.2010.05.002
18. signal|Definition of signal in English by Oxford Dictionaries. https://en.oxforddictionaries.com/definition/signal#dictionary
19. entity type|Microsoft Docs. https://docs.microsoft.com/en-us/dotnet/framework/data/adonet/entity-type
20. The Open Group. (2014). *Technical Standard for Future Airborne Capability Environment (FACETM), Edition 2.1*. Burlington, MA, USA: Author.
21. Friedenthal S, Moore A, Steiner R. (2008). A practical guide to SysML. https://doi.org/10.1016/B978-0-12-374379-4.X0001-X
22. No Magic, Inc. New and Noteworthy. https://www.nomagic.com/new-and-noteworthy-sections?start=200
23. Karagiannis, D., & Kühn, H. (2002). Metamodelling platforms. In *Proc Third International Conference EC-Web 2002* (p. 182). https://doi.org/10.1007/3-540-45705-4
24. Team O 2 E. (2017). Meta-modeling and the mof meta-modeling and the OMG meta object facility (MOF).
25. Omg. (2012). Unified profile for DoDAF and MODAF (UPDM). *Management* (2.1).
26. OMG. (2012, March). Service oriented architecture modeling language (SoaML) specification. *Language (Baltim)* (pp. 1–144).
27. Lafleur JM, Lantoine G, Hensley AL, et al. (2008). A systematic concept exploration methodology applied to venus in situ explorer (p. 404).
28. Tzeng, G.-H., & Huang, J.-J. (2011). *Multiple attribute decision making, methods and applications*. Boca Raton, FL, USA: CRC Press.
29. IBM. (2010). Rational software architect version 8.0.0, UML profile-based integrated architecture (UPIA).
30. Consortium A. (2010). EAST-ADL domain model specification. *Proj Deliv 1*(218). http://scholar.google.com/scholar?hl=en&btnG=Search&q=intitle:EAST-ADL+Domain+Model+Specification#0

Chapter 21
The Data-Ontology-Rule Footing: A Building Block for Knowledge-Based Development and Event-Driven Execution of Multi-domain Systems

Maria Coelho, Mark A. Austin, and Mark R. Blackburn

21.1 Introduction

21.1.1 Problem Statement

Our work is concerned with the comprehensive development of knowledge-based representations and event-driven behavior models for engineering systems spanning a multiplicity (e.g., physical, cyber, human, natural environment) of domains. We simply refer to such entities as multi-domain systems. Multi-domain systems in both the civilian and military sectors are of interest, as are those that have benefited from remarkable advances in technology—particularly, computing, communications, and materials technology—over the past three decades. From a model-based systems engineering (MBSE) perspective, the latter opens doors to the development of systems having (1) new forms of functionality, (2) superior levels of performance and agility in the face of pre-planned and unforeseen disruptions, and (3) economical operations over extended time horizons. While end users applaud the benefits that these technological advances afford, systems engineers are faced with a multitude of new design challenges that can be traced to the presence of heterogeneous content; system-level behaviors that are distributed and concurrent; network structures that are spatial, interwoven, and dynamic; and design conflicts that inevitably occur when a multiplicity of stakeholders from separate domains have competing objectives and concerns.

In a decentralized system structure, no decision-maker knows all of the information known to all of the other decision-makers, yet, as a group, they must cooperate

M. Coelho · M. A. Austin (✉)
Department of Civil and Environmental Engineering, University of Maryland, College Park, MD, USA
e-mail: austin@isr.umd.edu

M. R. Blackburn
Stevens Institute of Technology, Hoboken, NJ, USA

S. Adams et al. (eds.), *Systems Engineering in Context*,
https://doi.org/10.1007/978-3-030-00114-8_21

to achieve system-wide objectives. Communication and information exchange are important to the decision-makers because communication establishes common knowledge among the decision-makers, which, in turn, enhances the ability of decision-makers to make decisions appropriate to their understanding, or situational awareness, of the system state and its goals and objectives. And even if the resulting cross-domain relationships are only weakly linked, they are, nonetheless, still linked. When part of a system fails, there exists a possibility that the failure will cascade across interdisciplinary boundaries, thus making connected systems more vulnerable to various kinds of disturbances than their independent counterparts [1, 2]. The introduction of automation into a system's operation, perhaps as a replacement for human-centered control, expands the range of design concerns that need to be addressed. For example, a new fundamental question is: How do we know that an automated management system will always do the right thing? A second important question is: If part of the system fails unexpectedly, what assurances do we have that the system will handle and recover from disruptions in a manner that is both sensible and timely? Our research is motivated by the belief that as systems become progressively complex, good answers to these questions will be unobtainable unless we have an ability to model and formally reason with the semantics and data in multi-domain systems.

21.1.2 Objectives and Scope

State-of-the-art approaches [3–5] to semantic modeling of engineering systems focus on the capture and representation of knowledge within one or more domains. A common objective is development of ontologies for the comprehensive representation of knowledge within a domain (e.g., sensors and sensor networks, satellites), with far less effort going to the development of rules for the validation, use, and interaction of the ontology with other ontologies. Two further problems include (1) a lack of discipline in the development of ontologies for system development and (2) a lack of computational support for evolution of semantic graphs in response to events. The first factor is one of the reasons why formal representations of ontologies have a reputation of being difficult to develop and use. As a case in point, within the integrated model-centric engineering ontologies (IMCE) developed at JPL (Jet Propulsion Laboratory) during the 2000–2010 era [5], the electrical engineering ontology (i.e., electrical.owl) imports the mechanical engineering ontology (i.e., mechanical.owl). Both the electrical and mechanical engineering ontologies import a multitude of foundation ontologies (e.g., analysis.owl, mission.owl, base.owl, project.owl, time.owl) and make extensive use of multiple inheritance mechanisms in the development of new classes. The result is ontologies containing more than several hundred classes, with some classes containing three or four dozen data and object properties. Notions of simplicity in system design through modularity of semantic models (e.g., bundling of ontologies and rules) do not seem to exist. SySML [6] steps around these difficulties by providing a variety of diagram types (e.g., use case, requirements, block, activity, sequence) that an engineer can use to visually describe what a proposed system will do and how it will do it.

This paper proposes a framework for knowledge-based development and event-driven execution of multi-domain systems where the complementary rolls of data, ontologies, and rules are highlighted and have equal importance. We call the building block for this framework the data-ontology-rule footing. The remainder of this paper proceeds as follows: Sect. 2 introduces our approach to semantic modeling and rule-based decision-making. Section 3 describes the data-ontology-rule footing, its use in the construction of multi-domain semantic models, and formalisms for the modeling and visualization of domain ontologies. The case study (see Sect. 4) exercises these ideas through an examination of fault tolerance of drone operations in a military mission.

21.2 Semantic Modeling and Rule-Based Decision-Making

21.2.1 Framework for Semantic Modeling

Model-based systems engineering development is an approach to system-level development in which the focus and primary artefact of work is models, as opposed to documents. A tenet of our work, which is motivated by the steady trend toward engineering systems becoming increasingly complex, is that methodologies for strategic approaches to design will employ semantic descriptions of application domains, and use ontologies and rule-based reasoning to enable validation of requirements, automated synthesis of potentially good design solutions, and real-time management of interactions among participating domains.

Figure 21.1 is adapted from our recent work [3] and shows the essential elements of a framework for semantic modeling of multi-domain systems. On the left-hand side, textual requirements are defined in terms of mathematical and logical rule expressions for design rule checking. Engineering models will correspond to a

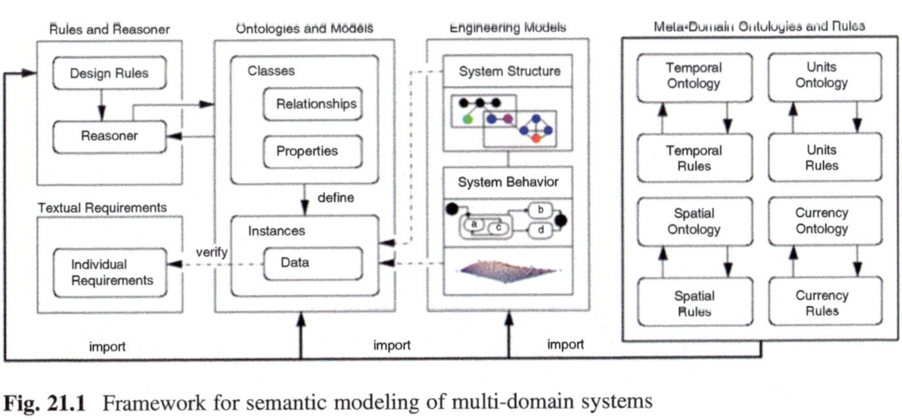

Fig. 21.1 Framework for semantic modeling of multi-domain systems

multitude of network and hierarchy graph structures for the system structure and behavior. Behaviors will be associated with components, with discrete behavior being modeled with finite-state machines (FSM) and statecharts. System-level behavior will correspond to networks of FSM/statechart behaviors. Ontology models and rules glue the requirements to the engineering models and provide a platform for the development of system structures, adjustments to system structure over time, and system behavior. Collections of ontologies and rules will be developed for specific domains as well as meta-domains (e.g., space, time, physical units) that are an integral part of all engineering domains.

21.2.2 Use of Semantic Web Technologies for Rule Checking

Semantic models (see Fig. 21.1) consist of ontologies, graphs of individuals (specific instances of classes), and inference-based rules. An ontology is a formal and explicit representation of the concepts, referred to as classes, and data and object properties. Support for relationships among classes is provided by object properties. Individuals are instances of ontology concepts, and their purpose is to represent the data in a domain. Inference rules and their associated reasoning mechanisms provide a way to derive new information based on the existing data stored in the ontology in the form of *if <condition> then <consequences>*. A key benefit of this formalism is that ontologies and rules are human readable, yet they can also be compiled into code that is executable on machines.

21.2.3 Working with Jena and Jena Rules

Our experimental software prototypes employ Apache Jena and Jena Rules. Apache Jena [7] is an open-source Java framework for building Semantic Web and linked data applications. Jena provides the interfaces for code development and the construction of resource description framework (RDF) graphs and semantic descriptions of domains in OWL (Web Ontology Language). The Jena rule-based inference subsystem allows for a range of inference engines to be plugged into Jena, and Jena Rules is one such engine. Jena Rules uses facts and assertions described in OWL to infer additional facts from instance data and class (ontology) descriptions. As we will soon see in the case study prototype, domain-specific ontologies can import and use multi-domain (or crosscutting) ontologies. And rules can be distributed among domains and can be programmed to respond to events that involve (or affect) reasoning among a multiplicity of domains. These inferences result in event-driven structural transformations of the semantic graph model.

21.3 Proposed Methodology

In state-of-the-art development of semantic models, two common strategies are as follows: (1) provide classes and data properties for all possible configurations within a domain, as well as linkages to related domains, and (2) maximize generality of the knowledge representation through extensive use of multiple inheritance mechanisms. In practical engineering settings, the result can be ontologies containing hundreds of classes, each having dozens of data and object properties defined through inheritance mechanisms. In our view, notions of "simplicity of system design" through modularity of semantic models (e.g., binding of ontologies and rules) do not seem to exist. This practice makes MBSE of multi-domain systems more complicated than it needs to be and misses opportunities for adding value to the system development process.

21.3.1 The Data-Ontology-Rule Footing

In a step toward mitigating these complexities, we propose a semantic modeling framework (see Fig. 21.2) that supports (1) concurrent data-driven development of domain data models, ontologies, and rules and (2) executable processing of incoming events.

Instead of creating a few (very large) ontologies and a few rules (perhaps for validation of semantic graph properties), our goal is to put development of the data models, ontologies, and rules on an equal footing. We call this arrangement the data-ontology-rule footing. Each row of the footprint contains rules, ontologies, and data models for a specific domain or design concern. And as we will soon see in the case study problem, domains interact through rule interactions. Figure 21.2 shows, for example, data-ontology-rule footings for the semantic modeling of drone operations in a military mission. The overall system representation is highly multidisciplinary covering concerns for drone operations and communications, the military mission, and the potential impact environmental factors may play in affecting the feasibility of proposed drone operations. Our second strategy is to minimize the use of multiple inheritances in the specification of OWL ontologies and, instead, explore opportunities for replacing them with object property relations.

A key advantage of this approach is that it forces designers to provide data that are needed for decision-making and increases the likelihood that data not needed for decision-making will be left out. The latter occurs because co-development of rules, ontologies and data sources forces developers to think about the chain of dependency relationships that allow the rules will work—rules require data and object properties from the ontologies, which in turn, require data values from the data models shown along the right-hand side of Fig. 21.2. Rules will be developed for the verification of semantic properties (e.g., has a specific data property been initialized?) and for reasoning with data sources and incoming events, possibly from a

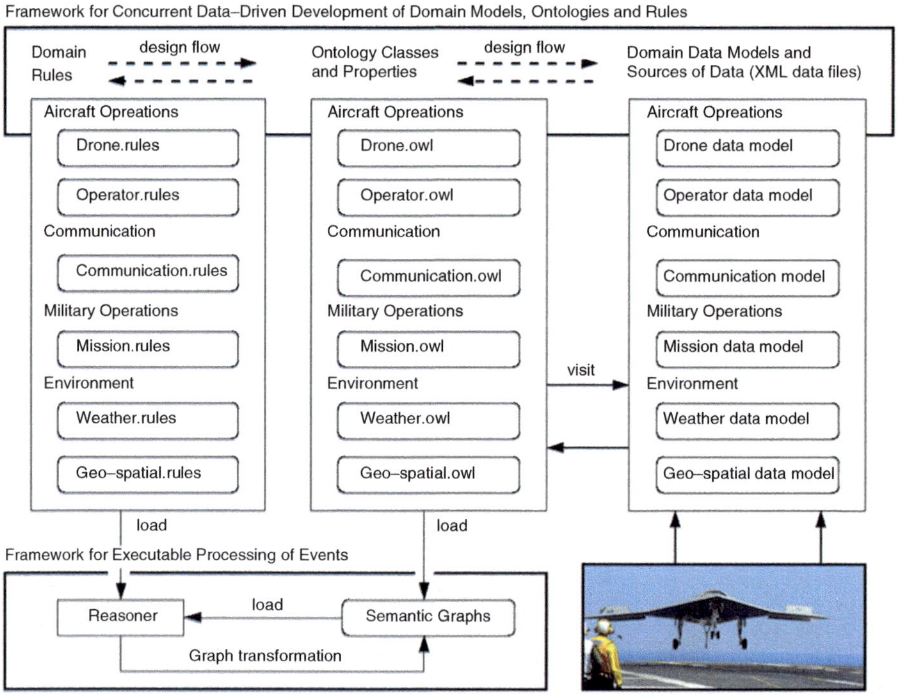

Fig. 21.2 Framework for (1) data-driven development of multi-domain systems, ontologies, and rules and (2) executable processing of events

multiplicity of domains. Implementation of the latter leads to semantic graphs that can dynamically adapt to the consequences of incoming data and events (e.g., a weather event) acting on the system.

21.3.2 Data-Driven Approach to Generation of Individuals in Semantic Graphs

By themselves, ontologies and rules provide a framework for the representation and transformation of semantic graphs, but they do not contain actual data (i.e., individuals in the semantic graph). In a practical implementation, the data will come from a variety of sources and be highly heterogeneous (see the right-hand side of Fig. 21.2).

We address this challenge with a data-driven approach to generation of semantic models, where, as illustrated in Fig. 21.3a, semantic models obtain data on individuals by systematically visiting the data models that are designed to host visitors. This is an implementation of the visitor software design pattern. Generally speaking, data models obtain their data from data files and a variety of online sources (e.g., weather and geographic information servers) and can be highly heterogeneous. SysML

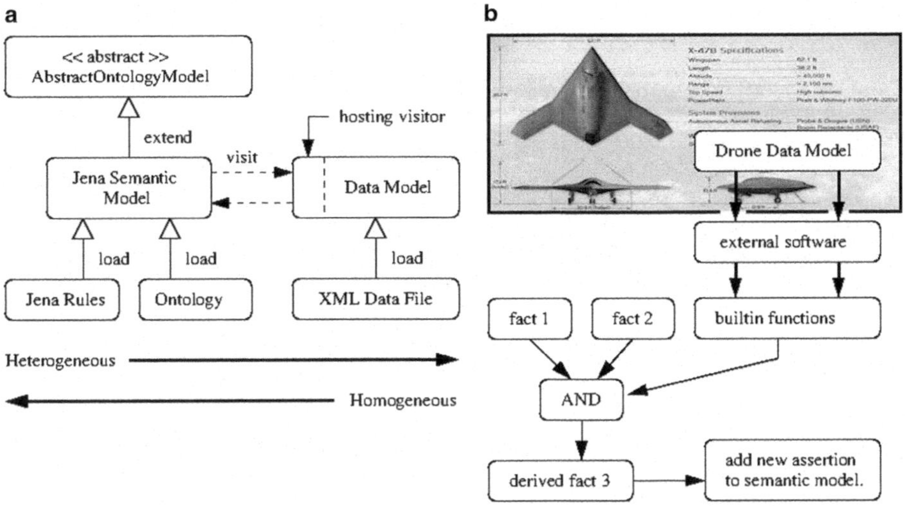

Fig. 21.3 Data-driven approach to semantic modeling: (**a**) generation of individuals, (**b**) forward chaining facts and results of built-in function evaluations

models of system structure and behavior can also be viewed as data. The challenge in parsing these data sources can be vastly simplified if the XML data files and corresponding data models are designed to work with JAXB (the XML binding for Java). We are currently exploring opportunities for using JAXB in the description of general-purpose components, their finite-state machine (or statechart) behaviors, and various specifications for performance and functionality. This aspect of our research is a work in progress. It is important to note that when a semantic model visits a data model in search for data, the results have to fit into one of eight basic data types supported in Jena (i.e., integer and floating point numbers, true and false, and character strings). While this transformation simplifies the systems integration problem and makes the overall data problem more homogeneous, it forces designers to think carefully about how the data will be used by the ontologies and rules. This leads to novel solutions on the semantic level—for example, it is much easier to store sets of geo-spatial coordinate data as character strings than as individual data points. Backend software routines associated with the rule-checking procedures (see Fig. 21.3b) can parse these strings and work with the individual data values. Once the data has been transferred to the Jena Semantic Model and used to create an ontology instance, the rules are applied. It is important to note that while Fig. 21.2 implies a one-to-one association relationship between semantic graphs and data, in practice a semantic graph model might visit multiple data models to gather data on individuals.

21.3.3 Organization and Visualization of Domain Ontologies

Domain-specific footings are conveniently organized into groups along the lines of their contribution to the system-level functionality. In Fig. 21.2, for example, data-ontology-rule footings are provided for aircraft operations, communications, military operations, and the environment. A footprint model to support fault detection and diagnostic analysis of equipment in buildings would contain data, ontologies, and rules for the building geometry and operations, the building occupants, sensors, equipment, and procedures to be followed in the fault detection and diagnostic procedure.

Visualization of domain ontologies and their connection to rules is an essential component of verifying our models make sense. Figure 21.4 shows a simplified semantic model for the structure and operation of a drone having behavior that can be described by a statechart (i.e., states, transitions, guard conditions). The limitations of drone operations (e.g., maximum payload) can be represented by a collection of specifications. We employ red rectangles with heavy dashed edges to highlight classes that participate in rule checking.

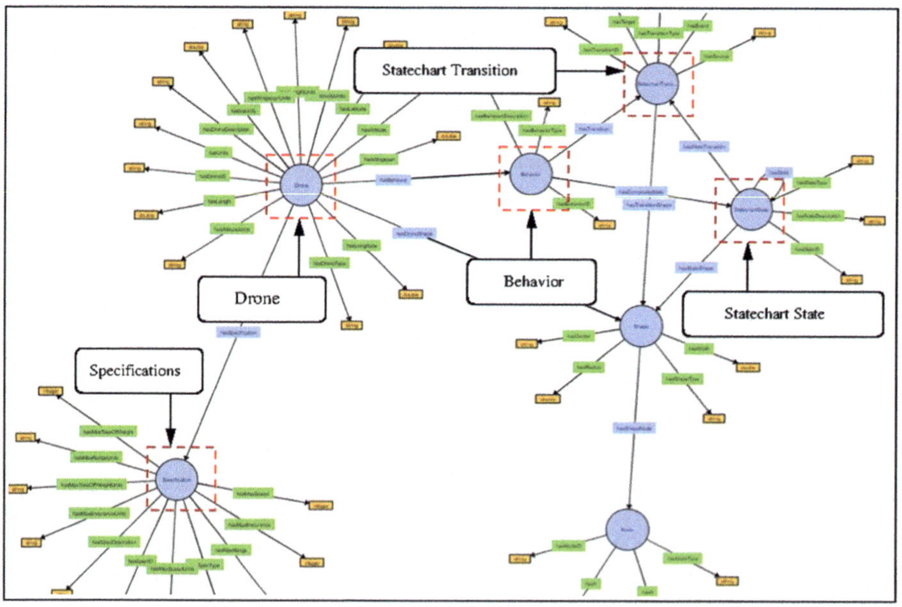

Fig. 21.4 Simplified semantic model for the structure, behavior, and performance specification of an unmanned aircraft (drone)

21.4 Case Study Simulation

21.4.1 Problem Description

Our case study simulation corresponds to a simple scenario where a drone (an unmanned air vehicle) is involved in a reconnaissance operation of a geographical area of military interest. The modeling objective is to obtain a high-level description of the participating processes—a drone, a human operator, a communications channel between the operator and the drone—and their interactions in response to an unexpected disruption.

The participating domains—rules, ontologies, and data models—are shown along the rows of Fig. 21.2. We assume (see Fig. 21.5) that the drone, pilot operator, and communication system behaviors are concurrent and can be represented by a collection of statechart behavior models. The flight operations will follow a set of states and transitions as the drone travels to and from the mission area. Two modes of flight operation and control are supported: manual (or human-centered) control occurs when the drone is piloted from a remote location. Autonomous operations and control are independent of the remotely positioned pilot and state of the drone-operator communications system and offer advantages for missions of long duration, and/or the area of military interest is large.

At this point, statechart behavior models are hardcoded in Java, which requires developers to work at a very low level of detail. To make the development process less tedious and hopefully more scalable, we are exploring opportunities for embedding a variant of the statechart XML language specification [8] within our systems

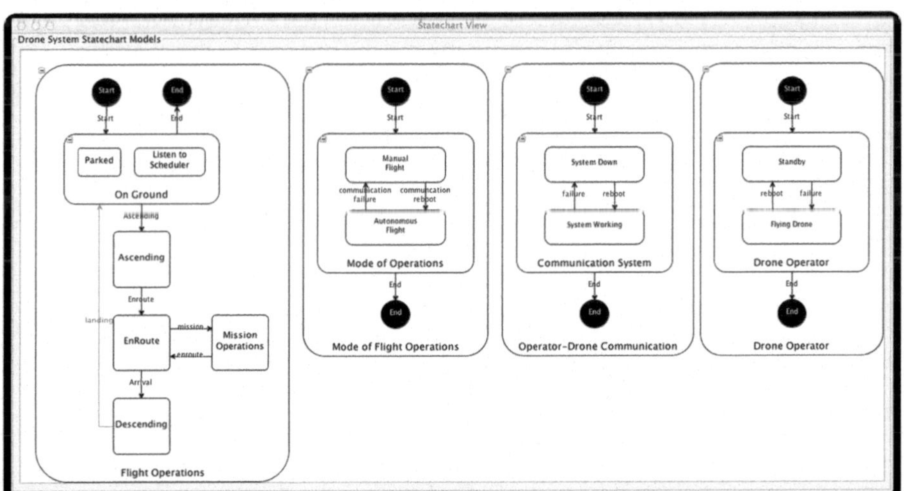

Fig. 21.5 Statechart models for drone, communication system, and human operator behavior

data model specification for components—a drone will be modeled as a component—and then automatically generating the model and visualization for statechart behaviors. We are also extending the Whistle scripting environment [9] so that it can handle the specification and integration of data-ontology-rule models.

21.4.2 Failure of the Communication System

Now suppose that the ontology-drone communication channel is active but then suddenly fails. This event triggers a chain of rules and actions to adjust the flight operations from manual to autonomous operation. Rules and actions act across a multiplicity of domains.

Figure 21.6 shows details of the Jena Rules responsible for the transformation of drone, human operator, and communications semantic models in response to the communications system failure. Figure 21.7 is a snapshot of the corresponding multi-domain evaluation and forward chaining of rules. Rule 01 gathers data from the operator and drone domains and updates the semantic graph to acknowledge the communication channel is active (F3). Then, Rule 02 set parameters in the drone and operator domains for manual operation of the drone. Rule 03 is triggered when the communication system fails—manual operation of the drone is no longer possible—the operator status switches to standby (F9), and the mode of drone operation switches from manual to autonomous (F8). The systems integration of data and rules acting across multiple domains is possible because Jena Rules are capable of working across multiple name spaces (see the prefix declarations at the top of Fig. 21.6). Finally, the actions associated with the activation of these rules can also be pushed to the visualization of statechart behavior, thereby, providing visual feedback and verification for how the system is responding to the communication network failure.

```
@prefix dr:  <http://austin.org/drone#>.
@prefix op:  <http://austin.org/operator#>.
@prefix cm:  <http://austin.org/communication#>.
@prefix rdf: <http://www.w3.org/1999/02/22-rdf-syntax-ns#>.

// Rule 01: Active Communication ....

[ ActiveCommunication: (?x rdf:type dr:Drone) (?y rdf:type op:Operator) (?z rdf:type cm:Communication)
                       (?y op:hasConnection ?z) (?x dr:hasConnection ?z) -> (?z cm:isActive af:True)

// Rule 02: Manual Operations ....

[ ManualOperations: (?x rdf:type dr:Drone) (?y rdf:type op:Operator) (?z rdf:type cm:Communication)
                    (?z cm:isActive af:True) -> (?y op:isFlyingDrone af:True) (?x dr:isInAutonomousFlight af:False)]

// Rule 03: Drone Communication Failure ....

[ NoCommunication: (?x rdf:type dr:Drone) (?y rdf:type op:Operator) (?z rdf:type cm:Communication)
                   (?y op:hasConnection ?z) (?x dr:lostConnection ?z) -> (?z cm:isActive af:False)

// Rule 04: Autonomous Operations ....

[ AutonomousOperations: (?x rdf:type dr:Drone) (?y rdf:type op:Operator) (?z rdf:type cm:Communication)
                        (?z cm:isActive af:False) -> (?y op:isFlyingDrone af:False) (?x dr:isInAutonomousFlight af:True)]
```

Fig. 21.6 Jena Rules for transformation of drone, human pilot, and communication system semantic models

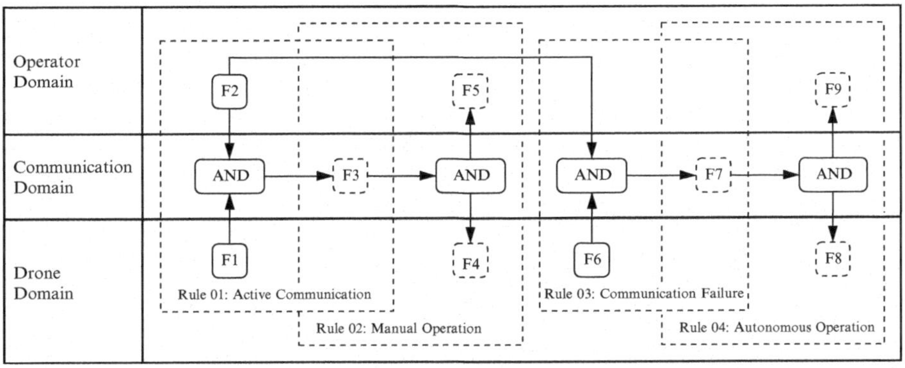

Fig. 21.7 Snapshot of multi-domain evaluation and forward chaining of rules. Legend: F1: Drone001 hasConnection Channel001. F2: Operator001 hasConnection Channel001. F3: Channel001 isActive true. F4: Drone001 isInAutonomousFlight false. F5: Operator001 isFlyingDrone true. F6: Drone001 lostConnection Channel001. F7: Channel001 isActive false. F8: Drone001 isInAutonomousFlight true. F9: Operator001 isFlyingDrone false

21.5 Conclusions and Future Work

The data-ontology-rule footing is a new approach to the development and event-driven execution of semantic models spanning multiple domains. Support for the forward chaining of rules across domains is a powerful mechanism that allows one to develop software that "thinks like a human." Our ongoing research will include development of tools to streamline the specification of data models (structure and behavior) and families of domain and cross-domain ontologies and rules that will be organized into data-ontology-rule modules and test programs.

Acknowledgments This work was supported by grants RT170 and RT195 from the NAVAIR/SERC/DoD on Systems Engineering Transformation.

References

1. Austin, M. A., Delgoshaei, P., & Nguyen, A. (2015). Distributed systems behavior modeling with ontologies, rules, and message passing mechanisms. In *13th Annual Conference on Systems Engineering Research (CSER 2015)*, Hoboken, New Jersey, March 17–19, 2015.
2. Coelho, M., Austin, M. A., & Blackburn, M. (2017). Distributed system behavior modeling of urban systems with ontologies, rules and many-to-many association relationships. In *The 12th International Conference on Systems (ICONS 2017)* (pp. 10–15), Venice, Italy, April 23–27, 2017.
3. Petnga, L., & Austin, M. A. (2016). An ontological framework for knowledge modeling and decision support in cyber-physical systems. *Advanced Engineering Informatics, 30*(1), 77–94.
4. Russomanno, D. J., Kothari, C., & Thomas, O. (2005). Sensor ontologies: from shallow to deep models. In *37th Southeastern Symposium on System Theory* (pp. 107–112). IEEE.

5. Wagner, D. A., Rouquette, N., Bennett, M. B., Jenkins, S., Karban, R., & Ingham, M. (2012). An ontology for state analysis: formalizing the mapping to SysML. In *Proceedings of 2012 IEEE Aerospace Conference*, Big Sky, Montana, March 2012.
6. Fridenthal, S., Moore, A., & Steiner, R. (2008). A Practical Guide to SysML, MK-OMG.
7. Apache Jena. (2016). An Open Source Java framework for building Semantic Web and Linked Data Applications. For details, see https://jena.apache.org/
8. State Chart XML: State Machine Notation for Control Abstraction (SCXML). *Wikipedia*. Retrieved April 13, 2018, from https://en.wikipedia.org/wiki/SCXML
9. Delgoshaei, P., Austin, M. A., & Pertzborn, A. (2014). A semantic framework for modeling and simulation of cyber-physical systems. *International Journal on Advances in Systems and Measurements, 7*(3–4), 223–238.

Chapter 22
Why Not Teaching Systems Architecture as a Studio Art Class?

Alejandro Salado, Tom McDermott, Kirsten Davis, and Alejandra Moral

22.1 Introduction

Systems architecture is both an art and a science [1, 2]. Its scientific side deals with producing actual designs; its artistic one drives the value of the system architecture [3]. Strong ability in both sides is exhibited in effective systems engineers, who possess a good balance of hard and soft skills [1, 4]. However, most of the development and training of systems engineers focuses in the analytical and methodological side. Yet, there is a growing demand for promoting the artistic side and pursing elegance when architecting engineering systems [5–7]. We know how to teach the science, but how should we teach the art? We contend that systems architecture should be taught in a way similar to how the arts are taught, rather than the traditional instructional approaches employed when teaching engineering sciences. In support of this idea, we propose to teach systems architecture mimicking a studio art class.

A. Salado (✉) · K. Davis
Virginia Tech, Blacksburg, VA, USA
e-mail: asalado@vt.edu

T. McDermott
Stevens Instiute of Technology, Hoboken, NJ, USA

A. Moral
Radford University, Radford, VA, USA

S. Adams et al. (eds.), *Systems Engineering in Context*,
https://doi.org/10.1007/978-3-030-00114-8_22

22.2 The Importance of Teaching the Art of Systems Architecting

Teaching systems engineering concepts to undergraduate students and to working professionals provides an interesting insight. Although working professionals nod and share *aha* moments by contextualizing the concepts to their previous experiences, undergraduate students usually react with a *so what?* attitude. For undergraduate students, systems engineering concepts seem obvious. In fact, some have even argued that systems engineering is just good engineering or even just commonsense engineering [8]. We do not go that far necessarily but suggest that systems engineering concepts are generally easy to grasp. Systems engineering becomes difficult (very difficult) when diving into its practice. This is a departure from traditional engineering sciences. In professional practice, applying electrical theory or doing structural calculations gets *simpler* by applying codes, substituting model sophistication with safety margins, or using software (among other methods). In systems engineering however, the student is usually lost when trying to apply the concepts that appeared to be so obvious earlier. As complexity grows, they can often lose interest in the broader purpose of the system and its architecture, leading to poor decisions.

Although in this paper we focus on systems architecture, this applies to various aspects of systems engineering. For example, verification and validation have been defined in a fantastic and simple way. Verification indicates if the product is built right and validation if the right product was built. These concepts are easy and deep at the same time. Yet most systems engineers are unable to distinguish between verification testing and validation testing. One only needs to look at V&V standards in industry [9, 10]. Engineers exhibit similar behavior when writing engineering requirements. The guidelines for writing good requirement seem obvious and easy to follow. A requirement must be quantified, must enforce solutions, etc. Yet most requirement sets out there are of very poor quality [11–13].

We have seen the same pattern when teaching systems architecture. For sure, heuristics are easy to grasp, for example, "simplify, combine, and eliminate," "system structure should resemble functional structure," and "group elements that are strongly related; separate elements that are unrelated" [2]. Concepts are not sophisticated, but a heuristic without appropriate foundational knowledge and practice is just a set of words. We preach "simplify" but the "art" of simplification is seldom taught in engineering. We discuss abstraction, interfaces, function/component, and assessment, among others. Yet, students have trouble in the creational aspect. In particular, they struggle with applying those concepts and using those heuristics to architect effective solutions to problems. Therefore, we suggest that teaching systems architecture must focus more strongly on conveying its art.

22.3 Why Mimic a Studio Art Class

Our proposal to mimic a studio art class to teach systems architecting builds on three key ideas. First, artists and systems architects need to develop similar competencies. Second, systems architectures are created for others. Third, the art of systems architecture must be developed through experience.

22.3.1 Similar Competencies for Architects and Systems Architects

Previous work has explored the relationship between systems engineering and artistic endeavors. The principles and practices of systems engineering to cope with complexity and achieve harmony in the solution seem to be exhibited in the creation of original film scores [14]. The techniques that master painters use to convey beauty seem to match those that systems architects use to architect effective systems [15]. This has led the first two authors of this paper to map competencies that students need to learn in the arts and in systems architecting [16]. They are listed in Table 22.1.

When evaluating the necessary competencies in the arts, we reach a similar conclusion to our analysis of systems architecting. That is, several concepts are simple, yet their application to create something meaningful is very difficult. For example, for understanding how different colors interact, one *only* needs to use a color wheel. Similarly, there are several rules that help in composing the frame or scenario. Common ones include the golden ratio, the rule of thirds, or the leading lines. However, just a few people are able to use them in a way to create something beautiful, let alone something that transcends.

As a result, practice is a key educational method in the arts. In fact, a person would hardly be presented as an artist because he/she knows composition rules and color patterns. Someone will be called an artist when showing a portfolio of their work, their originality, and their art. We believe that a similar stance should be taken for architects. Knowing systems architecture heuristics and concepts should not be sufficient. The quality of an architect must be measured on the architectures he/she creates. Therefore, this similarity informs our suggestion to teach systems architecture in a studio art setting as opposed to how it is traditionally taught.

22.3.2 Architecting Is for Others

Several of the competencies listed in Table 22.1 can be gained through traditional instruction. To the best of our knowledge, we (as a community) teach systems engineering through lectures, case studies, individual assignments, and group projects. For sure, they are all helpful. Case studies can help assimilate failures and

Table 22.1 Common learning concepts spanning art and systems architecture [16]

Learning concepts	Art	Systems architecting
Abstraction	Understanding symmetry between internal context and observed system	Understanding symmetry between business context and observed system
Precedence	Interpreting the masters' unprecedented work	Studying unprecedented architectures
Time	Representing movement	Attributes of change
Heuristics and patterns	Learning across the arts	Learning across management and engineering
Decomposition, recomposition	Lists and groups, split, subtract details, symmetries, aesthetics	Elements and interfaces, attributes, objectives (requirements) flow down and derivation, integration sequences and verification coverage, purpose
Boundary setting	Observing, framing, scene-setting	Heuristics, context, external interfaces, problem formulation, ConOps
Simplifying	Subtract details, split	Determining abstraction levels, lumping or splitting components
Synthesizing	Remix and reconnect	Emergence, partitioning, integration, generic architectures, architectural frameworks
Focusing	Emphasize, power of the center, contrast and balance	Views and viewpoints, centralization/networked
Communicating	Aesthetics, color/musical palette, frame/viewpoint, title/description, patterns	Domain knowledge, standards, views and viewpoints, documentation/metadata
Analytical competencies	Domain-driven use of materials and media	Domain-driven technical and business analysis
Methodological and executional competencies	Creativity, use of processes and patterns, experimentation and risk taking, ability to engage and keep audience focus	Creativity, use of processes and patterns, use of frameworks, planning, managing, ability to engage and facilitate agreement
Learning outcomes	Understanding of composition, inventive design, risk taking, creating expression, breadth of work, individualized transformation of concepts	Composition of operational and technical design, methods and execution, planning, breadth and depth, group communication of transformation

successes of using certain heuristics and approaches to architecting. Lectures may be effective to convey key concepts and share the experiences of the instructors. However, these instructional approaches do not let the student apply that knowledge to create architectures. They must be complemented.

A traditional way to do so is to assign individual and group projects to the students. With individual assignments, students are forced to apply the concepts they learnt to a number of problems. With group assignments, students collectively create an architecture. Both approaches suffer from two weaknesses, which we discuss here.

The first one has to do with exposure. For sure, trying to architect systems is a necessary effort to become a systems architect. However, the student receives feedback from the instructor on his/her work. The student loses the opportunity to receive feedback from and provide feedback to his/her peers. Most importantly, we believe, students lose the opportunity to compare and contrast how other engineers may face and solve the same problem differently. We suggest that understanding strengths and weaknesses of different architectures for solving the same problem is key to grow as a systems architect.

The second key weakness relates to the success criteria associated with the assignment. Lacking an authentic context, the student solves the problem for him-self/herself and for the instructor. That is, the student's intent is to create a solution that the architect believes successful and in such a way that the instructor will be able to interpret it. However, a key mission of systems architecture is to stimulate communication, align mental models, and seek feedback that informs a new iteration on the architecture [17]. In fact, that feedback is essential to become a successful systems architect. Specifically, it allows for gaining experience in understanding multiple perspectives to a problem. We suggest that the student loses again an opportunity, with traditional instructional approaches, to build up an ability to think about the architecture he/she creates from diverse perspectives.

Note that being exposed to multiple perspectives, as discussed in relation to the second weakness of traditional instruction, differs from the exposure to different solutions we discussed for the first one. Diversity of perspectives has to do with how different stakeholders may look at or use the architecture. For sure, group assignments offer the possibility of generating feedback. However, all team members work toward the same purpose and under the same role/viewpoint: they are all architects. This difference in perspective or viewpoint is essential to effective architecting. For example, one of the outcomes of creating a systems architecture is the set of objectives or requirements for the lower-level components. Defining objectives or writing requirements for self-understanding is easy. Criticality in flowing down and deriving objectives and requirements lays on the understanding and interpretation that other people will do with them. Without experiencing those interesting interactions (e.g., unconscious misinterpretation, deliberate misinterpretation, misalignment of assumptions), it is harder for the student to learn and deeply understand (and feel) the reach of the consequences of the architecture that has been created for the system under development.

We found in the competencies required for an artist that similar needs with respect to interpretation by external observers are key to become successful. In fact, the purpose of art is to convey experiences and emotions. It is understood as a bi-directional communication from the artist to the consumer of art. This similarity informs our suggestion to design the instructional approach for systems architecture around the studio art concept. Systems architecting might be considered a process of aligning the architects' internal viewpoints, external stakeholder viewpoints, and the business and technical viewpoints of the architecture itself [15]. Without the ability to recognize and practice all three, the quality of the architecture will likely suffer.

22.3.3 To the Art Through Experience

Creativity and innovative thinking are increasingly recognized as skills that engineering students need to develop [18–20]. Engineering education researchers have identified specific sub-skills that contribute to creativity in engineering work, including observation, big-picture thinking, problem finding, divergent thinking, and iterative experimentation [19, 21–26]. They have further explored methods for developing these skills in engineering students, with variable results. Competitions [24], collaborative activities [23–25], open-ended projects [19, 23, 27], and feedback from peers and mentors [26, 28] have all been suggested as methods for improving student creativity. A limited amount of research has indicated that teaching ideation strategies [27, 29], incorporating reflection [26, 28, 29], and interdisciplinary work [24] can help in developing these skills. On the other hand, some research has indicated that the traditional experience in engineering education might actually reduce the level of creativity in designs produced by students. Two studies found that designs by senior students were less innovative (although more practical) than those of first-year students, particularly on more open-ended tasks [27, 30]. Industry feedback has also suggested that engineering students are not graduating with sufficient creativity [19].

There are limited examples of using the approach we are describing in an engineering context. There is an increasing acknowledgment that engineering students could benefit from a more artistic perspective [31, 32]. Some schools have responded to this call by introducing interdisciplinary courses where engineering and art students work together on a project (e.g., [31, 33]). Such classes are reported to help students think outside the bounds of their own discipline but can often be a challenge to set up administratively and coordinate teaching across multiple departments in a way that places all disciplines on equal footing [31]. Another approach has been to require upper-level engineering students to work on a studio art project with art faculty members and reflect on how their art experience connects to their engineering work [32]. In this case, the engineering students began to see clear connections between the creative processes used in their art projects to their engineering work. Closer to our proposal, software engineering has begun to introduce "studio" courses where students work on their projects in class and have opportunities to give and receive feedback with peers [34, 35]. They report that such a course improves students' reflective practice and ability to give and receive feedback. Similar approaches are now being adopted in other disciplines as well (e.g., [36, 37]). However, the courses reported so far have attempted to align only their lesson structure and (in a few cases) classroom setting to that used in traditional studio art courses. In our proposal, we are looking to take the idea of a studio further to consider not only the lesson structure but also the classroom setting and course structure as a whole.

In previous work, we presented a general framework for engineering competency development relating the analytical and methodological skills traditionally taught in the sciences and engineering to development of conceptual abilities (heuristics, precedence, abstraction) linked to systems thinking and compositional capabilities (simplification, synthesis, focus, communication) linked to design [16, 17]. Participation in the arts should enhance the compositional abilities of the engineer, as many of the fundamental composition rules are transferrable across different domains of the arts, architecture, engineering, and computer science. Incorporation of these composition rules in student project artifacts focused on communicating architectures (such as operational views) helps to improve their communication skills and even their systems thinking abilities [17]. However, more is needed to establish the foundations and competencies of architects as "composers"—we should teach systems architecture as a studio art class.

22.4 A Concept to Teach Systems Architecture as a Studio Art Class

22.4.1 A Studio Art Class in Art

We describe in this section the typical setting and a structure for a studio art class. We build the example on elements of two semester-long undergraduate courses in arts at Radford University, one in drawing and one in figure drawing. Both courses employed traditional instructional approaches in studio art settings. We discuss three aspects of the course: the classroom setting, the structure of the lessons, and the structure of the course.

22.4.1.1 Classroom Setting

The class takes place in a large space. The objects or models to be drawn are placed in the middle of the room, and the students sit surrounding them. This implies that while all students look at the same point, they look at it from different perspectives (viewpoints) physically, as well as emotionally by convoluting their inner self with the scenario they observe. The room has walls that are set to hang the drawings made by the students. All drawings can be hanged on the same surface.

22.4.1.2 Lesson Structure

The lesson is structured in three blocks:

1. *Lecture*: The instructor explains techniques, ideas, and guidelines as they relate to the objective of the lesson.
2. *Practice*: Students spend time applying the techniques explained in class on a model the instructor defines. The instructor provides feedback to students individually as they practice their drawing. This phase allows for a deeper interaction

between the instructor and the student in relation to the concepts and techniques explained in class. Furthermore, the frequent feedback allows for iterating the student's work as it is created.
3. *Exposition and critique*: Toward the end of the lesson, the works of all students are exhibited together. Students then critique each piece of work, comparing them holistically and reflecting on what each person captured from the model, as well as how well they accomplished it.

Exposition is actually a key element in the studio art setting. By making students share what they see in each other's work, the student becomes aware of how other people interpret his/her own work. In other words, students are enabled to understand the level of alignment between what they wanted to convey through their work and what other people interpreted from it.

22.4.1.3 Course Structure

The overall course is structured in such a way that techniques and skill build step by step. Specifically:

1. *Draw basic forms*: The student is able to capture basic forms such as cubes, pyramids, or spheres.
2. *Draw objects*: The student is able to capture objects that combine multiple basic forms, such as vessels, caps, or fruits.
3. *Draw a person from a model*: The student is able to draw the human body. While one could think that there is no difference between drawing objects and drawing a human conceptually (both are divisible to basic forms), there is a fundamental difference wired to our brains. Apparently, the human brain has a high level of accuracy and precision in differentiating human faces and eyes. Hence, while a regular person would not be able to detect a given variation between a fruit and a drawing of it, they would be able to distinguish between a face and its drawing. Thus, there is a level of added granularity in the way a face needs to be captured that is sufficient to warrant a dedicated learning step.
4. *Draw a person from imagination*: The student is able to create the drawing of a person from imagination, without looking at a model. This step is divided in two sub-steps:
   (a) *Learn patterns*: The student learns certain patterns to abstract reality, such as those expressed through a mannequin made of geometric figures.
   (b) *Apply imagination to patterns*: The student embellishes the basic figures in the patterns until the work captures the detailed aspects that make the pattern vanish and the imagination emerge.
5. *Draw a story from imagination*: The student is not only able to create a person from imagination but to place the person within an imaginary context to create a story, a narrative. This is the ultimate purpose of art: it tells a story to convey emotions.

22.4.2 Our Proposal: A Studio Art Class in Systems Architecting

We describe in this section our planned approach to organize a course in systems architecture within a studio art class setting. Using the previous section as a paradigm, we present the classroom setting, the structure of the lessons, and the structure of the course.

22.4.2.1 Classroom Setting

We propose to tailor the studio art class setting to capture the specific characteristics of systems architecting problems. The object of study is the problem for which students need to create an architecture. Therefore, lacking a physical form, placing the problem at the center of the room may not seem especially important. However, using a center in the class enables placing each student at the same distance from the problem, which, at least in the traditional studio art class, facilitates student engagement with the object. Therefore, we suggest that such a structure, the problem at the center and students surrounding it, is maintained when teaching systems architecture. Furthermore, we also suggest enabling the simultaneous visualization of all solutions (architectures). This has two caveats. The first one is that architecture views are created in the computer and, therefore, the room needs the capability to display all architectures at a sufficiently large size. The second one is that architectures have multiple dimensions. Evaluating an architecture is a dynamic activity, not a static one like the artistic drawings we have used as a paradigm in the previous section. This creates a different dynamic of interaction between students and their solutions. Hence, an ideal classroom setting should count with multiple large displays that enable interaction.

22.4.2.2 Lesson Structure

We plan to use the same lesson structure as in the arts class:

1. *Lesson*: The instructor explains basic concepts such as architecting heuristics, abstraction, or interfaces.
2. *Practice*: Students spend time applying the concepts to architect a solution to a given problem. The instructor provides feedback to students individually as they practice their architecting skills. This phase allows for a deeper interaction between the instructor and the student in relation to the concepts and techniques explained in class. For example, the instructor can tailor his/her experience much more to the specific effort of the student in learning the material. Furthermore, the frequent feedback that allows for iterating the student's architecture is developed, increasing the student's exposure and understanding of perspectives/viewpoints.

3. *Expositions*: Toward the end of the lesson, students share their work and compare it holistically. Students are then asked to find strengths and weaknesses in the architectures of their peers. This enables students to gain exposure on how different solutions could have worked. It also helps them learn about different interpretations from diverse viewpoints about the solution idea that he/she tried to convey through the models employed to capture the architecture.

As was the case for the arts class, we also find the exposition to be a key aspect of teaching architecting. Only by facilitating those exchanges, students can gain experience in aligning mental models and can become aware of the gaps their mental models can create when architecting systems.

22.4.2.3 Course Structure

When defining the course structure, we aimed at capturing the staged development of skills present in the drawing course described in the previous section. In particular, we plan for four main developmental steps:

1. *Learn forms*: The student is able to capture basic systems architecture elements such as functions, behaviors, or interfaces, in various architectural frameworks.
2. *Learn of objects*: The student is able to capture basic architectural blocks, such as a command and control functionality/mechanism, a communication functionality/mechanism, or a security protection functionality/mechanism.
3. *Capture the architecture of an existing solution*: The student is able to capture the underlying architecture of an existing system.
4. *Create an architecture*: The student is able to create an effective architecture for a given problem. We have divided this step into three sub-steps:
    (a) *Generic architectures*: The student is able to recognize traditional solutions to common problems, such as current architectures for automobiles, computers, or airplanes.
    (b) *Architectural patterns and heuristics*: The student is able to recognize architectural patterns and heuristics as a function of the problem characteristics and desired solution effectiveness.
    (c) *From blank to solution*: The student is able to apply and tailor all techniques to create effective architectures to unprecedented problems.
5. *Depict a system principle representing a person and an engineered system*: The student is able to depict an architecture with intent. Some examples include "represent both centering and scale," "represent both focus and diversity," "visualize boundaries," "visualize flow (movement)," "visualize emergence (time)," and "visualize nonlinearity (feedback)."

22.5 Future Plans

We have presented in this paper a concept to teach a systems engineering course in a studio art format. The idea builds upon previous work that has identified parallelisms between the practices and learning outcomes of systems architects and artists.

We plan to pilot test the proposed class format during the Fall 2018 semester in an introductory graduate course in systems engineering at Virginia Tech. We additionally plan to study the impact of this change in pedagogy on the outcomes of the course. The instructor who will be teaching the course in Fall 2018 has taught the course in two previous semesters to around 60 students using a traditional format. We will use the final projects from the previous semesters as a baseline to which we can compare the final work of the students in the newly designed course. To do a formal comparison, we propose to develop rubrics to connect to the developmental steps of systems architecting described above and use these rubrics to score final projects from both the previous courses and the Fall 2018 course. This will allow us to identify developmental differences in students' systems architecting competence that may be connected to our proposed changes in pedagogy.

References

1. Ryschkewitsch, M., Shaible, D., & Larson, W. J. (2009). The art and science of systems engineering. *Systems Research Forum, 03*(02), 81–100.
2. Maier, M. W., & Rechtin, E. (2009). *The art of system architecting*. Boca Raton, FL: CRC Press.
3. Muirhead, B. K., & Thomas, D. (2010). The art and science of systems engineering tightly coupled programs. *SAE International Journal of Passenger Cars—Electronic and Electrical Systems, 3*(2), 117–130.
4. Jansma, P. A. T. (2012). Exploring the art and science of systems engineering. In *IEEE aerospace conference*. Big Sky, MT, USA.
5. Griffin, M. D. (2010). How do we fix systems engineering? In *61st international astronautical congress*. Prague, Czech Republic.
6. Salado, A., & Nilchiani, R. (2013). Using Maslow's hierarchy of needs to define elegance in system architecture. *Procedia Computer Science, 16*, 927–936.
7. Madni, A. M. Elegant system design: Creative fusion of simplicity and power. *Systems Engineering, 2013, 15*, 347–354.
8. Akeel, U. U., & Bell, S. J. (2013). Discourses of systems engineering. *Engineering Studies, 5* (2), 160–173.
9. ECSS. (2009). *Space engineering - verification*. Noordwijk, The Netherlands: European Cooperation for Space Standardization.
10. NASA. (2007). *Systems engineering handbook.*
11. Salado, A., & Nilchiani, R. (2014). A categorization model of requirements based on max-neef's model of human needs. *Systems Engineering, 17*(3), 348–360.
12. Salado, A., & Nilchiani, R. (2017). Reducing excess requirements through orthogonal categorizations during problem formulation: Results of a factorial experiment. *IEEE Transactions on Systems, Man, and Cybernetics: Systems, 47*(3), 405–415.
13. Hooks, I. F. (2010). Why Johnny still can't write requirements. In *22nd annual SSTC conference*. Salt Lake City, UT, USA.

14. Salado, A., & Salado, C. (2015). Systems engineering practices exhibited in the creation of a film original score. *INCOSE International Symposium, 25*(1), 1147–1158.
15. Salado, A., Iandoli, L., & Zollo, G. (2016). Painting systems: From art to systems architecting. *INCOSE International Symposium, 26*(1), 773–787.
16. McDermott, T., & Salado, A. (2017). Improving the systems thinking skills of the systems architect via aesthetic interpretation of art. *INCOSE International Symposium, 27*(1), 1340–1354.
17. McDermott, T., & Salado, A. (2018). Art and architecture: Effectively communicating models of systems. In *2018 Annual IEEE international systems conference (SysCon)*. Vancouver, Canada.
18. National Academy of Engineering. (2004). *The engineer of 2020: Visions of engineering in the new century*. Washington, DC: National Academies Press.
19. Cropley, D. H. (2015). Promoting creativity and innovation in engineering education. *Psychology of Aesthetics, Creativity, and the Arts, 9*(2), 161–171.
20. Rover, D. T. (2005). New economy, new engineer. *Journal of Engineering Education, 94*(4), 427–428.
21. Charyton, C., & Merrill, J. A. (2009). Assessing general creativity and creative engineering design in first year engineering students. *Journal of Engineering Education, 98*(2), 145–156.
22. Duval-Couetil, N., & Dyrenfurth, M. (2012). Teaching students to be innovators: Examining competencies and approaches across disciplines. *International Journal of Innovation Science, 4* (3), 143–154.
23. Raviv, D. (2008). Innovative thinking: Desired skills and related activities. In *2008 ASEE annual conference and exposition*.
24. Raviv, D., Barak, M., & VanEpps, T. (2009). Teaching innovative thinking: Future directions. In *2009 ASEE annual conference and exposition*.
25. Raviv, D., & Barbe, D. (2010). Ideation to innovation workshop. In *2010 ASEE annual conference and exposition*.
26. Kershaw, T. C., Hölltä-Otto, K., & Lee, Y. S. (2011). The effect of prototyping and critical feedback on fixation in engineering design. In *CogSci '11*.
27. Genco, N., Hölltä-Otto, K., & Seepersad, C. C. (2012). An experimental investigation of the innovation capabilities of undergraduate engineering students. *Journal of Engineering Education, 101*(1), 60–81.
28. Green, G., & Kennedy, P. (2001). Redefining engineering education: The reflective practice of product design engineering. *International Journal of Engineering Education, 17*(1), 3–9.
29. Pappas, E. (2009). Cognitive-processes instruction in an undergraduate engineering design course sequence. In *2009 ASEE annual conference and exposition*.
30. Lai, J. Y., Roan, E. T., Greenberg, H. C., & Yang, M. C. (2008). Prompt versus problem: Helping students learn to frame problems and think creatively. In *Third international conference on design computing and cognition*.
31. Sochacka, N. W., Guyotte, K. W., Walther, J., & Kellam, N. N. (2013). Faculty reflections on a STEAM-inspired interdisciplinary studio course. In *2013 ASEE annual conference and exposition*.
32. Beams, D. M., Gullings, K., & Ross, C. E. (2016). Seeking new perspectives: Engineers experiencing design through creative arts. In *2016 ASEE annual conference and exposition*.
33. Shooter, S. B., & Orsborn, S. (2013). "Impact! Exploring innovation across disciplines" - Engaging the university innovation ecosystem through a university-wide course. In *2013 ASEE annual conference and exposition*.
34. Bull, C. N., & Whittle, J. (2014). Observations of a software engineering studio: Reflecting with the studio framework. In *IEEE conference on software engineering education and training*.
35. Bull, C. N., & Whittle, J. (2014). Supporting reflective practice in software engineering education through a studio-based approach. *IEEE Software, 31*(4), 44–50.
36. Thompson, B. E. (2002). Studio pedagogy for engineering design. *International Journal of Engineering Education, 18*(1), 39–49.
37. Chance, S. M., Marshall, J., & Duffy, G. (2016). Using architecture design studio pedagogies to enhance engineering education. *International Journal of Engineering Education, 32*(1), 364–383.

Chapter 23
A Research Plan to Discover Patterns of Unknown Vulnerabilities Associated with Adopting SysML

Paul Wach and Alejandro Salado

23.1 Introduction

Terms such as model-based systems engineering (MBSE), digital thread, or digital twin are gaining momentum in the research community, government, and industry [1–3]. All these initiatives are built under a central idea: system development is supported by a single model that represents the system of interest [1, 4]. Several benefits are anticipated: improved communication and understanding among engineers (and with other stakeholders) [5], faster verification and validation cycles [6], and predictive analytics to anticipate and overcome unforeseen operational problems [7], among others. However, converging to a single model may have detrimental unintended consequences that must be understood and characterized.

Models are abstractions of systems to be developed and realized [8]. Models are inherently unable to capture all characteristics of the system they model, which has led to research to reduce model approximation [9]. In fact, even a sophisticated digital twin will omit certain aspects that its twin real system will realize. These omissions may become sources of potential system vulnerability: you cannot protect against what you do not know; and you do not know what you have not modeled, such as an unplanned system feature [10]. We hypothesize that converging to a central, core model results in specific design patterns (inherent to the underlying modeling language), which are not able to capture certain realization patterns that enable the emergence of unknown vulnerabilities. In other words, converging to a single model may reduce coverage in the identification of system vulnerabilities, even if the model is sophisticated.

In this paper, we present a research plan to study if SysML inherently creates patterns of unknown vulnerabilities, once the systems it models are realized. We

P. Wach (✉) · A. Salado
Virginia Tech, Blacksburg, VA, USA
e-mail: paulw86@vt.edu

S. Adams et al. (eds.), *Systems Engineering in Context*,
https://doi.org/10.1007/978-3-030-00114-8_23

have chosen SysML due to its widespread and high level of adoption within academia and the community of practice.

23.2 Inherent Limitations to Modeling

23.2.1 All Models Are Wrong, Some Are Useful

It is widely accepted that all models contain errors [11], as previously introduced. The purpose of modeling is hence to capture only some aspects of reality sufficiently well, so that the model can predict certain aspects of the modeled system, with certain level of confidence [1]. Therefore, every model presents two main limitations. One, resulting from the error associated to the model sophistication, that is, to the fidelity that can be achieved with the particular modeling effort. The second one, resulting from the purpose to which the model is created. That is, there are always certain elements of reality that are purposefully not modeled.

23.2.2 Limitations Imposed by the Modeling Language

The limitations in the previous section are controlled by the modeler. However, the modeling language adds an inherent limitation to the model. Essentially, modeling languages are unable to capture reality completely, irrespective of the amount of effort and sophistication that is invested in the modeling effort. We use *language* to further strengthen this argument and articulate the emergence of patterns of features inherent to the modeling languages.

Language, understood as the system that enables communication between humans, facilitates the creation of models of our thoughts and feelings [12]. Research in linguistics has shown that there are ideas, concepts, and thoughts that can be expressed (captured, modeled) in one particular language, but cannot be translated to other ones. More specifically, there is a loss of information and intent when they are translated. For example, there is not an exact translation for the English word "disgust" in Hindi or Malayalam [13]. Such limitation is not controllable by the speaker (the modeler), but inherent to the language that is employed.

In addition, research in linguistics has shown that language is also responsible for the emergence of communication patterns [14]. For example, the syntax of the German language is such that the verb (the action) is often placed at the end of the sentence. As a result, communication in German tends to be slow (a speaker needs to find the specific terms that capture his/her ideas and thoughts) and without interruption (the intent of the sentence is only really unveiled at the end of the sentence) [15]. Spanish, on the contrary, leverages flexibility in syntax and its richness in synonyms to capture differences in thought and intention [16]. Therefore, given our understanding that language is *just* another modeling language, we contend that any modeling language used in engineering will also exhibit certain modeling patterns that are inherent to the language, beyond the control of the modeler.

23.3 Potential Vulnerability Threats of Adopting a Core, Central Model

23.3.1 Every Solution Creates New Problems

The notion that every solution creates new problems is an important element of the systems thinking folklore, although we hope the new solution is more useful [17]. There are indeed several instances where the solution to a problem has created worse problems than the original problem they were trying to solve. An example includes the unintended consequences found with health information technology (HIT), which have been noted to create technology-induced errors [18]. While there are positive intended consequences of HIT, an overreliance on the technology has created the unintended consequences of errors of both omission and commission such as ordering of malapropos medication [18].

However, the aspect of unintended consequences is not only limited to those that are worse than the initial problem. Some solutions create unintended consequences that, although detrimental, are still less severe than the causes that led to needing a solution in the first place. For example, there have been aircraft failures associated with autopilot, yet improvements have been made on prevention such as found with Loss of Control study conducted by the National Aeronautics and Space Administration (NASA) Aviation Safety Program [19]. At any rate, the main purpose is that the solution does not create worse problems than the original ones, not necessarily that it does not create them.

As with any system, methodological advances also come at a *price*. Consider, for example, the advent of sophisticated computation with microelectronics. There is no question that the benefits of such computation ability have been unprecedented—we are not going back to slide rulers. Yet, beneath the impressive things we can achieve thanks to contemporary computers, their wide adoption has also come with a few unintended drawbacks. Discussing with some of our more senior colleagues, it seems that junior engineers have largely lost the feeling for understanding precision. When computing was costly, engineers needed to decide carefully if it was worthy to get one more decimal. Now, at a click you get an amazing precision free. And this leads to engineers not being able to contextualize results. For example, a civil engineer designs a road to millimeter precision. Have they seen the construction trucks working on the road?

These experiences lead us to suggest that in transitioning to MBSE, integrated modeling, or digital twins, we should not be simply convinced by their anticipated benefits. We believe that they will surpass their possible drawbacks. However, we also contend that we must actively understand the potential unintended consequences they may generate so that we can take action to control them. As discussed, we focus this paper on the emergence of unknown system vulnerabilities.

23.3.2 You Cannot Protect What You Don't Know

In essence, vulnerability assessments consist in looking at potential security holes in the system under development. This should be done before the system exists, so that protection mechanisms to those security holes can be implemented to mitigate the vulnerabilities, remove them, or monitor them. Models are necessary to accomplish such effort. Three main dimensions are worth considering. First, the functional model or architecture. That is, an attacker may affect a function to change or alter the behavior of the system. Second, the physical model or architecture. The attacker may directly affect the physical form of the system; or may use it to access the functionality to be affected. Third, the contractual or organizational model. The attacker may affect the system by accessing the interactions between the different organizations that are developing the system.

Whether the vulnerability extends from an intended attack or an unintended consequence of the way the model was constructed, the importance of evaluating the risk is paramount. As an example of unintended consequences, the NASA Office of Safety and Mission Assurance has sponsored research on model-based assurance. With the use of diagnostic tree verification (DTV) to test procedure for lighting failure in the NASA Habitat Demonstration Unit (HDU), a wide range of vulnerabilities was discovered in their model. The vulnerabilities in the model included missing procedure, inconsistent level of detail; and undocumented, missing, or invalid assumptions [20].

For malicious-intended attacks, the Pacific Northwest National Laboratory (PNNL) has worked to understand the vulnerability of blended attack; between physical only, cyber only, physical-enabled cyber, and cyber-enabled physical attacks. The study relied on three levels of assessment, one by subject matter expert (SME) review, a high-level quantitative modeling, and finally detailed modeling. The SME review was found to be useful for immature, newly developed systems. However, even the more in-depth modeling requires expansion of the research to protect against vulnerabilities [21].

Hence, the fidelity of the models is essential to guarantee an adequate identification, and hence protection, of the vulnerabilities of a system. However, as discussed before, models have inherent limitations. Therefore, uncontrolled interfaces, characteristics, behaviors, or interactions will emerge when realizing a real-world system out of a model. Therefore, by definition, every system has a set of unknown vulnerabilities that result from the abstractions that models create.

23.3.3 A Single Model May Reduce Coverage of Vulnerabilities

Consider the assumption we made previously that any modeling language inherently yields modeling patterns. Since the realization of a system is a transformation from a model into a real-world system, we contend that the real-world system will exhibit patterns that are caused by the modeling patterns from which it has been realized. Some of those patterns may be of unknown vulnerabilities. Therefore, there is potentially a linkage between a modeling language and patterns of unknown vulnerabilities in systems that are realized from such modeling language. We believe that this does not occur when multiple modeling languages are used, because they act as a kind of randomizer.

This linkage yields a critical security threat for systems that rely on an integrated modeling approach or a digital twin. Imagine an attacker focuses on *hacking* or understanding the vulnerabilities that stem from the modeling language. Then, an attacker can envision a vector attack that is *independent* of the system to be attacked. In other words, all systems developed with a given modeling language may be vulnerable to such attack vector. Furthermore, because the modeling language is open, the vector may be generated without knowing the specifics of the system under attack. Protection against this type of vulnerabilities needs an understanding of the fundamental behavior of the modeling language. This has seeded our interest to pursue the work planned in this paper.

23.4 A Research Plan for Discovering Patterns of Unknown Vulnerabilities in SysML

23.4.1 Research Goal and Overview

The research for which a plan is presented here has the goal to discover patterns of unknown vulnerabilities that are inherent to modeling using SysML. The research consists of two main activities. First, we will explore the emergence of unknown vulnerabilities during the realization of systems using a formal mathematical framework. Second, we will use the mathematical framework to identify patterns of unknown vulnerabilities associated to the modeling patterns inherent to SysML.

It should be noted that we present a general plan of the research approach and that several questions about its actual implementation remain open at this time.

23.4.2 Formal Exploration

We will use Wymore's mathematical framework [22] to explore the mathematical properties associated to transforming models into real-world systems. The basic structure for the mathematical elaborations is formed on the definition of system and the concept of homomorphism. Adapted definitions are given below.

Definition 1 (Adapted) A discrete system is a quintuple$z = (SZ, IZ, OZ, NZ, RZ)$, where z is the name of the system, SZ is the set of its states, IZ is the set of its inputs,

OZ is the set of its outputs, NZ is its next state function, and RZ is its readout function that specifies the outputs for each state.

Definition 2 (Adapted) The system z_1 is a homomorphic image of the system z_2 with respect to a set of inputs $I_2 \subseteq IZ_2$, a set of outputs $O_2 \subseteq OZ_2$, and a set of states $Q_2 \subseteq SZ_2$ if and only if:

1. There exists a surjection $hi : I_2 \rightarrow I_1$, where $I_1 \subseteq IZ_1$,
2. There exists a surjection $ho : O_2 \rightarrow O_1$, where $O_1 \subseteq OZ_1$,
3. There exists a surjection $hq : Q_2 \rightarrow Q_1$, where $Q_1 \subseteq SZ_1$,
4. $hq(NZ_2(x, i)) = NZ_1(hq(x), hi(i)), \forall x \in Q_2, i \in I_2$,
5. $ho(RZ_2(x)) = RZ_1(hs(x)), \forall x \in Q_2$.

A model of a real-world system is a homomorphic image of the latter. As indicated in the definition of the concept, a model captures a subset of the transformations that the real-world system will perform. Therefore, a model is actually a homomorphic image of an infinite number of systems. We will therefore explore this problem by looking at divergent patterns that derive from the model itself, as well as convergent patterns that result from a large amount of different real-world systems. Once the patterns have been found, we will classify them as vulnerable or free of vulnerabilities. Vulnerable patterns will be those in which the real-world system allows for an input trajectory that results in a behavior that is different from the one predicted by the model. Patterns free of vulnerabilities will be those in which, although they may exhibit behavior in addition to that predicted by the model, the behavior predicted of the model is also exhibited for the system for all input trajectories.

23.4.3 *Identify Vulnerability Patterns*

We expect the previous activity to provide insights on the behaviors in relation to vulnerability patterns to which modeling languages may lead. This activity centers on identifying those specific to SysML. We will collect various SysML models from different modelers in order to randomize the effect of the modeler. We plan to verify their consistency and adequacy by an independent panel of experts in SysML. We propose the following process.

23.4.3.1 Transform SysML to Wymore's Framework

The SysML model will be transformed into Wymore's mathematical models. In particular:

- *Sequence diagrams* will be transformed into sets of inputs and input trajectories, sets of outputs and output trajectories, and potentially modes.

- *State machine diagrams* will be transformed into sets of inputs, sets of outputs, states, and next state and readout functions.
- *Activity diagrams* will be transformed into sets of inputs and input trajectories, sets of outputs and output trajectories, and potentially modes.
- *Block definition diagrams* will be transformed into input and output ports.
- *Internal block diagrams* will be transformed into input and output ports, and into system couplings (specifically coupling recipes).

The resulting mathematical description of the system will capture at once all relationships and aspects of the different models in SysML. This enables us to explore patterns emerging not only from the different types of models that SysML can produce, but also those emerging from their connections. The remaining SysML diagrams do not contribute to the definition of the system structure. Therefore, they are not considered relevant for vulnerability assessment, with the exception of the *Parametric diagram*, which we believe may be worth of study in this regard.

Naming will be converted to a predefined dictionary. This is intended to facilitate identification of patterns across the models of different types of systems. In a way, this step can be considered an abstraction that enables focusing attention to the structural relationships of the models and systems.

23.4.3.2 Generate Realizations

Using the insights gathered in the initial exploration activity with respect to the generation of homomorphisms, we plan to create several potential real-world system realizations from the original model. Because the amount of potential real-world realizations is infinite, a subset will be considered for this research. The subset will be created automatically using a predefined algorithm. The algorithm will incorporate randomized selection to limit the extent to which we induce a pattern into the selection of the homomorphisms.

23.4.3.3 Filter Inherent Limitations from Lack of Sophistication

We anticipate that several of the potential real-world systems generated in the previous step may be the result of the lack of sophistication of the source SysML models, not necessarily artifacts of the inherent limitations of SysML as a modeling language. Therefore, we will attempt in this step to distinguish between and separate them. The process is depicted in Fig. 23.1.

Using the reverse process as when transforming SysML models to Wymore's framework, we will now generate sequence diagrams, state machine diagrams, activity diagrams, block definition diagrams, and internal block diagrams for all real-world realizations. Because of the potential inherent limitations of SysML, some of the elements in Wymore's models will not be captured in the newly generated SysML models. Therefore, we will transform back these new SysML

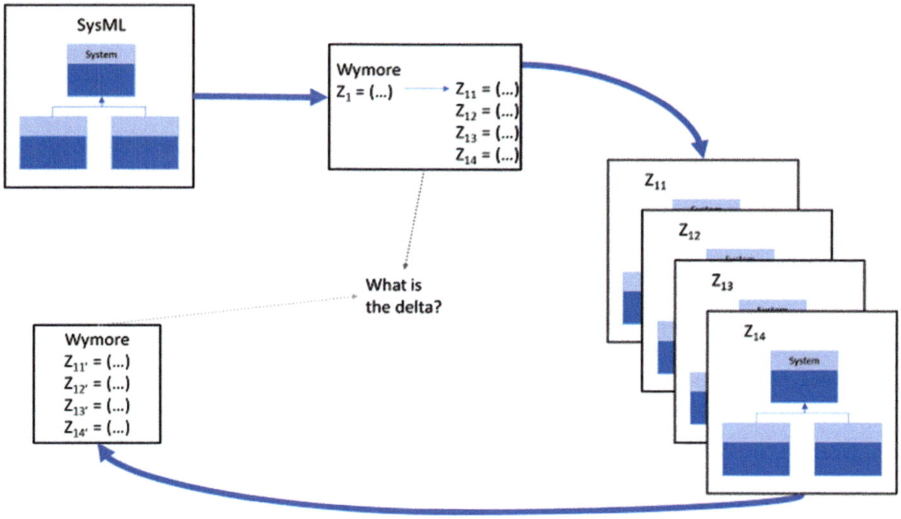

Fig. 23.1 Process to disambiguate between artifacts of the model and artifacts of the modeling language

models to Wymore's notation, which will not be identical to the original Wymore's models. Hence, by comparing the similarity we believe we will be able to identify precisely the artifacts associated to the inherent limitations of SysML as a modeling language.

23.4.3.4 Identify New Vulnerabilities and Patterns

In a final task, we will study the results of all initial SysML models collectively, aiming at identifying patterns of unknown vulnerabilities. The patterns will be identified by exploring the common aspects to all surrogates that could not be transferred from Wymore's models to SysML.

Specific approaches and methods need to be defined once we have more insights on how the data will look. Our initial intention is to base the comparison of patterns using mathematics. Yet, we recognize that we may need to employ executable models at some point. If this would be the case, it should be noted that it would suffice to create executable models out of Wymore's notation, not from SysML. This is because the identification of patterns is made in the models defined within Wymore's framework, since we are concerned about aspects of the system that SysML is not able to capture. To our knowledge, Wymore's framework enables the development of executable models, so we do not anticipate major obstacles to pursue this path, should it be required.

23.4.4 *Potential Limitations*

We anticipate four key obstacles in the proposed plan. First, instead of actually creating real-world systems, we are creating mathematical models of them. Effectively, we are exploring the inherent limitations of a modeling language by overlaying another modeling language. Therefore, any discovered pattern may be in principle the result of confounding both languages; hence, not necessarily inherent to SysML. Furthermore, even if we could isolate the effects of Wymore's modeling language, the discovered patterns will be limited only to those that Wymore's modeling language can discover. As with many other problems, completeness cannot be guaranteed in this case. Therefore, part of the preliminary work will consist in developing mechanisms with the research approach that allows us to distinguishing if patterns are generated because of SysML or because of Wymore's mathematical framework.

The second problem lays on the creation of homomorphisms. As previously discussed, the amount of real-world systems that are a homomorphic image of a given model is infinite. We will have to execute a selection mechanism to reduce the set of resulting homomorphisms to a practically manageable amount. Hence, both the creation and selection mechanisms may induce specific patterns in the solutions that we will use to explore the patterns associated with SysML. Therefore, we also plan to investigate specific creation and selection processes that guarantee randomization in the definition of homomorphisms.

The third issue is related to Wymore's definition for system, in particular with regards the uses for *state*, which is described as a behavioral aspect of the system. For example, a car with an input of pressing the acceleration peddle will produce an output of a force. In this case, the car is in a behavioral state of acceleration. We foresee that the translation of SysML to behavioral states may not be as simple as it first appears, in particular regarding non-functional system characteristics such as physical attributes. Therefore, a translation medium may be necessary. If this is the case, the application of Category Theory will be explored to perform this translation.

The last key issue concerns the meaning of vulnerability. In particular, there may be unknown aspects of a system that do not imply a system vulnerability. Therefore, it is necessary to incorporate machinery that enables differentiating between unknown aspects that lead to potential vulnerabilities and those that do not. Ideally, this machinery also facilitates computing the extent to which the vulnerability may be harmful to the system.

23.5 Conclusions

We have suggested in this paper that convergence of MBSE to a single modeling language could result in patterns of unknown vulnerabilities. We base this assertion on the inherent limitations of modeling languages to capture reality fully. If MBSE is going to be widely adopted, we believe that the systems engineering community should explore the unintended consequences that it may create with respect to current practice.

Aiming at studying the existence of such vulnerability patterns in MBSE, we have outlined a research plan for exploring and gaining a fundamental understanding of the vulnerabilities that exist inherently to modeling using SysML. We plan to use the Wymore's mathematical framework to create surrogates of real-world systems, using SysML models as a starting point. We expect the surrogates to facilitate identifying patterns of unknown vulnerabilities that are inherent to SysML. These patterns would be seen as aspects common to all surrogates that cannot be captured in SysML when trying to convert the surrogates back to SysML.

We have also identified four key potential limitations of the proposed research plan that need to be overcome for it to be successful. First, using Wymore's framework to create surrogate realizations of SysML model may create problems related to confounding of modeling patterns. Second, there are potentially infinite homomorphisms for each SysML model. Selection of a subset can yield an artificial selection pattern that also confounds with the modeling patterns. Third, effort is needed to investigate the extent to which Wymore's framework can capture non-behavioral system aspects. Fourth, an unknown aspect not captured by a model may not necessarily imply a vulnerability. Hence, it will be necessary to separate unknown aspects that lead to vulnerabilities and those that do not.

References

1. Friedenthal, S., Moore, A., & Steiner, R. (2015). *A practical guide to SysML: The systems modeling language* (3rd ed.). Amsterdam: Elsevier.
2. Hedberg, T., Lubell, J., Fischer, L., Maggiano, L., Barnard Feeney, A. et al. (2016). Testing the digital thread in support of model-based manufacturing and inspection. *Journal of Computing and Information Science in Engineering, 16*(2), 021001.
3. Miller, M. Z., & Wheaton, M. J. (2017). Interest grows in digital twins, digital engineering. *Aerospace America, 55*(11), 15.
4. Long, D., & Scott, Z. (2011). *A primer for model-based systems engineering* (2nd ed.). Blacksburg: Vitech Corporation.
5. Piaszczyk, C. (2011). Model based systems engineering with department of defense architectural framework. *Systems Engineering, 14*(3), 305–326.
6. Khan, M. O., et al. (2013). Model-based verification and validation of the SMAP uplink processes. In *IEEE aerospace conference aerospace conference*. IEEE.
7. Kovarik Jr., V. J., & Muralidharan, R. (2017). Model-based systems engineering: Lessons learned from the joint tactical radio system. *Journal of Signal Processing Systems, 89*(1), 97–106.
8. Saitta, L., Zucker, J.-D., & Ebrary, I. (2013). *Abstraction in artificial intelligence and complex systems* (1st ed.). New York: Springer.
9. Avitabile, P., Marinone, T., & Dardeno, T. (2018). Reduced model approximation approach using model updating methodologies. *Journal of Engineering Mechanics, 144*(3), 04018005.
10. Fithen, W. L., et al. (2004). Formal modeling of vulnerability. *Bell Labs Technical Journal, 8* (4), 173–186.

11. Box, G. E. P. (1976). Science and statistics. *Journal of the American Statistical Association, 71* (356), 791–799.
12. Lucy, J. A. (1992). *Language diversity and thought: A reformulation of the linguistic relativity hypothesis* (Vol. 12). Cambridge: Cambridge University Press.
13. Kollareth, D., & Russell, J. A. (2017). The English word disgust has no exact translation in Hindi or Malayalam. *Cognition and Emotion, 31*(6), 1169.
14. Torre, I. G., et al. (2017). Emergence of linguistic laws in human voice. *Scientific Reports (Nature Publisher Group), 7*, 43862.
15. Boase-Beier, J., Lodge, K. R., & Ebrary, I. (2003). *The German language: A linguistic introduction* (1st ed.). Malden: Blackwell Publishing.
16. Zagona, K. T. (2002). *The syntax of Spanish*. New York: Cambridge University Press.
17. Field, E. H. (2015). All models are wrong, but some are useful. *Seismological Research Letters, 86*(2A), 291–293.
18. Lehmann, C. U., Séroussi, B., & Jaulent, M. C. (2016). Troubled waters: Navigating unintended consequences of health information technology. *IMIA Yearbook, 1*, 5–6.
19. Jacobson, S. R. (2010). Aircraft loss of control causal factors and mitigation challenges. In *AIAA GNC conference*. Toronto: NASA Center for Aerospace Information (CASI).
20. Lutz, R., Johnson, J., & Patterson-Hine, A. (2011). Using Model-based Assurance to strengthen diagnostic procedures. In *IEEE/ACM international conference on automated software engineering (ASE 2011) automated software engineering (ASE)*. IEEE Computer Society.
21. MacDonald, D., et al. (2013). Cyber/physical security vulnerability assessment integration. In *IEEE PES innovative smart grid technologies conference (ISGT)*. IEEE.
22. Wymore, A. W. (1993). *Model-based systems engineering*. Boca Raton, FL: CRC Press.

Chapter 24
Optimizing Fleet Life Cycle Management Decisions Through Graphical Dominance Analysis

Tobias Lemerande

Abbreviations

AMS	Asset management system
DF	Dominance factor
FLCM	Fleet life cycle management
FLCO	Fleet life cycle objective
GDA	Graphical dominance analysis
LID	Line of increasing dominance
PEW	Point of equal weighting
PoD	Point of dominance
RAN	Royal Australian Navy
SRI	Stakeholder region of interest
WC	Weighting criteria

24.1 Introduction

Australia's Chief of Navy has repeatedly urged the national naval enterprise to embrace asset management, specifically calling for the nation's maritime defense industry to deliver more cost-effective sustainment methods and practices by enhancing readiness levels through innovative techniques and improving affordability and operational availability across the fleet [1]. Separately, the country's Head of Navy Capability acknowledged that future success in the Royal Australian Navy (RAN) requires a smart and balanced approach to proper asset management that can

T. Lemerande (✉)
Department of Engineering, University of South Australia, Mawson Lakes, SA, Australia
e-mail: toby.lemerande@unisa.edu.au

S. Adams et al. (eds.), *Systems Engineering in Context*,
https://doi.org/10.1007/978-3-030-00114-8_24

maximize capability for each ship across its 30–40-year life cycle [2]. A proposed fleet life cycle management (FLCM) concept applies asset management to the Australian naval fleet [3]. An asset management system (AMS) and high-level life cycle objectives are fundamental framework components that support FLCM with decision support being a key function within the knowledge management component of an ISO 55001 compliant AMS [4]. In FLCM, decisions must be made with a clear understanding of the repercussions on achieving primary objectives while subordinating others. Decision-making in multi-objective FLCM scenarios presents unique challenges when trying to optimize a ubiquitous master solution applicable to every situation. In some cases, software code could be generated to find the optimal solution [5], but in many cases visual representation can help naval enterprise stakeholders and decision-makers formulate better courses of action and discard others without expensive and complicated software solutions.

This paper applies the mixing triangle as a method to bound weighting criteria (WC) and graphically display those boundaries for naval enterprise stakeholders to consider when making decisions related to high-level FLCM objectives. A brief explanation of each fleet life cycle objective (FLCO) is followed by a short summary of the mixing triangle and its usage. A longer discussion and analysis illustrates how graphical dominance analysis (GDA) identifies appropriate WC for each objective when considering asset-related decisions. Mathematically derived exclusion zones and preference areas are presented as methods to categorize regions of interest by establishing WC limitations that prevent excessive dominance by a single objective. The paper concludes with recommendations for incorporating GDA into an AMS leading to enhanced knowledge management within a FLCM framework.

24.2 Asset-Related Decisions in Fleet Life Cycle Management

24.2.1 Fleet Life Cycle Objectives

FLCM seeks to optimize high-level objectives across the RAN fleet [6]. Figure 24.1a graphically represents the ideal mixes of availability, capability, and affordability in a Venn diagram displaying all desirable results from the interaction of these sets [7]. The green-shaded region signifies the concurrent achievement of FLCOs, previously identified as a fundamental goal of asset management.

Availability, in a technical sense, is the probability a system or system of systems (e.g., a warship or submarine) will be in a satisfactory condition to execute its assigned tasking when needed [8]. Affordability acknowledges the possibility to execute and sustain a program with the fiscal resources likely to be available in the future [9]. Capability refers to the nature of operational effects and how, when, where, and for how long those effects can be achieved or produced [10]; naval capability is how these operational effects are achieved in the maritime battlespace. In the Australian context, this means the RAN can generate a strategic effect by

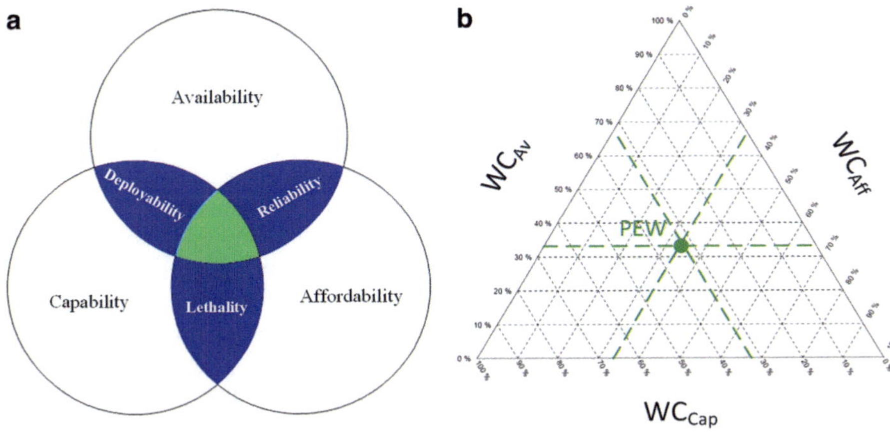

Fig. 24.1 (**a**) Fleet life cycle objectives; (**b**) the mixing triangle and the point of equal weighting

conducting offensive operations that strike at an adversary's key capabilities, destroying them through lethal means [11]. Further analysis shows additional objectives—lethality, reliability, and "deployability" in the blue-shaded regions—result from the concurrent achievement of two FLCOs. Reliability is "the probability that a system (or system of systems) will perform its intended functions without failure, within design parameters, under specific operating conditions, and for a specific period of time." [12] Lethality is the RAN's ultimate purpose and manifests itself as the ability to "generate decisive outcomes in conflict." [13] Deployability simply means a seagoing naval asset (i.e., warship, submarine, replenishment oiler, amphibious ship, etc.) has the ability to conduct and complete deployed missions supporting Australia's maritime doctrine and operations [14]. These six objectives provide the basis for making asset-related decisions in the FLCM framework and should be considered the main criteria for any decision support tool within Australia's naval enterprise.

24.2.2 The Mixing Triangle

The mixing triangle in Fig. 24.1b uses WC for any three parameters under consideration. The sum of all relative weights must always equal 100%. Each point inside the triangle represents a weighting combination with all possible combinations of relative weights existing inside the triangle. When all criteria have the same value (33 1/3%), the three weighting lines cross at the triangle's center, designated as the point of equal weighting (PEW). This point represents ideal cases where each criterion holds the same sway among stakeholders and decisions are considered holistically with no criteria being considered more important than another. At the triangle's corners, only one criterion is weighted at a value of 100%. On the borders, only two criteria are considered with the third criteria having no weight. Mixing triangles are used extensively in chemistry and metallurgy to evaluate the effects of different elemental compositions. Life cycle sustainment efforts in environmental and ecological studies use them to evaluate effects from different potentially hazardous products. Hofstetter et al. adapted the mixing triangle as a means to analyze and correlate multiple criteria to evaluate environmental impacts when conducting life cycle impact assessments [15]. Finkbeiner, Schau, Lehmann, and Traverso applied the mixing triangle to weigh social, economic, and environmental effects in a life cycle sustainability assessment [16]. A leading company in sustainability and environmental impact assessments offers a simple and free tool on its web site that can evaluate any three criteria [17]. The mixing triangle supports decision-making by providing a visual representation of the weighting factors [18]. If four criteria are considered, a mixing tetrahedron can be used. The advantage of the mixing triangle is its simplicity and easily understood visual representation. It is conceivable that an *n-sided* three-dimensional object could be used to represent more than four weighting criteria, but the results would be much harder to visualize and discern. The mixing triangle offers a simple representation that not only conveys unambiguous results but also provides limits to possibilities decision-makers and stakeholders can consider. The mixing triangle's simple design, use, and approach are keys to simplifying the decision-making process.

24.2.3 Decision Support in Asset Management

Asset management relies on a set of fundamentals, one of which contends that an asset's primary purpose is to add value to not just the organization but to internal and external stakeholders as well. This is achieved through identifying stakeholders' needs and organizational objectives to define what value really means. Decision-making processes play a key role in establishing rules for organizations to follow that can deliver results that best meet the identified needs and increase assets' value [19]. Within an organization, several factors influence the criteria for making asset management decisions. Asset management policy, attitudes to risk, and stakeholders' needs rank chiefly among the top influences. A firm understanding of asset-related decisions and how they are made is an important enabling function within asset management [20]. Internationally recognized standards require organizations to establish criteria for making asset management-related decisions [21]. A common asset management problem is that seemingly unacceptable solutions for one criterion may actually be acceptable when considered in a larger holistic context. Conversely, optimal decisions for a single criterion can be irrelevant when considered against all other criteria [22]. Making decisions with multiple and conflicting objectives is complex and challenging. When objectives are in direct competition, a single alternative or course of action cannot maximize all the objectives [23].

Applying the mixing triangle to FLCOs provides the needed visual representation and decision aid to help stakeholders better understand the implications of decisions when considered within the context of FLCM. A graphical display that addresses multiple criteria may not formally lead to problem resolution or definitive alternatives but does provide a method for decision-makers or decision-making groups with the ability to apply trade-offs through subjective means to understand the implications when compared against other criteria [24]. Decision-making can be easier if the problem space can be visually represented in a simple format [25]. GDA supports this method by using weighting sets where one or two alternatives dominate the other(s) as boundary conditions when considering multiple options. It provides informational views that show the effects of different weighting sets and the dominance over other inferior criteria [18]. Decision support for FLCM must consider the full spectrum of solutions to optimize concurrent achievement of FLCOs in a multi-criteria approach. Using the mixing triangle in FLCM deliberations, decisions can be made by weighting the different objectives to quantify the importance and understand the trade-offs. To make these trades and assign appropriate discrete weighting values, decision-makers must understand the interactions and relationships between the different weighting factors [26].

24.3 Graphical Dominance Analysis

24.3.1 Conditions of Dominance

In addition to equal weighting, this paper considers six instances of dominance (see Fig. 24.2) where concurrent achievement of all FLCOs is not accomplished. Venn diagrams, in this instance, show the relationships between different objectives. The overlapped regions are not scaled to show the amount of concurrent achievement or dominance, but merely to show the relationships between objectives. The six scenarios represent boundary conditions that exist at the triangle's corners and midpoints on each side of the mixing triangle. Letter designations (i.e., A, B, C, D, E, and F) correlate to a specific point of dominance (PoD) with defined WC against each objective.

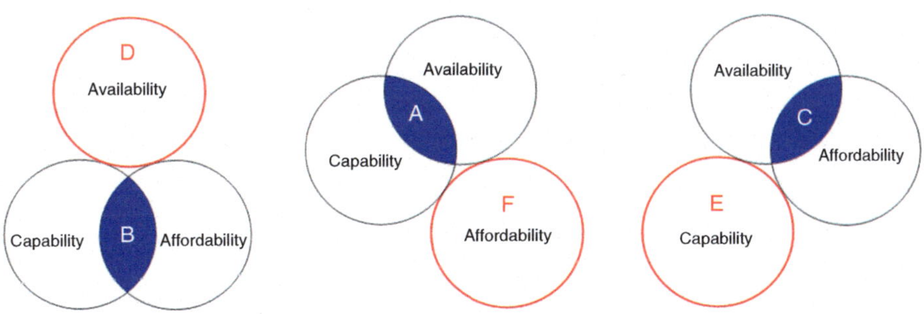

Fig. 24.2 Six instances of dominance that set the boundary conditions

Table 24.1 Dominance weighting criteria

Point of dominance	WC_{Av}	WC_{Cap}	WC_{Aff}	Dominant FLCM attribute
A	50%	50%	–	Deployability
B	–	50%	50%	Lethality
C	50%	–	50%	Reliability
D	100%	–	–	Availability
E	–	100%	–	Capability
F	–	–	100%	Affordability

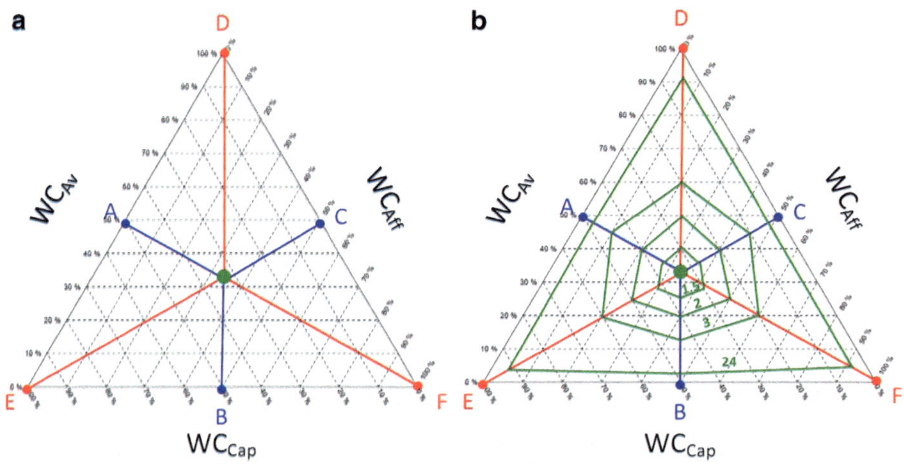

Fig. 24.3 (**a**) Points of dominance and lines of increasing dominance; (**b**) dominance factor defined regions

Table 24.1 shows the values assigned to availability weighting criteria (WC_{Av}), capability weighting criteria (WC_{Cap}), and affordability weighting criteria (WC_{Aff}) associated with each PoD. A dominant objective receives greater weighting than the others; an inferior weighting is given the least important weighting.

Figure 24.3a plots the six PoDs on the mixing triangle. A, B, and C signify instances when only two of the objectives are given weights in the decision-making process. D, E, and F represent cases where only a single objective is given any weight. A line of increasing dominance (LID) radiates from PEW outward to each PoD. WC on a LID becomes more dominant as it approaches a PoD and gets further from PEW. As this happens, the inferior objective(s) receive less weighting. At a PoD, the inferior objective(s) is not considered and has no weighted value.

24.3.2 Dominance Factors, Exclusion Zones, and Preference Areas

A dominance factor (DF), in this paper, is the ratio of the dominant (or non-inferior) WC to the inferior WC along each LID. DF = 1 at the mixing triangle's center and is

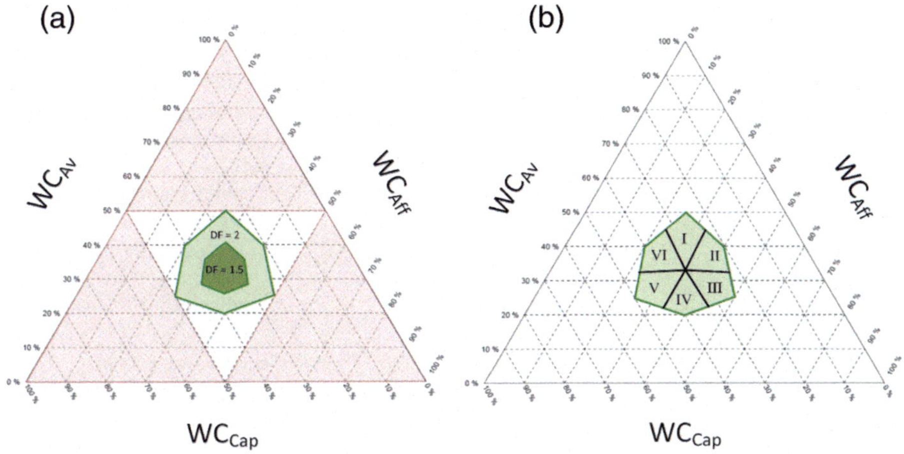

Fig. 24.4 (**a**) Exclusion zones and preference areas; (**b**) stakeholder regions of interest

collocated with PEW and thus has no discernible area or region. Values of DF > 1 plotted and connected on the six LIDs in Fig. 24.3b quantify the range of dominance between FLCOs. These points provide a means to discretize points on LIDs within the mixing triangle that, when connected, yield defined regions about PEW. Enclosed areas for DF values equal to 1.5, 2, 3, and 24 show the regions' areas increase exponentially as DF values increase.

In the FLCM concept, each objective must be allocated a reasonable weighting with no single WC being more than 50%. Exclusion zones are areas within the mixing triangle where a single criterion is allocated a majority (i.e., >50%) of the total weighting. Decisions that utilize WC in these areas heavily favor one objective over the others. This approach does not align to asset management principles and should not be employed. The red-shaded areas in Fig. 24.4a signify exclusion zones that visually identify regions with WC to be avoided. Exclusion zones represent 75% of the mixing triangle's original area. Figure 24.4b shows preference areas with WC regions outside the identified exclusion zones while also allocating WC that support decisions that can simultaneously represent naval enterprise stakeholders' interests and support FLCM principles and practices in the short, medium, and long terms. Where DF > 2, WC will encroach the exclusion zones. Thus, the preference area is bounded by DF $\leq$ 2. This area, equal to 10% of the mixing triangle, indicates WC boundaries that best support asset-related FLCM decisions.

24.3.3 *Stakeholder Regions of Interest*

The preference area is divided into six discrete regions that favor one set of naval enterprise stakeholders over the others. Each stakeholder region of interest (SRI) exists entirely within the preference area, but its dominant objective(s) does not

Table 24.2 Interested stakeholders by region of interest

SRI	Dominant objective	Stakeholders[a] with the greatest interest	Life cycle periods
I	Availability	• Operational commanders • Maintenance providers • Warship/submarine class managers • Fleet support staff	• Operational running periods • Post-maintenance certification • Training work-ups • Design gate reviews
II	Reliability	• Authorized engineering organizations • System designers • Logisticians	• Maintenance planning • Design gate reviews • Modernization/update planning
III	Affordability	• Financial managers • Budget submission authorities • System program offices	• Budget preparation • Maintenance planning • Acquisition gate reviews
IV	Lethality	• Naval headquarters staff • Coalition warfare planners • Warfare community sponsors	• Strategic planning • Operational analysis • Design gate reviews
V	Capability	• Strategic maritime planners • Intelligence community • R & D laboratories	• Analysis of alternatives gate reviews • National strategy planning • Coalition and allied war-gaming
VI	Deployability	• Joint operations staff planners • Warship/submarine class managers • Squadron commanders • Warship/submarine commanding officers	• Pre-deployment work-ups • Mission preparation periods • Exercise preparation periods

[a]Does not include contracted service providers. Contracted service providers are incentivized through performance-based contracting methods and thus will respond to identified performance indicators specified within the identified contracted statement of work. Through performance-based contracting, the naval enterprise can ensure contracted service providers always weight FLCOs with WC within predetermined preference areas that support the FLCM concept

maintain an inordinately high or lopsided WC compared to other objectives. SRIs provide visibility to decision-makers where WC can be adjusted to satisfy interested stakeholders' needs while not sacrificing fundamental tenets of FLCM. Military operations and the defense environment operate on planned schedules but also respond to emergencies and contingencies. Crises often require short-term actions that can extend into medium-term activities. Regardless of the tasking, asset management ensures long-term considerations are not sacrificed in favor of immediate needs and manages assets across the entire life cycle. Recognizing these SRIs, decision-makers can dynamically adjust assigned WC to best meet naval enterprise needs according to specific scenarios, operational contingencies, real-world emergencies, or routine asset management activities while still adhering to FLCM principles and practices. Table 24.2 summarizes different cases stakeholders can consider across the life cycle and how those situations can help influence WC assignment according to naval enterprise needs.

24.4 Conclusion

The mixing triangle can be applied to FLCOs and illustrate how GDA can assist Australia's national naval enterprise stakeholders and decision-makers when considering asset-related decisions to better support the FLCM concept. Internationally recognized asset management standards specifically identify decision-making criteria as a key component to an AMS. GDA can show WC boundaries for different scenarios that can be incorporated into a decision support tool within an AMS. WC limits of multiple objectives can be predetermined to establish constraints and restrict the relative importance among different FLCOs. These limits graphically display the relative weightings of availability, capability, and affordability to ensure naval enterprise decision-makers uphold fundamental tenets of asset management and the FLCM concept. DFs in basic scenarios identify exclusion areas to limit maximum WC values. FLCM should incorporate the identified preference areas into decision support tools used in an AMS to guide the decision-making process for Australia's naval enterprise stakeholders. Some basic and fundamental rules should be followed to standardize the concept and application of GDA to support FLCM:

- Decisions with any WC $= 0\%$ do not support FLCM and preclude concurrent achievement of FLCOs. Scenarios that assign WC $= 0\%$ for any objective will fall on the mixing triangle's border and must not be allowed.
- WC that give a single objective a majority of the weighting (i.e., inside the defined exclusion zones) should be disallowed as it seriously detracts from the ability to reasonably achieve all three objectives.
- Good decisions that support FLCM have WC that fall inside the defined preference area where DF ≤ 2 and ensure each objective receives at least 20% WC in any decision.
- Stakeholders and decision-makers must ensure WC are assigned to FLCOs in the region where $1 \leq$ DF ≤ 2.
- Stakeholders should match real-world situations with the appropriate SRI and WC within defined preference areas.

A fundamental tenet to the FLCM concept is the concurrent achievement of availability, capability, and affordability. The mixing triangle offers a method for visually showing the emphasis placed on each objective during the decision-making process. GDA, as an optimization technique, does not discriminate or favor one objective over another. The importance of each criteria is left for users to decide. The dynamic nature of maritime operations and the 30–40-year life span of naval ships requires agility in decision-making. The mixing triangle offers users a flexibility to apply weighting criteria to best meet circumstances as they arise. This should not be considered a constraint on decision theory but rather an advantage stakeholders and decision-makers can exploit when considering options.

24.5 Recommendations

GDA allows standardized WC limits to be formally identified and established. This allows senior policy-makers to define boundaries and constraints within the naval enterprise. Experiential knowledge applied in decision support mechanisms has been shown to be invaluable [27], but it should be expected that recognized experts, stakeholder groups, and decision-makers will place greater emphasis and weighting on the criteria to which they are most attracted or for which they have the most responsibility [28]. Regardless of the subjectivity, it is imperative that naval enterprise stakeholders identify and agree on when and how SRIs will be applied during different scenarios and when exceptions will be allowed to support unplanned maritime contingencies within Australia's designated areas of operations. Establishment of WC for FLCOs is an enterprise activity and should be agreed within the naval enterprise and included in FLCM policy and strategy. This should also be included in performance-based contracts to ensure service providers support asset management practices and principles. Once preference areas have been identified and agreed, stakeholders must accept these boundaries as limitations to decisions and guidance that will help ensure the best asset-related decisions are being made. Further research is needed to codify, establish, and automate WC into a formal decision support tool within an AMS. This includes the need to digitally integrate the WC as constraints within a decision support tool that will maintain the integrity of asset management and ensure alignment to ISO 55001 practices, principles, and methods. The naval enterprise should pursue incorporating GDA and automated WC restrictions into an AMS that encompasses the entire Australian naval fleet.

References

1. Barrett, T. (2017). *The navy and the nation*. Melbourne: Melbourne University Press.
2. Laird, R. (2017). Rear Admiral Jonathan Mead focuses on the way ahead for the Royal Australian Navy. In *Second Line of Defence: Delivering Capabilities to the Warfighter*. Retrieved October 11, 2017, from http://www.sldinfo.com/rear-admiral-jonathan-mead-focuses-on-the-way-ahead-for-the-royal-australian-navy/
3. Lemerande, T. J. (2017). Life cycle management applied to major fleet units. In *Defence Assets Lead Speaker presentation at 12th Annual World Congress on Engineering Asset Management*. Brisbane, 4 August 2017.
4. Lemerande, T. J. (2017). Fleet life cycle management: Applying asset management to sustain a multi-class submarine Fleet. In *Proceedings of the submarine science technology and engineering conference Adelaide*, 5–7 November 2017.
5. Azapagic, A., & Clift, R. (1999). The application of life cycle assessment to process optimisation. *Computers & Chemical Engineering, 23*(10), 1509–1526.
6. Lemerande, T. J. (2017). Optimising availability, capability, and affordability across the fleet: A total life cycle management approach for improving seaworthiness. In *e-Proceedings of World Congress on Engineering Asset Management (WCEAM)*, Brisbane, 2–4 August 2017.
7. Chen, H., & Boutros, P. C. (2011). Venn diagram: A package for the generation of highly-customizable Venn and Euler diagrams in R. *BMC Bioinformatics, 12*(1), 35.

8. Wagner, D. H., & Mylander, W. C. (Eds.). (1999). *Naval operations analysis*. Annapolis, MD: Naval Institute Press.
9. Porter, G. H., Conley, K. M., Gordon, C. V., Kneece Jr., R. R., Rieksts, B. Q., Shaw, A. H., et al. (2015). *Affordability of defense acquisition programs (No. IDA-PAPER-P-5243-REDACTED)*. Alexandria, VA: Institute for Defense Analyses.
10. Capability Development Group. (2014). *Defence capability development handbook*. Canberra, ACT: Commonwealth of Australia.
11. Barrett, T. (2015). Speech to Industry Leaders' Luncheon. In *Royal Australian Navy Sea Power Conference*, 7 October 2015. Retrieved from http://www.navy.gov.au/media-room/publications/chief-navy-speeches
12. Institute of Electrical and Electronics Engineering. (2000). *The authoritative dictionary of IEEE standard terms* (7th ed.). New York: Standards Information Network IEEE Press.
13. Barrett, T. (2015). *Chief of Navy speech to Naval Warfare Officers Association*, 30 October 2015 http://www.navy.gov.au/media-room/publications/chief-navy-speeches
14. Commonwealth of Australia. (2010). *Australian Maritime Doctrine: RAN Doctrine 1–2010*. Canberra, ACT: Sea Power Centre.
15. Hofstetter, P., Braunschweig, A., Mettier, T., Müller-Wenk, R., & Tietje, O. (1999). The mixing triangle: Correlation and graphical decision support for LCA-based comparisons. *Journal of Industrial Ecology, 3*(4), 97–115.
16. Finkbeiner, M., Schau, E. M., Lehmann, A., & Traverso, M. (2010). Towards life cycle sustainability assessment. *Sustainability, 2*(10), 3309–3322.
17. See Pre's web site at https://www.pre-sustainability.com/triangle-tool
18. Hofstetter, P. (1998). *Perspectives in life cycle impact assessment: A structure approach to combine models of the technosphere, ecosphere and valuesphere*. New York: Springer Science + Business Media LLC.
19. Standards Australia. (2014). *Asset management – Overview, principles and terminology (AS ISO 55000: 2014)*. Sydney: Council of Standards Australia.
20. Standards Australia. (2014). *Asset management – Management systems – Guidelines for the application of ISO 55001 (AS ISO 55002: 2014)*. Sydney: Council of Standards Australia.
21. Standards Australia. (2014). *Asset Management – Management systems – Requirements (AS ISO 55001: 2014)*. Sydney: Council of Standards Australia.
22. Thomas, E., Levrat, E., & Iung, B. (2008). Maintenance decision making tool reaching a compromise between maintainability and reliability performances. In *17th IFAC World Congress, Jul 2008, Seoul, South Korea*. pp. 907–912.
23. Brownley, C. W. (2013). *Multi-objective decision analysis: Managing trade-offs and uncertainty*. New York: Business Expert Press.
24. Blanchard, B. S., & Fabrycky, W. J. (2005). *Systems engineering and analysis* (4th ed.). Englewood Cliffs, NJ: Prentice Hall.
25. Azapagic, A. (1999). Life cycle assessment and its application to process selection, design and optimisation. *Chemical Engineering Journal, 73*(1), 1–21.
26. Seppälä, J., Basson, L., & Norris, G. A. (2001). Decision analysis frameworks for life-cycle impact assessment. *Journal of Industrial Ecology, 5*(4), 45–68.
27. Teah, H. Y., Fukushima, Y., & Onuki, M. (2015). Experiential knowledge complements an LCA-based decision support framework. *Sustainability, 7*(9), 12386–12401.
28. Finnveden, G., Hauschild, M. Z., Ekvall, T., Guinée, J., Heijungs, R., Hellweg, S., et al. (2009). Recent developments in life cycle assessment. *Journal of Environmental Management, 91*(1), 1–21.

Chapter 25
Comparing Frequentist and Bayesian Approaches for Forecasting Binary Inference Performance

Sean D. Vermillion, Jordan L. Thomas, David P. Brown, and Dennis M. Buede

Nomenclature

DT	Decision tree classifier
f	False-positive rate
FN	False-negative count
FP	False-positive count
IE	Inference enterprise
IEM	Inference enterprise model
p	Precision
r	Recall
SVM	Support vector machine classifier
TN	True negative count
TP	True positive count

25.1 Introduction

In this paper, we compare two approaches for forecasting inference enterprise model performance: (1) a frequentist approach and (2) a Bayesian approach. An inference enterprise (IE) is an organizational entity that uses data, tools, people, and processes to make mission-focused inferences [1]. For example, airport security organizations use scanning equipment with trained agents' judgment to infer whether or not a passenger is a security threat. An inference enterprise model (IEM) is a model that uses available organizational information to forecast inference quality given changes to an IE. In many cases, organizations are reluctant to release representative data or

S. D. Vermillion (✉) · J. L. Thomas · D. P. Brown · D. M. Buede
Innovative Decisions, Inc., Vienna, VA, USA
e-mail: svermillion@innovativedecisions.com

S. Adams et al. (eds.), *Systems Engineering in Context*,
https://doi.org/10.1007/978-3-030-00114-8_25

lack the data needed for IEM activities [2]. Thus, the quality of inferences made by a proposed or hypothetical IE architecture is uncertain. The question becomes: *how do we best forecast the performance of an IE's processes and architecture?*

For forecast generation, there are two primary schools of thought: the frequentist approach and Bayesian approach. The frequentist approach to forecast generation is generally seen as the more quantitative approach to probability, as it does not necessitate prior knowledge [3]. However, the primary limitation to the frequentist forecast is that all uncertainty contained in the forecast is due to randomness in the sample—i.e., aleatory uncertainty [4]. This means that the frequentist-generated forecast is dependent on the system being random and repeatable. An additional limitation of the frequentist approach is that it does not condition on observed data [5]. Bayesian probability on the other hand is generally thought to be the more subjective approach to probability because it requires information about a prior [3]. Since Bayesian probability uses prior knowledge, it is able to incorporate uncertainty due to lack of knowledge—i.e., epistemic uncertainty [4]—in addition to the uncertainty due to randomness.

While many researchers have sought to find agreement between frequentist and Bayesian methods [6–8], significantly less research has been conducted on comparing the ability to predict performance using the two schools of thought. In this paper we aim to compare the frequentist and Bayesian approaches to forecasting inference quality. Specifically, we generate 18 forecasts using both approaches and score the forecasts against ground truth answers. Additionally, we qualitatively compare the properties of the forecasts generated using both approaches.

This paper is organized as follows: Sect. 2 outlines the foundations relevant to this research including a background on the metrics used to measure classifier performance and a summary the Bayesian probabilistic framework used in the Bayesian forecast approach. In Sect. 3, we provide details of the methodology, and in Sect. 4 we discuss our results. We conclude the paper with a summary of the research and directions for future work.

25.2 Foundations

25.2.1 Binary Classifier Performance

In this paper, we only consider IEs that make binary inferences. Such IEs are analogous to, or even incorporate, binary classifiers, and thus we characterize inference quality similar to how a binary classifier's performance is characterized. Binary classification experiment results can be summarized as a confusion matrix; see Table 25.1. Each data point used in an experiment is binned into one of the four possible states in the confusion matrix, and TP represents the total number of true positive data points, FN represents the total number of false-negative data points, etc., so that TP+FN+FP+TN $= N$, the total number of data points used in an experiment.

Table 25.1 Binary classification confusion matrix

	Predicted label	
Actual label	Target class	Nontarget class
Target class	True positive (TP)	False negative (FN)
Nontarget class	False positive (FP)	True negative (TN)

Classifier performance metrics are functions of elements in the confusion matrix [9]. Three such performance metrics are recall (r), precision (p), and false-positive rate (f), which are defined as the following:

$$r = \frac{\mathrm{TP}}{\mathrm{TP} + \mathrm{FN}}, p = \frac{\mathrm{TP}}{\mathrm{TP} + \mathrm{FP}}, f = \frac{\mathrm{FP}}{\mathrm{FP} + \mathrm{TN}} \tag{25.1}$$

Recall is a classifier's effectiveness to identify target labels, precision measures the agreement of the positive labels given by the classifier with the true labels, and false-positive rate measures a classifier's failure to identify nontarget labels. Recall and precision are to be maximized while false-positive rate minimized.

When we have uncertainty in our data, or we test a classifier using multiple datasets, we generate a confusion matrix for each experiment. Therefore, we compute a recall, precision, and false-positive rate for each experiment. From a frequentist perspective, we can fit distributions to the set of recall, precision, and false-positive rate values generated through experiments, from which we make performance forecasts. In the next section, we discuss an alternative, Bayesian approach to generating distributions for recall, precision, and false-positive rate from empirical data.

25.2.2 Probabilistic Framework for Classifier Performance Prediction

Goutte and Gaussier introduce a probabilistic framework for inferring recall and precision distributions given empirical results in the form of a confusion matrix as in Table 25.1 [10]. We extend this framework to inferring a distribution for false-positive rate as well. For tractability, we summarize the framework in terms of inferring a distribution for recall and then discuss translation to precision and false-positive rate. Through Bayes' theorem, the probability of recall, r, given empirical observations, D, is the following:

$$\Pr(r|D) \propto \Pr(D|r) \times \Pr(r) \tag{25.2}$$

where $D = [\mathrm{TP}, \mathrm{FP}, \mathrm{FN}, \mathrm{TN}]$, $\Pr(D|r)$ is the likelihood distribution for r, and $\Pr(r)$ is the prior belief distribution over r. Since recall takes a value between zero and one, we intuitively model our prior belief using a beta distribution such that

$$r\tilde{}\text{Beta}(\alpha_r, \beta_r) \Rightarrow \Pr(r) \propto r^{\alpha_r - 1} \times (1 - r)^{\beta_r - 1} \tag{25.3}$$

Goutte and Gaussier model the distribution of D as a multinomial distribution, whose marginal and conditional distributions are binomial. Using this property, we derive the likelihood distribution, Pr(D| r), through the following:

$$\Pr(D) = \frac{N!}{\text{TP}! \times \text{FN}! \times \text{FP}! \times \text{TN}!} \pi_{\text{TP}}^{\text{TP}} \times \pi_{\text{FN}}^{\text{FN}} \times \pi_{\text{FP}}^{\text{FP}} \times \pi_{\text{TN}}^{\text{TN}} \Rightarrow \Pr(D|r) \propto r^{\text{TP}} \times (1 - r)^{\text{FN}} \tag{25.4}$$

Combining Eqs. (25.2–25.4), the posterior distribution for recall is the following:

$$\Pr(r|D) \propto r^{\text{TP}+\alpha_r - 1} \times (1 - r)^{\text{FN}+\beta_r - 1} \Rightarrow r \mid D\tilde{}\text{Beta}(\text{TP} + \alpha_r, \text{FN} + \beta_r) \tag{25.5}$$

Using the same logic, we generate a distribution for precision, Pr(p| D), by replacing FN in Eq. (25.5) with FP, and we generate a distribution for false-positive rate, Pr(f| D), by replacing TP and FN in Eq. (25.5) with FP and TN, respectively.

The above probability distribution definitions incorporate only a single observation of a confusion matrix. However, we can use several observations to update recall, precision, and false-positive rates. Consider we perform one experiment and generate D_1; our recall distribution is computed as in Eq. (25.5). Then we perform a second experiment and generate D_2. The posterior distribution generated from D_1, Pr (r| D_1), becomes our new prior distribution, and our new posterior is $r \mid D_1, D_2 \sim$Beta ($\text{TP}_1+\text{TP}_2+\alpha_r$, $\text{FN}_1+\text{FN}_2+\beta_r$). Extending this procedure to n empirical observations, r is modeled as the following:

$$r \mid D_1, D_2, \ldots, D_n\tilde{}\text{Beta}\left(\sum_j^n \text{TP}_j + \alpha_r, \sum_j^n \text{FN}_j + \beta_r\right) \tag{25.6}$$

Precision and false-positive rate distributions are updated similarly.

25.3 Methodology

25.3.1 General Approach

To compare forecasts generated from frequentist and Bayesian perspectives, we use both approaches to generate 18 performance forecasts using the same empirical observations. We then score the quality of two forecast sets against ground truth answers using three scoring metrics. In the remainder of this section, we discuss the specific forecast questions, IE structures under consideration, data simulation, and scoring metrics used for comparing the two forecasting approaches.

25.3.2 Forecast Questions

In this paper, we use the 18 forecast questions asked in the 15th challenge problem in the Intelligence Advanced Research Program Activity-sponsored SCITE research program. This challenge problem is motivated by a particular IE structure use case and narrative in which an organization wishes to identify line managers within their organization of approximately 3500 individuals using data on their employees' online activity. This data potentially includes their web proxy data, email habits, VPN logs, and human resource (HR) data. The organization contracts an independent modeling team to estimate the recall, precision, and false-positive rate of a decision tree (DT) classifier and a support vector machine (SVM) classifier for detecting line managers based on (1) only web proxy data, (2) only web proxy and email data, and (3) all available data. The organization provides the modeling team with data between 4 October 2015 and 24 September 2016 and asks the modeling team to predict performance for 2 October 2016 to 25 February 2017. The combination of three performance metrics, two classifiers, and three data structures yields a total of 18 forecasts the independent modeling team is asked to provide the organization. Forecasts include a point estimate for the mean recall, precision, and false-positive rate with a 60% certainty interval for each classifier and data structure under consideration.

25.3.3 Data Simulation

For training and testing the DT and SVM classifiers, we are only provided aggregated statistics of the organization's approximately 3500 employees' online activity. These statistics include means, standard deviations, correlations, histograms, and weekly autocorrelation for 142 variables for each week in the training period: 4 October 2015 to 24 September 2016. The 142 variables are grouped as being one of the following categories:

1. Web proxy detectors, e.g., total number of unblocked connections, total number of blocked connections, etc.
2. Email detectors, e.g., total number of emails sent, total number of emails received, etc.
3. VPN detectors, e.g., average time of day of first connection to the VPN each day, maximum time of day of first connection to the VPN each day, etc.
4. Human resources detectors, e.g., years at the organization, number of unique IP addresses associated with a user, etc.

Separate sets of statistics are provided for line managers and non-line managers during this time period.

The focus of this paper is not on describing data simulation procedures in depth, but we summarize our procedure here. An overview of our Monte Carlo approach for

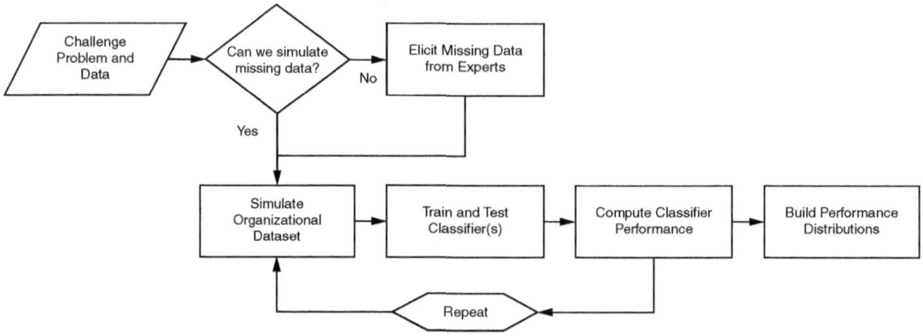

Fig. 25.1 Inference enterprise modeling metamodel

simulating datasets from these aggregate statistics is presented in Fig. 25.1. For each week in the training period, we sample from a 142-dimensional Gaussian copula to generate a correlated sample set using the correlation matrices supplied. We transform this sample using the supplied autocorrelations and properties of a conditional multivariate Gaussian distribution to generate the data sample for another week in the training period. We then transform the uniform marginals of each week's Gaussian copula dataset to match the supplied histograms for each of the 142 variables. We extrapolate this procedure to the test period—2 October 2016 to 25 February 2017—by allowing for variability in correlations and autocorrelations. With each simulated dataset, we train and test the DT and SVM classifiers. We repeat this procedure 30 times.

25.3.4 Forecast Generation

The primary focus of this paper is on the final step shown in Fig. 25.1: build performance distributions. For each simulated dataset and each classifier-data structure pair, we compute a confusion matrix as in Table 25.1. The frequentist and Bayesian inference approaches use confusion matrix results in different ways to build performance distributions and ultimately performance forecasts. With the frequentist approach, forecast mean performances are computed using the arithmetic means, so, for instance, the mean for a recall forecast is computed as the following:

$$\bar{r} = \frac{1}{n} \times \sum_{i=1}^{n} r_j = \frac{1}{n} \times \sum_{i=1}^{n} \frac{\mathrm{TP}_j}{\mathrm{TP}_j + \mathrm{FN}_j} \tag{25.7}$$

where n is the number of simulation runs. See Eq. (25.1) for computing precision and false-positive rate. We compute the upper and lower 60% certainty bounds for the frequentist approach by finding the 80% and 20% percentile performance metric values, respectively.

With the Bayesian updating approach, we are explicitly building a beta distribution and can thus use the mean of a beta distribution and a beta distribution's cumulative distribution function (CDF) to compute our forecast. For a recall forecast, the distribution for recall given empirical results, D, is $r \mid D$~Beta ($\sum_j \mathrm{TP}_j+\alpha_r$, $\sum_j \mathrm{FN}_j+\beta_r$) where α_r and β_r are the parameters for the prior distribution. Therefore, the forecast mean estimate is computed as the mean of this distribution:

$$\bar{r} = \frac{\sum_j^n \mathrm{TP}_j + \alpha_r}{\sum_j^n \mathrm{TP}_j + \alpha_r + \sum_j^n \mathrm{FN}_j + \beta_r} \tag{25.8}$$

We compute the 60% certainty interval for the Bayesian approach using the generated beta distribution's CDF and finding the values where the CDF is 0.8 and 0.2, respectively, to serve as the upper and lower bounds. In this paper, we assume all prior distributions are uniform so that $\alpha = \beta = 1$ for all forecasts. The uniform prior reflects our naivety in that we have no prior experience for where recall, precision, or false-positive rate lie.

25.3.5 Forecast Scoring Metrics

We score forecasts using available ground truth answers, a_i, for each of the 18 forecast questions using three metrics: (1) mean squared error (MSE), (2) certainty interval calibration (CIC), and (3) interval scoring rule (ISR). MSE measures the average error in our forecasts' central tendencies and is defined as the following:

$$y_1 = \frac{1}{T} \times \sum_{i=1}^{T} (a_i - m_i)^2 \tag{25.9}$$

where m_i is the mean of the ith forecast. A smaller MSE is preferable to a larger one to reflect the desire to mitigate forecast likelihood error. CIC measures the proportion of forecasts where the ground truth answer is within the forecast's certainty interval and is defined as the following:

$$y_2 = \frac{1}{T} \times \sum_{i=1}^{T} I(l_i \leq a_i \leq u_i) \tag{25.10}$$

where l_i and u_i are the lower and upper certainty interval bounds for the ith forecast and $I(\cdot)$ is an indicator function giving a value of 1 if the argument is true and 0 if the argument is false. Since we compute certainty intervals at 60% confidence, the target CIC score is 0.6 indicating that 60% of our certainty intervals contain the ground truth answer. ISR simultaneously measures certainty interval width and the difference between the ground truth answer and certainty interval bound should the ground truth answer lie outside the certainty interval:

$$y_3 = \frac{1}{T} \times \sum_{i=1}^{T} [(u_i - l_i) + 5 \times (l_i - a_i) \times I(a_i < l_i) + 5 \times (a_i - u_i) \times I(a_i > u_i)] \tag{25.11}$$

If the ground truth answer lies within the certainty bounds, the interval score for that prediction is simply the interval width. If the answer lies outside the certainty bounds, then the score is the interval width plus five times the difference between the ground truth and the upper bound or lower bound (whichever is closer). A smaller ISR is preferable to a large one to reflect the desire to provide informative forecasts without overly wide intervals.

25.4 Results and Discussion

The frequentist and Bayesian forecasts to the 18 forecast questions are shown in Fig. 25.2, and the forecast scores are listed in Table 25.2. From Table 25.2, we see the frequentist approach outscores the Bayesian-generated forecasts in two of the three aggregate scoring metrics: CIC and ISR. The two approaches produce similar MSE scores. The frequentist approach systematically produces forecasts with wider certainty intervals, and thus it is unsurprising that these forecasts score higher in CIC. Additionally, due to wider intervals, the distance between a frequentist-generated bound and a ground truth answer is lessened even if the corresponding mean is far from the ground truth answer contributing to a lower ISR. While the Bayesian-generated forecasts have small interval widths, they are penalized in ISR for having greater distance between a bound and the ground truth.

For all forecast questions, the Bayesian-generated forecast certainty intervals are consistently narrower than those generated through the frequentist approach. This observation is unsurprising for the context in which we are generating forecasts. The variance of a random variable distributed with a beta distribution such that $X \sim \text{Beta}(\alpha, \beta)$ is the following:

$$\text{var}(X) = \frac{\alpha \times \beta}{(\alpha + \beta)^2 \times (\alpha + \beta + 1)} \tag{25.12}$$

As α and β increase, the denominator in Eq. (25.8) increases much more quickly than the numerator; thus variance decreases. In this paper, we are dealing with datasets with thousands of data points so that the TP, FP, FN, and TN values in the confusion matrix of each simulation run are potentially large numbers. Since the

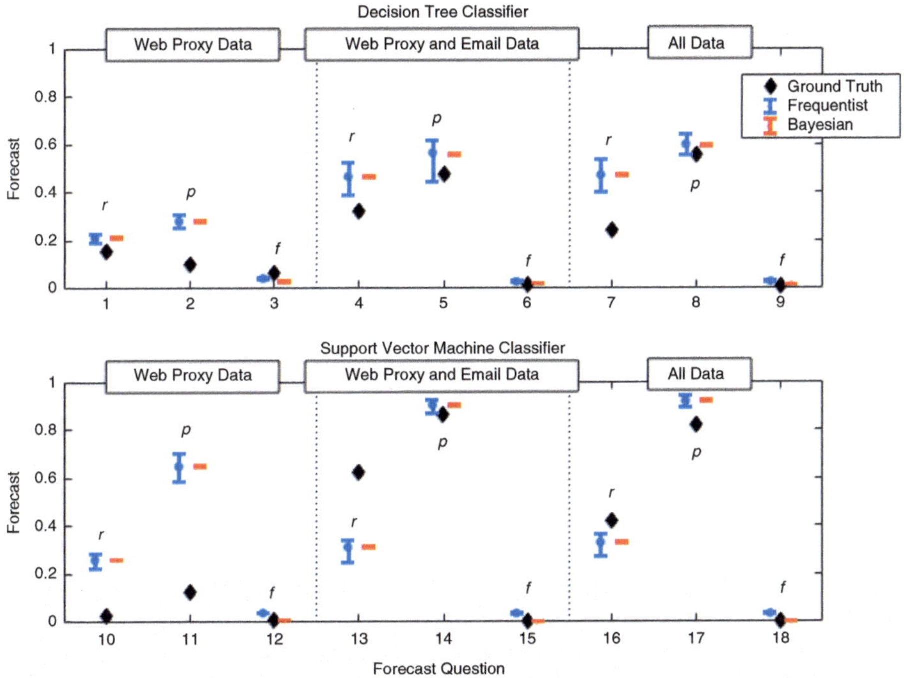

Fig. 25.2 Ground truth answers and forecasted answers to the 18 performance questions

Table 25.2 Forecast scores with bold values signifying optimal scores

Metric	Frequentist score	Bayesian score
MSE	0.0311	0.0311
CIC	**0.1667**	0.0000
ISR	**0.4833**	0.5823

parameters for our posterior beta distributions for recall, precision, and false-positive rate use these large confusion matrix elements, the variance for our posterior distributions is smaller resulting in small certainty intervals.

25.5 Summary

In this paper, we compared performance forecasts generated using a frequentist inference approach and a Bayesian inference approach on a representative binary classification task. While the two approaches generally produced forecasts that align with each other around their means, the Bayesian approach consistently produced narrower certainty intervals due to the large number of data points we use in our comparison study. Therefore, the Bayesian approach arguably produces overly confident forecasts. If the prior distributions allow for a lot of performance uncertainty, the data is the driving element in defining the performance posterior distributions' parameter values so that the priors have little impact on the posteriors. This is to say that if the parameters of the prior distribution, α and β, are much less than the elements in a confusion matrix as in Table 25.1, the empirically generated elements of posterior performance distributions drown out prior beliefs so that information from the priors have little influence on the posterior distributions.

Bayesian inference is an attractive framework for producing performance forecasts since we can incorporate our beliefs over inference quality. However, there are hurdles to overcome while employing this framework to making reasonable forecasts. Immediate future work includes the following:

- Generalize the differences between the frequentist and Bayesian forecasting approaches beyond the binary classification task presented in this paper.
- Conduct a parameter search to determine if there exist particular performance prior beliefs that would produce posterior forecasts that would outscore the frequentist approach and investigate how reasonable those prior beliefs are.

Additionally, we seek to extend the Bayesian framework presented in this paper by allowing users to constrain or limit the ability of empirical results to overwhelm the prior distributions, potentially yielding better balance between the contributions of the priors and empirical results to the posteriors.

Acknowledgments Research reported here was supported under IARPA contract 2016-16031400006. The content is solely the responsibility of the authors and does not necessarily represent the official views of the US government.

References

1. Huang, E., Zaidi, A. K., & Laskey, K. B. (2018). Inference enterprise multimodeling for insider threat detection systems. In A. M. Madni, B. Boehm, R. G. Ghanem, D. Erwin, & M. J. Wheaton (Eds.), *Disciplinary convergence in systems engineering research* (pp. 175–186). Cham, Switzerland: Springer.
2. Turcotte, M. J., Kent, A. D., & Hash, C. (2017). Unified host and network data set. *arXiv preprint arXiv:170807518*.
3. Efron, B. (1986). Why isn't everyone a Bayesian? *The American Statistician, 40*, 1–5.
4. O'Hagan, T. (2004). Dicing with the unknown. *Significance, 1*, 132–133.
5. Wagenmakers, E.-J., Lee, M., Lodewyckx, T., & Iverson, G. J. (2008). *Bayesian versus frequentist inference. Bayesian evaluation of informative hypotheses* (pp. 181–207). Berlin, Germany: Springer.
6. Bartholomew, D., & Bassett, E. (1966). A comparison of some bayesian and frequentist inferences. II. *Biometrika, 53*, 262–264.
7. Jeffreys, H. (1998). *The theory of probability*. Oxford, UK: OUP.
8. Tiao, G. C., & Box, G. E. (1973). Some comments on "Bayes" estimators. *The American Statistician, 27*, 12–14.
9. Sokolova, M., & Lapalme, G. (2009). A systematic analysis of performance measures for classification tasks. *Information Processing & Management, 45*, 427–437.
10. Goutte, C., & Gaussier, E. (2005). A probabilistic interpretation of precision, recall and F-score, with implication for evaluation. In *European Conference on Information Retrieval* (pp. 345–359). Berlin, Germany: Springer.

Chapter 26
The Danger of Using Ratio Performance Metrics in System Evaluations

William T. Scherer and Stephen Adams

26.1 Introduction

As we delve further into the data science/big data era, we create more and more forecasts or predictions for the evaluation of systems. Many practitioners use ratio variables as performance metrics, one of the most common being the benefit/cost (B/C) ratio [1]; however, one of the subtler is the "percent change" metric. Other ratio metrics are used throughout disparate industries, including financial ratios [2–4] and television performance indices. There is significant risk associated with using ratio metrics [5], and in this paper we focus on one simple trap that can result from forecasting a ratio metric. We illustrate and motivate the issues with a stylized television-viewing index forecasting problem, where the goal is to forecast the index of future television shows. This is similar to forecasting, for example, the percentage increase in a college enrollment for a new program designed to increase enrollments, where we forecast the original enrollment (O) and forecast the new enrollment (N), the end goal being to forecast O/N.

Such an idea is best demonstrated with a fabricated and simple numerical example involving television viewership, where the total viewers (T) and behaviorally targeted (BT) viewers (a subset of the total viewers) are estimated from a single variable—the number of panelists (P) that we expect to watch a show. In practice, there might be many factors (variables) that figure into the viewership estimates; however, the purpose of this example is to illustrate the concept with an elemental example. Consider that we desire to forecast the T/BT of a new television show and that for any (independent) previously broadcast television shows, we can measure via our data the number of panelists (a small subset of viewers that we carefully monitor) that watched the show. We can also measure the T and the BT for any

W. T. Scherer · S. Adams (✉)
Department of Systems and Information Engineering, University of Virginia, Charlottesville, VA, USA
e-mail: sca2c@eservices.virginia.edu

S. Adams et al. (eds.), *Systems Engineering in Context*,
https://doi.org/10.1007/978-3-030-00114-8_26

Table 26.1 Sample historical data

Show #	Panelists	Total viewers	Behavioral target	*T*/BT
1	431	1766	460	3.84
2	22	81	21	3.88
3	458	1837	492	3.74
4	260	932	248	3.76
5	115	440	105	4.21
6	104	402	108	3.72
7	98	375	109	3.42
8	421	1510	454	3.33
9	407	1465	470	3.12
10	322	995	319	3.12

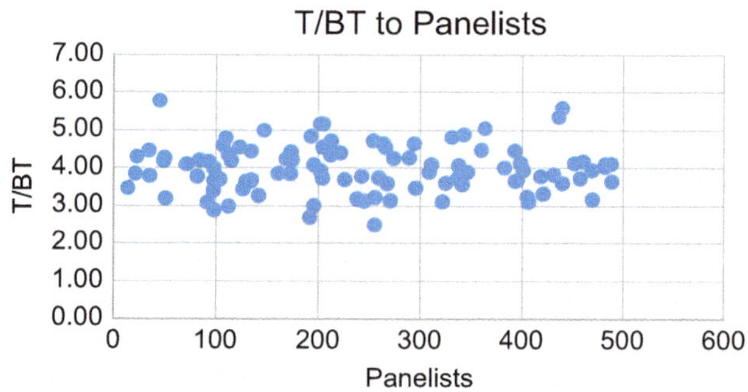

Fig. 26.1 Relationship of panelists to *T*/BT ratio

historical show and, therefore, calculate the *T*/BT ratio for each previously broadcast television show. Note that the goal is typically to have as many as possible behavioral viewers fall into the total viewership, so the goal is a lower *T*/BT ratio, where 1 is the best case (i.e., all viewership was part of the desired behavioral target). So, the previously broadcast show (e.g., last television season) data would look like the data in Table 26.1 (viewership's behavioral target in 1000 s).

For any proposed new show, we can estimate the number of panelists that would watch the show, and given that panelist number, we would like to forecast the *T*/BT for the show to make the decision regarding pricing of the show. A natural approach to this problem is to build a model that predicts the *T*/BT given the panelists expected; however, a quick look (see Fig. 26.1) at the data (we have 100 samples in the data and only show ten samples in the data above for space issues; data available from the author) shows little relationship between variables.

This does not look promising, and the R^2 from a simple linear model is approximately 0 as observation would suggest, i.e., it appears that there is no relationship between the panelists and the *T*/BT. As we dig deeper into the data, however, we can see some very interesting things. For example, the BT versus panelists is shown on the left side of Fig. 26.2 where we see a clear relationship between the two variables,

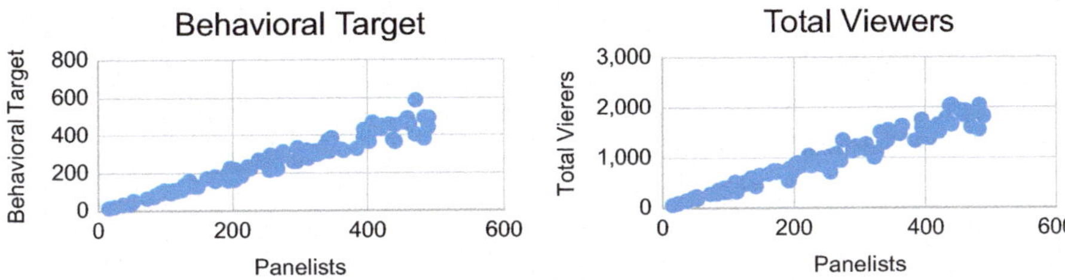

Fig. 26.2 Left—linear relationship of panelists to behavioral target (behavioral target in 1000 s). Right—linear relationship of panelists to total viewers (total viewership in 1000 s)

Fig. 26.3 No relationship of forecasted *T*/BT to actual *T*/BT

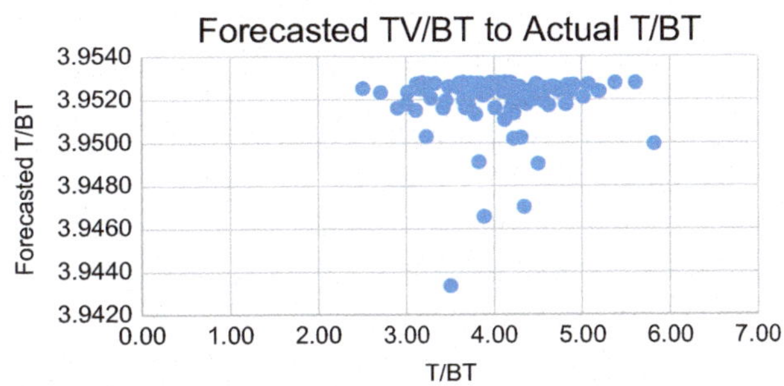

with an R^2 greater than 0.95 for a simple linear regression. The more panelists we expect, the more the BT viewers we would also expect. In a similar fashion, we show on the right side of Fig. 26.2 the same predictability exists where panelist also correlates with total viewership (again $R^2 > 0.95$).

It makes sense that since we can generate good predictions of the BT and the *T* values individually, we should be able to predict the *T*/BT ratio for future television shows. Applying the linear BT and the linear *T* models from above to calculate the predicted *T*/BT, we get the following result of the actual *T*/BT (x-axis) versus the forecasted *T*/BT (y-axis) in Fig. 26.3. We clearly see no relationship between the forecasted *T*/BT ratios and the actual *T*/BT ratios!

Thus, we can predict the *T* and BT values reasonably accurately ($R^2 > 0.95$), but the forecasted TV/BT ratio from these models has no value whatsoever (we've ignored for this simple example a holdout sample, etc.)! What happened? At a 100-thousand-foot view, the forecasted *T*/BT is actually noise about the ratio of the *T*/BT means—it has virtually no modeling or predictive value! This appears counterintuitive since we are able to get reliable viewership and behavioral predictions. Below we look under the hood to see how the numbers were generated and why the result occurred in this stylized example; however, this result can be found throughout the actual practice.

26.2 Some Technical Details

In our above example, the total viewership and behaviorally targeted data were calculated as follows:

$$T = 4N(\mu_{\mathrm{VT}}, \sigma_{\mathrm{VT}})\ P \text{ and } \mathrm{BT} = N(\mu_{\mathrm{BT}}, \sigma_{\mathrm{BT}})\ P,$$

where we assume P is the number of panelists, $\mu = \mu_{\mathrm{TV}} = \mu_{\mathrm{BT}}$, $\sigma = \sigma_{\mathrm{TV}} = \sigma_{\mathrm{BT}}$, $\mu = 1$, $\sigma = 0.1$, $N(\mu,\sigma)$ is the normal distribution, and the two normal variates are independent. Thus, one variable was basically a linear transformation of the other with a small amount of noise introduced by the normal variables. We should note that these settings for the mean and variance assure a near-zero probability of the random variables assuming a nonpositive value. So, the *T*/BT ratio is *T*/BT = 4 $N(\mu,\sigma)/N(\mu,\sigma)$, a constant times the ratio of two normal random variables. The distribution of the ratio of two normal random variables is a "ratio distribution" and is quite complex to evaluate (e.g., see Cauchy distributions, where we assume zero means—often there is no mean value existing with ratio distributions, etc.) [6], but we can see that resulting *T*/BT is a function of the two noise random variables and is not predictive of the actual TV/BT ratio (*P is not in the T/BT forecast ratio above*!).

Going further, we can take a Taylor series approximation of the ratio of the two normal random variables above [7, 8] with the assumption of no mass at 0 (notation: for random variables z and y, $E[z]$ = expectation z, Var$[z]$ = variance z, Cov $[z, y]$ = covariance between z and y, and $E^n[]$ is the expectation raised to the nth power). If we assume that z and y are independent, and both with mean $\mu = 1$, and standard deviation $\sigma = 0.1$, then:

$$E[z/y] \cong E[z]/E[y] - \mathrm{Cov}[z,y]/E^2[y] + \mathrm{Var}(y)\mathrm{E}[z]/E^3[y],$$

and since $E[z] = E[y] = 1$, Cov$[z,y] = 0$, and $E^3[y] = 1$,

$$E[z/y] \cong 1 + \mathrm{Var}[y],$$
$$E[z/y] = 1 + (\sigma)^2 = 1.01 \text{ (for example)}.$$

Note that the simplest (first-order expansion) estimate for $E[z/y]$ is simply $E[z]/E[y]$. Also, we can estimate the variance of the ratio as:

$$\mathrm{Var}[z/y] \cong {\mu_z}^2/{\mu_y}^2\ \left[{\sigma_z}^2/{\mu_y}^2 - 2\mathrm{Cov}[z,y]/(\mu_z\mu_y) + {\sigma_y}^2/{\mu_y}^2\right]$$
$$= 1\ \left[{\sigma_z}^2 + {\sigma_y}^2\right] = 2\,\sigma^2 = 2\ (0.1)^2 = 0.02 \text{ (for the example)}.$$

So we know the estimate of the ratio of the mean = $1 + \sigma^2$ and the variance of the ratio = $2\sigma^2$. Going back to our original problem,

$$T/\mathrm{BT} = 4[N(1,0.01)/N(1,0.01)],$$
$$E[T/\mathrm{BT}] = 4E[N(1,0.01)/N(1,0.01)] = 4 \times (1.01),$$
$$\mathrm{Var}[T/\mathrm{BT}] = 4^2\mathrm{Var}[N(1,0.01)/N(1,0.01)] = 16 \times (0.02).$$

So the estimated mean forecasted *T*/BT = 4 × 1.01 = 4.04 (or very close to using the simplest estimate of *E*[*T*/BT] = 4 $\mu_{\mathrm{TV}}/\mu_{\mathrm{BT}}$ = 4), and the variance is estimated as 16 × 0.02 = 0.32, while the actual *T*/BT ratios were in the range of 2.5 to almost 6. A simple back-of-the-envelope (BoE) validation experiment generating 20,000*T* and BT samples (from above normal distributions) gets a mean *T*/BT of 4.0456 and a variance of 0.3391—so Taylor expansion estimate appears in BoE evaluation to be reasonable.

Note that if the two normal variables (noise errors in the above example) in the ratio are correlated (correlation coefficient $\rho > 0$), then the variance of the ratio will be smaller, and as ρ approaches 1, the variance will also go to 0 (see equation above for Var [*z*/*y*]; since Cov[*z*,*y*] = σ^2 if $\rho = 1$). This makes sense since a ρ of 1 means that the two normal variables will be varying identically and the ratio, therefore, always the same. This would occur if there were a reason that the errors of the numerator and denominator were moving in unison, which could occur in practice if there was some underlying issue in the predictive model of the forecasts.

What causes this overall problem of accurately forecasting the components of the ratio and not the ratio itself? One major issue is that the underlying estimates of the ratio components, e.g., total viewership and behavioral target, are highly correlated (in the stylized example the correlation is almost 1!); however, this is typical in the earlier mentioned B/C exercises where we typically *expect* there to be a relationship between the benefits and costs of a project (and not just the noise illustrated above). In the percentage example, we also expect that the predicted enrollments between the two programs would be correlated at some high level as a function of the total enrollment.

So, what's the solution? Apply advanced machine learning (SVM, neural nets, etc. [9]) to the data to forecast the *T*/BT as a function of *P*? No—we could easily overfit a model and generate meaningless predictions where a relationship does not exist! Alternatively, we could not use the *T*/BT ratio unless we can *directly* predict it, or we could look at the two dimensions separately and do Pareto analysis or apply a value/utility function to the two non-commensurate attributes [10]. Using Pareto analysis, for example, Fig. 26.4 shows in the gray box all of the television shows that are dominated by the show in the upper left corner of the box—these shows in the gray box could (possibly) be eliminated from consideration if the chart was the set of candidate or proposed shows.

We have seen in our experiences multiple examples where the performance metric being used was a ratio of ratios—imagine the potential for spurious predictions in those cases. The bottom line is it is easy to start predicting and generating models that appear to make sense but that are, in reality, predicting noise. So, always do a thorough walkthrough of the data, and avoid ratios unless it is very clear that they make sense.

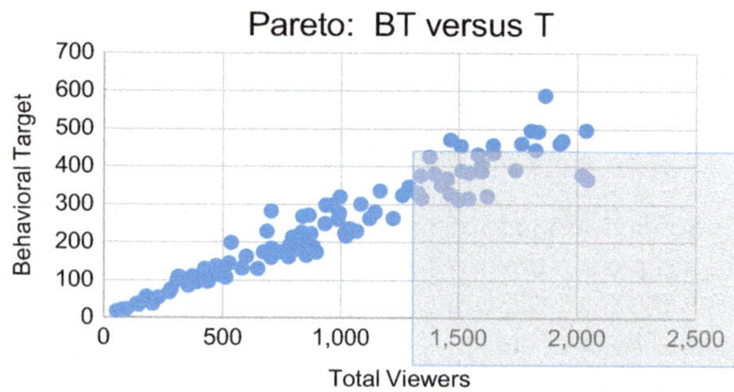

Fig. 26.4 Pareto analysis of projects illustrating dominated shows (total viewers, behavioral target in 1000 s)

26.3 An Example: Television Index Forecasting

Consider the following example, which is based on data used in television index modeling. The question posed was: "Can we use 'our' data in lieu of the equivalent data being purchased from a third party provider (XYZ data)?" Using the sample data in Table 26.2, we see what the data looks like.

Looking at the ABC Late Fringe row, for example, in the ABC network Late Fringe (termed "daypart"), we see a rating 1.29—thus 1.29% of the adults 18+ watched this network/daypart out of the possible 100%. From the next column, we see there was a rating of 1.72 for the demographic of interest (Rx); therefore, 1.72% of the possible Rx patients in the demo (18+) watched this ABC/Late Fringe. As a result, the "XYZ – Rx Patient Index" is 1.72/1.29, or 133%, meaning that there were 33% more of the people watching the time/network (daypart Late Fringe/ABC) that were in the demo of interest, Rx, compared to the overall viewership. To explain further, if the same portion of Rx people watched as the average overall population, the index would be 100. Obviously, advertisers want large indices meaning that the people viewing the network/daypart were more concentrated in the Rx population of interest.

Alternatively, the equivalent data that the company owned (our data) is shown in the last three columns—the question being can we use the data in the last three columns to predict the XYZ values, where the main number of interest is the index, the number of value to advertisers? Note that the equivalent columns are not of the same scale due to different data providers using different methods of estimating the desired index.

If we use our data (e.g., "Universe Rating") to model XYZ data (e.g., "XYZ Universe Rating") with a simple regression, we get an R^2 of about 78%, appearing that we may be able to forecast the XYZ rating data from our rating data, especially if we add more independent variables (e.g., day of week, etc.). This is illustrated in Fig. 26.5, where the correlation is easy to observe and would likely be higher with some outlier elimination. Likewise, if we model XYZ Rx patient data versus our data Rx patient data, we get an R^2 of 79% with the same conclusion.

It appears that we can forecast the numerator and the denominator of the index and thus likely forecast the XYZ index as well. However, as we saw earlier, the ratio

Table 26.2 Television data sample

Network	Daypart	XYZ universe rating (A18+)	XYZ—Rx patient rating (A18+)	XYZ—Rx patient index	Universe rating	Rx patient rating	Rx patient index
AEN	Early fringe	0.23	0.17	74	0.15	0.08	54
AEN	Late fringe	0.19	0.17	90	0.12	0.12	95
AEN	Prime time	0.39	0.32	83	0.26	0.22	85
AEN	Weekday daytime	0.26	0.19	73	0.11	0.07	64
AEN	Weekday morning	0.20	0.17	88	0.09	0.04	42
AEN	Weekend daytime	0.28	0.22	79	0.12	0.11	90
ABC	Early fringe	4.84	5.57	115	2.45	4.43	180
ABC	Late fringe	1.29	1.72	133	0.44	0.62	141

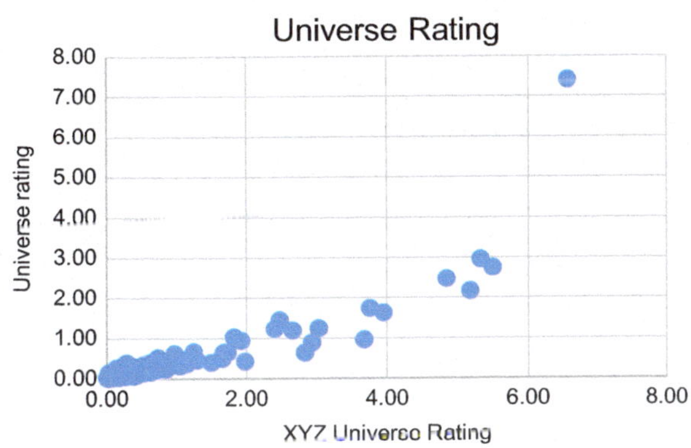

Fig. 26.5 Near-linear relationship between "Universe Rating" and "XYZ Universe Rating"

of these two could be very difficult to forecast if the numerator and denominator are highly correlated—in this case the "XYZ Universe Rating" and the "XYZ – Rx Patient Rating" are correlated at 0.99. The actual data looks very much like the stylized example earlier, where one variable is basically a linear transformation of the other and the ratio forecast is just noise ratios! Figure 26.6 shows a plot of the two indices (XYZ versus ours), where we see the effect and the resulting non-correlation ($R^2 \sim 0.00$) and the inability to forecast one from the other.

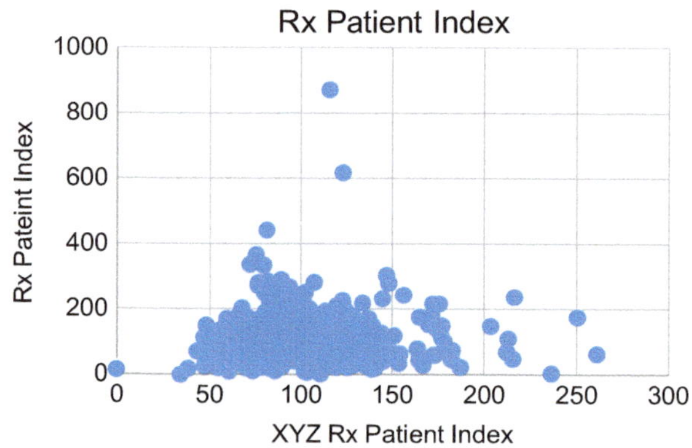

Fig. 26.6 No relationship between our patient index and the XYZ patient index

26.4 Another Example: Forecasting Viewership for Political Campaigns

As another example, we look at data on forecasting the index (as above), but this time for the viewership of a political Internet video campaign, in the identical vein as the example above where the denominator is the audience universe ratings and the numerator is the targeted audience ratings, and we want to predict the index (ratio). The three components and their correlations are illustrated in Fig. 26.7 (left) the denominator (audience) and Fig. 26.7 (right) the numerator (demo). The correlation of the ratio variables in Fig. 26.8 (left) and in Fig. 26.8 (right) illustrates the inability to forecast the ratio of demo to audience (where we include the regression fits). From the equations in the figures, the R^2 are in the range of 0.90–0.95, and once again the indices are not significantly correlated ($R^2 = 0.11$).

26.5 Conclusion

We have demonstrated the dangers of using ratio metrics, especially in light of the frequent ability to forecast the components of the ratio. We see many cases, illustrated by digital marketing, where we often use indices that are the ratios of ratio, something like index = [(A/B)/(C/D)], where A and B are highly correlated as are C and D. We often build forecasting models for the four components and combine the four forecasts into an index, where the forecast is fraught with the issues discussed in this paper. Our goal in this paper is to make practitioners aware of these pitfalls.

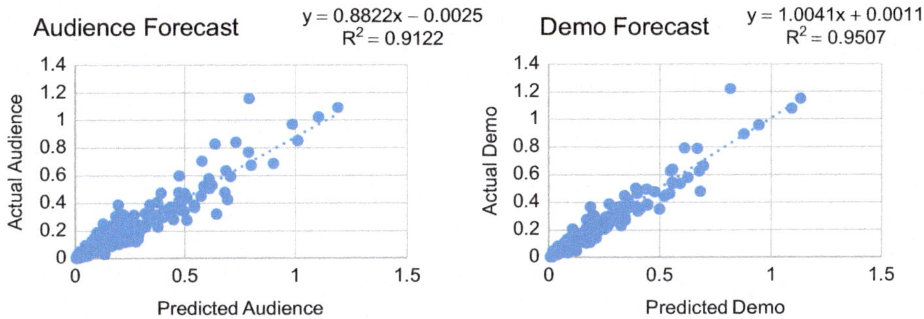

Fig. 26.7 Left—ability to accurately forecast denominator. Right—ability to accurately forecast numerator

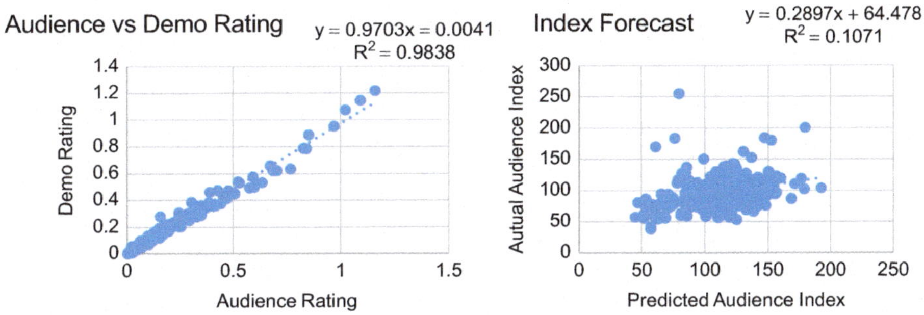

Fig. 26.8 Left—correlation of two ratio variables,—audience and demo. Right—lack of ability to forecast index ratio

References

1. Blank, L., & Tarquin, A. (2005). *Engineering economy*. New York: McGraw-Hill.
2. Chen, K. H., & Shimerda, T. A. (1981). An empirical analysis of useful financial ratios. *Financial Management*, 51–60.
3. Shirata, C. Y. (1998). Financial ratios as predictors of bankruptcy in Japan: an empirical research. In *Proceedings of the second Asian Pacific interdisciplinary research in accounting conference* (pp. 437–445).
4. Yeh, Q. J. (1996). The application of data envelopment analysis in conjunction with financial ratios for bank performance evaluation. *Journal of the Operational Research Society, 47*(8), 980–988.
5. Gibson, J. E., Scherer, W. T., Gibson, W. F., & Smith, M. C. (2016). *How to do systems analysis: Primer and casebook*. New York: Wiley.
6. Casella, G., & Berger, R. L. (2002). *Statistical inference* (Vol. 2). Pacific Grove, CA: Duxbury.
7. Kendall, M., Stuart, A., Ord, J. K., & O'Hagan, A. (1994). *Kendall's advanced theory of statistics, volume 1: Distribution theory* (6th ed.). Arnold.
8. Seltman, H. (2017). *Approximations for mean and variance of a ratio*. www.stat.cmu.edu/~hseltman/files/ratio.pdf
9. Michalski, R. S., Carbonell, J. G., & Mitchell, T. M. (2013). *Machine learning: An artificial intelligence approach*. Berlin, Germany: Springer.
10. Clemen, R. T., & Reilly, T. (2013). *Making hard decisions with decision tools*. Boston: Cengage Learning.

Chapter 27
Uncertain Requirements in the Design Selection Problem

Somayeh Moazeni, Roshanak Nilchiani, and Antonio Pugliese

27.1 Introduction

The selection of a specific design or architecture for a system is an important component of the systems engineering process. Given a set of requirements, capturing objectives and constraints, there may be several designs that will satisfy them at various levels. More precisely, the designs that fully satisfy the constraints, the suitable designs, perform differently in achieving the objectives. The purpose of the selection process is then to identify a design, or designs, that better achieve the objectives, thus maximizing the utility for the stakeholders.

When the constraints are well defined and the suitability of a design can be precisely measured, the search for an optimal configuration is about finding the maximum of the objective function. However, when the constraints are subject to uncertainty, both the modeling and optimization should take into account these uncertainties. One of these modeling and solution strategies is the robust approach, which tries to select a design with the optimum of the objective function under the worst-case scenario when all the uncertainties act upon the constraints in the direction of shrinking the design space. Selecting a design with the robust optimization approach means to build a system that will work under all scenarios with a performance at least as expected, and likely better than expected, when not all of the uncertain events happen simultaneously. This approach is preferred for the design of those systems in which failing a requirement cannot be tolerated and acceptable under any circumstances.

In this paper, two concerns regarding the classical robust optimization, due to its significant reliance on pre-specified sets for uncertainties, are rigorously discussed. First, we show that the robust counterpart problem may become infeasible, even

S. Moazeni (✉) · R. Nilchiani · A. Pugliese
Stevens Institute of Technology, Hoboken, NJ, USA
e-mail: smoazeni@stevens.edu

S. Adams et al. (eds.), *Systems Engineering in Context*,
https://doi.org/10.1007/978-3-030-00114-8_27

when a feasible design exists for many realized values of the uncertain inputs. Thus, in such cases, models explaining uncertainties in the requirements must be readjusted and re-specified in order to attain a feasible design space. In addition, even when the robust counterpart problem is feasible and a robust solution can be found, the system's performance at the robust solution might not be satisfactory and may be significantly worse than the objective function value at the nominal design. This conservativeness of the approach has been previously noted in the literature [1, 2]. Whence to reduce conservativeness of the robust solution, similar to the previous case, the uncertainty sets need to be redefined.

Second, we provide an argument that for an effective adjustment, uncertainty sets of all constraints and the objective function must be taken into account simultaneously, in accordance with the claim that pairwise requirement comparison does not help in conflict resolution [3]. Specifying the parameters individually might be ineffective. Highlighting these uncertainty set-related shortcomings of the classical robust optimization can motivate further research to develop more favorable approaches in the systems engineering process to obtain a robust solution.

27.2 Literature Review

The problem of design selection has been studied extensively in the recent years. This problem is often referred to as *tradespace exploration* [4]. The tradespace exploration paradigm allows for the search of an optimal configuration or, to be more precise, a set of Pareto optimal configurations that take into account the utility of the system and its cost [5]. This approach is based on the consideration that typically space systems are evaluated using four metrics: cost, time, quality, and risk [6]. A different approach considers also the market potential of the design, taking into account competitor products [7], and the possibility of generating a product line out of a certain design, thus looking for flexibility in the design [8]. Both these methods focus mainly on the evaluation of design alternatives, and the selection of the best ones, without thinking whether the design space in which they are searching is in fact an optimal one.

The shape of the design space is defined by requirements representing constraints that the system needs to satisfy. The work on conflicting requirements [9] explains how some sets of requirements can define design spaces that have no solution and that it is necessary to consider all the requirements at the same time, thus showing that pairwise comparison can hide the conflict. This model assumes that requirements are exact, and there is no uncertainty about their formulation. The problem of conflict in requirements in more general settings has been discussed in [10] and references therein.

Uncertainties are often considered at early phases of the system design [11]. Several efforts have been made toward a standardization of the type of uncertainties to be considered [12]. When uncertainties are considered, one of the attractive approaches to address optimization under uncertainty is the (classical) robust optimization [13, 14]. In robust optimization, uncertainty is defined deterministically through uncertainty sets, which include all or most possible realizations of the uncertain inputs. Robust optimization, then, offers a solution which has the best worst performance when inputs belong in the given uncertainty sets. Most of the current literature on robust optimization is devoted to the applications of robust optimization or to develop tractable algorithms to solve the minimax robust counterpart problem, both under the assumption that prior uncertainty sets are given. This critical assumption about the decision environment is, however, a large burden on the decision-maker. Determining uncertainty sets is a difficult task. Firstly, there are not many statistical techniques and studies available on constructing the uncertainty sets. Secondly, as it is shown in [10], the robust solution and its objective function value can be very sensitive to changes in parameters of the uncertainty sets.

27.3 Methodology

In this section, we outline the model of data uncertainty considered in this paper. The design goals are represented as an objective function that will be maximized. Suitability (or feasibility) of designs is defined by a set of constraints. For the purpose of illustration, we concentrate on constraints and objectives which are linear in design parameters with (well-defined) deterministic coefficient matrix. Here, we assume that solely the coefficients of the objective function and the entries of the right-hand-side vector are subject to uncertainty. In reality, not all the system requirements will be linear, but this assumption simplifies the model and allows us to show the effects of uncertainty.

27.3.1 Robust Optimization and Design Selection Problem

An uncertain linear programming problem with a deterministic coefficient matrix $A \in \mathbb{R}^{m \times n}$ is of the form

$$\max_{x \in \mathbb{R}^n} \quad \tilde{c}^{\,T} x \text{ s.t.} Ax \leq \tilde{b} \tag{27.1}$$

where $\tilde{c}$ and $\tilde{b}$ are n-vector and an m-vector, whose entries are subject to uncertainty and represent, respectively, the importance that stakeholders give to the various variables and the values of the constraints. Throughout, superscript ^{T} denotes transpose of a vector or a matrix. Without loss of generality in the following discussion, we assume that all of the $\tilde{b}_i$s are uncertain.

Following the common practice in robust optimization [13], we assume that the uncertain inputs, $\tilde{c}$ and $\tilde{b}$, depend on a set of primitive independent uncertainties $\{\tilde{z}_l\}_{l=1}^{N}$, where $(\tilde{z}_1, \cdots, \tilde{z}_N)$ belongs in a set defined through a vector norm $\|\cdot\|$. The choice of norm may depend on the statistical distribution assumption of uncertain parameters or a subjective opinion of the decision-maker or the model user.

$$\mathcal{U}_{b_i}(\Omega_i)\underline{\underline{\text{def}}}\left\{\tilde{b}_i = b_i^{(0)} + \sum_{l=1}^{N} \Delta b_i^{(l)} \cdot \tilde{z}_l : \tilde{z} = (\tilde{z}_1, \cdots, \tilde{z}_N) \in \mathbb{R}^N, \|\tilde{z}\| \le \Omega_i \right\},$$
$$i = 1, \cdots, m$$
$$\mathcal{U}_c(\Omega_c)\underline{\underline{\text{def}}}\left\{\tilde{c} = c^{(0)} + \sum_{l=1}^{N} \Delta c^{(l)} \cdot \tilde{z}_l : \tilde{z} = (\tilde{z}_1, \cdots, \tilde{z}_N) \in \mathbb{R}^N, \|\tilde{z}\| \le \Omega_c \right\} \tag{27.2}$$

Here, $c^{(0)}$ and $b^{(0)} = \left(b_1^{(0)}, \cdots, b_m^{(0)}\right)^T$ are the nominal values of the data. Denote $\Delta b_i = \left(b_i^{(1)}, \cdots, b_i^{(N)}\right)$ and $\Delta c = (\Delta c^{(1)}, \cdots, \Delta c^{(N)})$. We assume that the directions of data perturbation are nonnegative, $\Delta b_i \ge 0$ and $\Delta c \ge 0$. The parameters Ω_c and Ω_i are referred to as the *budgets of uncertainty* for the objective function and the constraints, respectively.

Let a_i be the ith column of A^T. It is easy to see that the robust counterpart of the constraint $a_i^T x \le \tilde{b}_i$with respect to the uncertainty set $\mathcal{U}_{b_i}(\Omega_i)$ will be

$$a_i^T x \le b_i^{(0)} - \Omega_i \|\Delta b_i\|^* \tag{27.3}$$

where $\|\cdot\|^*$ denotes the dual norm given by $\|u\|^* = \sup \{u^T x : \|x\| \le 1\}$.

Budget of uncertainty parameters is usually specified subjectively by the decision-maker or affected by several elements of institutional, social, physical, economical, and environmental limitations. These are the design margins that are used by the engineers to take into account manufacturing or integration uncertainties or the monetary budget margins that program managers use as a buffer in order to overcome funding uncertainties. The parameter Ω_c is mainly to control the degree of conservatism of the robust solution. As Ω_c increases, one expects that the robust optimal value decreases. The role of the budgets of uncertainty, Ω_is, is to adjust robustness of the proposed solution. Bertsimas and Sim [2] suggest to select these parameters by establishing probabilistic guarantee for feasibility under reasonable probabilistic assumptions on uncertain inputs. Some attempts to determine uncertainty sets through risk measures have been made [15, 16]. However, these methods assume that the support of the risk measure is given and is still performed constraint-wise.

Determining proper values for the parameters Ω_c and Ω_i is often a challenging task. An improper assignment of these parameters may result in an infeasible robust problem; whence the model user remains with no solution. Even when the robust counterpart problem is feasible, the stakeholders may be unsatisfied by the level of optimality or robustness of the proposed solution. Thus, to cure infeasibility or to reduce conservativeness of an obtained robust solution, these parameters must be modified jointly. These issues are addressed further in Sects. 3.2 and 3.3.

27.3.2 Measure of Infeasibility

We are now going to introduce a measure to quantify infeasibility of a system of linear inequalities. The quantification of the infeasibility in the design selection problem, based only on its requirements, has the goal of measuring by how much the constraints need to be adjusted in order to have a non-empty design space. Define the *measure of infeasibility* of the system $Ax \leq \tilde{b}$, as below:

$$D(A, \tilde{b}) \overset{\text{def}}{=} \max\left\{0, \min_{x \in \mathbb{R}^n} \max_{i=1,\cdots,m} \frac{a_i^T x - b_i^{(0)} + \Omega_i ||\Delta b_i||^*}{||\Delta b_i||^*}\right\}$$

This measure is scale-independent, meaning that for every positive $\lambda \in \mathbb{R}^m$, we have $D(\mathrm{Diag}(\lambda)A, \mathrm{Diag}(\lambda)\tilde{b}) = D(A, \tilde{b})$. Furthermore, $0 \leq D(A, \tilde{b}) \leq \max_{i=1,\cdots,m} \Omega_i$. Moreover, $D(A, \tilde{b}) = 0$ if and only if the feasible region of the robust counterpart problem, associated with the uncertainty sets $\mathcal{U}_{b_i}(\Omega_i)$, is non-empty, meaning that a robust solution exists. However, when $D(A, \tilde{b}) > 0$ there is no $x \in \mathbb{R}^n$ which satisfies inequalities (27.3), for all $i = 1, \cdots, m$. In other words, there is no design that satisfies all the constraints simultaneously, together with their margins.

The min-max problem in the definition of $D(A, \tilde{b})$ can be computed efficiently by solving the following linear programming problem:

$$\begin{aligned} \min_{x \in \mathbb{R}^n, \psi \in \mathbb{R}} \quad & \psi \\ \text{s.t.} \quad & a_i^T x - ||\Delta b_i||^* \psi \leq b_i^{(0)} - \Omega_i ||\Delta b_i||^*, i = 1, \cdots, m \end{aligned}$$

Thus, if $D(A, \tilde{b}) > 0$, $D(A, \tilde{b})$ is the minimum decrease in the budgets of uncertainty which guarantees a feasible robust counterpart problem. More precisely, the system

$$a_i^T x \leq b_i^{(0)} - (\Omega_i - \psi)||\Delta b_i||^*, i = 1, \cdots, m$$

is always feasible. It is worth mentioning that the definition of $D(A, \tilde{b})$ can be extended to the case when A is also subject to uncertainty, by replacing $a_i^T x$ with $\max_{a_i \in \mathcal{U}_{a_i}} a_i^T x$, where $\mathcal{U}_{a_i}$ is the uncertainty set of the elements in a_i.

27.3.3 Design Space and a Non-zero Measure of Infeasibility

When $D(A, \tilde{b}) > 0$, the classical robust optimization would not propose any solution to the model user leaving the problem as an infeasible one. However, consider an uncertain problem in which the number of primitive uncertainties equals one, i.e.,

$N = 1$, and $\tilde{z}$ is randomly uniformly distributed. Thus each $\tilde{b}_i$ is uniformly distributed in the interval $\left[b_i^{(0)} - \Omega_i \Delta b_i, b_i^{(0)} + \Omega_i \Delta b_i\right]$. Whence, we have

$$\begin{aligned}
&\Pr(Ax \le \tilde{b} \text{ is feasible}) \\
&= \Pr\left(Ax \le \tilde{b} \text{ is feasible and } b_i^{(0)} - \left(\Omega_i - D(A, \tilde{b})\right)\Delta b_i \le \tilde{b}_i, \text{ for all } i\right) \\
&\quad + \Pr\left(Ax \le \tilde{b} \text{ is feasible and } b_i^{(0)} - \left(\Omega_i - D(A, \tilde{b})\right)\Delta b_i > \tilde{b}_i, \text{ for some } i\right) \\
&\ge \Pr\left(Ax \le \tilde{b} \text{ is feasible and } b_i^{(0)} - \left(\Omega_i - D(A, \tilde{b})\right)\Delta b_i \le \tilde{b}_i, \text{ for all } i\right).
\end{aligned} \tag{27.4}$$

Using the notion of conditional probability, the probability in (27.4) can be rewritten as

$$\begin{aligned}
&\Pr\left(b_i^{(0)} - (\Omega_i - D(A, \tilde{b}))\Delta b_i \le \tilde{b}_i, \text{for } i = 1, \cdots, m\right) \times \\
&\Pr\left(Ax \le \tilde{b} \text{ is feasible} \mid b_i^{(0)} - (\Omega_i - D(A, \tilde{b}))\Delta b_i \le \tilde{b}_i, \text{for } i = 1, \cdots, m\right).
\end{aligned} \tag{27.5}$$

Since the system $a_i^T x \le b_i^{(0)} - (\Omega_i - D(A, \tilde{b}))\Delta b_i \le \tilde{b}_i, i = 1, \cdots, m$ is always feasible, the conditional probability (27.5) equals 1. Hence,

$$\begin{aligned}
&\Pr\left(Ax \le \tilde{b} \text{ is feasible and } b_i^{(0)} - (\Omega_i - D(A, \tilde{b}))\Delta b_i \le \tilde{b}_i, \text{for } i = 1, \cdots, m\right) = \\
&\qquad \Pr\left(b_i^{(0)} - (\Omega_i - D(A, \tilde{b}))\Delta b_i \le \tilde{b}_i, \text{for } i = 1, \cdots, m\right).
\end{aligned}$$

Using this equality along with inequality (27.4), we arrive at

$$\begin{aligned}
\Pr(Ax \le \tilde{b} \text{ is feasible}) &\ge \Pr\left(b_i^{(0)} - (\Omega_i - D(A, \tilde{b}))\Delta b_i \le \tilde{b}_i, \text{for } i = 1, \cdots, m\right) \\
&= \prod_{i=1}^{m} \Pr\left(b_i^{(0)} - (\Omega_i - D(A, \tilde{b}))\Delta b_i \le \tilde{b}_i\right) \\
&= \prod_{i=1}^{m} \left(1 - \frac{D(A, \tilde{b})}{2\Omega_i}\right)
\end{aligned}$$

where the last equality comes from the assumption that $\tilde{b}_i$ is uniformly distributed with support $\left[b_i^{(0)} - \Omega_i \Delta b_i, b_i^{(0)} + \Omega_i \Delta b_i\right]$. Now assume $\Omega_i = 1$ for all $i = 1, \cdots, m$. Thus, we get

Table 27.1 Lower bounds on the feasibility probability of the system $Ax \leq \tilde{b}$ *for normally distributed* $\tilde{b}$. Here, $\sigma = 0.25$, $\Omega_i = 1$ for $i = 1, \cdots, m$, and $D(A, \tilde{b}) = 0.001$

m	$\prod_{i=1}^{m} \Pr\left(b_i^{(0)} - (\Omega_i - D(A, \tilde{b}))\Delta b_i \leq \tilde{b}_i\right)$
1500	0.95283
500	0.98402
100	0.99678
25	0.99919
15	0.99951

$$\Pr\left(Ax \leq \tilde{b} \text{ is feasible}\right) \geq \left(1 - \frac{D(A, \tilde{b})}{2}\right)^m. \tag{27.6}$$

Let $D(A, \tilde{b}) = 0.001$. Thus when the optimization problem (27.1) has 20 constraints, i.e., $m = 20$, inequality (27.6) yields to

$$\Pr\left(Ax \leq \tilde{b} \text{ is feasible}\right) \geq 0.99.$$

This indicates that although the classical robust methodology considers the problem infeasible and proposes no robust solution, the underlying system is feasible with probability more than 99%; in other words, the problem is feasible for many realizations of the uncertain inputs. This is where robust optimization fails in the search of a suitable design. The assumption that all the constraints need to be satisfied at the same time, in order to have a robust system, will reject a large amount of designs that can easily satisfy $m - 1$ uncertain requirements or even m requirements in case a correction to the margin is applied.

A similar result can be observed when the entries of $\tilde{b}$ are independently and *normally* distributed. Let $\tilde{b}_i$ be normally distributed with mean $b_i^{(0)}$ and standard deviation $\sigma\Delta b_i$. Table 27.1 presents lower bounds on the feasibility probability of the system $Ax \leq \tilde{b}$, when $\sigma = 0.25$. This table shows that the system remains feasible with a high probability even for a fairly large number of constraints.

In such cases that the problem is feasible for many realizations of the uncertain parameters, one expects that the chosen budgets of uncertainty are capable to offer a robust solution. This expectation, however, may not be fulfilled unless these parameters are determined jointly and based on a systematic method. In the next section, we discuss that the parameters determining the uncertainty sets must be reassigned (to make the robust counterpart problem feasible) simultaneously. This is in particular important, when we are dealing with large number of constraints in the optimization problem.

27.3.4 Adjustment of Parameters Defining the Uncertainty Sets

One of the shortcomings of the classical robust optimization, frequently reported in the literature, is its conservative nature, in the sense that too much of optimality for the nominal problem may be lost in order to ensure robustness [1]. One of the remedies suggested in the literature to find a less conservative robust solution has been to use different vector norms for a fixed budget of uncertainty. Since norms are related to each other:

$$\|z\|_\infty \leq \|z\|_2 \leq \|z\|_1 \leq \sqrt{N}\|z\|_2 \leq N\|z\|_\infty$$

a smaller norm results in a bigger uncertainty set and (most likely) a more conservative robust solution. For example, Ben-Tal and Nemirovski [1] suggest to use ellipsoidal norm instead of $\|\cdot\|_\infty$. Bertsimas and Sim [2] propose to use $\|z\|_\Gamma = \max\{\frac{1}{\Gamma}\|z\|_1, \|z\|_\infty\}$, to control the level of conservatism in the robust solution and to obtain less conservative solutions compared to Soyster's method [17].

However, when the coefficient matrix A is not subject to uncertainty, inequality (27.3) shows that for a chosen norm $\|\cdot\|_s$ and the budget of uncertainty Ω_i, there exists some budget of uncertainty $\widehat{\Omega}_i$, which, along with another norm $\|\cdot\|_t$, describes the same robust feasible region. Indeed, it is enough to set $\widehat{\Omega}_i = \Omega_i \frac{\|\Delta b_i\|_s^*}{\|\Delta b_i\|_t^*}$, where $\|\Delta b_i\|_s^*$ and $\|\Delta b_i\|_t^*$ are dual norms of $\|\cdot\|_s$ and $\|\cdot\|_t$, respectively. For example, if $\|\cdot\|_s$ is $\|\cdot\|_\Gamma$and $\|\cdot\|_t$ is $\|\cdot\|_2$, we should assign

$$\widehat{\Omega}_i = \Omega_i \frac{\|\Delta b_i\|_s^*}{\|\Delta b_i\|_t^*} = \Omega_i \frac{\max\left(\frac{1}{\Gamma}\|\Delta b\|_1, \|\Delta b\|_\infty\right)}{\|\Delta b\|_2}$$

Hence, when the coefficient matrix defining the feasible region, A, is deterministic, the norm describing the uncertainty can be fixed and the conservativeness can be solely controlled through the budgets of uncertainty. An inappropriate selection of these parameters, then, tends to result in an over conservative solution, if feasible at all.

To obtain a less conservative solution, the budgets of uncertainty, for both the objective function and constraints, need to be updated. However, this readjustment may not necessarily be effective, when it is done individually. As an example, consider a linear programming problem with four constraints and two decision variables (x, y):

$$\max_{x,y \in \mathbb{R}} \quad \tilde{c}_1 x + \tilde{c}_2 y \text{ s.t. } \tilde{b}_1 \leq x \leq 1, \tilde{b}_2 \leq y \leq 1$$

where $\tilde{b}_1 \in [-2, 0]$ and $\tilde{b}_2 \in [-2, 0]$ with nominal values $b_1^{(0)} = -1$ and $b_2^{(0)} = -1$. Further,

$$\mathcal{U}_c(\Omega_c) = \left\{ \begin{pmatrix} \tilde{c}_1 \\ \tilde{c}_2 \end{pmatrix} = \begin{pmatrix} c_1^{(0)} \\ c_2^{(0)} \end{pmatrix} + \begin{pmatrix} 1 \\ 0 \end{pmatrix} \tilde{z}_1 + \begin{pmatrix} 0 \\ 1 \end{pmatrix} \tilde{z}_2 : \left\| (\tilde{z}_1, \tilde{z}_2) \right\|_1 \le \Omega_c \right\}.$$

Here, $\left(c_1^{(0)}, c_2^{(0)}\right) = (1, 1)$. The nominal problem will be

$$\begin{array}{cc} \max\limits_{x, y \in \mathbb{R}} & x + y \\ \text{s.t.} & \begin{array}{c} -1 \le x \le 1, \\ -1 \le y \le 1. \end{array} \end{array}$$

Thus the unique nominal solution is $x = -1$ and $y = -1$, in which case the nominal optimal objective function value equals -2. The robust counterpart problem is

$$\min_{x, y \in \mathbb{R}} \max_{\left(\tilde{c}_1, \tilde{c}_2\right)^T \in \mathcal{U}_c(\Omega_c)} \tilde{c}_1 x + \tilde{c}_2 y \text{ s.t. } \tilde{b}_1 \le x \le 1, \tilde{b}_2 \le y \le 1.$$

Thus the unique robust solution is $x = 0$ and $y = 0$, and the robust optimal objective value will be 0. This solution does not depend on Ω_c. Thus the robust optimal objective function value cannot get improved even when the model user updates Ω_c, unless the interval uncertainty sets for $\tilde{b}_1$ and $\tilde{b}_2$ also get adjusted. However, if in this example we update both the uncertainty set for $\tilde{b}_1$ and the budget of uncertainty Ω_c to $\tilde{b}_1 \in [-2, -0.5]$ and $\Omega_c = 0.5$, then the unique robust solution will be $x = -0.5$ and $y = 0$. Thus the robust optimal objective function value is decreased to -0.25. Therefore, improvement can only be achieved when the budgets of uncertainty are updated simultaneously.

This example shows the need for an approach to balancing the optimality and robustness, in that parameters of the uncertainty sets for the constraints as well as for the objective function are updated jointly and not individually. Translating this result in systems engineering words, we see that the value delivered to the stakeholders cannot be maximized by only solving the uncertainties related to their utility functions, which means that working only on defining more precise system performances, and getting the stakeholders to sign the most precise contracts in order to cancel any uncertainty on what are their real needs, is not going to help. There is a need to work on the uncertainties that relate to the constraints of the problem; otherwise the optimal design will not be reached. Therefore, a robust optimization approach should be equipped with some preprocessing, to infer as much useful information as possible about the structure of the feasible region polytope and to detect contributing constraints and factors to the robustness of the problem.

27.4 Conclusions

The classical robust optimization approach to deal with parameter uncertainty in optimization problems heavily relies on the description of the given uncertainty sets. In this paper, we address several shortcomings of the approach due to this dependence. Improper values for the parameters specifying the uncertainty sets may make the robust counterpart problem infeasible, while the problem has feasible points for many realizations of the uncertain inputs. Furthermore, individual adjustment of these parameters, to make the problem feasible or less conservative, can be ineffective. Thus updating the budgets of uncertainty must be done simultaneously. The discussion also highlights the importance of developing efficient techniques for data preprocessing before adopting the classical robust optimization framework, devising methods for determining uncertainty sets jointly, and proposing approaches less dependent on pre-specified uncertainty sets.

References

1. Ben-Tal, A., & Nemirovski, A. (2000). Robust solutions of linear programming problems contaminated with uncertain data. *Mathematical Programming, 88*, 411–424.
2. Bertsimas, D., & Sim, M. (2004). The price of robustness. *Operations Research, 52*(1), 35–53.
3. Salado, A., & Nilchiani, R. (2013). *The concept of order of conflict in requirements engineering*. Piscataway, NJ: IEEE.
4. Ross, A. M., Diller, N. P., Hastings, D. E., & Warmkessel, J. M. (2002). *Multiattribute tradespace exploration in space system design*. Houston, TX: International Astronautical Congress.
5. Ross, A. M., & Hastings, D. E. (2005). *The tradespace exploration paradigm*. INCOSE, pp 1–13.
6. Ross, A. M., Hastings, D. E., & Diller, N. P. (2003). Multi-attribute tradespace exploration with concurrent design for space system conceptual design. In *Aerospace Sciences Meeting*, AIAA2003–1328. Reno, NV: January, pp 6–9.
7. Li, H., & Azarm, S. (2000). Product design selection under uncertainty and with competitive advantage. *Journal of Mechanical Design, 122*(4), 411–418.
8. Li, H., & Azarm, S. (2002). An approach for product line design selection under uncertainty and competition. *Journal of Mechanical Design, 124*(3), 385–392.
9. Salado, A., & Nilchiani, R. (2014). The concept of problem complexity. *Procedia Computer Science, 28*, 539–546.
10. Moazeni, S., Coleman, T. F., & Li, Y. (2013). Regularized robust optimization for optimal portfolio execution. *Computational Optimization and Application, 55*(2), 341–377.
11. de Weck, O., Eckert, C., & Clarkson, J. (2007). A classification of uncertainty for early product and system design. *International Conference on Engineering Design (ICED)*, pp. 1–12.
12. McManus, H., & Hastings, D. (2005). A framework for understanding uncertainty and its mitigation and exploitation in complex systems. In *INCOSE International Symposium* (Vol. 15, pp. 484–503). Hoboken, NJ: Wiley Online Library.
13. Ben-Tal, A., Ghaoui, L. E., & Nemirovski, A. (2009). *Robust optimization*. Princeton, NJ: Princeton University Press.
14. Bertsimas, D., Brown, D. B., & Caramanis, C. (2011). Theory and applications of robust optimization. *SIAM Review, 53*(3), 464–501.

15. Chen, X., Sim, M., & Sun, P. (2007). A robust optimization perspective on stochastic programming. *Operations Research, 55*(6), 1058–1071.
16. Bertsimas, D., & Brown, D. B. (2009). Constructing uncertainty sets for robust linear optimization. *Operations Research, 57*(6), 1483–1495.
17. Soyster, A. L. (1973). Convex programming with set-inclusive constraints and applications to inexact linear programming. *Operations Research, 21*, 1154–1157.

Chapter 28
Applying Model-Based Validation to Inference Enterprise System Architecture Selection

Sean D. Vermillion, David P. Brown, and Dennis M. Buede

Nomenclature

IARPA	Intelligence Advanced Research Projects Activity
IE	Inference enterprise
IEM	Inference enterprise model
MBV	Model-based validation
NB	Naïve Bayesian network classifier
RF	Random forest classifier
SCITE	Scientific Advances to Continuous Insider Threat Evaluation
TN	True negative count
TP	True positive count
VM	Voting machine
vNM	von Neumann-Morgenstern

28.1 Introduction

In this paper, we describe the application of a model-based approach to comparing and selecting inference enterprise system architectures. An inference enterprise (IE) is an organizational entity that uses data, tools, people, and processes to make mission-focused inferences [1]. For example, airport security organizations use body and baggage screening equipment in conjunction with expert judgment to infer if a passenger is attempting to smuggle contraband into a flight terminal. Missile defense agencies use radar, infrared sensors, and aircraft flight plans to

S. D. Vermillion (✉) · D. P. Brown · D. M. Buede
Innovative Decisions, Inc., Vienna, VA, USA
e-mail: svermillion@innovativedecisions.com

S. Adams et al. (eds.), *Systems Engineering in Context*,
https://doi.org/10.1007/978-3-030-00114-8_28

infer whether or not an airborne object is an enemy missile. In both examples, there are conceivably several IE system architectures—e.g., architectures differentiated by the data models, tools, people, and processes incorporated therein—that organizations can deploy to make the same inference. Intuitively, organizations desire to deploy the best performing alternative, so the question becomes: *How do we rank order alternatives?*

In this paper, we approach the issue of IE system architecture alternative selection in the context of the Intelligence Advanced Research Projects Activity (IARPA)-sponsored Scientific Advances to Continuous Insider Threat Evaluation (SCITE) project. SCITE is a research program for developing fundamental knowledge on inference enterprise modeling for insider threat detection applications. Inference enterprise modeling (IEM) involves using available information about organization activities, characteristics of observables on those activates, and other relevant organizational data to predict IE inference performance—e.g., recall, precision, etc.—given proposed changes in the IE's data, tools, people, or processes. At present, we build such models using expert judgment and statistics aggregated from raw organizational data, which reflects practical situations where an organization might have missing data or be reluctant to disclose personally identifiable data [2]. Therefore, IEM inherently involves dealing with uncertainty and forecasting performance in terms of probabilities rather than deterministic values. While we use IEM to evaluate changes to IEs, we lack a framework for *rigorously* comparing these alternatives in the face of uncertainty to ultimately support IE architecture selection.

To leverage IEM advances made in the SCITE program, a model-based approach is intuitive for comparing candidate IE architectures. In this research, we refine model-based validation (MBV) to a framework for IE comparison to support our selection problem. Originally applied to comparing engineering design methodologies, MBV is a validation technique that combines normative decision theory with inductive inference to compare methodologies [3, 4]. Specifically, MBV applies decision analysis to empirical results to make inductive inferences on alternatives' rank ordering with regard to a given preference structure. In our application of MBV to IEs, we formulate mission-focused preference to rank order Monte Carlo simulation results generated using IEM and use statistical tests to infer a rank ordering beyond the simulations with a given confidence.

This paper is organized as follows. The next section discusses the foundations for this research, particularly IEM, vNM expected utility theory, and MBV. In Sect. 28.3, we describe the MBV framework for comparing IE architectures. In Sect. 28.4, we apply the MBV framework to comparing three proposed IE architectures for detecting *exfil* threats, i.e., individuals in an organization that transfer sensitive information outside of the organization. We conclude with a summary of the research and directions for future work.

28.2 Foundations

28.2.1 Inference Enterprise Modeling

In this section, we simply provide a high-level description of IEM as used in the SCITE program. IE systems considered in the SCITE program largely consist of two components: (1) organizational data on the activities, e.g., email habits or computer access histories, of individuals in that organization and (2) machine learning algorithms to predict if an individual is a threat given his observed activities. Since insider threat IE systems rely on data directly tied to an individual, it is likely organizations outsourcing IEM development would be reluctant to provide such data [2]. Alternatively, an organization might not have collected data to provide for IEM development. Since IEM developers and evaluators are likely to encounter both of these scenarios, IEM research efforts largely focus on simulating an organization's data in order to train machine learning algorithms and evaluate their performance.

Figure 28.1 describes the IEM process developed for SCITE challenge problems. In a challenge problem, we are given a particular behavior, which defines threats, to predict and data *describing* individuals' activities. With the problem structure and data received, we assess if we can simulate the organization's data. If necessary data elements are missing, e.g., detector correlations or time series autocorrelations, we elicit this information from domain experts. We then simulate an organization's data using stochastic processes. For example, we often use copula methods [5] to generate samples correlated according to our provided information or information we elicit from experts and then fit the samples to given or elicited detector, e.g., number of outgoing emails or marginal distributions. Alternatively, we may use a discrete event approach of probabilistically assigning detector values to artificial individuals according to detector value probability distributions and correlations.

Once we simulate an organization's, i.e., each individual's, data, we train a machine learning classifier, e.g., random forest classifier, naïve Bayesian network, support vector machine, etc., using a subset of the simulated data. Then we evaluate

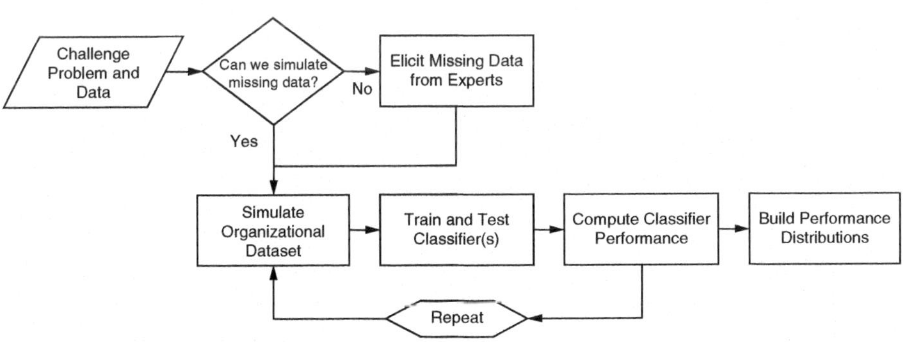

Fig. 28.1 Inference enterprise modeling meta-model

the classifier's performance on the rest of the simulated data. Since we are simulating data using stochastic processes, we generate additional data sets and repeat the classifier evaluation. Therefore, classifier performance metrics such as recall, precision, and false positive rate are stochastic, and we build distributions for these metrics through Monte Carlo simulations.

28.2.2 Expected Utility Theory

von Neumann-Morgenstern (vNM) expected utility theory is an axiomatic decision-making framework for decisions under uncertainty [6]. This framework is framed in terms of *lotteries*. Lotteries comprise of a set of outcomes, $\mathbf{Z}$, and a probability of occurrence, p_i, for each outcome, $z_i \in \mathbf{Z}$. Lotteries are comprehensive in that an outcome in $\mathbf{Z}$ will occur and so $\sum_i p_i = 1$. The axiomatic framework culminates in the following theorem about modeling rational decision-making:

Theorem 1 (*vNM expected utility theorem*). For any vNM rational decision-maker, there exists a function $u : \mathbf{Z} \rightarrow R$, unique up to a positive affine transformation, which assigns a real number to each lottery outcome $z \in \mathbf{Z}$ such that for any two lotteries, $L_1 \prec L_2$ if and only if $E[u(L_1)] < E[u(L_2)]$.

This theorem states that we can mathematically describe a preference function, and the goal of a rational decision-maker is to select an alternative that maximizes her expected utility. We use this theorem as the basis for rank ordering IEM candidates.

28.2.3 Model-Based Validation

Frey et al. formulated MBV, originally to compare engineering design methods, based on the methodology for validation in the medical sector [3, 4]. In the medical sector, candidate medical treatments are validated on human analogs, e.g., mice, in order to build confidence in their effectiveness. Treatments that are proven effective in the human analogs, i.e., human models, are then tested and developed further. MBV likewise uses experiments based on models that are representative of a given context to build confidence in one method's effectiveness over another's effectiveness. For example, Frey et al. use MBV to compare Pugh controlled convergence to a Borda count method for concept selection by simulating a team of people involved in iterated use of the methods [7]. The validation terminology stems from the notion that MBV seeks to validate that one method is superior to another in a series of experiments, thus inferring that the contextually superior method is generally superior beyond the experiment results.

MBV incorporates vNM expected utility theory as the means to rank order the empirical results generated by different alternatives. Models used in experiments output meaningful metrics or figures of merit over which decision-makers formulate preferences. By grouping the empirical results associated with a particular alternative, we effectively create a lottery as described. In a strict sense, this application of vNM expected utility constitutes ex post decision analysis, as opposed to ex ante decision analysis, since we are looking at empirical results and determining which alternative to choose after the fact. In a loose sense, empirical results inform decision-maker's beliefs over the outcomes associated with a given alternative, and then the decision-maker rank orders alternatives in an ex ante sense. In this paper, we follow the strict sense and thus recognize that we make an inductive inference on the generalizability of alternative rank orders generated from empirical results.

28.3 Model-Based Validation of Inference Enterprises

28.3.1 Model Outputs

In this section, we apply the foundations discussed in the previous section. In the SCITE program, we are largely concerned with IEs that make binary inferences, i.e., an individual is classified as either a threat or nonthreat. During IEM, we assign a true label to a simulated individual, and the IE system architecture candidates each predict the label. From a simulated population's true and predicted labels, we can create a 2×2 confusion matrix that informs IEM performance; see Table 28.1. The confusion matrix separates and lists (1) the number of correctly labeled threats (true positives), (2) the number of correctly labeled nonthreats (true negatives), (3) the number of actual threats labeled as nonthreats (false negatives), and (4) the number of actual nonthreats labeled as threats (false positives). Each individual in a simulated organization falls into one of these outcomes, and the sum of these outcomes is the number of individuals in the data test set.

Let us refer to generating a confusion matrix using a candidate and simulated organization combination as an experiment. Each experiment potentially populates the confusion matrix with different values. In the remainder of this section, we describe the three fundamental steps to the MBV framework: (1) formulating a preference model based on the confusion matrix, (2) conducting experiments to generate confusion matrices, and (3) performing statistical inference tests on experimental results.

Table 28.1 Binary classification confusion matrix

	Predicted label	
Actual label	Threat	Nonthreat
Threat	True positive (TP)	False negative (FN)
Nonthreat	False positive (FP)	True negative (TN)

28.3.2 Preference Model

Since IEs make mission-focused inferences, organizations act and make resource allocations according to these inferences. Therefore, we aggregate the counts in the confusion matrix to value to the organization. In other words, each inference the IE makes results in a value to the organization, including negative value. Allowing for each outcome in the confusion matrix to yield a different value, the value generated by experiment i is the following:

$$V_i = \mathrm{TP}_i \times V_{\mathrm{TP}} + \mathrm{FN}_i \times V_{\mathrm{FN}} + \mathrm{FP}_i + V_{\mathrm{FP}} + \mathrm{TN}_i \times V_{\mathrm{TN}} \tag{28.1}$$

In experiment i, the IE makes TP_i inferences that each yield V_{TP}, FN_i inferences that each yield V_{FN}, and so forth. This approach is similar to cost-sensitive classification and learning, which often only considers the costs of making incorrect predictions, i.e., costs or penalties generated by false negatives and false positives [8, 9]. We characterize a decision-maker's risk attitude toward uncertain value by defining a utility function, $u(\cdot)$. Therefore, for each experiment i, we compute V_i to feed into a utility function to yield $U_i = u(V_i)$. We do not prescribe a particular function form for $u(\cdot)$ since risk preference is subjective and dependent on the decision-maker.

28.3.3 Monte Carlo Experiments and Statistical Inference

Since we are uncertain in our organization's data, we use IEM to simulate this data, as discussed in Sect. 28.2. To propagate our data uncertainty to the counts in the confusion matrix, we employ a Monte Carlo approach. We simulate the organization's data through MIEM n times. For each simulated data set and for each IEM candidate, we compute a confusion matrix. From these confusion matrices, we compute the value, V_i, generated in the experiment and its corresponding utility, $U_i = u(V_i)$. See Fig. 28.2.

Previously, we defined an experiment as comprising of a simulated organization data set and an IEM candidate. Experiment i comprises of data set k and IEM candidate j such that $i = \langle k, j \rangle$. Using this notation, we estimate the expected utility of candidate j as the following:

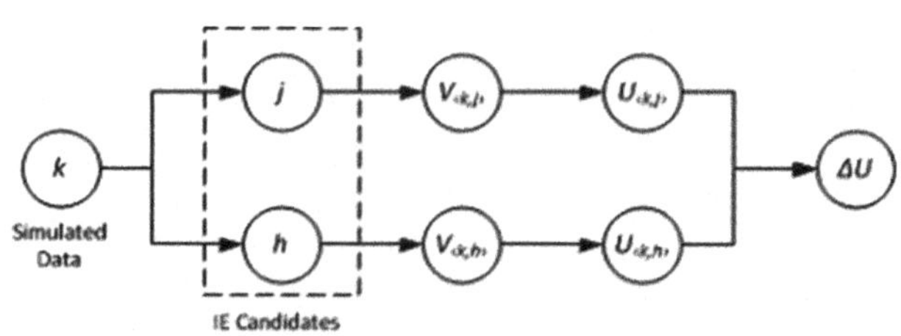

Fig. 28.2 A decision-theoretic approach to using model-based validation to compare two methods

$$E[U]_j \approx \frac{1}{n} \sum\nolimits_{k=1}^{n} U_{\langle k,j \rangle} \tag{28.2}$$

From Theorem 1, we know that candidate j is more preferable, or ranked higher, than candidate h if $E[U]_j > E[U]_h \Rightarrow E[U]_j - E[U]_h > 0$. To make the inference that candidate j is superior to candidate h with a given confidence, we compute the paired t-interval of $E[U]_j - E[U]_h$ using our estimate in Eq. (28.2):

$$\frac{1}{n} \sum\nolimits_{k=1}^{n} \left[U_{\langle k,j \rangle} - U_{\langle k,h \rangle} \right] \pm t \times \frac{s_d}{\sqrt{n}} \tag{28.3}$$

where t is determined from a Student's t table and s_d is the standard deviation of the paired utility differences. If every point in the confidence interval is positive, then we infer that candidate j is preferable over candidate h. We repeat this pairwise procedure for all pairs of IEM candidates to eliminate those we are confident are dominated. Cases where we cannot eliminate an alternative intuitively suggest we need to simulate more data sets to reduce interval width.

28.4 Case Study: Exfil Threat Inference Enterprises

28.4.1 Problem Description

We demonstrate the framework described in Sect. 28.3 by applying it to the 16th SCITE challenge problem. In the narrative of this challenge problem, a fictional organization purchased a threat alerting system that provides two classifier options, a random forest (RF) and naïve Bayesian network (NB), to detect potential cases where employees transport or transfer the organization's intellectual property or other sensitive information outside of the organization, i.e., an *exfil threat*. The organization employs independent consultants to make a recommendation as to whether the organization should configure the threat alerting system to use RF, NB, or a combination of the two. For the purposes of this demonstration, we model the organization's risk attitude using the following utility function:

$$U = u(V) = \frac{\exp(-c \times V_{\min}) - \exp(-c \times V)}{\exp(-c \times V_{\min}) - 1} \tag{28.4}$$

Parameters for Eqs. (28.1) and (28.4) are given in Table 28.2. This utility model characterizes constant absolute risk aversion and is scaled so that $0 \leq U \leq 1$. The state values listed in Table 28.2 are notional and represent a case where an organization expends \$1000 just to investigate an individual, but the risk value for not detecting a threat is \$1000,000.

Table 28.2 Case study preference parameters

Parameter	Description	Value
V_{TP}	Value generated by true positives	–\$1000
V_{FN}	Value generated by false negatives	–\$1000,000
V_{FP}	Value generated by false positives	–\$1000
V_{TN}	Value generated by true negatives	0
c	Risk coefficient	\$2E-9
V_{min}	Minimum value (maximum cost)	–\$1E9

28.4.2 Available Data

The organization makes available 10 days'—two working weeks'—worth of data for approximately 4000 individuals within the organization. This data set contains information measured from 26 detectors about an individual's daily email habits, USB transfer habits, website access habits, and machine access habits. Additionally, the organization provides expert-elicited labels for each individual indicating whether or not the individual is an exfil threat. We use the provided data set in IEM activities to simulate alternate data sets to account for variation in an individual's habits. At a high level, we aggregate, for each label, daily detector values and generate statistics such as their means, standard deviations, detector marginal distributions, correlations between detectors, and detector autocorrelations from day to day. Then, we use a Gaussian copula to generate correlated data sets that we then transform to have similar marginal distributions and autocorrelations across days as our original data set. We use this procedure to simulate 14 data sets, i.e., 10 days' worth of data for each individual, in addition to the provided data set to use for IEM performance comparison.

28.4.3 IE System Candidates

As mentioned, the organization can deploy an RF, an NB, or a combination of the two. For this demonstration, we use the RF and NB classifiers implemented in scikit-learn v0.19.1 with default parameter settings [10]. Each classifier is trained on 5 days' worth of detector values and makes predictions using 5 days' worth of detector values to provide weekly threat reports. Additionally, we consider two voting machines that combine the results of the RF and NB. One voting machine, VM1, labels an individual as a threat if at least one classifier labels the individual a threat. The second voting machine, VM2, labels an individual as a threat if both RF and NB label the individual as a threat.

28.4.4 Results

Pairwise comparison results are shown in Fig. 28.3. In addition to pairwise analysis with expected utility, we do the same analysis using the F_1 score in place of expected utility. The F_1 score is computed using the confusion matrix in Table 28.1 as the following:

$$F_1 = \frac{2 \times \text{TP}}{2 \times \text{TP} + \text{FN} + \text{FP}} \tag{28.5}$$

The F_1 score is a commonly used metric for binary classification performance that seeks to balance type I and type II errors [11]. From Fig. 28.3, figure of merit choice impacts IE system choice. Using expected utility, the best choice is VM1. Using F_1, the best choice is RF. However, the expected utility approach explicitly considers value information and risk preferences, while expected F_1 score does not. Using the cost information in Table 28.2, deploying the RF-based IE system would lead to a 24% increase in expected cost over using the VM1-based IE system.

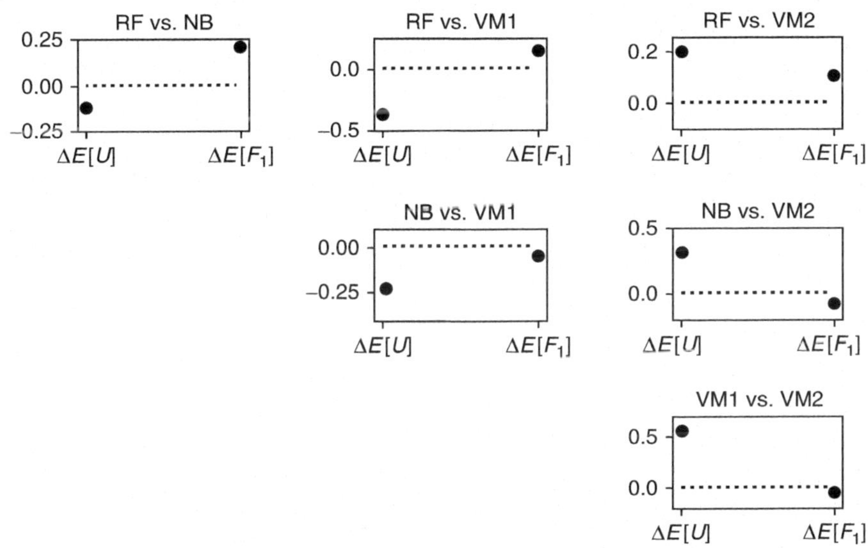

Fig. 28.3 Pairwise comparison results using expected utility and expected F_1 score. A value above zero (dashed line) indicates the candidate listed first in the subplot title is ranked higher than the second; the opposite is true for values below zero. Uncertainty intervals are computed at 95% confidence and 14 degrees of freedom. Intervals are small due to consistent metric differences across simulation runs

28.5 Discussion and Summary

In this paper, we describe a model-based framework for making inference enterprise model selection decisions. This framework leverages several disciplines: (1) inference enterprise multi-modeling, (2) normative decision theory, and (3) statistical inference. We use inference enterprise modeling and normative decision theory to evaluate and rank order inference enterprise models in the context of the organization deploying the inference enterprise system. While demonstrating the framework, we compare using expected utility as the performance metric with the generic, context-independent F_1 score, which is commonly used for binary classification performance. Selecting an inference enterprise based on the F_1 score would be approximately 24% more costly than the inference enterprise chosen by expected utility with respect to the cost information in Table 28.2. This observation strengthens the argument for contextual information and preference structures when making inference enterprise model selection decisions.

Future work on inference enterprise system architecture selection is twofold. First, we seek to incorporate stochastic dominance rules to eliminated dominated inference enterprise model candidates. The expected utility approach discussed in this paper requires detailed value and risk attitude information. Using stochastic dominance rule allows us to potentially reduce the number of considered inference enterprise model candidates with minimal preference structure knowledge [12]. Second, we seek to incorporate prior beliefs over inference enterprise model performance. Currently, we use Monte Carlo simulations to propagate data uncertainty to inference enterprise model performance. Through Bayesian updating, we can potentially update beliefs using Monte Carlo simulation results. Goutte et al. provide a probabilistic framework for modeling binary classifier performance [13].

Acknowledgment Research reported here was supported under IARPA contract 2016-16031400006. The content is solely the responsibility of the authors and does not necessarily represent the official views of the US government.

References

1. Huang, E., Zaidi, A. K., & Laskey, K. B. (2018). Inference enterprise multimodeling for insider threat detection systems. In A. M. Madni, B. Boehm, R. G. Ghanem, D. Erwin, & M. J. Wheaton (Eds.), *Disciplinary convergence in systems engineering research* (pp. 175–186). Cham: Springer International Publishing.
2. Turcotte, M. J., Kent, A. D., & Hash, C. (2017). Unified host and network data set. arXiv preprint arXiv:170807518.
3. Frey, D. D., & Dym, C. L. (2006). Validation of design methods: lessons from medicine. *Research in Engineering Design, 17*, 45–57.
4. Frey, D., & Li, X. (2006). Model-based validation of design methods. In K. E. Lewis, W. Chen, & L. C. Schmidt (Eds.), *Decision making in engineering design* (pp. 315–323). New York: ASME Press.

5. Nelsen, R. B. (2007). *An introduction to copulas*. Berlin: Springer Science & Business Media.
6. Von Neumann, J., & Morgenstern, O. (2007). *Theory of games and economic behavior* (60th Anniversary ed). Princeton, NJ: Princeton University Press.
7. Frey, D., Herder, P., Wijnia, Y., Subrahmanian, E., Katsikopoulos, K., & Clausing, D. (2009). The Pugh Controlled Convergence method: model-based evaluation and implications for design theory. *Research in Engineering Design, 20*, 41–58.
8. Domingos, P. (1999). Metacost: A general method for making classifiers cost-sensitive. In *Proceedings of the Fifth ACM SIGKDD International Conference on Knowledge Discovery and Data Mining* (pp. 155–164). New York: ACM.
9. Elkan, C. (2001). The foundations of cost-sensitive learning. In *International Joint Conference on Artificial Intelligence* (pp. 973–978). Mahwah, NJ: Lawrence Erlbaum Associates Ltd.
10. Pedregosa, F., Varoquaux, G., Gramfort, A., et al. (2011). Scikit-learn: Machine learning in Python. *Journal of Machine Learning Research, 12*, 2825–2830.
11. Sokolova, M., & Lapalme, G. (2009). A systematic analysis of performance measures for classification tasks. *Information Processing & Management., 45*, 427–437.
12. Hadar, J., & Russell, W. R. (1969). Rules for ordering uncertain prospects. *The American Economic Review, 59*(1), 25–34.
13. Goutte, C., & Gaussier, E. (2005). A probabilistic interpretation of precision, recall and F-score, with implication for evaluation. In *European Conference on Information Retrieval* (pp. 345–359). Berlin: Springer.

Chapter 29
A Pilot Study on Ranking the Critical Components of a System

Satish M. Srinivasan, Raghvinder S. Sangwan, Colin J. Neill, and Nil Kilicay-Ergin

29.1 Introduction

Components of complex systems are interdependent in various dimensions including physical, cyber, social, functional, and organizational. As systems grow, complex interactions between the components that compose, develop, operate, or interact with them create challenges for system developers in evolving, maintaining, and using these systems—these various networks, which evolve either intentionally or organically, mean that changes to the system must be considered carefully with regard to their impact. Considerable effort is required when a system undergoes portability or a major upgrade, for example. Such processes could require a change to the system's architectural components which poses a risk to the structural stability of the system. In addition, socio-technical aspects such as complex relationships within an organization, between business processes, and across functions mean that we may have to consider more than just the physical components of a system when making the change—will it impact the team structures in the development organization, for example? It is useful in such scenarios, therefore, if we can identify the most important, critical, or significant elements within these various systems (technical, social, organizational, etc.) so that we can prioritize activities, anticipate risks, or design for future considerations effectively.

The question, however, is how do we identify the significant components of a large or ultra-large-scale system? Recent interest in graph-theoretic approaches that have emerged in the network analytic community has found favor in this regard [1–3]. There exist several different network algorithms for ranking components that can be used to determine their significance: closeness centrality, betweenness centrality, and eigenvector centrality being the most commonly used [4–6]. Each of these algorithms, however, takes a different approach to ranking components within

S. M. Srinivasan (✉) · R. S. Sangwan · C. J. Neill · N. Kilicay-Ergin
School of Graduate Professional Studies, Penn State University, Malvern, PA, USA
e-mail: sus64@psu.edu; rxs69@psu.edu; cjn6@psu.edu; nhe2@psu.edu

S. Adams et al. (eds.), *Systems Engineering in Context*,
https://doi.org/10.1007/978-3-030-00114-8_29

a system and, therefore, results in different ranking orders, reflecting the structural properties of the network measured by each approach. Each has value but obviously with respect to different system attributes or characteristics.

In this paper, we will examine the utility of each algorithm along with that of a new ranking algorithm based upon discrete-time Markov chain that has shown promise in ranking highly those elements of complex systems deemed important by experts [7]. We then compare these four algorithms using three system engineering examples explored in detail by Eppinger and Browning [8].

29.2 Component Ranking

Using a simple graph (Fig. 29.1) as an example, we provide a brief demonstration of component rankings from three mainstream algorithms, namely, betweenness centrality, closeness centrality, and eigenvector centrality along with discrete-time Markov chain that we have developed. For brevity, we refer to these algorithms as betweenness, closeness, eigenvector, and Markov chain, respectively, throughout the remainder of this paper. The details of how betweenness, closeness, and eigenvector algorithms work can be found in [4–6]. We describe the mechanics of the Markov chain in more detail since this is an algorithm that we have developed.

Markov chain constructs a component graph that represents the dependencies among the components of a system and treats the resultant graph as a discrete-time Markov chain (DTMC), a representation used for the stochastic modeling of complex systems [9]. The resulting component graph is a directed graph consisting of a

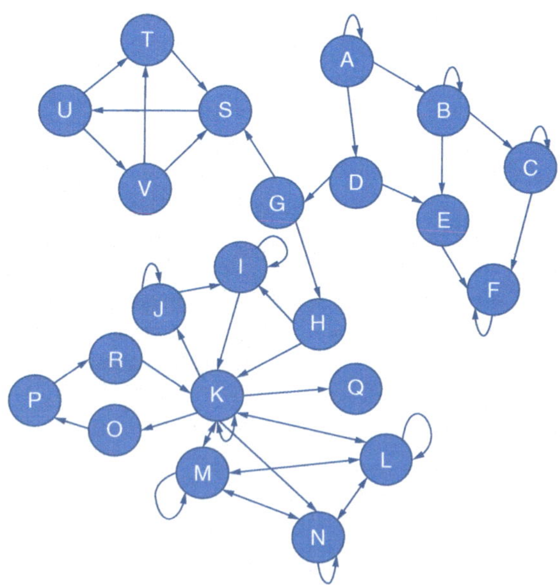

Fig. 29.1 A component graph

set of nodes representing the components and a set of links representing relationships (physical/energy-information interaction, dependency, information flow, inputs, and outputs to parameter decision) between those components. Using DTMC as an input, the Markov chain algorithm determines the weights of each component and uses these weights to determine their ranking and, therefore, their relative significance within the system under study.

If we visualize the component graph as a communication network where each node communicates a value to its neighbors along the connected edges and represent it as a probabilistic state transition with a transition probability, then this component graph is a DTMC [1]. The transition probabilities between the nodes are represented using a transition probability matrix (TPM) with each row summing up to one. Assuming all the nodes have an equal weight to begin with, we can determine the weight of the link by computing the product of the transition probability of an edge with the weight of the propagating node.

If the underlying component graph of any system is irreducible (strongly connected), recurrent (it is possible to start and eventually return back to the same state), and aperiodic (all nodes have a self-loop), then the component graph converges or attains a stationary convergence [10]. On such a component graph, the Markov chain algorithm computes the final weight of the components by multiplying the initial weights of the components against the transpose of the TPM, in multiple rounds. In each round the weights of the components are updated accordingly. This computation is continued until the component graph attains a stationary convergence. Using the final weight of each component, the Markov chain ranks the components of the system in the decreasing order of its weight.

Betweenness highly ranks components that are frequently located on as many of the shortest path between pairs of other components. Closeness identifies and highly ranks components that are closest to the cluster containing high-profile components in a system. Eigenvector aims at highly ranking components that are well connected to the components that hold a prestige of being a high-profile component in a system. Markov chain on the other hand ranks those components highly that are high profile or those that lie on as many shortest paths between pairs of other components within a system.

With this background, we now return to Fig. 29.1 to show the relative performance of each of these algorithms. The graph in Fig. 29.1 is composed of 22 nodes and 48 edges. To better understand the ranking strategies of the ranking algorithms, we have orchestrated the design of the graph as follows: it has a completely connected subgraph comprised of nodes K, N, M, and L; a clique that includes nodes S, U, T, and V; and a bridge, node G.

Figure 29.2 shows the rankings of the nodes in this graph from these respective algorithms. We can see that the Markov chain ranked the nodes of the completely connected subgraph highly, the clique in the middle, and the bridge toward the bottom. A degree distribution analysis of the nodes in the graph reveals that nodes with higher in-degrees were ranked higher in the list with some exceptions. Node I with a higher in-degree than nodes J, O, and Q was ranked lower. This is because nodes J, O, and Q have an incoming link from node K (a higher ranked node), thus

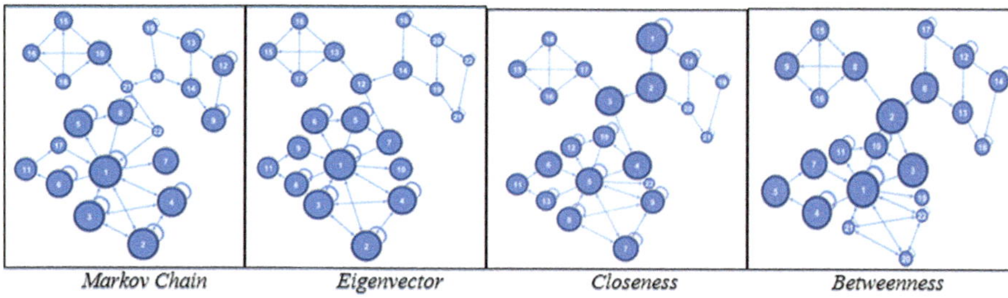

Fig. 29.2 Ranking of components for the component graph of Fig. 29.1

receiving a majority of weights in contrast to node I which has only an outgoing link to node K. Therefore, the ranking strategy of Markov chain is mutifaceted. Primarily, the nodes with a larger in-degree tend to receive higher weighting. If such a node also lies in the shortest path between other nodes in the graph, then they move up higher in the ranking. As an exception to having higher in-degree, nodes that lie frequently in the shortest path between other nodes (that were ranked highly) in the graph get highly ranked even though they have a significantly lower in-degree. This is evident from node J ranked higher than node I. Another interesting observation is that nodes with self-loops generally have a tendency to be ranked higher. We can see that node A is ranked higher than node D irrespective of the fact that both have the same in- and out-degree. We also observe that node T has an in-degree from node U and node V which are again ranked below node T. This is because nodes that have a direct incoming and outgoing link from and to highly ranked nodes get ranked higher. This also explains why the nodes N, M, and L are ranked together below the top-ranked node K.

Figure 29.2 also shows that, similar to Markov chain, eigenvector ranked the nodes of the completely connected subgraph highly. The node K has been consistently ranked higher by Markov chain, eigenvector, and betweenness but not by closeness. However, between closeness and betweenness, the ranking of nodes N, M, and L was higher in closeness than in betweenness. Secondly, we were interested to see how Markov chain compares with other ranking algorithms on assigning a rank to the node G. Betweenness and closeness have highly ranked the node G, i.e., second and third respectively. In contrast, eigenvector and Markov chain ranked the node G lower in 12th and 21st position, respectively. All the four algorithms consistently ranked the nodes of the clique lower. Overall, the ranking of nodes in the graph by each of these algorithms highlights different perspectives. So which ranking should one consider?

To make this determination, we chose to rank the components of three different systems discussed in [8]. The systems analyzed here include the Pathfinder, Super Hornet Organization, and the NASA ISS Organization. For these systems, we were able to determine an ideal list of ranking for the components from the descriptions of the system provided in [8]. We wanted to see how the rankings from the four ranking algorithms discussed above compare with the ideal lists for these systems.

29.3 Systems and Their Components

This section provides a short description of the systems analyzed in this study and provides an ideal rank order for critical components within these systems. We chose systems described by Eppinger and Browning [8] since they have explored each in detail, and their analysis provides the basis for the ideal ranking—the ranking an expert in a given system would provide. Furthermore, their use of design structure matrices (DSMs) provides us the necessary data to construct the directed graphs used by network metrics. Thus, to prepare an ideal rank for a system, we examined its corresponding DSM and the accompanying description of the system provided in [8]. Three systems (Pathfinder, NASA ISS Organization, and the Super Hornet Organization) were selected for analysis since the descriptions of their DSMs included criteria that could be readily used for ranking components objectively.

The Pathfinder mission system was designed to demonstrate new technologies that could help reduce the cost of delivering scientific instruments to Mars. Table 29.1 enumerates the most critical components in a rank order based on their technology risk scores determined from the DSM [8].

The second system we looked at was the NASA International Space Station (ISS). This organization supports various functionalities related to on-orbit operations and space shuttles. In total there are 36 critical functions performed by various teams within the ISS that are required to support the on-orbit operations. Interdependency between these functions can be grouped into three categories, namely, no, moderate, and high. Table 29.2 ranks the functionalities of the ISS sustaining engineering operations based on how many of their interdependencies were categorized as high [8].

Table 29.1 Ideal rank for Pathfinder

Rank	Components	Notes
1	Telecommunications boards VME I/F	Components with high technology risk scores [8]
2	Backshell pyro switching and pyros	
3	Parachute	
4	1553 data bus	
5	Flight computer and I/F boards	
6	VME bus	
7	Solid rocket assisted decelerator	
8	Aeroshell (heat shield and backplane)	
9	Telecommunications boards TEL I/F	
10	Lander remoter eng. Unit	
11	UHF modem	
12	Airbags	

Table 29.2 Ideal rank for NASA ISS Organization

Rank	Components	Notes
1	Assure engineering safety	Critical functions performed by various teams within the ISS ranked based on how many of their interdependencies are categorized as high [8]
2	Assure engineering quality	
3	Assess operations safety	
4	Assess and manage subsystem risk	
5	Flight control off nominal ops	
6	Close anomalies (CA RC)	
7	Maintain program requirements	
8	Flight control normal ops	
9	Develop/maintain mission plans	
10	Perform real-time engineering support	
11	Perform systems resource analysis and integration	
12	Perform logistics planning analysis	
13	Perform anomaly investigation	
14	Perform subsystem analysis and integration	
15	Develop and verify hardware modifications	
16	Maintain subsystem analytical models	
17	Support certification of flight readiness process	
18	Perform anomaly engineering triage	
19	Perform mission integration	
20	Analyze integrated systems performance	
21	Maintain system technical data	
22	Test repair overhaul and procure hardware	
23	Perform subsystem management	
24	Maintain logistics and maintenance tech databases	
25	Perform on-orbit maintenance repair	
26	Maintain system configuration	
27	Perform problem trend analysis	
28	Analyze and trend subsystem performance	
29	Develop and verify software modifications	
30	Approve change requests	
31	Maintain software	
32	Maintain personnel skills	
33	Maintain program engineering tools	
34	Maintain support facilities	

Third, Super Hornet Organization, represents the interactions among the teams that make up this organization that undertook major redesign of the Boeing F/A-18E/F Super Hornet, a fighter/attack aircraft originally developed by McDonnell Douglas for the US Navy. The DSM for this system describes the integrative mechanisms among the program's cross-functional development teams and their interactions during the engineering manufacturing development (EMD) phase. Table 29.3 ranks 21 teams in the descending order of their frequency of interactions on a daily basis.

Table 29.3 Ideal rank for the Super Hornet Organization

Rank	Components	Notes
1	Assembly AT	Ranking of the different teams in the super hornet organization hierarchy based on their frequency of interactions on a daily basis [8]
2	LE flaps/horizontal tail	
3	Structural dev and test	
4	Inner/outer wing	
5	Composite center AT	
6	TE flap/aileron	
7	Armament	
8	Flt ctrls computer soft	
9	Flying qual/control laws	
10	Flt ctrls syst integ testing	
11	Stability and control	
12	Main landing gear	
13	Maneuvering loads	
14	E and B loads	
15	Structural integrity	
16	Weapons separation	
17	High-speed drag and perf	
18	NLG/doors/hooks	
19	Mechanisms/flt controls	
20	ECS	
21	Electrical	

In subsequent sections, we will refer to the ranking of components for these three systems (Tables 29.1, 29.2, and 29.3) as the ideal rank. In the next section, we present a comparison of the performance of the different ranking algorithms in ranking the components of the three different systems against the ideal ranks.

29.4 Ranking System Components

To further explore the ranking capabilities of all the four ranking metrics, we designed and performed three experiments. *In each experiment, we constructed a component graph with its nodes representing the components of the system and the edges representing the relationship between the components as indicated in the DSM and modeled the graph as a DTMC.* Using the constructed DTMC, we computed the final weights of the components in the system and ranked them as discussed in Sect. 2. Next, we performed a pairwise comparison of the rankings of the components by each of the metrics against the ideal ranks. We only show this for Pathfinder in Table 29.4 due to space limitations.

Table 29.4 Ranking of the components for Pathfinder by different ranking algorithms

Ideal rank	Components	Markov chain	Betweenness	Closeness	Eigenvector
1	Telecommunications boards VME I/F	1	1	1	2
2	Backshell pyro switching and pyros	2	6	6	3
3	Parachute	3	4	5	7
4	1553 data bus	5	5	4	8
5	Flight computer and I/F boards	4	3	9	12
6	VME bus	6	2	10	1
7	Solid rocket assisted decelerator	7	10	11	10
8	Aeroshell (heatshield and backplane)	8	8	3	4
9	Telecommunications boards TEL I/F	9	7	2	5
10	Lander remoter eng. Unit	10	12	7	6
11	UHF modem	11	9	8	9
12	Airbags	12	11	12	11

In order to evaluate the different ranking algorithms, we employed three different statistical techniques: the Spearman's rho correlation, Kendall's tau-b correlation coefficients, and an information retrieval metric MAP (mean average precision). Since we are interested in understanding whether there is a statistical dependence between the ranking metric and ideal rank, we employed two nonparametric rank correlations: Spearman's rho and the Kendall's tau-b. These measure the strength of relationship between two ordered variables. A high value of associativity (close to 1) indicates a strong relationship between the two variables. In addition we drew upon information retrieval evaluation statistics, specifically precision@k. This is a metric that corresponds to the number of relevant results on the first search. Here precision@k measures the ratio of the total count of the number of components that are ranked in the exact same order by the ranking algorithm when compared with the ranking in the ideal rank to the total number of ranked components. So, while the ranking correlations assess the similarity of the rankings with the desired ideal rank, the precision@k metric assesses the relative positions of the components in each ranking. Finally, the MAP is computed as the mean of the precision@k. All the statistical computations conducted in this study were performed in R, a language for statistical computing.

The results of the pairwise comparison are summarized in Table 29.5.

For the Pathfinder system, both eigenvector and closeness resulted in the p-value (Kendall) of greater than 0.05, thus indicating that the null hypothesis cannot be rejected. Therefore, we conclude that the ranking for eigenvector and closeness is uncorrelated with the ranking of the Pathfinder components. The ranking of the

Table 29.5 Correlation coefficients for different ranking strategies across three different systems

	Correlation coefficient		p-value	
	Spearman rho	Kendall tau	Spearman rho	Kendall tau
	Pathfinder			
Markov chain	0.993007	0.969697	**2.20E-16**	**2.51E-08**
Betweenness	0.7902098	0.5454545	**0.001808**	**0.006885**
Closeness	0.4405594	0.3333333	0.07711	0.0763
Eigenvector -	0.4055944	0.3636364	0.09631	0.05798
	Super Hornet Organization			
Markov chain	0.9883117	0.952381	**2.20E-06**	**1.33E-15**
Betweenness	0.8987013	0.8095238	**1.49E-06**	**1.61E-09**
Closeness	0.7220779	0.5714286	**0.0001625**	**7.76E-05**
Eigenvector	0.725974	0.5714286	**0.0001454**	**7.76E-05**
	NASA ISS Organization			
Markov chain	0.9101604	0.7575758	**2.20E-16**	**2.76E-13**
Betweenness	0.4649351	0.3404635	**0.003035**	**0.002142**
Closeness	−0.3876241	−0.2869875	0.988	0.9922
Eigenvector	0.6213904	0.4688057	**6.14E-05**	**2.83E-05**

Pathfinder components by Markov chain and betweenness is concordant with the ranking of the components in the ideal rank with coefficients of 0.96 and 0.54 for the Kendall's statistic. However, between Markov chain and betweenness, the rankings of the components by Markov chain were more consistent with the rankings in the ideal rank. Based on the p-value of 2.50×10^{-8} at $\alpha = 0.05$, we conclude that the ranking of the Pathfinder components by Markov chain is positively correlated with the ranking of the Pathfinder components in Table 29.1. *It is worth noting, of course, that this level of precision for the p-value is remarkable given the sample size, but this is an anomaly of statistical tests. We would not anticipate this level of significance in general.*

In the Super Hornet Organization, all four ranking strategies resulted in a p-value <0.05. The Markov chain algorithm resulted in the highest correlation coefficient of 0.95 followed by betweenness (0.80) and 0.57 for both the eigenvector and closeness.

For the NASA ISS Organization, Markov chain resulted in a correlation coefficient of 0.76. Eigenvector and betweenness demonstrated the second and third best performance, respectively. Closeness on the other hand resulted in negative correlation, i.e., the ranking of the components in the NASA ISS Organization by closeness is discordant with the ranking of the components in the ideal rank.

So, while most of the algorithms performed well across the three systems, Markov chain consistently scored the highest correlation coefficients against the ideal ranks.

Table 29.6 records the precision@k ($k = 5, 10, 15$) and MAP for each of the ranking strategies across the three systems. Across all the systems, our Markov chain algorithm exhibited a superior performance, i.e., the MAP is consistently high for

Table 29.6 Precision@k and MAP computation for different ranking strategies across three different systems

	Markov chain	Betweenness	Closeness	Eigenvector
	Pathfinder			
Precision@5	1	0.8	0.6	0.4
Precision@10	1	0.9	0.9	0.9
MAP	1	0.85	0.75	0.65
	Super Hornet Organization			
Precision@5	1	0.6	0.6	0.8
Precision@10	0.9	0.7	0.7	0.7
Precision@15	1	1	0.8	0.8
MAP	0.966666667	0.766666667	0.7	0.766666667
	NASA ISS Organization			
Precision@5	0.6	0.2	0	0.4
Precision@10	0.9	0.6	0.2	0.6
Precision@15	0.866666667	0.533333333	0.4	0.666666667
MAP	0.788888889	0.444444444	0.2	0.555555556

Markov chain when compared against the other ranking strategies. Overall, across all the three systems analyzed in this study, our Markov chain algorithm consistently outperformed the other three component ranking metrics.

29.5 Discussion and Conclusion

Highly ranked components can make up the critical core of a system that is perhaps safety critical and needs to be highly reliable. The project team may need to allocate more resources for testing such components and also improve their testability for ease of diagnosis and troubleshooting. As systems evolve over time and get larger, ranking algorithms can provide an automated way of discovering these critical components using a dependency structure of the system under consideration.

In this paper, we used four such algorithms on three distinct systems analyzing component graphs that were created from DSMs for these systems. While distinct from each other, each algorithm specifically targets an important characteristic of the component graph. Betweenness highly ranks components that are frequently located on as many of the shortest path between pairs of other components. From the perspective of reliability, these components could be gateways or bridges that control access from one part of the system to another. They could become potential bottlenecks or single points of failure. Closeness identifies and highly ranks components that are closest to the cluster containing significant components. The cluster, for instance, could be a core element of a system providing services to the components highly ranked by closeness. Eigenvector ranks components highly if they are well connected to the core significant components. Similar to closeness, components ranked highly by eigenvector are consumers of the services provided by core significant components of the system. Markov chain, on the other hand, highly ranks components that demonstrate structural significance. These include nodes with high in-degree that many other parts of the system depend on. Making these components highly reliable and localizing the impact of their change is critical to prevent faults or changes from propagating to other parts of the system causing undesirable side effects.

References

1. Inoue, K., Yokomori, R., Fujiwara, H., Yomamoto, T., Matsushita, M., & Kusumoto, S. (2005). Ranking significance of software components based on use relations. *IEEE Transactions on Software Engineering, 31*(3), 213–225.
2. Lowe, W., Ludwig, A., & Schwind, A. (2001). Understanding software – static and dynamic aspects. In *The 17th international conference on advanced science and technology*.
3. Winter, V., Perry, J., Siy, H., Srinivasan, S. M., Farkas, B., & McCoy, J. (2011). The tyranny of the vital few: The Pareto principle in language design. *Journal of Software Engineering and Applications (JSEA), 4*, 146–155.
4. Landherr, B., Friedl, J., & Heidemann, J. (2010). A critical review of centrality measures in social networks. *Business and Information Systems Engineering, 2*(6), 371–385 discussion paper WI-282.
5. Matthews, L., & Richard, P. (2014). *Who is central to a social network? It depends on your centrality measure*. http://www.activatenetworks.net/who-is-central-to-a-social-network-it-depends-on-your-centrality-measure/
6. Zimmermann, T., & Nagappan, N. (2008). Predicting defects using network analysis on dependency graphs. In *Proceedings of the 30th international conference on software engineering* (pp. 531–540). New York: ACM.
7. Srinivasan, S. M., Sangwan, R. S., & Neill, C. J. (2017). On the measures for ranking software components. *Innovations in Systems and Software Engineering, 13*(2–3), 161–175.
8. Eppinger, S. D., & Browning, T. R. (2012). *Design structure matrix methods and applications*. Cambridge, MA: MIT Press.
9. Prowell, S. J. (2005). Using Markov chain usage modes to test complex systems. In *Proceedings of the 38th Hawaii international conference on system sciences*.
10. Srinivasan, S. M., & Azadmanesh, A. H. (2007). Exploiting Markov chains to reach approximate agreement in partially connected networks. In *International symposium on performance evaluation of computer and telecommunication systems (SPECTS 2007)*.

Chapter 30
A Method for Robust Requirements Definition

Marc Gethers and Dale Thomas

30.1 Engineering Errors

It is sufficiently difficult to realize a complex engineered system that performs any nontrivial mission well; most engineered systems are specified to perform a specific mission within a prescribed environment. Unfortunately for system developers, both missions and environments are inherently dynamic owing to a multitude of factors, from evolving technologies to competition. Further complicating the realization of a properly functioning complex engineered system are the engineers. Human faults due to miscommunication or lack of communication are the root cause of most failures during system development and operations [1]. These faults amplify a system's conditional and scenario requirements, which are often overlooked, ignored, or considered highly improbable [2]. Off-nominal requirements are not typically included in system performance specifications (SPSs) due to the lack of knowledge of how to organize these requirements [2]. Omission of these nonstandard requirements results in systems which, as realized, yield systems which perform narrowly prescribed missions in projected environments. Robust systems are characterized as "systems that have acceptable behavior over a larger class of situations than was anticipated by their designers" [3]. The focus of this paper is the development of robust requirements to enable the realization of robust systems—systems subjectively characterized by terms such as evolvable, adaptable, and graceful degradation.

To fully understand the context of robust systems, insight can be gained from looking at systems with a demonstrated lack of robustness. The Demonstration of Autonomous Rendezvous Technology (DART) was an autonomous satellite that had a mission of meeting up with another satellite but instead collided. This failure was attributed to the advanced video guidance sensor (AVGS), which failed to activate

M. Gethers · D. Thomas (✉)
The University of Alabama in Huntsville, Huntsville, AL, USA
e-mail: mg0045@uah.edu

S. Adams et al. (eds.), *Systems Engineering in Context*,
https://doi.org/10.1007/978-3-030-00114-8_30

when the DART missed a waypoint. Although ground operators recognized the missed waypoint, they were unable to take corrective action since the DART was not designed to take ground commands [4]. DART exemplifies a non-robust system since no consideration was given to the scenario requirement of what to do if the system misses a navigational waypoint. The Mars Climate Orbiter (MCO) offers another example. MCO failed upon entering Mars' orbit with a trajectory of 170 km below the planned insertion point. It is widely known that this error was due to the fact that the operations navigation team's Angular Momentum Desaturation (AMD) file modeled the trajectory of the spacecraft in English units, while the Software Interface Specification (SIS) called for SI units [5]. Less widely known is that the AMD file called for the MCO to execute a daily 180-degree flip maneuver. This daily flip had been deemed unnecessary by analysts, but that result was never passed to the operation navigation team [5]. The series of daily flips progressively worsened the wrong unit's impact even more. Once the unit issue was recognized, a possible mission-saving maneuver called the TCM-5 could have been attempted to maneuver the spacecraft to a safer altitude. However, analysis, test, and procedures had not been completed nor attempted, so the maneuver was not performed [5]. Though the TCM-5 was listed as a contingency maneuver, it was of no use since the team had not adequately prepared for the possibility of executing the maneuver. MCO epitomizes the lack of communication between project teams which contributed to a system failure when faced with an off-nominal operational scenario.

30.2 Robust Requirement Definition

Robustness is not a quantifiable measure, and as such it is hard to determine at what point a system may be considered to be robust. As illustrated with the DART and MCO examples, it is easy to see when a system is not considered robust in retrospect. System engineers must acknowledge the difficultly in creating robust systems and formulate a framework for robust system development—a framework which will include the development of robust requirements. In addition to nominal system behaviors, robust requirements consider off-nominal behaviors due to degradation, failure, or anomalies and include calculated responses. While this characterization of robust requirements may fuel fears of constraining the solution space with too many requirements, that would be counterproductive to robust system realization. Likewise, broadening requirements unnecessarily will lead to "gold plating" the system such that it will not adequately meet the overall customer needs within constraints of performance, budget, and schedule. Rather, the key concept is one of coverage which can appropriately respond to situations not anticipated by system designers which will yield a robust system. The traditional method for requirements development fails to produce robust requirements since, as previously stated, it fails to consider conditional, scenario, and off-nominal requirements. The method proposed for robust requirements development is a top-down hierarchal decomposition of system goals called the goal-function tree, which will be described and illustrated in the following sections.

30.3 Goal-Function Tree

The goal-function tree derived from complications that arose from defining abort conditions for the Ares I [6]. The Ares I used engineering methods such as the ARA, hazards analysis, fault trees, event sequence diagrams, and failure modes and effects analysis. Those engineering methods entered a classification process to generate the Abort Condition List (ACL) that was documented in the Abort Conditions Report (ACR) for the Ares I [7]. However, issues arose from using that methodology to obtain the Abort Condition List. The ACL did not contain all possible abort conditions—a shortcoming of the classification process. The classification process was an abort condition that would pass through the set of criteria that would define the condition as monitored or non-monitored. Non-monitored conditions were accepted as program risk. The problem with the non-monitored classification was some conditions even though they fit the criteria for non-monitored were still technically monitored downstream. Thus the condition was technically monitored but not directly by a sensor assigned to that abort condition. For example, the Thrust Vector Control (TVC) Hardover Abort Condition was classified as a non-monitored condition since the gimbal angle position sensor was not capable enough to detect the condition. However, the GN&C abort triggers such as attitude error and attitude rate error could detect the abort condition; therefore, even though the TVC Hardover Abort Condition was listed as non-monitored condition, it was actually monitored [7]. Since it is possible that one abort trigger can monitor multiple abort conditions, there should be documentation of all the different possible abort conditions one trigger could cover. Furthermore, the program risk calculation would not be reliable considering that some non-monitored conditions were actually monitored downstream. Another issue identified was determining if the ACL was truly complete and could all of the potential abort triggers detect all the possible ways of failures which could cause loss of crew [6].

The goal-function tree addresses those questions by the top-down approach of the success space which focuses on what is needed to occur for mission success [6]. If lower-level goals cannot be accomplished, then the overall higher-level goals are compromised as well. To address the completeness and link the goal representation to the system design, the GFT employs state variables which define the success of the goal. The state variable enters an equation $y = f(x)$ with a set range for the success of the function; anything outside of the intended range would be off-nominal [8]. Quantifiable values, for instance, the mass flow of air, along with other attributes are included in the GFT equation to determine if they are able to satisfy the function's requirements. The values are required to be measurable in real life such that they can be detected by a sensor; otherwise the GFT is not modeling a realistic environment. State variables link the physics and behavior logic of the system.

Exemplified by the inability to achieve system goals moving from the bottom of the tree to the top will mirror the effect of a real-life system failure [6]. The net coverage of safety was addressed by the application of off-nominal functions on each goal that down-streamed to other goals which all had off-nominal functions as well to protect the crew [6].

To illustrate the application of the GFT a system, for example, a printer will be analyzed with the main overall goal of printing paper. Printers often have ink estimates which give the user an estimated amount of ink remaining in the cartridge. After continued use, the estimate will decrease, and the user will be given an alert that the ink is running out as it approaches zero. This alert to the user in the GFT context is an off-nominal warning that the system's state variables are no longer within the nominal bounds. The system will experience a degradation of performance until complete failure. Failure to print due to no ink will occur unless an off-nominal goal is implemented such as replacing the cartridge. The early warning by the system to the operator allows new off-nominal goals to be implemented before system failure. This example proves the identification of the degradation of system performance through state variables before failure in conjunction with planned actions for off-nominal status can prevent some system failures. This concept will be applied to a UAV through the creation of robust requirements by the goal-function tree.

30.4 Goal-Function Tree Application

30.4.1 Operational Scenario

An operational scenario is how the system will operate during a mission phase. The focus of this paper is on the mission operation phase for the UAV system of flying to a location and back. To facilitate the designing of the functional flow of the mission, the use case diagram is implemented to illustrate the behaviors the system will perform for an external actor or system, illustrated in Fig. 30.1 [9]. The creator of the diagram will use a use case specification that entails the conditions for the use cases, triggers, actors/stakeholders involved in the invoking of the use case, main success scenario, and extensions for the use cases. The main success scenario is the nominal case for the mission with the extensions acting as the off-nominal functions, that is, the deviation from the base use case. Triggers will be fully illustrated later in the paper in the function diagram shown in Fig. 30.3 which are the guards between the transitions of nominal to off-nominal states and vice versa. The advantages that the use case diagram has are it removes ambiguity about the functionality of the UAV by highlighting behaviors the system shall perform and accounts for possible deviations from the traditional path.

The main overall use case for the UAV is "Fly UAV" which includes "Send and Receive Data" and "Allocate Power Allowances." Both of those have extensions act as the off-nominal events that could occur with those types of system behaviors. The

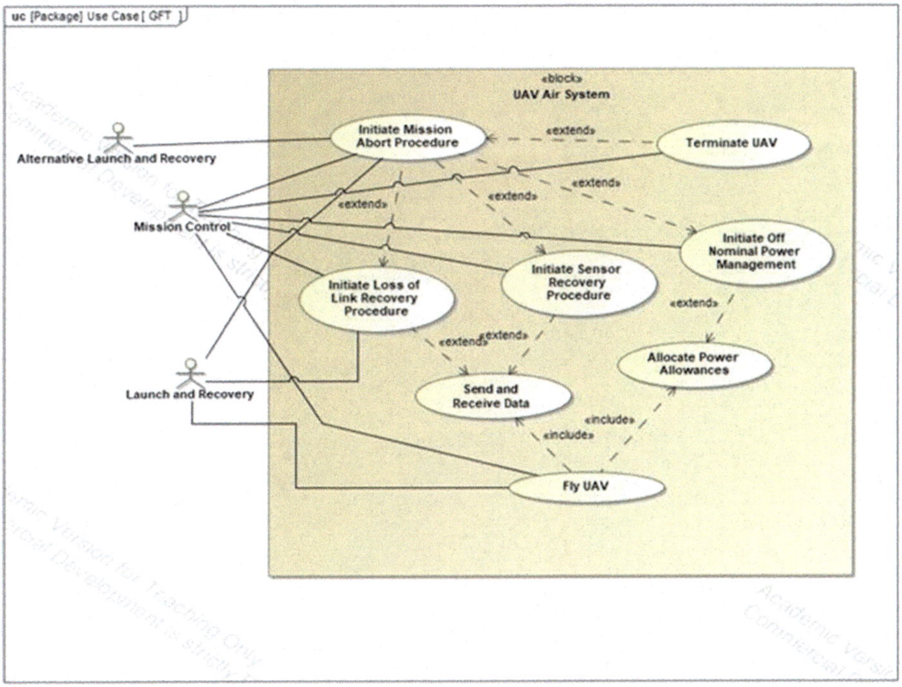

Fig. 30.1 Use case diagram

three identified actors for the mission was Launch and Recovery, Mission Control, and Alternative Launch and Recovery. These actors are responsible for invoking and/or involved with the service that the use case will perform.

30.4.2 Goal Diagram

The Goal Breakdown View Goal Diagram aspect of the GFT defines the primary goal of the mission and the secondary goals which must be met in order to accomplish the primary goal. To illustrate the GFT methodology, consider the mission goal for a UAV to have a "successful flight from point A to point B and return." The mission sequence is:

- Take off from point A under launch and recovery.
- Fly to point B under mission control guidance.
- Land under launch and recovery at point A.

The goals and secondary goals are listed below:

- Successful data exchange between ground station and UAV—Maintain connectivity.

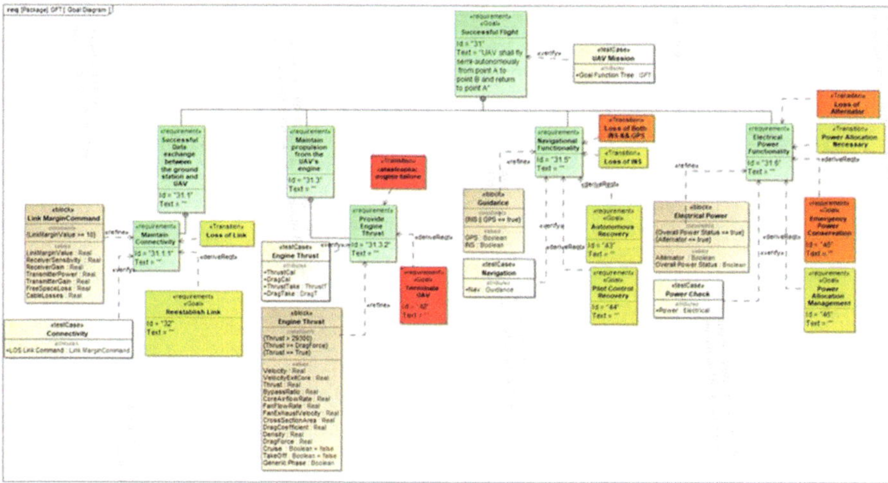

Fig. 30.2 Goal diagram

- Maintain propulsion—Provide engine thrust.
- Navigational functionality—GPS and INS functionality.
- Electrical power functionality.

The goal diagram illustrates the goals and secondary goals as shown in Fig. 30.2. Those secondary goals must be completed, or the UAV will not be able to complete its mission. The goals that have variables that will pass through a function to be monitored are maintenance goals which must stay within the ranges of success, or they will go into the off-nominal function. Initially, there were two types of off-nominal functions determined for the UAV mission—off-nominal recoverable and mission abort.

- Off-nominal recoverable: Activate management of subsystem with the new goal of mitigating off-nominal event to be able to return to nominal status and continue its mission.
- Off-nominal mission abort: Noncommon mission critical components go off-nominal causing new goals of aborting the mission and returning to base.

After building the simulation and conducting further research on fault analysis of UAVs, a third off-nominal goal was discovered for the simulation. The third function was off-nominal flight termination which causes the UAV to intentionally crash, and the crash site can then be targeted by air or ground forces which can salvage or dispose of the UAV.

- Off-nominal flight termination: Triggered by an off-nominal mission abort if the UAV is unable to make it to base a or an alternative base, so the flight computer or operator triggers the new goal of a forced crash in an ideal location on the way back to base.

A more ideal location is called out for that factor stated above and to avoid crashing in areas which secondary disposal may be problematic due to diplomatic or collateral damage. The off-nominal mission abort function has four factors that ultimately cause the triggering of the off-nominal flight termination function. Next the significance of each of the secondary goals must be assessed to determine possible quantifiable variables which can be monitored for nominal or off-nominal status. These are the main factors which determine success for the secondary goals which then allows for the success of main achievement goal.

30.4.3 Function Diagram

After creating the goal diagram, the simulation process flow is then created from it, as seen in Fig. 30.3. The simulation process flow for this mission is not a sequence of events for the mission but is a sampling of a moment of time during the mission. Reasoning for this decision is that a UAV has about ten phases which have very similar functions in each of the phases, so the simulation would become extremely redundant. It would be a better practice of creating a test plan of simulating one phase at a time and recording the results accordingly. A sequence of events for the simulation process flow would be good for a space telescope slewing mission which has five major steps but only one mission phase. The simulation process flow is created from a SysML state machine diagram due to its ability to clearly illustrate the transition from nominal to off-nominal. These transitions are extremely important to identify since these triggers represent real-life changes in the mission status. The state machine will run on instance blocks which are user input values which represent instrument values which will be the parameters for which the UAV will be tested on to determine nominal or off-nominal status.

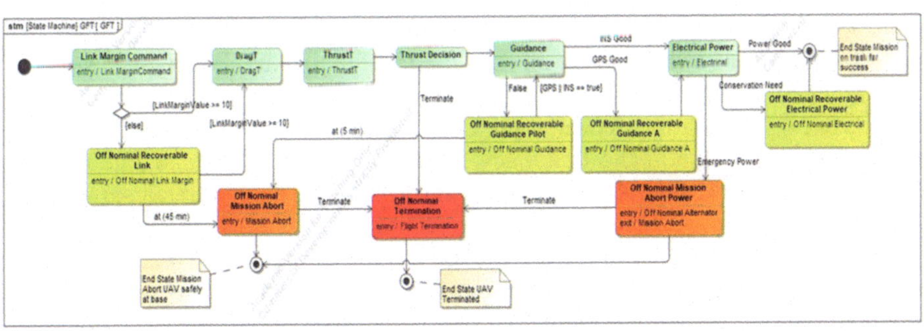

Fig. 30.3 Function diagram

30.4.4 Data Connectivity Portion of UAV Mission

To illustrate the process of the GFT fully, this section will walk through the "successful data exchange between the ground station and UAV" branch of the GFT. Data uplink will be command and control, and voice and data downlink will be ATC voice, telemetry data, and payload data. Most of the flight will be automated besides takeoff and landing, but the connection to ground is still required for communication between ATC and the operator which alerts ATC of the current flight path of the UAV so it can be clear of other air vehicles or munitions [10]. Successful data exchange was further elaborated to maintain connectivity. Maintain connectivity was selected since in order to have a successful data exchange, the UAV must have connectivity or it will not be able to send or receive data. The variable was link margin which is the margin required to maintain the link in the presence of signal fade [11]. Various factors such as antenna sensitivity, antenna gain, receiver sensitivity, transmitter power, cable loss, and free space loss will contribute to the link margin calculation [11]. The link margin required for the mission was 10 dB to accommodate excess path losses, and 10 dB is a typical target value used in link budgets for aeronautical systems [12]. Link margin less than 10 dB will not be sufficient for the mission and will be considered a loss of link which triggers off-nominal status. Shown in Fig. 30.2 on the far left side of the diagram, the Link Margin Command block has a constraint of "LinkMarginValue > = 10" which refines the requirement "Maintain Connectivity." This refine action is a stereotype which gives context that a value of 10 or greater is needed to satisfy that requirement. Also shown in Fig. 30.2 is the test case "Connectivity" which verifies the requirement by an activity diagram which contains the link margin function.

$$\begin{aligned}\text{Link Margin} = {} & \text{Receiver Sensitivity} + \text{Receiver Gain} \\ & + \text{Transmitter Power} + \text{Transmitter Gain} \\ & - \text{Free Space Loss} - \text{Cable Loss} \end{aligned} \tag{30.1}$$

In the event that the link margin is <10, it will be sent to off-nominal recoverable goal. Figure 30.2 shows the blocks "Loss of Link" which is the trigger for the off-nominal recoverable goal as well as "Loss of Link time limit exceeded" which is the trigger for off-nominal mission abort if the off-nominal recoverable goal is not successful. In a loss of link scenario, the UAV would likely be under mission control during this event due to the frequency of loss of link in a beyond line of sight scenario. As seen in Fig. 30.4, mission control is unable to communicate with the UAV, so the primary control module identifies loss of link and starts the preprogramed loss of link process. The loss of link process is for the UAV to fly to a safe zone and loiter there until connection is reestablished to the UAV and mission control or a time limit is reached. If the connection has not been reestablished to the UAV within a certain time interval, the UAV will transition to off-nominal mission abort by a time trigger. Under mission abort, the UAV will enter a criteria to determine if terminate UAV is warranted. If not, the UAV will fly back

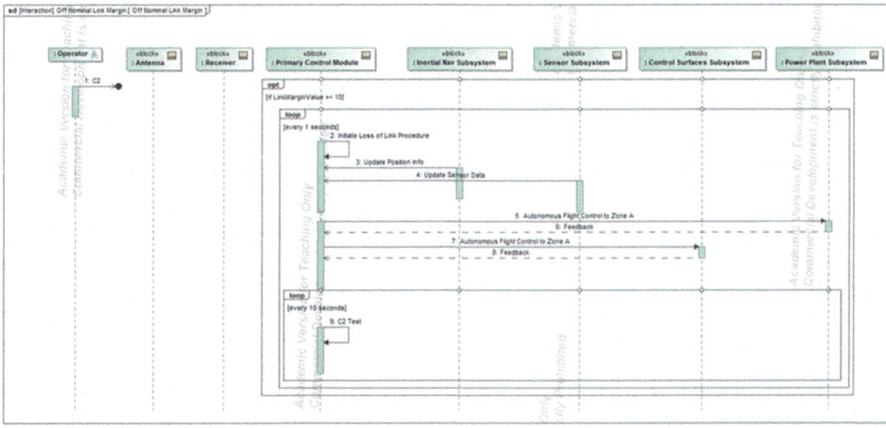

Fig. 30.4 Off-nominal link margin

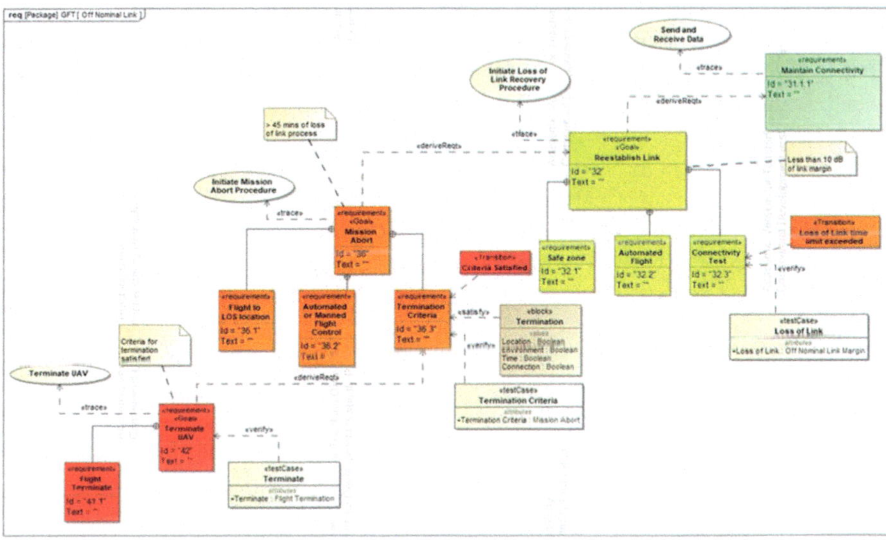

Fig. 30.5 Maintain data connectivity leaf of goal diagram

under autonomous control until it can connect with Launch and Recovery which will land the UAV. The goal diagram shown in Fig. 30.2, can be followed for the remaining mission requirements in a similar process of the linkage between constraint blocks, test cases, and off-nominal functions for the mission requirements.

Shown in Fig. 30.5 is the nominal Maintain Connectivity linkage to the off-nominal process previously described throughout this section. This diagram expands on the goal diagram shown in Fig. 30.2 by connecting off-nominal actions the UAV takes to the off-nominal requirements for robust system realization. This method of clear unambiguous communication between the system and user is necessary to prevent the propagation of the fault to a system-wide level by having a method of response as soon as a fault is detected.

Furthermore, the goal diagram can be visually inspected to ensure that each level of the Goal Tree has functions which correspond to possible faults. This net of coverage strengthens the requirement development process by early identification of system constraints which equate to faults that can be corrected or mitigated through off-nominal systems, as well as the identification of assigning sensor coverage which is responsible for monitoring the real-life inputs into the system which have been modeled by the GFT.

30.5 Future Work

More research will be invested into developing a framework for when the GFT should be implemented in the system lifecycle. Ideally the GFT would be implemented at the start of the system lifecycle defining the capabilities required for the system given the solution space and identification of possible alternatives given less than optimal inputs. Then as the analysis matures, the model can achieve a higher fidelity that could provide insight into the sensitivity of the systems as they relate to each other. Future models will focus on increasing fidelity through implementing more numerical analysis into the simulation. For example, the overall electrical power allocation could be rolled up to determine if the allocation meets the nominal amount. Currently the model represents some numerical analysis such as power allocation as Boolean values to determine nominal or off-nominal status.

30.6 Conclusion

It is hard to define the extent of requirements needed to achieve the goal of a robust system. Systems lacking robustness if tested will fail under situations outside of their prescribed environment as exemplified by the DART or Mars Climate Orbiter. A framework which enables the development of robust systems is clearly needed. Methods such as the robust requirements conceived by the goal-function tree in MBSE facilitate the identification of goals needed to complete a mission. The identification of those goals facilitates the identification of constraints of the mission functions. When constraints are not met, a set of procedures are put into place to mitigate mission failure with the attempt to return to nominal status. Much can be learned about a system while going through the methodology of building the goal-function tree. System designers can learn from mission constraints to build margins or systems that overcome mission constraints, facilitating the development of robust requirements for the system, as well as mission planners can learn from the GFT the limitations of their systems to more optimally plan future operations. The GFT provides a path forward in the event of system degradation, the goal of robust requirements.

References

1. Johnson, S. (2011). *System health management: With aerospace applications*. Hoboken, NJ: Wiley.
2. Sage, A. (2016). *Systems engineering: Analysis, design, and development*. Hoboken, NJ: Wiley.
3. Sussman, G. J. (2007). *Building robust systems an essay*. Cambridge, MA: Massachusetts Institute of Technology.
4. Thomas, D. (2007). Selected systems engineering process deficiencies and their consequences. *Acta Astronautica, 61*(1–6), 406–415.
5. National Aeronautics and Space Administration. (1999). *Mars climate orbiter mishap investigation board phase I report*. Washington, DC: NASA.
6. National Aeronautics and Space Administration. (2013). *Space launch system program (SLSP) integrated mission and fault management (M&FM) design analysis and performance assessment volume 2: SLS goal tree/success tree (GT/ST)*. Washington, DC: NASA.
7. Greg, P., Anupa, B., & Dwight, S. (2007). *The Ares I abort failure detection, notification, and response system: An overview of the development process*. Piscataway, NJ: IEEE.
8. Johnson, S. B. (2013). *Goal-function tree Modeling for systems engineering and fault management*. Reston, VA: AIAA.
9. Sanford, F., Alan, M., & Steiner, R. (2015). *A practical guide to SysML: The systems Modeling language*. Needham, MA: Object Management Group.
10. *Data link basics: The link budget*. L3 Technologies, New York.
11. Medina-Pazmiño, W., Jara-Olmedo, A., & Valencia-Redrován, D. (2016). Analysis and determination of minimum requirements for a data link communication system for unmanned aerial vehicles. In *2016 IEEE Ecuador technical chapters meeting* (ETCM).
12. Steve, H. (2008). *Unmanned aircraft system control and ATC communications bandwidth requirements*. Washington, DC: National Aeronautics and Space Administration.

Chapter 31
Identifying Barriers to Agile Technology Roadmapping

Lars Ivansen, Jonas Andersson, and Dag Bergsjö

31.1 Introduction

Product developing companies must defend their position in the forefront of technology by investing large resources in technology development. These investments include developing or acquiring new technologies as well as refining existing ones in order to close technology gaps necessary to provide competitive products to the market. To deliver a return on these investments in terms of selling more products, or reaching higher profit margins, they must be carefully planned and executed. Moreover, as expenditure increases, so does also the risk of potentially failing to develop a technology altogether or falling behind competition and thereby missing revenue. Such investments, or technology gaps, are investigated in terms of failure to predict technology maturity and the time it takes to develop a mature technology. The studies are based on the assumption that in order to manage the increasing complexity and risk related to the development of new technology, a technology roadmap aligned with a planning process and supporting methodology needs to be implemented to plan and to govern an organization's entire technology portfolio. A roadmap is a strategy and foresight tool supporting organizations to reach their long-term objectives and goals in manageable steps optimizing risk, time, and resources. Roadmaps typically align multiple views over time such as scenarios, corporate

L. Ivansen (✉)
Chalmers University of Technology, Göteborg, Sweden

Mycronic AB, Täby, Sweden
e-mail: ilars@chalmers.se

J. Andersson
University College of South East Norway, Kongsberg, Norway

D. Bergsjö
Chalmers University of Technology, Göteborg, Sweden

University College of South East Norway, Kongsberg, Norway

S. Adams et al. (eds.), *Systems Engineering in Context*,
https://doi.org/10.1007/978-3-030-00114-8_31

strategy, products, and technology [1]. One attractive aspect of roadmaps is their potential to make explicit unforeseen technology gaps or upcoming opportunities, thereby promoting that early and agile action can be taken. One important reason for this is change due to competition. This paper focuses on the use of technology roadmaps in high-tech industries and their potential to support innovation without compromising the need of mature technology in time for product development. The goal is to identify important aspects that need to be managed within the roadmap, in order to design predictable models where risk and revenue as well as important technology decisions and alternatives can be visualized.

31.2 Research Approach

The case study research presented in this paper is retrospective case studies covering a time approximately 5 years initiated to identify factors involved in technology reuse. Initial finding called for a need of a deeper understanding of the engineering context in order to identify potential significant factors affecting technology roadmapping. The nature of the topic of interest leads to the selection a qualitative case study research, seeking to generate hypotheses [2, 3] for further research. Furthermore, the organization of the research project as a partnership with the case company also gave access to detailed information regarding its technology development projects. This access to data was the main rationale for conducting the case study, i.e., an opportunity to study a situation otherwise inaccessible to researchers, which Yin [2] refers to as a "revelatory case." The case company can be described as an organization typical of the high-tech electronics development industry, which provides rationale for regarding the case as "typical" [3]. Reviews of existing literature were conducted primarily from the academic fields of *technology management* and *systems engineering*, which helped in positioning the empirical findings and observations in a larger context.

31.3 Frame of Reference

The findings in the conducted research have been linked to three main domains: technology roadmapping (long-term planning), systems theory (complexity), and product development (systems engineering). After providing a general overview on technology roadmapping, we will point out important barriers to successfully implement in an organization in terms of complexity and technology development. The aspect of complexity is believed to indicate the magnitude of risk undertaken when maturing and developing a new technology. The technology development is included specifically for its ability to estimate status in the development process and hence approximating the potential risk relating to internal development decisions or external purchasing decisions.

Technology roadmapping has been adopted by many organizations to support innovation, strategy and policy development, and deployment [1, 4, 5]. Roadmapping provides a structured approach for mapping the evolution and development of complex systems by visualizing and analyzing the linkage between technology capability, corporate strategy, and external contextual variables [1]. By integrating business modeling, strategic roadmapping, and technology roadmapping, the methodology improves decision-making, discovery of innovation opportunities, and quality of business plans [6]. Business models and roadmaps are used as communication tools where gaps and problems are made visible [7]. The graphical time-based framework for linking the multiple views together is one of the key features of technology roadmapping. Roadmapping stimulates cross-functional interactions where people share knowledge about specific issues. It thereby becomes an instrument for setting the direction for the whole firm [1]. Product proposals in roadmaps may be compendious, and requirements are subject to change, especially in projects with a high degree of uncertainty about their goals [8, 9]. By estimating the maturity of a technology, the risk and cost of further development can be estimated to support the decision of when a handover from the *technology* development process to the *product* development processes is viable [10]. The most widely adopted metric for assessing technology maturity is the technology readiness levels (TRLs) developed by NASA [11]. The scale ranges from one to nine whereof TRL levels up to five or six are often considered *technology* development.

Systems engineering is an interdisciplinary approach supporting the realization of successful systems [12]. It considers the whole life cycle of the system and is supported by processes and life cycle stages, e.g., as described in the international standard ISO/IEC/IEEE 15288 [13]. In the field of systems engineering, a system is regarded as a purposeful whole consisting of interacting parts [14]. Another definition is offered by Webster's Dictionary stating that a system is a complex unity formed of many often diverse parts subject to a common plan or serving a common purpose [15]. The International Council on Systems Engineering (INCOSE) states that a system becomes complex when local interactions between parts of a system give rise to a novel, nonlocal emergent pattern and is thereby characterized by its interdependencies [12]. Even systems consisting of relatively few parts can therefore exhibit a high level of complexity.

Moreover, the complexity of a system can mean different things to different people, and there is no unanimous definition in science [16]. The scientific notion of complexity has therefore often been conveyed using examples of real-world systems which are believed to be complex and typically beyond the scope of individual comprehension. The field of complexity theory acknowledges the interrelated nature of things, as well as emergent behavior, and where the whole is greater than the sum of its parts [17]. The many interrelationships of the parts of complex systems cause dynamic behaviors and self-organizations that make them hard to predict. It is only possible to understand matters we are immediately involved in and that are not too far into the future [18]. Within the field of engineering design, several methods have been promoted to better understand and quantify system complexity, such as product function [19, 20], and structural complexity, encompassing the constituent number of parts and their interconnections [21, 22]. Understanding the level of system complexity provides insights when estimating system cost or managing change or other product-related matters. The degree of complexity has implications on product development projects. Shenhar and Dvir [8] suggest that a common theme of unsuccessful projects is failure to appreciate up front the extent of uncertainty and complexity involved and neglecting to adapt the project management style to the situation. Based on their findings, they suggest and approach to more clearly describe projects and their outcome intending to improve success rate of the projects. This approach, the Project Diamond approach, is a simple yet powerful framework to classify the style of a project based on four main classification variables: novelty of the product, technological uncertainty, complexity of the project, and pace based on its criticality. The framework exerts project complexity as a function of system scope, i.e., the number of hierarchical layers of the system.

31.4 Mycronic Case Studies

Mycronic is a Swedish high-tech company engaged in the developing, manufacturing, and marketing of production solutions for the electronics and display industry. Technology development was typically not separated from the normal product development and consequently brought forward uncertainty and technical risk. This occasionally led to missed milestones and cost overruns rather late in the product life cycle. In order to deal with these issues, the company was aware that it needed to implement a more structured approach for technology development and management.

In this study, we defined milestones on time in product development projects as an indicator of project success. Conversely, we regarded missed milestones as an indicator of potential project failure. One can argue that the criterion for project success is far more complex than simply achieving milestones on time. However, it is our experience that missed milestones due to technical problems often impact not only directly the schedule but also indirectly the scope for mitigating the problems, thereby increasing the cost. A software tool was created to extract statistical information from the project execution log at Mycronic. Analyzing a full year, we could see that of a total of 142 milestones across all running projects, 87 milestones were completed on time, while 55 milestones missed their deadlines by 1 day or more. That corresponds to a success rate of 61% and a failure rate of 39%. This was regarded to be a poor result considering the target for successfully meeting milestones was set to 80% or better.

There are many potential reasons for missing a milestone. To learn about the impact of technical problems on the projects, another data extraction was performed where delays were correlated to root causes. Six types of root causes were used in the analysis: (1) resources not available, (2) technical problems, (3) lack of decisions, (4) late deliveries from suppliers, (5) change in requirements, and (6) others. Together with resources not being available, technical problems top the list of root causes for missed milestones. From the analysis we could see that out of 55 missed milestones, 27% were caused by technical problems. The observation indicates that technical problems in product development were caused by a lack of technical readiness. With support from the technology readiness levels (TRL) for assessing technology maturity [11], we shall now discuss these samples in two illustrative cases.

31.4.1 Hydra Mount Head for 01005 Components (Hydra)

Pick and place equipment are robotic machines used for placing surface-mount devices onto a printed circuit boards, aka PCB. Central to these machines is the mount head that picks up components from tape feeders by means of vacuum nozzles, moves the components to the proper position over the board, spins them to their desired angular orientation, and places them on the PCB surface in wet solder paste. The components come in many shapes and sizes, and one of the smallest currently in use today is the type 01005 which puts very high demands on assembly and interconnection processes. The company already provided a mount head with 01005 capabilities for medium speed mounting. However, there was a need to integrate this capability into the high-speed Hydra mount head to increase the throughput; see Fig. 31.1. The challenge in this particular project was to solve the tiny components' tendency to stick to the nozzles.

We found in hindsight that the actual technology maturity level at project start was rather low. In the TRL scale, it was about 2, defined as "practical applications of basic principles identified and technology/application concept formulated." However, the project assumed a higher maturity of 5, "component and/or breadboard validation in relevant environment or by simulation with a thoroughly validated model." Basically the organization took for granted that it possessed the needed technology capability and was ready for product development. The cause for this estimation error was partly confusion with the existing precision mount head, partly due to the rating of the technology maturity outside of the organization. Other manufacturers offered support for 01005 components at high speed, and the team made the inference that this design could be remade in-house but failed to recognize that the environment where this capability was originally achieved was different. The recontextualization caused the actual technology maturity to drop. When the estimate was originally made, the team did not explicitly distinguish between internal and external technology maturity which may have contributed to the confusion.

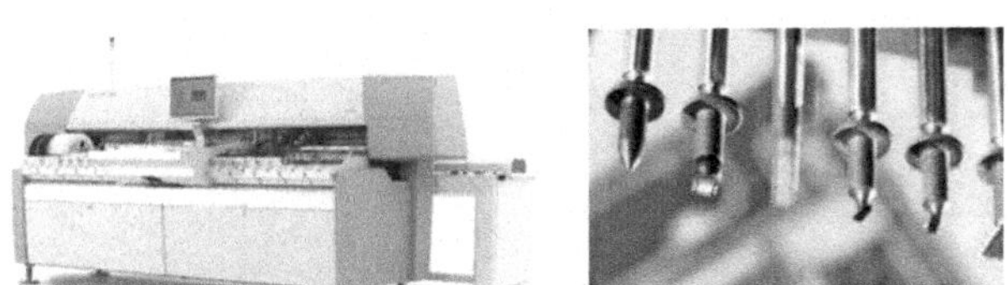

Fig. 31.1 MY200 pick and place machine (left) and nozzles of the Hydra mount head (right)

During project execution, the team discovered multiple challenges related to complexity. Unforeseen technical complexity was materialized, mainly caused by interference between system parts spanning across multiple technical domains. This in turn extended the number of collaborative engineering disciplines in the project. They were originally limited to mechanics, electronics, and software, and hence the project was planned with this limited scope in mind. Sticky residues from multiple sources on the nozzle tips called for chemistry expertise, and the delicate nature of the 01005 component required a modified mount cycle to be executed by the servo system. The issues that originated as technical complexity became organizational complexity. Using the terminology from the diamond framework [8], the Hydra project ran at regular pace, mainly as a consequence of the original misjudgment of the in-house technology maturity and the level of system complexity. When the actual situation became apparent, resources and expertise were added, but it was not possible to catch up the delay. In retrospect, it is likely that the project would have run at a higher pace from start if the actual internal technology maturity and system complexity had been appreciated up front. The data shows that the Hydra project missed its milestones mainly due to technical problems. The detailed examination revealed a significant discrepancy between the actual and required technology maturity. Other samples from the analysis of missed milestones in the project logs revealed similar results. We consider that failure to appreciate the actual technology gap a root cause candidate to why these projects failed to meet their targets.

However, the number of samples was limited, and further studies are needed to learn if the case is representative for a wider range of projects.

31.4.2 Electrical Heater for Dispenser System (Heater)

A new heater feature was needed for the Mycronic MY600JD jet dispenser platform, a system that jet fluids on-the-fly at high speed; see Fig. 31.2. One application is to dispense adhesives for the purpose of underfill on a printed circuit board, PCB. Underfill is glue under the component/flip-chip used primarily to decrease forces on soldered joints. With the use of the optional integrated heating unit, the material

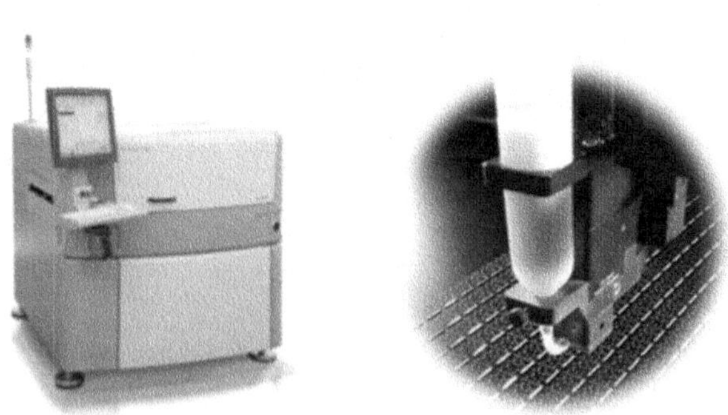

Fig. 31.2 MY600JD jet dispensing machine (left) and glue dispenser head (right)

viscosity of the glue can be adjusted for best dispense results. The general principle of heating the PCB to improve the flow of the fluid under the component was in use in the industry, so high-level conceptualization could be made by observing the competition. Since Mycronic had no prior experience of this technology, it was concluded in the assessment that the heating feature represented a technology gap. When comparing the moment of gap identification with the planned start of the product development, it was discovered that the project was already behind schedule. The gap suggested that 2 years were needed, but the roadmap granted only 1. Some work on the underfill feature had luckily been initiated a year earlier, but it was performed at low intensity. The actual TRL of this technology was estimated to 2. The required internal technology readiness was set to TRL 5 corresponding to the gate criterion for entering product development. Closing the gap would encompass maturing the heating technology from TRL 2 to TRL 5, advancing through three maturity levels mainly in the research domain.

Heaters for this type of application have been around since the 1960s, so in the external world, the technology was already at its maximum maturity of TRL 9. This observation suggests that the best strategy for closing the gap would be to acquire the technology from outside the firm. This highlights an interesting dilemma, being that other companies possessed the capability to make heaters for dispensing equipment, but it was considered unlikely that this capability could be transferred unless there was an alliance facilitating exchange of knowledge between competitors. The remaining options were to do research and development in-house and as a backup initiate a search for neutral partners with applicable knowledge. Another aspect of the external technology assessment was the risk for it being protected by patents. While the general principle of heaters was well-known in the art, it was possible that specific design choices were not. It became evident that the strategy of in-sourcing an established technology had to be combined with a meticulous IP clearance effort. If the external technology maturity had been low and in-sourcing would be the natural choice, the emphasis would likely have been on IP protection.

The project had been running for about a year when a reassessment identified the heater as a larger technology gap than originally anticipated. The scope represented by the number of domains being investigated from start was limited to one: the heater element itself. The technical complexity of the system was regarded low. However, detailed investigations of the design exposed additional interfaces to the rest of the system and researching the heater itself would not suffice. In addition to providing heat (the main function), the scope was extended to include positioning errors caused by temperature expansion, impact on vision and cameras due to air turbulence, accelerated material deterioration in cables and lubrication from heat exposure, and finally the change of atmosphere and outgassing in system affecting sensors. The total number of required technologies was now five, significantly increasing the level of technical complexity. This finding substantiated that the introverted research was extended to span across far more disciplines than was anticipated, thereby affecting project process and coordination.

The remaining time until the capability was needed was just a year, and the pace was too low to meet the deadline. To catch up the project would have to be upgraded to time-critical. Executing in time-critical mode has downsides, mainly in terms of impact to competing projects. The organization would have to lock staff dedicated to this particular project, thereby limiting the flexibility of the workforce. The personnel would be selected for specific skills and experiences, sometimes to the potential detriment of other projects. The schedule would be put under tight monitoring by management for early detection of deviations from the plan, typically at the cost of increase overhead. Exposed to the consequences of rushing the project, the management decided to postpone the product development.

31.5 Discussion

The identified barriers discussed in this chapter are based on observations from the two case studies. These observations will be discussed as barriers for implementing a structured technology roadmap.

31.5.1 Internal and External Technology Readiness

Technology development focuses on increasing the technology maturity of a specific function or aspect of a system. When the projects in the study were executed, the informal TRL assessment gave false indications of the maturity of the technology. In the Hydra case, we identified two types of confusions. Firstly, by referring to proven performance of another in-house product as evidence of mature technology thus failing to recognize that it reached its performance in a different context. Secondly, the fact that competition offered solutions was seen as conviction for technology maturity in general and the team underestimated the associated uncertainty and risk in the project. In the heater case, the technology gap was correctly identified but only after running the project for some time. The completion time was originally not seen as critical. Similar to the Hydra case, the team recognized that the maturity of heaters among competition was high, which may have influenced the team's confidence in their design. When the magnitude of the gap was realized, the project was running late for to the planned product introduction. This suggests that both internal and external technology maturities need to be monitored in order to successfully set up, model, and simulate a technology roadmap. It also indicates that the assessments would benefit from proof to strengthen their validity and suppress personal bias of the assessor. The findings can be induced by the following hypotheses:

1. Evidence-based assessment of technology maturity increases validity of the estimate.
2. Distinguishing internal and external technology maturity clarifies recontextualization implications.

31.5.2 Complexity

Both cases revealed interactions and dependencies between system elements initially not accounted for. Therefore, a greater number of interfaces across different technologies had to be explored. The increased degree of technical complexity resulted in the need of introducing additional engineering disciplines and specialist knowledge. The Hydra project started with experts on mechanics, electronics, and software but was later expanded to include chemistry and mechatronics as well. The Heater Project launched with the presumable simple task of integrating a heater element into an existing system but did not account for other interfaces related to thermal expansion, air turbulence affecting interferometers, accelerated material deterioration, and contamination. In order to explore these emerging side effects, the projects extended the number of engineering disciplines which in turn increased the project complexity. Interviews with team members suggest that the system context in both cases were originally defined too narrowly. There were, for instance, no checklist of affected areas of expertise to walk through and no visualization of the system to use as a basis for stakeholder communication. Instead, the input was limited to a set of textual requirements which did not provide enough overview. Instead, the side effects were subsequently discovered when the system exhibited emergent behavior and failed to reach the targeted performance. When the project broadened its scope and explored these effects, a change from technical complexity into organizational complexity could be observed, similar to *project complexity* described in the Project Diamond approach [8]. The complexity problem is, as demonstrated in the two case studies, also a major barrier toward implementing successful technology roadmapping. In the cases, complexity is presented as unforeseen interactions between system parts and the involvement of several engineering fields within the technology development project. Hence, complexity is identified as a potential barrier to successful development. Quantification of the technical and organizational complexity up front has the potential to improve the overall scoping quality of the technology development initiatives. The findings can be induced by the following hypotheses:

3. Quantification of technical and organizational complexity increases scoping precision of technology development projects.

31.5.3 Implications for Technology Roadmapping

The two cases studies revealed a reactive management style where the organization struggled to provide technology assets in time to satisfy its product roadmaps. The scope of commitments gradually expanded as the development teams uncovered new interactions between the system elements and realizing that the actual maturity of the technology was lower than first anticipated. It was also found that the degree of system complexity was difficult to estimate early in the projects. Complexity had direct impact on the scope, which in turned expanded the width of expertise needed to cover all aspects of the system. At the time for the study, the technology roadmapping did not rely on a structured process for analyzing and describing the technology development efforts. The technology view was populated, but there were no explicit assessments of gaps between current capability and derived product requirements up front. Once the gaps were understood in terms of technology maturity and system complexity, the roadmap did not provide ample time slack to close the gap before the capability was expected.

We propose that in order to design and maintain future technology roadmapping initiatives, it is important to identify and maintain knowledge regarding the *internal* and *external* technology readiness. Process formality, in terms of providing actual evidence of the particular maturity, plays a crucial role for the quality of the assessment. Informal assessments entail risk for confusing internal with external maturity, compromising the ability to properly recognize impact from recontextualizing of technology. In addition, an unbiased evidence-based approach to correctly assess the technical and organizational complexity is needed in technology projects. This poses the question how the two presented projects would have benefited from such an approach. While they represent past events that cannot be replicated for validation of methods, we still believe that explicit definition and visualization of gaps in the technology roadmap would have likely facilitated important discussions among subject-matter experts earlier on, which in turn would have improved the quality of subsequent planning and proactivity in the management. Proactivity would in turn entail decoupling of technology development from product development and provide the slack needed to bring the technology through each level of maturity. A structured method for defining the gaps, in terms of external and internal maturity as well as their inherent complexity, would likely encourage objectivity and suppress some of the rushed assessments observed in the study. The findings can be induced by the following hypothesis:

4. Visualization of technology gaps facilitates communication which enables proactivity.

31.6 Conclusion

The findings of the presented case study suggest that long-term technology planning to a large extent is driven by management. To improve the alignment with short-term product development, we suggest that technology roadmapping deserves more attention at the engineering level. The study also draws attention to the need for high-tech organizations to introducing strategies for the systematically mapping, quantifying, and prediction of long-term technology development to leverage their investments in research and development. This paper has attempted to identify barriers involved when designing a technology portfolio and presents ideas toward efficient technology development within the studied organization. In particular, it demonstrates challenges related to correctly assessing gaps of technology maturity, both for internally developed technology and technology acquired externally. The case studies reveal the need for a better understanding of the technical and organizational complexity that needs to be mastered when introducing new or improved technologies. Especially complexity causing unforeseen scope changes was found difficult to mitigate. Finally, the nature of the technology gaps was not well understood and communicated in the organization which led to technical risks related to product development projects. As a platform for our continued research, we suggest four hypotheses that will be used for testing a revised technology roadmapping approach.

Acknowledgment The project was supported by Mycronic AB and instigated by the Chalmers University of Technology and the University College of South East Norway. The financial support by Vinnova (Swedish Governmental Agency for Innovation Systems) is greatly acknowledged. This work was conducted in the Vinnova Vinnex Excellence Centre for Product Realization and the Production Area of Advance at Chalmers University.

References

1. Phaal, R., Farrukh, C. J., & Probert, D. R. (2004). Technology roadmapping—A planning framework for evolution and revolution. *Technological Forecasting and Social Change, 71*(1), 5–26.
2. Yin, R. K. (1994). *Case study research: Design and methods*. Thousand Oaks, CA: SAGE.
3. Gerring, J. (2006). *Case study research: Principles and practices*. Cambridge, UK: Cambridge University Press.
4. Barker, D., & Smith, D. J. (1995). Technology foresight using roadmaps. *Long Range Planning, 28*(2), 21–28.
5. Groenveld, P. (1997). Roadmapping integrates business and technology. *Research-Technology Management, 40*(5), 48–55.
6. Abe, H., Ashiki, T., Suzuki, A., Jinno, F., & Sakuma, H. (2007). *Integration studies of business modeling and roadmapping methods for innovation support technology (IST) and its practical application to real-world-cases*. Portland International Center for Management of Engineering and Technology.
7. Abe, H., Ashiki, T., Suzuki, A., Jinno, F., & Sakuma, H. (2009). Integrating business modeling and roadmapping methods–The Innovation Support Technology (IST) approach. *Technological Forecasting and Social Change, 76*(1), 80–90.
8. Shenhar, A. J., & Dvir, D. (2007). *Reinventing project management: The diamond approach to successful growth and innovation*. Boston, MA: Harvard Business Review Press.
9. Larman, C., & Basili, V. R. (2003). Iterative and incremental developments: A brief history. *Computer, 36*(6), 47–56.
10. Nolte, W. L. (2008). *Did I ever tell you about the whale?: Or measuring technology maturity*. Charlotte, NA: IAP.
11. Mankins, J. C. (1995). *Technology readiness levels*. White Paper.
12. Walden, D. D., Roedler, G. J., Forsberg, K. J., Douglas Hamelin, R., & Shortell, T. M. (2015). *Systems engineering handbook: A guide for system life cycle processes and activities*. Hoboken, NJ: Wiley.

13. ISO/IEC/IEEE 15288: 2015. (2015). *Systems and software engineering-content of systems and software life cycle process information products (documentation).* Geneva, Switzerland: International Organization for Standardization/International Electrotechnical Commission.
14. von Bertalanffy, L. (1968). *General system theory* (p. 40). New York: George Braziller.
15. Grove, P. (2002). *Webster's third new international dictionary.* Merriam-Webster Incorporated.
16. Johnson, N. (2009). *Simply complexity: A clear guide to complexity theory.* United Kingdom: Oneworld Publications.
17. Senge, P. M. (2006). *The fifth discipline: The art and practice of the learning organization.* New York: Broadway Business.
18. Flood, R. L. (1999). *Rethinking the fifth discipline: Learning within the unknowable.* London: Psychology Press.
19. Griffin, A. (1997). The effect of project and process characteristics on product development cycle time. *Journal of Marketing Research, 34*, 24–35.
20. Bashir, H. A., & Thomson, V. (1999). Estimating design complexity. *Journal of Engineering Design, 10*(3), 247–257.
21. Meyer, M. H., & Lehnerd, A. P. (1997). *The power of product platforms.* New York: Simon and Schuster.
22. Sinha, K. (2014). *Structural complexity and its implications for design of cyber-physical systems.* Cambridge, MA: Massachusetts Institute of Technology.

Chapter 32
A System of Knowledge Briefs to Support Decision-Making and Knowledge Reuse in Product Development

Linn Sandvold, Daniel Stenholm, Henrik Mathiesen, and Dag Bergsjö

32.1 Introduction

Industries today face inefficiencies in all types of processes; one such challenge is addressed within this paper: the creation and potential reuse of knowledge (by others) within the organization. Approximately 20% of a designer's time, and even more for technical specialists, is spent searching for information. Even though more suitable information from other sources exist, around 40% of all design information stem from personal storages [1]. The main challenge with knowledge management (KM) is known to be the transition of implicit knowledge to explicit knowledge [2].

Reuse can be said to be the use of an existing asset to satisfy a need [2]. According to Pfleeger and Bollinger, anything can be reused [3]. *Knowledge* reuse can consequently be the process where an entity is able to use shared knowledge. It can also be the idea of implementing previously developed capabilities into a new project for the purposes of positively influencing the development cycle [3, 4]. Successful reuse can be defined as applications where the benefits outweigh the cost [2]. When it comes to knowledge reuse in engineering, several challenges exist in planning for reuse,

---

L. Sandvold
Kongsberg Automotive, Kongsberg, Norway

University College of South East Norway, Kongsberg, Norway

D. Stenholm (✉)
Chalmers University of Technology, Göteborg, Sweden
e-mail: Daniel.stenholm@chalmers.se

H. Mathiesen
Kongsberg Automotive, Kongsberg, Norway

D. Bergsjö
University College of South East Norway, Kongsberg, Norway

Chalmers University of Technology, Göteborg, Sweden

S. Adams et al. (eds.), *Systems Engineering in Context*,
https://doi.org/10.1007/978-3-030-00114-8_32

implementing reuse, and measuring reuse [2]. Quantifiable measurements of knowledge reuse are difficult to obtain, thus making it challenging to express its impact. Reuse can assist in problem-solving, during replacement of employees, implementation of improvements, and optimization of product characteristics [5]. Reusing previously developed capabilities can be very advantageous. Decisions made early in the development life cycle have high impact, as they determine up to 80% of the cost at later stages [6]. Hence, it is beneficial to initiate reuse early in the project. Attempting to implement reuse at later stages in the project increases the risk of inadvertently affecting other aspects of the system under development [2].

Three factors have been identified during successful knowledge reuse in the systems engineering domain: technology, organization, and people. Fundamentally, the knowledge product must be capable of being reused, and optimally, the knowledge product is designed with intent of reuse. The people involved in knowledge work must have sufficient familiarity with the products, and in addition, they must be familiar with the challenges of reusing knowledge and its process. Process steps involve, e.g., identifying critical knowledge, capturing the knowledge in a reusable manner, and disseminating and making the knowledge available to people that potentially can gain value by reuse.

In addition, knowledge is dynamically and continuously updated. The stored knowledge must be adapted to new factors, data, inventions, and problems. Knowledge must not be reduced to an object but rather evolve through the skills, understanding, and relationship of the employees, as well as through the tools, documents, and processes that withhold this knowledge [7]. KM directly improve organizational processes, such as innovation, decision-making, and both individual and collective learning. Improving these organizational processes leads to improved organizational performance [8].

Many methods and principles have been introduced to adapt to the evolution of the automotive industry. Common to them is that they provide guidance to obtain higher efficiency during the development process. The guidance subsequently leads to time and cost reductions. An approach to obtain the above effects has been the implementation of lean methods in systems engineering which comprise creating the environment that ensures the right information is available to the right person (s) when needed [9]. In lean product development, A3-reports are commonly used to facilitate that environment. A3-reports originally refer to Toyota's form to communicate complex information and solve problems, which are commonly created on a single sheet of paper [10, 11]. These A3-reports are commonly referred to as knowledge briefs or K-briefs. The name "A3" originates from a paper size (297 $\times$ 420 mm), which has proved to be a good size that limits the report space available to the creator, therefore called briefs. When the K-brief report is done, it is usually stored digitally on the company server [11, 12]. To increase the understanding and enable thorough information, visual information is recommended in the largest possible degree. In the KM literature, different types and purposes for K-briefs are suggested, problem-solving K-briefs being the most common type [10]. Problem-solving K-briefs encourage systematic problem-solving, including problem formulation and experimental design, which address high-quality solutions to immediate local problems [13].

Challenges when performing activities to enhance effective knowledge flow to the right person are highlighted by many authors [14–18]. The challenges can be separated into several steps along the organizational KM cycle, starting with the beginning of (1) creation of knowledge, which involves learning to extend or replace existing knowledge. The creation is triggered when new knowledge is needed, and past knowledge was not sufficient to close the existing knowledge gap [19]. This is followed by (2) identifying possessed and valuable knowledge that is of interest for the organization to make available [17] and then (3) encapsulating the knowledge into something that can be (4) transferred and shared, which includes preparing for availability and accessibility for future potential users [20, 21]. Followed by the situation when need for existing knowledge is identified by (5) searching and understanding knowledge [22], (6) evaluate the value of the gained knowledge to the current problem or decision [23, 24], and finally (7) applying the knowledge to the current context [25]. Other activities that are mentioned in the literature also regard (8) ensuring the validity over time, (9) requesting new knowledge to be codified [26, 27], and (10) owning the knowledge asset [28].

32.2 Research Methodology

The research in this paper has been conducted in collaboration with Kongsberg Automotive (KA), in Kongsberg, Norway, who provides products to the global vehicle industry. The general perception in the company is that the implementation of knowledge-based development (KBD) has proven to be somewhat inadequate, depriving the company of reaching its full potential. A way to rapidly amend the situation has traditionally been to introduce alternative means to solve the problem of knowledge reuse. The result, so far, is to some extent an even more complicated development process to navigate through. It leaves behind confusion, document disorder, and waste in terms of time and resources.

The rationale behind the untapped potential of knowledge reuse has yet to be understood in its entirety. Only when fully understood can the most suitable improvement strategies be identified. The problem statement is thus:

"We perceive the strategy for knowledge reuse in the company as suboptimal and that several challenges exist that limit the implementation of KBD."

The chosen approach to understanding the complex matter is to focus on examining how knowledge reuse can be better utilized by investigating the following research question:

"What are the main barriers toward effective knowledge reuse in the KBD process?"

With this question, we specifically will target the current codification strategy involving K-briefs, in order to identify problems and potential future improvements

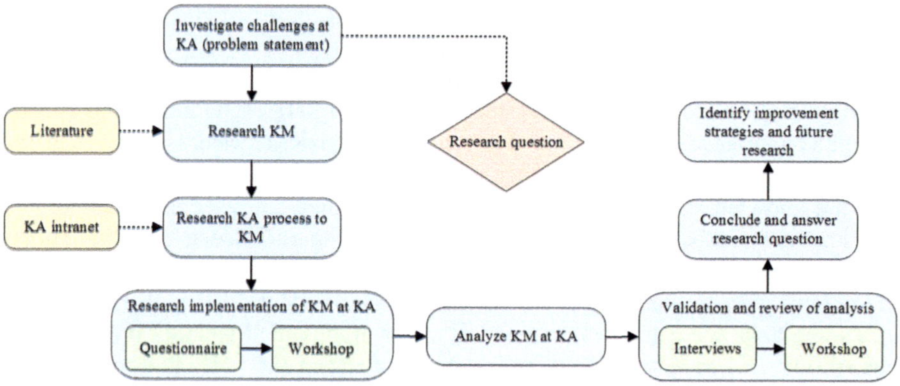

Fig. 32.1 Research methodology

in the methodology. Further, we will investigate the current personalization strategy centered around the COP with the same type of questions.

To tackle the challenge, a literature review was performed on the topic knowledge reuse and systems engineering in order to gather insights along with studying the company's KM process.

The findings were used to shape a questionnaire to gather the employees' personal experience and viewpoints. The age, seniority, experience, and positions within the company were varying among the 17 attendants. The questionnaire consisted of descriptive and Likert scale questions [29].

An analysis of the questionnaire and workshop followed, and the literature on KM was evaluated to benchmark theory toward findings in the company. To validate the analysis and to refine conclusions, open interviews were conducted with knowledge experts in the company, as well as a second workshop.

The flow diagram of the research methodology in this study is shown in Fig. 32.1.

32.3 Study Case: Kongsberg Automotive Practical Process of Knowledge Management

KA has adopted a multitude of different methods and principles in their KBD process. The main objectives are to reduce repetitive problems and time to market, as well as improve quality by better exploit knowledge and thus optimize the KBD process [30].

KBD was adopted as a mandatory process at KA in 2012, to assist and drive KM. The principles derive from lean product development and Kennedy's knowledge value stream. These principles are today widespread and well-known methodologies in system engineering [9, 31]. KBD consists of two value streams, knowledge value stream and product value stream, as shown in Fig. 32.2a.

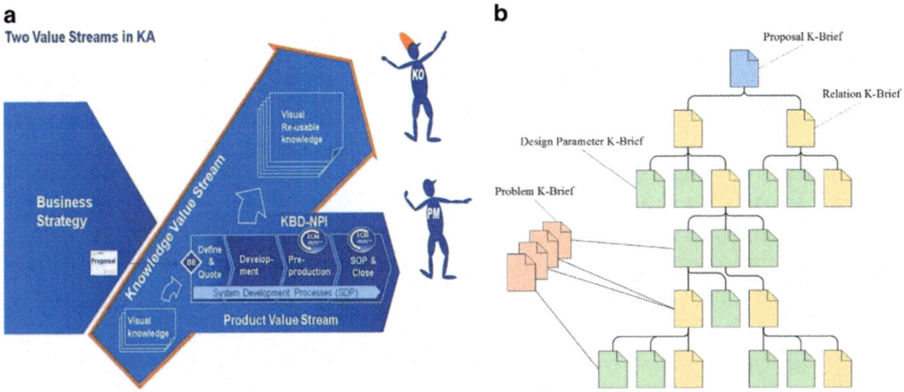

Fig. 32.2 (**a**) Model of KA KBD process; (**b**) KA knowledge standard, a network of documents

The knowledge value stream withholds the company's knowledge and supports the product value stream with reusable knowledge throughout development projects. The product value stream comprises the knowledge-based development-new product introduction (KBD-NPI), which consists of four phases: define and quote, development, preproduction, and start of production (SOP) and close. It is through these phases that new knowledge is generated to the knowledge value stream, and it is subsequently pushed out to the product value stream, i.e., the streams are dynamic flows, nourishing each other.

The high-level purpose of the flows is to facilitate continuous growth of organizational knowledge. This knowledge results in increased value in the products released to market, both short term and long term. The need for knowledge reuse on a global level is brought by over 10,000 employees located in 18 countries.

32.3.1 *Knowledge Standard: A Network of Documents*

The knowledge value stream is focused around a network of documents, referred to as knowledge standards, illustrated in Fig. 32.2b. The knowledge standard documents are preferably one-pagers and within KA known as K-briefs. The knowledge standards are utilized for knowledge gap closing, problem-solving, and optimizing design during the KBD-NPI and to avoid former mistakes. The knowledge standard documents are stored in the product life cycle management system which is based on a project/product methodology, meaning that the projects are partitioned in the conceptual framework of the product development.

Furthermore, our research discloses that the general perception of the employees is that approximately 20% of the problem K-briefs result in future changes of best practices, processes, standards, or the like. During our workshops, it was mentioned that one of the reasons for the low number of reused knowledge standards might be due to the fact that the captured knowledge is highly connected to its specific context. This means that there is a need for an effort to make it more general, which in many cases is not conducted. However, the final workshop indicated that improvement and attempts have been seen during the last years on this matter. The example brought up was that after conducting a comprehensive project, a thorough "lessons learned" process was performed, resulting in a list of identified and closed knowledge gaps during the project. This list was formalized into K-briefs and presented with support of presentation slides. One slide for each work area such as mechanical, electric, test, project management, and so forth. The purpose of this process was to identify the problem, explore the root cause, elaborate on solution space, and select the most appropriate solution based on gained insights. Finally, the knowledge was spread within the organization. The employee stressed that this was "telling the whole story" and was a great source of knowledge. However, the subsequent effort to make it reusable and available to the rest of the organization fell short. In the end, the employee felt that he pleased the top management more than actually bringing reusable knowledge into the organization. On the other hand, the project outcome of 169 K-briefs, comprising 106 problem K-briefs, is an indication that the use of K-briefs is highly adopted inside the company.

The employees state that the knowledge they find when searching is many times insufficient, too specific, not applicable in their context, etc. The K-brief format is known for its short and concise format; however, some employees argue that the mantra "keep it short" doesn't always allow capture of the full potential of the learning. For example, one employee argued that he had worked on something for 4 years and was then asked to summarize it on one single K-brief. The employees also state that it is challenging to transform implicit knowledge into explicit knowledge, involving discussions that explicit knowledge is more suitable for reuse than implicit knowledge due to documentation difficulties. Research states that the main challenge with KM is knowledge conversion and transition of implicit knowledge to explicit [32]. However, that does not imply that implicit knowledge is less important.

The current process to support knowledge value stream does not offer any information in regard to which team members are responsible to initiate and take ownership of the K-briefs throughout the process. It appears that guidance on how to perform KM on a daily basis is insufficient in the existing structure. One employee stated that during the last year, only a few relational K-briefs had been produced and uploaded to the knowledge standard. The relational K-briefs are combined information with the purpose to be more general in order to be spread inside the company. However, during the workshop, it was noticed that more than double the amount of relational K-briefs had been created and shared. This highlighted that there is no formal approach to create awareness when new knowledge that could be of value for relevant people has been captured. On the other hand, one initiative that resulted in improved knowledge sharing from documented knowledge is the lately started *design forums* which work as a COP in the company. It has shown to be a valuable forum for sharing knowledge and references to existing documents within and across disciplines.

32.3.2 People's Responsibilities for Knowledge Management

KA has initiated a structure to assign people to different roles in relation to KM. The process is in early stage and is today not spread throughout the company in a formal manner. The main roles so far are knowledge owner and knowledge expert. They manage the knowledge value stream by being responsible for the following KM activities:

- Transforming knowledge captured throughout the project team into knowledge standards to satisfy current and future customer interests.
- Distributing the knowledge to all relevant parties.
- Coaching problem owners in problem-solving/knowledge building.
- Managing reuse of previously captured knowledge.
- Supporting or rejecting design decisions in integrating events, where the project team describes what was decided and on which ground it was decided.

The knowledge owners' dedicated responsibility is to create and maintain the knowledge for system functionality such as a complete product. The PLM system is structured in certain areas where the knowledge owners store all the gathered knowledge. It is paramount that the knowledge owners monitor the completeness of their respective areas. This implies that they are responsible of identifying knowledge gaps in their specific areas and must strive to close them.

The knowledge experts' dedicated responsibility is to create knowledge standards for systems, sub-systems, processes, and materials that are important in order to achieve the product functionality. The knowledge owner is normally a functional manager, and knowledge expert is normally a senior expert.

An overview of the responsibilities related to KA's practical process can be seen in Table 32.1.

32.4 Analysis

Our study shows that the knowledge value stream in the company mainly focuses on K-briefs as knowledge storage format, even if various formats are used. The employees express confusion in regard to what format to apply to different contexts and state that the use often differs among authors. We did not obtain any procedure on how to store knowledge that is not suited to be stored in K-briefs, such as computer-aided design models.

In relation to the organizational aspects, the knowledge value stream describes it as the knowledge owner's and knowledge expert's responsibility to extrapolate the knowledge within the expert area, relying solely on their involvement to store and handle the knowledge. The process expresses focus on reusing design decisions but illustrate less focus on the road that leads there, such as properly documenting the rationale behind the decisions and the KM system that was used in the process. We

Table 32.1 Summary of the responsibilities in KA's practical process of knowledge management with numbering connecting to literature

KM activities	Responsibilities of knowledge owner	Responsibilities of knowledge expert
Transforming knowledge to knowledge standards	Outmost responsible that knowledge is documented and approve documented knowledge (8)	Elicit and extract knowledge from team members, and together capture useful learnings (2 and 3)
Distributing knowledge	Distributing knowledge between projects and being a contact person for global sharing in assigned knowledge field. Store and manage knowledge in PLM system (4)	Share knowledge in involved projects. Search for new knowledge regarding specific sub-system both externally and internally (4)
Coaching problem owners	Be aware of knowledge gaps on system level; initiate steps to close them. Provide coaching during problem-solving	Be aware of knowledge gaps for lower sub-system; initiate steps to close them. Provide support during problem-solving
Managing reuse of knowledge	Participate in project start-up, share experiences (4), and highlight potential valuable knowledge (support 5). Making sure that knowledge is available for team members (4)	Guide throughout the project, and especially support during knowledge gap closure and key decisions (support 7)
Evaluate design decisions during IE	Support or question key decisions (8)	Summarize background information and sub-system status

could not obtain a formal process from the company addressing the issue on what to store, leaving it up to the employees to evaluate the value of the knowledge under consideration. What one employee see as unimportant might be essential for another, and employees state that the individual interpretation to what the documents shall contain leads to interpretation issues when attempting to reuse.

Findings from the workshop and interviews show that knowledge is not sufficiently accessible in the existing technology structure, causing difficulties when attempting to find knowledge. This may be the reason why our research discloses that knowledge reuse is conducted mainly through direct contact with other employees or through the employees' own mind or personal system. Literature support this as a common situation [1]. A negative NPS (-14 and -13) to questions regarding the level of success in the company's approach to capture, store, and reuse knowledge further supports this claim. More specifically, the workshop disclosed that the database structure is not satisfactory and the various other databases make it difficult for the employees to know what to store and where. However, this isn't solely a challenge of the technology itself. The workshop highlighted a people domain challenge, as the employees admit to not consult the database before they are convinced that the knowledge they seek exist. To be sufficiently convinced, they must have entered the knowledge in the database themselves, been told that it exist, or be aware that someone have had a similar problem before. Our research also concludes that the naming convention for each K-brief is undefined, which strengthens the challenge to find the K-brief in the database structure. Even K-brief and A3 are used interchangeably in the name of the documents.

Hence, the people aspect of KM cannot be neglected. In the workshops, scepticism was expressed toward knowledge reuse in fear of preventing new thinking and innovation in the projects. Other employees contradicted this statement claiming that reuse could free up resources which could be used for innovative thinking, making the company more efficient. A clear strategy will reduce the level of confusion at this point. The employees admitted that unused potential would still exist even with an optimized database for KM. The many tools and methods that the company has implemented throughout the years may also be a contributor to confusion and demotivation because the employees seem to perceive the storage of knowledge as a cumbersome and tiring process. Half-hearted effort and dissimilar entry to the knowledge base may be yet another reason to why the employees express that the knowledge is difficult to evaluate, trust, and interpret. However, the employees express a willingness to improve what to store and how to reuse knowledge.

In an attempt to ease the process, the employees have made their own systems for KM, such as spreadsheets, which only escalate the problem with reuse. The people domain challenges, in combination with a cumbersome process, may be a contributor to why the employees state that time deprivation causes KM to yield for other, more time-pressing tasks.

It is apparent that KA has invested much effort in improving their KM, hereby proving that the willingness exists. The study has enlightened several initiatives aimed at increasing the value flow of knowledge. Although the initiatives are performed on both individual and group level, the study shows a low degree of synchronization among local initiatives but also between the local and global level. This study and analysis indicate that the company and employees are gaining new knowledge in high speed but have yet to exploit the full potential in regard to documenting and reusing knowledge.

Figure 32.3 shows the perceived challenges during knowledge reuse in the systems engineering domain at KA.

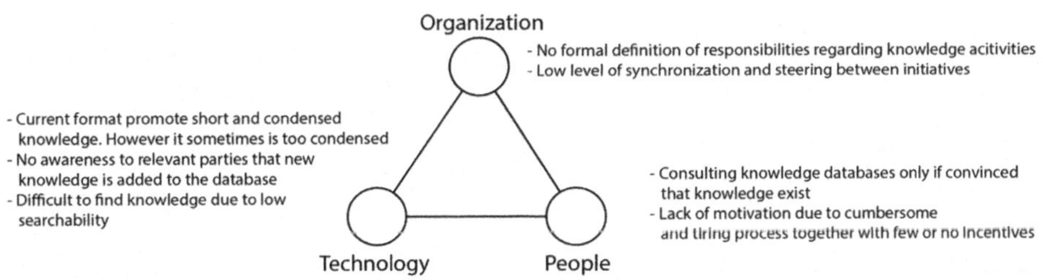

Fig. 32.3 Highlighted perceived challenges in knowledge flow at KA

32.5 Conclusion

The case company has experienced significant progression when investing a major effort in the implementation of KBD. They have accomplished this parallel to experiencing an expanding business. However, potential improvements will always exist. Our findings in this research verify that the case company exercises a suboptimal process and structure for KM, limiting the implementation of KBD. We conclude that knowledge reuse is intricate and consists of multiple, interconnected challenges. Several challenges toward knowledge reuse exist, depriving the company of exploiting the full potential of KBD.

The employment of standard K-briefs for problem-solving is also regarded as the main codification strategy to achieve future knowledge reuse. One main challenge, connected to this strategy, is in visualizing the network structure of the knowledge standards in the company database. The database presents as a flat collection of a multitude of formats. Furthermore, difficulties referencing to the K-briefs in the system lead to poor traceability, and no summarization exists of all K-briefs, depriving the company of quantifying the total knowledge gap closed. A continuously expanding structure further complicates the visualization. The structure also limits KM because it does not facilitate storage of other formats than K-briefs. In summary, this means that the intention of the K-briefs, as a format to focus on core and summarized knowledge and decisions, is not enough, since the company is currently experiencing an overload of documents in relation to how difficult the captured knowledge is to find.

As we see it, consistent KM currently heavily relies on the personal interaction of knowledge owners, this being the personalization strategy employed. This results in a fragile process that demands excessive time resources from specific employees. These individuals must have the ability to withhold a holistic view at all times. This challenge inadvertently results in lost knowledge, inconsistent KM, and outdated content in the database.

32.6 Future Research

To continue this research in the future, we see a combination of codification and personalization strategies. This relates both to the format, how knowledge is structured and how the core knowledge from the K-briefs is to be effectively communicated and reused in future projects. The K-brief format has been shown to be ideal for collecting knowledge giving a condensed rationalization regarding the knowledge and decisions taken in the specific instance. However, it has shown to be an overburden in reusing knowledge, as the K-briefs include much of irrelevant information for the reuse scenario. Hence, a structure is needed that only focuses on the relevant knowledge available in the K-briefs and lessons learned. This may lead to a higher adoption of the new knowledge documentation which may in turn decrease the burden of the knowledge owners. A suitable strategy for reducing K-briefs to dedicated knowledge documentation can be found in the thin-slicing methodology, to some extent also demonstrated within the case company where knowledge is captured in condensed and actionable pieces of experience-based knowledge.

Acknowledgment The financial support by Vinnova (Swedish Governmental Agency for Innovation Systems) and Kongsberg Automotive is greatly acknowledged. This study was conducted within the Vinnova Vinnex Excellence Centre for Product Realization and the Production Area of Advance at Chalmers University of Technology in collaboration with University College of South East Norway.

References

1. Alistair, L., Chris, M., & Steve, C. (2004). Information access, storage and use by engineering designers, part 1 and 2. *The Journal of the Institution of Engineering Designers, 30*, 23–26.
2. Fortune, J., & Valerdi, R. (2008). Considerations for successful reuse in systems engineering. In *AIAA Space 2008 Conference & Exposition 2008-7758*.
3. Pfleeger, S. L., & Bollinger, T. B. (1994). The economics of reuse: New approaches to modelling and assessing cost. *Information and Software Technology, 36*, 475–484.
4. Satyadas, A., Harigopal, U., & Cassaigne, N. P. (2001). Knowledge management tutorial: An editorial overview. *IEEE Transactions on Systems, Man, and Cybernetics, Part C (Applications and Reviews), 31*, 429–437.
5. Kurt, S. (2014). Knowledge reuse: Survey of existing techniques and classification approach. In *European Business Intelligence Summer School* (pp. 126–148). Berlin, Germany: Springer.
6. Boothroyd, G. (1994). Product design for manufacture and assembly. *Computer-Aided Design, 26*, 505–520.
7. Etienne, W., Arnold, M. R., & William, S. (2002). *Cultivating communities of practice: A guide to managing knowledge*. Boston, MA: Harvard Business Press.
8. King, W. R. (2009). *Knowledge management and organizational learning*, *Annals of Information Systems* (Vol. 4, pp. 3–13). Berlin, Germany: Springer.
9. Welo, T., Tonning, O. R. B., & Rølvåg, T. (2013). Lean systems engineering (LSE): Hands-on experiences in applying LSE to a student eco-car build project. *Procedia Computer Science, 16*, 492–501.
10. Sobek II, D. K., & Art, S. (2011). *Understanding A3 thinking: A critical component of Toyota's PDCA management system*. Boca Raton, FL: CRC Press.
11. Morgan, J. M., & Liker, J. K. (2006). *The Toyota product development system*. Seattle, WA: Productivity Press.
12. John, S. (2008). *Managing to learn: Using the A3 management process to solve problems, gain agreement, mentor and lead*. Cambridge, MA: Lean Enterprise Institute.
13. Raudberget, D., & Bjursell, C. (2014). A3 reports for knowledge codification, transfer and creation in research and development organisations. *International Journal of Product Development, 19*, 413–431.
14. Riege, A. (2005). Three-dozen knowledge-sharing barriers managers must consider. *Journal of Knowledge Management, 9*, 18–35.
15. Miles, J. C., & Moore, C. J. (1991). An expert system for the conceptual design of bridges. *Computers & Structures, 40*, 101–105.
16. Moore, C. J., & Miles, J. C. (1991). Knowledge elicitation using more than one expert to cover the same domain. *Artificial Intelligence Review, 5*, 255–271.

17. Gavrilova, T., & Andreeva, T. (2012). Knowledge elicitation techniques in a knowledge management context. *Journal of Knowledge Management, 16*, 523–537.
18. Connelly, C. E., Ford, D. P., Turel, O., Gallupe, B., & Zweig, D. (2014). 'I'm busy (and competitive)!'antecedents of knowledge sharing under pressure. *Knowledge Management Research and Practice, 12*, 74–85.
19. Nonaka, I. (1994). A dynamic theory of organizational knowledge creation. *Organization Science, 5*, 14–37.
20. Argote, L., & Ingram, P. (2000). Knowledge transfer: A basis for competitive advantage in firms. *Organizational Behavior and Human Decision Processes, 82*, 150–169.
21. Bock, G.-W., Zmud, R. W., Kim, Y.-G., & Lee, J.-N. (2005). Behavioral intention formation in knowledge sharing: Examining the roles of extrinsic motivators, social-psychological forces, and organizational climate. *MIS Quarterly, 29*(1), 87–111.
22. Menon, T., & Pfeffer, J. (2003). Valuing internal vs. External knowledge: Explaining the preference for outsiders. *Management Science, 49*, 497–513.
23. Mankins, J. C. (1995). *Technology readiness levels.*
24. Corin Stig, D., Isaksson, O., Högman, U., & Bergsjö, D. (2015). Tera—An assessment of technology reuse feasibility. *Procedia Computer Science, 44*, 507–516.
25. Duffy, S. M., Duffy, A. H. B., & Maccallum, K. J. (1995). *A design reuse model.*
26. Appleyard, M. M. (1996). How does knowledge flow? Interfirm patterns in the semiconductor industry. *Strategic Management Journal, 17*, 137–154.
27. Galup, S. D., Dattero, R., & Hicks, R. C. (2002). Knowledge management systems: An architecture for active and passive knowledge. *Information Resources Management Journal, 15*, 22.
28. Jarvenpaa, S. L., & Staples, D. S. (2001). Exploring perceptions of organizational ownership of information and expertise. *Journal of Management Information Systems, 18*, 151–184.
29. Jamieson, S. (2004). Likert scales: How to (ab)use them. *Medical Education, 38*, 1217–1218.
30. Stenholm, D., Mathiesen, H., & Bergsjo, D. (2015). Knowledge based development in automotive industry guided by lean enablers for system engineering. *Procedia Computer Science, 44*, 244–253.
31. Michael, K. (2008). *Ready, set, dominate: Implement Toyota's set-based learning for developing products and nobody can catch you.* Richmond, VA: Oaklea Press.
32. Nonaka, I., & von Krogh, G. (2009). Perspective—Tacit knowledge and knowledge conversion: Controversy and advancement in organizational knowledge creation theory. *Organization Science, 20*, 635–652.

Chapter 33
Ontology-Driven Requirements Engineering in the Responsive and Formal Design Process

Nadew Kibret, William Edmonson, and Solomon Gebreyohannes

33.1 Introduction

Requirements that are understood and shared by all stakeholders as well as system engineers are highly critical in the development of a successful system [1]. Requirements specification is commonly generated in natural language during the process of requirements engineering (RE) [2]. For software systems, for example, an overwhelming 71.8% of requirements specifications are generated using a common natural language, while only 15.3% use structured language and 5.3% use formal specification languages [3]. However, the inherent ambiguity and impression of natural languages restrain one from generating accurate and unambiguous requirements specifications.

A formal representation of requirements is necessary to overcome inherent problems such as ambiguity associated with the natural language representation of requirements. In this work, the use of ontologies for representing domain knowledge from its requirements will be explored for the responsive and formal design (RFD) process. Ontology languages such as ologs (ontology logs) [4] can be used to build domain ontologies from requirements in the RFD process. Ologs constitute a formal technique for information presentation, storage, and sharing.

The RFD process is a systems engineering methodology introduced in [5, 6]. It integrates formal methods with model-based systems engineering (MBSE) to achieve two main goals: introduce formal reasoning from early stage of system construction and effectively manage design complexity using MBSE. The RFD process is composed of horizontal mappings and vertical abstractions; the former is captured through the design parameters that include requirements in natural language form, set of logical formulas (theory), and system of interconnected models together with their associated simulations. These design parameters are intelligently

N. Kibret (✉) · W. Edmonson · S. Gebreyohannes
North Carolina A & T State University, Greensboro, NC, USA
e-mail: nskibret@aggies.ncat.edu; wwedmons@ncat.edu; shgebrey@aggies.ncat.edu

S. Adams et al. (eds.), *Systems Engineering in Context*,
https://doi.org/10.1007/978-3-030-00114-8_33

integrated in such a way that they facilitate the seamless integration of formal verification and validation activities in the construction process from the onset. Vertically, the RFD process is divided into hierarchically arranged layers that are linked through *refinement* and *abstraction* functions. The two functions enable the formalization of traceability as well as preservation of consistency between abstraction layers.

Following the inputs from our work on the RFD process [5–9], this work presents a modification to the RFD abstraction layers. The main goal of the reformulated RFD process is to enable seamless integration of verification based on formal methods and validation based on modeling and simulation. The contributions of this work include reformulating the RFD process and building a framework for ontology-based RE process in the RFD.

Ontologies are used to formally represent entities and their relationship in a domain. A domain knowledge captured in ontology constitutes a shared conceptualization that can be understood by different stakeholders. Due to this, the use of ontologies in requirements engineering has been explored in various works [10–14]. In [14], an ontology is used as an intermediate representation of system behavior from natural language requirements. From the ontology representation, an automated transition to logic-based modeling using Maude language is provided, but the scope of the work is limited to smart space domain (evolution from ubiquitous computing) and functional requirements only.

The remainder of the paper is presented as follows: Sect. 33.2 presents background material on ontology and methodologies for developing ontologies. In Sect. 33.3, the reformulated RFD process is introduced. Ontology-based requirements engineering for the RFD is presented in Sect. 33.4. A case study to demonstrate the use of ontologies in the requirements engineering activities of the RFD will be developed in Sect. 33.5. Section 33.6 concludes the paper.

33.2 Background

33.2.1 What Is Ontology?

Based on [15] "an ontology is a formal explicit specification of a shared conceptualization of a domain of interest." *Conceptualization* is understood as an abstraction of a domain. The term *explicit* refers to the idea of concepts and their relationships being unambiguously defined. In addition, the term *shared* in the definition refers to a consensus reached among experts in the domain.

The use of ontologies for requirements engineering is an active research area [13, 14]. Ontology languages such as OWL have been used to represent requirements formally and perform requirements analysis [13]. In [12], a detailed description of the use of ontologies for software engineering and technology is presented. Ontologies can assist in the development process of software systems or during their operation (run time). Some of the uses in the development process include facilitating communication among project stakeholders, avoiding ambiguities of natural language, and filtering of knowledge when defining models and metamodels [12].

33.2.2 Ontology Development

There are different methodologies that describe the steps to construct ontologies [16, 17]. The most important steps are outlined in [16] for building ontologies in the domain of autonomous systems. The methodology is referred to as OASys (ontology for autonomous systems) and provides general guidelines to build ontology. The steps are as follows:

1. Enumerate important items—identify nouns and verbs that make up the glossary of terms in the domain.
2. Define concept taxonomies— create concept hierarchy using top-down (from general to specific), bottom-up, or a combination of both.
3. Define relations—some of the verbs identified belong to relations between concepts.
4. Define attributes—some of the nouns identified describe other nouns (concepts).
5. Define instances—individuals belonging to concepts are created by giving them names and concept they are associated to.
6. Define axioms, rules, and functions—give additional structure to the ontology.

33.2.3 Ontology Languages

Ontology languages are formal languages used in the construction of ontologies. In addition to allowing the encoding of knowledge in ontology, these languages provide reasoning mechanisms to infer knowledge from existing facts. In this work, olog is used to construct ontologies for our case study. Ontology log is introduced in [4]. The *olog* language is based on category theory and as such is highly expressive. Moreover, *ologs* can be presented graphically [4, 18] which makes knowledge communication effective between stakeholders and ontology modelers.

33.3 The Responsive and Formal Design Process

A systems engineering methodology called the responsive and formal design (RFD) is proposed in [5, 6]. The RFD process incorporates formal methods with model-based systems engineering (MBSE); formal methods provide the mathematical syntax and semantics required to specify and verify a system for correctness, and

MBSE ensure that modeling complexity grows manageably as the design progresses naturally by including detailed aspects of the system design. The RFD process consists of a set of layers. Adjacent layers are linked through *refinement* and *abstraction* functions. The *refinement* function involves the breakdown of an abstract model into smaller sub-models. The level of details in the sub-models is then enhanced by including granularities to their abstract version. *Abstraction* processes, on the other hand, involve first forming an abstract model by reducing the level of details in component models and eventually merging them together. The two functions provide assurance for consistency between the abstract and refined models and establish traceability between the different layers.

Based on the inputs obtained from our works on RFD [5–9], a modification is proposed to the RFD method. We model an abstraction layer (denoted as A_i) of the process using Eq. (33.1):

$$\begin{array}{lllll} A_i: & L_n & \Rightarrow & M & \Rightarrow & M_{S/P} \\ & & & \Downarrow & & \Downarrow \\ & & & T & & S_{S/P} \end{array} \tag{33.1}$$

where the design parameters represented are as follows:

$L_n \triangleq$ requirements written in natural language form
$M \triangleq$ logical/mathematical representation of a system
$T \triangleq$ theory based on the logical representation M
$M_{S/P} \triangleq$ system of interconnected models
$S_{S/P} \triangleq$ simulations based on the representation $M_{S/P}$ (subscript $_{S/P}$ denotes the shift from behavior to parametric representation in the later stages of the design)

Requirements in natural language form (L_n) are the starting points for each layer in our design methodology. Unfortunately requirements specified in natural language are ambiguous, incomplete, and inconsistent [2, 19, 20]. To address these challenges, there is a need to represent requirements formally. In the RFD, this is achieved by mapping L_n to M. The behavior of the system, its composition (parts and their relationship with each other), and operations (what each part does individually and together) are captured through a logic-based model (M) from L_n. Desired system properties are specified and evaluated on the model using techniques such as theorem proving or model checking (captured using T). The $M \Rightarrow T$ combination forms formal verification process part of the RFD.

Formal methods are used to establish system correctness by applying mathematical rigor in the verification process and are widely used in development of both hardware and software systems [21–25]. If a formal specification cannot be proved from the model, a modification is made on the model or requirements. This iterative process continues until all properties are correctly verified (i.e., $M \Rightarrow T$).

However, ensuring correctness through formal verification of a system construction is not enough to develop a "successful" system that is "fit for purpose" [1]. This is due to first the lack of formal methods that are expressive enough to be used to specify all types of system properties. Many of the formal methods developed have

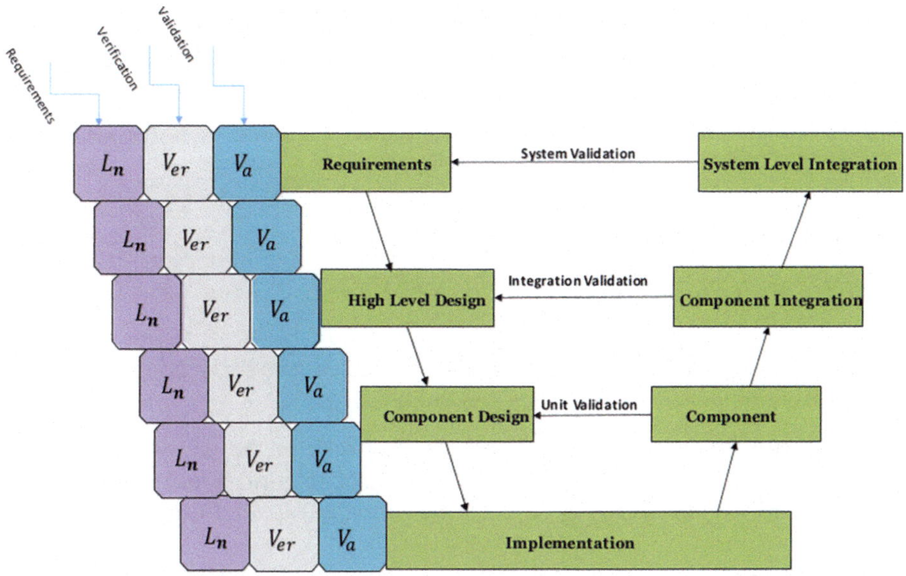

Fig. 33.1 The RFD process over the left side of the V-model

specific strength areas around which their development is anchored [26]. For example, the *Z* method based on Zermelo theory is developed to handle sequential processes and lacks the capability to describe timing and concurrency properties in its original form [26]. To address this issue, multiple formal techniques are used to verify different aspects of the system design. In this case, ensuring consistency among the different representations becomes necessary. Second, formal verification only ensures that the model built is logically sound. To validate if the system captures customer's needs, the RFD process provides a model-based simulation through the parameters M_S and S_S. The representation in M_S is consistent with the representation in M.

The same structure as Eq. (33.1) is adopted for each abstraction layer in the RFD except for two main changes; each stage adds granularity to the design, and as the design progresses toward the final stages, the modeling and simulation parameters change from capturing behavioral aspects of the system toward model-based parametric simulation of the system functionalities. This dynamism is captured by changing the design parameters M_S and S_S to their counterparts M_P and S_P, respectively.

Figure 33.1 shows the RFD process over the left side of the V-model. The RFD process supplements the left side of the V-model by utilizing rigorous verification and validation techniques. This in turn reduces the V-model's extreme reliance on post-implementation validation that is based on testing.

33.4 Requirements Engineering Using Ontology in the RFD Process

In this section, we present requirements engineering process for the RFD that incorporates ontology as one of its key components.

33.4.1 Building Ontology in the RFD Layer

The main concepts to build an ontology-driven requirements engineering and system development in the RFD process are presented in Fig. 33.2.

- **Ontology's Purpose Specification**: The first task in the development of an ontology is to identify the goals, the domain of interest, domain problems to solve, and requirements. These could be formally captured using competency questions.
- **Requirements in Natural Language (*L_n*)**: A document containing set of requirements in natural language are the starting point for every abstraction layer in the RFD process. In addition, the ontology modeler uses his knowledge of the domain, existing ontologies, and expertise in the domain to aid in the ontology development.
- **Term Extraction**: This stage corresponds to the enumeration of important terms in the OASys methodology. The glossary of terms that form the concepts, attributes, or relations in our ontology are extracted from our requirements document. Generally, nouns form either concepts or attributes and verbs constitute relations between concepts. The extraction process can be automated using natural language processing (NLP) techniques such as tokenization to separate each word in a sentence, POS (part-of-speech) tagging to identify the type of word in a sentence, and the use of proper parsing techniques to build a parse tree

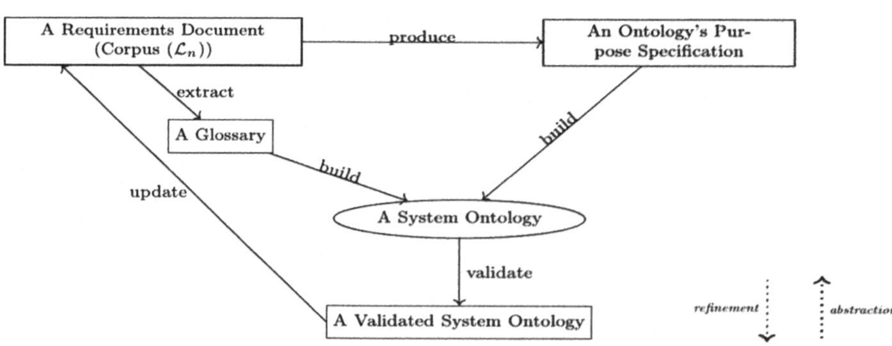

Fig. 33.2 A framework for incorporating an ontology-driven RE in the RFD

which helps to identify entities and their relations. The development of this approach is beyond the scope of our current work.

- **System Ontology**: Building the system ontology requires identifying the concepts, attributes, and relations, creating the taxonomy (concept hierarchy), identifying the domains and ranges of each relation, and identifying the concepts each attribute describes. Some of the concepts in the ontology correspond to components of the system, and the associated attributes describe the behavior of the component. The relations between components are described using the relations defined between them.
- **Ontological Properties (Axioms, Rules, and Functions)**: Ontological properties are defined as part of the system ontology. Axioms represent statements about the ontology which are always true. We populate these properties as part of the system ontology development. As the construction progresses through the refinement process, the ontological properties become domain-specific (sub-domain) with overlaps because they are extracted from the same requirements in natural language form.
- **Ontology Validation**: Among the benefits of domain modeling using ontologies is to aid communication between stakeholders and system engineers. This is even more so if graphical languages are employed in the modeling of the ontology. Since ontology developers are usually not domain experts, experts (customers and stakeholders) can be consulted to validate the constructed ontology, and comments can be obtained for further enhancement to the ontology.

33.4.2 Ensuring Consistency in the RFD Refinement Process

The RFD process starts from abstract level system description of a project, captured in a set of requirements. The subsequent stages in the RFD process produce refined set of requirements from which domain-specific system engineers produce design artifacts in their domain. Ensuring consistency among the domain-specific models with properties that depend on and are influenced by other domain model properties is critical for integration stages later in the development process.

Figure 33.3 shows RFD layers linked through refinement and abstraction processes. At the abstract level, requirements in natural language form (L_n^i) are used to construct abstract system ontology model and its ontological properties. Requirements in the refined level (L_n^{i+1}) are decomposed into domain-specific requirements representing the aspects of the system in each domain. Some of the ontological properties specified for each system aspect overlap, as can be seen from the overlapping inner ellipses in the refined layer in Fig. 33.3. These overlapping properties can be used as a basis for ensuring consistency among the different domains.

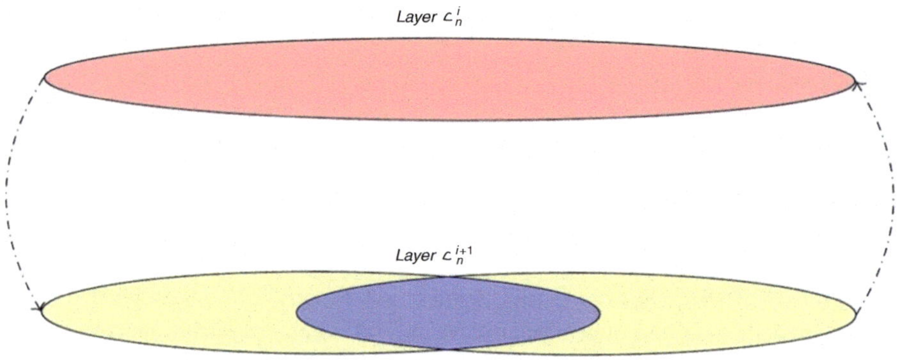

Fig. 33.3 Refinement over system ontology

33.5 Case Study

An example of a small satellite mission taken from [27] is used in this section to illustrate the ontology-driven RE approach presented for the RFD process. We first present the system description informally.

33.5.1 System Description

The mission consists of three formation flying satellites situated at low Earth orbit (LEO) [27] (see Fig. 33.4). Two of these satellites are from the U-class spacecraft (CubeSat) categories (1U), and they are deployed at equatorial LEO, while the remaining small satellite is allowed to orbit in a sun-synchronous LEO. The satellites are equipped with visible light communication (VLC) capabilities to enable efficient and fast communications. In addition, the small satellites will have a radio-frequency (RF) communication capability which will furnish the initial communication link needed with Earth prior to orbital insertion and which will function as a backup communication system in the case of a fault or failure of the primary visible light relay transponder system.

33.5.2 Inter-satellite Communication System Ontology

- **ISC Ontology's Purpose Specification**: We are interested to develop an ontology for the domain of inter-satellite communication (ISC) for small satellites. Table 33.1 shows some of the uses the ISC ontology can provide, among which enabling shared understanding among different stakeholders and enabling requirements analysis in the early stages are the main ones.

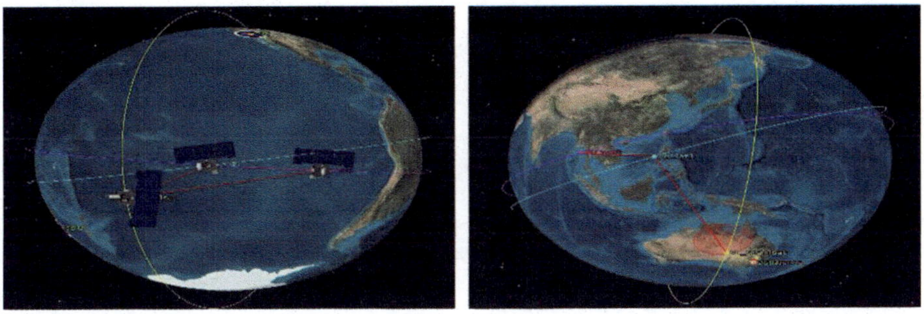

Fig. 33.4 Mission concept [27]

Table 33.1 Determining the scope and purpose of the inter-satellite communication ontology

Inter-satellite communication system ontology	
What is domain of interest?	Inter-satellite communication
What the ontology is used for?	• Understanding and analysis of requirements of an inter-satellite communication system for small satellite • Enabling shared understanding between stakeholders and system engineers • Enabling automated mapping to logic-based modeling of system behavior
Who will use the ontology?	Requirements engineers, domain experts, stakeholders, system designers

- **ISC Ontology**: To build the ISC ontology, glossary of terms (corresponding to concepts, attributes, and relations) are elicited from both a requirements document prepared for MBSE requirements analysis and design of inter-satellite communication for small satellites and domain knowledge. Figure 33.5 shows partial presentation of the ISC system ontology which is obtained from a partial list of requirements. The ontology construction uses the *olog* language. Ten concepts (shown in brown) and nine relations are defined as shown in the figure. In addition, six attributes (shown in olive) are identified. The concepts and attributes correspond to *types* in the *olog* language, and the relations correspond to *aspects*. To preserve space, some of the terms are abbreviated (RF corresponds to radio frequency; Tx and Rx for transmitter and receiver, respectively; VLX for visible light/RF hybrid system; and C&DH for control and data handling). The graphical presentation of the olog ontology improves communication among stakeholders. Besides the precise conceptualization obtained through the encoding of requirements in the ontology, ologs provide the means to capture equivalent paths (commutative diagrams) as facts, and through *limit* and *colimit* constructions, we can encode more structures.

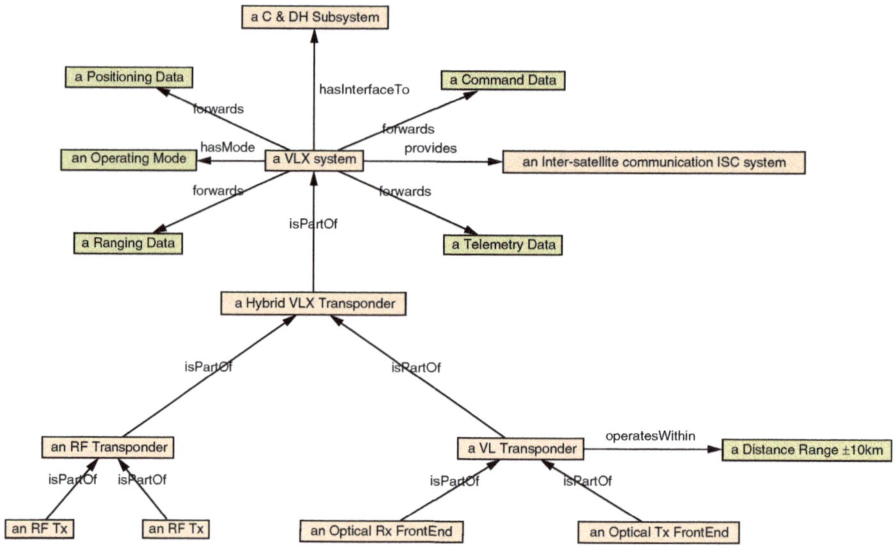

Fig. 33.5 A partial presentation of inter-satellite communication system ontology

33.6 Conclusion and Future Work

In this work, the responsive and formal design process is reformulated to enable the seamless integration of formal verification and validation based on modeling and simulation. The process method defines a hierarchical approach by introducing refinement and abstraction activities to manage complexity of the design at different abstraction stages. In addition, ontology-driven requirements engineering approach is introduced. The use of ontologies enables requirements engineers to identify incomplete and inconsistent requirements early on.

As a future work, automated transition from the ontology model to logic-based model and the formalization of this transition will be investigated. The transition will allow safety critical aspects of the system under construction to be verified using the closed-world assumption of formal methods.

References

1. Holt, J., Perry, S. A., & Brownsword, M. (2012). *Model-based requirements engineering*. London: IET.
2. Halifax, N. S. (2016). *Leveraging natural language processing in requirements analysis: How to eliminate over half of all design errors before they occur*. Halifax, NS: QRA Corp.
3. Berry, D. M., & Kamsties, E. (2004). Ambiguity in requirements specification. In *Perspectives on software requirements* (pp. 7–44). Berlin, Germany: Springer.
4. Spivak, D. I., & Kent, R. E. (2012). Ologs: A categorical framework for knowledge representation. *PLoS One, 7*(1), e24274.

5. Chenou, J., Edmonson, W. W., Esterline, A. C., & Neogi, N. A. (2014). *Formal framework for ensuring consistent system and component theories in the design of small satellite systems.* CSDM (Posters), CiteSeer.
6. Edmonson, W., Chenou, J., Neogi, N., & Herencia-Zapana, H. (2014). Small satellite systems design methodology: A formal and agile design process. In *8th Annual IEEE Systems Conference (SysCon).*
7. Gebreyohannes, S., Edmonson, W., Chenou, J., Neogi, N., & Esterline, A. (2015). Formal requirement management for the responsive and formal design process. In *2015 IEEE International Symposium on Systems Engineering (ISSE).*
8. Gebreyohannes, S., Edmonson, W., & Esterline, A. (2017). Formal behavioral requirements management. *IEEE Systems Journal, 12*(3), 3006–3017.
9. Gebreyohannes, S., Edmonson, W., Esterline, A., Homaifar, A., & Kibret, N. (2017). Fuzzy classification context for the responsive and formal design process. In *2017 Annual IEEE International Systems Conference (SysCon).*
10. Breitman, K. K., & Hinchey, M. G. (2008). The use of formal ontology to specify context in ubiquitous computing. In *International Symposium on Leveraging Applications of Formal Methods, Verification and Validation.* Springer.
11. Dermeval, D., Vilela, J., Bittencourt, I. I., Castro, J., Isotani, S., & Brito, P. (2014). A systematic review on the use of ontologies in requirements engineering. In *2014 Brazilian Symposium on Software Engineering (SBES).*
12. Ruiz, F., & Hilera, J. R. (2006). Using ontologies in software engineering and technology. In *Ontologies for software engineering and software technology* (pp. 49–102). Berlin, Germany: Springer.
13. Sadoun, D., Dubois, C., Ghamri-Doudane, Y., & Grau, B. (2014). Formal rule representation and verification from natural language requirements using an ontology. In *International Workshop on Rules and Rule Markup Languages for the Semantic Web.* Springer.
14. Sadoun, D., Dubois, C., Ghamri-Doudane, Y., & Grau, B. (2013). From natural language requirements to formal specification using an ontology. In *2013 IEEE 25th International Conference on Tools with Artificial Intelligence (ICTAI).*
15. Gruber, T. R. (1993). A translation approach to portable ontology specifications. *Knowledge Acquisition, 5*(2), 199–220.
16. Bermejo, J. (2007). *A simplified guide to create an ontology.*
17. Jones, D., Bench-Capon, T., & Visser, P. (1998). *Methodologies for ontology development.*
18. Spivak, D. I. (2014). *Category theory for the sciences.* Cambridge, MA: MIT Press.
19. Boehm, B. W. (1981). *Software engineering economics.* Englewood Cliffs, NJ: Prentice-Hall.
20. Stecklein, J. M., Dabney, J., Dick, B., Haskins, B., Lovell, R., & Moroney, G. (2004). Error cost escalation through the project life cycle. In *INCOSE International Symposium* (Vol. 14, pp. 1723–1737). https://doi.org/10.1002/j.2334-5837.2004.tb00608.x.
21. Baier, C., Katoen, J.-P., & Larsen, K. G. (2008). *Principles of model checking.* Cambridge, MA: MIT Press.
22. Clarke, E. M., Grumberg, O., & Peled, D. (1999). *Model checking.* Cambridge, MA: MIT Press.
23. Clarke, E. M., & Wing, J. M. (1996). Formal methods: State of the art and future directions. *ACM Computing Surveys (CSUR), 28*(4), 626–643.
24. Easterbrook, S., Lutz, R., Covington, R., Kelly, J., Ampo, Y., & Hamilton, D. (1996). *Experiences using formal methods for requirements modeling.*
25. Hall, A. (1990). Seven myths of formal methods. *IEEE Software, 7*(5), 11–19.
26. Rouff, C. A., Truszkowski, W. F., Rash, J., & Hinchey, M. G. (2005). *A survey of formal methods for intelligent swarms.* Greenbelt, MD: NASA Goddard Space Flight Center.
27. Awele, E. (2018). An MBSE conceptual design phase model of a VLC system for inter-satellite communication. In *2018 Annual IEEE International Stems Conference (SysCon).*

Chapter 34
A Review of Value Modeling in the NASA Systems Engineering Research Consortium

Joseph Clerkin and Bryan Mesmer

34.1 Introduction

The NASA Systems Engineering Research Consortium consists of multiple universities, companies, and centers striving to understand the theory of systems engineering (SE). A primary output of the consortium is an annual report entitled "Engineering Elegant Systems: Theory of Systems Engineering." In this report, 7 postulates and 12 principles have been proposed. A topic that spans multiple principles—i.e., principles 1–6, 8, and 10—is "the need to represent stakeholder preferences." One way to represent stakeholder preferences is mathematically through value models. Significant work has been performed on value models in the consortium: examining how value models are created and how they can be used to aid in decision-making.

This paper will summarize and review the previous value modeling work performed at the University of Alabama in Huntsville (UAH) for the NASA consortium. There are four main areas of research that have been pursued. The first is a lunar mining mission value model formed to demonstrate what a value model is and how it can be applied to a NASA project. The second is a congressional stakeholder value model for allocation of heavy launch vehicle funding to explore how a high-level stakeholder's preference can influence design decisions. The third is an extension of the congressional stakeholder value model with a focus on technologies beyond launch vehicles. The final topic examines how to form value models through the use of content analysis of NASA documentation. The models being covered by this paper have been restricted to these four most recent UAH value models—some of which are still under development. This summarization and review will not contain value models created by other organizations in the NASA consortium.

J. Clerkin (✉) · B. Mesmer
University of Alabama in Huntsville, Huntsville, AL, USA
e-mail: jhc0006@uah.edu; bryan.mesmer@uah.edu

S. Adams et al. (eds.), *Systems Engineering in Context*,
https://doi.org/10.1007/978-3-030-00114-8_34

This paper will first discuss what value models are, how they are used, and how they are formed. Each value model will then be discussed, focusing on their strengths and weaknesses, assumptions and results, and what the future might hold for the model.

34.2 Background

To understand value models, there must first be a brief discussion on where value models come from and what value models are based on. The simplest progression is that value models find their first identifiable ancestor in decision-based design [1–4] (DBD). DBD incorporated decision analysis and utility theory into the design process. Value-driven design [5] (VDD)—utilizing decision analysis and economics with a scorecard approach—could be argued to be an offshoot of DBD. An offshoot of VDD is value-based design [6–8] (VBD)—utilizing multidisciplinary design optimization along with the value approach from VDD. The value models reviewed in this paper primarily utilize VBD, but due to its ancestry, elements of DBD and VDD will also be apparent. This background section will describe briefly DBD, VDD, and VBD.

34.2.1 Decision-Based Design

DBD arose from the idea that, given the advances in computer processing power, engineers would primarily be decision-makers [2]. There are problems that must be overcome when making decisions in an engineering context. The primary problem is making decisions under the uncertainty that is common in engineering. This problem can be overcome by utilizing decision theory where normative decision analysis can be used—utilizing a utility function—to make rational decisions under uncertainty [3]. A utility function combines a preference on measurement with a preference concerning risk. The measurement element is similar to that of an objective function in optimization, enabling a rank ordering of alternatives.

34.2.2 Value-Driven Design

VDD finds its roots in DBD. While VDD is related to DBD, VDD derived its creative impetus in the cost overruns of modern large-scale complex engineered systems [5]. Therefore, VDD utilizes value, generally in monetary units, to represent the preferences of the stakeholders in a value model [5]. The value of a design can then be maximized by examining the local derivatives of the value function to determine what changes in the attributes or behavior variables of the system will help improve the value of the overall system [5]. Since local derivatives are used, care must be taken to move in small steps through the design space. VDD, while very useful for making decisions, may also arrive on infeasible solutions, i.e., the highest value changes to the system may be contradictory due to the focus on systems attributes (behavior variables) and not systems design variables.

34.2.3 Value-Based Design

VBD inherits the value functions of VDD but utilizes a multidisciplinary design optimization (MDO) approach to use total derivatives rather than partials. This use of total derivatives allows the decision-maker to understand how changes to the design variables affect the entire system. This allows for optimization to be performed on the entire system [6]. This approach also enables coupling analysis from MDO.

Once a value model is constructed, the model can be distributed to engineers throughout the design organization. This dissemination enables consistent decisions to be made. This consistency refers to all decisions being made to improve the value of the system, represented by a single value model. The framework for VBD is seen in Fig. 34.1. While value models offer many benefits, the models themselves are complex to produce.

Value models are part of the value-based engineering (VBE) paradigm. VBE relies on the ability to create value models in order to find optimal solutions. Value models are used to find optimal solutions by being used as objective functions in optimization approaches. Value models determine, in economic units, the value of a design given that the design is typified by a set of attributes that are both specific and feasible. Conceptually, the designers could then select a set of design variables that maximize value. Not only does this approach allow the systems engineer to make better decisions, those based on the stakeholders' preferences, but it can also allow the systems engineering a means for succinctly disseminating stakeholder preferences throughout the organization.

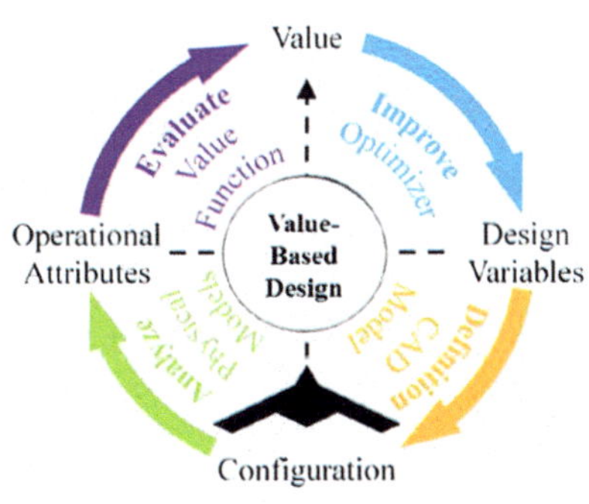

Fig. 34.1 VBD framework

Fig. 34.2 Research tasks

34.3 UAH Value Model Research

The UAH is the lead university in the NASA Systems Engineering Research Consortium. With leaders in the field of value-based engineering, the UAH has taken the role of principle value model researcher for the consortium. The value model research has included simple NASA preference representation, simple and complex congressional preference representation, and formation of value models using content analysis. Figure 34.2 highlights the research tasks.

34.4 Research Task 1: Lunar Mining Mission Value Model

34.4.1 Introduction

This research task was to form a value model for a lunar mining mission. The model allows a comparison between lunar mining approaches. There was also a certain intent when creating the value model to be able to transition some of the content to mining opportunities for planets and asteroids, although this was not a primary motive within the creation of the value model. As the first research task on value models, the model does not incorporate diverse technologies. The value model is created for a NASA stakeholder utilizing net present value (NPV) as the measure of value.

34.4.2 Basis

The model was built from a monetary basis, looking at the profits and costs of a lunar mining mission. Important mission attributes that are reflected in the value model include manufacturing cost (facilities, payloads, etc.), launch cost, development cost, payload mass, launch frequency, and reliability (probability of a successful launch). The NPV basis is calculated for the mission per Eq. (34.1). Necessary, decomposed equations for the value model are seen in Eqs. (34.2), (34.3), (34.4), and (34.5).

$$\mathrm{NPV} = \sum_{t=0}^{n} \left[(P_t - C_t) \times (1+i)^{-t} \right] \tag{34.1}$$

$$\mathrm{NPV} = P_{\mathrm{perlaunch}} \times \left(\frac{\mathrm{LaunchFreq}}{0.07} \right) - C_{\mathrm{Dev}} \times (1 + .07)^{t_{\mathrm{dev}} - t_{\mathrm{offset}}} \tag{34.2}$$

$$P_{\mathrm{perlaunch}} = (R_{\mathrm{perlaunch}} \times \mathrm{Reliability}) - C_{\mathrm{Manufacturing}} - C_{\mathrm{Launch}} - C_{\mathrm{Loss}} \times (1 - \mathrm{Reliability}) \tag{34.3}$$

$$R_{\mathrm{perlaunch}} = \frac{C_{\mathrm{UnitManuf}}}{1 - M_{\mathrm{Profit}}} \times \frac{\mathrm{MaxPayloadMass}}{\mathrm{NomPayloadMass}} \tag{34.4}$$

$$C_{\mathrm{Loss}} = P(\mathrm{Crew}) \times C_{\mathrm{CrewLoss}} + (1 - P(\mathrm{Crew})) \times C_{\mathrm{CargoLoss}} \tag{34.5}$$

where:

t = Time
C = Cost
P = Profit
R = Revenue
$P()$ = Probability
M = Profit Margin

34.4.3 Assumptions and Results

This model makes assumptions on the time from the start of development to the centroid of development, the profit margin of the project, the unit manufacturing cost, the payload mass, the percentage of launches that will be crew launches, the cost of losing a crew, and the cost of losing a payload. There are other basic assumptions as well such as the economy will continue to function in a similar fashion as it currently does. With these assumptions, the model was able to output a numerical result. Thc model indicates that there would be a positive NPV ($0.5425 Billion) for a lunar mining mission. A sensitivity analysis was conducted on the value model, and it indicated the model has some veracity.

34.4.4 Future of the Model

This model was used to influence some of the ideas for Research Task 2: Value Model of NASA Funding Allocations with a Congressional Stakeholder. This influence can be seen most notably in the way that the congressional value model deals with extraterrestrial mining. The lunar mining value model, while somewhat simplistic, was a demonstration of the capabilities of value models to aid in decision-making. The value model leads to Research Tasks 2–4.

34.5 Research Task 2: Value Model of NASA Funding Allocations with a Congressional Stakeholder [9]

34.5.1 Introduction

This research task's research question was: Which type of launch system will give a congressional stakeholder the most value? This model attempts to answer that question by comparing a NASA heavy launch system to private launch capabilities—e.g., SpaceX and United Launch Alliance. This model uses one stakeholder, so both macro- and microeconomic concepts can be applied without a profusion of assumptions about what the stakeholder might desire. There are a large number of assumptions, and the model is very high level. Additionally, there was an intent for this model to be used as a basis for Research Task 3: Broader-Spectrum Congressional Stakeholder Value Model for NASA.

34.5.2 Basis

This model was built using a Cobb-Douglas production function (Eq. 34.6), which is used to describe the interaction of capital and labor [9]. Equation (34.7) shows the Cobb-Douglas function with the elasticity of substitution (σ)—which describes the marginal utilities of two products—and the distribution parameter (δ) set to 0.5 for this model [9]; additionally, there are the production factor (A_p), the values of the products being compared (V_1,V_2), and the production output (Y_T) [9].

$$Y_T = A_R \times K_T^{\alpha} \times L_T^{\beta} \tag{34.6}$$

$$Y_T = A_p \times \left[\delta \times V_1^{\frac{\sigma-1}{\sigma}} + (1-\delta) \times V_2^{\frac{\sigma-1}{\sigma}}\right]^{\frac{\sigma-1}{\sigma}} \tag{34.7}$$

These equations were then used with the properties of the launch vehicles to form the value function. The model incorporated the benefits of extraterrestrial mining, job creation, tax income, patent development, and stock market value [9]. This gave the model some basis to assess how the choice between launch vehicles would impact the overall economy. In this way, NASA can make decisions consistent with how the congressional stakeholder would make decisions. The model utilizes a single design variable to simplify calculations. The design variable is the percentage (X_p) of the NASA budget for launch capability development, where (X_p) is the percentage designated for heavy launch systems and one minus (X_p) is the percentage allocation for commercial launch [9]. The model changed the allocation between heavy launch systems and commercial launch depending on how much value each generated the previous year.

34.5.3 Assumptions and Results

This model makes a large number of assumptions. The foundational assumption of this model is that the Congress primarily cares about the economy [9]. This assumption is made because of an evidence that the voting public primarily cares about the economy [9]. It is also assumed that mass of equipment needed to create extraterrestrial mines is equivalent to the amount needed for a terrestrial mine; that the creation of product patents leads to new businesses; that the creation of process patents accounts for the learning curve of producing a launch device; that there will be no astronomical shifts in the launch capability paradigm; that new businesses are of a standard size; that taxes—both individual and corporate—follow a flat tax scheme; that minerals will continue to increase in value linearly; that the stock market will not react overly positively or crash due to an increase in minerals and space-associated businesses; and that the latency function reasonably delays the full benefits of new businesses and project completion. With the assumptions covered, a discussion of the results can be undertaken [9].

The model, after optimization was run on the initial funding allocation, showed that the vast majority of the launch capability budget should go to the heavy launch system. Interestingly, for the first 7 years, before the heavy launch system has reached full launch capability, the model steadily increased the funding to commercial launch; after the heavy launch is at full capability, the model then decreases the funding for commercial launch for the rest of the program (Fig. 34.3).

34.5.4 Future of the Model

This model significantly influences Research Task 3: Broader-Spectrum Congressional Stakeholder Value Model for NASA. Research Task 3 challenges assumptions made to simplify the value model of Research Task 2.

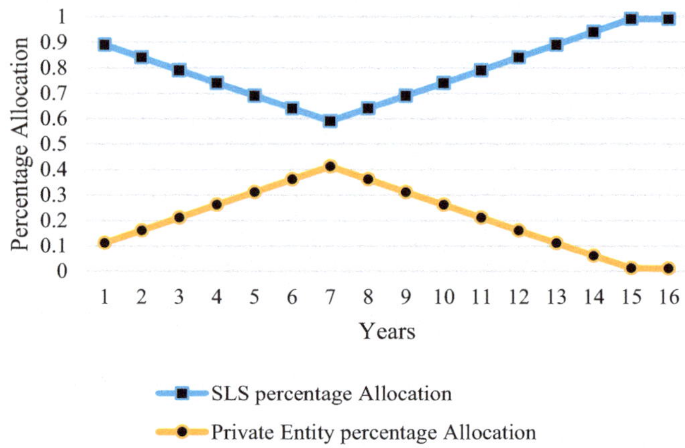

Fig. 34.3 Results from Research Task 2

34.6 Research Task 3: Broader-Spectrum Congressional Stakeholder Value Model for NASA

34.6.1 Introduction

This ongoing research task is developing a comprehensive value model for NASA projects with the value being set for a congressional stakeholder. This value model is an expansion of the Research Task 2 value model. Where the previous value model is focused on just launch systems, this value model is intended to incorporate multiple technologies. This value model also intends to allow other stakeholder value models and physics-based models to be "plugged" into the bottom end—enabling broader and more detailed representations. This ability to "plug in" lower-level models is intended to eventually allow NASA to optimize technologies with consideration to congressional preferences.

34.6.2 Basis

While this value model is mostly an expansion of a previous value model, the Cobb-Douglas function is not being carried over; instead, the value model will be using NPV. In addition to being based on the previously discussed value model, this value model utilizes NASA's 2014 "Strategic Plan" [10]. This was done to fully determine what NASA's goals are. The "Strategic Plan" was pared down to remove instances of repetition and areas where objectives overlapped. This paring down leads to a set of distinct objectives; the objectives are grouped into objectives and sub-objectives. These objectives are being used to develop value. For instance, NASA's goal of traveling into space indicates that there is a value to traveling in space. The

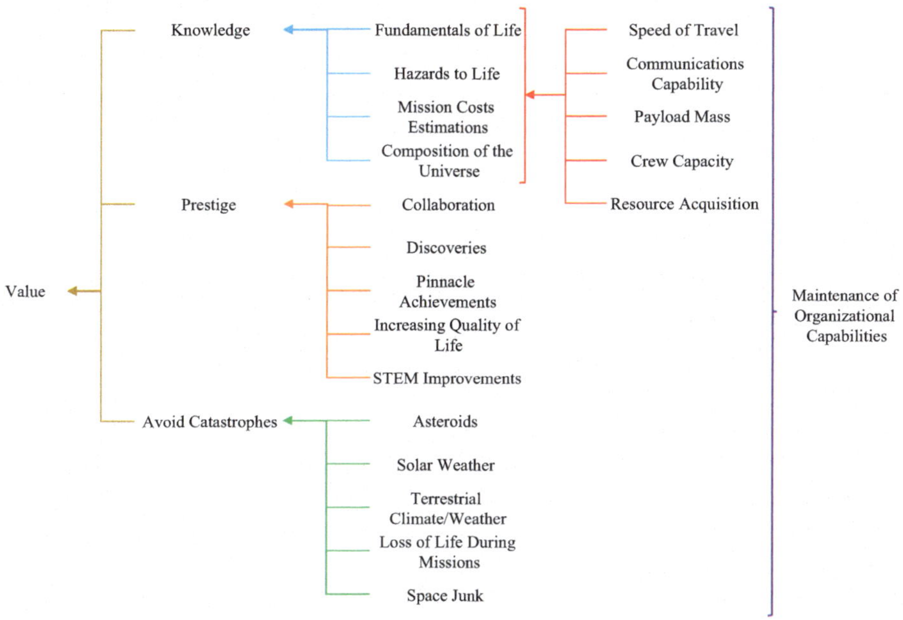

Fig. 34.4 The objectives and how they generally flow through the Broader Spectrum Congressional Stakeholder Value Model for NASA

objectives and the general way they connected are shown in Fig. 34.4. While most of objectives are somewhat self-explanatory, the "Maintenance of Organizational Capabilities" represents the increased costs that will be incurred as time passes between developments of a type of technology and influences all of the attributes.

34.6.3 Assumptions and Results

In addition to the many assumptions of the parent model, this model has a few unique assumptions of its own. These assumptions are mostly related to the complicated questions the model is asking about the value of attributes in a technology. While the answer to questions like "How valuable it is to travel a certain distance in space?" can be very complicated. This model is assuming that simple answers will suffice, and more complicated answers can be worked into the model over time. Essentially, meaningful placeholders are being developed for future disciplinary teams to elaborate at a later time.

34.6.4 Future of the Model

This is an ongoing research task. At the time of this writing, the model is still preliminary with the objectives being expanded into meaningful attributes and design variables.

34.7 Research Task 4: Value Model of Cislunar Habitats [11]

34.7.1 Introduction

This ongoing research task approaches the value model process differently than the first three tasks. The previous tasks assumed a top-level preference of the stakeholder, specifically maximize NPV and maximize the value of the Cobb-Douglas function. This research task does not make that assumption and instead uses content analysis to investigate how to create a value model. With that said, it will still be referred to here as a value model as the intent is to build a value model. This value model focuses on the value that can be derived from a cislunar space habitation capability [11]. The capability of the system is discussed in NASA's "Nextstep-2" habitat documentation. This model is focusing on capturing and modeling the NASA preferences that are being communicated to the competing commercial contractors [11] in the documentation. When complete, the value model will represent NASA's preferences and enable optimization of the cislunar habitat.

34.7.2 Basis

This model is being built by analyzing NASA documents on the Habitat project. The documents are being analyzed by a process of coding the documents and through the use of the NVivo 11 software [11]. Coding documents is a qualitative process for analyzing the meaning of phrases, sentences, and paragraphs [11]. The NVivo 11 software is a computer program that can be used for finding word frequencies and trend or patterns in text, audio, and videos [11]. These patterns and frequencies can be used to determine what the document is communicating [11]. This can be done by looking at a set of correspondences to determine how what NASA describes is being reflected in descriptions of the products the commercial contractors are designing. With an understanding of what NASA is communicating and by observing the responses of contractors, the value model can be precisely tuned to NASA's preferences.

Fig. 34.5 A word cloud depicting the most common words found in NASA habitat documentation

34.7.3 Assumptions and Results

The foundational assumptions here are that NASA's preferences are not changing throughout the selection process and that NASA is releasing documents that contain true NASA preferences. Additionally, there are assumptions that NASA's preferences can be reliably determined through content analysis and that there will be no shift in the need for a cislunar habitat. As this model progresses, there will be more assumptions that need to be made. Despite the early stage, there are some results that can be discussed about the model.

An initial result of this value model is a word frequency cloud of 81 words that are significant to the NASA habitat documentation, seen in Fig. 34.5. Keywords such as capabilities and exploration are seen to be repeated often in the documentation, perhaps leading to a value model that has value-influencing goals related to those words. As more information becomes available, more significant results are expected.

34.7.4 Future of the Model

This work is still under development. The research will turn the qualitative analysis into a quantitative mathematical representation of the preferences of NASA for a cislunar habitat. The value model formed will eventually be used in conjunction with and potentially to inform Research Task 3: Broader-Spectrum Congressional Stakeholder Value Model for NASA. The development of this value model will benefit

NASA by helping them understand what preferences they are communicating and benefit the contractors by helping them understand NASA's desires.

34.8 Discussion

This paper has discussed several value models. Each value model approaches the problem of value modeling in a slightly different way: from the Research Task 4: Value Model of Cislunar Habitats's method of analyzing the communications of the stakeholders to understand their true preferences to the Research Task 3: Broader-Spectrum Congressional Stakeholder Value Model for NASA's method of building on previous value models. The models also differed in their stakeholder perspectives: both congressional value models are looking at the problem from a very high-level viewpoint, while the "cislunar" and "lunar mining" models examine the problem from NASA's perspective. Value models can be built for any stakeholder. For instance, a NASA engineer working on a cooling pump for a crew capsule will likely have a different set of preferences from NASA itself. It is important to note how these variances in stakeholder preferences will affect the "value" in a value model. When multiple decision-makers interact and influence each other's decisions, game theory or other tools must be used to perform the analysis. Finally, value models are challenging to construct; the more precise the value model attempts to be, the harder it is to build. This challenge can be mitigated somewhat by examining value models on similar topics to see how they have dealt with a certain problem.

34.9 Conclusion

In summation, several value models were examined. There was a discussion of the value models, how they were formed, what they were intended to analyze, and how they could be improved. There was also a brief discussion of value modeling itself and the general challenges a potential value modeler faces. While each of these value model research tasks could be expanded, some of them are still unfinished. Overall, value modeling could give an interesting and informative perspective on engineering projects and force the engineer modeling the project to consider the inherent value of a system, attribute, or design variable and how it might be quantified. Therefore, value modeling might be useful for conceptualizing if something should be done as well as how it should be done—i.e., there is more value to a value model that simply determines the optimal design variables.

References

1. Lewis, K., & Mistree, F. (1998). Collaborative, sequential, and isolated decisions in design. *Journal of Mechanical Design, 120*(4), 643–652.
2. Shupe, J. A., Muster, D., Allen, J. K., & Mistree, F. (1988). Decision-based design: Some concepts and research issues. *Expert Systems, Strategies and Solutions in Manufacturing Design and Planning*, 3–37.
3. Lewis, K., Chen, W., Schmidt, L., & Chen, W. (2006). *Decision making in engineering design.* New York: ASME.
4. Hazelrigg, G. A. (1998). A framework for decision-based engineering design. *Journal of Mechanical Design, 120*(4), 653–658.
5. Collopy, P. D., & Hollingsworth, P. M. (2011). Value-driven design. *Journal of Aircraft, 48*(3), 749–759.
6. Topcu, T. G., & Mesmer, B. (2015). Customer, commercial, and government value functions for electric vehicle system design. In *IIE Annual Conference Proceedings Norcross* (pp. 959–968).
7. Bloebaum, C. L., & McGowan, A. M. R. (2012). The design of large-scale complex engineered systems: Present challenges and future promise. In *American Institute of Aeronautics and Astronautics* [cited 2017 Oct 24]. Available from http://arc.aiaa.org/doi/10.2514/6.2012-5571
8. Lee, B. D., & Paredis, C. J. J. (2014). A conceptual framework for value-driven design and systems engineering. *Procedia CIRP, 21*, 10–17.
9. Dyas, J., Clerkin, J., & Mesmer, B. (2018). A congress-based value model for NASA funding allocation. In *AIAA SPACE and Astronautics Forum and Exposition* [Internet]. American Institute of Aeronautics and Astronautics [cited 2018 Jan 30]. Available from https://arc.aiaa.org/doi/abs/10.2514/6.2017-5327
10. Palma, G. E., & Mesmer, B. (2018). A preliminary content analysis of NASA's nextstep-2 habitat documentation for preference representation. In *2018 AIAA Aerospace Sciences Meeting* [Internet]. American Institute of Aeronautics and Astronautics [cited 2018 Jan 30]. Available from https://arc.aiaa.org/doi/abs/10.2514/6.2018-0709
11. NASA Strategic Plan 2014 [Internet]. (2014). National Aeronautics and Space Administration. Available from https://www.nasa.gov/sites/default/files/files/FY2014_NASA_SP_508c.pdf

Chapter 35
Risk Analysis and Prognostics and Health Management for Smart Manufacturing

Michael Malinowski, Stephen Adams, and Peter A. Beling

35.1 Introduction

Manufacturing industries are expanding in both size and complexity. Many companies are undergoing major transformations and adapting new technologies to provide competitive, high-quality products to their customers. Examples of these new technologies include various types of sensors, automation, advanced machining, and robotics. As manufacturing industries continue to grow in complexity, there is an emerging need to identify, characterize, and communicate essential metrics to decision-makers regarding the health of the overall system. Specifically, managers are seeking to enhance their ability to gather timely and actionable intelligence about their system to achieve their major objectives (e.g., minimizing cost, minimizing production/maintenance time, maximizing system health, maximizing product quality, and maximizing product output) subject to maintaining acceptable risks to human health and to the environment.

Many systems and methodologies have already been developed to provide this type of intelligence, especially within the field of prognostics and health management (PHM) [1]. In particular, PHM systems seek to provide tools for predicting remaining usable life (RUL) of equipment, for diagnosing errors or failures, and for developing maintenance or mitigation strategies. However, there are multiple challenges to implementing PHM across all manufacturing sectors. Manufacturing systems are inherently complex and can involve multiple subsystems, processes, and components that are all reasonable targets for the development of PHM systems. This scoping problem is made more difficult because the potential costs and benefits of those PHM systems are subject to random and known uncertainties [2, 3]. Thus,

M. Malinowski · S. Adams (✉) · P. A. Beling
Department of Systems and Information Engineering, University of Virginia, Charlottesville, VA, USA
e-mail: sca2c@virginia.edu

S. Adams et al. (eds.), *Systems Engineering in Context*,
https://doi.org/10.1007/978-3-030-00114-8_35

manufacturers require a way to determine which subsystems or components thereof are most critical and thus would most benefit from an established PHM system.

The field of risk analysis (modeling, assessment, management, and communication) provides systematic and well-established approaches for identifying the major risk scenarios that could impair a system from achieving its major objectives. Through slight modification, these risk methodologies, first proposed by the authors in [4], can be customized to identify targets/components within a manufacturing process that would most benefit from a PHM system. A previously developed methodology, called the WEAR methodology, utilizes risk analysis tools in its first step to determine where to focus PHM efforts and to scope the manufacturing system [5]. The remaining three steps of the WEAR methodology focus on evaluating PHM systems using simulation and building the business case for a PHM system.

This paper outlines a modified risk-based methodological approach for scoping the boundaries of a manufacturing system and determining targets for a PHM system. The methodology incorporates hierarchical holographic modeling (HHM) [6]; risk filtering, ranking, and management (RFRM) [6]; and fault tree analysis (FTA) [7] to provide a scope and direction for the PHM system designer. The method presented in this paper is an extension of the method used in the first step of the WEAR methodology. The primary difference is the addition of the FTA step. To validate the developed methodology, this paper walks through a case study with a US manufacturer. This is a different manufacturer than the one presented by Adams et al. [5] and solely focuses on the risk-based methodology for scoping a manufacturing system and determining targets for a PHM system.

35.2 Background

Smart manufacturing is the integration of information, technology, and human ingenuity to promote rapid revolution in manufacturing intelligence across all business sectors [8]. In the context of an industrial facility, smart manufacturing incorporates the available data across all systems, machines, and components in order to facilitate improvements toward achieving the predefined objectives of the organization (e.g., cost, safety, production, or environmental sustainability). Many industrial facilities are undergoing major transformations and adapting new technologies, such as sensing, automation, machining, and robotics capabilities, to provide competitive, high-quality products to their customers. As the complexity of these manufacturing processes increases, so too must their system models expand. Most organizational and technology-based manufacturing systems are hierarchical in nature [9, 10]; thus their models must reflect this reality. In the context of smart manufacturing, the natural structure of these systems of systems can be represented as follows: (1) system (e.g., factory), (2) subsystem (e.g., workstation), (3) machine (e.g., lathe), (4) component (e.g., spindle), and (5) primitive (e.g., cutting insert). Due to their inherent, complex hierarchical nature, smart manufacturing systems require advanced technologies to facilitate widespread information flow. This information can include the health, performance, and risk of the system failing to meet a critical objective [11].

PHM is an engineering discipline that divides into two major portions. The prognostics aspect refers to the prediction of the future status, health, or performance of components and systems. The major metric used in prognostics is the RUL, and it is often calculated from detection modes (e.g., sensors), maintenance schedules, speed of operations, signs of wear, failure mode analyses, and precursors to system failure. Health management on the other hand refers to the process of making maintenance and logistics decisions from the prognostics information, available resources, and operational demand [12]. The purpose of health management is to compare the trade-offs between available decisions, policies, or schedules in order to meet the objectives established by the facility (usually to minimize cost, minimize production/maintenance time, maximize system health, maximize product quality, and maximize product output) [13].

The use of PHM systems to improve manufacturing performance has been demonstrated in numerous case studies within the automotive [14], aerospace [15], machine tool [1], and power generation [16] industries. Further, PHM information can be used to inform large-scale control problems in a smart manufacturing system [17, 18]. However, as the manufacturing processes increase in size and complexity, it becomes increasingly difficult to determine which components or subsystems could most benefit from a PHM system. Even data-driven approaches, which rely on collected data and mathematical models, lose accuracy and become less predictive as complexity increases [19]. When available resources for PHM efforts are limited, designers and implementers of PHM systems face a difficult problem in deciding where to deploy scarce resources to maximize benefit. A smart manufacturing system may involve multiple subsystems or processes that present reasonable targets for the development of PHM systems [12]. This selection problem is made more difficult because the potential costs and benefits of those potential PHM systems are subject to random and known uncertainty [2, 3].

35.3 The Synergy of Risk Analysis with PHM

The fields of risk analysis and PHM are harmonious and compatible with respect to their theory, methodology, and practice for ensuring the "health" (safety and optimal operability and performance) of a smart manufacturing system of systems. Both schools of thought embrace the need to constantly adapt in order to remain relevant and true to the system of interest. They also follow similar guidelines when designing and implementing a PHM system or a risk mitigation plan. Both fields have a design phase where the high-value critical systems must be identified to provide a scope for the designer. There is also a modeling stage where available data and expert knowledge should be integrated. Then, both risk analysis and PHM implement a data collection phase that collects health and performance metrics from the components of interest. Finally, both fields embrace the need to iterate back to the initial phase, collect new information, learn from the process results, and make improvements as needed.

When comparing risk analysis with PHM, both can be viewed as nearly identical at the core of their fundamental classifications. Risk analysis can be divided into two major parts: (1) risk assessment and (2) risk management (which may include risk communication). Similarly, PHM can be split into two parts: (1) prognostics and (2) health management. Comparing the first parts and the second parts of both fields, respectively, it becomes apparent that the two fields are truly synergistic and can benefit from sharing their methodologies, theories, and lessons learned.

The risk assessment part of risk analysis can be articulated by asking four major questions: (1) What can go wrong? (2) What is the likelihood of that happening? (3) What are the consequences? [20] (4) What is the time frame? [6, 21]. Similarly, the prognostics stage of PHM refers to the prediction of the future status, health, or performance of the components of the manufacturing systems. In terms of the four questions of risk analysis, prognostics might offer the following alternative questions: (1) Which components or subsystems can become nonoperational? (2) What is the probability of that happening? (3) Which systems will be affected? (4) What is the RUL of those components or subsystems? A quick comparison shows that prognostics and risk assessment are seeking answers to the same types of questions. The main difference is that the field of prognostics (with respect to smart manufacturing) specifically focuses on components and subsystems in an industrial environment, as well as the main negative outcome being an expired RUL. On the other hand, risk analysis tends to be broader in its scope of relevant systems and possible risk scenarios.

In a similar fashion as risk assessment, risk management can be simplified by seeking the answers to three major questions: (1) What can be done, and what options are available? (2) What are the associated trade-offs (costs, benefits, risks)? (3) What are the impacts of current decisions on future options? [6, 21]. Health management of PHM is extremely similar to risk management. The central difference between these two definitions is that health management in PHM is directly related to decisions in a manufacturing environment, such as operational strategies or maintenance schedules. The above is effectively risk management but with a more narrow scope. In similar terms as the three risk management questions, health management might ask these questions: (1) What are the possible operational and maintenance strategies? (2) What are the trade-offs (costs, productivity, risks)? (3) How will the decision impact future operational and maintenance strategies?

By examining some of the fundamental properties, theories, and philosophies of risk analysis and PHM, we note that these seemingly different schools of thought are actually extremely compatible and similar. A noticeable difference is that PHM specifically narrows the scope of risk analysis, usually for data-based solutions and systems for a manufacturing process. By realizing the similarities between these two fields, we can begin to share benefits that each has to offer. Risk analysis provides many tools, resources, methodologies, and theories that can advance the fast-growing field of prognostics and health management in smart manufacturing.

35.4 Methodology

In this paper, we harmonize risk analysis with PHM by presenting a methodology that builds on risk analysis methods and by modeling and analyzing a manufacturing process as interconnected and interdependent systems of systems (SoS). The proposed methodology can be divided into three stages: modeling the system, identifying risks, and identifying critical components. Each of these stages relies on an established method from risk analysis.

To model a complex manufacturing environment, we utilize some of the existing tools and methods established by the field of risk analysis. Hierarchical holographic modeling (HHM) [6, 21] is a methodology in the field of risk analysis that considers all possible sources of risk (e.g., hardware, software, organizational, human, cultural, and external). In a PHM setting, sources of risk are failures of subsystems or system components. HHM aims to represent and display through a visual model the inherent diverse qualities and attributes of a system, including its multiple aspects, perspectives, views, dimensions, and hierarchies. A model using the HHM methodology is created by forming an organized team of system experts containing varied experiences and knowledge bases and having them discuss and organize the various characteristics and components of the system under study. The actual models are visually represented by *head topics* (representing different categories for sources of risk). Each head topic will have one or many *subtopics* which contain the details and components of these potential risks. If more detail is required, each head topic and subtopic has the potential to be expanded into an HHM model in and of itself. These expanded HHMs can lead to even more insight into the system's intrinsic hierarchy.

After modeling the manufacturing systems using HHM, we seek to identify the major sources of risk that would most benefit from PHM. Risk filtering, ranking, and management (RFRM) [6] is another risk analysis methodology developed that encompasses the *identification*, *prioritization*, *assessment*, and *management* of potential risk scenarios. In the RFRM methodology, decision-makers and stakeholders for large-scale systems must consider the likelihoods and consequences of each risk. The objective is to produce a reduced list of possible risk scenarios from the full list of risk scenarios for a system so that acceptable mitigation options can be developed and implemented. RFRM often reduces the list from hundreds of risk scenarios to a manageable number, which can vary in number based on the complexity of the system and resources available to the stakeholders and decision-makers. RFRM heavily relies on the results from the HHM methodology, and these two methods are often paired. The RFRM methodology offers eight major phases to guide risk management: (1) scenario identification, (2) scenario filtering, (3) bi-criteria filtering, (4) quantitative ranking, (5) multi-criteria evaluation, (6) risk management, (7) safeguarding against missing critical items, and (8) operational feedback. For the purpose of identifying major subsystems for PHM, only phases 1 through 5 were utilized.

Finally after identifying the major sources of risk to a system, it is essential to understand the critical components involved in that specific risk event. Therefore, fault tree analysis (FTA) [7] is the next step in the risk analysis methodology that provides scope for a PHM system designer. It should be used after the identification of high-priority PHM subsystems from the HHM, RFRM, and PHM-specific RFRM methodologies. FTA helps to determine exactly how a risk scenario could occur and more specifically which components might be involved in the failure of a manufacturing machine.

A fault tree is created using standard FTA components. The top event represents the risk scenario of interest, which was determined from the RFRM methodology. This top event then breaks down into multiple levels of intermediate events, until finally ending at the bottom with an undeveloped event. In theory, these undeveloped events should represent components in a machine hierarchy that would most benefit from PHM. The events in a fault tree are connected by AND gates and OR gates. If events are connected with AND gates, they are in parallel, and all of the events must occur in order for the above event to trigger. If events are connected with OR gates, they are in series, so the occurrence of any of the events would cause the above event to trigger.

35.5 Case Study

The case study with a manufacturing company validates our premise that there exists a harmony and synergy between the seemingly two different fields of risk analysis and PHM. Additionally, this case study has enabled us to further streamline the theory and methodology with the actual implementation of PHM systems and management strategies. By collaborating with an industry, we have refined the developed and established methodologies of identifying major precursors to risks. The case study provides a unique perspective of a system of systems and unique takeaway lessons. Due to the competitive nature of the specific manufacturing facility with which we collaborated, the identity of the company has been omitted from this paper and shall henceforth be referred to as *company*. The presented case study is an extension of the study in [4].

The company is a relatively small-scale manufacturer in the food and food supplement industries. It has developed a complex chemical process to convert raw, organic material into two major products. The main objectives for the company are to maximize production of their food product and to minimize the risk of a system failure (production shutdown or delay) at acceptable quality and safety levels. To help achieve these objectives, the plant managers desire to implement a PHM system into their product packing and bagging process. However, they currently have limited monetary resources allocated toward this project. Therefore, the company has requested a risk analysis consultation in addressing the company's components/machines/subsystems so as to benefit from a PHM system.

The case study benefitted from the new aggregated methodology for the design phase of a PHM system in manufacturing by (1) building upon HHM, RFRM, and FTA; (2) providing a scope and direction for a PHM system by identifying a

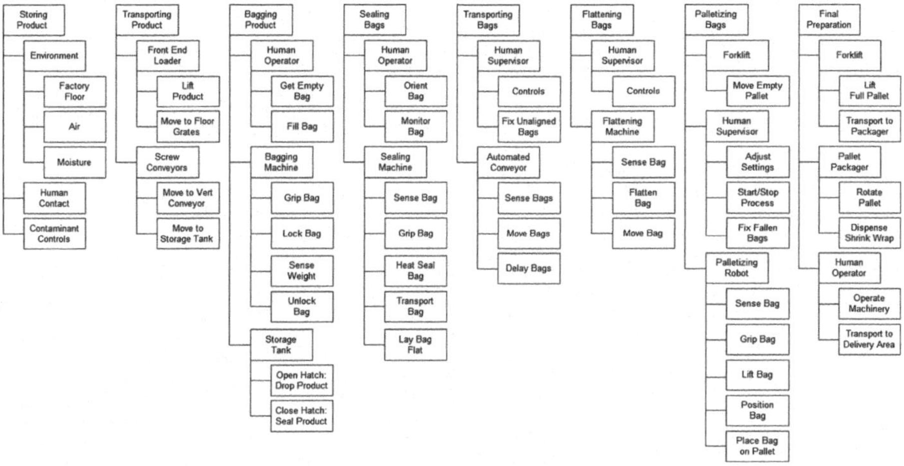

Fig. 35.1 HHM for a company—bagging system process

prioritized set of targets that would most benefit from PHM capabilities; and (3) testing the outcome of the developed methodology with the manufacturing company. Currently, there are multiple methods to determine the major failure modes of a system following an accident or a catastrophe. However, the developed methodology in this paper allows for a thorough analysis to be conducted even prior to the occurrence of a failure in a manufacturing environment.

35.5.1 Modeling the Complexity of a Smart Manufacturing System

First, an HHM model was developed from multiple company experts with the perspective/decomposition of the different *processes* involved in the food bagging system. The head topics for this model were (1) storing product, (2) transporting product, (3) bagging product, (4) sealing bags, (5) transporting bags, (6) flattening bags, (7) palletizing bags, and (8) final preparation for delivery. The complete HHM model for the bagging system processes at the company can be seen in Fig. 35.1.

35.5.2 Identifying the Major Sources of Risk to a Manufacturing System

Phase 1 (scenario identification) of RFRM is to identify possible failure scenarios for the system of interest. This can be a daunting task, but the process may be expedited by utilizing a comprehensive risk scenario structuring model, such as one developed using HHM. From an HHM model, each head topic can represent a major category

Table 35.1 Top risk scenarios of interest for a food company

Risk ID	Subsystem/process	Risk description
1.a	Storing product	Environment failure
2.a	Transporting product	Front-end loader failure
2.b	Transporting product	Screw conveyor failure
3.a	Multiple processes	Human operator failure
3.b	Bagging product	Bagging machine failure
3.c	Bagging product	Storage tank failure
4.b	Sealing bags	Sealing machine failure
5.b	Transporting bags	Automated conveyor failure
6.b	Flattening bags	Flattening machine failure
7.a	Palletizing bags	Forklift failure
7.c	Palletizing bags	Palletizing robot failure
8.b	Final preparation	Pallet packager failure

or subsystem that can branch out into multiple failure scenarios. Subject matter experts and stakeholders are required to help with this failure identification and definition process. Phase 2 (scenario filtering) of RFRM is often combined with the first phase. This step involves a crude, preliminary filtering of the previously identified risk scenarios on the basis of scope, temporal domain, and level of decision-making. The purpose of phase 2 is to reduce the original number of risk scenarios (often in the hundreds) down to a much more manageable number (10–20) before continuing with the complete RFRM process. Table 35.1 contains the results of RFRM phases 1 and 2 for the company. Each risk scenario was assigned a unique risk ID based on its location from the original HHM model. The subsystem/process category refers to the main workstation or business unit where the failure occurs. The risk description has been abbreviated in these tables but is associated with a detailed scenario defined by the subject matter experts.

Phase 3 (bi-criteria filtering) of RFRM is to filter and rank the list of risk scenarios based on two criteria: probability (likelihood) and severity (impact). This involves placing the scenarios on a qualitative matrix with two axes and then selecting the risks that have met a predetermined threshold. The matrix used in this process has been adapted from the Military Standard 882 of the US Department of Defense [22]. The impact levels are selected by the decision-makers for the system and in this case ranged as follows: (1) minor or no impact, (2) minor equipment degradation, (3) reduced production speed, (4) loss of product, and (5) complete production shutdown. The likelihood values are defined as (1) unlikely, (2) seldom, (3) occasional, (4) likely, and (5) frequent. Phase 4 of RFRM is to assign specific probability ranges to each of the likelihood terms. Again the probability ranges are subject to change by the decision-makers and analysts, but in this case they were assigned as follows: unlikely (0–1%), seldom (1–5%), occasional (5–10%), likely (10–50%), and frequent (50–100%). Table 35.2 contains the analysis for RFRM phases 3 and 4 for the company. Table 35.3 contains the new list of the top identified risks for a food company, which appeared in the "high risk" and "extremely high risk"

Table 35.2 Risk matrix for a food company

Impact	Likelihood/Probability				
	0 – 1%	1 – 5%	5 – 10%	10 – 50%	50 – 100%
Production Shutdown		7.c			
Loss of Product	3.c	2.b	3.a, 3.b		
Reduced Speed	2.a, 7.a	5.b, 8.b		6.b	
Minor Degradation	1.a, 4.b				
Minor or No Effect					

Low Risk	Moderate Risk	High Risk	Extremely High Risk

Table 35.3 Reduced list of risk scenarios for a food company

Risk ID	Subsystem/process	Risk description
3.a	Multiple processes	Human operator failure
3.b	Bagging product	Bagging machine failure
6.b	Flattening bags	Flattening machine failure
7.c	Palletizing bags	Palletizing robot failure

categories. The risks with the highest impact and likelihood for a food company were a palletizing robot failure, human operator failure, bagging machine failure, and flattening machine failure.

Phase 5 (multi-criteria evaluation) of RFRM is to evaluate the reduced list of risk scenarios by three major criteria: redundancy, resilience, and robustness. *Redundancy* refers to the capability of auxiliary components or subsystems to be substituted for failed components or subsystems and perform the same function. *Resilience* refers to the ability of a system to recover after a failure and can be evaluated based on the time to recover and the resources needed to recover. *Robustness* is the insensitivity of the system's performance to external stresses, i.e., the ability of the system to oppose potential risks. These three categories have been broken down into 13 subcategories to assist the ranking process. The top risk scenarios must be evaluated by each of the 13 criteria and assigned a rating of "high," "medium," or "low." As a generalization, the scenarios with more "high" ratings will be ranked above those with more "low" ratings. It is up to the stakeholders to decide exactly how to characterize and assign these ratings. Additional scales and filtering criteria may also be used in order to further reduce the number of scenarios under consideration.

The RFRM process is essential because it limits the number of risk scenarios for a manufacturing facility that ought to be followed and studied further to a manageable quantity. However, the process must be modified in order to identify the risks that are applicable to realistic and practical PHM strategies. When adapting RFRM to the process of scoping and designing a PHM system, the goal is to identify failures that result from routine use of equipment or the passage of time and can be mitigated by the PHM system. While outside influences (e.g., power surges, operator error, or natural disasters) may be the cause of a system failure, PHM addresses failures that occur during routine operation. Therefore, any outside influences must be factored out during the modified RFRM process. The modifications to the standard RFRM filtering process as described in Adams et al. [5] are as follows:

M1. The categories *undetectability*, *uncontrollability*, and *irreversibility* are considered to be "tier 1" risks, while the other categories are considered to be of "tier 2." Scenarios shall firstly be ranked by number of high, medium, and low ratings in tier 1 and secondly by number of high, medium, and low ratings in tier 2.
M2. A new category shall be added to the list called *human*, referring to the effect of human error on the risk. If the risk is completely classified as *human*, then it shall be filtered out during RFRM.
M3. A new category shall be added to the list called *organizational*, referring to the effect of facility's organizational structure on the risk. If the risk is completely classified as *organizational*, then it shall be filtered out during RFRM.
M4. Risks that are rated "high" for *undetectability* shall be filtered out during RFRM, unless it is possible to add a detection method (such as a sensor to a machine).
M5. Risks that are rated "high" for *uncontrollability* shall be filtered out during RFRM, unless it is possible to insert control modes to the process that could mitigate the risk.
M6. Risks that are directly classified as an *operating environment* risk, and thus cannot be mitigated on a day-to-day basis shall be filtered out during the RFRM.
M7. Risks that are related to hardware, software, or programs that are rated "high" in the category of *design immaturity* and that have never been used, tested, or verified shall be filtered out during RFRM.

The modified RFRM methodology for PHM in manufacturing was performed to help identify the major subsystems that would most benefit from PHM system capabilities. The result of this process was a list of the top risk scenarios for each company, which aligned with the subsystems that would most benefit from PHM. The top areas are in order, palletizing robot failure (7.c), bagging machine failure (3.b), and flattening machine failure (6.b).

35.5.3 Identifying Critical Components Involved in a Subsystem Failure

A FTA was performed in the case study to provide additional scope for the PHM system designer. This process helped to determine exactly how a risk scenario could occur and more specifically which components might be involved in the failure of a manufacturing machine. The result of FTA is a list of the smallest combination of

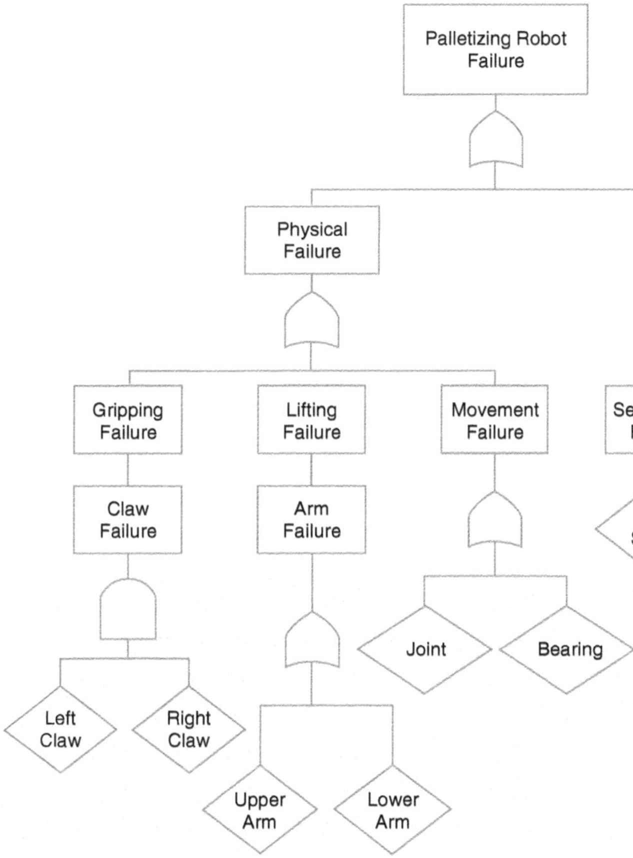

Fig. 35.2 Food company—palletizing robot failure fault tree

component failures that would lead to a subsystem or system level failure. This combination is referred to as a "minimal cut set." A single failure from any of the components in the minimal cut set will propagate to failure of the top event [7]. A PHM system designer can use knowledge of these components to provide guidance for locations of sensors or health monitoring devices. Since the top risk scenario from the RFRM methodology was the palletizing robot failure, this was used as the top event for the fault tree analysis. The fault tree for the company can be seen in Fig. 35.2.

From this model, the system analyst can understand every combination of component-level failures that would cause a domino effect into a system-level failure, in this case the failure of the palletizing robot. The minimal cut set for the palletizing robot failure is comprised of the following component failures: (1) left claw and right claw, (2) upper arm, (3) lower arm, (4) joint, (5) bearing, and (6) claw sensor. These six components all represent viable targets for PHM sensors within the palletizing robot.

35.6 Conclusion

The major findings presented in this paper were the results from the case study with a manufacturing company, where the analysis successfully identified critical subsystems and components that would most benefit from PHM system capabilities. The HHM models were developed to provide hierarchical system views to help the subject matter experts develop critical risk scenarios. Next the risk scenarios were filtered and ranked based on their likelihood and severity. Finally, a FTA revealed the critical component-level paths to failure for the highest-ranked subsystems.

For the company, the top three subsystems for PHM were the palletizing robot, the bagging machine, and the flattening machine. The critical paths to failure for the palletizing robot were the following six component-level combinations: (1) left claw and right claw, (2) upper arm, (3) lower arm, (4) joint, (5) bearing, and (6) claw sensor. These results were presented to the decision-makers at the company, who confirmed that they would use the information when scoping and designing a PHM system, as well as when developing their risk management strategies.

The case study outlined in this paper has demonstrated that risk analysis methodologies, (1) HMM, (2) RFRM, and (3) FTA can be modified to provide scope and direction for a PHM system designer. This combination of tools represents a unique contribution to the PHM literature for the benefit of practitioners contemplating a PHM system implementation. This methodological approach can be adapted and conducted by other manufacturing companies. It has been shown that the methodology is compatible between very different industries and of varying scales of production. Another major contribution of this paper is the demonstration of the synergy between the fields of risk analysis and PHM. The case study demonstrated that theories and methodologies originally used for risk analysis could be modified and applied toward PHM system design projects. This demonstration of compatibility opens the door for additional collaboration efforts toward adapting and modifying methodologies between these two fields.

References

1. Biehl, S., Staufenbiel, S., Recknagel, S., Denkena, B., & Bertram, O. (2012). Thin film sensors for condition monitoring in ball screw drives. In *1st Joint International Symposium on System-Integrated Intelligence: New Challenges for Product and Production Engineering*.
2. Feldman, K., Jazouli, T., & Sandborn, P. (2009). A methodology for determining the return on investment associated with prognostics and health management. *IEEE Transactions on Reliability, 58*(2), 305–316.
3. Hou-bo, H., & Jian-min, Z. (2011). Cost-benefit model for PHM. *Procedia Environmental Sciences, 10*, 759–764.
4. Malinowski, M., Beling, P.A., Haimes, Y., Laviers, A., Marvel, J., Weiss, B. (2015). System interdependency modeling in the design of prognostic and health management systems in smart manufacturing. In *Annual conference of the prognostics and health management society* 2015.

5. Adams, S., Malinowski, M., Heddy, G., Choo, B., & Beling, P. A. (2017). The WEAR methodology for prognostics and health management implementation in manufacturing. *Journal of Manufacturing Systems, 45*, 82–96.
6. Haimes, Y. Y. (2016). *Risk modeling, assessment, and management* (4th ed.). Hoboken, NJ: Wiley.
7. US Nuclear Regulatory Commission (NRC). (1974). *Reactor safety study: An assessment of accident risks in US commercial nuclear power plants* (Vol. 2). Rockville, MD: NRC.
8. Chand, S., & Davis, J. F. (2010). What is smart manufacturing. *Time Magazine Wrapper, 7*, 28–33.
9. Alvandi, S., Bienert, G., Li, W., & Kara, S. (2015). Hierarchical modelling of complex material and energy flow in manufacturing systems. *Procedia CIRP, 29*, 92–97.
10. He, N., Zhang, D. Z., & Li, Q. (2014). Agent-based hierarchical production planning and scheduling in make-to-order manufacturing system. *International Journal of Production Economics, 149*, 117–130.
11. Jung, K., Morris, K. C., Lyons, K. W., Leong, S., & Cho, H. (2015). Mapping strategic goals and operational performance metrics for smart manufacturing systems. *Procedia Computer Science, 44*, 184–193. https://doi.org/10.1016/j.procs.2015.03.051
12. Barajas, L. G., & Srinivasa, N. (2008). Real-time diagnostics, prognostics and health management for large-scale manufacturing maintenance systems. In *ASME 2008 International Manufacturing Science and Engineering Conference collocated with the 3rd JSME/ASME International Conference on Materials and Processing* (pp. 85–94). New York: American Society of Mechanical Engineers.
13. Lee, J., Wu, F., Zhao, W., Ghaffari, M., Liao, L., & Siegel, D. (2014). Prognostics and health management design for rotary machinery systems – Reviews, methodology and applications. *Mechanical Systems and Signal Processing, 42*(1–2), 314–334. https://doi.org/10.1016/j.ymssp.2013.06.004
14. Holland, S. W., Barajas, L. G., Salman, M., & Zhang, Y. (2010). PHM for automotive manufacturing & vehicle applications. In *Prognostics & Health Management Conference Fielded Systems Session*, October 14, Portland, Oregon.
15. Batzel, T. D., & Swanson, D. C. (2009). Prognostic health management of aircraft power generators. *IEEE Transactions on Aerospace and Electronic Systems, 45*(2), 473–482.
16. Hofmeister, J. P., Wagoner, R. S., & Goodman, D. L. (2013). Prognostic health management (PHM) of electrical systems using condition based data for anomaly and prognostic reasoning. *Italian Association of Chemical Engineering, 33*, 991–996. https://doi.org/10.3303/CET1333166
17. Choo, B. Y., Adams, S. C., Weiss, B. A., Marvel, J. A., & Beling, P. A. (2016). Adaptive multiscale prognostics and health management for smart manufacturing systems. *International Journal of Prognostics and Health Management, 7*, 14.
18. Choo, B. Y., Adams, S., & Beling, P. (2017, June). Health-aware hierarchical control for smart manufacturing using reinforcement learning. In *2017 IEEE International Conference on Prognostics and Health Management (ICPHM)* (pp. 40–47). Piscataway, NJ: IEEE.
19. Bai, G., Wang, P., & Hu, C. (2015). A self-cognizant dynamic system approach for prognostics and health management. *Journal of Power Sources, 278*, 163–174. https://doi.org/10.1016/j.jpowsour.2014.12.050
20. Kaplan, S., & Garrick, B. (1981). On the quantitative definition of risk. *Risk Analysis, 1*(1), 11–27.
21. Haimes, Y. Y. (1991). Total risk management. *Risk Analysis, 11*, 169–171. https://doi.org/10.1111/j.1539-6924.1991.tb00589.x
22. Roland, H. E., & Moriarty, B. (1990). *System safety engineering and management* (2nd ed.). New York: Wiley.

Chapter 36
Technology Maturity in NASA's Small Business Innovation Research Program

Andrea Belz, Aleksandar Giga, Richard Terrile, Michael Kawas, and Fernando Zapatero

Nomenclature

$mTRL_i$	PI-measured initial TRL
$aTRL_f$	PI-anticipated final TRL
$mTRL_f$	PI-measured final TRL
$mTRL'_f$	NASA-measured final TRL

36.1 Introduction

Technology readiness levels (TRLs) have been used in the aerospace industry for many years as a domain-agnostic management process; the usefulness of a standardized scale has been demonstrated by recent extensions to manufacturing, systems, and innovation. This last activity is of particular importance on a national scale and is the focus of the early-stage Small Business Innovation Research (SBIR) program, funding roughly \$2 B in research and development (R&D). While the SBIR program has been studied extensively in terms of its economic outcomes, such as revenues

A. Belz (✉) · A. Giga
Viterbi School of Engineering, University of Southern California, Los Angeles, CA, USA
e-mail: abelz@usc.edu

R. Terrile
Jet Propulsion Laboratory, California Institute of Technology, Pasadena, CA, USA

M. Kawas
Department of Economics, University of Southern California, Los Angeles, CA, USA

F. Zapatero
Marshall School of Business, University of Southern California, Los Angeles, CA, USA

S. Adams et al. (eds.), *Systems Engineering in Context*,
https://doi.org/10.1007/978-3-030-00114-8_36

generated, subsequent venture capital, and patent generation, little is known regarding the actual technical advances of the research funded with this vehicle.

Another area in which data are sparse is the study of behavioral effects in decision-making and reporting, such as biases and overconfidence, in natural entrepreneurial settings. While behavioral economics scholars have explored biases in many laboratory experiments, few studies exist in the realm of innovation and early-stage technology management. A study of entrepreneurial biases during the course of ordinary operations would provide much needed insight into early-stage technology management.

In this work, we extract information reported by principal investigators (PIs) in their SBIR proposals to assess both NASA's funding strategy as well as the PIs' behavior. In the first study, we segment proposals and awards by the company headcount, denoting "microfirms" and "standard small businesses" (SSBs) as those with 1–5 employees and 6–499, respectively. We measure the distributions of technology maturities to understand the risk profile of the NASA SBIR portfolio managers, looking for "sweet spots" and the agency's risk appetite. In the study's second part, we search for differences reported by both NASA's representative and the PI to indicate biases or other behaviors.

This work contributes to the literature on systems engineering by linking funding to technology advancement. We offer new data for the discussion of innovation policy and subsidies. Finally, this work supplements experimental reports by reporting on observations of biases "in the wild."

36.2 Literature Review

36.2.1 Aerospace Industry and TRL Measurements

The aerospace industry is typically viewed as a nationally strategic industry with strong links between defense and civil markets [1]. The industry experiences high development costs and cyclical cash flow, with systems characterized by heterogeneity; low-volume, high-performance, and reliable requirements; and extreme ambient operating conditions[2, 3]. The high degree of customization leads to extended product life cycles and often to launching variations of existing models rather than new products [4–6]. Uncertainty in technology maturation has been recognized as a major constraint in developing broader system-level schemes to integrate various technologies [7].

To address these concerns, the aerospace industry adopted the technology readiness level ("TRL") scale ranging from 1 (idea) to 9 (flown successfully) to compare technologies broadly and assess their readiness for infusion into flight projects [8] (Table 36.1). The general concept of ordinal scales to describe readiness has been extended to automotive manufacturing [9], Integration Readiness Level and System Readiness Level [10], and the Innovation Readiness Levels [11]. While their use is limited to that of a simple ordinal scale [12], these tools provide important general insight.

Table 36.1 Technology readiness level (TRL) definitions

TRL	Definition
1	Basic principles observed and reported
2	Technology concept and/or application formulated
3	Analytical and experimental critical function and/or characteristic proof of concept
4	Component and/or breadboard validation in laboratory environment
5	Component and/or breadboard validation in relevant environment
6	System/subsystem model or prototype demonstration in a relevant environment
7	System prototype demonstration in a space environment
8	Actual system completed and "flight qualified" through test and demonstration
9	Actual system "flight proven" through successful mission operations

Many factors inhibit the adoption of new technologies at NASA; the agency experiences cost and schedule slippages from new technology infusion [13], and organizational factors may affect program advancement [14]. At NASA, a TRL of 6 is required for a new technology to be considered for infusion into a flight project [15]. Even with this requirement, slippages will occur with typical schedule modeling practices, and more sophisticated analyses linking schedule and risk may be needed [16]. At the systems level, low TRL and TRL heterogeneity strongly influence schedule slippage [17], and technology interaction coupling effects can complicate high- and low-TRL systems [18]. As development costs increase nonlinearly with TRL [19], demonstration of technology performance becomes progressively more difficult and may limit further funding for development [20], which may already be constrained by the limited frequency and number of mission opportunities [21]. Thus, advanced development funding becomes difficult to attract; ranking projects by the TRL will prioritize proven heritage solutions over new, riskier technologies potentially offering value. Several new frameworks have been discussed to incorporate prospective benefits in technology assessments [22–25], but these are not commonly used.

36.2.2 *Entrepreneurship Policy and Behavior*

In the United States, one mechanism to support R&D in small companies is the SBIR program [26], originally justified in part with research showing that small companies are more innovative than large ones [27]. Start-ups contribute disproportionately to economic growth [28–30], but young, small technology firms may struggle to attract resources from private capital due to high risk, especially in less active investment periods [31]. To remedy this, subsidy programs have developed around the world [32–34]. Although funding could in principle "crowd out" or displace private investment [35], the subsidy instead enables the firm to pursue research that would be otherwise discontinued [36, 37], and private investments may follow [38–40]. The SBIR program has grown in importance because basic research is declining in large companies as it increases in universities and small businesses [41]. SBIR awards have been associated with increases in entrepreneurial activity, venture capital, company growth, high-tech entrepreneurship, patent generation, and externally generated patents [42–47].

Early-stage ventures have also been studied by behavioral economics scholars, with surveys indicating that entrepreneurs overweight their odds of success [48] and that the beliefs persist over time [49]. Those with higher optimism tend toward self-employment, after which their optimism grows [50], although they show worse outcomes than realists [51, 52]. An entrepreneur's optimism has been negatively linked to venture performance in terms of revenue and employment growth [53]. In more mature companies, overconfident CEOs are more likely to conduct acquisitions without generating value [54]; however, in innovative industries, they are successful in conducting more R&D and generating higher patent and citation rates [55].

While TRL schemes have been studied extensively in large projects managed with traditional systems engineering, the research on the TRL advancement in entrepreneurial settings is sparse. TRL may be used to characterize the technologies proposed by small businesses, but the portfolio has not been studied from this perspective. We contribute to the currently sparse literature analyzing the relationship between early-stage technology subsidies and the funding agency's mission [56, 57]. In addition, we provide an early view into entrepreneurial behavior measured by comparing the PI's assessment of technology readiness with that of NASA.

36.3 Data Overview and Analysis Methods

Applications are publicly solicited for SBIR funding in two consecutive tranches or phases. The award stipulates no rights for the agencies nor a demand for equity or intellectual property. Only Phase I awardees may apply for Phase II funding; over 96% of eligible firms elect to do so. The entire award portfolio for the year is selected at once; this is true for both phases. At NASA in the last decade, awards have ranged from \$70 k to \$125 k for Phase I and an estimated \$500–750 k for Phase II, with award terms of 6 months and 2 years, respectively. Further funding may follow from other programs at NASA if the likelihood of using the technology in a flight project is sufficiently high.

Since 2009, NASA SBIR applications at each phase have required that the principal investigator (PI) estimate the proposed technology's current status as an initial measured TRL ($mTRL_i$) and the anticipated final TRL ($aTRL_f$) upon the award's conclusion. In the final report at each phase, PIs estimate the final measured TRL ($mTRL_f$) as does a NASA representative $\left(mTRL_f'\right)$. These measurements repeat for both phases (Fig. 36.1).

Data used for this analysis were compiled from the Electronic Handbook (EHB) for the NASA SBIR program. This data set is for restricted use but is searchable over

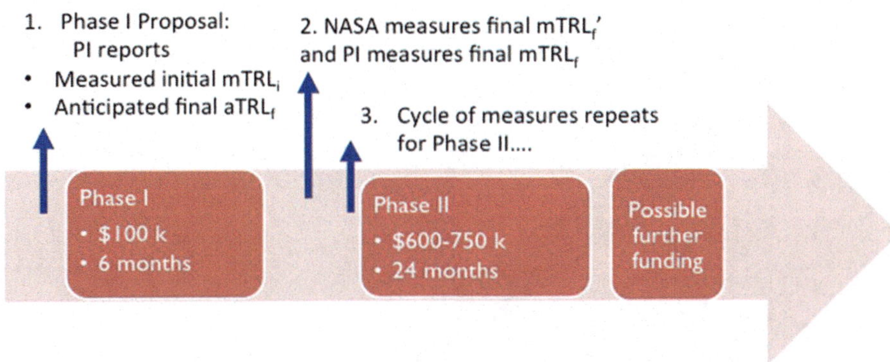

Fig. 36.1 Schematic of program and associated measurements

Table 36.2 Data set of NASA SBIR proposal pool 2009–2016

	Phase I		Phase II	
Population	Submissions	Selection rate	Submissions	Selection rate
All	10,204	22%	2176	41%
Microfirms (1–5 employees)	2640	17%	374	36%
Standard small businesses (6–499 employees)	7564	24%	1802	42%

many years of the program. To protect the procurement sensitive nature, we compiled data from 2009 to 2016 over all NASA Mission Directorates and present aggregated results only to prevent identification of individual companies. From the database, we extracted the firm's headcount, the proposal measures ($mTRL_i$ and $aTRL_f$), and the final evaluation measures ($mTRL_f$ and $mTRL_f'$). We divide the pool into two populations, the microfirms with 1–5 employees and SSBs with 6–499 employees because roughly one-fourth of the submitted proposals come from microfirms; the median employment in applicant firms is only 12 employees [58]. In general, the headcount does not vary over the time period; only 4% of the firms change categories. In this group, 887 companies progress fully from Phase I applications to Phase II awards (Table 36.2). These data were used to analyze the selection probabilities and the evolution of the TRL.

For the behavioral studies, data were available only for 2009–2012, with 502 proposals for analysis. We tabulate the distributions of Phase I final reports with a given $mTRL_f$/$mTRL_f'$ combination and sort them into three categories: accurate ($mTRL_f = mTRL_f'$), humble ($mTRL_f < mTRL_f'$), and arrogant ($mTRL_f > mTRL_f'$) (Fig. 36.2). Proposals in those groups were then tracked to identify if Phase II awards followed.

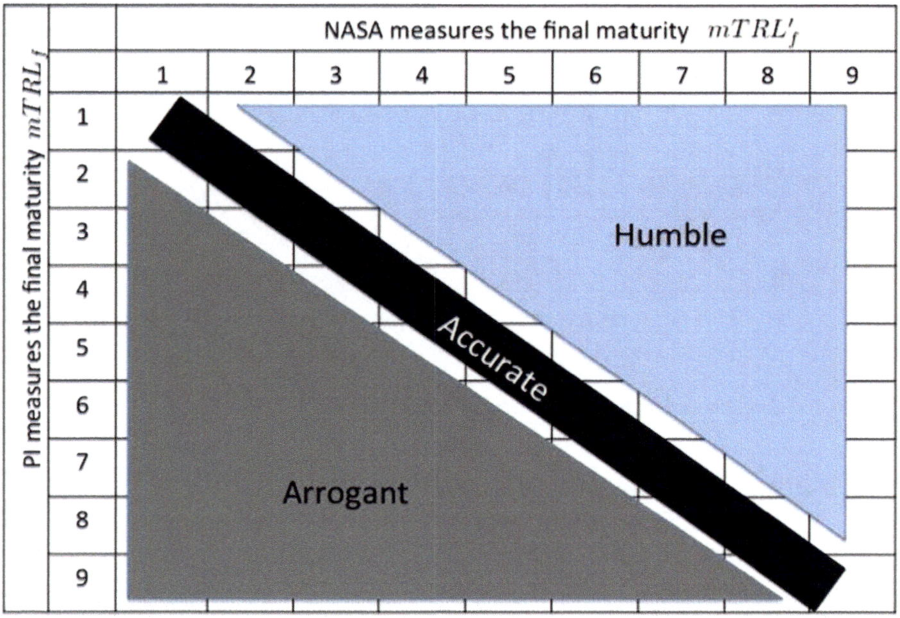

Fig. 36.2 Framework for bias analysis comparing the final technology readiness and the end of Phase I as measured by the PI (rows) and NASA (columns)

36.4 Results

36.4.1 Aggregated Results for Technology Advancement

To estimate the progress anticipated in the proposal pool, $mTRL_i$ and $aTRL_f$ distributions were studied for the awardees at each phase (Fig. 36.3). Consistent with an early-stage technology program, Phase I anticipates evolving technologies from TRL 2 to 4, whereas Phase II technologies are expected to migrate from TRL 3 to 6. These distributions are highly skewed, particularly for technologies entering the program.

Descriptive statistics indicate that selection preserves both mean and skew in Phases I and II (Table 36.3)—i.e., the selection process is largely TRL-independent. As the SSBs comprise the majority of the proposal and award pools, they dominate the aggregated numbers. Kurtosis analysis suggests that major outliers tend to be removed in award selection, particularly in the microfirms. Because of the high degree of skew and the low mean, the kurtosis must result primarily from high TRL; evidently, these high $mTRL_i$ and $aTRL_f$ values, particularly from microfirms, are excluded in Phase I, whereas SSBs do not experience the reduction in outliers. At Phase II, this effect is much smaller.

We directly examine the dependence of the selection process on TRL (Fig. 36.4). For each population (microfirms and SSBs), selection is largely independent of both

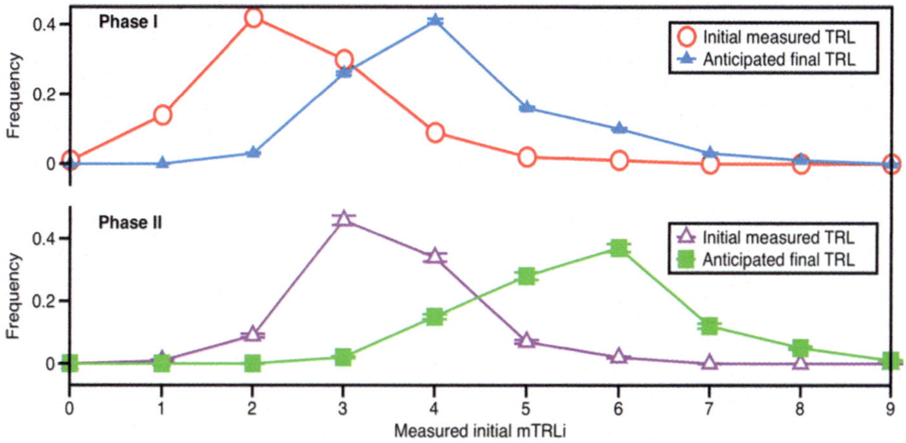

Fig. 36.3 Measured and anticipated final TRL for (top) Phase I awardees and (bottom) Phase II awardees

the initial and anticipated final TRL for values lower than 7. In Phase I, SSBs consistently show approximately 20% higher likelihood of selection at all TRL levels, whereas in Phase II, that difference is virtually eliminated.

Finally, we examine reporting tendencies in the PI population and specifically asymmetries between the PI and NASA's representative. We compare the measurements at the end of Phase I, when both parties estimate the final TRL prior to the Phase II evaluation. Sixteen percent of the population appears to overestimate the status of the technology (Table 36.4), and roughly 5% underestimate it, with the microfirms showing considerably higher levels of arrogance. These tendencies do not affect the selection probability; proposals and winners show statistically equivalent levels of these behavioral characteristics.

36.5 Discussion

We have examined two interesting facets of the NASA SBIR program: [1] NASA's portfolio management strategy and [2] the PI's behavior in assessing the technologies. First, we find that in contrast to concerns that heritage technologies would be prioritized [20], NASA's selection process is generally independent of both the technologies' initial maturity and the envisioned final state. Evidently, NASA is prepared to accept technical risks throughout the program. On the other hand, NASA appears to reduce business risk at Phase I with roughly 20% increased selection of firms with six or more employees, even at modest TRLs. This could result from poorer proposals from microfirms, which is certainly possible with inexperienced proposers or from deliberate decisions by NASA program managers. The lack of

Table 36.3 Data set of NASA SBIR proposal pool 2009–2016

		Initial TRL			Anticipated final TRL		
Stage	Population	Mean	Skew	Kurtosis	Mean	Skew	Kurtosis
Phase I	All						
	Proposals	2.42 ± 0.01	0.67 ± 0.02	1.39 ± 0.05	4.17 ± 0.01	0.89 ± 0.02	1.18 ± 0.25
	Winners	2.46 ± 0.02	0.67 ± 0.05	1.32 ± 0.10	4.13 ± 0.02	0.84 ± 0.05	0.79 ± 0.10
	Microfirms (1–5 employees)						
	Proposals	2.42 ± 0.02	0.72 ± 0.05	1.09 ± 0.10	4.22 ± 0.03	0.83 ± 0.05	0.83 ± 0.10
	Winners	2.46 ± 0.05	0.56 ± 0.11	0.48 ± 0.23	4.18 ± 0.06	0.70 ± 0.11	0.10 ± 0.23
	Standard small businesses (6–500 employees)						
	Proposals	2.42 ± 0.01	0.64 ± 0.03	1.60 ± 0.06	4.15 ± 0.01	0.90 ± 0.03	1.21 ± 0.06
	Winners	2.46 ± 0.02	0.71 ± 0.06	1.32 ± 0.11	4.12 ± 0.02	0.88 ± 0.06	1.01 ± 0.11
Phase II	All						
	Proposals	3.46 ± 0.02	0.46 ± 0.05	1.12 ± 0.11	5.62 ± 0.02	0.18 ± 0.05	0.27 ± 0.11
	Winners	3.53 ± 0.03	0.55 ± 0.08	1.32 ± 0.16	5.63 ± 0.04	0.20 ± 0.08	0.14 ± 0.16
	Microfirms (1–5 employees)						
	Proposals	3.47 ± 0.05	0.53 ± 0.13	0.71 ± 0.25	5.67 ± 0.06	0.11 ± 0.13	−0.02 ± 0.25
	Winners	3.62 ± 0.09	0.54 ± 0.21	0.45 ± 0.42	5.74 ± 0.12	0.26 ± 0.21	−0.11 ± 0.42
	Standard small businesses (6–500 employees)						
	Proposals	3.46 ± 0.02	0.44 ± 0.06	1.18 ± 0.12	5.61 ± 0.03	0.19 ± 0.06	0.33 ± 0.12
	Winners	3.51 ± 0.03	0.53 ± 0.09	1.49 ± 0.18	5.61 ± 0.04	0.16 ± 0.09	0.13 ± 0.18

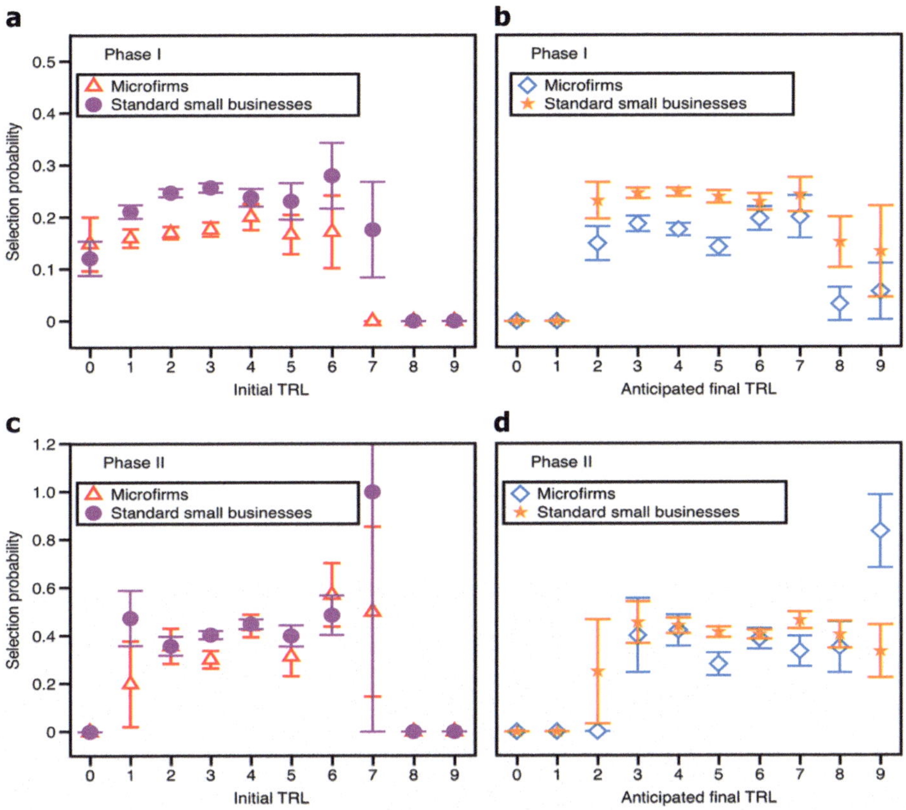

Fig. 36.4 Selection probabilities for (**a**) Phase I initial TRL, (**b**) Phase I final TRL, (**c**) Phase II initial TRL, and (**d**) Phase II final TRL

experience may be consistent with the outliers rejected in the Phase I selection process. These issues are virtually eliminated at Phase II.

This study offers another view into the NASA strategy by analyzing the full portfolio's distribution. Based on the information provided in the proposals, the program advances in Phase I from roughly 2.5 to 4, and in Phase II from 3.5 to 5.5. The initial TRL at Phase II is slightly lower than the ending TRL in Phase I and may be related to the later findings in PI behavior. Indeed, we see that roughly 1 in 5 microfirm PIs overestimates the status of the technology at the end of Phase I, compared with 1 in 8 PIs from SSBs.

This study points to other avenues of future research. For instance, we may measure the aggregate progress through the program as a technology transition matrix. This would have tremendous value on its own to report the actual technology advance financed by NASA, particularly in conjunction with an optimization model to describe the funding strategy. Another approach would be to consider this from a game theoretic perspective, given that the presence of inaccuracies in Table 36.3 suggests a possible asymmetry of information at the proposal stage.

Table 36.4 Descriptive statistics of TRL distributions for Phase II, based on assessments after Phase I

	Accurate		Humble		Arrogant	
	($mTRL_f = mTRL'_f$)		($mTRL_f < mTRL'_f$)		($mTRL_f > mTRL'_f$)	
Population	Proposals	Winners	Proposals	Winners	Proposals	Winners
Total pool	78.5 ± 0.9%	80.3 ± 1.4%	5.3 ± 0.5%	5.5 ± 0.8%	16.2 ± 0.8%	14.2 ± 1.2%
Microfirms (1–5 employees)	77.8 ± 2.2%	76.3 ± 3.7%	3.1 ± 0.9%	2.3 ± 1.3%	19.2 ± 2.1%	21.4 ± 3.6%
Standard small businesses (6–499 employees)	78.7 ± 1.0%	81.0 ± 1.5%	5.7 ± 0.6%	6.1 ± 0.9%	15.6 ± 0.9%	12.9 ± 1.2%

Other possible extensions to this work include segmenting the data by technology type to determine the investment efficiency. Private venture capital has migrated strongly to software [41] precisely because of this advantage; i.e., a dollar invested in software buys more advancement than one invested in hardware. Indeed, with the advent of advanced computation, particularly machine learning, technology can potentially advance itself and thus would require minimal capital investment. The role of SBIR in financing hardware solutions becomes more important to the national industrial landscape.

Finally, we contribute to the literature on behavioral economics and give objective measurements of inaccuracies in human estimation of technology maturity. Additional measures of inconsistency are possible, such as those comparing the technology's maturity at the end of Phase I with that at the beginning of Phase II; in principle, they should coincide exactly. This is an exciting opportunity to link systems engineering with cutting-edge behavioral studies.

Acknowledgment This work was conducted under JPL award 1550874 and National Science Foundation award 1440080.

References

1. Ecorys. (2009). *Sector competitiveness studies - Competitiveness of the EU aerospace industry.* Rotterdam: Author.
2. Corallo, A., Laubacher, R., Margherita, A., & Turrisi, G. (2009). Enhancing product development through knowledge based engineering (KBE). *Journal of Manufacturing Technology Management, 20*, 1070–1083.
3. Tamaskar, S., Neema, K., & DeLaurentis, D. (2014). Framework for measuring complexity of aerospace systems. *Research in Engineering Design, 25*, 125–137.
4. McAdam, R., O'Hare, T., & Moffett, S. (2008). Collaborative knowledge sharing in composite new product development: An aerospace study. *Technovation, 28*, 245–256.
5. Corallo, A., Lazoi, M., & Secundo, G. (2012). Inter-organizational knowledge integration in collaborative NPD projects: Evidence from the aerospace industry. *Knowledge Management Research & Practice, 10*, 354–367.
6. Corallo, A., Errico, F., & Lazoi, M. (2014). The aerospace "networked" business model: Evidences and suggestions. *Universal Journal of Management, 2*, 151–159.
7. Sauser, B., Ramirez-Marquez, J. E., Magnaye, R., & Tan, W. (2008). A systems approach to expanding technology readiness within defense acquisition. *International Journal of Defense Acquisition Management, 1*, 39–58.
8. Mankins, J. C. (1995). *Technology readiness levels: A white paper.* Washington, DC: NASA.
9. Ward, M. J., Halliday, S. T., & Foden, J. (2012). A readiness level approach to manufacturing technology development in the aerospace sector: An industrial approach. *Proceedings of the Institution of Mechanical Engineers Part B: Journal of Engineering Manufacture, 226*, 547–552.
10. Magnaye, R. B., Sauser, B. J., & Ramirez-Marquez, J. E. (2010). System development planning using readiness levels in a cost of development minimization model. *Systems Engineering, 13*, 311–323.
11. Evans, J. D., & Johnson, R. O. (2013). Tools for managing early-stage business model innovation. *Research-Technology Management, 56*, 52–56.
12. Kujawski, E. (2013). Analysis and critique of the system readiness level. *IEEE Transactions on Systems, Man, and Cybernetics, 43*, 979–987.

13. GAO. (2017). *NASA: Assessments of major projects, GAO-17-303SP*. Washington, DC: United States Government Accountability Office.
14. Dwyer, M., Cameron, B., & Szajnfarber, Z. (2015). A framework for studying cost growth on complex acquisition programs. *Systems Engineering, 18*, 568–583.
15. National Aeronautics and Space Administration. (2007). *NASA systems engineering handbook, NASA/SP-2007-6105*. Washington, DC: Author.
16. Dubos, G. F., Saleh, J. H., & Braun, R. (2008). Technology readiness level, schedule risk, and slippage in spacecraft design. *Journal of Spacecraft and Rockets, 45*, 836–842.
17. Dubos, G. F., & Saleh, J. H. (2011). Spacecraft technology portfolio: Probabilistic modeling and implications for responsiveness and schedule slippage. *Acta Astronautica, 68*, 1126–1146.
18. Battat, J. A., Cameron, B., Rudat, A., & Crawley, E. F. (2014). Technology decisions under architectural uncertainty: Informing investment decisions through tradespace exploration. *Journal of Spacecraft and Rockets, 51*, 521–532.
19. Terrile, R. J., Doumani, F. G., Ho, G. Y., & Jackson, B. L. (2015). Calibrating the technology readiness level (TRL) scale using NASA mission data. In *2015 IEEE Aerospace Conference*. Big Sky, MT: IEEE.
20. Terrile, R. J. (2013). Balancing innovation with commercialization in NASA's science mission directorate SBIR program. In *2013 IEEE Aerospace Conference*. Big Sky, MT: IEEE.
21. Szajnfarber, Z. (2014). Space science innovation: How mission sequencing interacts with technology policy. *Space Policy, 30*, 83–90.
22. Mankins, J. C. (2002). Approaches to strategic research and technology (R&T) analysis and road mapping. *Acta Astronautica, 51*, 3–21.
23. Mankins, J. C. (2009). Technology readiness and risk assessments: A new approach. *Acta Astronautica, 65*, 1208–1215.
24. Terrile, R. J., Jackson, B. L., & Belz, A. P. (2014). Consideration of risk and reward in balancing technology portfolios. In *IEEE Aerospace Conference*. Big Sky, MT: IEEE.
25. Browning, T. R. (2014). A quantitative framework for managing project value, risk, and opportunity. *IEEE Transactions on Engineering Management, 61*, 583–598.
26. Wessner, C. W. (2008). An assessment of the small business innovation research program. In *Committee for capitalizing on science, technology, and innovation*. Washington, DC: National Academies Press.
27. Edwards, K. L., & Gordon, T. (1982). *Characterization of innovations introduced on the U.S. market in 1982*. Washington, DC: Small Business Administration.
28. Audretsch, D. B., Keilbach, M. C., & Lehmann, E. E. (2006). *Entrepreneurship and economic growth*. New York: Oxford University Press.
29. Akcigit, U., & Kerr, W. R. (2010). *Growth through heterogeneous innovations*. Philadelphia, PA: National Bureau of Economic Research.
30. Haltiwanger, J., Jarmin, R. S., & Miranda, J. (2013). Who creates jobs? Small versus large versus young. *Review of Economics and Statistics, 95*, 347–361.
31. Nanda, R., & Rhodes-Kropf, M. (2013). Investment cycles and startup innovation. *Journal of Financial Economics, 110*, 403–418.
32. González, X., & Pazó, C. (2008). Do public subsidies stimulate private R&D spending? *Research Policy, 37*, 371–389.
33. Hall, B. H., & Lerner, J. (2010). The financing of R&D and innovation. *Handbook of the Economics of Innovation, 1*, 609–639.
34. Zúñiga-Vicente, J. A., Alonso-Borrego, C., Forcadell, F. J., & Galán, J. I. (2014). Assessing the effect of public subsidies on firm R&D investment: A survey. *Journal of Economic Surveys, 28*, 36–67.
35. Wallsten, S. J. (2000). The effects of government-industry R&D programs on private R&D: The case of the small business innovation research program. *The Rand Journal of Economics, 31*, 82–100.
36. Feldman, M. P., & Kelley, M. R. (2006). The ex ante assessment of knowledge spillovers: Government R&D policy, economic incentives and private firm behavior. *Research Policy, 35*, 1509–1521.

37. Belz, A. (2017). Real options valuation of a federally funded small business portfolio. In *Proceedings of the IEEE Technology Engineering Management Society*. San Jose, CA: IEEE.
38. Lach, S. (2002). Do R&D subsidies stimulate or displace private R&D? Evidence from Israel. *Journal of Industrial Economics, 50*, 369–390.
39. Almus, M., & Czarnitzki, D. (2003). The effects of public R&D subsidies on firms' innovation activities: The case of eastern Germany. *Journal of Business & Economic Statistics, 21*, 226–236.
40. Bronzini, R., & Iachini, E. (2014). Are incentives for R&D effective? Evidence from a regression discontinuity approach. *American Economic Journal: Economic Policy, 6*, 100–134.
41. Belz, A. (2016). Trends in industry-university relationships. *A vision for the future of center-based, multidisciplinary engineering research.* National Academy of Engineering. Retrieved July, 2017, from https://www.nae.edu/File.aspx?id=169641
42. Lerner, J. (1999). He government as venture capitalist: The long-run impact of the SBIR program. *Journal of Business, 72*, 285–318.
43. Toole, A. A., & Turvey, C. (2009). How does initial public financing influence private incentives for follow-on investment in early-stage technologies? *Journal of Technology Transfer, 34*, 43–58.
44. Cumming, D., & Li, D. (2013). Public policy, entrepreneurship, and venture capital in the United States. *Journal of Corporate Finance, 23*, 345–367.
45. Qian, H., & Haynes, K. E. (2014). Beyond innovation: The small business innovation research program as entrepreneurship policy. *Journal of Technology Transfer, 39*, 524–543.
46. Galope, R. V. (2016). A different certification effect of the small business innovation research (SBIR) program: Evidence from the Kauffman firm survey. *Economic Development Quarterly, 30*, 371–383.
47. Howell, S. (2017). Financing innovation: Evidence from R&D Grants. *American Economic Review, 107*, 1136–1164.
48. Cassar, G. (2010). Are individuals entering self-employment overly optimistic? An empirical test of plans and projections on nascent entrepreneur expectations. *Strategic Management Journal, 31*, 822–840.
49. Landier, A., & Thesmar, D. (2009). Financial contracting with optimistic entrepreneurs. *The Review of Financial Studies, 22*, 117–150.
50. Dawson, C., de Meza, D., Henley, A., & Arabsheibani, G. R. (2014). Entrepreneurship: Cause and consequence of financial optimism. *Journal of Economics and Management Strategy., 23*, 717–742.
51. Arabsheibani, G., de Meza, D., Maloney, J., & Pearson, B. (2000). And a vision appeared unto them of a great profit: Evidence of self-deception among the self-employed. *Economics Letters, 67*, 35–41.
52. Dawson, C., de Meza, D., Henley, A., & Arabsheibani, G. R. (2015). *The power of (non) positive thinking: Self-employed pessimists earn more than optimists.* (IZA DP No. 9242). Bonn: IZA.
53. Hmieleski, K. M., & Baron, R. A. (2009). Entrepreneurs' optimism and new venture performance: A social cognitive perspective. *Academy of Management Journal, 52*, 473–488.
54. Malmendier, U., & Tate, G. (2008). Who makes acquisitions? CEO overconfidence and the market's reaction. *Journal of Financial Economics, 89*, 20–43.
55. Hirshleifer, D., Low, A., & Teoh, S. H. (2012). Are overconfident CEOs better innovators? *Journal of Finance, 67*, 1457–1498.
56. Edler, J., & Fagerberg, J. (2017). Innovation policy: What, why, and how. *Oxford Review of Economic Policy, 33*, 2–23.
57. Mazzucato, M., & Semieniuk, G. (2017). Public financing of innovation: New questions. *Oxford Review of Economic Policy, 1*, 24–48.
58. Giga, A., Terrile, R. J., Belz, A. P., & Zapatero, F. (2016). The impact of NASA's small business innovation research program on invention and innovation. In *2016 IEEE Aerospace Conference*. Big Sky, MT: IEEE.

Chapter 37
Visualizing Communication Patterns in Design Teams Over Time

Connor Forsythe, Nikolai Joseph, Zoe Szajnfarber, and Erica Gralla

37.1 Introduction

Communication is integral to the design of complex systems, which are typically composed of several smaller subsystems that interact with and depend upon one another in many ways. There are different engineering teams responsible for the design of the respective subsystems, which often correspond to different engineering disciplines. These teams or "disciplines" are often not collocated and have limited channels of communication. Because of dependencies present between subsystems, information must be shared between the respective disciplines throughout the design process. Understanding the communication that happens throughout the design process is important for several reasons. First, it leads to an understanding of the design process and in particular the interdependencies among subsystems; this type of knowledge has been used for a variety of types of research in design, systems engineering, and management [1–3]. Second, observed communication patterns can enable organizations to improve their "communication infrastructure" by guiding decisions about when standing meetings are needed or which subsystem teams should be colocated [4–6]. Taken together, these efforts have the potential to improve the efficiency and effectiveness of the design process.

Representing organizational communication is challenging because the data are rich, multidimensional, and complex. Tools for representing these data must enable both intuitive understanding and also analysis. A commonly used tool for visualizing communication and interdependencies in design processes is a Design Structure Matrix (DSM) [7]. The DSM is useful in part because it is a compact, easily visualized, and readily manipulated representation of communication patterns. However, the DSM does not address some important aspects of communication:

C. Forsythe (✉) · N. Joseph · Z. Szajnfarber · E. Gralla
Engineering Management and Systems Engineering, The George Washington University, Washington, DC, USA
e-mail: connorforsythe@gwu.edu; njoseph37@gwu.edu; zsajnfa@gwu.edu; egralla@gwu.edu

S. Adams et al. (eds.), *Systems Engineering in Context*,
https://doi.org/10.1007/978-3-030-00114-8_37

multilateral communication and how communication changes through time. Yet, over time, different events, challenges, and participant dynamics emerge and can change communication throughout the design process. Multiple disciplines can be in communication at any one time. By providing methods for identifying dynamic and multilateral relationships throughout the design process, as we do in this research, communication during the design process can be better understood. Our research goal is to develop visualization tools and data structures for better understanding and analysis of communication during a design process, including changes over time and multilateral communication.

This paper proposes visualizations and representations that build on the DSM to allow analysis of dynamic and multilateral communication patterns but retain the DSM's ability to represent such rich data in a compact, tractable format. The representations are illustrated using data collected from observing design sessions in a concurrent engineering environment at the NASA Goddard Space Flight Center, to provide examples of the types of insights that can be achieved.

37.2 Background and Data

37.2.1 Background

A popular visualization of communication within an organization is the Design Structure Matrix (DSM). Eppinger and Browning have defined an "Organizational Architecture DSM" as a matrix where both the *X*- and *Y*-axes represent the same set of disciplines involved in a design process and the elements of the matrix represent the communication between the two respective entities [7]. The elements of this matrix may be binary representations [3, 8], categorical representations indicating different levels of communication [9], or representations of the extent of communication among subsystems [10]. A DSM provides a simple and effective way of describing the static relationships present among all disciplines contributing to the design process and has many valuable uses including optimization of organizational structure to determine, for example, how best to group disciplines or design tasks into modules [8, 9].

37.2.2 Data Collection

Our development of new representations that build upon the DSM was motivated by a larger study of communication throughout a concurrent design process. We use data collected from this larger study in order to provide an illustration of the new representations, understand the viability of the necessary data collection, and demonstrate the types of insights that can be achieved. Therefore, this paper draws on observations of NASA's Goddard Space Flight Center's (GSFC) Mission Design

Laboratory (MDL), a concurrent design facility that provides high-level mission designs for many different types of space mission concepts. These design processes involve a team of approximately 19 disciplinary experts who are each responsible for either a spacecraft subsystem (e.g., propulsion, communication) or an integrating function (e.g., systems engineer, team lead). The team meets in one room for a single workweek and produces a spacecraft conceptual design. Products include discipline block diagrams, a solid model of the spacecraft, a mission plan and concept operations, and preliminary cost, mass, and power budgets.

Our data collection involved having one or more observers make note of all visible communication between disciplinary experts in the design facility at all times throughout the week-long studies. The observers sat in the back of the design facility and were also able to move around and follow groups if they left the room for sidebar conversations. It was feasible for one observer to note all conversations that occurred in this manner, since there were rarely more than a few conversations happening at once. The data indicates when each communication began and ended and who was involved in it. Data sheets containing diagrams of the concurrent design facility were used to denote which disciplines were communicating, at what times, when people joined and/or left the conversation, where the conversation took place, and whether any props (such as whiteboards or "Tinkertoy" models) were used. These data were input into a database described below. The data recording procedures were verified and validated by calculating an inter-coder reliability score for two observers.

Data records were structured so that temporal and multilateral aspects of communications were maintained. Recording communications was complicated because some conversations changed over time: for example, two people might begin talking and a third might join later, or a larger conversation might break into two separate conversations. Therefore, communication records were stored in a database using two entities: "conversations" and "interactions." An interaction contains a start time, end time, and a list of participants (along with a location and other characteristics). A conversation is a set of interactions that are centered around a certain topic or purpose. This allows for the data to reflect when personnel changes occur throughout a conversation. Each study included hundreds of interactions, which provides a rich basis for developing new ways to visualize and interpret design process communication.

37.2.3 *Applicability*

Although the work presented in this paper centers around concurrent design, the techniques utilized could be applied to any design organization. The only necessity is that the data collected for the organization includes all parties participating in as well as the beginning and end time of each communication. Such data could be obtained using records of emails, phone calls, and meetings, for example. Future research can examine the feasibility of obtaining such data in other contexts.

37.3 Data Structures and Visualization

To address limitations of the DSM, a three-dimensional data structure was developed: the "conversation matrix." This structure was then manipulated in three ways so as to, first, replicate a standard DSM; second, represent dynamic relationships throughout all disciplines participating in the design process; and third, visualize multilateral conversations. These manipulations are named "Standard DSM," "Time-Discretized DSM," and "Egocentric Communication Dynamics View," respectively.

37.3.1 Conversation Matrix

Although the data can be stored and collected in a database that tracks both conversations and interactions along with their associated characteristics, it is not easily manipulated in this form. The conversation matrix was developed to be manipulated so that different visualizations can be created.

The conversation matrix is a three-dimensional matrix where the *X*- and *Y*-axes represent the disciplines present in the design process, the *Z*-axis represents time, and the elements of the matrix contain unique identifiers for each conversation. The *Z*-axis is defined in discrete time increments where the time between indices is smaller than the shortest interaction length recorded. Conversation IDs are placed in all cells (x,y,z) where entities x and y are interacting between times z and $z + 1$, to indicate between whom and when communication occurs. Figure 37.1 provides a generic representation of the conversation matrix. It maintains all of the information collected throughout the observation.

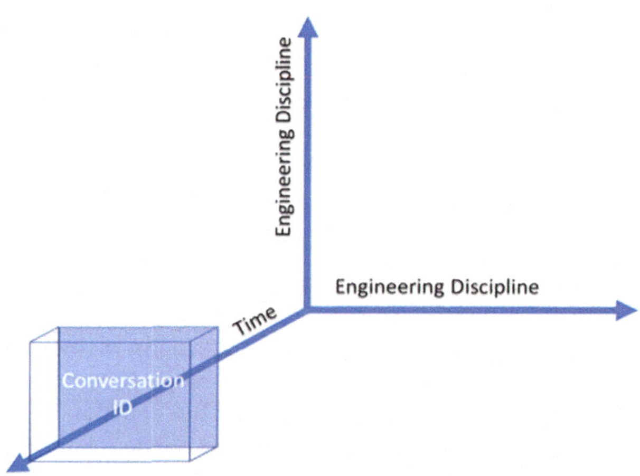

Fig. 37.1 Generic representation of conversation matrix

Unfortunately, the conversation matrix is difficult to visualize and interpret. However, manipulations of this matrix provide easily interpreted visualizations which are described in turn below.

37.3.2 Standard DSM

The conversation matrix can be manipulated to replicate a standard DSM. As discussed, the DSM provides an effective way of visualizing static relationships in the design process and is an important check on the consistency of our representation.

In order to create the Standard DSM, the dimension of time in the conversation matrix must now be compressed. To do so, the count of non-blank entries present in all x–y pairs of coordinates along the Z-axis is multiplied by the time step (Δt) in the conversation matrix. The resultant two-dimensional matrix represents the disciplines on the X- and Y-axes, and the elements contain the total time in communication between every discipline pair throughout the entirety of the design process (Fig. 37.2a). These values can then easily be transformed into binary or categorical representations; a sample categorical representation is shown in Fig. 37.2b (categories indicate communication less than the 33rd percentile of nonzero matrix entries, between the 33rd (inclusive) and 67th (exclusive), and greater or equal to the 67th percentile).

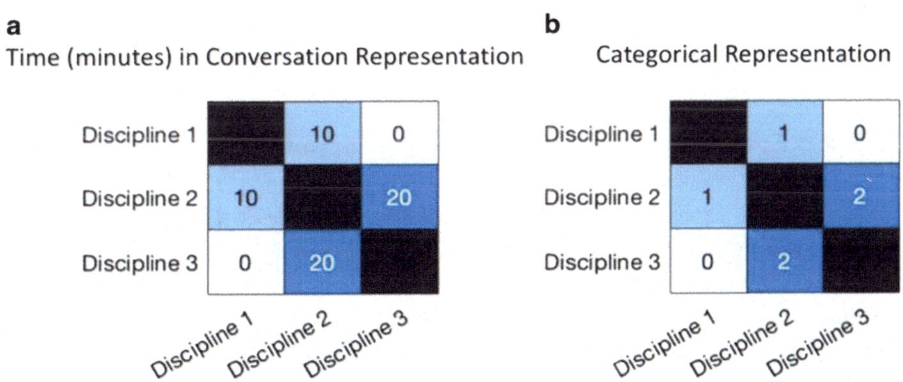

Fig. 37.2 Standard DSM showing the total number of minutes of communication between each pair of disciplines (**a**, left) and a "categorical standard DSM" (**b**, right) showing low (0), medium (1), and high (2) communication

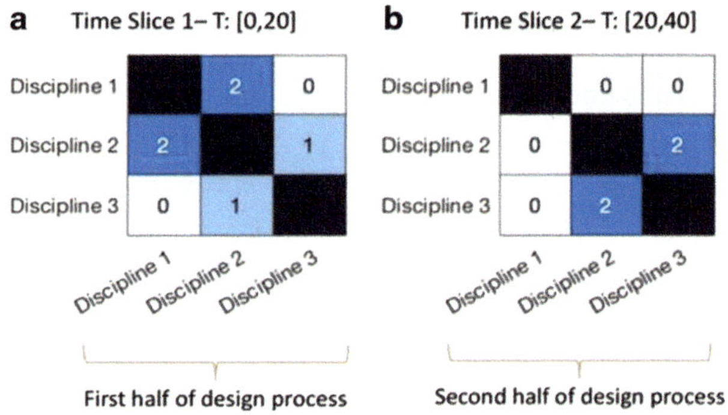

Fig. 37.3 (**a**) "Early" time slice where *T*: [0,20]. (**b**) "Late" time slice where *T*: [20,40]

37.3.3 Time-Discretized DSM

Dynamic relationships are those relationships that change throughout the design process. Analyzing specific segments of time rather than the entire time spent, designing can bring to light how relationships are changing as a result of scope changes, major design trades, identified problems, and other causes.

The procedure to create "Time-Discretized DSM" is comparable to what is done to create the "Standard DSM" as described above. It involves summing communication times along the *Z*-axis for each *x*–*y* pair value. However, this summation is over specific intervals of time throughout the design process, yielding the amount of time any two disciplines were in conversation during the specified interval. For example, should the design take place over a total of 40 min, the "conversation matrix" could be summed over two intervals of 20 min (Fig. 37.3). In the example, the time slices show that all of the communication between Disciplines 1 and 2 occurred in the first half of the design process, and none occurred in the second. Just as for the "Standard DSM," the "Time-Discretized DSM" can be further manipulated to represent discipline interaction in either binary form or categorically.

37.3.4 Egocentric Communication Dynamics View

While the "Time-Discretized DSM" reveals some aspects of the changes in communication over time, they do not reveal multilateral communication nor lengths of individual conversations. Moreover, they do not easily visualize discipline-specific communication patterns: for example, disciplines may show a tendency toward short or long, bi- or multilateral, or early or late communication. Egocentric Communication Dynamics (ECD) view provides a way of identifying these characteristics for each discipline.

Suppose we are interested in all the conversations that involved the propulsion team. From the conversation matrix, we pull out the "slice" along the *X*-axis for the propulsion team. We are left with a two-dimensional matrix in which the vertical

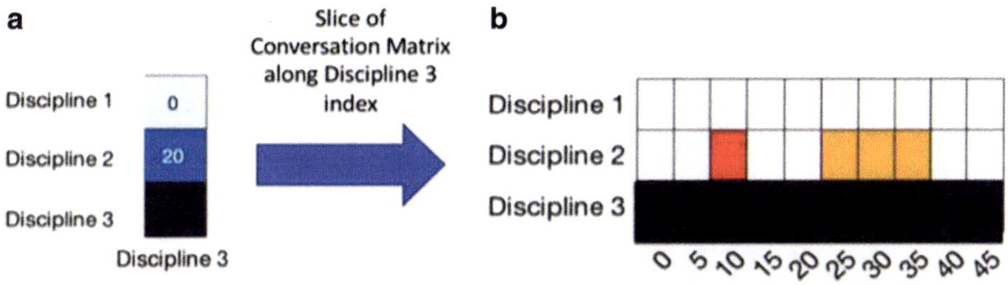

Fig. 37.4 Egocentric slice for Unit 3 (where **(a)** is the respective Standard DSM column for **(b)** the egocentric slice)

axis represents the disciplines and the horizontal axis represents time (formerly the *Z*-axis of the conversation matrix); each element contains the conversation ID for any interaction between the propulsion team and the other discipline on the *Y*-axis, between times z and $z + 1$. If a particular conversation involves multiple other disciplines, the same conversation ID would appear in multiple rows. The horizontal axis for the ECD view can be on any range of times between z_0 and z_n. This visualization allows for each discipline's communication patterns to be viewed. In an example (Fig. 37.4), each time interval is 5 min. We see that "Discipline 3" has two conversations, both with "Discipline 2": the first conversation (orange box) lasts for 5 min and the second (yellow boxes) lasts for 15 min.

37.4 Application and Insights

In this section, we illustrate the data structures and visualizations just described using the data from the NASA Goddard Mission Design Lab (MDL), to demonstrate the types of insights that can be achieved.

37.4.1 Conversation Matrix

The data collected from the MDL allowed for the construction of a single conversation matrix collating all 4 days of design. Each day of data is bound [9:00,17:00] with a Δt of 6 s for each step, and the *X*- and *Y*-axes represent the disciplines involved in the study.

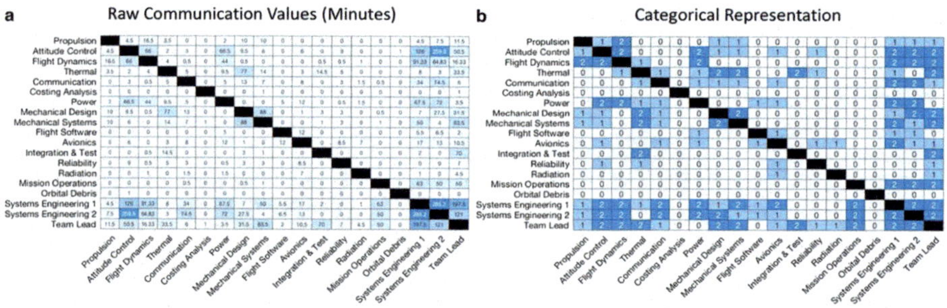

Fig. 37.5 Standard DSM with true conversation values (**a**) and categorical representation (**b**)

37.4.2 Standard DSM

The standard DSM with elements containing the total true amount of conversation between any two disciplines (Fig. 37.5a) and a categorical representation of the same data (Fig. 37.5b) was constructed.

Just as in the literature, static relationships can be identified through the resultant DSMs in Fig. 37.5a, b. For example, we see clearly that the systems engineers and team lead have meaningful amounts of communication with many disciplines. This is to be expected since these are the "integrators" for the project and should be communicating to ensure the respective disciplines are interfaced properly. In addition, there are other pockets of conversation that are visible within the DSM. For example, Mechanical Design has large amounts of communication with Mechanical Systems and Thermal. Mission Operations has a similar level of communication with all three integrators: Systems Engineering 1–2 and Team Lead. These groupings represent important interdependencies in the spacecraft design.

Potential Applications. The information yielded from the DSM clearly identifies bilateral channels of communication that are present within the MDL in this study. By analyzing this type of data across many studies, one could identify which disciplines most often need to communicate with one another and find common patterns of interaction in the spacecraft design process.

37.4.3 Time-Discretized DSM

Figure 37.6a, b present a Time-Discretized DSM with categorical representation for the intervals Monday–Tuesday and Wednesday–Friday, respectively.

This time-sliced view allows us to examine time dynamics. Recall that the Standard DSM (Fig. 37.5a) identified several relationships that required relatively high amounts of communications (discussed in Sect. 37.4.2). We can see how these relationships change throughout the design process by analyzing the time slices in Fig. 37.6. For example, in the standard DSM (Fig. 37.5), the Mechanical Design to

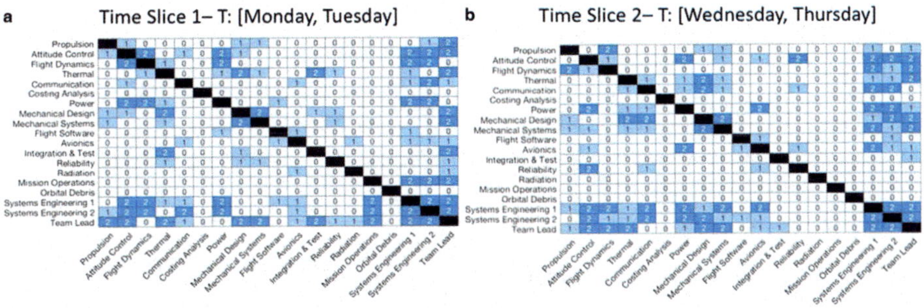

Fig. 37.6 Time-discretized DSM depicting communication Monday–Tuesday (**a**) and Wednesday –Thursday (**b**)

Thermal cell and Team Lead to multiple discipline cells all show high levels of communication. When viewed in the time slices in Fig. 37.6a, b, the Mechanical Design to Thermal cells maintain consistent and high levels of communication. However, the set of disciplines that the Team Lead is talking to is different in time slices 1 and 2. All of this information is lost in a standard DSM.

Potential Applications. Although the results described in these sections only describe time slices applied to the beginning and end of the relevant study, the possible applications of this technique are numerous. The periods of time for respective time slices can be tailored to surround specific events within the design process, such as major changes in design or requirements. With this information, one can identify the communication patterns that surround these events. While these transient communication patterns might not show up clearly on a standard DSM which aggregates communication across the entire study, supporting this type of "problem-solving" communication around design or requirements changes is critical to improving the design process.

37.4.4 Egocentric Communication Dynamics View

The ECD view allows us to examine the multilateral nature of conversations as they evolve over time. In order to demonstrate the information that can be yielded from the ECD view, examples were constructed for two different disciplines each of which were in communication with multiple disciplines for fairly large periods of time. Figure 37.7a, b show the conversations that each of the Team Lead and Systems Engineer 2, respectively, had over the course of the study. (It is important to note that the role of Systems Engineer 2 was fulfilled by two individuals, which allows for multiple conversations to occur simultaneously.) Where the standard DSM can only tell us that they spoke to many people for a long time, and the time slices can only tell us in which time periods these high levels of communication occurred, the egocentric slices also show us which group of disciplines were involved in each conversation.

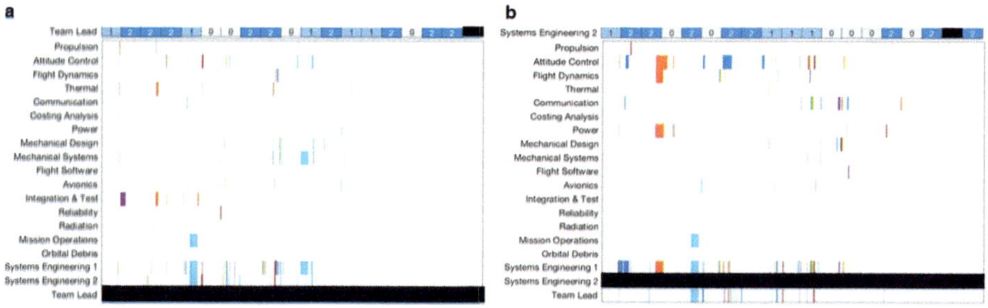

Fig. 37.7 Egocentric communication dynamics view for team lead (**a**) and systems engineer 2 (**b**)

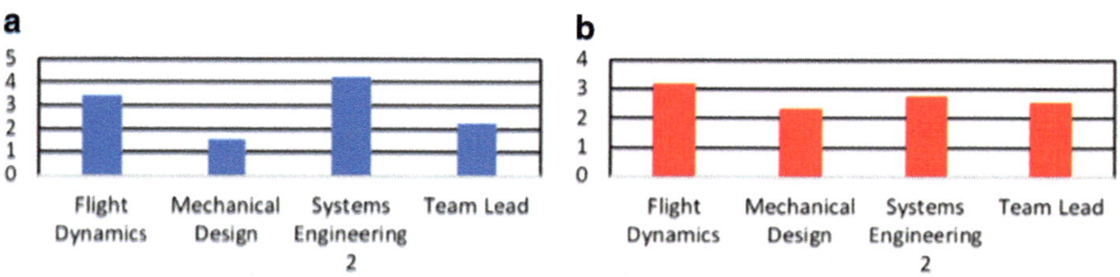

Fig. 37.8 Average length of interactions (**a**) and weighted average of disciplines in interactions (**b**) by discipline

Figure 37.7b shows the egocentric plot for Systems Engineering 2 over the entire study. From this plot, it is easily seen that this discipline participated in rather long conversations (shown by the wide lines present). Most of the longer conversation were earlier in the week. As is expected, conversation comes to an almost complete standstill at the end of the week. Lastly, we can see that Systems Engineering 2 participated in communication with small and large groups throughout the entirety of the design process (shown by the large number of same-colored lines at the same point in time, i.e., position on the *X*-axis) and the conversations with the larger groups are more concentrated at the beginning of the study. These characteristics can be further verified through sample statistics calculated for the disciplines (Fig. 37.8).

The egocentric plot for Team Lead (Fig. 37.7a) shows a different communication style than Systems Engineering 2. Team lead can clearly be seen to participate in shorter conversations throughout the entirety of the study, except for an occasional longer conversation. Just as with Systems Engineering 2, Team Lead all but ceases communication at the end of the study.

The ECD view suggests further ways to analyze the data. Building on the differences identified above by examining the egocentric visualizations, we computed the average length of interactions and weighted average number of disciplines participating per interaction for all the disciplines (partially shown in Fig. 37.8). These statistics complement the ECD view by providing quantitative insights into the types of communicators present in the study.

Potential Applications. With these insights, it is possible to identify "archetypes" of communicators. Two possible archetypes are the "talker" and the "checker." A "talker" talks to many people at once for longer periods of time (flight dynamics). For this archetype, you would expect to see egocentric plots with many instances of multilateral conversation over large swaths of time. A "checker," rather, participates in interactions with one to two people for short amounts of time (Mechanical Design). This is indicative of someone verifying some small piece of information. A checker's egocentric plot would have many small conversation blocks. These two archetypes are simple examples, but more archetypes focused on different communication styles could be identified through further data analysis.

Just as with the results provided for the time slices, the presented ECD view results only scratch the surface of the possibilities provided by the ECD view. One can identify communication archetypes (like the two identified) and patterns surrounding events, which can inform organization structure. For example, those who prefer long periods of multilateral communication can plan for consistent meetings with relevant disciplines. For those who prefer quick conversations, a constant line of communication such as an online chat capability might best support their communication needs. The egocentric analysis and accompanying quantitative approaches can identify such dynamics in many design situations.

37.5 Conclusion

The design of complex systems necessitates communication between many different disciplines. While current techniques being used to understand this communication, such as DSMs, only allow for the identification of static, bilateral relationships, the design process contains multilateral communication dynamics that change through time. The utilization of the data structures and visualizations proposed in this paper allow for the same high-level insights one can gain from using tools currently available but also enable visualization and analysis of dynamically changing and multilateral communication patterns. These capabilities were demonstrated through the examples provided.

The insights from data collected at NASA's MDL demonstrate the kind of valuable information that can be gained. In this case, the insights focus on the concurrent design process. Our analysis found that long, multilateral communications were commonplace, especially early on in the study. The communication characteristics we identified therefore justify the importance of concurrent design, since it provides an environment that supports the required lengthy multilateral communication. The tools developed here can also be used to understand how more traditional design environments facilitate communication and to redesign environments to support more efficient and effective design processes.

The work presented in this paper provides the tools for visual analysis of design communication and interdependencies but also provides the infrastructure necessary to complete quantitative analysis of the same data. Combining visualizations with quantitative measurements of patterns identified visually is a powerful tool for understanding the design process. Moreover, it provides a manner of comparison between many design processes carried out by the same or comparable organizations. Features such as conversation lengths and number of disciplines in communication can be compared and analyzed to support organizational redesign of communication infrastructure, such as information systems, colocation of teams, etc. In addition, these analyses could support future research on varied topics, including how differences in technical product architectures affect communication during the design process.

The key contribution of this paper is to suggest data structures for capturing rich, complex data on communication during the design process in a tractable format, enabling the data to be both visualized for intuitive understanding and quantified for analysis and comparison. By building on the successful Design Structure Matrix, we have developed tools that capture dynamic, multilateral communication within the design process, to lay a foundation for further research into more effective and efficient design processes.

Acknowledgment This work was supported in part by the National Science Foundation under Grant CMMI—1563408.

References

1. Pimmler, T. U., & Eppinger, S. D. (1994). Integration analysis of product decompositions. In *Proceeding of the ASME Design Theory and Methodology Conference*. Minneapolis, MN. Retrieved February 15, 2018, from https://dspace.mit.edu/bitstream/handle/1721.1/2514/1/SWP-3690-30681755.pdf
2. Suh, E. S., De Weck, O. L., & Chang, D. (2007). Flexible product platforms: Framework and case study. *Research in Engineering Design, 18*(2), 67–89. https://doi.org/10.1007/s00163-007-0032-z
3. Sosa, M. E., Eppinger, S. D., & Rowles, C. M. (2003). Identifying modular and integrative systems and their impact on design team interactions. *Journal of Mechanical Design, 125*(2), 240. https://doi.org/10.1115/1.1564074
4. Sosa, M. E., Eppinger, S. D., & Rowles, C. M. (2004). The misalignment of product architecture and organizational structure in complex product development. *Management Science, 50* (12), 1674–1689. https://doi.org/10.1287/mnsc.1040.0289
5. Feng, W., Crawley, E. F., de Weck, O., Keller, R., & Robinson, B. (2010). Dependency structure matrix modelling for stakeholder value networks. In *Proceedings of the 12th International DSM Conference* (pp. 3–16). Cambridge.
6. Browning, T. R. (1998). Integrative mechanisms for multiteam integration: Findings from five case studies. *Systems Engineering, 1*(2), 95–112. https://doi.org/10.1002/(SICI)1520-6858(1998)1:2<95::AID-SYS1>3.0.CO;2-3
7. Eppinger, S. D., & Browning, T. R. (2012). *Design structure matrix methods and applications*. MIT Press http://app.knovel.com/hotlink/toc/id:kpDSMMA003/design-structure-matrix/design-structure-matrix

8. Batallas, D. A., & Yassine, A. A. (2006). Information leaders in product development organizational networks: Social network analysis of the design structure matrix. *IEEE Transactions on Engineering Management, 53*(4), 570–582. https://doi.org/10.1109/TEM.2006.883706
9. Browning, T. R. (2001). Applying the design structure matrix to system decomposition and integration problems: A review and new directions. *IEEE Transactions on Engineering Management, 48*(3), 292–306. https://doi.org/10.1109/17.946528
10. Gokpinar, B., Hopp, W. J., Iravani, S. M. R., & Ross, S. M. (2010). The impact of misalignment of organizational structure and product architecture on quality in complex product development. *Management Science, 56*(3), 468–484. https://doi.org/10.1287/mnsc.1090.1117

Chapter 38
A Value-Centric Tradespace Approach to Target System Modularization

Adam M. Ross and Hank H. Roark III

38.1 Introduction

During conceptual design of a system, much of the lifecycle cost is committed. One of the architecture choices that can be committed in this phase is how a system's components are interconnected and hence the level of modularity of the components. This modularity level can have long-term impacts on the changeability of a system over its life span. One of the difficulties with the previous statement is that modularity has many definitions, introducing ambiguity into how to measure and value modularity.

With so many definitions, potential lifecycle benefits, and potential trade-offs, design engineers are faced with many objectives and challenges when deciding how to modularize components within a system. Some work has been done on this front, including the development of clustering algorithms that take into account the endogenous system structure to determine component clusters [1] and the engineering system matrix that uses qualitative knowledge of endogenous and certain exogenous system factors as a screening for potential areas of high system change [2]. One of the author's industry experiences, however, is that the decision on organizing system components into modules during conceptual design seems to be based on expert experience, rules of thumb, heuristics, and iteration. Stakeholders and system designers can benefit from practical tools to aid in making design decisions on where to focus modularity efforts during conceptual and early design.

A dynamic Tradespace-based approach can help make design decisions regarding incorporating changeability into a system, based on changes in customer needs over time and physics-based models of system performance [3]. This paper describes an approach that uses dynamic Tradespaces coupled with a network representation of a

A. M. Ross (✉) · H. H. Roark III
Massachusetts Institute of Technology, Cambridge, MA, USA
e-mail: adamross@mit.edu

S. Adams et al. (eds.), *Systems Engineering in Context*,
https://doi.org/10.1007/978-3-030-00114-8_38

system's components to aid in value-based decisions about if, and where, to modularize a system to make it more robust to stakeholder requirements changes [4]. This approach will provide design engineers with tools to make modularization decisions of a system in support of product variety and evolvability.

38.1.1 On Modularity

One of the more difficult aspects of this research was to settle on an operational definition of modules, modularity, and modularization. MIT's Engineering System Division defines a module as "a part of a system that is constructed to have minimal, standardized interactions with the rest of the system" and modularity as "the degree to which the components of a system can be designed, made, operated, and changed independently of each other" [5]. Several authors have defined modularity as something along the lines of the one-to-one mapping of a system function to the system form and including certain component interface characteristics [6], while others have defined modularity as a series of design rules that enable various operations, called modularity operators, on the system [7]. While these definitions suffice as conceptual definitions, this research requires a definition with an operational measure of modularity.

This research is primarily concerned with increasing the changeability of a system based on how well certain components are connected as modules and their connectivity to the other elements in the system. With this desire, and the need for an operational measure, the definition provided by Ref. [8] will be adopted by this research: modularity is "a measure of the lack of technical interface connectivity between components of a system." Based on network analysis and graph theory, there are three metrics for measuring component modularity and their impacts on component redesign: degree modularity $M(D)$, distance modularity $M(T)$, and bridge modularity $M(B)$, which are a normalization of existing network centrality metrics (degree, Freeman closeness, and Freeman betweenness) to a range of [0, 1], where 1 is the highest level of modularity for each metric [8]. From these metrics there is a correlation between the set of outdegree modularity and outdistance modularity and the likelihood of planned redesign of a component to change the system's performance level. These metrics and this correlation finding will be used for the modularity analysis of components needed to change a system's performance.

38.1.1.1 Benefits and Challenges to Modularity

Modular systems have been proposed to provide many lifecycle benefits to systems and products, including, but not limited to, increasing overall system economic value [7], aiding in system flexibility and evolvability [9], increasing product variety [10], and as an aid to complexity management [11]. Ref. [12] proposed modularity as a mitigation strategy for certain types of uncertainties. In this regard, modularity can be viewed as a means to achieve desired goals and may be critical for achieving other life cycle properties [9, 13].

While modular architectures have many potential benefits, these benefits do not come without costs: for systems with relatively high energy density, modularity increases weight, increases cost, and/or results in lower performance [14]. For systems with business and performance constraints where stakeholders value lower weight, smaller size, or higher performance, one tends to find more integral architectures; conversely when stakeholders value commonality and reuse across products in order to achieve cost savings, one tends to find more modular architectures [15].

38.1.1.2 Existing Modularization Approaches

The primary methods to aid a design engineer seem to be focused on clustering or heuristics. A clustering approach, using design structure matrix (DSM) representations of a system, can be used to cluster a proposed system into blocks where the interactions within a module cluster are maximized and between clusters are minimized [1]. A shortcoming of this approach is that it only looks endogenously at the system and lacks tools necessary to identify areas where modularization may aid in achieving desired lifecycle properties and system variability across decision makers or through time. Function cluster was proposed as a way to consider potential changes in mass, energy, and information flow between components and to cluster a system into modules and suggested system cleavage points to introduce interfaces between modules that will minimize the likelihood of propagation of changes should a module need to change [16]. This approach takes into consideration exogenous factors that could require changes in the system but leaves it up to the system architect to recognize potential changes that may emerge due to changing or different stakeholder needs.

On the other end of the modularization approach spectrum are heuristics [17, 18]. The shortcoming with these approaches is that they suggest extensive experience is required by the system architect to make modularity choices, leaving the developing organization to rely on tacit knowledge and the designer with little experience to blindly trust in the heuristic's applicability.

38.2 Approach

The scope of this work includes the development of an approach to focus system modularization efforts and its application to a case. The Responsive Systems Comparison (RSC) method, based on Multi-Attribute Tradespace Exploration (MATE), will be used, as it develops value-focused designs and indicators of potential design changes [19]. DSMs will be used to model the system architecture and the connectivity between the components, and network centrality measures based on the DSM

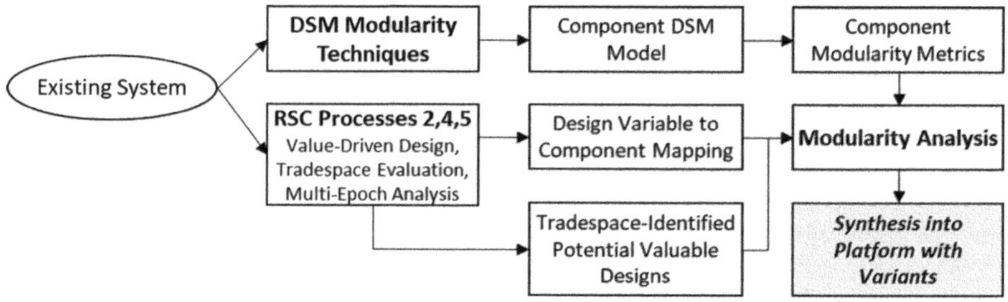

Fig. 38.1 Approach for considering modularity with RSC

will be used to measure modularity of the components. These two different techniques, RSC and DSM, will be combined into one approach (Fig. 38.1). To demonstrate the approach, the case study will be a mobile terrestrial machine (LifeTrac), which was chosen because it has multiple use scenarios and is available as open source. The goals of the approach are as follows:

1. Provide a mechanism to understand potentially desired changes to design variables, based on decision-maker preferences, and link those changes to components that may need to be altered in response. The purpose of this goal is to provide designers a value-centric approach to target modularization efforts.
2. Build upon the Responsive Systems Comparison (RSC) method by making connections to RSC's existing process steps. The purpose of this goal is to use the decision analysis tools from RSC to make trade-off decisions in terms of decision-maker utility, system cost, and modularization.
3. Utilize the component modularity metrics to evaluate the component modularity of a proposed system architecture [7]. The purpose of this is to be able to measure component modularity during early-stage architecture synthesis or to be able to reverse engineer existing systems and quantify the modularity of components in that existing system.

Component modularity analysis consists of two activities: quantification of the modularity of the components of a system and linkage of the components to the design variables used in RSC. Quantification of the components is done by representing the system in a way that is susceptible to network centrality measures, such as a DSM representation, and then calculating the component modularity metrics [7]. It should be noted that constructing the form-form interaction network requires either development of a DSM or network representation of the form-form interactions of the system. Because new systems are often an evolution of an existing system, for many development efforts, this assumption is achievable. There are other scenarios, such as greenfield design, where representation of the form-form interactions of the underlying system cannot be built due to lack of necessary inputs.

Another needed activity is to provide a linkage between components of the system and design variables. This linkage is an indication of the system components that are likely to need to change because of a change in level of a design variable.

This linkage is represented as a multiple domain mapping table between the components and the design variable.

If a designer chooses to have design(s) able to be changed readily over its life, then the designer can place appropriate modularity requirements on the components to reduce the time and/or costs to execute the change. The component modularity analysis is used to derive a table that links the design variables to the components that are likely to change because of a design variable changing. This linkage would only be in place if there is a system architecture being proposed at this stage, as may be the case in changes to existing products or systems. Modularity requirements should also be carried into downstream architecture synthesis activities.

38.3 Case Example: LifeTrac Tool

The case study system is the OpenEcology Project's LifeTrac tool, which was designed for two functions: acting as a simple wheeled skid steer and as a simple agricultural tractor [20]. The purpose of the skid steer function is to push, lift, and move material around a work site. The purpose of the agricultural tractor function is to provide tractive energy for pulling agricultural implements, such as tillage or seeding equipment, through a field. It is interesting to note that the LifeTrac tool is described as a "modular" design, but little is given on the website to backup that claim. The LifeTrac tool was chosen for this case study for the following reasons:

- The LifeTrac tool is targeted for use by two types of users. The first type is the farmer who desires a tool to aid in field operations; the second type is a construction work site operator who desires a tool for moving material around a job location. With these two types of users and use cases, one might expect two different measures of utility that might result in a different desirable system.
- The LifeTrac tool is an existing system, demonstrating the use of the proposed modularization design approach in the evolution of a system. This is considered an acceptable starting scenario as many design efforts are incremental in nature.
- While a simple design, LifeTrac is sufficiently complex to demonstrate the proposed modularization design approach. Complexity was determined by the number of components ($n = 47$), and interactions ($i = 218$) between the components, based on the analysis of the component DSM model of the system. By keeping with a simple design, it is hoped that the research is more approachable.
- The design and costing information is covered under an open-source license, allowing ease of research access [21].

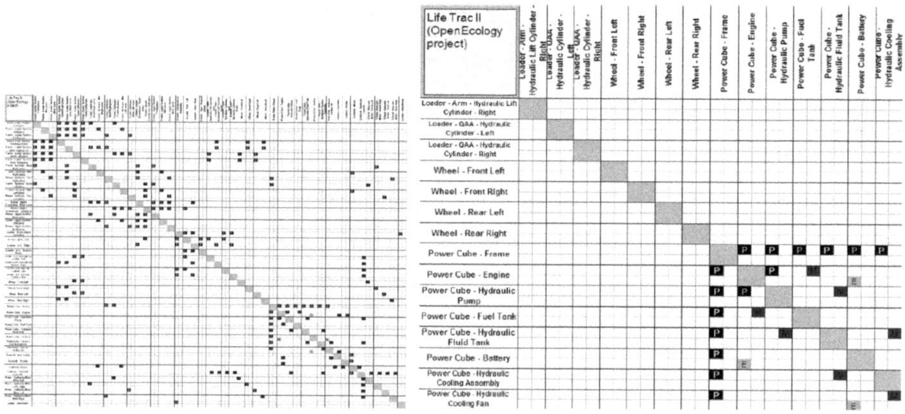

Fig. 38.2 Component DSM for the LifeTrac system; full (left), lower-right quadrant zoom (right)

38.3.1 DSM Modularity Techniques

The first activity in the proposed approach is to construct a component DSM model of the underlying system. This was done for the LifeTrac tool by reverse engineering open-source computer-aided design (CAD) models of LifeTrac [22]. The component DSM model in Fig. 38.2 considers four types of dependencies: spatial (P), mass flow (M), information flow (I), and energy flow (E). Overall, there are 47 DSM elements in the DSM, with 218 element-to-element dependencies, leading to an interaction density of 0.10.

These interaction types were simplified to binary, single-type (there is either an interaction between two elements or there is none) interactions for the remainder of the case study. The decision to use binary, single-type interactions is a trade-off in model fidelity versus execution time for later analysis. Further, physical and information dependency may be more easily determined than energy or mass flow dependencies [8]; by collapsing to one type, it is suggested that at least the connectivity between components will be more completely recorded for use in the subsequent network centrality analysis. Many alternatives are available for increasing the fidelity of the model of interactions, including types of interactions as well as strength of interactions [8].

At this point in analysis it is not uncommon to apply clustering algorithms to the DSM to group components together into modules. However, that step is not necessary here as we are not interested in the grouping of components into modules (clustering does not change the connectivity between the components). Instead, we want to explore how connected each component is to all other components in the system as an indication of the "cost" of propagation of change.

The next activity in the approach is to calculate the metrics for degree, distance, and bridge centrality [8]. Each of these metrics is in the range of [0, 1], with higher values corresponding to a higher level of modularity for that metric. These metrics, which have been calculated for all LifeTrac components, provide the following insights into the connectivity of a component to other components:

Degree modularity: the number of other components that have direct dependencies with a given component. The less direct dependencies component i has with other components, the higher the value of $M(D)_i$. This is the simplest of the three metrics.

Distance modularity: how far away (or how close) a given component is to other components. This is built on the concept of farness (or its inverse, closeness) in network theory. This measure captures the concept that design changes may propagate not just to/from immediate neighbors (as measured in degree modularity) but also through the network of the design dependencies from components. If component i has high distance modularity $M(T)_i$, then changes to that component would have a longer distance to traverse to reach other components in the system.

Bridge modularity: tells one about how many design dependency paths a component lives on between other components. This is built on the network theory concept of centrality. The idea with this measure is to capture the degree to which a component is on design dependency paths between other components. If component i has high bridge modularity $M(B)_i$, then it lies on fewer design dependency paths between all other components.

38.3.2 RSC Processes 2, 4, and 5: Value-Driven Design, Tradespace Evaluation, and Multi-Epoch Analysis

For this case, two decision-makers are considered for the LifeTrac: one is a consumer that will use the LifeTrac as a skid steer to move materials around a work site (i.e., construction usage) and another is a farmer that will use the LifeTrac for tractor field work (i.e., agricultural usage). In practice, this can be represented as four use scenarios ("epochs"): two for construction ((1) demanding needs and (2) balanced needs) and two for agricultural (row spacing (3) 24 in. and (4) 30 in.). For the construction use, there are three attributes of interest, material capacity, maneuverability, and lifting capacity, whereas in the agricultural use, there is only one attribute of interest, efficiency, along with a row spacing constraint. Table 38.1 lists the measurements, units, acceptance range, and utility weights for each of the attributes across these two uses, including the minimum ($U = 0$) and maximally desired ($U = 1$) levels. These attributes are now described:

Material capacity—maximum load is a measure of max weight the vehicle has capability to vertically lift and carry around.

Maneuverability—vehicle width is a measure of ability of the vehicle to fit through openings and passages.

Lifting capacity—breakout force is a measure of force to break material apart (e.g., pulling an embedded stone out of ground).

Table 38.1 Attributes for both users: construction (material capacity, maneuverability, lifting capacity) and agriculture (efficiency, subject to row spacing constraint)

Attribute	Measurement	Units	$U=0$	$U=1$	Epoch 1 Wt	Epoch 2 Wt	Epoch 3 Wt	Epoch 4 Wt
Material capacity	Maximum load	Pounds	1200	2000	0.0	0.5	N/a	N/a
Maneuverability	Vehicle width	Inches	96	72	0.0	0.2	N/a	N/a
Lifting capacity	Breakout force	Pounds	1500	2500	0.0	0.3	N/a	N/a
Efficiency	Work rate	Acres/hr	2	6	N/a	N/a	1	1
Row spacing	Constraint	Inches	N/a	N/a	N/a	N/a	24	30

Efficiency—vehicle work rate is a measure of efficiency to perform key agricultural tasks quickly over an area.

For both agricultural and construction usage, lifecycle cost is measured as the acquisition cost of the underlying system plus the fuel usage over the period of use. The period of use is 5 years with 300 h of engine time per year, typical for usage patterns and the life span of products of this type and size. The acquisition cost model for LifeTrac was built from online information [23, 24]. For operating expenses, costs were assumed to be USD1 per horsepower-hour.

Next, we conduct design-value mapping (DVM), with the goal to ensure design variables (factors in our control) related to achieving decision-makers' goals and is considered in subsequent Tradespace development. The DVM is in Table 38.2.

For each design variable, ranges were determined by the availability of off-the-shelf components; in a clean sheet design, these dependencies would be relaxed. A description and justification for each design vector element are as follows:

Bucket width represents the width of the attachment on the front of the machine. Width of the attachment bucket minimum was set in a range typically available from commercial suppliers (based on a review of attachment bucket sizes from Deere, Bobcat, and Caterpillar websites). For the sake of simplicity, only bucket width is determined to affect maneuverability because width of the vehicle is fixed in this epoch and is considered not tradeable.

Available hydraulic power represents the amount of hydraulic power, as measured by the brake power of the engine. The base design of the LifeTrac has an engine power of 28 horsepower (HP). The minimum and maximum power of this design variable is set at 4 HP and 40 HP, as it is the limit of available air-cooled internal combustion engines available off the shelf [24] (same source the LifeTrac team sourced the current engine).

Engine power represents the amount of engine power available, as measured by the brake power of the engine, that will be translated to tractive force. The base design of the LifeTrac has an engine power of 28 HP. The ranges were selected for the same reasons as stated in the description of hydraulic power.

Table 38.2 Tractor and skid steer attributes to design vector (design-value) mapping

Design vector		Attributes				
Variable	Range	Material capacity	Maneuverability	Lifting capacity	Efficiency	Row spacing
Bucket width	[56–84] in.	X	X			
Available hydraulic power	[4–40] HP	X		X		
Engine power	[4–40] HP				X	
Vehicle width	72, 90 in.					X

Vehicle width represents the tire-center to tire-center spacing. Because this is a constraint, depending on the epoch, only one of the two levels will provide a feasible design in each epoch (i.e., if considering the 24 in. row spacing epoch, only the 72 in. vehicle width will result in feasible designs because $72 \bmod 24 = 0$ and $90 \bmod 24 \neq 0$). Since the variable will need to have different values in different epochs to produce feasible designs, it could affect the design of components within the system, making those components potential candidates for modularity analysis.

Now that we have a list of value-driving design variables, the next step is to map these to the DSM components to identify which components would be affected by alternative choices for the design variables. A description of the components identified with each design variable now follows, along with a mapping to components in Table 38.3.

Bucket width: The main component affected by this design variable is the "loader attachment" component, the DSM entity that is for the attachment element of the LifeTrac.

Hydraulic power: All components associated with generating and transmitting hydraulic power to the loader arm are determined by the setting of this design variable. Also affected is the length of the loader arm as it is a lever that transmits the hydraulic force to the attachment.

Engine power: This list contains all the components that are responsible for generating and transmitting power to the wheels. These largely determine the ability to create tractive force required for pulling implements through the ground.

Vehicle width: These are the main frame structural elements that determine the overall width of the vehicle.

Next, we generate our Tradespaces. The alternative LifeTrac designs are generated by varying the design variables across their allowed ranges. Combinations of design variables at particular levels result in a unique design, which was then evaluated in terms of the attributes via a physics-based performance model and a cost model. Due to the row spacing constraints in the agricultural epochs (3 and 4), half of the designs were infeasible (i.e., not usable at all). Table 38.4 lists the number of designs in each of the four epochs.

Table 38.3 LifeTrac design variable to DSM component mapping

Design variable		DSM component i	Component names
Skid steer (construction)	Bucket width	47	Loader - Attachment
	Hydraulic power	20, 21, 23, 24, 25, 26, 32, 33, 42	Loader - Arm - Left, Right; Loader - Arm - Hydr. Lift Cyl. - Left, Right; Loader - QAA - Hydr. Cyl. - Left, Right; Power Cube – Engine; Power Cube - Hydr. Pump; Controls - Hydr. Controls
Tractor (agriculture)	Engine power	32, 33, 42, 43, 44, 45, 46	Power Cube – Engine; Power Cube - Hydr. Pump; Controls - Hydr. Controls; Wheel - Hydr. Motor - Front Left, Right; Wheel - Hydr. Motor - Rear Left, Right
	Vehicle width	1, 2, 3	Frame - Lower section - Front piece, Mid piece, Back piece

Table 38.4 Basic metrics from LifeTrac Tradespaces (designs considered $n = 592$)

Tradespace	Feasible designs	Designs $U(X) \geq 0$
Skid steer, full soln. (epoch 1)	592	256
Skid steer, partial soln. (epoch 2)	592	256
Tractor, 30 in. (epoch 3)	296	256
Tractor, 24 in. (epoch 4)	296	272

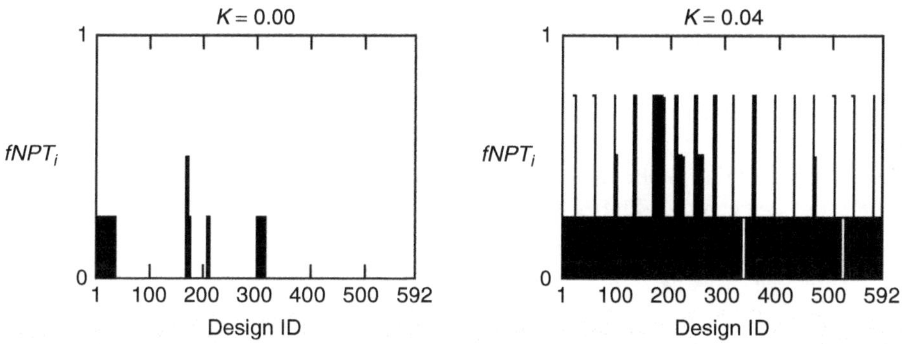

Fig. 38.3 Fuzzy normalized Pareto trace by design for $K = 0$ and $K = 0.04$

A large number of Tradespace analyses were performed to identify design variables that drive value. Screening metrics for looking across epochs are useful for identifying valuable designs, that is, designs that are most efficient in utility for cost (i.e., Pareto efficient). Fuzzy normalized Pareto trace (fNPT) can be used to identify designs that, while not strictly non-dominated, could be valuable across epochs. The factor K will be used to denote a tolerance to uncertainty and was varied from 0.00 to 0.10 in steps of 0.02 (the situation of $K = 0$ is the same as the strict Pareto set) [25].

The results for $K = 0$ and 0.04 are shown in Fig. 38.3. We would expect to find no designs achieving a $\text{fNPT}_i = 1$ because of the constraint on width across epochs 3 and 4 (i.e., one half of designs in those epochs are not feasible). As K is increased, we see more and more designs within being "close" to Pareto efficient. Upon inspection, it is when $K = (0.02, 0.04)$ that we see designs that, across the epoch 3 and epoch 4 constraints, have an fNPT of 0.75; this means that a design is good in both construction use epochs and one agricultural use epoch. These designs are now selected as "passively value robust" and provide the reduced set of potential candidates for modularity.

Another way to view the previous analysis is as a means to layer changeability onto the most passively robust designs identified through the fuzzy Pareto tracing. The goal of layering changeability onto the most passively robust designs is to achieve a high effective value robustness, one that leverages both passive and active robustness. Modularity is an enabler for changeability, with the implication being that one can to link together the mostly passively robust designs into a completely robust (passive and active) design through component modularity. This is the same as attempting to find or synthesize, through component modularity, designs with an effective fuzzy normalized Pareto trace, efNPT equal to one [26].

38.3.3 Modularity Analysis

At this point in the approach, we have a set of potentially valuable designs, as well as insights into which design variables exhibit variance across these designs. If this were a product for commercial markets, the providing organization might want to offer all the designs to the market as options for different consumers (e.g., "good, better, best" product options [27]). Or, the scenario in question could be for a single decision maker that desires a level of skid steer functionality at some time $t = 0$ but could foresee needing a higher level of functionality at $t > 0$. An alternative scenario could be that a single tractor decision maker is in epoch 1 (30 in. rows) and may want to switch to epoch 2 (24 in. rows) in the future. With each of these change scenarios, we want to understand the relative effort of supporting different designs and design changes inside of the given product architecture. Since the design variables may need to be changed, the next step focuses on answering the question, "What might component modularity tell the designer about the design's ability to support these changes?"

Returning to our modularity metrics, we can inspect the modularity of the selected components associated with particular design variables (Fig. 38.4). Those components that are more modular can support easier change. Inspecting all four design variables resulted in identification of two types of modules: *attachment modularity*—readily changed and executable during use or production—and *frame modularity*, harder to change and executable during production only. These modularities enable change paths in the Tradespace, allowing a design to "change" into another. As an example, allow us to start with design 22. With modularity path enablers in place for changing the attachment width and changing the width of the product, one could transition from design 22 to the remainder of the designs in Table 38.5 through executing the modularity path enablers.

From this shorter list of designs, we will inspect the utilities to determine what designs could be offered into the target markets. Beginning with the construction epochs, the utilities are given in Table 38.6 for each of the designs from Table 38.5. This list of designs is sorted in descending order of utility in epoch 1, to make the "good, better, best" offerings easier to discern. The "good, better, best" determination was done via inspection as an example, but other determinations may be made.

The same information can be used for determining the offerings in the agricultural sector. The list of designs is shown in Table 38.7 with the utilities for the agricultural epochs. Of note is that in each of epoch 3 and epoch 4, the acceptable designs have the same utilities. Upon inspecting the Tradespaces for these epochs, we find that

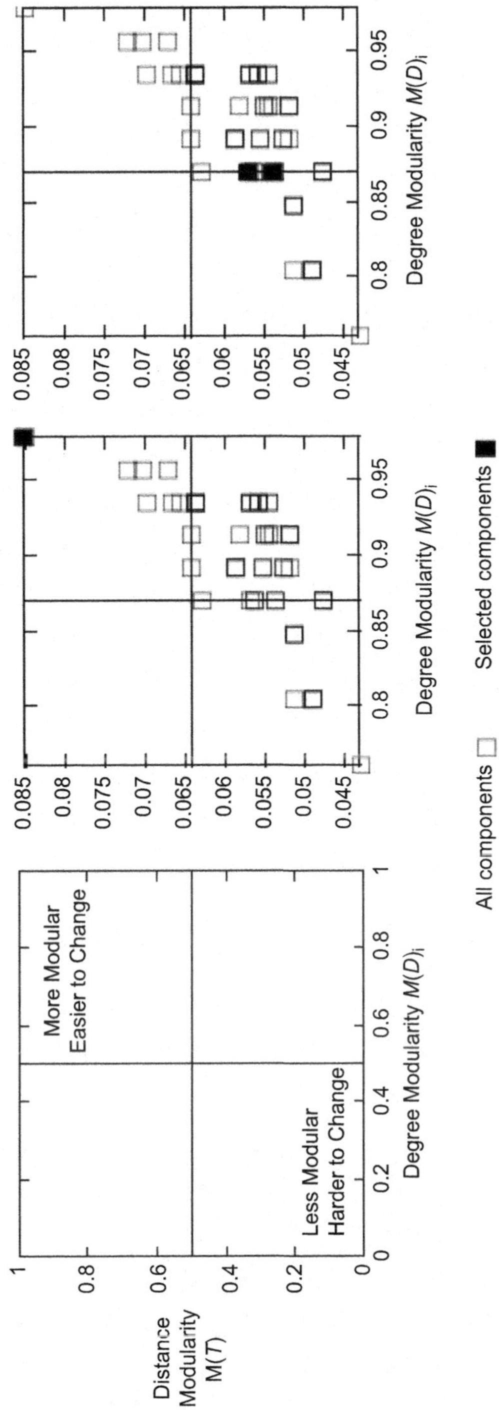

Fig. 38.4 Description of $M(T)$ vs. $M(D)$ (left), modularity of attachment modules (e.g., bucket) (center), and frame modules (e.g., tractor width) (right)

Table 38.5 Example designs reachable through modularity path enablers

Design ID	Hydraulic/engine power (HP)	Vehicle width (in.)	Bucket width (in.)	Reachable via	Reachable phase
22	25	72	56	Baseline	
170	25	72	72	Attachment mod	Production or use
281	25	72	84	Attachment mod	Production or use
318	25	90	56	Frame mod	Production only
466	25	90	72	Attachment and frame mods	Production or use
577	25	90	84	Attachment and frame mods	Production or use

Table 38.6 Product offerings for construction epochs (epochs 1 and 2)

Design ID	Hydraulic/engine power (HP)	Vehicle width (in.)	Bucket width (in.)	Utility: construction		Product offering
				Epoch 1	Epoch 2	
577	25	90	84	0.079	0.645	
170	25	72	72	0.060	0.662	Best
281	25	72	84	0.031	0.579	Better
22	25	72	56	0.016	0.341	Good
466	25	90	72	0.015	0.512	
318	25	90	56	0.004	0.191	

Table 38.7 Product offerings for agriculture epochs (epochs 3 and 4)

Design ID	Hydraulic/engine power (HP)	Vehicle width (in.)	Utility: agriculture		Product offering
			Epoch 3	Epoch 4	
22, 170, 281	25	72	Unacceptable	0.809091	Yes, all three
318, 466, 577	25	90	1	Unacceptable	Just 318

utility is a function of the overall engine power. Because the engine power is not varied in the considered designs, and a modularity path enabler for engine power was not in place, there is only one product available to users for each of epochs 3 and 4. The producing entity might now decide to make the cost/time trade-off decision to put an engine power modularity path enabler in place so that it could offer a wider selection of products to target markets.

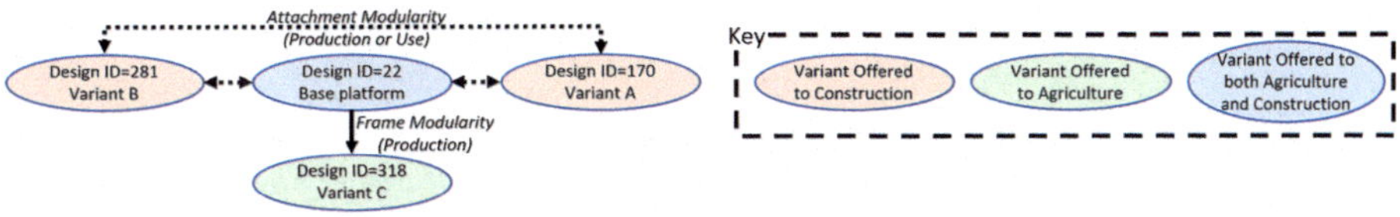

Fig. 38.5 Design variants for different target users connected via modularity-enabled paths

38.3.4 Synthesis into Platform with Variants

Now that 4 designs, 22, 170, 281, and 318, have been selected as offerings to two different markets, they can be tied together into an overall product line. Design 22 could be considered the base “platform” by the supplier; design 22 could then be transitioned to the other designs via modularity in production or use of either the frame or the attachment. The family offering of products, based on design 22 as the base platform, is shown graphically in Fig. 38.5. Also shown is an indication of which variants would be offered to which use types: construction, agriculture, or both. In the case of designs 22, 170, and 281, if a construction customer acquired any of these three designs, then they could transition to any of the other two designs via modularity in attachment during the use phase. In the case of designs 22 and 318, if an agricultural customer acquired either of these, they could not transition to the other design because modularity in the frame is only available at the time of production; design 22 would be acquired for epoch 4 needs, and design 318 would be acquired for epoch 3 needs.

38.4 Discussion

While only qualitatively considered, a connection was made between component modularity and system path enablers and transition paths. If one is starting with an existing design, the modularity metrics could be used to indicate which modularity path enablers are present and to what degree, providing higher fidelity as compared to binary existence of modularity. Additionally, these opportunities for targeted modularity can be used to identify families of systems that are, together, value robust across multiple epochs. As an example of the benefits of the proposed approach, the results of the Tradespace study and modularity analysis were combined to demonstrate how a complete product line could be developed and offered to the market based on the results of the approach. Finally, this approach has utilized RSC to provide a value-centric focus to modularity efforts. Instead of relying on experiential and tacit knowledge, this approach can focus designer’s modularization efforts on elements of the system that may need to change to accommodate changes in decision-makers’ needs or use contexts, whether the system is an evolution or entirely new.

References

1. Yu, T., Yassine, A. A., & Goldberg, D. E. (2007). An information theoretic method for developing modular architectures using genetic algorithms. *Research in Engineering Design., 18*, 91–109.
2. Bartolomei, J. E. (2007). *Qualitative knowledge construction for engineering systems: Extending the design structure matrix methodology in scope and procedure* (PhD dissertation). Cambridge, MA: MIT.
3. Ross, A. M. (2006). *Managing unarticulated value: Changeability in multi-attribute tradespace exploration* (PhD dissertation). Cambridge, MA: MIT.
4. Roark, H. H. III. (2012). *Value centric approach to target system modularization using multi-attribute tradespace exploration and network measures of component modularity* (SDMthesis). MIT, Cambridge, MA
5. Allen, T., Mcgowan, D., Moses, J., Magee, C., Hastings, D. E., Moavenzadeh, F., et al. (2002). *ESD terms and definitions*. Cambridge, MA: MIT. http://esd.mit.edu/WPS/esd-wp-2002-01.pdf
6. Ulrich, K. (1995). The role of product architecture in the manufacturing firm. *Research Policy, 24*, 419–440.
7. Baldwin, C. Y., & Clark, K. B. (2000). *Design rules, The power of modularity* (Vol. 1). Cambridge, MA: MIT.
8. Sosa, M. E., Eppinger, S. D., & Rowles, C. M. (2007). A network approach to define modularity of components in complex products. *Journal of Mechanical Design, 129*(11).
9. de Weck, O., Roos, D., & Magee, C. (2011). *Engineering systems : Meeting human needs in a complex technological world*. Cambridge, MA: MIT.
10. Johnson, H. T., & Broms, A. (2000). *Profit beyond measure: Extraordinary results through attention to work and people*. New York: Free Press.
11. Browning, T. R., & Eppinger, S. D. (2002). Modeling impacts of process architecture on cost and schedule risk in product development. *IEEE Transactions on Engineering Management, 49* (4), 428–442.
12. McManus, H. L., & Hastings, D. E. (2006). A framework for understanding uncertainty and its mitigation and exploitation in complex systems. *IEEE Engineering Management Review, 34*(3), 81–94.
13. Ross, A. M. (2008). *Defining and using the new "ilities"* (Working Paper 2008-4-1). MIT: Cambridge, MA.
14. Whitney, D. (2003). Physical limits to modularity. In *MIT ESD Symposium 2003*, Cambridge, MA.
15. Holtta-Otto, K., & de Weck, O. (2007). Degree of modularity in engineering systems and products with technical and business constraints. *Concurrent Engineering, 15*(2), 113–126.
16. Holtta-Otto, K. (2005). *Modular product platform design* (PhD dissertation). Helsinki, Finland: Helsinki University of Technology.
17. Maier MW and Rechtin E. The art of systems architecting. CRC Press: Boca Raton, FL, 3rd, 2009.
18. Stone, R. B., Wood, K. L., & Crawford, R. H. (2000). A heuristic method for identifying modules for product architectures. *Design Studies, 21*(1), 5–31.
19. Ross, A. M., McManus, H. L., Rhodes, D. H., Hastings, D. E., & Long, A. M. (2009). Responsive systems comparison method: Dynamic insights into designing a satellite radar system. In *AIAA Space 2009 (SPACE09)*. Pasadena, CA.
20. Open-Ecology-Project. (2011). "LifeTrac". http://opensourceecology.org/wiki/LifeTrac
21. Open-Ecology-Project. (2011). "Open Source Ecology License". http://opensourceecology.org/wiki/License
22. Open-Ecology-Project. (2011). "LifeTrac III (SolidWorks parts & assembly)". http://openpario.mime.oregonstate.edu/attachments/download/4304/LifeTrac_III_SolidWorks_partsassembly_.zip

23. Open-Ecology-Project. (2011). "LifeTrac Bill of Materials". http://opensourceecology.org/wiki/LifeTrac/Bill_of_Materials
24. SmallEngineSuppliers.com. (2011). *Briggs and stratton vertical engines retail prices*. http://smallenginesuppliers.com/shop/html/pages/briggs_vertical_shaft_engines.html
25. Ross, A. M., Rhodes, D. H., & Hastings, D. E. (2009). Using pareto trace to determine system passive value robustness. In *3rd IEEE Sys Conf. (SysCon09)*. Vancouver.
26. Fitzgerald, M. E., Ross, A. M., & Rhodes, D. H. (2012). Mitigating contextual uncertainties with valuable changeability analysis in the multi-epoch domain. In *6th IEEE Sys Conf. (SysCon12)*. Vancouver.
27. Meyer, M. H., & Lehnerd, A. P. (1997). *The power of product platforms*. New York: Free Press.

Chapter 39
Framework for Design Exploration of Secure Embedded System Development

Paul A. Wortman and John A. Chandy

39.1 Introduction

As new technologies and techniques continuously become developed, there will be an equal eagerness to implement these new designs and methods. Complications of these developments will arise in the form of combining these various elements into a single functional design. The essence of this stems from the fact that no device or system operates independent from other individuals or systems. As such, when developing a new design solution, we need to recognize how each entity operates as well as what their scope of operation is. The focus of this paper is to propose a design framework for the exploration of early design choices to minimize risk, redesign costs, and ensure that all security requirements are met. For this purpose, we will heavily lean on an examination of the overall framework, the use of the Architecture Analysis and Design Language (AADL) as the descriptive language, as well as evaluation of generated designs. In this manner we are able to minimize the introduction of additional risk, security flaws, and policy violations when combining design elements. Early work on secure system development and modeling integrated security-related information, elements, and requirements into the unified modeling language (UML) and offshoots (e.g., UMLSec, SysML Sec) to include details of integrity and authenticity as well as security-aware architecture to improve secure system design [1–3]. Sadeghi et al. worked to prove narrow forward privacy for RFID systems [4]. In a similar vein, Abidin et al. worked to provide better forward security for improving a biometric authentication protocol [5]. By contrast Dong and Kiefer worked on new security policies for improved server client privacy [6]. However, this prior work does not adequately capture quantifiable metrics to enable a systematic approach to secure system design.

P. A. Wortman (✉) · J. A. Chandy
Department of Electrical and Computer Engineering, University of Connecticut, Storrs, CT, USA
e-mail: paul.wortman@uconn.edu

S. Adams et al. (eds.), *Systems Engineering in Context*,
https://doi.org/10.1007/978-3-030-00114-8_39

In this paper, we present a security design framework that uses risk and impact as quantifiable metrics for a security design that goes beyond just access control. The contributions of our work are as follows:

- A design exploration framework for ensuring security-based functional and architectural needs is met.
- Examples of AADL code for the definition and verification of various security properties to show the ease of writing security evaluations.
- Showing a clear path for development of a mapping system to combine secure component models with embedded systems requirements.

AADL, like most modeling languages, must be able to describe not only the requirements of a system but also the constraints, capabilities, and costs of various implementations and methods for the purpose of modeling a gamut of different designs. Coupling this already large space with security causes the considerations and influences on the problem to grow considerably. Our paper chooses to examine the field of modeling embedded system security through the lens of the Architecture Analysis and Design Language (AADL). One needs to develop a verification and selection process for taking all of the cataloged information and possible solutions to then compare and contrast solutions that meet user-defined security requirements while categorizing solutions that are maintaining within external constraints and the capabilities of the architectural components being used to produce a best fit, measured by this paper's verification and selection process. To begin this approach, we first need to define the constraints of the system, describe critical considerations, and standardize a method by which we can produce a comparable metric for generated embedded system security model designs.

39.2 Proposed Framework

We first introduce our larger security design methodology that can be applied to the larger process of embedded system security design. From this framework we show where the need for the modeling of functional security requirements that can be measured and validated. The method by which we develop our design-oriented examination of risk is reviewed and then applied to our adversarial model. The adversary influenced risk metric is shown to also be compatible with common monetary evaluations of embedded system design. We start our discussion of our proposed risk framework with a consideration of a methodology in which to explore the design space. We envision a design methodology built on platform-based design (PBD) [7] where one can take the functional space (including security requirements) and the architectural space (components and capabilities) in order to develop a mapping function that can produce solutions to a given design problem. These implementation solutions are then formally verified against the originating constraints of the system. Once the implementations are verified, code generation tools are used to produce a more detailed model that can be examined via

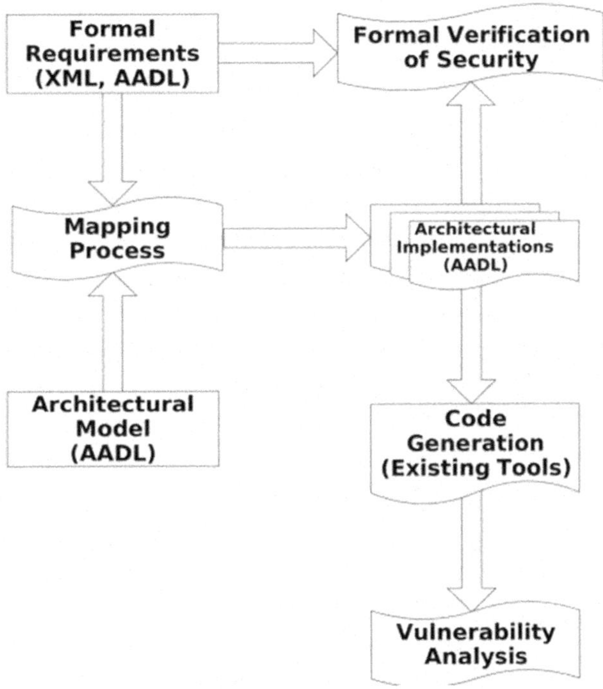

Fig. 39.1 Visualization of our Security Design Framework

vulnerability analysis. A visualization of the design process is shown in Fig. 39.1. In order to enable this framework, the following steps are required:

1. Creation of a low-level component library that would contain several implementations of each base component within the architectural space used for model generation. This library serves as the basis for the modeling capabilities and architectural knowledge of the design team. Each component would have various features and capabilities including security aspects. These details determine the functional capabilities and constraints of each component solution.
2. Formalized description and definition of higher-level security requirements that may come from user-defined needs or from the experience and knowledge of security experts. Currently, high-level needs are manifest in a vague form of "my data must be secure." From a design perspective, this is not very useful. We favor metric-based requirements that capture risk and the value of assets to both the defender and adversary. Formalized definitions can be used to abstract these higher-level quantifiable constraints into architectural implementations.
3. Creation of a mapping process by which security requirements and secure component specifications can be uniformly verified and selected to allow for the generation of potential secure architectural system model solutions to the given inputs. The mapping process is responsible for producing various system design implementations that meet the functional requirements of the system while maintaining realistic architectural constraints. It is in this step that having an

evaluable security metric allows for the generation of different security-based embedded system design models.

4. Verification tools to validate mapping implementation solutions. This step is where each produced design implementation is compared and contrasted to the original design specifications (i.e., functional and architectural constraints) to determine the best solutions of those presented. These instances can then be used as part of a genetic algorithm to thoroughly explore a given design space. Through the use of automated tools, we can move to exhaust this design exploration through the aid of computers.

We have proposed a vision of the overall design process, but in this paper, we particularly focus on steps 1 and 2—modeling components and requirements. Specifically, we focus on how to model security goals and requirements in such a way that they can be measured and validated.

39.3 AADL as Security Modeling Language

The Architecture Analysis and Design Language (AADL) is traditionally used for the specification, analysis, automated integration, and code generation of real-time performance-critical distributed computer systems [8, 9]. The "performance-critical" aspects of a system include timing, safety, schedulability, fault tolerance, and security. AADL works to provide a new language vehicle for allowing analysis of system designs prior to development. The advantage being that the language and associated tools allow for developers to model, analyze, and generate embedded real-time systems while supporting model-based engineering concepts. Due to AADL's wide adoption and use in real-time embedded system design, there is a large scale of available tools already developed by the user community [10]. Although security was not an initial concern in the development of the AADL language, there has been recent work made to extend the language to include behavioral and security-centric capabilities. These developments have occurred in a variety of annexes, most recently the security annex extensions, alterations, and expansion of security analysis tools that have been occurring throughout the past 2 years; starting in 2016 [11–15].

Current language standards used to describe security concepts, requirements, and constraints have not been developed well enough to be "all-encompassing." The motivations of security-based expansion of AADL center around three core purposes [14]. First is to detect security policy violations early. Second is to assure that the system implementation correctly enforces policies and that no security risks are introduced by the runtime architecture. Lastly is to automate the execution of security assurance plans. Recent work by Ellison et al. [11]. has extended the security-centric properties of AADL to include user-defined "Access Group" and "Access Mode" properties that are used for evaluating and validating an existing system model based on user expectations of functionality. The extension adds different security types (e.g., unclassified, confidential, secret, top secret). When implementing these extended security properties, one must determine the viability of a system given confidentiality requirements of data objects and security clearance by users. These security extensions allow for analysis of the modeled system to ensure processes and threads are mapped to appropriate hardware, communicate over secured channels, and store data in protected memory as well as deriving minimum security requirements for a given software architecture. While these two properties are effective at validating that a system will retain some arbitrary level of security, this is still not the level of detail that effectively allows for modeling of component specifications.

There are certain improvements of AADL that we believe are a step in the right direction. Including a numerical value to "security levels" allows for quantitative analysis of this property through a system's design. The concept of exposure (e.g., how exposed a component or system is) as a numerical value allows one to determine the possibility of attack within a design. Both of these additional properties open the door to formalize a verification of potential security hazards and complications that may arise further down the development cycle or even once out in the field. One problem with the approach, however, is the move to describe the security aspect of "trust" as a non-binary value, as one cannot say that they trust a component, or system, 80% of the time, whereas reliability can be easily described as a percentage. While the released work [12] states that this scale is representative of proof of a component's correctness, we believe that this merging of concepts does not accurately reflect security concerns and does not allow for accurate verification and validation of security requirements and behavior. Having the language to describe and define the correctness of a component is advantageous; this is not the same as trust but should rather be seen as a combination of trust and reliability.

39.4 Security Risk Modeling and Evaluation

In this section we discuss how components and security properties are modeled in the AADL language, explore how security requirements are defined, and speak upon the evaluation of mapped designs constructed from the architecture and functional needs. These aspects form the crux of our proposed secure embedded system design framework and allow for its automated use and execution in system design and development.

39.4.1 Component Security Models

As part of the proposed design framework, we need to have defined component security models, so that the design process can make selections that optimize not only functional requirements but also security requirements. In order to formalize

Table 39.1 Illustration of different component variations

Component types table					
	Types				
Elements	I	II	III	IV	V
Memory	Unprotected	Protected	Encrypted	Obfuscated	Combo
Bus	Unprotected	Encrypted	Non-sniffable		
Processor	Simple	Embedded Encryption	HW Encryption		
Data	Plaintext	Encrypted	Protected		
Port	Normal	Encrypted	Protected		

this framework, we need to produce a form of component library that represents the possible combinations, or solutions, that a designer or developer can pull from to generate new implementation models. Listing 39.1 shows an example of how we use AADL to define properties on a processor and an encrypted transmitter. A system design will have a combination of components to meet the desired functional and security requirements. Table 39.1 shows defined potential component types that would populate the architectural space of this new security framework. Each element in the table has distinct variations that have their own unique security properties. For example, the method by which a designer implements memory is reliant on the technology available as well as any cost constraints. A developer could decide to protect memory by creating physical boundaries for access (e.g., use a trusted platform module (TPM)) or perhaps ensure that memory is encrypted to prevent an attacker from being able to leverage access due to an exploit. Furthermore, one can choose to obfuscate memory operations (e.g., ORAM) to thwart side-channel attacks or implement a combination of all these techniques in an effort to improve security of a given system component. Depending on the combination of component types, one will need to define these distinctive characteristics of the system. A system is defined as a grouping of elements that serve a given purpose that are created via the combination of different components to perform a specific task or function. Furthermore, this new system can be abstracted into a "single-layer" device object model. Variations in the models come from mixtures of different design decisions to form larger systems; each expressed with unique properties and functions. A regular or simple system has no encryption, no data access control, and no assurance of "security level maintained." In this implementation one can interpret everything is at the "default" lowest level of "security" that can be assured. An encrypted system has at least security implementation at some level; on the input or output, bus, memory, or processor. Any mixture of the aforementioned "levels" of encryption is possible but must be verifiable, due to the need of verifying security. A protected system contains elements that either interact or are responsible for access control to "secure" data, information, or services.

Listing 39.1 Example of User-Defined Lower Level Components

```
system implementation transmitter.encrypt_i
  -- Subcomponents of the transmitter
  subcomponents
    ant_in : system recv_antenna.normal_i;
    ant_out : system trns_antenna.encrypt_i;
    encrproc : processor procbase.encryptembedded_i;
  -- Connection definitions of the transmitter
  connections
   c0 : port ant_in.wired_out -> encrproc.input_port;
   c1 : port encrproc.output_port -> ant_out.wired_in;
  -- Flow path definition for the transmitter
  flows
  f0 : end to end flow ant_in.f0 -> c0 -> encrproc -> c1 -> ant_out.f0;
  -- Additional properties of the transmitter
  properties
  securityspecs::has_encryption => true;
end transmitter.encrypt_i;

processor implementation procbase.encryptembedded_i
  properties
    securityspecs::has_encryption => true;
    securityspecs::encryptmodule_class => embedded;
    securityspecs::encryption_class => AES;
    securityspecs::encryption_variation => b256;
    securityspecs::has_PUF => false;
    securityspecs::has_TPM => false;
    securityspecs::has_encryptedflash => false;
    securityspecs::isTamperProof => false;
end procbase.encryptembedded_i;
```

39.4.2 Security Requirements Model

Security requirements are traditionally very vague and undefined. For the most part, they define that either "data must be secure" or certain access controls—for example, which users have access to which data. We require a quantifiable metric against which to measure the quality of a design. Ferrante et al. came up with a notion of a security level (SL) which arbitrarily assigned a 0–100 value to a component [16]. A more satisfying metric is the notion of risk—i.e., the combination of probability of attack and the impact/cost of the attack. While traditional views of risk only deal with a single source of probability, the examination of security risk is more involved due to multiple sources of probability. This starts with an understanding of attacks from the adversary's perspective. Each possible attack vector upon a system is dependent on a series of exploits, meaning that as multiple steps are taken to perform a successful data exfiltration attack, one needs to accurately aggregate the individual costs of each event in the process; having its own event probability and cost. Beyond the notion of an attack vector, the traditional view of risk also does not account for an attacker's motivations. Therefore, the risk metric must also capture the level of embedded system security risk influenced by both the potential of attack as well as impacted loss. Security risk (SR) of a given solution is influenced by both the monetary value that an attacker places upon the presented system as well as what is required by the defender to recover from a successful attack. Through the use of the resulting metric, a developer can tailor their design to better protect systems while also presenting a higher cost to an adversary willing to expend resources. The *SR* value allows for the approximation of a monetary value of the security strength of any potential design solution. However, while the calculated metric does begin to influence implementation decisions, it still requires incorporation into the overall cost of manufacture, operation, and maintenance.

Thus, as part of the security requirements, the system specifications must also specify the attack impact and value on a system's service. Listing 39.2 shows an example of how AADL could be used to specify security requirements on a wireless sensor service. Security prerequisites are specified in terms of value of a service feature to an attacker and impact to the system user if the service was to be compromised. For example, if the wireless sensor leaks data, the system user will lose $1000. At the same time, the leaked data is worth only $800 to an attacker. In a more detailed requirements listing, these values may be more granular and also specify different values to different attackers. Unfortunately obtaining these attack values can be a tedious process as it is difficult to obtain data relating to value and impact of attacks on critical systems from larger corporations; understandably due to the sensitive nature of the information. To properly arrive at meaningful values for these security properties, one must recognize and pinpoint the impact and costs of attacks upon a given system. These costs include damage to one's reputation, time required to repair or even fathom the extent of the impact, loss of proprietary knowledge, or even additional expenditure of resources (e.g., power consumption). Furthermore, depending on the system being examined or the operational parameters of its purpose, these considerations can greatly vary. For instance, the concerns placed upon the design of a secure embedded system are going to be remarkably different than those for a secure network architectural configuration. The amount of exposure of each component, ease of access for the system at large, power consumption of individual elements, even the methods and costs that an attacker would have to pay to successfully exploit vulnerabilities will differ. Without exception the economics vary based on the situation and require detailed knowledge to accurately represent costs and impacts.

One should note that as with any of these properties, the AADL code block shown in Listing 39.2 shows a simplified single-pay cost but it could easily incorporate aggregate costs due to multiple attackers or defenders of a given system or even temporal centric value based on run time and resource usage. Bereft of these considerations, a designer or developer cannot accurately ensure that the security requirements of a produced design have been met. Depending on the value of data elements to different attackers (e.g., hacker, information broker), an aggregate cost of these impacts must be considered. The same can be said of the defensive security considerations. For example, a company may be defending the personally identifiable information (PII) of several customers. In the scenario where that information is stolen, there are two considerations for impact. The first is the cost, in terms of reputation and recovering for the company hosting this service, while the second is the individual costs to each customer. In the case of the latter, these values can greatly range depending on the sensitivity of the information as well as the degree of damage that a malicious actor can cause (e.g., loss of reputation, increase in debt).

Listing 39.2 User-Defined Higher Level Security Requirement

```
abstract implementation sysreq.wireless_sensor_i
  subcomponents
    serv_ADConv: abstract sysserv.ADConv_i {
         servatrb::dynamicRange => 0..5 V;
         secatrb::integrity::atkImpact => 300;
         };
    serv_wrlsTrans: abstract sysserv.wrlsTrans_i {
         servatrb::distance => 100 m;
         secatrb::authentication::atkValue => 600;
         secatrb::authentication::atkImpact => 400;
         secatrb::authorization::atkImpact => 1200;
         secatrb::dataleakage::atkImpact: => 1000;
         secatrb::dataleakage::atkValue: => 800;
         };
    fnc_data: abstract security_props.data_i {
         dataatrb::data_class => Sensor;
         secatrb::atkImpact => 800;
  properties
    secatrb::hasProtection => false;
    secatrb::AuthGroup => Employees;
end sysreq.wireless_sensor_i;
```

39.4.3 Design Mapping and Evaluation Model

The mapping process is a difficult and intricate element of the model that requires automation. In its current state the mapping step requires that a system expert produces a design, or series of designs, for a given secure embedded system. From the produced models, the next step is to derive the attack tree and impacts that correspond to the vulnerabilities and security properties of each scheme. The final step is to then compare and contrast the gathered information to determine if the original requirements are still met. In this section we examine the mapping step of our proposed framework as well as the additional steps of evaluating the produced designs for refinement of the mapping function.

The security requirements of a system are used by a designer to select components from the component library described in Sect. 4.1 to optimize the functional and security requirements. It is possible to develop a mapping process that will not only account for the needs of the modeled system but also take into account the constraints of embedded systems. These constraints originate both from the user-defined requirements and needs of the system but also from the architectural limitations of the components available to form the base of the existing design space. To accurately represent these costs, one has to write in the corresponding properties into the AADL language; which will then be used in the mapping of these various architectural schemes to produce a series of final designs. The other half of this mapping process is to incorporate any number of functional requirements imposed upon the system. As stated before in Sect. 2, these can be vague and poorly defined. In this scenario the developer will need to make decisions on how to best implement these functional needs in more concrete methods. Using the platform-based design method, we abstract these functional aspects into more detailed implementations. Once an appropriate level of tangible functional implementation has been reached, this is combined with the architectural models as input to the mapping process. The mapping element then combines these two aspects using special consideration toward the original constraints of the system in order to produce a series of final design mappings. These are then evaluated in terms of the original functional and architectural needs as well as the potential security risk cost and impacts. From this evaluation we then further refine the constrains of the mapping function to produce a more polished series of potential design solutions. However, in its current form, we still require the use of a security expert and their knowledge.

With respect to security, as the designer evaluates a potential system design (combination of components), a possible attack tree will be constructed against all potential valuable assets and services in that system. For each asset and service under consideration, we evaluate the potential attack vectors in the tree with respect to their probabilities and costs. A determination can be made if these design/implementation costs are greater than that the attack impact or vice versa the attack costs are less than the value to the adversary. In other words, we construct the attack value and impacts as quantifiable constraints in our design selection strategy. The security constraints are structured from a requirements specification (as in Listing 39.2) that assigns potential impact or value costs for every asset. Thus, the mapping process for our Listing 39.2 example would have constraint as follows:

$$C_d(data) < \$1000$$

$$C_a(data) > \$800$$

where $C_d(data)$ is the cost of the defensive measures needed to protect against data leakage and $C_a(data)$ is the cost (to the attacker) to exploit vulnerabilities in these defensive measures. The first constraint says that the cost of the defensive measures should be less than the value of the asset (data). In other words, it makes no sense to spend more on defensive measures than the value of the asset. Likewise, the second constraint says that while there may still be vulnerabilities in the designed system, the cost for an attacker to exploit these vulnerabilities will be higher than the value of the attack—thus making the attack needless. For example, the defensive measures may have locked down on all "standard" exploits thus leaving only vulnerabilities in hardware that can be exploited with expensive IC decapping and probing.

Determining these C_a and C_d costs is dependent on the design being evaluated. Using actual costs is much more concrete than a 0–100 scale, because it allows for each potential design to be evaluated individually rather than all together in a single scale. The defensive costs are the aggregation of amortized system design costs, implementation costs, ongoing costs, etc. allocated to defense of a particular asset. The attack costs are determined by evaluating the attack tree and using probabilistic weights and estimates of exploits of each node in the attack tree. With these probabilities and costs (both attack and defensive), one can then arrive at risk values for a system design under consideration. The various costs (e.g., risk, operational, maintenance) produce a monetary metric that effectively compares each design in terms of security requirements met. These metrics are then used to further refine the constraints of the mapping process based on the favorability of each design to the original functional and architectural requirements. Additional constraints take the form of architectural limitations due to the types of components available to a given designer, functional restrictions due to the type of system or operational scenario of a proposed scheme, or to the cost of materials due to the budget allotted to the design project. Furthermore, these constraints can be expanded to include the type of users that are interacting with the system, assumptions made based on the accessibility of each developed device (e.g., sensors in a network), or even the roles of multiple attackers or defenders of a system.

All of these security considerations carry a weight and cost that must be fed back into the mapping process to better refine the constraints and produce an improved design model. While currently this mapping process requires human input to produce each mapping of functional and architectural needs, our evaluation scheme allows for a method of ensuring that requirements are met as well as determining the risk and impact of each design. The next steps in further developing this aspect of our framework depends upon automating the process. Through automation we will be able to more thoroughly explore the design space of any given secure embedded system development project.

39.5 Verification and Validation Approaches

While the focus of our paper is not on verification of a design, it is worthwhile to discuss some approaches to verify that a design has met the security requirements outlined in Sect. 4. The core of the existing Automated Assurance of Security Policy Enforcement (AASPE) work centers around the use of security levels/types, checks that are written using Resolute, as well as simple use of different security properties (e.g., exposure) for a given system. For the purpose of our proposed framework, we steer the conversation toward the use of Resolute as an evaluation and validation tool. When examining the security-based extensions to AADL, Resolute would make an effective method of verification and validation that a produced security model meets all of the required user-defined security assurances. These could range from networking connectivity requirements to dependencies on correctly implemented access control parameters. The disadvantage of this tool, in its current state, is that the assurance checks are relatively simple, and there has been little to no documentation on the user community's efforts to improve the tool for security-centric applications of modeling. An example of one such security check is shown by Listing 39.3 examining that all connections between given components maintain an expectation of encryption.

Listing 39.3 User-Defined Resolute Checks

```
package arch_resolute_checks
public

annex resolute {**
check_encryption_flow(comp : component) <=
    ** " Check the encryption property of connections of features on
component " comp **
    forall (conn : connections(comp)) . has_property(source(conn),
securityspecs::has_encryption) and
   has_property(destination(conn), securityspecs::has_encryption) =>
  (property(source(conn), securityspecs::has_encryption)) =
(property(destination(conn), securityspecs::has_encryption))
**};

end arch_resolute_checks;
```

As one can see from the above example, the language of the security check is relatively simple. While Listing 39.3 does improve the detection of security architecture violations, it is limited in terms of the type of verification it can perform. This stems from the method by which properties can be compared and contrasted. Resolute is an excellent tool for ensuring security properties match and that architectural requirements are met. However, there are aspects of a secure embedded system (e.g., security policy and behavior) that cannot be modeled using Resolute alone. In these cases, other tools such as Architecture-led Incremental System Assurance (ALISA) [17], ASSPE [15], and Behavior Language for Embedded Systems with Software (BLESS) must be implemented to ensure detection in security functional and behavioral violations as early as possible. Further work is needed to develop Resolute, or other annex tools, to account for security assurances through verification and validation solutions. Furthermore, standardization of the dialect must take place to allow for ease of communication of ideas, specifications, requirements, and expected behavior. Otherwise the advantages of this new security framework would not outweigh the cost and time spend learning the syntax and tools required. Through BLESS we can improve each model to evaluate not only the functional and architectural properties but also ensure secure behavior of the system. An example of such a linguistic implementation of BLESS properties is shown in

Listing 39.4. The main point to illustrate is that depending on the data model used by a given device design, the device should act differently. Limitations of this BLESS expansion are that a knowledge base of formal verification is required to write the necessary proofs.

Overall the incorporation of these tools does allow meaningful expansion of security modeling. However, there are concerns that stem from limitations of the tools, complications that arise in learning to use it, or even formalization of practices and implementation. As with any model, if it is made too complex, then a great deal of effort is placed in a design that cannot be very flexible. On the other hand, without enough detail the model is far too generic and cannot properly represent the developed system. In this scenario the focus of the model could be around policy specification and enforcement. In summary, we require the development of a formal method to design and develop different levels of security properties and functionality for improving system design and understanding.

Listing 39.4 AADL BLESS Definition Example

```
data auth_method
    properties
       Data_Model::Data_Representation => Enum;
       Data_Model::Enumerators => ("None","shared","user","key");
       BLESS::Types => "enumeration (None, shared, user, key);
end auth_method
...
device authenticator
    features
       auth : in data port auth_method;
       auth_status : out data port auth_rating;
    ...
end authenticator
...
device implementation authenticator
    ...
    annex BLESS {***
    ...
       if (auth=None)~> auth_status!(Fail) []
       if (auth=shared)~> auth_status!(Weak) []
       if (auth=user)~> auth_status!(Average) []
       if (auth=key)~> auth_status!(Strong) []
    ...
    ***};
end authenticator.impl;
```

39.6 Conclusions and Future Work

In this paper we explore our proposed framework for effectively modeling secure systems using AADL as the defining language. Through our work, one can easily extend any model to not only work at the scope of a network or individual proprietary system but also expand it to contain desired security properties. Our proposed framework is very customizable to any situation that requires modeling. Our future work includes development of behavioral checks for security behavior, maturation of the mapping process for generating design solutions, and formalization of security verification and evaluation functions.

References

1. Jürjens, J. (2005). *Secure systems development with UML*. Berlin: Springer Science & Business Media.
2. SysML Sec. http://sysml-sec.telecom-paristech.fr
3. Jürjens, J. (2002). *UMLsec: Extending UML for secure systems development, UML 2002 the unified modeling language* (pp. 412–425). Berlin: Springer Publishing.
4. Sadeghi, A., Visconti, I., & Wachsmann, C. (2009). Anonymizer-enabled security and privacy for RFID. In *International Conference on Cryptology and Network Security* (pp. 134–153). Berlin: Springer.
5. Abidin, A., Matsuura, K., & Mitrokotsa, A. (2014). Security of a privacy-preserving biometric authentication protocol revisited. In *International Conference on Cryptology and Network Security* (pp. 290–304). Cham: Springer.
6. Dong, C., & Kiefer, F. (2015). Secure set-based policy checking and its application to password registration. In *International Conference on Cryptology and Network Security* (pp. 59–74). Cham: Springer.
7. Sangiovanni-Vincentelli, A. (2007). Quo Vadis, SLD? Reasoning about the trends and challenges of system level design. *Proceedings of the IEEE, 95*(3), 467–506.
8. AADL. http://www.aadl.info/aadl/currentsite
9. Feiler, P. (2010). *SAE AADL V2: An overview*. Pittsburgh: Carnegie Mellon University.
10. AADL Tools. https://wiki.sei.cmu.edu/aadl/index.php/AADL\tools
11. Ellison, R., Householder, A., Hudak, J., Kazman, R., & Woody, C. (2015). *Extending AADL for security design assurance of cyber-physical systems*. Pittsburgh: Software Engineering Institute, Carnegie Mellon University CMU/SEI-2015-TR-014.
12. Delange, J., Feiler, P., Klieber, W., Nam, M., & Seibel, J. AADL security annex. https://github.com/saeaadl/userdays/blob/master/UserDays/May2016/security-annex-May2016.pdf
13. Delange, J., Nam, M., & Seibel, J. AADL security analysis tools. https://github.com/saeaadl/userdays/blob/master/UserDays/May2016/security-analysis-May2016.pdf
14. Feiler, P. Automated assurance of security-policy enforcement in critical systems. https://insights.sei.cmu.edu/seiblog/2018/02/automated-assurance-of-security-policy-enforcement-in-critical-systems.html
15. AASPE AADL Github. https://github.com/cmu-sei/AASPE
16. Ferrante, A., Milosevic, J., & Janjusévić, M. (2013). A security-enhanced design methodology for embedded systems. In *International Conference on Security and Cryptography (SECRYPT)*.
17. Alisa User Guide. https://rawgit.com/osate/alisa/develop/org.osate.alisa.help/contents/00-Main.html

Chapter 40
Engineering Elegant Systems: Postulates, Principles, and Hypotheses of Systems Engineering

Michael D. Watson, Bryan Mesmer, and Phillip Farrington

40.1 Introduction

There have been several approaches to the definition of systems engineering principles taken in the literature. INCOSE provided one of the earliest attempts to characterize a set of pragmatic systems engineering principles [1]. These principles are a good set of rules of thumb in the practice of systems engineering and were embodied in the systems engineering processes [2] and INCOSE handbook [3]. More recently, there have been several parallel threads related to systems engineering principles. Work on complex system governance has led to a set of metasystem functions [4, 5]. The metasystem functions have been applied to system acquisition and a system of systems context and seek to form an overarching framework for the governance of complex systems. An initial basis for systems engineering principles is presented as a set of elaborated points related to these metasystem functions. Another thread has developed a rich scientific basis for a set of seven axioms of systems theory [6, 7]. These axioms address the operation, or functioning, of a system and seek to advance systems theory in general. Another thread has taken a pathological view of systems theory [8]. This view seeks to define systems theory across many different approaches taken in the past. Systems science has also contributed a thread in this fabric. Systems science principles have been defined with work on the systemology and typology of system principles [9, 10]. System architecting has also contributed a thread bringing in the concepts behind

M. D. Watson (✉)
NASA Marshall Space Flight Center, Huntsville, AL, USA
e-mail: Michael.d.watson@nasa.gov

B. Mesmer
University of Alabama in Huntsville, Huntsville, AL, USA

P. Farrington
Trivector Services Inc., Huntsville, AL, USA

S. Adams et al. (eds.), *Systems Engineering in Context*,
https://doi.org/10.1007/978-3-030-00114-8_40

systems thinking [11]. This paper presents the work of another thread on systems engineering principles. The NASA Systems Engineering Research Consortium has developed a set of system engineering postulates, principles, and hypothesis providing a basis for systems engineering as an engineering discipline and the application of systems engineering approaches and processes across many different system types. This work contributes a new dimension to the other threads and provides a framework that may potentially integrate these threads. Each of these threads has a rich literature basis contained in each of the references cited in this section.

40.2 Systems Engineering Framework

The NASA Systems Engineering Research Consortium has been studying the basic foundations of systems engineering since the fall of 2010. The consortium consists of various universities of government organizations that look at the basis of system engineering and approaches to conducting systems engineering in general [12]. This research has led to the identification of several fundamental basis of system engineering [13]. This basis has been captured as a set of systems engineering postulates, principles, and hypotheses.

Systems engineering as a discipline is comprised of two main thrusts: system integration and discipline integration. In this framework, these two thrusts encompass four components: mission context, system integrating physics, organizational structure and information flow, and policy and law (Fig. 40.1) [14].

Fig. 40.1 Systems engineering framework relationships

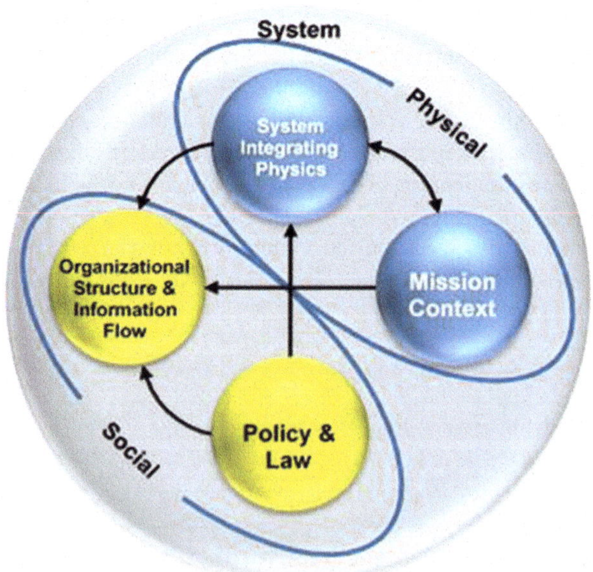

System integration consists of the physical and logical aspects of the system. System integrating physics includes the system integrating logic (for logical systems) as the control of many systems is based on logic (i.e., software). The software must have input on the system state to affect the intended system control and is coupled with the physical system. Environmental interactions such as thermal or radiation where hardware bit errors create logical anomalies in the operation of the system affect software. Also, included as part of system integrating physics are the human system integration aspects where the physical and logical functional design must consider human physiology and psychology. This provides a coupling of the user, operator, maintainer, and manufacturer to the system structure, and forms a bridge with the social systems that build, operate, and use the system. Mission context affects both the physical/logical system aspects as well as the social aspects. Mission context is part of system integration and mainly focuses on the definition of these aspects of the system. The social aspects of mission context are important, and the physical/logical choices made for the system can emphasize or amplify these. For example, when a planetary satellite is intended for Neptune, the social perturbations are small. When the physics determines that a nuclear-powered satellite is necessary for this distance from the sun, much greater social concern is generated due to potential interaction of the nuclear device with the Earth's environment in the unlikely occurrence of an accident during launch. In this example mission context influence of the physical system on the social response can be seen.

The social aspects are a major thrust defined by the organizational structure and information flow and in the application of policy and law. Organizational structure and information flow deal with the maintenance and flow of system information within the organization. This brings in the aspects of sociology in the functioning of the organization. Information flow is a key element in designing and operating an elegant system. Systems engineering assures that the organizational structure supports the necessary flow of information among the system disciplines and assures the design captures this information flow. Gaps, barriers, and organizational reservoirs of information in the flow of information through the organization particularly concern systems engineers. The system design and operations represent the knowledge of the system residing in the organizational structure.

Policy and law are generally social influences on the system. Policy and law certainly influence the physical/logical aspects of the system (e.g., requiring a crash-proof casing for the nuclear power cell for launch for the Neptune mission) but are included with the social aspects of the system due to their social considerations.

40.3 Systems Engineering Postulates, Principles, and Hypotheses

The Systems Engineering Consortium has identified a set of postulates, principles, and hypotheses to articulate the basic concepts that guide systems engineering. These postulates and hypotheses emerged looking at the work of Ludwig Boltzmann and his postulates on gas distributions as an early example of how to characterize the interactions of complex systems. This led us to articulate a set of underlying postulates and hypotheses underlying systems engineering, leading to the seven postulates and four hypotheses stated in this section. These postulates define the domain of systems engineering as well as the system aspects and influences that are of concern to the systems engineer. The hypothesis contains the seeds of a holistic mathematical basis for systems engineering. In addition, the system postulates define a set of systems engineering principles. The principles serve as an extension of the postulates and are listed after them.

40.4 Systems Engineering Postulates

A postulate is something assumed without proof to be true, real, or necessary [15]. The postulates of systems engineering identify the basis for the discipline. These are further expanded by a set of principles in Sect. 3 below.

Postulate 1 Systems engineering is system- and environment-specific and context dependent.

Description: This is the first and foundational statement on systems engineering. The product (i.e., the system) and its operational environment drive systems engineering and the system integrating physics, logic, social, and cognitive relationships (i.e., context) that are foundational to the specific product or system. Essential to this is the understanding of the mission or use of the product as formulated by the product goals. This includes the aspects of the system needed to operate in an elegant manner and thus considers the entire product life cycle.

Evidence: The ubiquitous tailoring of systems engineering approaches provides strong support for this postulate. Systems engineering must be consistent with the system being developed or operated. Our research surveying the "NASA 17 Systems Engineering Processes" provides support for this postulate indicating 72% of companies interviewed have systems engineering processes unique to their product. More than 7% of the respondents [16] do not follow a standard process.

Implications: This postulate states that any application of systems engineering should be organized based on consideration of the system being developed or operated. The systems engineering methods applied to a product will and should vary in emphasis and application based on the nature of that product, its environment, and its context.

Postulate 2 The systems engineering domain consists of subsystems, their interactions among themselves, and their interactions with the system environment.

Description: From a physical, logical, and structural sense, a system is not a single mechanical or electrical or chemical entity; it encompasses a set of interacting subsystems. Systems engineering is concerned with combining multiple subsystems, of various physical and logical types, into a best-balanced functional whole to accomplish the mission goals. This postulate addresses the system integration aspects of systems engineering. Postulate 3 addresses the discipline integration aspects below.

Evidence: The individual engineering disciplines deal with the development of their specific functions extremely well. When these functions are integrated with each other and with the environment, the interrelationships drive the final system performance including emergent properties not evident from the individual system functions. Thus, the engineering of the individual functions is well addressed, while the integration of the engineering functions is what makes these functions a system. The domain of systems engineering is the set of these integrated relationships.

Implications: The systems engineer focuses on the interaction of these subsystems, not as a design engineer focused on the details but as a well-versed integrator. These system interactions, including interactions with the system environment, can drive the design as strongly as the subsystem functions themselves and, when coupled, can potentially create unexpected system responses. The systems engineer must predict and manage these responses.

Postulate 3 The function of systems engineering is to integrate engineering disciplines in an elegant manner.

Description: The systems engineering discipline is its own engineering discipline, but it is not independent from other engineering and social disciplines. Systems engineering seeks to integrate and incorporate the other engineering and social disciplines in an elegant manner to produce an elegant system throughout the system life cycle. This postulate addresses the discipline integration aspects of systems engineering. Postulate 2 above addresses the system integration aspects.

Evidence: Any complex system is developed by multiple engineering disciplines with many social aspects influencing the integration. These engineering disciplines with social influences work in an integrated fashion, formerly and informally, to produce these systems.

Implications: The interaction of the disciplines is the focus of the systems engineering domain. The objective is a basic understanding of each discipline with a detailed understanding of their interactions. This incorporates various organizational integration aspects. The systems engineer must be cognizant of the organizational and sociological influences on the system development and operations. The systems engineer must also "engineer" these relationships.

Postulate 4 Systems engineering influences and is influenced by organizational structure and culture.

Description: The technical aspects of the system are not the only focus of systems engineering. The system under development drives the development process which has a corresponding influence on the structure of the system's developmental and operational organizations. Similarly, the structure of the organization has an influence on the engineering of the system. These factors also impact the culture of the organization.

Evidence: Organizational mirroring provides examples where the organization maps to system functions. Our current research in "Biased Information Sharing" also shows that system margin is maintained by the organization and not always clearly identifiable in the system design.

Implications: The systems engineer must be cognizant of the culture, the organizational interactions, and their potential impact on the design of the system. The systems engineer must understand how information flows through the organization, is filtered and interpreted by the organization, and is captured by the system design or operational procedures. The systems engineer should work with project management and line management to address issues in organizational information flow and culture to improve the elegance of the system.

Postulate 5 Systems engineering influences and is influenced by budget, schedule, policy, and law.

Description: Every project has overarching constraints that extend beyond the physical and environmental. Specifically, most (if not all) projects have a limited budget and schedule. In addition, all systems must conform to established organizational and government policy and laws. These policies and laws put additional constraints on budgets, schedules, and technical solutions. These factors provide a context in which the system is developed and operated. In addition, the system design choices also influence these factors. Government policy and law is based on the understanding of legislators on what systems can actually achieve their intents. Similarly, corporate/company policy is influenced by the types of systems the corporation or company chooses to develop.

Evidence: Every project has these constraints. Infinite budgets or schedule do not exist. Policy and law issues and constraints pervade our systems. Government policy and law are based on the legislators understanding of solutions needed to accomplish their intents. Similarly, corporate/company budgets and schedules are based on the executives' understanding of the budget and time frame necessary to develop a system. This understanding can be seen in budget and schedule allocations, which encompass both a total funding and a time frame understanding, that are provided by the government or corporate/company executives.

Implications: Social choices drive the establishment of these constraints. People make choices to define budget limits, schedule limits, policies, and laws, whether at the national or organizational level. Thus, physical and logical solutions through these constraints link social choice theory. These choices are based on an understanding of system's abilities to achieve the government and corporate/company executive's intents. This understanding drives the budget and schedule allocations and the policies put in place. Similarly, the available budget, available expected duration, and existing policy and law can influence choices in the development of a system.

Postulate 6 Systems engineering spans the entire system life cycle.

Description: Systems engineering is not just a development phase activity but continues throughout system operation, decommissioning, and disposal. The organizational relationships and goals change as the system progresses through these phases, but systems engineering continues to integrate the system functions and the system disciplines throughout all phases of the system life cycle. Operations engineering is responsible for the operation of the system. Systems engineering is responsible for the various changes/upgrades to the system capabilities.

Evidence: Systems engineering during the development phases is well understood. During the operational phases, systems engineering is still essential as the system goes through maintenance upgrades, new application adaptations, obsolescence-driven redesigns, etc. In addition, during decommissioning and disposal, systems engineering is essential to deal with the proper decoupling of the system and ensuring conformance with policy and laws affecting the system disposal.

Implications: As the system progresses through its life cycle, the need for systems engineering changes. A shift takes place from development to operations in terms of the scope of changes and organizational responsibility. Operations engineering is responsible for operating the system, while systems engineering is responsible for the system changes/upgrades. The baseline operational system, then, becomes the medium in which operational phase system changes take place. The organization changes significantly as the system transitions from development to operations. Organizational relationships and needs are different. Culture can be very different. All of this affects the system and must be dealt with in systems engineering. Another organizational change and culture shift occurs during decommissioning and disposal.

Postulate 7 Understanding of the system evolves as the system development or operation progresses.

Postulate 7 Corollary Understanding of the system degrades during operations if system understanding is not maintained.

Description: A deeper understanding of the system as a whole is gained as the system progresses through development and operations. As the system progresses through development, more detailed decisions are needed and as understanding deepens these detailed decisions can be made. Understanding of the system could also regress, if organizational changes occur due to inactivity of an organizational element (loss of experience), retirement of key experienced individuals, or closure of suppliers.

Evidence: This deepening of understanding is seen in any system development. The technical assessment process shows this as systems progress from concept review to requirements, review to design, and review to acceptance review. Lessons learned from the operations phase are abundant for any system. This deepening of understanding of the system and its application drives commercial product upgrades or new models. Regression of system understanding can be seen in some life cycle extension activities. When system understanding is not maintained, the basis of systems specification becomes unclear, and some systems have been found not to perform (either underperform or over perform) to their system specifications. In addition, operational procedures can lose their basis and be difficult to determine when they should be retired or maintained as the system ages.

Implications: Requirements are derived as the system design progresses. Thus, while mission requirements (i.e., part of understanding the mission context) are defined at the beginning of development, the system requirements cannot be established up front. They are a function of the design choices made and are understood progressively throughout the development phase. This also applies to cost and schedules, particularly for new systems where the development or operations result in unexpected changes. Similarly, systems engineers develop models to predict system capabilities, and then refine these models as testing and operational experience is achieved. System models gain fidelity as the design progresses, and the interaction between subsystem design maturity and system model maturity must be managed by the systems engineer. These system models become the basis of system operations, as discussed in "Engineering Elegant Systems: Theory of Systems Engineering", Draft 4, Section 4.9.2 [1]. If the system basis is not maintained, then the understanding of why certain procedures or specifications where defined can be lost. This becomes problematic for aging systems, particularly as they reach the generational gap for the workforce after 20 years of service.

40.5 Principles of Systems Engineering

Systems engineering postulates form the basis of the principles of systems engineering. Principles are accepted truths which apply throughout the discipline. These truths serve as a guide to the application of systems engineering.

Principle 1 Systems engineering integrates the system and the disciplines considering the budget and schedule constraints

This is the application of Postulate 5. Budget and schedule constrains the integration of the system and the integration of the disciplines developing or operating the system. Note that budget is the amount allocated to execute the system development or operation and is not the actual cost. The focus of systems engineering is to keep the cost within the budget or recommend when the solution space defined by budget and schedule does not meet the intended system application.

Principle 2 Complex systems build complex systems

This principle is fundamental to the execution of systems engineering. The systems engineer must deal with both the complex system (the organization) that develops the system and the complex system itself. This dual focus forms the basis of the systems engineering framework [i.e., [1] mission context and systems integrating physics and [2] organization structure and information flow]. Postulates 4 and 5 also capture this duality when the systems engineer is responsible for both integration of the systems discipline functions defined in Postulate 2 and the development organization disciplines defined in Postulate 3.

Principle 3 The focus of systems engineering during the development phase is a progressively deeper understanding of the interactions, sensitivities, and behaviors of the system

This principle is the application of Postulate 7. What you do up front does not confine systems engineering, and it does not fade as one progresses through the system development. Instead, the knowledge captured, maintained, and improved by systems engineering deepens as the discipline organizations complete their development work and the system functions are integrated. This deepening of understanding enables the systems engineering decisions necessary to produce an elegant system. The focus of systems engineering is on understanding the interactions of the system, many of which are not apparent until system integration (e.g., physical integration, logical integration), as current systems engineering tools do not allow sufficiently deep understanding of system interactions (which we are addressing with new tools discussed in "Engineering Elegant Systems: Theory of Systems Engineering", Draft 4, Sect. 5 [1]). This leads to a continuous reduction in system uncertainties and identification of system sensitivities. The systems engineer should understand the behavior of the system, including the emergent behaviors, prior to the operational phase. As the development progresses, the systems engineer seeks the best balance of performance, cost, schedule, and risk.

There are several sub-principles to this progressively deeper understanding of the system interactions, sensitivities, and behaviors.

Sub-Principle 3(a) Requirements and models reflect the understanding of the system.

The accuracy and completeness of system requirements and system models reflect the understanding of the system. A system that is not well-understood leads to poorly stated requirements, requirement gaps, and inaccurate system models and representations. The objective of system engineering is to understand the system (Principle 4(a)) which then produces the proper specification of requirements and proper representation of the system in the system models.

Sub-Principle 3(b) Requirements are specific, agreed-to preferences by the developing organization.

Preferences are an individual attribute. The organization as a whole, however, must at some point consolidate these individual preferences and agree on specific values (i.e., performance, cost, schedule) that the system will achieve. These agreed-to preferences along with some agreement on the uncertainty in their measure are the system requirements. These are specific to the system being developed, and the requirements (agreements) that are necessary for the successful completion of the system should be carefully defined as part of systems engineering. Integration of the disciplines is dependent on these requirements (agreements) between the different disciplines developing or operating the system. Configuration management is an important systems engineering function in maintaining these requirements (agreements) and managing their change in a consistent and coherent manner.

Sub-Principle 3(c) Requirements and design are progressively defined as the development progresses.

Mission requirements are defined early in the understanding of the system as a part of mission context. The remaining technical requirements are derived based on system design decisions that progress throughout the development phase. Subsystem requirements are not defined completely until PDR and component requirements may not be fully defined until CDR.

Sub-Principle 3(d) Hierarchical structures are not sufficient to fully model system interactions and couplings.

System interactions and couplings are varied, involving serial, parallel, nested, and looping relationships. Often there are multiple peer relationships that provide connections among system functions and the environment. Looping, nested, and peer relationships support interactions and couplings not seen in hierarchical structures which generally only indicate parent/child relationships. In addition, hierarchical structures do not distinguish subtle interaction effects from strong interaction effects.

Sub-Principle 3(e) A product breakdown structure (PBS) provides a structure to integrate cost and schedule with system functions.

The PBS ties cost and schedule to the system functions and components. Cost and schedule are defining constraints (Postulate 5) on the system and must be clearly tied to the system functions and operations. The project manager is concerned with labor allocations through the work breakdown structure (WBS). The systems engineer is concerned with the system unit cost and driving cost components seen through the PBS.

Sub-Principle 3(f) As the system progresses through development, a deeper understanding of the organizational relationships needed to develop the system are gained.

As the organization works through the development activities, new relationships may be defined, and the magnitude of these relationships may change as the design matures. Organizational groups that do not share information in early development may be critical in sharing information late in the development. Similarly, organizational groups that may be critical at the concept development phase may complete the transfer of information, becoming less critical to information flow as the development matures.

Principle 4 Systems engineering spans the entire system life cycle.

This is the application of Postulate 6 through a set of sub-principles that are important throughout the system life cycle. Some of the roles of systems engineers are highlighted in the following sub-principles.

Sub-Principle 4(a) Systems engineering obtains an understanding of the system.

Understanding the system is essential to the successful development of any system. The level of understanding of the system possessed by the systems engineer underpins everything they do in terms of engineering the system.

Sub-Principle 4(b) Systems engineering models the system.

Systems engineering develops and maintains system-level models to aid in the design and analysis of the system. "Engineering Elegant Systems: Theory of Systems Engineering", Draft 4, Sect. 4 [1] describes the specific system-level modeling approaches.

Sub-Principle 4(c) Systems engineering designs and analyzes the system.

Systems engineering performs design and analysis at the system level. Ideally, this is not merely a cognitive integration of the results of various discipline models but rather uses system-level models to perform design at the system level. This then informs the system-level guidance to the discipline design to ensure the design closes at the system level as design analysis cycles are conducted. System analysis of the integrated results from the discipline analysis is then performed in a coherent level based on the system-level physics/logic.

Sub-Principle 4(d) Systems engineering tests the system.

System engineering is a critical aspect of system testing. The system engineer should define test objectives at the system level to ensure testing not only accomplishes specific discipline test objectives but also at the system level. This can involve separate system tests, modification of discipline tests for system-level objectives, or system-level analysis of test data to obtain a system-level understanding.

Sub-Principle 4(e) Systems engineering has an essential role in the assembly and manufacturing of the system.

The manufacturing of the system is an integrated activity between the system components and the tooling. In addition, changes during manufacturing often have system-level implications and can unexpectedly change system interactions. While this sub-phase is the purview of the manufacturing engineer, the systems engineer must stay involved to understand changes, update models, and perform analysis to ensure manufacturing changes are understood at the system level.

Sub-Principle 4(f) Systems engineering has an essential role during operations and decommissioning.

Systems engineering has a key role in system operations which are defined by system interactions. We obtain further understanding of the system interactions as the system operational experiences mature. These lead to updates of system models used for operations, and potential system maintenance upgrades or fixes. Similarly, systems engineering provides the understanding during decommissioning in how to de-integrate the system.

Principle 5 Systems engineering is based on a middle-range set of theories.

Systems engineering is comprised as a set of middle-range theories as discussed in "Engineering Elegant Systems: Theory of Systems Engineering", Draft 4, Section 1.2 [1]. Just as there is not a unified theory of physics, nor a unified theory of logic, nor a unified theory of sociology, then there is not yet a unified theory of systems engineering. Three possible theoretical bases are represented in the sub-principles below. These categories are broad systems engineering theoretical basis, system-specific physics/logic systems engineering theoretical basis, and sociological systems engineering theoretical basis.

Sub-Principle 5(a) Systems engineering has a physical/logical basis specific to the system.

Systems engineering incorporates the fundamental physical and logical mathematical concepts specific to the system. Thus, the mathematical basis of systems engineering incorporates the mathematical basis of the system physics/logic. The systems engineer must fully understand that this is different for different types of systems (Postulate 1).

Sub-Principle 5(b) Systems engineering has a mathematical basis.

There are several theories that are important to systems engineering, which enable a mathematical basis for the discipline. Systems engineers, in engineering the system, manage information about the system and its interactions as defined in Postulate 2, using this information to make development and operational decisions. The laws and relationships defined in Information Theory govern the information on the system. This also applies to the management of system information through the organization as contained in Postulate 3. Systems engineers use this information to control the system design or system operations which bring in control theory in a broad scope of controlling the information flow about the system and in defining the control methods to be used to control system states within relevant acceptable ranges over time. Statistical engineering is also a significant mathematical tool which allows for systems understanding and accounts for uncertainties and sensitivities as indicated by Postulate 2. Below are seven broad theoretical bases for systems engineering:

Systems Theory Basis: Postulate 2 derives this basis. Systems engineering uses key concepts such as the division between system and the environment and the recursive nature of systems engineering concepts as they apply to different "levels" of the system.

Decision and Value Theory Basis: Rational decision-making about the design of a system requires mapping of stakeholder preferences into a single scale of value. Hypothesis 3, below, states this is a feasible approach.

Model Basis: System information is represented and maintained in models, and exported to documents when needed. "Engineering Elegant Systems: Theory of Systems Engineering", Draft 4, Sect 4 and Sect. 5 [1] discuss specific system-level models.

State Basis: Systems representations maximize use of state variables, and functions are defined as mappings from input states to output states. "Engineering Elegant Systems: Theory of Systems Engineering", Draft 4, Section 4.4 [1] addresses this explicitly.

Goal Basis: Systems exist to achieve goals, which are represented as constraints on the output state variables of functions. "Engineering Elegant Systems: Theory of Systems Engineering", Draft 4, Section 4.4 [1] addresses this explicitly.

Control Basis: Constraints on function output state variables are achieved by using the physical laws to control those state variables within their ranges.

Knowledge Basis: Individuals and organizations construct and maintain knowledge of the system. Systems engineering takes advantage of existing knowledge structures and improve formation of new knowledge across them. Information Theory is an important part of this basis. This knowledge basis is a key aspect of Discipline Integration discussed in "Engineering Elegant Systems: Theory of Systems Engineering", Draft 4, Sect. 5 [1].

Predictive Basis: Knowledge of the system is inherently uncertain. Uncertainties must be modeled probabilistically to understand the level of confidence in system knowledge so as to enable proper decision-making.

Sub-Principle 5(c) Systems engineering has a sociological basis specific to the organization.

Systems engineering incorporates the fundamental sociological concepts specific to the development and operations organization. This is a result of Postulates 3 and 4.

Principle 6 Systems engineering maps and manages the discipline interactions within the organization.

This is an application of Postulates 3 and 4. Organizational mirroring, or the correspondence of the organization to the system, is an essential mapping activity in managing the information flow and engineering of the system. The maturity of the engineering organization establishes the need for organizational structure formality. Organizations inexperienced in a specific system will require more formal structure to successfully develop the system. Seasoned organizations with a specific system can operate successfully with little formal organization (driven more by culture than formal hierarchy). Note that project management and organizational line management are concerned with organizational unit responsibilities and personnel matters. A concern of the systems engineer is how these units interact as part of system knowledge and understanding (system information) flows through the organization. The systems engineer works with project management and line management to resolve identified system information gaps or barriers in the organizational structure as these gaps and barriers will lead to flaws in system design, manufacturing, and operation. System dynamics models provide an approach to this principle as discussed in "Engineering Elegant Systems: Theory of Systems Engineering", Draft 4, Section 5.6 [1].

Principle 7 Decision quality depends on the coverage of the system knowledge present in the decision-making process.

This principle derives from Postulate 2. Engineering organizations often create trade study or task teams to investigate and resolve specific problems, which is a process of organizational flattening.. Decision effectiveness depends on involving the right decision-makers with a sufficiently complete understanding of the decision context and the decision to be made. Decisions are process-dependent. Decision methods are directly driven by the information needed by the decision-makers.

Principle 8 Both policy and law must be properly understood to not overly constrain or under-constrain the system implementation.

This is the application of Postulate 5. Policy and law act as important constraints on the system. Requirements should not always contain policy and law though they are often written in a requirement-like format. The context for the policies and laws is much different, often being much looser than requirements and more likely reflecting high-level system expectations than specific system functional or operational choices. Often, most interpret policy as having more flexibility than law. The systems engineer should understand how much flexibility is acceptable by those who set the policy (whether government or organizational) and those who pass the laws.

Principle 9 Systems engineering decisions are made under uncertainty accounting for risk.

This principle derives from Postulates 2, 3, 4, and 7. Information about the system is progressively understood through the development process and through the operations process. There are several sources of uncertainty in the development and operations. Some of this is natural based on the progressive understanding of the system (Postulate 7). Uncertainty exists due to the inability to predict the future with certainty. Uncertainty arises from many aspects of systems engineering, including limited knowledge on system environments and social aspects of the organization which affects information maintenance, creation, and flow. Sensitivities must also be understood to ensure the proper focus is given to the different uncertainties. Uncertainty and sensitivities then should be modeled throughout the process. Systems engineering decisions need to be made with sufficient understanding of the system context and the knowledge that uncertainty does exist even as understanding is gained.

Principle 10 Verification is a demonstrated understanding of all the system functions and interactions in the operational environment.

Ideally requirements are level (i.e., at the same level of detail in the design) and balanced in their representation of system functions and interactions. In practice requirements are not level and balanced in their representation of system functions and interactions. Verification seeks to prove that the system will perform as the designers expect based on their requirements, models, and designs. This leads to the principle that the proper performance of the system functions (i.e., outputs are within required ranges for a given input state) is the focus of system verification. If requirements are truly level and balanced, then verification of the system functions will result although some redundancy of effort may be expended. If the requirements are not truly level and balanced, then the focus of system verification should be on the system functions. By focusing on the proper system functions, a verification approach can be defined for the system which focuses on its successful application.

Principle 11 Validation is a demonstrated understanding of the system's value to the system stakeholders.

System validation is based on the stakeholder's expectations, not on the system requirements, models, and design information. It melds the system as designed and as built with the system as expected by the stakeholders. It is often assumed that the requirements reflect the stakeholder expectations. This is difficult to accomplish in practice due to the melding of external stakeholder expectations with developer expectations. Thus, requirements do not clearly reflect the stakeholder (internal or external) expectations in many system developments. System value models appear to provide a mathematical basis to define and guide the system development with the stakeholder's expectations. "Engineering Elegant Systems: Theory of Systems Engineering", Draft 4, Sect. 5 [1] discusses this more.

Principle 12 Systems engineering solutions are constrained based on the decision timeframe for the system need.

This principle deals with the time-changing nature of systems based on when the decisions for the system are made. The systems engineering solution for a system is formed by the context of the current state of the art and emerging available technologies. For example, what formed the context for air passenger travel in 1935 was very different from the context found in 1965. With the pace of technological advancements, the available solution sets for a given system can change noticeably over as a little as 5–10 years such as seen in the electronics industry over the last five decades. Thus the decision time frame is an important aspect of the solution set available to the systems engineer.

Over time, the degree of consistency in stakeholder and user preferences tends to diminish due to environmental changes, emerging technologies, or changes in the makeup of stakeholder and user communities. For systems with long life cycle phases, these communities and their preferences can change significantly. This is seen primarily in the operations phase and can also occur in the development phase of long developments. This variation becomes more pronounced as the system life time increases. And with more variation in stakeholders and stakeholder preferences, changes can be introduced to the system which can impact the system's ability to adapt to these preferences or stretch out system long duration developments. A key to managing these social-driven changes is to recognize when these shifts indicate the need for a different system and the time for the current system to move into decommissioning.

40.6 Systems Engineering Hypotheses

The hypotheses are statements that the consortium members are debating and believe can be proven (or perhaps disproven) through research. These statements challenge some of the heuristic notions found in complexity theory and are set in a practical application context (i.e., with real boundaries and constraints) rather than in a theoretical infinite context. Each of the hypotheses is constrained by their time context as discussed by Principle 12 above.

Hypothesis 1 If a solution exists for a specific context, then there exists at least one ideal systems engineering solution for that specific context.

Description: For a given system context that has a system solution, there exists an ideal (optimal or best-balanced) design for the system to accomplish the mission. Budget, schedule, decision timeframe, policy, law, and organizational culture define the context.

Evidence: This hypothesis is stated to drive objective research into the question of an optimal system configuration (i.e., a best-balanced system). Hamilton's principle directly proves this through the relation:

$$\int_{t_1}^{t_2} (\delta T - \delta V + \delta W)dt = 0. \tag{40.1}$$

Exergy is an expansion of this principle, and our research on exergy efficiency of a rocket indicates that an optimal system with an objective of efficiency can be defined across multiple configurations. This is a result that has not previously been achievable in a quantifiable manner. In addition, the value model seems to offer the ability to define an objective function to optimize the system in each context.

Implications: This hypothesis makes no statement about a global optimum. Rather, this hypothesis states there is a local optimum within the confines of the specific developmental and operational context. Note, this means that if this context changes, the local optimum may also change. In the absence of the knowledge of a best balance, the system's development appears as a sociological balance of organizational preferences.

Hypothesis 2 System complexity is greater than or equal to the ideal system complexity necessary to fulfill all system outputs.

Description: In each operational context and decision time frame, the minimum system complexity required to fulfill all the system outputs is the optimal system complexity, and the complexity of alternative system designs are equal to or greater than the ideal (i.e., optimal). Note that this is not a simpler is better hypothesis. Minimal complexity involves all aspects of the system as defined by context in Hypothesis 1 description. Being simple in only one context is not necessarily the system with the minimal complexity. The minimal complexity solution involves a best balance of the system and may lead to some aspects being more complex than alternatives and other aspects being less complex. Systems engineers define the minimal complexity holistically and not based on a subset of system aspects. The definition of system complexity is a much-debated topic. Refer to Appendix B for a more detailed review of complexity.

Evidence: This is similar to the statement of Occam's razor. As Albert Einstein is reputed to have said, "everything should be made as simple as possible, but not simpler" (Einstein, n.d.), which underlines a powerful truth of system modeling and systems engineering.

Implications: This hypothesis asserts that less complexity is preferable for a given context. This also states that a more complex system solution than the optimum can fulfill the system application but not as elegantly. One must realize that the system complexity necessary to complete all intended outcomes of the system satisfies all its operational needs.

Hypothesis 3 Key stakeholders' preferences can be represented mathematically.

Description: Systems engineers must understand and mathematically represent the preferences of key stakeholders to make decisions that are consistent with the stakeholder's preferences and to accomplish system goals. This also provides a basis for the validation of the system performance. Making such representations provides a basis for understanding decisions made at any point in the system development.

Evidence: Several approaches have represented preferences in mathematical form including Game Theory and Decision Theory.

Implications: A system value model should be constructible for a given system and stakeholders.

Hypothesis 4 The real physical system is the perfect model of the system.

Description: This hypothesis provides a statement of the idea that has long been espoused among statistical modelers. The physical system is the only complete, full, or perfect model of the system.

Proof: Kullback-Leibler information provides a definition for "ideal" information. This information measure indicates how close a particular model matches the real physical system and is defined as:

$$I(f,g) = \int f(x)\log(f(x))dx - \int f(x)\log(g(x|\theta))dx \tag{40.2}$$

Setting this relationship to zero provides a relationship to define the differences in a given model to the real system. This provides a proof that the perfect model of the system is the system itself.

$$\int f(x)\log(f(x))dx - \int f(x)\log(g(x|\theta))dx = 0 \tag{40.3}$$

$$\int f(x)\log(f(x))dx = \int f(x)\log(g(x|\theta))dx \tag{40.4}$$

Note, also that copies of systems are not physically identical.

$$f_1(x) \neq f_2(x) \neq \ldots \neq f_n(x) \tag{40.5}$$

Thus, the physical system only represents itself identically and not other physical copies of the system.

Implications: This provides a mathematical proof of the idea that has long been espoused among statistical modelers. A perfect model, being the system itself, means all other models have limitations which must be recognized. There are various system models that can show various aspects of the system, but no system model can show the complete system. In addition, one copy of the physical system is not identical with another copy of the system. Thus, variation in copies of the same physical system is to be expected at various tolerance levels depending on the design and fabrication approaches.

40.7 Summary

The foundation of systems engineering as identified in the research of the NASA Systems Engineering Research Consortium have led to the statement of 7 postulates, 12 principles, and 4 hypotheses of systems engineering. These statements provide a clearer understanding of the foundations of systems engineering including both the physical and the social aspects of system integration and discipline integration. A mathematical definition of systems engineering has initially been defined leading to the distinction between system verification and system validation. The nature of systems engineering across the system life cycle has been identified by the principles. The aspects of information and flow and decision-making have also been captured as part of the principles. Future research in this area may expand or refine these principles. Areas of fruitful enquiry include further definition of the mathematical basis of systems engineering, principles and processes during the system operations phase, the incorporation of human factors principles, and proofs of the hypotheses.

Acknowledgements Special thanks to David Rousseau at the Centre for Systems Philosophy for an excellent bibliography used to guide our references.

References

1. Defoe, J. C. (1993). *National council on systems engineering: An identification of pragmatic principles*, Final Report. SE Practice Working Group. Subgroup on Pragmatic Principles. NCOSE WMA Chapter, 6600 Rockledge Dr., Bethesda, Maryland 20817.
2. "Systems and software engineering —System life cycle processes", ISO/IEC/IEEE 15288, first edition, 2015-05-15.
3. Walden, D. D., Roedler, G. J., Forsberg, K. J., Douglas Hamelin, R., & Shortell, T. M. (Eds.). (2015). *INCOSE systems engineering handbook: A guide for system life cycle processes and activities* (4th ed.). New York: Wiley.
4. Keating, C. B., Katina, P. F., Jaradat, R., Bradley, J. M., & Gheorghe, A. V. (2017). Acquisition system development: A complex system governance perspective. *INCOSE International Symposium, 27*, 811–825. https://doi.org/10.1002/j.2334-5837.2017.00395.x
5. Keating, C. B., Katina, P. F., Gheorghe, A. V., & Jaradat, R. (2017). Complex system governance: Advancing prospects for system of systems engineering applications.
6. Adams, K. M., Hester, P. T., Bradley, J. M., Meyers, T. J., & Keating, C. B. (2014). Systems theory as the foundation for understanding systems. *Systems Engineering, 17*(1), 112–123.
7. Whitney, K., Bradley, J. M., Baugh, D. E., & Chesterman Jr., C. W. (2015). Systems theory as a foundation for governance of complex systems. *International Journal of System of Systems Engineering, 6*(1–2), 15–32.
8. Katina, P. F. (2016). Systems theory as a foundation for discovery of pathologies for complex system problem formulation. In *Applications of systems thinking and soft operations research in managing complexity* (pp. 227–267). Cham: Springer.
9. Rousseau, D. (2018). Three general systems principles and their derivation: Insights from the philosophy of science applied to systems concepts. In *Disciplinary convergence in systems engineering research* (pp. 665–681). Cham: Springer.
10. Rousseau, D. (2018a). On the architecture of systemology and the typology of its principles. *Systems, 6*(1), 7.
11. Sillitto, H. (2014). *Architecting systems. Concepts, principles and practice* (pp. 33–38). London: College Publications (note: this list has been included in the Systems Engineering Handbook 4th edition (2015) (pp. 20–21)).
12. "Engineering Elegant Systems: Theory of Systems Engineering", Fourth Draft, NASA document, 9 February 2018.
13. Watson, M. D., & Farrington, P. A. (2016). *NASA Systems Engineering Research Consortium: Defining the path to elegance in systems*. Huntsville, AL: CSER.
14. Watson, M., Griffin, M. D., Farrington, P. A., Burns, L., Colley, W., Collopy, P., et al. (October 2014). Building a path to elegant design. In S. Long, E.-H. Ng, & C. Downing (Eds.), *Proceedings of the American Society for Engineering Management, 2014 International Annual Conference*. Virginia Beach, VA: American Society for Engineering Management.
15. V. Neufeldt, & D. B. Guralnik (Eds.). (1988). *Webster's New World Dictionary Third College Edition* (p. 1055). New York: Simon & Schuster.
16. Componation, P. J., Dorneich, M., Guiping, H., Farrington, P. A., & Hanson, J. L. (2013). Systems engineering and project success in government and commercial organizations. In *Proceeding of the American Society for Engineering Management, 2013 International Annual Conference*.

Chapter 41
Exergy Efficiency of Interplanetary Transfer Vehicles

Sean T. Owen, Michael D. Watson, and Mitchell A. Rodriguez

Nomenclature

a	Semi-major axis
F	Thrust
f	Final index
G	Universal gravitational constant
g_0	Standard acceleration due to gravity at Earth's surface
h_{prop}	Enthalpy of the propellant
I_{sp}	Specific impulse
i	Initial index
KE	Kinetic energy
m	Mass
M_E	Mass of the Earth
M_{planet}	Mass of the planet
M_{sun}	Mass of the Sun
m_0	Initial mass
$\dot{m}$	Mass flow rate
$M_{\text{vehicle,initial}}$	Mass of the vehicle on the pad
$M_{\text{vehicle, final}}$	Injected mass
PE	Potential energy
r	Distance, position, radius
S	Positive/negative sign

S. T. Owen
University of Alabama in Huntsville, Huntsville, AL, USA

M. D. Watson (✉)
NASA Marshall Space Flight Center, Huntsville, AL, USA
e-mail: Michael.d.watson@nasa.gov

M. A. Rodriguez
Jacobs Space Exploration Group, Huntsville, AL, USA

S. Adams et al. (eds.), *Systems Engineering in Context*,
https://doi.org/10.1007/978-3-030-00114-8_41

t	Time
T_{engine}	Engine thrust
V	Velocity
$\dot{V}$	Acceleration
V_e	Exhaust velocity
X	System exergy
X_{des}	Exergy destroyed
X_{exp}	Exergy expended
η_{exg}	Exergy efficiency
θ	True anomaly
μ	Gravitational parameter
φ	Horizon-relative flight angle

41.1 Introduction

Several space agencies, including NASA, are planning manned exploration of Mars in the upcoming decades. Many different mission architectures have been proposed for accomplishing this. It is the role of systems engineers to compare and optimize different space transportation systems and components, up to and including full mission architectures. To do this, some measure is needed that applies to all systems being compared, even though those systems may have considerable differences. Exergy efficiency, or how well a given system can use the work available to it, provides a measure to compare different interplanetary transfer systems.

41.2 Exergy Balance Relationship

Planetary transfer vehicles (i.e., satellites, planetary landers, and human and cargo transports as illustrated in Fig. 41.1) are integrated by system exergy. This includes their propulsion stages, electrical power systems (e.g., nuclear electric or solar electric), and crew volumes for transporting the crew. During propulsive trajectory changes, the exergy balance equation can be written for a spacecraft system as,

$$\begin{aligned}
&\sum\nolimits_{\text{stages}}\left[\Delta m_{\text{propellant}}\left(h_{\text{prop}}+\frac{V_e^2}{2}\right)\right]-X_{\text{des}}\\
&\quad=\sum\nolimits_{\text{stages}}\left[\left(M_{\text{vehicle, final}}\frac{V^2_{\text{vehicle, final}}}{2}-M_{\text{vehicle, initial}}\frac{V^2_{\text{vehicle, initial}}}{2}\right)\right.\\
&\quad\left.+\left(\frac{GM_E M_{\text{vehicle, initial}}}{r_{\text{altitude, initial}}}-\frac{GM_E M_{\text{vehicle, final}}}{r_{\text{altitude, final}}}\right)\right]
\end{aligned} \tag{41.1}$$

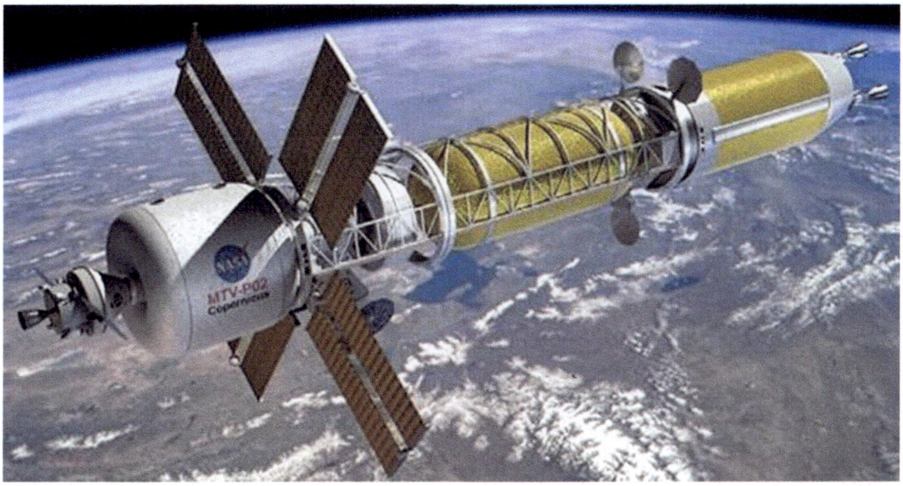

Fig. 41.1 Mars transfer vehicle

The propulsion engine (e.g., chemical, electric, nuclear thermal) characteristics (mass flow, enthalpy, exhaust velocity, and electrical power for electric propulsion) are all included on the left of the equation.

For coast phases of the flight trajectory, the exergy balance equation simplifies to the basic orbital mechanic's relationship for a balanced system. In this case the spacecraft energy (and exergy) is constant, and the kinetic and potential energies increase and decrease in opposite directions.

$$E_{\text{vehicle}} = \left(M_{\text{vehicle}} \frac{V_{\text{vehicle}}^2}{2} - \frac{GxM_E M_{\text{vehicle}}}{r_{\text{altitude}}} \right). \tag{41.2}$$

This creates an oscillatory relationship between the vehicle kinetic and potential energies with respect to the dominate body (typically the Sun in interplanetary space) [1].

Planetary and solar masses have a large effect on spacecraft exergy in interplanetary space. It is important to ensure an appropriate reference is used. A heliocentric reference is generally best for space travel within the solar system. When operating within a planetary body's sphere of influence (SOI), the sphere in which the planetary gravitational influence is greater than the Sun's influence then the solar influence can usually be ignored. In this case a planetary centric (geospatial reference system for the Earth) can be used. Equation (41.3), gives the general relationship for the planetary SOI [2].

$$r_{\text{SOI}} = r_{\text{sun, planet}} \left(\frac{m_{\text{planet}}}{m_{\text{sun}}} \right)^{2/5} \tag{41.3}$$

Planetary transfer uses a Hohmann transfer from Earth to Mars and back to Earth. The planetary stay is also important in calculating the possible trajectories. An 11-month stay on the planet is assumed with a total mission length on the order of 2–3 years. This trajectory contains four main burns: trans-Mars injection (TMI), Mars orbit insertion (MOI), trans-Earth injection (TEI), and Earth orbit insertion (EOI). Four different propulsion systems were analyzed using this basic course: low-enriched uranium (LEU) liquid hydrogen (LH2) nuclear thermal propulsion (NTP), high-enriched uranium (HEU) LH2 NTP, LEU CH4 (methane) NTP, and a chemical liquid oxygen (LO2)/LH2 system.

For the LEU CH4 NTP and CHM LOX-LH2 cases, the mass flow rate for the main engine can be calculated from I_{sp} by using Eq. (41.4).

$$\dot{m}_{\text{propellant}} = T_{\text{engine}} / \left(I_{\text{sp}} g_0 \right) \tag{41.4}$$

The mass flow rate of the reaction control system (RCS) thrusters is an important parameter in the maneuvers for the trajectory burns. For the calculations in this section, the mass flow rate for a typical RCS thruster of 7 kg/s with an I_{sp} of 291 s will be used.

Figure 41.2 shows the exergy efficiency of the LEU LH2 NTP case during the first 500 s of TMI and shows the decline in the efficiency during the RCS burn. Also visible in this plot is an efficiency drop just after the RCS burn; this corresponds to dropping an empty propellant tank. Exergy that was expended to accelerate the tank is lost when the tank is discarded, so dropping the tank registers as a decrease in efficiency.

Exergy calculations are sensitive to changes in position and velocity with respect to the departure and arrival planets, requiring a complete orbital trajectory to calculate exergy efficiency. A patched-conics trajectory is necessary to show the

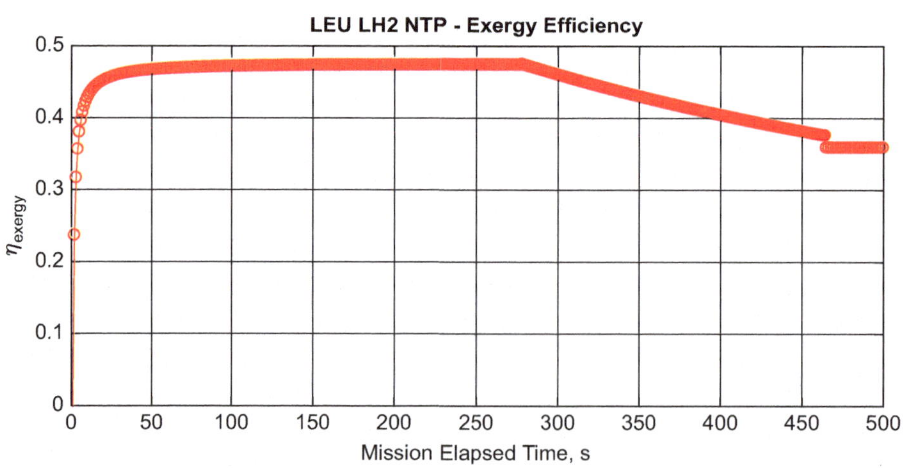

Fig. 41.2 Exergy efficiency during TMI

complete system and planetary environments within each planets SOI and in interplanetary space outside the planets SOI's.

41.3 Orbital Mechanics

For each leg of the mission, the departure planet's and arrival planet's position and velocity are important for the periods when the spacecraft is within the planets SOI [3]. Outside the planetary SOIs, the Sun is treated as the sole gravity source. Acceleration due to the Sun's gravity is broken up into vector components along the interplanetary trajectory path. Figure 41.3 shows the spacecraft trajectory path and planets orbital paths during the mission.

Using the planetary positions and the given position of the spacecraft at all points during the mission, the spacecraft's planet-relative distance, speed, and flight angle from the horizon are calculated for the days following the departure burns and leading up to the arrival burns using Eqs. (41.5)–(41.7).

$$\vec{r}_{\text{vehicle, planet}} = \vec{r}_{\text{vehicle, sun}} - \vec{r}_{\text{planet, sun}} \tag{41.5}$$

$$\vec{V}_{\text{vehicle, planet}} = \vec{V}_{\text{vehicle, sun}} - \vec{V}_{\text{planet, sun}} \tag{41.6}$$

$$\varphi_{\text{planetary horison}} = \frac{\pi}{2} - \operatorname{acos}\left(\frac{\vec{V}_{\text{vehicle, planet}}\,\vec{r}_{\text{vehicle, planet}}}{\left\|\vec{V}_{\text{vehicle, planet}}\right\|\left\|\vec{r}_{\text{vehicle, planet}}\right\|}\right) \tag{41.7}$$

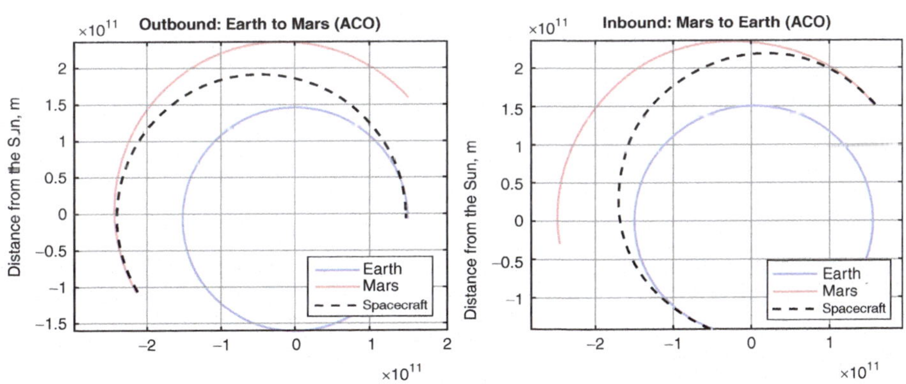

Fig. 41.3 Spacecraft interplanetary trajectory and planet trajectories during the outbound and inbound (return) phases

Using the spacecraft's distance from the planet over time, the exact time when it crosses the SOI boundary is interpolated with Eq. (41.8). The two points in time used for the interpolation are those just before and after crossing the SOI boundary, the radius of r_{SOI} defined in Eq. (41.3).

$$t_{\mathrm{SOI}} = t_i + \left(t_f - t_i\right) \frac{\vec{r}_{\mathrm{SOI}} - \left\|\vec{r}_{\text{vehicle, planet}, i}\right\|}{\left\|\vec{r}_{\text{vehicle, planet}, f}\right\| - \left\|\vec{r}_{\text{vehicle, planet}, i}\right\|} \tag{41.8}$$

With these values, the spacecraft's planet-relative velocity and flight angle from the horizon at that moment are similarly interpolated using Eqs. (41.9) and (41.10).

$$\vec{V}_{\text{vehicle, planet, SOI}} = \vec{V}_{\text{vehicle, planet}, i} + \left(t_{\mathrm{SOI}} - t_i\right)\left(\frac{\vec{V}_{\text{vehicle, planet}, f} - \vec{V}_{\text{vehicle, planet}, i}}{t_f - t_i}\right) \tag{41.9}$$

$$\varphi_{\text{horizon, SOI}} = \varphi_{\text{horizon}, i} + \left(t_{\mathrm{SOI}} - t_i\right)\left(\frac{\varphi_{\text{horizon}, f} - \varphi_{\text{horizon}, i}}{t_f - t_i}\right) \tag{41.10}$$

Additionally, a new reference frame is created based on the spacecraft's position and velocity while crossing the SOI boundary, using Eqs. (41.11)–(41.13). Planet-centric orbits within the SOI will be plotted in a 2D plane, and this reference frame will track the orientation of the plane relative to the solar ecliptic.

$$\widehat{i} = \frac{\vec{r}_{\text{vehicle, planet, SOI}}}{\left\|\vec{r}_{\text{vehicle, planet, SOI}}\right\|} \tag{41.11}$$

$$\widehat{k} = \frac{\widehat{i} x \vec{V}_{\text{vehicle, planet, SOI}}}{\left\|\widehat{i} x \vec{V}_{\text{vehicle, planet, SOI}}\right\|} \tag{41.12}$$

$$\widehat{j} = \frac{\widehat{k} x \widehat{i}}{\left\|\widehat{k} x \widehat{i}\right\|} \tag{41.13}$$

A transformation matrix is created using the new reference frame and Eq. (41.14) and will later be used to convert the SOI orbit back to a heliocentric reference frame.

$$T_{\text{Transform}} = \begin{bmatrix} \widehat{i}_X & \widehat{i}_Y & \widehat{i}_Z \\ \widehat{j}_X & \widehat{j}_Y & \widehat{j}_Z \\ \widehat{k}_X & \widehat{k}_Y & \widehat{k}_Z \end{bmatrix} \tag{41.14}$$

With conditions at the SOI intersection established, the planet-centric transfer and parking orbits within the SOI can be determined. First, the transfer orbit's semi-major axis is calculated using Eqs. (41.15) and (41.16).

$$V_{SOI} = \left\| \vec{V}_{ship, planet, SOI} \right\| \tag{41.15}$$

$$a_{transfer} = 1 / \left(\left(\frac{2}{r_{SOI}} \right) - \left(\frac{V_{SOI}^2}{\mu_{planet}} \right) \right) \tag{41.16}$$

The speed and flight angle of the spacecraft at the edge of the SOI is sufficient to define a hyperbolic orbit past the planet. The parking orbit periapsis is established (400 km above the planet's surface, roughly the altitude that the ISS orbits at over Earth in this example). This is the minimum shift that still puts the spacecraft's trajectory well above the atmosphere to avoid significant drag. Note, that aerobraking (not addressed here) requires an orbital altitude within the upper atmosphere with sufficient drag to reduce the spacecraft velocity (ΔV) to enter the prescribed parking orbit. Equations (41.17)–(41.22) are used to determine the apoapsis of the parking orbit for the listed ΔV at that periapsis.

$$e_{transfer} = 1 - \frac{r_{periapsis}}{a_{transfer}} \tag{41.17}$$

$$V_{periapsis, transfer} = \sqrt{\mu_{planet} \left(\frac{2}{r_{periapsis}} - \frac{1}{a_{transfer}} \right)} \tag{41.18}$$

$$V_{periapsis, parking} = V_{periapsis, transfer} - \Delta V \tag{41.19}$$

$$a_{parking} = 1 / \left(\left(\frac{2}{r_{periapsis}} \right) - \left(\frac{V_{periapsis, parking}^2}{\mu_{planet}} \right) \right) \tag{41.20}$$

$$e_{parking} = 1 - \frac{r_{periapsis, parking}}{a} \tag{41.21}$$

$$r_{apoapsis} = r_{periapsis} \left(\frac{1 + e_{parking}}{1 - e_{parking}} \right) \tag{41.22}$$

It is important for the apoapsis to remain within the planet's SOI and should be established to meet the parking orbit period necessary to meet mission objectives. Once the apoapsis and periapsis are established, the parking orbit periapsis is kept as the periapsis of the hyperbolic transfer orbit. This results in an extremely elliptical parking orbit with a very long period (particularly if it extends to the planetary SOI boundary). Equations (41.17)–(41.22) can be solved iteratively starting with an initial periapsis estimate and stepping in small increments (e.g., 100 mi periapsis altitude increases) until a reasonable apoapsis is found.

The eccentricity of the hyperbolic transfer orbit, the spacecraft's true anomaly at the SOI boundary, and its periapsis velocity can be calculated using Eqs. (41.17), (41.18), and (41.23).

$$\theta_{SOI} = \mathrm{acos}\left(\frac{a_{transfer}\left(1 - e_{transfer}^2\right) - r_{SOI}}{r_{SOI}e_{transfer}}\right) \tag{41.23}$$

By applying the listed ΔV at the new periapsis as a point-thrust burn, the shape of the parking orbit around the planet can be approximated using Eqs. (41.19)–(41.21). It is only an approximation because it assumes a point-thrust burn connects the transfer and parking orbit. As long as the chosen propulsion system is sufficiently high-thrust, the actual parking orbits will be quite close to the listed values here, as a sufficiently short burn time (on a timescale of minutes) will be negligible compared to the period of the parking orbit.

The parking orbits are only an approximation based on point-thrust burns. In order to properly calculate the exergy efficiency, plots of the spacecraft's position and velocity during each burn will be needed. To do this, Eqs. (41.24) and (41.25) can be used to track the spacecraft forward or backward in time from periapsis to establish its trajectory. Another acceleration vector from the spacecraft's engine is added, aimed directly opposite its velocity vector at any point in time for backward tracking. This new vector is split into $\widehat{i}$ and $\widehat{j}$ components for the calculations.

$$r_f = r_i + V_i\Delta t + \frac{1}{2}\dot{V}_i\Delta t^2 \tag{41.24}$$

$$V_f = V_i + \dot{V}_i\Delta t \tag{41.25}$$

At this point, a complete planet-centric course contains the spacecraft's position and velocity from engine start to SOI exit (or vice versa for entry scenarios). This course is then rotated such that the SOI exit/entry point lies directly on the $\widehat{i}$ axis of the planet-centric reference frame. Equations (41.26) and (41.27) are then used to plot the spacecraft's heliocentric position and velocity while it is inside the SOI.

$$\vec{r}_{vehicle,\,sun} = \vec{r}_{planet,\,sun} + \left(T_{Transform}\vec{r}_{vehicle,\,planet}\right) \tag{41.26}$$

$$\vec{V}_{vehicle,\,sun} = \vec{V}_{planet,\,sun} + \left(T_{Transform}\vec{V}_{vehicle,\,planet}\right) \tag{41.27}$$

41.4 Interplanetary Exergy Efficiency

With the modified mass data and orbital data in hand, the actual exergy calculations can begin. During each burn of the mission, changes in expended exergy are calculated using Eq. (41.28) which is taken from Eq. (41.1), with mass drops for each time step being calculated from the tank drops and consumable use schedules. These step changes are then summed to produce a plot of expended exergy that rises during burns but otherwise stays constant.

$$X_{\text{exp}} = \Delta m_{\text{propellant}}\left(h_{\text{prop}} + \frac{V_e^2}{2}\right) \tag{41.28}$$

In order to calculate destroyed exergy, changes in kinetic and potential energy must be tracked across the entire mission. To determine whether the change in kinetic or potential energy should be positive or negative during a given time step, the ruleset described below in Table 41.1 is applied based on Eqs. (41.29) and (41.30). Changes in the spacecraft's velocity and distance relative to the central body during that time step are taken into consideration when determining the sign. It should be noted that the values X, Y, and Z in the table are all greater than or equal to one.

$$KE : m_f V_f^2 - m_i V_i^2 = \begin{cases} > 0 \\ < 0 \end{cases} \tag{41.29}$$

$$PE : \frac{m_i}{r_i} - \frac{m_f}{r_f} = \begin{cases} > 0 \\ < 0 \end{cases} \tag{41.30}$$

Change in kinetic and potential energy during a given time step is then calculated using Eqs. (41.31) and (41.32), where S is the sign taken from the previous table, either 1 or $-$ 1.

$$\Delta KE_{\text{step}} = \frac{S}{2}\left|m_f V_f^2 \quad m_i V_i^2\right| \tag{41.31}$$

$$\Delta PE_{\text{step}} = S\mu\left|\frac{m_i}{r_i} - \frac{m_f}{r_f}\right| \tag{41.32}$$

Table 41.1 Sign convention for changes in kinetic and potential energy

Mass	Velocity	ΔKE_{step}	Distance	ΔPE_{step}
$M_f = M_i$	$V_f > V_i$	+	$r_f > r_i$	+
$M_f = M_i$	$V_f < V_i$	−	$r_f < r_i$	−
$\begin{cases} M_f > M_i \\ M_f = XM_i \end{cases}$	$\begin{cases} V_f > V_i \\ V_f = ZV_i \end{cases}$	+	$\begin{cases} r_f > r_i \\ r_f = Yr_i \end{cases}$	$\begin{cases} +(Y > X) \\ -(Y < X) \end{cases}$
$\begin{cases} M_f > M_i \\ M_f = XM_i \end{cases}$	$\begin{cases} V_f < V_i \\ V_i = ZV_f \end{cases}$	$\begin{cases} -(Z^2 > X) \\ +(Z^2 < X) \end{cases}$	$\begin{cases} r_f < r_i \\ r_i = Yr_f \end{cases}$	−
$\begin{cases} M_f < M_i \\ M_i = XM_f \end{cases}$	$\begin{cases} V_f > V_i \\ V_f = ZV_i \end{cases}$	$\begin{cases} +(Z^2 > X) \\ -(Z^2 < X) \end{cases}$	$\begin{cases} r_f > r_i \\ r_f = Yr_i \end{cases}$	+
$\begin{cases} M_f < M_i \\ M_i = XM_f \end{cases}$	$\begin{cases} V_f < V_i \\ V_i = ZV_f \end{cases}$	−	$\begin{cases} r_f < r_i \\ r_i = Yr_f \end{cases}$	$\begin{cases} -(Y > X) \\ +(Y < X) \end{cases}$

These step changes in kinetic and potential energy are summed over time to create a running total of energy changes. These sums are subtracted from the expended exergy using Eq. (41.33) to calculate the exergy destroyed, which then directly leads to the exergy efficiency, defined in Eq. (41.34), at that point in time.

$$X_{\text{des}} = X_{\text{exp}} - \sum \Delta KE_{\text{step}} - \sum \Delta PE_{\text{step}} \tag{41.33}$$

$$\eta_{\text{exergy}} = \frac{\Delta m_{\text{propellant}}\left(h_{\text{prop}} + \frac{V_e^2}{2}\right) - X_{\text{des}}}{\Delta m_{\text{propellant}}\left(h_{\text{prop}} + \frac{V_e^2}{2}\right)} = 1 - \frac{X_{\text{des}}}{\Delta m_{\text{propellant}}\left(h_{\text{prop}} + \frac{V_e^2}{2}\right)} \tag{41.34}$$

When the spacecraft is within a planet's SOI and not burning propellant, efficiency does not stay constant but fluctuates with the planetary gravity influences as the vehicle and planet both move along their respective trajectories. This is avoided by using a patched-conics model for the orbital modifications, where exergy calculations are applied to each SOI independently, not using the heliocentric portion of the trajectory. Whenever the spacecraft crosses into or out of a SOI, the most recent value for the total change in kinetic and potential energy is carried over to the next series of calculations. This ensures that exergy efficiency stays constant whenever the spacecraft's mass and velocity are constant, even across SOIs.

The final exergy efficiency plots over the whole mission for each propulsion system are given below in Figs. 41.4 and 41.5.

As seen previously in Fig. 41.2, exergy efficiency will sharply rise when using a main engine during a departure burn and then decrease during the following RCS burn. This is because of the RCS burn's lower I_{sp}, destroying more exergy for the same exergy expenditure, thus lowering the efficiency of that stage of the mission. Efficiency also drops when ejecting an empty propellant tank or spent consumables, as the exergy expended to move those components up to speed is lost when they are discarded.

Unlike the departure burns, braking burns when arriving at a planet show exergy efficiency decreasing during both the main burn and RCS burn. This is due to how sign conventions for kinetic and potential energy are defined, as well as what equation is being used for exergy efficiency. Both of the definitions used by the program were detailed in the previous section. However, Eq. (41.34) assumes that the vehicle is expending exergy to build up speed. By that definition, any exergy expenditure to shed speed, as in a braking burn, is detrimental to exergy efficiency.

It is possible that different sign conventions for kinetic and potential energy or a different definition of exergy efficiency, used only during braking burns, could show an increase in exergy efficiency during braking. That is future work to investigate the exergy efficiency relationships for braking maneuvers.

Notable efficiency values are given above in Table 41.2. The maximum exergy efficiency achieved (during the TMI burn) is shown in the top row. The second row shows the total exergy efficiency achieved from the TMI departure burn through parking orbit insertion (EOI) burn at the return to Earth.

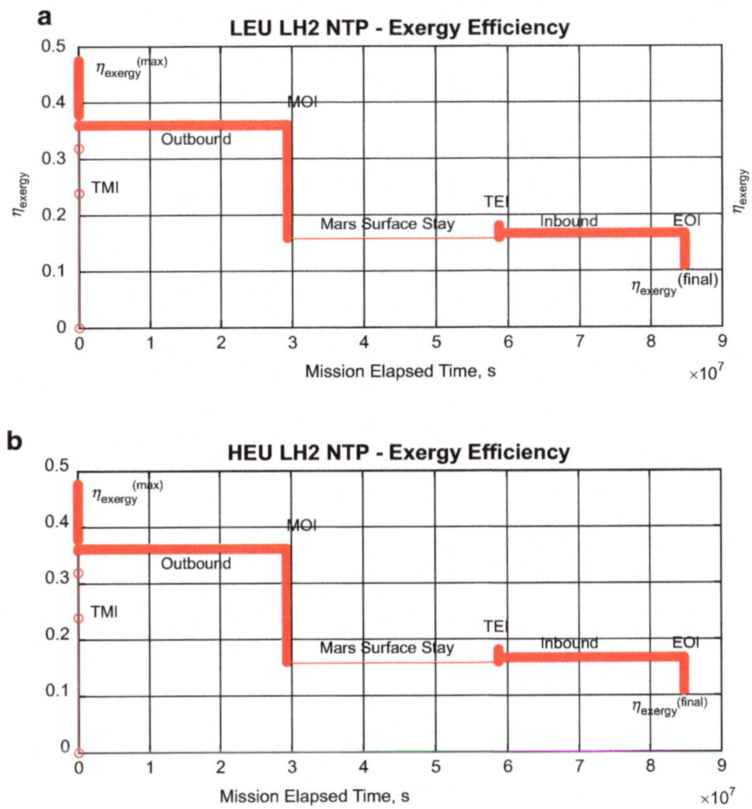

Fig. 41.4 Exergy efficiency throughout the mission using the LEU LH2 NTP system and the HEU LH2 NTP system

Overall, exergy efficiency roughly scales directly with I_{sp} and inversely with the total initial mass of the spacecraft. HEU LH2 NTP achieves the highest efficiencies, but only just barely, since it has the same I_{sp} as the LEU LH2 NTP case and is only minimally lighter due to reactor sizing to produce the same thrust. CHM LOX-LH2 has the lowest efficiencies by far, since its I_{sp} is considerably less than the other cases.

41.5 Summary

Exergy efficiency provides a mechanism to compare different system options in a clear and effective manner across the full system operational environment. This provides a measurement of performance (MoP) for systems engineers to use in comparing very different system options. This concept has been demonstrated for

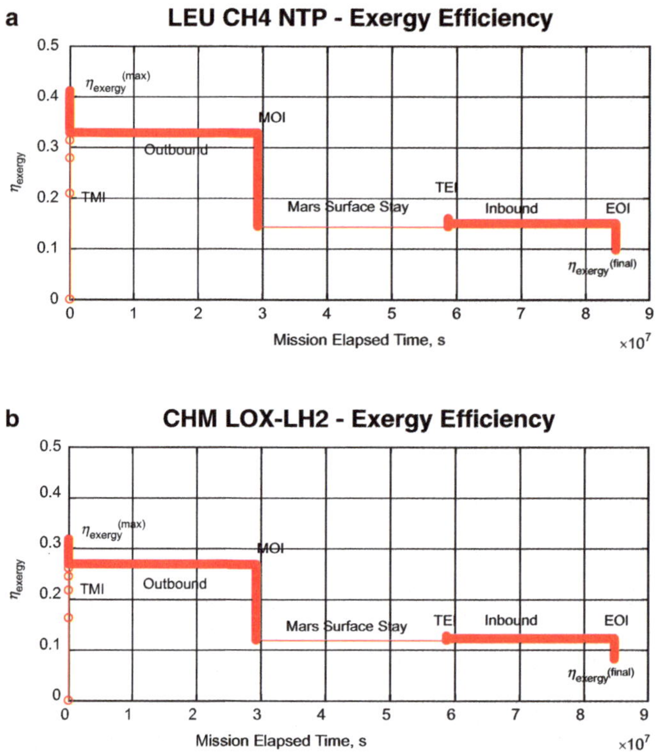

Fig. 41.5 Exergy efficiency throughout the mission using the LEU CH4 NTP system and the CHM LOX-LH2 system

Table 41.2 Final exergy efficiency results for all propulsion systems analyzed

	LEU LH2 NTP	HEU LH2 NTP	LEU CH4 NTP	CHM LOX-LH2
η_{exg} (max)	47.63%	47.68%	41.20%	31.83%
η_{exg} (total)	10.61%	10.62%	9.69%	8.18%

interplanetary transfer vehicles comparing four different system configurations. The result allows a balanced comparison between the options, informing system engineers of the best balanced system to select for the given application. The analysis of interplanetary transfer vehicles shows that NTP options are more efficient than chemical propulsion options, and NTP with liquid hydrogen as a fuel is more efficient than methane as a fuel. This also shows only very small difference between LEU NTP and HEU NTP. This provides key system characterization to support the selection of configuration for interplanetary transfer missions.

References

1. Watson, M. D. (2018). System exergy: System integrating physics of launch vehicles and spacecraft. *AIAA Journal of Spacecraft and Rockets, 55*(2), 451–461.
2. Curtis, H. D. (2014). *Orbital mechanics for engineering students* (3rd ed.). Boston: Elsevier.
3. NASA-SP-2009-566, Human Exploration of Mars Design Reference Architecture 5.0, Addendum #2, March 2014.

Chapter 42
Information Theory Applied to Decision-Making Structures

Michael D. Watson

Nomenclature

C	Cognitive function
f_n	Message or understanding
H	Information entropy and information uncertainty
I	Information
p_n	Probability that message is transmitted
q_n	Uncertainty
X_n, Y_n	Information contained in memory

42.1 Introduction

Information flow through an organization in the development or operation of a system is an important aspect of systems engineering. Systems engineering ensures that the correct information is provided to the correct engineers when or before it is needed. This is performed by understanding and managing the information about the system which resides in the design and in the organization. Information theory provides the tools to understand and manage this flow and the organizational decision structures which utilize this information.

Information theory has been applied to decision theory [1, 2] and provides a relevant structure to model a decision-making body (e.g., decision board). Webster's Dictionary defines information theory as "a theory that deals statistically with information and the measurement of its content in terms of its distinguishing

M. D. Watson (✉)
NASA Marshall Space Flight Center, Huntsville, AL, USA
e-mail: Michael.d.watson@nasa.gov

S. Adams et al. (eds.), *Systems Engineering in Context*,
https://doi.org/10.1007/978-3-030-00114-8_42

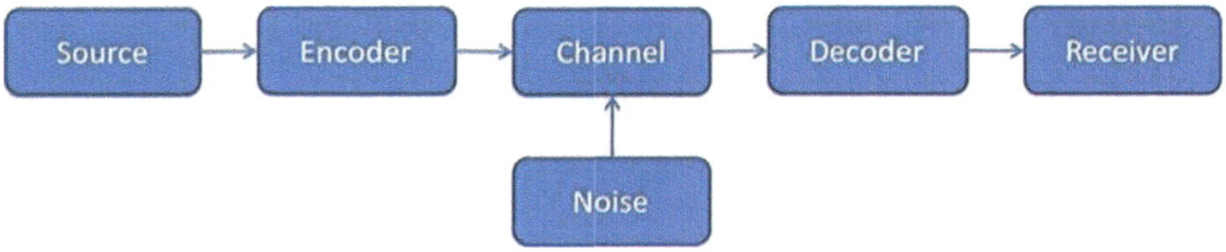

Fig. 42.1 Communication system model

essential characteristics or by the number of alternatives from which it makes a choice possible, and the efficiency of processes of communication between humans and machines." Expanding this definition to include human communication encompasses the organizational communications and hence decision-making bodies. The decision-making body essentially operates as a communication system where information is presented and shared in an open forum. Figure 42.1 illustrates a basic communication system model.

Information transmitted through this communication system model is a logarithmic function:

$$I = -\log p_n \tag{42.1}$$

where p_n is the probability that message f_n was sent. Taking the average, this is the measure of uncertainty that the transmitted information represents a specific event,

$$\bar{I} = H = -\sum_n p_n \log p_n \tag{42.2}$$

$\bar{I}$, representing the uncertainty that an event occurred, is also defined as the information entropy, H, of the communication system [3].

42.2 Single Board Structures

In the context of a board structure, each board member acts as a source and encoder, contributing information to the discussion. Each board member also acts as a decoder and receiver, receiving information and understanding (or interpreting) the meaning of the information. In this model, the board members include the board chairperson. The chairperson has the final decision authority in the board setting. In addition, subject matter experts (SME) often present information to the board or can be additional sources contributing information to the board discussions. The channel is the board meeting. Noise includes many factors including uncertainty in the information presented to the board, distractions (i.e., side conversations, board members working other issues on email, text, or side discussions), or physical noise in the room or on phone lines. Following this structure, a board can be modeled as illustrated in Fig. 42.2.

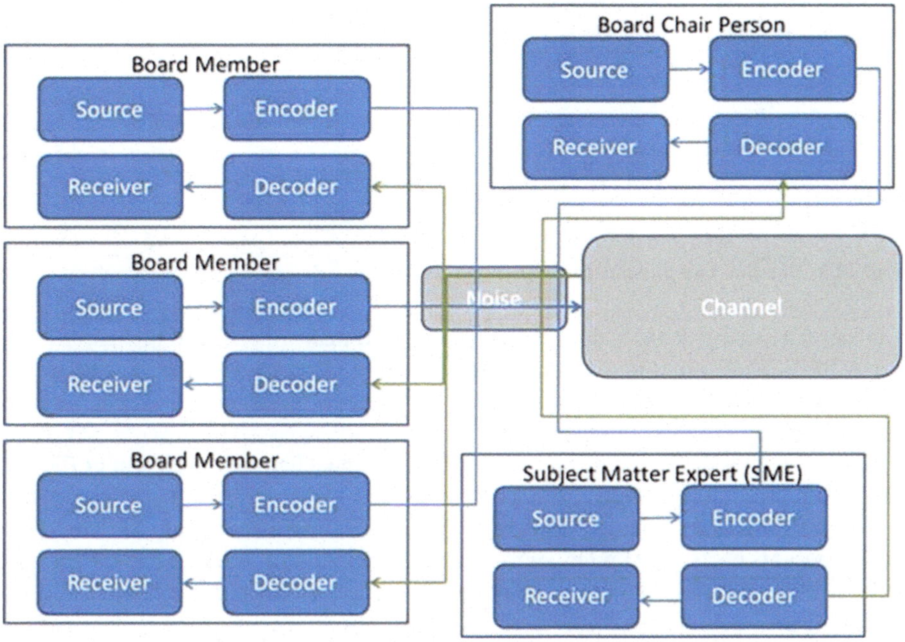

Fig. 42.2 Information theory board model

This model provides for the inclusion of the cognitive aspects of the board members. Each board member must present information in a clear and understandable manner. The extent of their skill in this is represented by the encoding of the knowledge that they possess. In addition, the decision to share or withhold information is a cognitive aspect of the board member. Similarly, the ability of each board member to understand what is being discussed is represented by the decoding of the information (understanding). Many cognitive factors influence the decoding (understanding) of the information including education background, experience, intuitive ability, etc. Cognitive science, then, can be used to establish the distribution functions for the knowledge, encoding, and decoding of each board member and SME.

In this simplest form, the board model assumes that all information needed for a decision is provided to the board and that the information is properly and completely understood. Therefore, the uncertainty in the decision is zero, and the information entropy $H = I = 0$. This does not mean that no information is conveyed by the board but that there is no uncertainty in the board decision. The information is fully sent and received with $p_n = 1$ such that the $\log_n p_n = 0$ leading to the average and uncertainty as zero. In this simple model, the uncertainty (or absence of) is absolute in the sense that the decision is fully understood and is not subjective. While there are simple decisions in practice, most decisions in practice involve various types of uncertainty in the decision-making. Understanding the decision outside the board is not addressed in this model and can lead to uncertainty in the larger context as well.

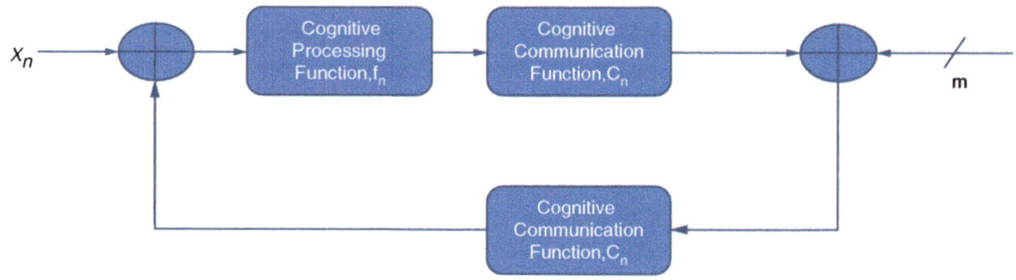

Fig. 42.3 Board member cognitive processing model

There are many sources of uncertainty in board decisions. These include hidden (or withheld) information, cultural biases (creating blind spots on certain topics or ignoring factors), ignorance (not understanding aspects of the topic) [4], and missing information in the board discussion.

Decision boards, as decision-making bodies, are chartered with controlling a particular program, project, system, etc. As such, control theory applies to the basic functions of a board. Boards can be modeled as a finite impulse response (FIR) system. Each board member comes to the board with information on a given topic. This information is cognitively processed forming preferences (i.e., weightings), relationships with other information, etc. These cognitive processing functions are quite complex. The board member then communicates with other board members during the board meeting and adds this information with their initial thoughts to create or modify their position. Thus, each board member thought processes can very simply be modeled as a cascade filter with feedback as shown in Fig. 42.3.

When all the board members and SMEs are combined, the board meeting then becomes a cascade filter model. In this representation, the information theory model relationship is clearly seen, where the addition of the information of the board members and SME during discussion is the channel, and noise is injected into the channel from external disturbances.

The board model can be updated with the board member decision-making model where the encoder is one form of cognitive-communication function, and the decoder is another form of cognitive-communication function. The source and receiver are combined as part of the cognitive processing function, and X_n is contained in memory. Figure 42.4 illustrates this model.

The equation represented in Fig. 42.4 can be written as

$$Y_n = X_n + \sum\nolimits_q f_{p,q+1}\left\{C_p\left[f_{p,q}(X_n) + \sum\nolimits_{m.m\neq p} C_m[f_m(X_n)] + \text{Noise}\right]\right\} \quad (42.3)$$

where X_n represents a specific piece of information, the subscript, p, represents a specific board member or SME, q is the number of iterations in the board discussion, and the sum is over m, the total board members and SMEs participating in the decision. Equation 42.3 then represents the decision, Y_n, reached by the decision-making board member with inputs from the other board members and SMEs. This

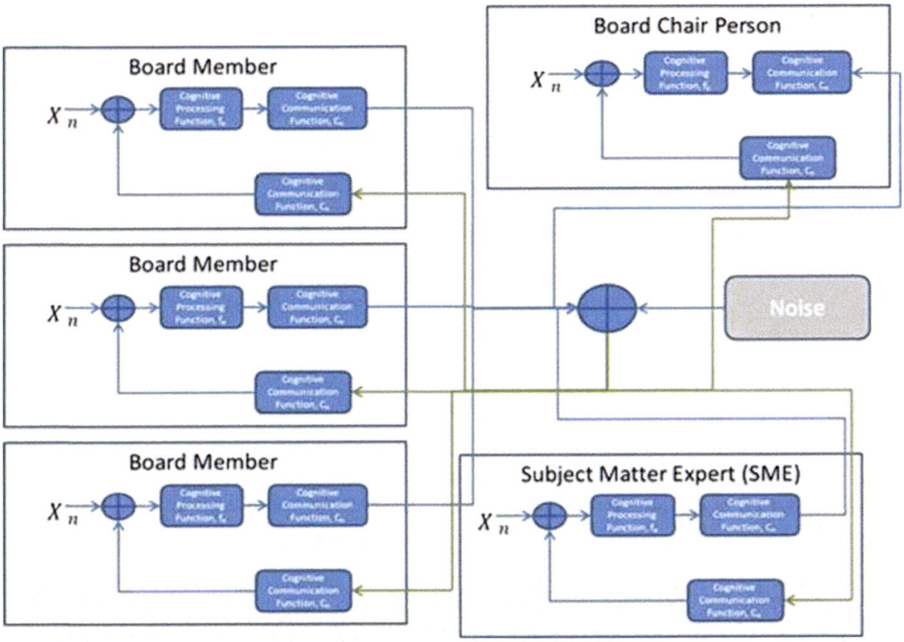

Fig. 42.4 Control theory board model with cognitive functions

model assumes all board members and SMEs start with the same basic information, X_n. It allows understanding of the information to vary among the board members represented by the function, $f_{p,q}$. In this model, if a board member or SME has no knowledge of the topic (i.e., ignorance of the subject), $f_{p,1}(X_n) = 0$. Similarly, if the board member withholds information on the topic, $C_p[f_{p,q}(X_n)] = 0$. Beyond this, the function, $f_{p,q}$ represents the level of understanding on a subject. Similarly, a decision to not share information is represented by this function as well. This function also encompasses preconceived ideas on the given information, preferences (personal or shared), intuition, deductive reasoning, and inductive reasoning. Clearly, the form of this function is complex. Some recent work in cognitive science [5, 6] may provide improvements in this representation.

Using this control theory-based model, the transfer function of the board can be represented as the ratio of the initial understanding of the information divided by the final decision as shown in Eq. (42.4):

$$T_n = \frac{X_n + f_{p,q}(X_n)}{X_n + \sum_q f_{p,q+1}\left\{ C_p \left[f_{p,q}(X_n) + \sum\limits_{m.m \neq p} C_m[f_m(X_n)] + \text{Noise} \right] \right\}} \tag{42.4}$$

This transfer function provides a model of the cognitive information processes as a beginning point to incorporate cognitive science models.

There are other information sources which can contribute to this model similar to SME inputs. These include text messages and emails to board members and personal side discussions (which also contributes to noise and affects the intake of other information). Since these inputs do not go to the whole board, but rather to individual members, and the external SME (particularly in electronic communication) may not be receiving all the board discussion, they have a function $C_s[f_s(X_n)]$, where s represents a specific SME, and there is no iteration with the board discussion, q. The inputs are single events since the external SME is not part of the board discussion.

42.3 Multiple Board Structures

A question often asked is what is the most efficient board structure? Will a single board suffice or are multiple boards more efficient? This has been a difficult question to answer. The set theory view of information theory provides the answer to this question.

A range that is too small (missing expertise) cannot be properly mapped to a decision domain for the intended outcome of the system. If this range can be mapped, then the missing expertise is not necessary for the decision in the context of the system. This immediately tells us that our board must have the right distribution of expertise for the system context and is therefore system specific. In addition, the information uncertainty increases as the information is distributed among separate boards as shown by the relation:

$$H(S, D, X, Y, Z) \leq H(S) + H(D) + H(X) + H(Y) + H(Z) \tag{42.5}$$

Information theory provides additional keys to understanding the board membership. Partitioning of information entropy, H, can only increase the uncertainty in the system by the relation:

$$H(p_1, p_2, \ldots, p_n, q_1, q_2, \ldots q_m) \geq H(p_1, p_2, \ldots, p_n). \tag{42.6}$$

Thus, as more members are added to the board (additional members are represented by q), more uncertainty is created in the decision. This is balanced by range mapping being complete. Thus, the board structure needs to have only those members necessary for the system decisions (satisfying the mapping condition) and no more (minimizing H). This applies to a single board structure or to a multiple board structure.

Within the set theory view of information theory, the board can be split (or delegated) if the information needed in one board is different than that needed in any other board. Then,

$$I_A \not\subset I_B \text{ and } I_B \not\subset I_A \tag{42.7}$$

so, there is no intersection of the information needed by the board and the board's domain (scope) can be different for each board.

When there is scope overlap, then I_A∩I_B, and the boards cannot be separated. In this case

$$I_A \subset I_B \text{ and/or } I_B \subset I_A. \tag{42.8}$$

42.4 Statistical Properties of Boards

Splitting a board into multiple boards where there is significant overlap greatly increases the information uncertainty, *H*, in the board structure as shown in Eq. (42.5). To examine this, we need to start with the characteristics of the uncertainty, or entropy, function itself. There are four (4) axioms the information entropy must meet:

42.4.1 Continuity

$$H(p_1, p_2, \ldots, p_n) \tag{42.9}$$

is continuous in all p_n. Thus, there are no discontinuities in the information probabilities. This means, as noted earlier, that the range maps completely to the domain within the board. Discontinuities lead to highly uncertain, or in some cases blind, decisions. A robust board has all disciplines (i.e., affected or contributing parties) represented. This satisfies the range to domain mapping criteria and the related continuity property.

42.4.2 Symmetry

$$H(p_1, p_2, \ldots, p_n) = H(p_2, p_1, \ldots, p_n). \tag{42.10}$$

Thus, the order of uncertainty does not contribute to the uncertainty in the decisions. This must be distinguished from temporal order of information sharing leading to a momentary information void on a subject until all aspects are explained for understanding. The process of understanding is always assumed to be complete in this model, and symmetry holds for a complete understanding of a subject. The order in which you discuss or think of a subject does not matter if you fully understand the subject.

42.4.3 Extrema

$$\mathrm{Max}[\boldsymbol{H}(\boldsymbol{p}_1,\boldsymbol{p}_2,\ldots,\boldsymbol{p}_n)] = \boldsymbol{H}\left(\frac{\mathbf{1}}{\boldsymbol{n}},\frac{\mathbf{1}}{\boldsymbol{n}},\ldots,\frac{\mathbf{1}}{\boldsymbol{n}}\right). \tag{42.11}$$

The maximum uncertainty arrives when all decisions are equally uncertain. If any single decision can be distinguished from the others, then the uncertainty to choose or not choose that option is smaller. Similarly, if no options satisfy the decision criteria, then the board has no information on which to base a decision leading to

$$\mathrm{Min}[\boldsymbol{H}(\boldsymbol{p}_1,\boldsymbol{p}_2,\ldots,\boldsymbol{p}_n)] = \boldsymbol{H}(\mathbf{0},\mathbf{0},\ldots,\mathbf{0}) = \mathbf{0}. \tag{42.12}$$

42.4.4 Additivity

If a probability of occurrence, p_n, can be subdivided into smaller segments, q_k, then the uncertainty can be represented as

$$H(p_1,p_2,\ldots,p_{n-1},q_1,q_2,\ldots q_k) = H(p_1,p_2,\ldots,p_n) + p_nH\left(\frac{q_1}{p_n},\frac{q_2}{p_n},\ldots\frac{q_k}{p_n}\right). \tag{42.13}$$

42.4.5 Principles

Following from these four properties, information can be subdivided during discussions if all the information is presented (i.e., all q_k is present in the discussion) without affecting the uncertainty of the decision. Note that this requires all information to be present. Subdividing boards and segmenting the information do not meet these criteria and result in higher uncertainty.

These four properties and five equations (Eqs. (42.9)–(42.13)) from information theory provide guidance in the structuring of boards. These relationships indicate that impossible solutions do not affect the information entropy. These solutions do not fit in the domain of the solution and cannot be mapped from the range of the original decision question. In addition, the continuity of H requires all information to be present for a decision. While a decision may be made with missing information, the decision is not actually addressing the original question. The question essentially changes, when all information is not present, and the decision addresses a different question than the one intended.

Information theory assumes a statistical basis of the information. Before we proceed further, we need to establish the statistical nature of boards, not that they are predictable but that their underlying operations can be represented statistically.

There are four principles that establish the statistical nature of a decision board:

1. Uncertainty exists in complex decisions. In these cases, simplifying assumptions lead to a lower understanding of the decision intricacies and a higher uncertainty (not always recognized) in the decision process. Interactions among differing factors in complex decisions have dependencies that are not recognized (ignorance) [4] or not well understood. Missing information is not always easily recognized. Factors not considered important in the decision can end up driving the system. Missing information comes from events (physical, chronological, or fiscal) not recognized as relating to the decision, unknown environments in which a system operates, unrecognized dependencies, and cultural biases (e.g., politics).
2. The uncertainty of which option is best collectively, and in some cases individually, leads to a statistical representation of which answer is best. In a board decision, the board vote is a statistical event with a distribution of yes and no positions. This is tied back to the cognitive functions. This statistical function is then combined with other statistical functions (i.e., other board members and SMEs) to produce a decision based on these functions.
3. The potential for misunderstanding (i.e., error) is also statistical. This includes miscommunication (not stating clearly what is meant or not understanding clearly what is stated (and therefore meant). These lead to unintended consequences in the decision-making process. These unintended consequences can be social, physical, chronological, fiscal, or environmental.
4. Cultural and historical bias lead to suboptimal decisions. Large social population actions form the basis for these biases and the effects on a person's cognitive information processing function, f_n, are statistical in nature.

Decisions can be represented statistically with various distribution functions depending on the individual preferences, biases, knowledge, and experience with the subject as discussed in the control theory model above. The cognitive processing functions, based on the properties of H, should fulfill continuity, symmetry, extrema, and additivity.

42.5 Information Bounds in the Board Context

In the board context, the board discussion forms the information channel as discussed above. The board members and SMEs are both information sources and sinks as modeled in Fig. 42.4. Information theory treats communication as the transmission of symbols. Natural language, where letters form words, words form sentences, and the order of the symbols and words are important in interpretation fits this model perfectly. And the board discussion is the channel where this information is transmitted between the board members and SMEs.

Information theory models the transfer of information through the board channel very well. A definition of terms is convenient at this point.

$H(X_n)$ is the average information shared by a single board member or SME as defined in Eq. (42.2).

$H(Y_n)$ is the average information received by a single board member or SME also following the definition in Eq. (42.2).

$H(X_n,Y_n)$ is the joint probability that what was shared by one member and heard by another (the average uncertainty in the total transmission through the board channel).

$H(Y_n|X_n)$ is the probability that one member actually heard what was stated by another. This brings in the effects of noise (and misunderstanding) in the channel. This focus is on the receiver of the information.

$H(X_n|Y_n)$ is the equivocation probability that one member actually stated what was heard by another. This brings in the effects of recovery (or proper understanding) of the information sent and is a measure of how well the information is understood by the receiving member.

If the board discussion is clear, and no misunderstanding is present, then the information provided by the speaker is accurately received by the listener (receiver). The information is perfectly transferred and information theory tells us that

$$I(X;Y) = H(X,Y) = H(X) = H(Y). \tag{42.14}$$

Now, if there is complete confusion, then what is stated is not related to what is heard. This is the case where the received information is independent of the transmitted information, and

$$I(X;Y) = 0. \tag{42.15}$$

In this case, no information is transmitted through the channel (i.e., discussion). These two extremes, perfect transmission and no transmission, provide bounds on the information sharing in a board meeting. Typically, neither of these conditions are achieved, and there is always some noise or misunderstanding during the discussion that limits the amount of information transferred among the board members.

42.6 Information Theory Representation of a Board

Set theory provides the mathematical basis for information theory which fits the board structure well. Information shared in a board discussion is the sum of all the information provided by the individual board members. This is illustrated in Fig. 42.5 for the example board structure used in Fig. 42.4.

This picture is somewhat complex in that there are many different areas of shared information. Note that the symbol, |, is read as "and not" so that $I(S;D,X|Y,Z)$ is the information shared between the SME, S; the board decision-maker or chair, D; board member, X; and not board members Y and Z. For a decision to be fully informed, the information for the decision must be contained in the center most ellipsoid, $I(S;D,X,$

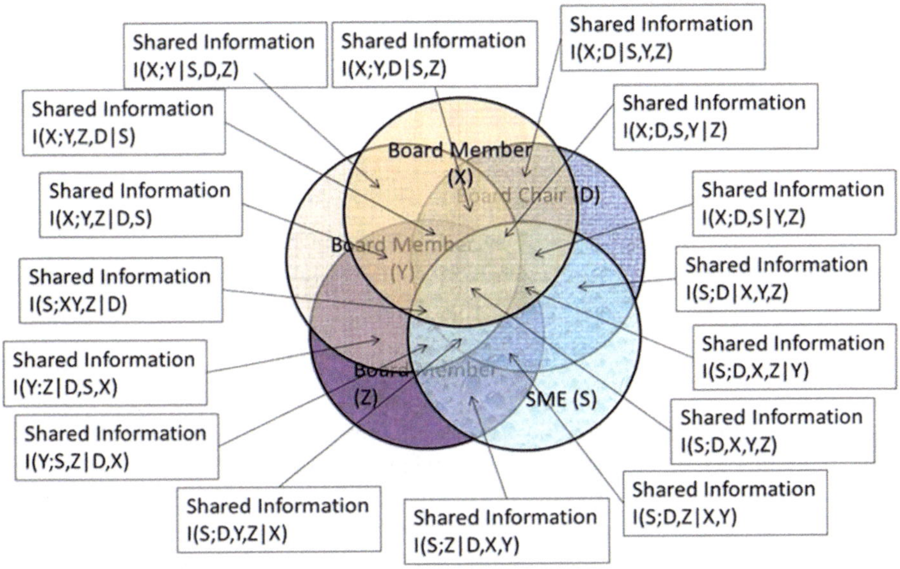

Fig. 42.5 Set theory representation of board

Y,Z). This represents the set of all information shared and received in the board discussion. Other information is shared based on the knowledge of individual board members and the SME, the ability of each to understand the information, and individual distractions. This can lead to board discussions which do not fully incorporate all board member knowledge. All permutations of this case are represented in the figure except for $I(Y{:}D,Z|S,X)$, $I(S{:}X,Z|D,Y)$, $I(S;X,Y|D,Z)$, $I(X{:}D,Z|Y,S)$, $I(S;X,Z|Y,D)$, which is an artifact of the figure geometry (where nonadjacent sets cannot be shown as excluded).

Information theory represents this as shown by Eq. (42.5) above:

$$H(S,D,X,Y,Z) \leq H(S) + H(D) + H(X) + H(Y) + H(Z)$$

where $H(S)$, $H(D)$, $H(X)$, $H(Y)$, and $H(Z)$ are how well the board members and SME communicate their information. This indicates that the sum of information can be no more than that provided by each of the members. Noise (distractions, misunderstanding, poorly stated (poor transmission)) and information not shared (intentional, unintentional, missing board member) invoke the inequality in the relationship. Following the work of Reza [3], set theory can relate the rules for information. This yields the following relationships:

$$I(X;Y) = f(X \cap Y) \qquad (42.16)$$

which is the expected value of mutual information shared in the discussion. In set theory, this is a function of the intersection of the information held by X and Y.

$$H(X, Y) = f(X \cup Y) \tag{42.17}$$

which is the average uncertainty of the discussion. This is a function of the union of the information available.

$$H(X|Y) = f(XY') \tag{42.18}$$

which is the information received by X given the information that Y shared. This is the probability that the board understood the information shared by Y. Note, in set theory this is a function of the information X has that Y does not.

$$H(Y|X) = f(YX') \tag{42.19}$$

which is the information shared by Y given the information that X heard. This is the probability that the board understanding is what was shared by Y. Note, in set theory this is a function of the information Y has that X does not.

From these relationships, then, perfect understanding occurs when $f(X) = f(Y)$ and both parties understand the information fully. When there is, no information shared $I(X; Y) = f(X \cap Y) = 0$. Thus, there is no intersection of the information sets and no common understanding. In the board example used above, $I(S; D, X, Y, Z). = f(S \cap D \cap X \cap Y \cap Z)$, and the shared information is represented in Fig. 42.5 by the intersection of the five circles representing the knowledge to share for each decision.

In these representations, $H(X)$, etc. represent the uncertainty in the information shared by board member X. This uncertainty stems from the board members understanding (or knowledge) of the decision requested and the associated decision factors, cultural bias (which indicates if information will be shared or withheld), and personal comfort in sharing specific information or engaging in debate about the information. Channel capacity (i.e., board capacity) in information theory is

$$C = \max(I(X; Y)) = \max(f(X \cap Y)). \tag{42.20}$$

Thus, channel capacity (i.e., the board capacity) for a decision is defined by the mutual information or the intersection of information, shared in the board discussion. The maximum board capacity then is based on the intersection of knowledge held by each board participant. The intersection represents the integration of individual board participant's knowledge to form a decision. Note, this indicates that if a board is segmented, and required knowledge for a decision is not present, then the board does not have the information necessary to decide. A decision can be made, but the scope of the decision does not address the actual question being considered. This results in unintended consequences for the decision because the board does not have all the facts.

One implication of this capacity is that a board with a missing member(s) will have a lower capacity since mutual information for the topic will be reduced.

Similarly, adding a member that has largely overlapping knowledge can create disjoint relationships where the two members approach the topic differently (based on their differing cognitive functions) and do not overlap in their understanding, and $I(S;D,X,Y,Z) = 0$ blocking the board decision.

42.7 Summary

Information theory provides a rich mathematical structure to model and understand decision-making structures within systems engineering and project management. The models produced by information theory provide guidance in the structuring of boards, both their membership and their relationship to other board structures. Since boards are established to control the system or project, control theory provides a model of the individual decision-making process. This provides a starting point for the future addition of cognitive processes in the decision-making model. Set theory understanding of information theory is also helpful in understanding the relationships between board members and the information that they share among the board.

References

1. Tishby, N., & Polani, D. *Information theory of decisions and actions*. Retrieved February 16, 2018, from http://www.cs.huji.ac.il/labs/learning/Papers/IT-PAC.pdf
2. Dorpinghaus, M., Roldant, E., Nerit, I., Mery, H., & Julicher, F. (2017). An information theoretic analysis of sequential decision-making. In *IEEE International Symposium on Information Theory (ISIT)*, Aachen, Germany.
3. Reza, F. M. (1994). *An introduction to information theory*. New York: Dover.
4. Merton, R. K. (1996). The unanticipated consequences of social action (1936). In P. Sztompka (Ed.), *Social structure and science* (pp. 173–182). Chicago: University of Chicago Press.
5. Fan, J. (2014). An information theory account of cognitive control. *Frontiers in Human Neuroscience, 8*, 680.
6. Thornton, C. *A new way of linking information theory with cognitive science*. Retrieved February 16, 2018, from https://pdfs.semanticscholar.org/c944/20f28388bc053fa9e1170e547e51aa4fc6c2.pdf

Chapter 43
Using MBSE in Satellite Architecture Trade Studies: A Practical Example

Daniel Cipera, David Jacques, and Thomas Ford

43.1 Introduction

Historically, operational space missions have been performed by large, highly capable satellite systems. While these traditional satellite designs meet or exceed very demanding performance and reliability requirements, their drawbacks of expense and potential vulnerability have incentivized consideration of alternative architecture types. The trade space for alternative architectures is limited only by creativity; architectures for a given mission could vary by satellite size, number, orbit, payload types, design life, or any other number of parameters or concepts. Evaluation of concepts and options within this trade space is helped by a structured and systematic approach; model-based systems engineering provides such an approach.

This paper explores the practical usage of model-based systems engineering to assess a possible alternative to traditional satellite architectures. Since the inception of the concept in 1999, CubeSats have been used for education and technology demonstration [1]. More recently, commercial companies have begun to deploy constellations of CubeSats for Earth imaging and other purposes [2]. CubeSats have the potential for being both cheaper and potentially less vulnerable, though with sacrifices in capability or performance. As an example of how MBSE can be used to assess solutions from a vast space architecture design trade space, this paper will step through the setup and results of a recent Master's thesis simulating traditional and CubeSat architecture performance in a disaster response imagery scenario. Given the size of this project, it was helpful to break the setup into four basic questions: what to design, what to analyze, how to design, and how to analyze.

D. Cipera · D. Jacques (✉)
The Air Force Institute of Technology, Wright-Patterson Air Force Base, OH, USA
e-mail: David.Jacques@afit.edu

T. Ford
Integrity Applications Incorporated, Beavercreek, OH, USA

S. Adams et al. (eds.), *Systems Engineering in Context*,
https://doi.org/10.1007/978-3-030-00114-8_43

This paper concludes with suggestions on how this example can scale toward analyzing other architecture solutions as well.

43.2 The "Whats": What to Design and What to Analyze

This project required constructing two conceptual models of electro-optical (EO) imagery satellite architectures: one "traditional" and one CubeSat. In this project, a "traditional" satellite architecture refers to satellites exceeding 1000 kg, the upper limit being the capacity of modern launch vehicles [3]. A CubeSat architecture refers to a satellite design solution that conforms to the CubeSat standard developed in 1999 [4]. Performance analysis to compare the two architectures required identifying a mission and a scenario to simulate. EO imagery in support of disaster response was chosen as a mission, with 2017s Hurricane Maria providing a realistic setting for the scenario. In such a disaster response scenario, responders require information that is high-quality and timely and addresses the entire affected area [5, 6]. Attributes of an imagery system that address these needs include spatial resolution, timeliness, and geographic coverage. These attributes form the basis for the measures of effectiveness (MOEs) used to quantitatively compare architecture performance.

Spatial resolution is a measure of whether an object of a given size is distinguishable from other nearby objects. It helps answers the question, "how useful is this image to an analyst?" A spatial resolution of 10-m ground sample distance (GSD) is sufficient for building identification and location, while discerning building damage requires a GSD on the order of 1 m [7]. Timeliness, for this scenario, refers to the amount of time between the natural disaster event (i.e., a hurricane making landfall) and the time a given image collected by the system is available for an analyst to exploit. Coverage refers to the amount of affected area that can be imaged at a nominal spatial resolution in a given timeline. In addition to MOEs, a set of notional performance requirements were established. These requirements are shown in Table 43.1.

Table 43.1 Mission requirements

Attribute	Threshold	Objective	Units
Spatial resolution	10	1	Meters
Timeliness	72	24	Hours
Coverage	11,522	Same	Square kilometers in a 72-h period
Access	37°	All	Degrees latitude
Change detection	Must be capable	N/A	N/A

43.3 The "Hows": How to Design and How to Analyze

A comparison of architectures could be as straightforward as manually creating and executing a simulation in STK. However, an MBSE approach provides a structured and disciplined, yet flexible, means of architecture development. Models provide a centralized source of traceability from scenario and mission requirements, to system requirements, and finally to physical designs themselves. Setting up an MBSE project requires three up-front decisions: choice of a modeling language, choice of method, and choice of a modeling tool [8]. Systems Modeling Language (SysML) was chosen as the language. The modeling method followed the system specification and design process from the Object-Oriented System Engineering Method (OOSEM). Described by Friedenthal, Moore, and Steiner [9], this method had a pedigree in space system modeling, having been used by Thompson [10]. This method outlines a design process of four steps:

1. Analyze stakeholder needs
2. Analyze system requirements
3. Define logical architecture
4. Synthesize candidate physical architectures

Figure 43.1 shows a mapping of steps taken for this project to Thompson's modified OOSEM process [10]. The bottom arrow from the "Architecture Performance/Cost Analysis" block back to the "Analyze System Requirements" block and "Synthesize Candidate Architecture" block illustrates the iterative nature of this process. This method was modified as needed; for example, the process in Fig. 43.1 does not describe the development of a "logical architecture" but rather

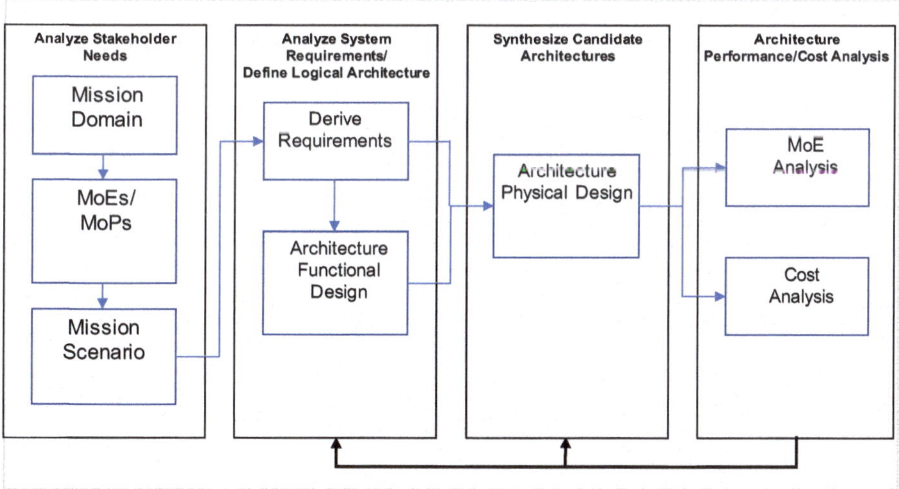

Fig. 43.1 Project steps mapped to OOSEM system specification and design process

describes the identification and analysis of functions to synthesize physical architectures.

The modeling tool for the project was Cameo Systems Modeler, version 18.5, by No Magic MagicDraw, from which Cameo is based and has been used for CubeSat modeling by Kaslow et al. [11]. Cameo was used to generate models consisting primarily of three SysML diagram types: use case diagrams, requirement diagrams, and block diagrams. These three diagrams enabled the designs to go from a general solution trade space to a model of sufficient fidelity for analysis. Details of the models developed for this project are described in depth in Cipera [12]. For architecture performance modeling and simulation, Systems Toolkit (STK) was used.

Comparing the system models requires development of measures of effectiveness (MOEs) from the relevant system attributes of spatial resolution, timeliness, and geographic coverage. In disaster response literature, spatial resolution is often measured by ground sample distance (GSD). GSD is the instantaneous field of view in one linear dimension for one pixel for a given sensor [13]. GSD depends on range and elevation, as well as the design parameters of focal length and detector pitch. The equation used by STK [14] to determine GSD is defined in Eq. 43.1.

$$\text{GSD} = \frac{\text{Detector Pitch} \times \text{Range}}{\text{Focal Length} \times \sqrt{\sin(\text{elevation})}}. \tag{43.1}$$

Slant range and elevation describe the geometry between the satellite and target at a certain point in time; these variables are outputs of the STK simulation for a given image collect.

Overall timeliness is a combination of the length of time (ΔT) of each contributing function of the architecture, from the time a target is affected to the time the imagery is available to an analyst. This relationship is defined in Eq. 43.2.

$$\text{Timeliness} = \Delta T_{\text{Target}} + \Delta T_{\text{Ground}} + \Delta T_{\text{Process}} \tag{43.2}$$

where:

ΔT_{Target} = time between hurricane landfall and image collect, with target access windows occurring at night disregarded.
ΔT_{Ground} = time between image collect and downlink to ground station.
$\Delta T_{\text{Process}}$ = time to process an image, from data download until a softcopy image is available to a user.

The two dominant terms in this equation are ΔT_{Target} and ΔT_{Ground}. ΔT_{Target} is a function of event timing and orbital mechanics. Event timing is a random variable accounted for by sampling 360 possible satellite locations at a given event start time. ΔT_{Ground} is a function of ground station placement and orbital mechanics. Ground station placement, including number of stations and their locations, is a design parameter; careful consideration of ground station placement in a regional scenario can help minimize ΔT_{Ground}. The time of ground station access for each satellite pass is an output of the STK simulation. In this scenario, $\Delta T_{\text{Process}}$ is assumed to have only minor variation between architectures and is assumed to be negligible compared to ΔT_{Target} and ΔT_{Ground}; this term was not calculated as part of the analysis.

Coverage, the overall amount of area imaged, for a simulated scenario is defined in Eq. 43.3.

$$\text{Coverage} = \text{A} \times \text{I} \times \text{P} \times \Delta T_{\text{scenario}} \times N_{\text{sat}} \tag{43.3}$$

where:

A = area/image
I = images/satellite pass
P = satellite passes/unit time
$\Delta T_{\text{scenario}}$ = time of interest between hurricane landfall and scenario end
N_{sat} = number of satellites.

With the assumption of a scanning sensor, the area imaged A is equal to swath width times the distance scanned on the ground. Images per pass are equal to the available target access time in a given pass divided by the amount of imaging time for a single image. This equation assumes no overlap between consecutive images.

Performance modeling and analysis were accomplished using a loosely integrated combination of Cameo, Python scripts, and Systems Toolkit (STK). Relevant model design parameters were transferred either manually or via Excel table into a Python script, which generated and ran the scenario in STK via STK's Connect interface. Outputs of the STK simulation, including target and ground station access, azimuth/elevation/range during each access, and target lighting time reports, were exported in Excel tables and ingested by post-simulation Python scripts, which calculated the measures of effectiveness. Modeling assumptions included:

- No cloud cover
- Nighttime satellite passes over target area do not count toward requirement satisfaction
- Regional scenario with no competing targets
- No real-time data relay

43.4 Performance Simulation Results

Performance simulation results for each architecture are in Table 43.2. Green indicates a value that meets an objective requirement. Yellow indicates a value that meets a threshold requirement but not the objective requirement. Red indicates a value that fails to meet the threshold requirement.

Table 43.2 Performance simulation results

	GSD (Meters)		Timeliness (Hours)		Coverage (Sq Km)	
	Traditional	CubeSat	Traditional	CubeSat	Traditional	CubeSat
Minimum Value	0.31	4.11	10.73	10.73	30938	16692
Maximum Value	2.51	32.74	21.78	23.34	46751	38601
Mean	1.78	23.21	15.44	16.59	35451	19697
Standard Deviation	0.63	8.26	3.28	3.50	3373	3580
% of Images Meeting Threshold	100%	10.21%				
% of Images Meeting Objective	14.44%	0%				

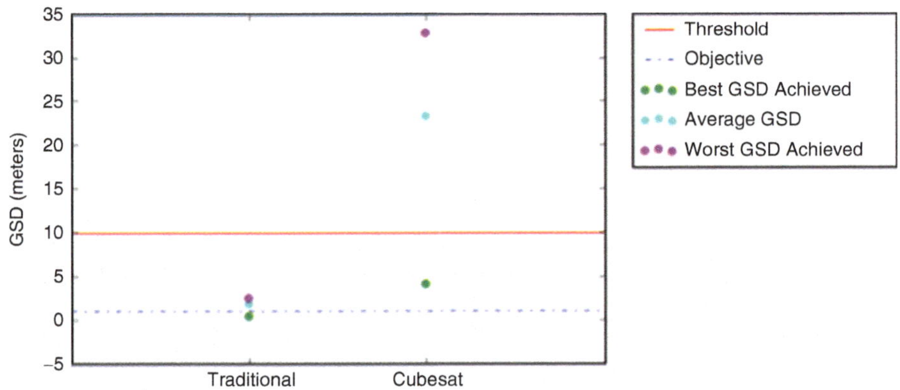

Fig. 43.2 GSD performance comparison

43.4.1 Spatial Resolution

Both architectures are capable of meeting the threshold GSD requirement of 10 m, indicating that both architectures would provide at least some useful imagery in response to a hurricane disaster scenario. The objective requirement of 1-m GSD is more challenging; only the traditional architecture is capable of meeting this value and only 14.44% of the time. A visual comparison of the best, worst, and average resolution of both architectures is provided in Fig. 43.3.

Compared to the traditional architecture, the CubeSat architecture has a much greater range of GSD values. As both the average value and Fig. 43.2 show, much of this range is above the 10-m threshold; most collects from the CubeSat design do not meet resolution requirements. In this simulation, 10.21% of collects meet the threshold requirement, and 0% meet the objective. The fact that most CubeSat collects would not meet the threshold requirement is accounted for when calculating timeliness and coverage.

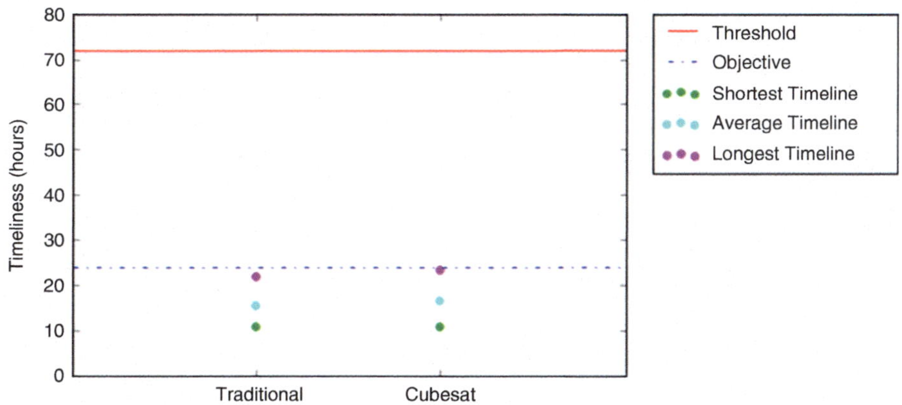

Fig. 43.3 Timeliness performance comparison: maximum, minimum, and mean

43.4.2 Timeliness

Timeliness is measured from the time of the event to the time the first collect is downlinked to a ground station. Both architectures meet the 24-h timeliness objective requirement, meaning that both architectures can provide imagery on a timeline relevant to users. Performance of both architectures is similar, with the average values being within 8% of each other. A visual comparison of these values is provided in Fig. 43.3.

43.4.3 Coverage

Both architectures meet the coverage requirement of 11,522 square km within 72 h of the disaster event, with some margin; this is visible in Fig. 43.4, where the solid line is the requirement and the lowest dots are the minimum achieved coverage values.

The results for coverage show that both architectures are capable of providing users with images of any location in this given affected area on a meaningful timeline. Recall that only collects meeting the threshold GSD requirement were counted toward meeting this requirement; this means that for the CubeSat architecture, even with only 10% of geometries/accesses yielding usable imagery, given enough satellites, this design solution is viable. Once again, however, the traditional architecture has more favorable performance, owing to a much higher percentage of images meeting threshold GSD requirements.

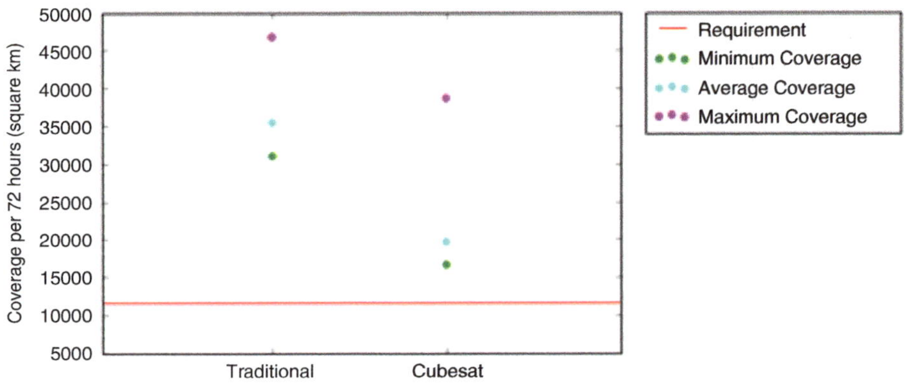

Fig. 43.4 Coverage performance comparison: maximum, minimum, and mean

43.5 Conclusions and Suggestions for Future Research

While it is intuitive that CubeSat-sized satellites would not directly match the performance of a larger traditional satellite architecture, this project demonstrated that the utility of CubeSats is not all that diminished compared to traditional architectures. Ground resolution is the most significant disparity between the two solutions in this simulation. A CubeSat architecture can provide useful EO imagery in the sub-10-m range for a portion of collects, but cannot meet a sub-meter requirement; a traditional architecture easily meets a sub-10-m GSD requirement and can meet a sub-meter requirement for a portion of collects. In terms of user needs in a disaster scenario, these results mean that CubeSat architecture imagery would be useful for identifying broader phenomena such as areas of flooding, but could not identify features such as individual structure damage. Imagery from the traditional architecture would be useful in addressing all user needs, but higher-resolution imagery would be less frequent.

For coverage, both architectures are capable of providing sub-10-m GSD imagery covering the entire island within 72 h. For the traditional architecture, this requires one satellite. The CubeSat architecture requires three satellites, owing partially to the fact that only a percentage of CubeSat collects meet the sub-10-m GSD threshold. For this scenario and set of design solutions, timeliness between architectures is comparable, with timelines meeting user needs for both architectures.

In using MBSE for this trade study, rather than a domain-specific, stand-alone analysis method, several advantages emerged:

- The OOSEM system specification and design process provided a structured and traceable means of defining and constraining trade space based on realistic user needs.
- After the initial work of developing one model instance, it was relatively easy to modify that base model to represent a different architecture.

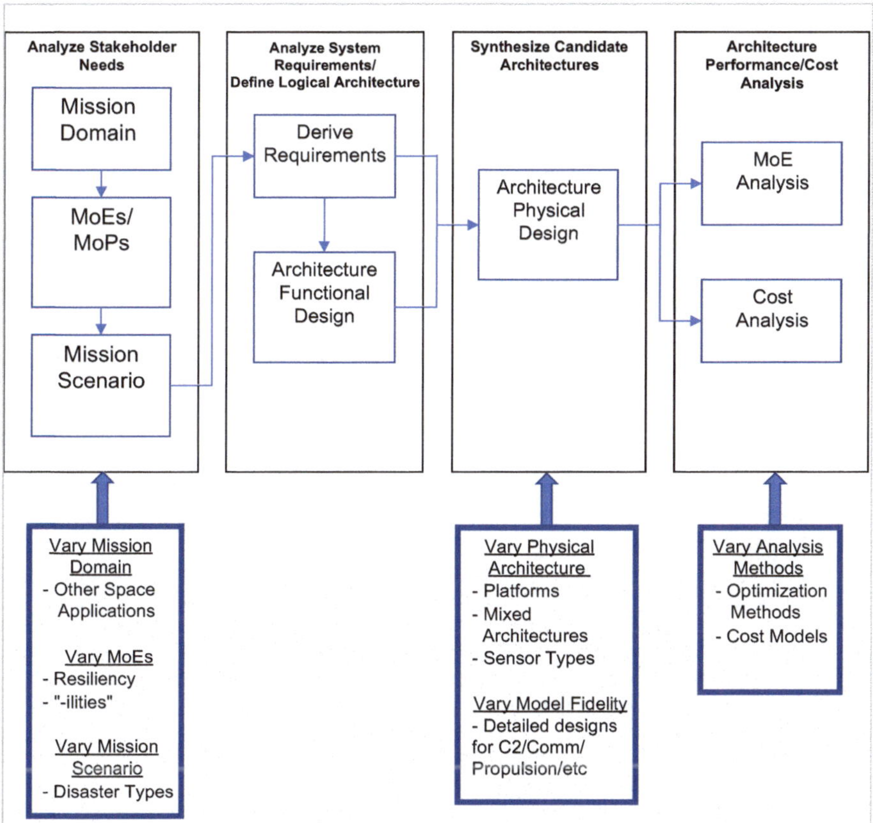

Fig. 43.5 Ideas for variation and further investigation in a traditional vs CubeSat trade study

- Using MBSE, along with system modeling tools like Cameo, prevents unnecessarily limited focus; Cameo enabled the modeling of all segments of the system, including ground functions such as command and control and image processing.
- Where more domain-specific analysis was desired, it was not difficult to export key model parameters for integration with a more domain-specific analysis tool such as STK.

The model developed for the traditional architecture, and its derivative CubeSat model, was used for one limited application. As alluded to in the introduction, these or similar models could be used to do much more. Figure 43.5 illustrates where "tap points" for varying trade space exist in the four steps of the OOSEM system specification and design process. It would be relatively straightforward to incorporate SysML models of other design solutions for this scenario, such as UAVs or a mixture of different sensors and satellites. Specific to the CubeSat/traditional architecture comparison, this paper focused solely on performance, without addressing measures of cost or vulnerability. As a result, this paper highlighted CubeSat limitations, leaving investigation of quantified advantages such as lower cost, launch flexibility, or increased resiliency for future research. Work is ongoing to integrate SysML-based models with cost modeling tools such as COSYSMO [15]; future investigation could involve tailoring this approach to space systems. Comparison of vulnerability, and its related concept of resiliency, requires the development of quantifiable measures of effectiveness/measures of performance.

References

1. National Academy of Sciences. (2016). *Achieving science with cubesats: Thinking inside the box*. Washington DC: The National Academies Press.
2. Boshuizen, C. R., Mason, J., Klupar, P., & Spanhake, S. (2014). Results from the planet labs flock constellation. In *28th Annual AIAA/USU Conference on Small Satellites*. Reston, VA: AIAA.
3. Konecny, G. (2004). Small satellites - A tool for earth observation? In *XXth ISPRS Congress Technical Commission IV* (pp. 580–582). Istanbul: International Society for Photogrammetry and Remote Sensing (ISPRS).
4. European Space Agency. (n.d.). *CubeSat concept*. Retrieved from eoPortal Directory: https://directory.eoportal.org/web/satellite-missions/c-missions/cubesat-concept
5. Battersby, S. E., Hodgson, M. E., & Wang, J. (2012). Spatial resolution imagery requirements for identifying structure damage in a Hurricane disaster: A cognitive approach. In *Photogrammetric engineering & remote sensing* (pp. 625–635). Fairfax, VA: American Society for Photogrammetry and Remote Sensing.
6. Department of Homeland Security. (2013, February 22). *Remote sensing and incident support*. 32. Retrieved from www.napsgfoundation.org/wp-content/uploads/2013/02/NAPSG-Remote-Sensing-Webcast-022213.pdf
7. Womble, J., Ghosh, S., Adams, B. J., & Friedland, C. J. (2006). *Advanced damage detection for Hurricane Katrina: Integrating remote sensing and VIEWS field reconnaissance*. Buffalo, NY: MCEER.
8. Delligatti, L. (2014). *SysML distilled*. London: Pearson Education.
9. Friedenthal, S., Moore, A., & Steiner, R. (2015). *A practical guide to SysML* (3rd ed.). Waltham, MA: Elsevier.
10. Thompson, R. E. (2015). *A methodology for the optimization of disaggregated space system conceptual designs*. PhD Dissertation, Air Force Institute of Technology, Department of Systems Engineering and Management, Wright-Patterson AFB.
11. Kaslow, D., Soremuken, G., Hongman, K., & Spangelo, S. (2014). Integrated Model-Based Systems Engineering (MBSE) applied to the simulation of a CubeSat mission. In *IEEE Aerospace Conference* (pp. 1–14). Big Sky, MT: IEEE Service Center.
12. Cipera, D. (2018). *Comparison of traditional versus cubeSat remote sensing: A model-based systems engineering approach*. Master's Thesis, Air Force Institute of Technology, Department of Systems Engineering and Management, Wright-Patterson AFB.
13. Evans, H., Lange, J., & Schmitz, J. (2014). *The phenomenology of intelligence-focused remote sensing*. New York: Riverside Research.
14. Analytical Graphics Incorporated. (2017, March). *Sensor resolution*. Retrieved from STK Programming Help: help.agi.com/stk/11.1/#stk/sn-13.htm?Highlight=GSD.
15. Edwards, D. J. (2016). *Exploring the integration of COSYSMO with a model-based systems engineering methodology in early tradespace analytics and decisions*. Masters Thesis, Naval Postgraduate School, Monterey, CA.

Chapter 44
Systems of Systems Thinking

Jo Ann Lane and Barry Boehm

44.1 Introduction

As more of our systems interoperate in one or more independently evolving systems of systems (SoS), more SoSs are overlapping while independently evolving, and user needs focus more on "new capabilities" rather than a "new system," so too do our systems thinking processes need to evolve to SoS thinking. Examples of multi-SoSs are crisis management SoSs for overlapping types of crises (hurricanes, floods, fires, earthquakes, terrorist attacks, etc.), international SoSs that operate differently in different nations, and defense SoSs that serve different defense-service needs. The coordination of multiple overlapping SoSs also requires more socio-technical thought and investment in trend analysis, potential conflict identification, and conflict-resolution negotiation. This paper presents a framework for defining and guiding SoS thinking.

44.2 Background

Much thought has been given to systems thinking. Early applications of systems thinking include the design and development of the Egyptian pyramids, with more rigorous thought and elaboration in recent times. Many definitions have been provided in an attempt to capture the essence of systems thinking from a broad point of view. Arnold and Wade [1] surveyed many of these definitions and used

J. A. Lane (✉)
San Diego State University, San Diego, CA, USA
e-mail: jalane@sdsu.edu

B. Boehm
University of Southern California, Los Angeles, CA, USA

S. Adams et al. (eds.), *Systems Engineering in Context*,
https://doi.org/10.1007/978-3-030-00114-8_44

them to create a model of systems thinking. This model has three constructs: purpose, elements, and the relationships between the elements of systems thinking.

The Systems Engineering Body of Knowledge (SEBoK) [2] in its description of "system of systems" indicates that SoS is a relatively new conceptual area of systems engineering and little thought has been given to extending systems thinking to SoS thinking. However, a literature review shows that several have discussed aspects of SoS thinking from various perspectives or within a specific domain. For example, Rebovich [3] discusses systems thinking from the enterprise perspective. He defines an enterprise as "people, processes, organizations, technology, funding" that interact with each other to achieve enterprise goals, with technology often comprised of multiple interacting systems. Within an enterprise, systems engineering and thinking expands to include "working across and bridging organizational cultures, agendas, and socio-political-economic differences" [3]. Others have discussed SoS thinking in policy development [4], improved transportation [5], and product development [6]. Lastly, the systems resilience literature provides further examples and guidance on socio-technical thought and investment in trend analysis, potential conflict identification, and conflict-resolution negotiation in rapidly evolving complex environments [7, 8].

Chen and Unewisse [9] have also given some thought to SoS thinking as applied to the conceptualization and understanding of military SoS. While their paper describes a few important aspects of SoS thinking, it primarily focuses on assessing operation and technical status of the "as is" SoS using four parameters: constituents' involvement, cooperation and interoperability, maturity of emergent behaviors, and cooperation uncertainty/disorder. Chen and Unewisse also discuss the development cycle of the SoS, but not of the constituent systems (CSs) that work together to perform SoS capabilities or the development cycle of a given capability. Their focus certainly informs some aspects of SoSE but is quite limited when compared to the breadth and depth of SoS thinking required to evolve the SoS to meet new needs and performance levels over time.

This paper presents a more comprehensive model of SoS thinking than Chen and Unewisse using the systems thinking constructs identified by Arnold and Wade [1] and elaborated using the SoS engineering (SoSE) framework described in the US Department of Defense (DoD) SoSE Guidebook [10]. The SoSE framework was developed through extensive interviews with and feedback from engineers responsible for the performance and evolution of 18 major US DoD SoS and reviewed with engineers with experiences in enterprises outside of the DoD. The SoSE framework elements identify the high-level activities performed by SoS engineers to manage and evolve an SoS. They can be summarized as follows, where elements 1–7 are from the DoD SoSE Guidebook [10] and element 8 is added to extend the DoD SoSE framework to the multi-SoS environment:

1. Translating SoS capability needs and objectives into SoS requirements
2. Assessing SoS performance with respect to SoS capability objectives
3. Monitoring and assessing the impact of environmental and CS changes on the SoS

4. Understanding the systems which contribute to the SoS and their relationships, strengths, and weaknesses with respect to SoS objectives
5. Defining an SoS architecture (or capturing the "as is" SoS architecture from a set of systems operating as an SoS) and evolving it to meet performance and capability needs
6. Analyzing SoS requirements and evaluating solution options
7. Orchestrating upgrades to the SoS through coordination, monitoring, and negotiation with the CSs
8. Participating in negotiating and evolving CS and SoS changes for CSs involved in multiple independently evolving SoSs.

This view of SoSE is not limited to DoD SoS. Rather, studies of SoS in other domains such as supply-chain management [11], healthcare [12], transportation [5], and utilities [5] show that the goals, concerns, and engineering activities at the SoS level are indeed similar.

Embedded in the DoD guidance for SoSE activities are aspects of SoS thinking necessary to successfully maintain and evolve SoS capabilities. These have been extracted to develop an SoS thinking model discussed in more detail in this paper. But before the SoS thinking model is presented, it is important to understand key differences and relationships between systems and SoS as well as systems engineering and SoSE as summarized in the next section.

44.3 Differences Between Systems and SoS and Between Systems Engineering and SoSE

For the purposes of SoS thinking, a single system can be viewed as physical entity with its own purpose, sponsorship (that includes funding), management, and engineering team to support and evolve the "physical entity" and the functions that it performs. This "physical entity," which may be referred to as a "system of interest," is often comprised of multiple components that may include hardware, software, and firmware. In addition, the system may contain aspects (hardware, software, or firmware features) that allow it to be tailored, configured, or adapted to a variety of environments. Systems engineering for this single system is focused on the system internals as well as the system interfaces to the outside world or user.

An SoS is a much larger "system of interest" that is comprised of a set of independent systems (i.e., single systems) that interoperate to perform cross-cutting capabilities. As a result, SoSE is focused on how the CSs currently interoperate to perform the desired SoS capabilities and the strengths and weaknesses of the constituents to continue to support the existing SoS capabilities as well as improved or new capabilities. SoSE is capability-focused with emphasis on options for flexibility, reconfigurability, adaptability, and evolvability of the SoS. While the technical aspects of the CSs are important to the SoS needs, so are the administrative aspects of the CSs. This includes compatible goals at both the system and SoS level,

CS funding to continue to upgrade and evolve the CS, as well as CS leadership commitment to support SoS needs. Also of concern to SoSE is how many SoS does a given CS participate in and are the evolutionary directions of all of these SoS compatible.

Successful SoSE means that the SoS both performs at least adequately today and continues to evolve to meet changing needs in a changing environment. Therefore, SoSE is concerned with aspects of the physical systems such as the long-term technical viability of SoS CSs; CS interoperability with other constituents; capability tradespace; current vs. desired capability performance and efficiency; overall SoS performance; CS ability to meet desired SoS performance; both SoS and CS dependability, safety, resilience, and security; compatibility of CS security approaches; CS options and limitations for meeting future needs; and the evolution of standards and conventions across the technical space. "Options and limitations for meeting future needs" also bring in less technical concerns such as political, managerial, and feasibility issues related to SoS stakeholder needs and commitment vs. CS and other SoS owner commitments, willingness to collaborate, and cross-agency or organization policies.

44.4 SoS Thinking Model

As mentioned earlier, the SoS thinking model presented here is based on the Arnold/Wade systems thinking modeling approach [1] elaborated for SoSE. The Arnold/Wade systems thinking constructs are model purpose, model elements, and relationships between model elements. The SoS thinking model elements and the relationships between the model elements are illustrated in Fig. 44.1 and described in the following sections.

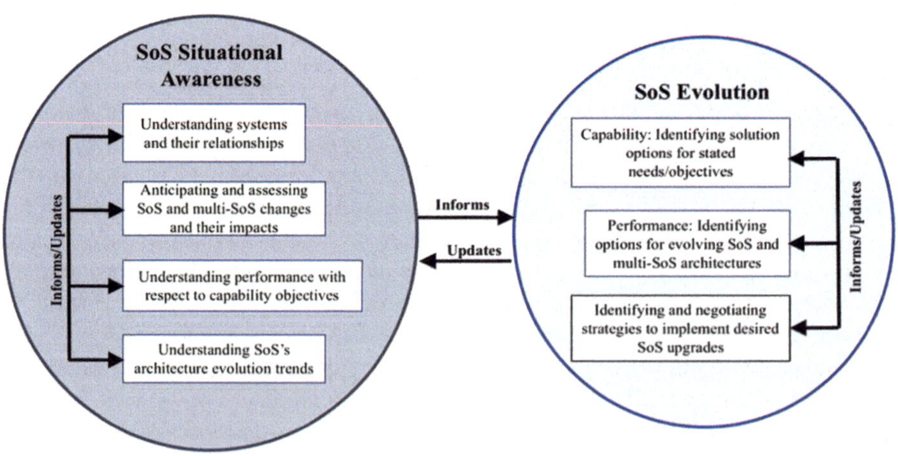

Fig. 44.1 SoS thinking model

44.4.1 Purpose of SoS Thinking

The first step in developing the SoS thinking model is to define the purpose of the model by extending the Arnold/Wade model purpose to the SoS and multi-SoS domains. Whereas "systems thinking" focuses on a single system, the SoS thinking model primarily focuses on a group of interacting systems or SoSs that perform capabilities that no single system can perform. In addition, SoS thinking must pay attention to the strengths and weaknesses of the SoS constituents to determine when upgrades are needed to constituents or constituents are becoming too fragile to support the SoSs they participate in.

Therefore, the purpose of SoS thinking can be described as follows: SoS thinking is a set of synergistic analytic skills used to improve the ability to identify and understand a set of related systems that function as an SoS, predicting their behaviors and devising modifications to or replacements for them in order to produce desired effects. These skills work together as a system of systems, providing technical, administrative, and strategic insights to CSs, SoS, and multi-SoS capabilities and the ability to evolve these capabilities.

44.4.2 Elements of SoS Thinking

An analysis of the SoSE elements [10] shows that many of the engineering activities of SoSE include SoS thinking. In fact, many of the SoSE components are described as activities required to understand current SoS and multi-SoS capabilities, the systems that contribute to the SoS capabilities, how well the CSs and SoSs interoperate with each other, the ability to modify the existing CSs both technically and politically, as well as the strengths and weaknesses of the SoS component systems that contribute to existing SoS and multi-SoS capabilities. The components of SoS thinking, illustrated in Fig. 44.1, can be organized into two categories: SoS situational awareness and SoS evolution.

44.4.2.1 SoS Situational Awareness

The SoS situational awareness elements are those that strive to understand the CSs of the SoS, how they contribute to SoS or multi-SoS capabilities, the performance of the existing capabilities, the ability of the architecture and CSs to support desired capability performance, the strengths and weaknesses of the CSs to support current capabilities, as well as the ability to evolve to support future capability enhancements and new capabilities. A CS's "ability to evolve" includes both the technical ability and the political ability to evolve, e.g., whether the CS leadership is willing to support current and future SoS needs. These SoS thinking elements are described in more detail in the following sections along with examples of questions that guide SoS thinking activities.

Understanding Systems and Their Relationships

The goal of this SoSE thinking element is to understand the CSs which contribute to the SoS or multi-SoSs and the relationships between the CSs over time. These SoSE thinking activities identify and monitor the evolution of CSs that can participate in the SoS of interest. The focus is on CS boundaries, interfaces, the SoS capabilities that the CS currently supports, as well as the functions and services available through the CS interfaces that might support future needs or improvements. Also, this activity monitors the life expectancies of the CSs: Is a given CS a new system with a long life expectancy, or is it close to retirement or replacement? Other questions that guide these SoS thinking activities include:

- Who are the owners and stakeholders associated with each CS?
- What are the goals/objectives of the current CS owners and stakeholders?
- Are these goals/objectives aligned with the SoS or multi-SoS goals/objectives?
- How compatible are the multi-SoS desired changes, and what needs to be done to harmonize them?
- What are near-term planned CS upgrades that may accommodate desired changes for SoS capabilities?
- When are CS upgrades going to be released (whether or not they include changes to support SoS needs), and will they be coordinated in a manner that does not adversely impact existing SoS capabilities?
- What are the key functions performed by each CS?
- How is funding managed at the CS and SoS levels?

A useful way to capture much of the technical information described above is in an SoS model or family of SoS models. One approach for developing this model uses Systems Modeling Language (SysML) notations to illustrate each SoS CS [13]. At the highest level, use cases can be used to capture the SoS capabilities and the external actors that interact with the SoS. For the SoS internal view, each CS is represented as an object class and the SoS CS functions and services as object attributes. Interface object classes are used to characterize each CS interface and the associated data attributes of each data item transferred over that interface. Finally, the relationships between the CSs are modeled using sequence diagrams that illustrate the flow of information/data between the CSs for each SoS capability of interest.

This aspect of SoS thinking can also identify several potential issues or problems that can adversely impact SoS capabilities if not sufficiently managed. Many of these are described as part of the Schedule Compliance Risk Assessment Method (SCRAM) for SoS [14] and include:

- CS organizational and technical issues.
- Slow or minimal progress on SoS capabilities already under development.
- CS limitations (e.g., CS priorities inconsistent with SoS or multi-SoS priorities, interoperability issues, fragile systems that are difficult to change).
- Immature technology or technology refresh coordination issues, especially those that may impact interoperability between CSs.
- Inaccessible CSs or incompatible data resulting from lack of planning for data/database conversions required for system upgrades.
- Capability deployments using an "all or nothing" approach rather than incremental rollout of capability parts.
- Asynchronous nature of CS changes to support capabilities and capability improvements when these related changes cross CS boundaries.
- Impacts related to any required SoS-level safety or security certifications.
- Existing CS technical debt that impacts the ability of the CS to support existing SoS or multi-SoS capabilities or evolve to support new or improved SoS capabilities.

Anticipating and Assessing SoS and Multi-SoS Changes and Their Impacts

Systems that participate in one or more SoS are typically independently owned and managed. This means that each CS is continually evolving based on the needs and objectives of its own stakeholders. In addition, each CS has its own engineering organization responsible for the evolution of the system. For a given SoS (or "SoS of interest"), this means that the SoS engineers must continually monitor CSs and other related SoS to anticipate changes that may impact the SoS of interest and, as those changes are better identified, assess the impact to the SoS of interest. Types of CS changes that might impact the SoS include technical changes such modernization/technical refresh of system components, implementation of new communications protocols or standards, as well as shifts in missions as the CS participates in other SoS. For CS changes that may adversely impact the SoS, SoS thinking can guide the SoS team identify solution options that will meet the needs of both the CS and the SoS. Likewise, there may be SoS external changes that may have implications for the SoS of interest. These might include changes in other SoS that the SoS of interest interoperates with, national or international events or opportunities, or game-changing technology breakthroughs. Questions that may help to guide these activities include:

- What changes are the CSs planning and which of these might impact SoS capabilities?
- How significant are the identified impacts? Which SoS capabilities might be impacted?
- Are fragile/nearly obsolete systems nearing retirement/replacement? Do the current needs/objectives depend on the "near retirement" system? If so, can the "near

retirement" system be updated? If not, do SoS capabilities requiring updates to the "near retirement" system need to be migrated to other systems within the SoS?

Understanding Performance with Respect to Capability Objectives

One of the key areas that SoSE pays attention to is the overall performance of the SoS with respect to the objectives of its SoS-level capabilities. In general, SoS performance often refers to single system performance features such as transaction speed/response time, number of users/sites supported, data security/protection, and safety at the SoS level. Over time, newer CSs can receive more attention and investment, while older systems closer to retirement can languish, often without thought on how these older systems may impact the performance of key SoS capabilities. To better understand SoS performance, SoSE sets up strategies to monitor performance for key capabilities. Because it can be difficult to measure SoS performance in developmental testing, key measures are more often captured in the operational environment and tracked over time. If performance starts straying from desired levels, SoSE performs causal analysis to determine why and what changes can be made to bring the SoS back to the desired level of performance. If the objective is to achieve better performance, SoSE evaluates the SoS to determine where strategic changes might be made that would provide this better performance. SoS thinking is key to both causal analysis of degraded performance as well as opportunities to make strategic changes to increase performance.

Another area of SoS performance that deserves attention is emergent behaviors. In some cases, SoS are designed to create desired emergent behaviors, and these can be monitored similarly to the performance characteristics listed above. However, over time in an SoS environment, unexpected (and sometimes undesirable) emergent behaviors arise under certain conditions. Sometimes these unexpected emergent behaviors are initially dismissed as an "unexpected anomaly" that might have happened due to an erroneous input. However, if it happens once, it will more than likely happen again. Through SoS thinking, tests can be developed to determine the cause(s) of these unexpected emergent behaviors and ensure that they do not become SoS vulnerabilities or problems.

Understanding SoSs Architecture Evolution Trends

The SoS architecture describes how the CSs are interconnected and how they share data and information to accomplish SoS capabilities. Data sharing includes data elements and protocols and how this is synchronized across the SoS. Questions to guide SoS thinking in this area include those that inform the "as is" architecture and how the users adapt it for various needs: How do the systems or SoSs work together; how do the users use/configure systems; how well does the architecture support current SoS functionality, nonfunctional requirements, data flow, communications, and multi-SoS interactions; what are the strengths and weaknesses of current architecture; and what are the impacts of the architecture weaknesses or limitations?

44.4.2.2 SoS Evolution

SoS evolution is typically more of a planning, negotiation, and oversight activity since the SoS leadership seldom has strong control or influence over the CSs that are independently owned and managed. In addition, CSs are more influenced and guided by their single system stakeholders. In some cases, a given CS may participate in multiple SoS. In these cases, the CSs must decide how to prioritize the needs of their own stakeholders as well as the needs of all the SoSs in which they participate. The SoS evolution process starts with translating new or improved SoS capability objectives into requirements which can then be used to identify options for further consideration.

Identifying Solution Options for Stated Needs/Objectives

Through engineering analyses and negotiations with CS engineers and leadership, options for meeting the objectives are identified and further analyzed. These tend to focus on providing new capabilities (or functions) or improving or expanding existing capabilities. In the early stages, this effort can be relatively independent of the CSs to open up and explore broader tradespace options. But as the options mature, they must be adjusted to reflect feasible approaches. Systems thinking plays a key role in these analyses which attempt to answer questions such as:

- What is driving each stated need or objective? Can stated needs/objectives be combined into a single solution?
- Will technology upgrades/refreshes address some of the needs/objectives? Which CSs will be included in the technology upgrades/refreshes? If not all can be upgraded/refreshed at the same time, which capabilities will be impacted?

Identifying Options for Evolving SoS and Multi-SoS Architectures

In addition to identifying solution options for technical needs/objectives, the SoSE team is also responsible for ensuring that the SoS architecture provides adequate performance as systems come and go from the SoS and as systems evolve over time. Areas of SoS architecture concern are similar to single system architecture concerns and often include aspects of configurability, adaptability, robustness, response time, network bandwidth, safety, and security, all at the SoS level. SoS thinking with respect to the SoS architecture is often guided by questions such as:

- What is required vs. desired change in performance/robustness/throughput/safety/security/etc.?

- What is the criticality of the required/desired changes (e.g., improved capability response times vs. security/safety vulnerabilities)?
- Are some of the CSs (or some of the CS components) no longer supported by vendors? If so, are there known replacement vendors? If so, what are the impacts of migrating to the replacement vendors? What is the actual impact if obsolete systems (or systems with obsolete components) are retired without replacement? Can any key functions be migrated to other CSs?
- Can incremental changes over time suffice?
- What are the current bottlenecks impacting performance/robustness?
- Are the bottlenecks within systems? Between systems? Between multi-SoSs?
- Will new system(s) planned for the SoS increase/decrease the need for added performance?
- Will planned technology refresh(es) resolve some/many of the performance issues?
- Will the SoS be prepared to interoperate within potential multi-SoSs?

Identifying and Negotiating Strategies to Implement Desired SoS Upgrades

At this step, SoS thinking is used to determine the set of options for unmet needs/objectives and architecture issues to implement in the next SoS evolutionary cycle. Options are reviewed along with their expected costs, risks, and return on investment. Opportunities within the CS upgrade cycles are re-evaluated and compared to the desired changes. Priorities are reassessed and funding strategies are evaluated to determine which options are most feasible. For example, if there are critical changes that may impact safety or security or upgrades that are needed to replace a system that is being retired, these may take precedence over new capabilities or improvements to new capabilities. There may also be situations where lower-priority needs can be included with little additional costs or schedule impacts if they can easily be combined with higher-priority options. Questions that guide these activities are used to assess the expected return on investment in the SoS CSs and might include:

- Where are the CSs in their current upgrade cycle?
- What opportunities (if any) are available to influence upcoming CS upgrades?
- What is/are the anticipated improvement(s) in SoS capabilities for each option?
- How soon is the capability improvement needed? (In the SoS environment, non-optimal solutions may be selected because they can provide new capabilities or improve existing capabilities sooner than the optimal solutions.)
- When will improvement(s) be available to the users if a given option is selected?
- What is the expected life remaining in the CSs targeted for change?

44.4.3 SoS Thinking Element Relationships

Most, if not all, of the CSs within an SoS are periodically upgraded on their own evolutionary paths to meet their own single system stakeholder needs. Much of SoS thinking is focused on understanding SoS technical, managerial, and political aspects in terms of CS constraints, goals, and plans for achieving those goals. As a result, much of the SoS thinking element activities utilize models of the SoS, collaboration with CS engineers to understand CS internals, and collaboration and negotiation with CS leadership to understand CS evolutionary directions and possibilities for changes to constituents to achieve the desired SoS capabilities and capability performance. Abstracting these activities to reflect the iterative and collaborative nature of SoS thinking at the SoS, multi-SoS, and CS levels, the relationships between the SoS elements become a flow of information that causes other elements to update their models, options, and plans for providing new and improved SoS capabilities. These update activities may include adjustments to current models, options, and/or plans for the current upgrade cycle; the selection of other alternatives for meeting needs when current plans are no longer viable; reassessment and reprioritization resulting from the deployment of the previous set of upgrades; and update of the SoS backlog of desired changes and upgrades.

44.5 Conclusion

We have proposed a model for SoS thinking for today's environment where most systems are part of one or more SoS and, in some cases, operate in multi-SoS environments. We have also shown how this model was designed to better inform SoSE activities and the selection of options to evolve SoS and SoS capabilities. This approach moved systems thinking to a new level where you can no longer focus on a single system and its environment but need to focus on an environment where "everything is connected to everything" and better understand the potential impacts of what seem to be minor changes but can seriously impact the interoperability of systems in an SoS or multi-SoS context.

References

1. Arnold, R., & Wade, J. (2015). A definition of systems thinking: A systems approach. In *Proceedings of the Conference on Systems Engineering Research, March 17–29, Hoboken, NJ.*
2. SEBoK Authors. " Systems of Systems." in BKCASE Editorial Board. (2017). *The Guide to the Systems Engineering Body of Knowledge (SEBoK), v. 1.9. R.D. Adcock (EIC).* Hoboken, NJ: The Trustees of the Stevens Institute of Technology. Retrieved February, 2018, from www.sebokwiki.org. BKCASE is managed and maintained by the Stevens Institute of Technology

Systems Engineering Research Center, the International Council on Systems Engineering, and the Institute of Electrical and Electronics Engineers Computer Society.
3. Rebovich, G. (2006). *Systems thinking for the enterprise: New and emerging perspectives*. Milton Keynes: MITRE publications. https://www.mitre.org/publications/technical-papers/systems-thinking-for-the-enterprise-new-and-emerging-perspectives
4. Hipel, K., Fang, L., & Bristow, M. *System of systems thinking in policy development: Challenges and opportunities*, published on-line April 2015. http://onlinelibrary.wiley.com/doi/10.1002/9781119036821.ch2/summary
5. Parker, J. *Applying a system of systems approach for improved transportation*, Sapiens Vol. 3, No. 2. http://journals.openedition.org/sapiens/1011
6. Belay, A. M., Helo, P., & Welo, T. SoS thinking in product development: A system dynamic approach, In *IEEE SoSE Conference 2012*. http://ieeexplore.ieee.org/document/6384193/
7. Hollnagel, E., Woods, D. D., & Leveson, N. C. (Eds.). (2006). *Resilience engineering: Concepts and precepts*. Aldershot: Ashgate.
8. Hollnagel, E., Paries, J., Woods, D. D., & Wreathall, J. (Eds.). (2011). *Resilience engineering perspectives volume 3: Resilience engineering in practices*. Farnham: Ashgage.
9. Chen, P., & Unewisse, M. (2017). SoS thinking: an approach to conceptualising and understanding military systems-of-systems. *International Journal of System of Systems Engineering, 8*(1), 74–101.
10. Department of Defense. (2008). *Systems engineering guide for system of systems, version 1.0*.
11. Friedman, T. (2005). *The world is flat: A brief history of the twenty-first century*. New York: Farrar, Straus and Giroux.
12. U.S. Government Accountability Office. (2013). Patient Protection and Affordable Care Act: Status of CMS efforts to establish federally facilitated health insurance exchanges, GAO-13-601: Published: Jun 19, 2013. Publicly Released: Jun 19, 2013.
13. Lane, J., & Bohn, T. (2013). Using SysML modeling to understand and evolve systems of systems. *Systems Engineering Journal, 16*(1), 87–98. (published on-line on November 2012).
14. Lane, J., Clark, B., Pitman, A., & Tuffley, A. (2015). SoS capability schedule prediction. In *Proceedings of the IEEE 10th Annual Systems of Systems Engineering Conference, May 17–20, San Antonio, Texas*.

Chapter 45
Information Entropy-Based Complexity Measurement for Systems Engineering and Trade-Off Analysis

Jean Charles Domerçant

45.1 Introduction

Over time, engineered systems have become more interdependent and interconnected due to technological advances in areas such as networking, embedded computing, and communications. While there are many technical and nontechnical factors that influence system design and development, complexity is often blamed for a host of problems including failed system development, cost and schedule overruns, unmaintainable and unrepairable systems, and obsolescence upon delivery [1]. Complexity is not inherently a bad property and helps meet evolving capability needs as more becomes demanded of systems.

While complexity adds value, affordability and complexity are two conflicting objectives that require compromise. Therefore, only value-added complexity must be incorporated into the system architecture and design. The impact of early architecting and design decisions ripple through the entire system life cycle, affecting downstream activities such as manufacture, maintenance, and operations. In some industries, 75–85% of life cycle production costs are determined during *conceptual design* [2]. Conceptual design is the first phase in the design process, where the least knowledge exists but also where decisions that are made have the greatest consequence on cost, schedule, and performance. With this in mind, it is imperative to manage complexity early during this phase of design.

Before system complexity can be managed, it must first be measured. However, a standard measure of complexity for systems engineering is not currently well defined. A recent, comprehensive study of complexity measures reaches the conclusion that "Most conceptions of complexity measures in the theoretical literature could not be applied to systems engineering easily, and many concepts of complexity within engineering do not have a close tie to theory" [3]. The goal of this research

J. C. Domerçant (✉)
Georgia Tech Research Institute, Atlanta, GA, USA
e-mail: jean.domercant@gtri.gatech.edu

S. Adams et al. (eds.), *Systems Engineering in Context*,
https://doi.org/10.1007/978-3-030-00114-8_45

is to develop a standard method of measuring complexity that addresses these shortcomings, enabling more informed trade-offs to meet design objectives.

45.2 Background: Causes of Complexity in Engineered Systems

The first step in defining a measure is understanding the causes of complexity in engineered systems. A system is considered an entity composed of separate yet interrelated system *nodes*. A node may refer to a subsystem or even an entire system that is part of a larger family or system of systems, depending on the scope. Nodes interact through the sharing and processing of resources such as information, energy, materials, and time to perform functions [4] and contribute to the overall system's behaviors, states, and processes. A complex system consists of interconnected parts that as a whole exhibit one or more properties (behaviors among the possible properties) not obvious from the properties of the individual parts [5]. Examples of non-obvious behavior include emergent and adaptive behavior, self-organization, and the difficult-to-predict interactions between elements of a system.

Many of today's systems have evolved from simpler arrangements into the more complex counterparts we see today. A prime example is avionics, which serve as the computing infrastructure aboard military and commercial aircraft. Avionics are responsible for aiding in navigation, flight control, and information display. Two to three decades ago, systems such as these were generally stand-alone, discrete systems or subsystems. Their function could be easily understood and their scope or purpose was very limited [1, 6]. Over time, information sharing between various components increased to reduce the number of black boxes required by each system. For example, "a single sensor providing heading and rate information could provide data to the navigation system, the weapons system, and the pilot's display system." The advent of digital technology and increased computational capability led to the following changes that also caused an increase in complexity [7]:

- Dramatic increases in functional integration
- Added functionality to meet expanding capability roles
- Increased performance requirements
- Increase in the number of subsystems
- Increased functional overlap/redundancies between subsystems and a blurring of functional boundaries
- Increase in the total amount of information/resources processed by the system
- Increased sharing of information/resource among existing subsystems
- Increased physical integration to meet more demanding size, weight, power, and other constraints

As a result, avionics have seen an increase in "performance, sensor types, functionality, cost, integration, complexity, supportability (reuse), software

Table 45.1 Six types of complexity [1, 3]

Complexity type	Example subtype
1. Structural	Size: number and types of elements, size of development process, total number of requirements
2. Structural	Connectivity: number, types, density, and strength of connections, connectivity of development process
2. Structural	Architecture: patterns, chunkiness of connections, inhomogeneity, boundaries
3. Dynamic	Short term: sudden rapid change in system behavior, development system behavior
4. Dynamic	Long term: changes in number and types of things and relationships, evolution of purpose
5. Sociopolitical	Social and political: human cognitive limitations, multiple stakeholders, global context, environmental sustainability, economics, "coopetition," supplier chain depth, distributed development

programs in terms of executable code, memory requirements, throughput, reliability, data handling, data links, and obsolescence" [6]. Simultaneously, there has been a reduction in size, weight, power consumption, and technology windows [6]. This is a pattern that applies to many other types of systems as well. The Defense Advanced Research Projects Agency (DARPA) has recently tried to address the significant growth in development time and cost with increasing complexity for aerospace defense systems [8, 25].

A detailed study of the complexity literature and existing complexity measures resulted in a framework for categorizing complexity types and their impact on system development efforts [1, 3]. A brief summary is provided in Table 45.1.

It is possible to draw lines of cause and effect between the previously mentioned architecture and design changes witnessed in avionics to the many complexity types listed in Table 45.1. Additionally, a complementary view on the causes of complexity in engineered systems is based on a mapping of interactions vs. coupling [9]. Systems such as aircraft (with their onboard avionics) possess both complex interactions and tight coupling. Even here, "the degree of coupling and interaction types have been inferred from a rough idea of the frequency of system accidents in the various systems, rather than derived from analysis of the properties of the systems independent of the nature of their failures" [9]. This further emphasizes the need for an objective quantification of independent, observable system properties that contribute to overall system complexity.

45.3 Technical Approach

45.3.1 Defining the Context

The next step is to understand the context in which systems are architected and designed. For military acquisition, this begins with the systems engineering process (SEP). The SEP transforms the customer's stated needs and requirements "into a set of system product and process descriptions" [10] while generating information for decision-makers. Similarly, a generalized, domain-based mapping of the design process has also been developed that can be applied to many different fields—software, hardware, systems, materials, organizations, and manufacturing systems [11]. The following is a description of the various domains:

- Customer domain: characterized by the attributes (CA) that the customer is looking for in a product or process or system or materials or organizations.
- Functional domain: the customer needs are specified in terms of *functional requirements* (FRs) and *constraints* (Cs). FRs are the minimum set of independent requirements that completely characterize the design objective based on CAs.
- Physical domain: in order to satisfy the specified FRs, *design parameters* (DPs) are conceived in the physical domain.
- Process domain: finally, to produce the product specified in terms of DPs, a process is developed that is characterized by *process variables* (PVs) in the process domain.

During the design process, architectures are generated to better describe and understand the system [12]. An architecture is defined as the structure of components, their relationships, and the principles and guidelines governing their design evolution over time [13]. The following is a generalization of the different architecture views available:

- Functional architecture details the complete set of functions to be performed and their sequence; it identifies and structures the allocated functional and performance requirements [11].
- Physical architecture details how the system is physically divided into subsystems and components [11].
- System architecture identifies all the products necessary to support the system [11].
- Data architecture defines the structure and meaning of data to ensure consistency and proper management; it also defines the approach toward the structure, semantics, redundancy, and storage of data [14].

In contrast to architecting, design is a decision-making process intended to produce technically feasible and economically viable solutions. Each solution represents an integration of system elements under both logical and physical constraints and is often a compromise of competing attributes and objectives according to the level of technology present at the time. Within this context, the overall approach to measuring complexity will be deemed useful if the causes of complexity can be captured at both the architecture and design levels of abstraction.

45.3.2 Uncertainty, Entropy, and Information Theory

Defining an absolute measure of complexity is not the goal of this research. The diversity of both natural and engineered systems makes defining an absolute measure of complexity difficult at best [15, 16]. Also, many existing complexity measures tend to be very domain specific or too theoretically abstract to usefully apply to real-world systems [15–17]. This research focuses on enabling informed trade-offs during the design process by developing a method to characterize system complexity as it relates to the functions, resources, and interactions between nodes. Information measures prove useful in this regard, as they "provide a precise method of dealing with trade-offs between knowledge and ignorance, and they supply a useful definition of complexity" [18]. In particular, *information entropy* [19] provides a statistical measure of information that relates the fundamental concepts of uncertainty, probability, and entropy, where *entropy* is a measure of unpredictability or the degree of randomness of a thermodynamic system [20]. Equation (45.1) is the mathematical formulation of the *information entropy, H*:

$$H = -K \sum_{i} p_i \log p_i \tag{45.1}$$

Information entropy "takes the concept of entropy out of the restricted thermodynamic setting in which it arose historically and lifts it to the higher domain of general probability theory" [20]. Maximum entropy occurs when all states are equiprobable. Furthermore, Eq. (45.1) has unique properties that qualify it formally as an information measure, including nonnegativity, symmetry, accumulation, and convexity [18]. The choice of a logarithmic base is arbitrary, as well as the value of the proportionality constant, *K*. When *K* is one and the base 2 logarithm is used, the unit of entropy is called a bit [19]. Therefore, a system that can be in two equiprobable states contains one bit of information. Information entropy provides a common basis of measurement across both the architecture and design levels of abstraction, making trade-off analysis easier. The challenge then lies in using information entropy, which is a measure of uncertainty, in a way that captures the fundamental interactions and relationships among nodes [2].

45.4 Architecture Complexity

In order to capture the different sources of system complexity at the architecture level of abstraction, a measurement framework [4] is used that characterizes an architecture according to two principal domains. The identified domains are the *functional domain* and the *resource domain*. Within each domain, there exists a *state* complexity measure, as well as a *processing* complexity measure. Each measure is defined as follows:

- Functional domain:
  - *Functional state complexity (FSC):* The allowable variation in either functional requirements or function outputs that determines the number of distinct functional states the system can inhabit and that must be accounted for.
  - *Functional processing complexity (FPC):* The accounting of all the potential independent process sequence paths that result during the execution of system functions.
- Resource domain:
  - *Resource state complexity (RSC)*: The allowable variation in resource properties that determines the number of distinct values or states these resources inhabit and that must be accounted for. Data/information, energy, time, and materials are examples of different categories or types of system resources that are exchanged and processed by the system.
  - *Resource processing complexity (RPC)*: A measure of the capacity to share, communicate, and process resources between system components.

Measurement within each domain makes use of the tools available to the architect or designer during the conceptual design phase. The following sections illustrate how this is accomplished.

45.4.1 Functional State Complexity

The role of a system function, as in mathematics, is the transformation of an input to a usable output. The functional architecture provides a description of the system in terms of what it does logically and in terms of the performance required when functions are executed [10]. A system will occupy certain functional states at any given time, based on if or how each function transforms given the resources. For example, navigation is a common avionics function where the system relies on information from sensors to determine the position and velocity of the center of mass of the aircraft. If the avionics were only able to compute a single value pair for position and velocity given a range of inputs, then the function would be a constant transform and thus highly predictable and simple. If, however, the position and velocity output are highly nonlinear and exhibit a wide range of values, this would lead to increased functional complexity, as there would be more distinct possible functional states describing the system.

It cannot be assumed that system functions are well defined during the conceptual design phase since the system components that will carry out these functions may not have been assigned or even exist. Thus, another method of specifying the number of possible states for a given function must be used. With this in mind, FSC is measured using the performance requirements defined in the functional architecture. FSC is calculated using Eq. (45.2) for each performance requirement:

$$\mathrm{FSC} = \sum_i log_2 \frac{| b_i - a_i |}{r_i} \tag{45.2}$$

The values of b_i and a_i in Eq. (45.1) represent the bounds of different subranges or "bins" within the overall *range* of a single performance requirement. This formulation assumes there is no constraint that bins within a performance requirement must be of uniform size or resolution. Meanwhile, r_i represents the associated *resolution* for each bin or the smallest significant/measurable interval. Further increases in resolution mean comparatively more fine-grained measurements are necessary in order to differentiate between significant functional states, adding to complexity. Intuitively, this matches the effect observed in chaotic systems, where the sensitivity of the parameters plays a key role. Table 45.2 provides a summary of example FSC calculations for a sample avionics terminal electrical requirement [7].

Table 45.2 illustrates how trade-offs in range and resolution of performance requirements affect architecture complexity. Also, the more functions added to the functional architecture, the greater the complexity. The FSC for each performance requirement is then summed to determine the total number of possible functional states.

Table 45.2 Functional state complexity for an avionics performance requirement

Function: changes in input level (transformer coupled)	Range (volts)	Resolution (volts)	Number of function states	FSC (bits)
1. Base requirement	**0.86–14.00**	**0.01**	**1314**	**10.36**
2. Increase in range	0.86–20.00	0.01	1914	10.90
3. Decrease in range	4.00–14.00	0.01	600	9.23
4. Increase in resolution	0.860–14.000	0.001	13,140	13.68
5. Decrease in resolution	0.9–14.0	0.1	131	7.03
6. Increase in both range and resolution	0.860–20.000	0.001	19,140	14.22

The bold values represent baseline values for comparison against increases or decreases in range and/or resolution

Table 45.3 Resource state complexity for various types of resources

Resource type	Range	Resolution	Number of resource states	RSC (bits)
A. Avionics bus Manchester II bit encoding (20-bit command word)	–	–	1,048,576	20.00
B1. Cold, dry cooling air (Mass flow rate – cubic feet per minute)	50–75	1	25	4.64
B2. Cold, dry cooling air (Temperature – °F)	40–60	1	20	4.32
B3. Cold, dry cooling air (Humidity – %)	6.00–10.00	0.01	400	8.64

45.4.2 Resource State Complexity

RSC is measured in a similar fashion to FSC; only now Eq. (45.2) is applied to a resource rather than a performance requirement. The RSC for the entire system is calculated through a simple summation, and RSC increases as more functions and therefore resources are included in the functional architecture. Table 45.3 provides sample RSC calculations for two different types of resources. Resource A is an avionics bus command word that specifies the function that a remote terminal is to perform [7]. Resource B is cold, dry air used to cool electronic components.

Table 45.3 also shows how RSC is determined via the decomposition of a resource into its fundamental physical properties. Digital resources, such as Resource A, are already encoded in bits, so there is no need to calculate the range and resolution.

45.4.3 Functional Processing Complexity

Systems are designed to execute their functionality within a programmatic sequence. The greater the number of possible program paths, the greater the complexity. For example, earlier versions of the Microsoft® Word program for word processing have been estimated to contain more than 2^{64}, or over 1.8 × 10 [19], separate program paths [2]. This large path size is caused by the presence of multiple feedback and feedforward loops, interdependent tasks that are coupled, and multiple branching points such as if-then statements or parallel pathways. Analyzing, testing, and maintaining software with this type of complexity are difficult. It is hard to trace the sequential program execution or to determine the impact of a change in one part of the program on the functioning of the rest of the program.

Software designers have long used control flow graphs (CFG) to illustrate functional sequences. A CFG is a graph-based visualization where nodes represent either a basic block of code or a branch point [21], and edges indicate the flow of program execution. Example CFG [2] can be seen in Fig. 45.1.

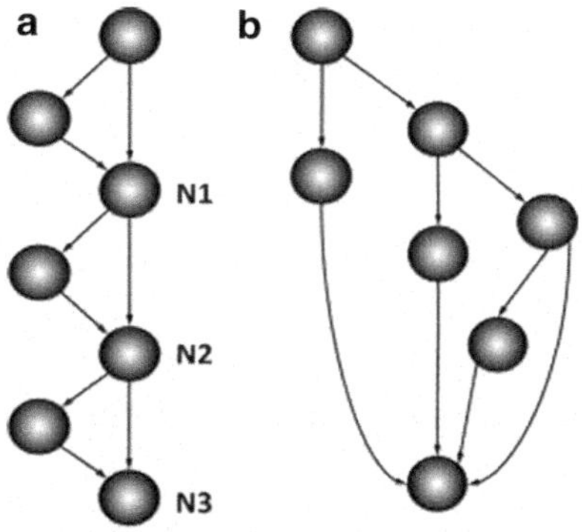

Fig. 45.1 (**a**) Less predictable control flow graph; (**b**) more predictable control flow graph

The CFG is a useful abstraction, and similar types of diagrams found in systems engineering include functional flow block diagrams [10], activity sequence diagrams, and task-based design structure matrices (DSMs) [22]. Any of these diagrams can be analyzed from a graph-theoretic perspective to determine the number of *acyclic paths* present. An acyclic path is defined as a path from program entry to exit that does not traverse an arc more than once. In Fig. 45.1, the left and right CFG possess the same number of nodes but different number of acyclic paths. The left CFG has eight acyclic paths, compared with four acyclic paths for the right CFG. The left CFG is therefore less predictable than the right CFG, and this is reflected in the *in-degree* of the nodes labeled N1, N2, and N3. In-degree is simply the number of edges entering a node. Nodes N1, N2, and N3 will be "the crossroad of more than one path and thus increase the uncertainty of the expected program behavior" [21].

FPC is determined by applying Eq. (45.1), which requires defining the probability or frequency of traversing each path. In the case that the probabilities are equiprobable, this means taking the logarithm of the number of acyclic paths. Under this assumption, the FPC of the left CFG is 3 bits, while the FPC of the right CFG is 2 bits. Various algorithms exist to automatically count acyclic paths for a given CFG, though care must be taken when features such as nesting or different types of conditional and iteration statements are present [21, 23].

45.4.4 Resource Processing Complexity

Nodes exchange resources with other nodes, systems, or the environment in order to function. At the most abstract level, this exchange of resources can be modeled as a communication, where messages originate at an information source, are encoded into a message using symbols, and then transmitted over a channel to a receiver which then decodes the message at the destination. This concept is applied to the exchange of resources between nodes, as different parts of the system effectively communicate via resources such as information, energy, materials, etc. in order for the system as a whole to function. More precisely, this communication is modeled as a stochastic "*m-order*" Markov chain where the probability of emission of a particular symbol depends on the preceding *m* number of symbols [20].

The first step in calculating RPC is to identify the system interfaces, which define the boundaries at which dependencies and coupling between nodes occur. Next, symbols must be identified for the resources exchanged at these boundaries. Table 45.3, shown previously, specifies the number of possible states for each resource in the avionics example. An arbitrary symbol is assigned to each possible resource state, and Table 45.4 lists example word/messages that are formed when the arbitrary symbols chosen correspond to actual resource values.

A word/message sequence (or even its absence) is a symbolic representation of the current state of a collection of resources being exchanged at a snapshot in time. To calculate RPC, each word/message is assigned a frequency $p(i)$, or the probability of that specific word/message occurring during the resource exchange. Transition probabilities $p_i(j)$ are also assigned to indicate the likelihood of a next word/message occurring in the sequence. Diagram probabilities $p(i,j)$ are then obtained with Eq. (45.3):

$$p(i,j) = p(i)p_i(j) \text{ where } \sum_j p_i(j) = \sum_i p(i) = \sum_{i,j} p(i,j) = 1 \quad (45.3)$$

RPC is then equal to the entropy of the $p(i,j)$ matrix. Maximum RPC occurs for equiprobable $p(i,j)$, signifying independence among the resource exchanges and great uncertainty in discerning the exact sequence of resource states as they are exchanged over time. Any deviations from equiprobability mean that correlations and dependencies exist in the transmission and processing of resources, helping to drive down the processing complexity. RPC requires the system architect to specify the transmission frequency of each resource, as different resources may be transmitted at different intervals compared to other resources. For example, the command word could be transmitted at twice the frequency as cold, dry air, as seen in example D of Table 45.4. Possible noise sources should also be considered along with the various architectural mechanisms and patterns (such as redundancy, backup systems, check sums, etc.) needed to help ensure error-free transmission of resources.

Table 45.4 Resource word message examples

Example resource message sequence (word/message)	Command word value	Cold, dry air values (CFM, °F, % humidity)
A. 1,001,111,000,001,110,010,165,400,701	10,011,110,000,011,100,101	65/ 40/ 7.01
B. 1,111,111,000,001,110,111,152,510,805	11,111,110,000,011,101,111	52/ 51/ 8.05
C. 0000101011001010010158530600	000010101100101001 01	58/ 53/ 6.00
D. 00001010110010100101000010101100 1010010158530600	000010101100101001 01	58/ 53/ 6.00

45.5 Design Complexity

45.5.1 Independence Axiom

In contrast to architecting, design is focused on integration, satisfying constraints, and ensuring the system is physically realizable. An axiomatic approach [11] to design seeks to define scientific foundations for the design of complex systems. The first axiom, the *independence axiom*, states that "when there are two or more FRs, the design solution must be such that each of the FRs can be satisfied without affecting any of the other FRs. This means that we have to choose a correct set of DPs to be able to satisfy the FRs and maintain their independence" [11]. This is represented mathematically in Eq. (45.4) using a design matrix $\boldsymbol{A}$ that provides the mapping between FRs and DPs.

$$\begin{Bmatrix} FR_1 \\ \cdot \\ \cdot \\ FR_m \end{Bmatrix} = \begin{bmatrix} A_{11} & 0 & \cdots & 0 \\ 0 & A_{22} & \cdots & 0 \\ \vdots & \vdots & \ddots & \vdots \\ 0 & 0 & \cdots & A_{mm} \end{bmatrix} \begin{Bmatrix} DP_1 \\ \cdot \\ \cdot \\ DP_m \end{Bmatrix} \tag{45.4}$$

where A_{ij} is a sensitivity coefficient and $A_{ij} = \delta FR_i/\delta DP_j$. Equation (45.4) represents an *uncoupled design* with a square design matrix (where there are m number of FRs and m number of DPs) with values only along the diagonals. A design that violates axiom 1 is either a *decoupled* design or a *coupled design.* A decoupled design has a square design matrix, but there is coupling between DPs due to nonzero off-diagonal elements. Finally, a coupled design has a rectangular matrix where p number of DP components exist with $m > p$. Axiom 1 can be maintained in a decoupled design when the DPs are controlled in some sequence [11, 24]. Equation (45.7) [24] provides an entropy-based formulation of the first axiom, assuming $\boldsymbol{A}$ is a square and nonsingular constant matrix:

$$\boldsymbol{h}(\boldsymbol{f}\{\boldsymbol{FR}\}) = \boldsymbol{h}(f_{DP}(\{\boldsymbol{DP}\}) + log_b|[\boldsymbol{A}]| \tag{45.7}$$

where $|[\boldsymbol{A}]|$ is the determinant of the design matrix $\boldsymbol{A}$. The variation inherent in the FRs, DPs, and PVs are typically continuous random variables [24]. If normal probability distribution functions (pdfs) are assumed with parameters (μ, σ^2), and the correlation between DPs is expressed as ρ, the total design complexity is captured by Eq. (45.8), assuming a natural logarithmic base:

$$h(\phi\{DP\}) = \sum_{l=1}^{p-1} \sum_{k=l+1}^{p} \ln\left(2\pi e \sqrt{(1-\rho_{kl}^2)\sigma_l \sigma_k}\right) + \ln\left((2\pi e)^p \prod_{l=1}^{p} \sigma_l^2\right)^{\frac{1}{2}} + \ln|[A]| \tag{45.8}$$

The following components of complexity are captured [24]:

- Variability: the measured entropy of the DPs, captured by the $(h(f_{DP}(\{DP\})$ term in Eq. (45.7) and the normal information sources, $\ln\left((2\pi e)^p \prod_{l=1}^{p} \sigma_l^2\right)^{\frac{1}{2}}$ in Eq. (45.8).
- Vulnerability: three factors related to the design matrix A influence the vulnerability, mapping, sensitivity, and dimension. Mapping is the topological structure of A corresponding to the position of the nonzero sensitivity coefficients $A_{ij} = \delta FR_i/\delta DP_j$. Sensitivity refers to the magnitude and sign of nonzero A_{ij}. Dimension refers to the size of the design problem itself, i.e., the number of the FRs, m. Vulnerability is associated with the $log_b|[A]|$ in Eq. (45.7) and $ln|[A]|$ in Eq. (45.8).
- Correlation: expressed in terms of a jointly distributed pdf between the correlated design variables or via a covariance matrix. Correlation (ρ_{kl} in Eq. (45.8)) is a causal relationship between DPs via some noise factors (e.g., common manufacturing variations).

Thus, the three components are separable and distinguishable from each other. This permits devising strategies for complexity reduction based on the component's significance. However, this might not be the case when the design is highly nonlinear with inseparable components due to the loss of additivity. Consequently, significant components of complexity may not be easily identified [24].

45.5.2 Information Axiom

The second axiom, the information axiom, states that the design with the highest probability of success is the best design [11], even though many different designs may be acceptable in terms of the independence axiom. In the general case of m FRs, the information content for the entire system is:

$$I_{\text{sys}} = -\log_b P_{\{m\}} \tag{45.9}$$

where $P_{\{m\}}$ is the joint probability that all m FRs are satisfied. The probability of success can be computed by specifying the design range for the FR and by determining the system range that the proposed design can provide to satisfy the FR. The area of overlap between the design range and system range indicates the probability of success.

45.6 Results and Conclusions

The complexity measures previously discussed directly address the list of the causes of increased complexity in systems based on the avionics system case study. This is illustrated in Table 45.5.

The aim of this research is to provide a useful measure of complexity for systems engineering that aids trade-off analysis. The approach taken is to identify observable system properties that lead to increased complexity. As part of the approach, notable causes of increased complexity are identified using avionics as a prime example. There is little doubt that systems such as avionics will continue to grow in

Table 45.5 System complexity factors captured by the architecture and design complexity measures

Measure	Complexity factor/trend/sensitive to	Required inputs
Functional state complexity (FSC)	Added functionality/subsystems, functional redundancy, and change in performance requirements	Functional architecture: defined set of atomic-level functions and performance requirements including range and resolution
Functional processing complexity (FPC)	Feedback and feedforward (process execution iterations), coupled tasks, and branching due to different factors: functional overlap, redundancies, function integration, and resource sharing	Functional flow block diagram, control flow graphs/program logic flow (software), design structure matrix, activity sequence diagram
Resource state complexity (RSC)	Added functionality/subsystems, changes in the amount and types (analog signal, digital signal, mechanical, energy, etc.) of resources being shared	Resource architecture: defined set of function resources including range and resolution
Resource processing complexity (RPC)	Increase in the amount of information/resources processed by the system, increased information and sharing among subsystems, dramatic increases in functional integration, increase in the number of subsystems, modularization and coupling and cohesion	Logical/objective architecture: function assignment to logical components and subsystems/modularization/logical interface boundaries, e.g., DSM with clustering applied, noise and loss functions; resource/data architecture: mutual information and resource coupling, redundancy, relative transmission frequencies, communication protocols
Axiomatic engineering design complexity	Increases in physical integration to meet more demanding size, weight, power, and other constraints; the size of the design problem along with coupling and correlation between design parameters; difficulty in meeting design goals and the impact of technology on probability of design success	Functional requirements, design parameters, system range, sensitivity of design parameters in meeting functional requirements, correlation between design parameters

complexity, and an objective measure of complexity provides the basis for evaluating different architectural patterns and design decisions. Future work will focus on applying the developed complexity measures to an avionics case study in order to further develop and assess relevant analyses and their correlation to emergent behavior, difficulty in modeling and simulation, and not to mention cost, schedule, and performance during acquisition. Ultimately, this should result in a useful trade-off environment for systems engineering and aid in the timely and affordable acquisition of systems moving forward.

References

1. Sheard, S. A., & Mostashari, A. (2010). 7.3. 1 A complexity typology for systems engineering. In *INCOSE International Symposium* (Vol. 20, No. 1, pp. 933–945).
2. Minai, A., & Braha, D. (2006). *Complex engineered systems: Science meets technology.*
3. Sheard, S. A., & Mostashari, A. (2013). 5.2. 2 Complexity measures to predict system development project outcomes. In *INCOSE International Symposium* (Vol. 23. No. 1).
4. Domercant, J. C. (2011). *ARC-VM: An architecture real options complexity-based valuation methodology for military systems-of-systems acquisitions.* Atlanta: Georgia Institute of Technology.
5. Joslyn, C., & Luis, R. (2000). Towards semiotic agent-based models of socio-technical organizations. In *Proc. AI, Simulation and Planning in High Autonomy Systems (AIS 2000) Conference, Tucson, Arizona.*
6. Moir, I., Seabridge, A. G., & Jukes, M. (2006). *Military avionics systems.* Hoboken, NJ: John Wiley & Sons.
7. Spitzer, C. R. (2001). *Avionics Handbook* (Vol. 200, p. 158). Boca Raton: CRC Press.
8. Eremenko, P. (2009). *META novel methods for design & verification of complex systems DARPA presentation, December* 22 (p. 10).
9. Perrow, C. (2011). *Normal accidents: Living with high risk technologies.* Princeton, NJ: Princeton University Press.
10. Lightsey, B. (2001). *Systems engineering fundamentals.* Fort Belvoir, VA: Defense Acquisition University.
11. Suh, N. P. (2005). *Complexity: Theory and applications.* Oxford: Oxford University Press on Demand.
12. Maier, M. W. (2009). *The art of systems architecting.* Boca Raton, FL: CRC Press.
13. IEEE Standard 1471 (IEEE 1471, 2000; Maier, Emery, & Hilliard, 2001).
14. Jamshidi, M. (2005). System-of-systems engineering-a definition. In *IEEE SMC* 2005 (pp. 10–12).
15. Kinsner, W. (2008). Complexity and its measures in cognitive and other complex systems. In *ICCI 2008. 7th IEEE International Conference on Cognitive Informatics.* Piscataway, NJ: IEEE.
16. Mitchell, M. (2009). *Complexity: A guided tour.* Oxford: Oxford University Press.
17. Alderson, D. L., & Doyle, J. C. (2010). Contrasting views of complexity and their implications for network-centric infrastructures. *IEEE Transactions on Systems, Man, and Cybernetics-Part A: Systems and Humans, 40*(4), 839–852.
18. Gell-Mann, M., & Lloyd, S. (1996). Information measures, effective complexity, and total information. *Complexity, 2*(1), 44–52.
19. Shannon, C. E. (1948). A mathematical theory of communication. *Bell System Technical Journal, 27*(3), 379–423.
20. Gatlin, L. L. (1972). *Information theory and the living system.*

21. Nejmeh, B. A. (1988). NPATH: A measure of execution path complexity and its applications. *Communications of the ACM, 31*(2), 188–200.
22. Steward, D. V. (1981). The design structure system: A method for managing the design of complex systems. *IEEE Transactions on Engineering Management, 3*, 71–74.
23. McCabe, T. J. (1976). A complexity measure. *IEEE Transactions on Software Engineering, 4*, 308–320.
24. El-Haik, B., & Yang, K. (1999). The components of complexity in engineering design. *IIE Transactions, 31*(10), 925–934.
25. Stuart, D., & Mattikalli, R. (2011). *META II complexity and adaptability*. St Louis, MO: Boeing Co.

Chapter 46
Capturing the Information Dependencies of Verification Activities with Bayesian Networks

Alejandro Salado, Hanumanthrao Kannan, and Farshad Farkhondehmaal

46.1 Introduction

Verification activities occur at various levels of a system's decomposition and at different times during its life cycle [1]. A verification strategy is established "aiming at maximizing confidence on verification coverage, which facilitates convincing a customer that contractual obligations have been met; minimizing risk of undetected problems, which is important for a manufacturer's reputation and to ensure customer satisfaction once the system is operational; and minimizing invested effort, which is related to manufacturer's profit" [2]. Essentially, verification activities are the vehicle by which contractors can collect evidence of contractual fulfillment in acquisition programs.

In current practice, a verification strategy is defined at the beginning of an acquisition program. As a result, the resources necessary to execute verification activities at various stages of the system development are allocated and committed at the beginning, when a small amount of knowledge about the system is available [1]. However, the necessity and value of a verification activity cannot be measured independently of the overall verification strategy [3, 4]. Instead, the necessity to perform a given verification activity depends on the results of all verification activities that have been previously performed. For example, testing the mass of a component is considered more necessary if a previous analysis had shown low margin with respect to the success criterion than if the analysis had shown ample margin. Essentially, the uncertain nature of system development may make verification activities that were not previously planned necessary and may make some of the planned ones unnecessary. However, existing techniques to optimize verification strategies consider the value of verification activities independent of each other (e.g., [1]).

A. Salado (✉) · H. Kannan · F. Farkhondehmaal
Virginia Tech, Blacksburg, VA, USA
e-mail: asalado@vt.edu

S. Adams et al. (eds.), *Systems Engineering in Context*,
https://doi.org/10.1007/978-3-030-00114-8_46

In order to address such limitation, we show in this paper how Bayesian networks can be an effective approach to capture the information dependencies of verification activities. A novel aspect in this paper is that the Bayesian network is used to capture how engineers' confidence or beliefs are shaped as a function of the results of verification activities, not just to compute the probabilistic relationships between traditional performance characteristics of verification activities such as efficiency and efficacy.

46.2 Using Bayesian Networks for Valuing Verification Strategies

46.2.1 The Value of a Verification Strategy

The purpose of executing verification activities is to achieve a desired level of confidence on the adequate functioning of the system under development [5]. Probabilistically, this confidence level can be defined as the complement of the probability of the system exhibiting an error or a problem during operation, after verification activities have been completed—we will use the term *belief* instead throughout the paper. Ideally, the confidence on the system being absent of errors should increase as successful results are received from the verification activities, and it should decrease if verification activities failed. If the confidence on the system being absent of errors does not change after the results of a given verification activity are received, such verification activity would not add any value to the system development. In other words, it would not provide any information on which new knowledge about the system can be created.

Under these terms, we use in this paper the notion that the value of a verification activity is a function of how much it changes the confidence on the system being absent of errors given its results.

46.2.2 Why Bayesian Networks for Defining Verification Strategies

Bayesian networks (BNs) or belief networks are directed acyclic graphs that capture probabilistic dependencies and enable computing probabilistic or belief update using Bayes' theorem [6]. A BN consists of nodes and arrows between the nodes. Nodes represent the random variables we are interested in gaining knowledge about, and arrows between nodes represent the probabilistic dependencies between the nodes [6]. They have been effectively used in different domains, including accounting, accident analysis, risk analysis, ecology, and cognitive sciences, among others [7]. For example, they have been effective in modeling soil erosion, by combining existing data and the complex dependencies between different factors that cause erosion [8]. Bayesian networks have also been employed in the field of systems engineering. For example, they have been used to validate expert judgment in evaluating technology readiness assessments [9, 10]. They have also been used to assess threats in air mission evaluation [11]. As previously discussed, executing a verification activity may result (ideally should) in an update of the confidence on the system being absent of errors (measured as a probability/belief). Hence, we suggest that BNs are adequate to model the information dependencies between verification activities in a verification strategy.

46.2.3 Mapping a Bayesian Network to a Verification Strategy

Consider a system z_0 built from components $z_{1,1},\ldots,z_{1,n}$ and a verification strategy $\Upsilon = (V, I)$, where V is the set of verification activities $\{v_1, \ldots, v_m\}$ and I is the set of tuples that capture information dependencies between the various verification activities $\{(v_i, v_j), \ldots, (v_l, v_k)\}$, with $n,m \in \mathbb{N}$ and $v_i,v_j,v_l,v_k \in V$. Information dependencies are defined as those that exist between two verification activities, in the sense that the results of one of them affect the confidence gained or lost by executing the other one. For example, consider the verification activities v_1 to run a thermal dissipation analysis and v_2 to perform a thermal test. The tuple (v_1, v_2) indicates that the information unveiled by the thermal dissipation analysis influences our confidence on the success of the thermal test.

A Bayesian network that models such verification strategy can be constructed by combining three graphs. The first one contains directly the graph of the verification strategy, $\Upsilon = (V, I)$. The nodes represent the possible outcomes of the verification activities. For this paper, a verification activity $v \in V$ can have two outcomes: v, which represents the verification activity being successful (no error is found), and $\bar{v}$, which represents the verification activity not being successful (an error is found). For each $v \in V$, there are beliefs associated with v and $\bar{v}$. The edges of the network capture the information dependencies between the activities. Conditional probability tables are associated to them.

The second graph captures the prior belief on the absence of errors in the various components that form the system and the system itself and the first verification activities executed on them. Mathematically, we can denote such graph as $S = (Z, A)$, where $Z = \{z_0, z_{1,1}, \ldots, z_{1,n}\}$ and $A = \{v \in V\colon\ \exists\, z \in Z \text{ such that } P(v|\, z) \neq P(v)\}$. This graph captures the dependency between the prior knowledge about the components forming the system, including the system itself, and the first verification activities that are carried out in the verification strategy. For example, consider two different components, one of them being a new development and the other one recurring. In such a case, the confidence on the proper functioning of the recurring component will probably be higher than the confidence on the new development. Therefore, the effect that a given verification activity will have on each of them will be different. Essentially, the maturity of the various components forming the system is captured in this graph.

Finally, the Bayesian network must contain the belief on the absence of error on the system z_0 as it relates to the belief on its components being absent of errors. Mathematically, we can denote such graph as $F = (Z, B)$, where $B = \{(z_i, z_j) : z_i, z_j \in Z, P(z_i | z_j) \neq P(z_i)\}$. This graph captures the coupling between the different components forming the system, that is, how they inform the confidence on the proper functioning of the system. For example, the mass of a system is a function of the mass of its components. Hence, our prior belief on the mass of the components shapes our prior belief on the mass of the system.

In summary, the resulting Bayesian network is given by $BN = (V \cup Z, I \cup A \cup B)$.

46.2.4 *Belief Update of a Bayesian Network*

We describe in this section how a Bayesian network can be used to update the belief on an event as a function of results. Consider a basic example of a system characteristic, such as mass, and two potential verification activities: a mass analysis and a mass test. The resulting Bayesian network is depicted in Fig. 46.1. Essentially, we hold a prior belief on the mass performance of the system before any verification activity is performed and assign confidence levels on the effectiveness of each verification activity to discover a deviation, should it be present. As can be seen, the network can then be used to update our confidence on the absence of deviation in

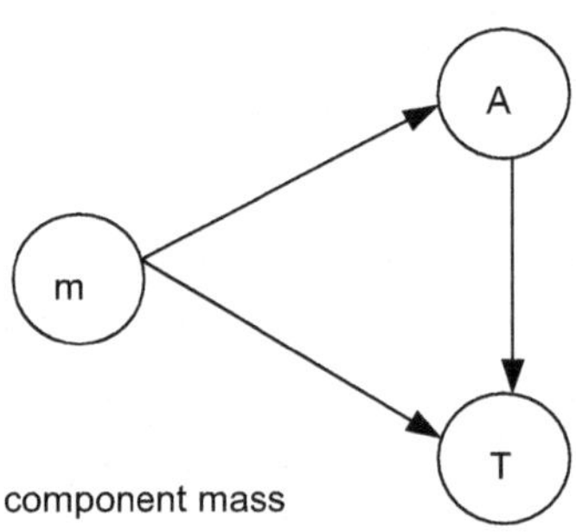

m: component mass
A: mass analysis
T: mass test

m	P(m)
d	0.4
/d	0.6

m	A	P(A|m)
d	a	0.75
d	/a	0.25
/d	a	0.00
/d	/a	1.00

m	A	T	P(T|m,A)
d	a	t	0.95
d	a	/t	0.05
d	/a	t	0.60
d	/a	/t	0.40
/d	a	t	0.00
/d	a	/t	1.00
/d	/a	t	0.00
/d	/a	/t	1.00

d: mass deviation; **/d:** no mass deviation
a: deviation found in A; **/a:** deviation not found in A
t: deviation found in T; **/t:** deviation not found in T

Fig. 46.1 Basic example of Bayesian update

case we get successful results in the mass test. In the case of the figure, knowing that the test indicates no error leads to a belief update of $P(/EPS|/t) = 0.91$.

46.3 Case Study

46.3.1 Problem Definition

In order to investigate the use of Bayesian networks to capture information dependencies between verification activities, we explore various scenarios that involve an observation about whether an error is found or not at some verification activities. This investigation provides insights about the entire sequence as well as about the individual verification activities.

We employ NASA's FireSat satellite as an example [12], because of its wide adoption for research in systems engineering. For demonstration purposes, we will specifically focus on the verification associated with a simplified version of its electric power system (EPS). We consider the EPS to be built from three components, namely, the power control and distribution unit (PCDU), the solar panels (SA), and the battery. A hierarchical breakdown is depicted in Fig. 46.2.

In order to investigate the impacts of various verification activities on the confidence on the absence of error (i.e., proper functioning), we assume that the EPS and PCDU need to be fully developed, the SA is based on an existing unit but needs some modifications, and the battery is recurring from a previous program. Essentially, we capture different levels of development maturity in the components that build the system [13]. Using varying levels of component maturity usually yields the need to use verification activities of varied fidelity levels. Hence, the proposed notional EPS enables a sensible definition of a notional verification strategy that incorporates diverse verification activities.

The notional verification strategy employed in this case study includes the following activities: analysis of SA (A), analysis of battery (B), analysis of PCDU (C), analysis of EPS (D), test of PCDU (E), and test of EPS (F). While notional, the proposed strategy is not arbitrary. The level of verification fidelity has been defined as a function of the maturity of the components according to the guidelines in [14]. It

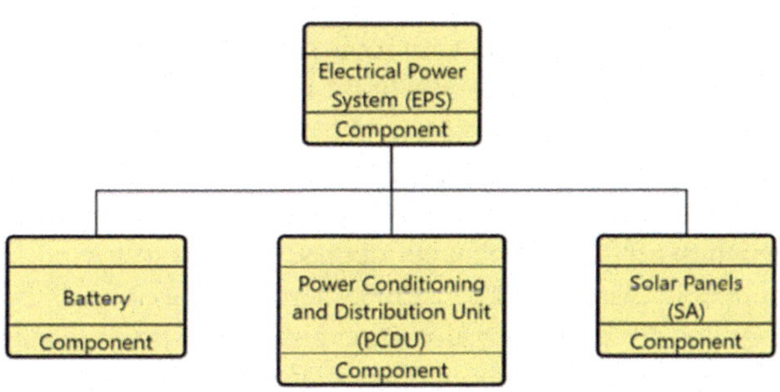

Fig. 46.2 Simplified FireSat EPS physical hierarchy

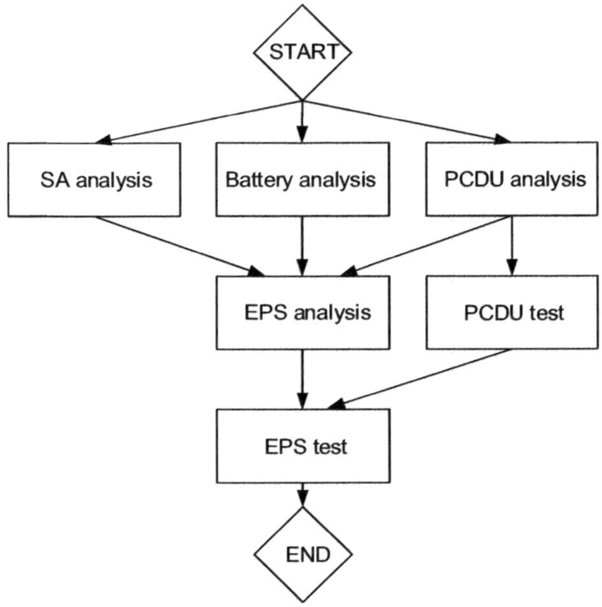

Fig. 46.3 Information dependencies between the various verification activities

should be noted that for generality, we have not identified any specific characteristic for verification. Instead, we just define generic verification activities. Figure 46.3 reflects the order in which verification activities are executed (from top to bottom), as well as the information dependencies between them in the form of arrows. We assume also in this case that errors, if found, are not corrected. This is done purposefully, so that we isolate confidence of proper functioning as a function of verification results as the only factors in the study.

46.3.2 *Bayesian Network of the Verification Strategy*

Figure 46.4 depicts the resulting Bayesian network. The nodes SA, battery, PCDU, and EPS represent the prior belief on each component and the system exhibiting an error, that is, the confidence on each component and system properly functioning before any verification activity is carried out on them. As discussed, that confidence is a function of the maturity of the components. Nodes A through F correspond to the verification activities presented in the previous section.

The probability values associated with the nodes are notional and have been defined under reasonable assumptions, in line with the maturity and coupling characteristics of the components defined previously. In addition, we have assumed for simplicity, and without loss of generality, that none of the verification activities yields false positives. When operationalizing the proposed approach in this paper, such values may be elicited using existing estimation techniques, such as from subject

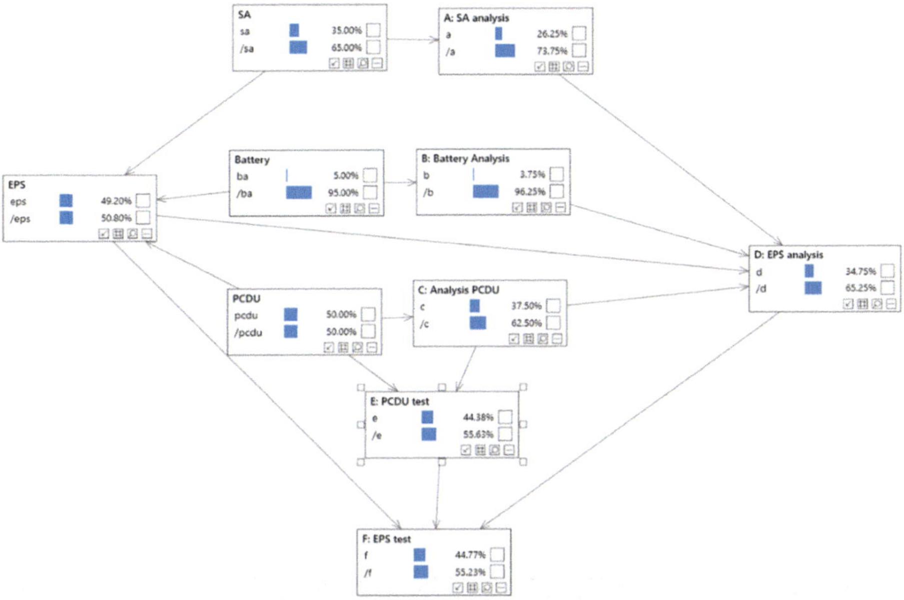

Fig. 46.4 Bayesian network for notional verification strategy

matter experts or historical datasets. In any case, the nature of the data in this example does not affect the purpose of this paper, which is to demonstrate the suitability of BN to capture information dependencies between verification strategies.

46.3.3 *Confidence on the System Being Absent of Errors (Proper Functioning) at the Beginning of the System Development*

The confidence on the system being absent of errors (proper functioning) before any verification activity is executed can be obtained by directly computing $1 - P(EPS)$ with the values in the BN. It yields a confidence of 0.51. We use this value as a reference for exploring how the confidence evolves as verification activities are executed.

46.3.4 *Evolution of the Confidence on the System Being Absent of Errors*

We explore four different cases of various predefined materializations of the notional verification strategy.

Table 46.1 Evolution of confidence on the system being absent of errors for Cases 1, 2, and 3

	Case 1		Case 2		Case 3	
Time	*v* executed	1 – P(EPS)	*v* executed	1 – P(EPS)	*v* executed	1 – P(EPS)
T1	A: successful	0.59	A: unsuccessful	0.27	A: unsuccessful	0.27
T2	B: successful	0.60	B: unsuccessful	0.13	B: unsuccessful	0.13
T3	C: successful	0.74	C: unsuccessful	0.10	C: unsuccessful	0.10
T4	D: successful	0.88	D: unsuccessful	0.00	D: successful	0.53
T5	E: successful	0.91	n/a	n/a	E: successful	0.53
T6	F: successful	0.98	n/a	n/a	H: successful	0.92

Case 1 We assume in this case that *all results are successful*. That is, no error has been found in any verification activity after carrying out all verification activities in the strategy. Table 46.1 shows the evolution of the confidence on the system being absent of errors for this case. The BN indicates that the confidence of the system to function properly increases as successful results are confirmed. This behavior is consistent with how verification shapes confidence. There are two aspects worth mentioning. First, note the impact of previous knowledge on the effect of verification activities. While an analysis on a mature component (the battery in this case) results in marginal increase in confidence, an analysis of the same characteristics on an immature component (the PCDU in this case) yields a significant increase. Second, it is worth noting how the confidence on the proper functioning of the system increases thanks to verifying the system. Specifically with the values in this example, the notional verification strategy manages to increase it from 0.51 to 0.98.

Case 2 We assume in this case that *all results are not successful*. That is, errors have been found but not corrected in every verification activity after carrying out all verification activities in the strategy. Table 46.1 shows the evolution of the confidence on the system being absent of errors for this case. The BN indicates that the confidence of the system to function properly decreases as unsuccessful results are confirmed. This behavior is also consistent with how verification shapes confidence. Because we have assumed no false positives in any verification activities, as well as no correction of errors, the strategy in this case reaches its end with activity D, since an error is found at the system level.

Case 3 We assume in this case *a combination of successful and unsuccessful results*. The combination has been defined arbitrarily. The results show that confidence updates as results become available, increasing with successful results and decreasing with unsuccessful ones. This insight will be key in future work to valuate verification activities. There are two aspects worth noting in this case. First, the Bayesian network captures the effects of margins at system level. While analyses

Table 46.2 Example of the value that verification activities provide as a function of prior knowledge

	Sequence 1		Sequence 2	
Time	*v*	1 − P(EPS)	*v*	1 − P(EPS)
T	E, F not performed	0.88	E, F not performed	0.88
T + 1	E: successful	0.91	F: successful	0.98
T + 2	F: successful	0.98	E: successful	0.98

have shown problems with the individual units, a successful result at system level indicates that they somehow compensate at system level. Yet, the final level of confidence that is achieved does not reach that one in Case 1.

Case 4 In this case we explore how the impact that a planned verification activity has on shaping the confidence on the system being absent of errors may change as results of prior verification activities become available. We define two different sequences. Other things being equal, in Sequence 1 activity E is performed before activity F. In Sequence 2, activity F is performed before activity E. That means that in Sequence 1 the results of activity E are known before executing activity F and vice versa in Sequence 2. Table 46.2 shows the BN prediction of how the confidence on the absence of errors in the EPS evolves in both cases. In Sequence 1, successful E increases the confidence from 0.88 to 0.91. However, the same activity in Sequence 2 only provides a marginal increase that is not even seeable with two-digit precision. Hence, the BN shows that the *value* of verification activities cannot be measured in absolute terms, but is always conditioned to the knowledge available at the time of its execution.

46.4 Discussion

The previous cases provide two key insights to the practice of systems engineering. The first one is that Bayesian networks can capture the information dependencies between verification activities that are carried out as part of a verification strategy. This is a point of departure with respect to existing techniques for optimizing verification strategies, which consider verification activities to act independently of each other in building confidence about the proper functioning of the system being verified. As an example, existing techniques consider different sequences of verification activities to be identical verification strategies. On the contrary, the sequence of execution of verification activities is a distinctive characteristic of a verification strategy [16]. We have shown in this paper that Bayesian networks have the potential to serve optimization techniques that leverage the value of verification activities as a function of the sequence in which they are executed.

The second key insight has to do with contracting of verification strategies. Verification activities provide the evidence of contractual fulfillment in the engineering of systems. In current practice, a verification strategy is defined at the beginning of an acquisition program and is agreed upon by customer and contractor at contract signature. Hence, the resources necessary to execute verification activities at various stages of the system development are allocated and committed at the beginning, when a small amount of knowledge about the system is available. However, the results of this paper infer that the uncertain nature of system development may make verification activities that were not previously planned necessary and some of the planned ones unnecessary, as results from previous activities become available. Under these terms, we contend that contractually committing to a fixed verification strategy at the beginning of an acquisition program fundamentally leads to suboptimal acquisition performance. We suggest that dynamic contracting of verification activities is necessary to guarantee optimality of acquisition programs in this area.

46.5 Conclusions and Future Work

We have shown in this paper that Bayesian networks provide the necessary machinery to capture information dependencies between verification activities. In particular, we have shown how they are able to capture how engineers' beliefs or confidence on the proper functioning of the system updates as the results of verification activities become available. This is a novel contribution of this paper. We have also discussed how valuable such belief is to reassess the necessity to execute planned verification activities later in time.

We suggest that the foundations set forth in this paper can seed future research. For example, in this paper we have only considered the knowledge discovered using a simplistic probability measure. Future work may focus on creating a utility function that captures the trade-offs between the increase in confidence, the cost associated with verification activities, and the cost associated with repairing identified errors. Furthermore, we plan to investigate the effect of prior beliefs on the value of verification strategies.

In addition, incorporating a utility function can also lead to formally integrating decisions related to verification strategies with design decisions, as proposed in [3]. Furthermore, the utility function can be used as an objective function to enable simultaneous optimization of design and verification, within a framework such as value-driven design [15]. This is the case in particular when considering the expected utility of a system, since the design aspects leverage the characteristics of the system and the verification aspects the confidence on achieving those characteristics.

Acknowledgment This material is based upon the work supported by the Naval Postgraduate School Acquisition Research Program under Grant No.N00244-17-1-0013. The views expressed in written materials or publications, and/or made by speakers, moderators, and presenters, do not necessarily reflect the official policies of the Naval Postgraduate School nor does mention of trade names, commercial practices, or organizations imply endorsement by the US Government.

References

1. Engel, A. (2010). In A. P. Sage (Ed.), *Verification, validation, and testing of engineered systems, Wiley series on systems engineering and management*. Hoboken: Wiley.
2. Salado, A. (2015). Defining better test strategies with tradespace exploration techniques and pareto fronts: Application in an industrial project. *Systems Engineering, 18*(6), 639–658.
3. Salado, A. (2016). Integrating design and verification decisions in value-driven design to increase system's expected value. In *Conference on Systems Engineering Research*. Huntsville.
4. Salado, A. (2016). Applying tradespace exploration to verification engineering: From practice to theory and back again. In *Conference on Systems Engineering Research (CSER)*. Huntsville.
5. INCOSE. (2015). *Systems engineering handbook: A guide for system life cycle processes and activities (version 4.0 ed.)*. Hoboken: Wiley.
6. Neapolitan, R. E. (2004). *Learning Bayesian networks*. Upper Saddle River: Pearson Prentice Hall.
7. Grover, J. (2013). A literature review of Bayes' theorem and Bayesian belief networks (BBN). In *Strategic economic decision-making* (pp. 11–27). New York: Springer.
8. Aalders, I., Hough, R., & Towers, W. (2011). Risk of erosion in peat soils–an investigation using Bayesian belief networks. *Soil Use and Management, 27*(4), 538–549.
9. Austin, M. F., Homberger, C., Ahalt, V., Doolittle, E., Polacek, G. A., & York, D. M. (2017) Applying Bayesian networks to TRL assessments: Innovation in systems engineering. In *INCOSE International Symposium* (Vol. 27, No. 1, pp. 1622–1634).
10. Austin, M. F., Homberger, C., Polacek, G. A., Doolittle, E., Ahalt, V., & York, D. M. (2018). *Using Bayesian networks to validate technology readiness assessments of systems*. Cham: Springer.
11. Costa, A. N., & Costa, P. C. G. (2018). *Simulation-based air mission evaluation with Bayesian threat assessment for opposing forces*. Cham: Springer.
12. Wertz, J. R., & Larson, W. J. (1999). *Space mission analysis and design*. Dordrecht: Microcosm.
13. ECSS. (2009). *Space engineering: Verification*. Noordwijk: European Cooperation for Space Standardization.
14. Larson, W. J., Kirkpatrick, D., Sellers, J., Thomas, L., & Verma, D. (2009). *Applied space systems engineering, Space technology series*. New York: McGraw Hills.
15. Collopy, P. D., & Hollingsworth, P. M. (2011). Value-driven design. *Journal of Aircraft, 48*(3), 749–759.
16. Salado, A., & Kannan, H. (2018). A mathematical model of verification strategies. *Systems Engineering 21*(6), 593–608.

Chapter 47
Towards Better Understanding of Software Maintainability Evolution

Pooyan Behnamghader and Barry Boehm

47.1 Introduction

Software developers can prevent failures and disasters and reduce the total cost of ownership by putting more emphasis on improving software maintainability in their software development process [1, 2]. One way to improve software maintainability is to produce clean code while changing the software and to continuously assess and monitor code quality while the software is evolving [3].

Prior research has been focusing on the analysis of official releases of software to understand how its code quality evolves [4–8]. This approach gives an insight on change in code quality over the major milestones of software development, rather than how code quality evolves during software development process. For example, a developer may unknowingly increase the amount of technical debt over a few commits. If that debt is not addressed quickly, it can impose extra cost. It gets even worse, if she leaves the team without paying that debt. In another example, a developer may simply commit broken code to the repository. This will break the code for other contributors and slows down the development. Since software developers always ship only compilable code, this detail may not be revealed in an analysis of official releases. Analyzing the impact of each commit on software quality can reveal a wealth of information because it holds fine-grained details in the software evolution, such as who commited each change and the time of that change.

Over the past couple of years, multiple tools and techniques have been introduced to study software evolution in commit level and visualizing the results [9–17]. Some mining algorithms are designed and implemented sequentially, and the execution of the study requires strong on-premise resources and takes multiple weeks [18]. Some mining algorithms are parallelized, but either the source code is skipped [10, 13], or a

P. Behnamghader (✉) · B. Boehm
University of Southern California, Los Angeles, CA, USA
e-mail: pbehnamg@usc.edu; boehm@usc.edu

S. Adams et al. (eds.), *Systems Engineering in Context*,
https://doi.org/10.1007/978-3-030-00114-8_47

lightweight static analysis technique is applied to each file, without considering the relationships between entities in a module or a body of software [9, 12].

There is also a relative scarcity of commit-level empirical study on using complex program analysis techniques (e.g., building the software, running FindBugs[1] on bytecodes, or running architecture recovery) on different modules, COTS tools with complex environments (e.g., SonarQube[2]), and dynamic analysis techniques (e.g., rendering webpages or test coverage). In addition, there is not much work on a multi-perspective analysis and visualization of software quality evolution from a module/system perspective.

We took steps toward addressing that scarcity by developing Software Quality Understanding by Analysis of Abundant Data (SQUAAD), a comprehensive framework including a cloud-based automated infrastructure accompanied by a data analytics toolset and web interfaces.

Our approach to conduct large-scale replicable empirical studies on software evolution has been to capitalize on cloud services to analyze full maintainability and technical debt commit histories of large families of open-source software systems available through GitHub.[3] SQUAAD automatically:

1. Retrieves a subject system's metadata (e.g., number of contributors) as well as its commit history from GitHub.
2. Distributes hundreds of revisions (i.e., official releases and/or revisions created by commits) on multiple cloud instances.
3. Compiles each revision and runs static/dynamic programming analysis techniques.
4. Collects and parses the artifacts generated by programming analysis techniques to extract quality attributes.
5. Runs various statistical analysis on software quality evolution.

The entire analysis workflow is automated. As soon as the framework is configured for a subject system, we can run the represented analysis on that system and study its evolution.

We have also developed web interfaces to illustrate the evolution of different quality attributes and the impact of each developer on software quality. Our integrated tool-based approach has been documented in multiple research publications [3, 19–21] empowering their empirical studies and is used by a major governmental entity.

[1]http://findbugs.sourceforge.net/.

[2]https://www.sonarqube.org/.

[3]https://github.com/.

47.2 Foundations

This section focuses on the foundational work that enables large-scale multi-perspective study of software quality evolution.

47.2.1 Cloud Computing

Cloud computing is a paradigm that enables worldwide access to shared pools of configurable system resources that can be rapidly provisioned with minimal management effort.

47.2.2 Version Control Systems

Version control systems (VCS), such as Subversion (SVN), Concurrent Versions System (CSV), and Git, are designed to track change over software evolution. SQUAAD analyzes systems that employ Git, a distributed VCS that facilitates collaborative software development. A Git "repository" contains all changes in the development history. Every developer has a "working directory" that is a copy of the repository. Developers can concurrently change the software in their working directory by a Git "commit." A commit produces a new revision of the software. It is an atomic entity containing the code that has changed and some metadata, such as who committed the change, when the change happened, and a message. Developers can synchronize their working repositories and fix potential conflicts. They can also tag some commits (e.g., as an official release).

47.2.3 Static and Dynamic Analysis

Software engineering community continuously produces program analysis techniques to assess different aspects of software quality. Researchers [22–24] have shown that issues identified by these automated techniques have a strong correlation with issues identified by manual inspection. Developers utilize program analysis tools to detect defects [25] and measure software quality [26] in the software development process. There are two types of program analysis: static and dynamic.

Static analysis does not need to run a body of software to analyze it and can be performed on the source code or on the bytecode. This analysis can help developers detect their mistakes in early stage of software development by revealing defects that are not visible to the compiler [27, 28]. These defects can be high-level to low-level problems, such as architectural decays [29] and code smells [30]. Static analysis can also produce artifacts (e.g., architectural facts [31] and control flow graphs [32]) to help developers better understand and maintain a complex software system.

Dynamic analysis executes a body of software to assess the properties related to its runtime behavior, such as visual representation of a web interface [33] or test coverage of a Java application [34]. This analysis has shown to be more accurate than static analysis for detecting runtime issues, such as concurrency faults [35]. Conducting dynamic analysis oftentimes depends on a specific execution environment, such as a sandbox [36].

47.2.4 Commit-Impact Analysis [3]

Every commit changes the software and introduces a new revision. The source code in that new revision may or may not be different from the previous one. If the source code is the same, the commit has changed other artifacts in the software repository (e.g., documentation). Even one line of code difference between the two revisions can affect almost all aspects of software quality. Commit-Impact Analysis is a lightweight mining software repository technique designed to identify "impactful commits" that change code. It can target a software module to understand its evolution by analyzing the software before and after impactful commits.

Focusing on a module helps in better understanding of heterogeneous projects that are developed in different programming languages and/or by different development teams. For example, Apache Avro[4] is implemented in more than ten programming languages in the same repository. Instead of evaluating the whole project, we can focus on its Java implementation by targeting the "lang/java" module. This also helps in analyzing heterogeneous projects even if the utilized program analysis technique does not support all languages. In another example, Apache Parquet-MR[5] contains different Java sub-projects. Each sub-project has its own set of developers and reviewers. Instead of focusing on the whole project, we can focus on each sub-project to evaluate the performance of each team.

An impactful commit can be:

- Orphan that introduces the module for the first time.
- A simple commit that introduces a new revision by changing its parent's revision.
- A merge that carries merge agent's combination of all developments over all branches leading to it.

[4]https://github.com/apache/avro.

[5]https://github.com/apache/parquet-mr.

47.3 Empirical Studies

In order to better understand the software maintainability, as well as conflicts and synergies among software quality attributes, we have conducted a series of empirical studies on open-source software systems.

47.3.1 Software Architecture Evolution

In a recent work, we conducted a large-scale empirical study of architectural evolution in open-source software systems [21]. The scope of our study is reflected in the total number of subject systems (23) and the examined official versions (931), as well as the cumulative amount of analyzed code (140 MSLOC). We employed three program analysis techniques to recover the architecture of each version from semantics- and structural-based architectural perspectives. We also designed two software change metrics to quantify the amount of architectural change between two versions. Our study resulted in several unexpected findings about the nature, scope, time, and frequency of architectural change. We found that the versioning scheme of a system is not necessarily reflecting the architectural change as the system may undergo major architectural modifications between minor releases and patches. We also found that the architecture may be relatively unstable during the pre-release phase.

47.3.2 Software Quality Evolution

Another recent maintainability trends analysis [3] involved a total of 19,580 examined commits from 38 Apache family systems across a time span from January 2002 through March 2017, comprising 586.33 million source lines of code. In this study, to obtain software quality, we used three widely used open-source static analysis tools: FindBugs, PMD,[6] and SonarQube. We selected a subset of quality attributes related to size (basic), code quality, and security. Table 47.1 shows metrics we used in our analysis. For each software system, we identified the module containing most of source code (the core module) and commits that change the core module (impactful commits). We compiled all revisions created by impactful commits and ran the static analysis techniques on those revisions. We analyzed software before and after each impactful commit to understand how each change impacts its quality.

We found that on average, the frequency of change in CS is two times of FN and four times of LC. Our result showed that although the security metrics do not change frequently, employing them together can reveal important data points where security

[6]https://pmd.github.io/.

Table 47.1 Quality metrics

Group	Abbr.	Tool	Description
Basic	LC	SonarQube	Physical lines
	FN	SonarQube	Functions
	CS	FindBugs	Classes
Code quality	CX	SonarQube	Complexity
	SM	SonarQube	Code smells
	PD	PMD	PMD code quality
Security	VL	SonarQube	Vulnerabilities
	SG	PMD	Security guidelines
	FG	FindBugs	Malicious code

problems are introduced. We identified multiple instances that the size of the software does not change, but at least another metric (code quality or security) changes. We calculated the probability for a metric to change, while another one does not change based on our empirical results.

47.3.3 Compilability Over Evolution

Some static and dynamic analysis techniques depend on the availability of bytecode. As a result, it is necessary to compile the software before conducting an analysis using those techniques. A high ratio of uncompilable commits over the commit history can result in an incomplete analysis. At the same time, committing uncompilable code is a symptom of careless development since a software revision is expected to be compilable. In our large-scale empirical studies on software evolution, we realized that developers tend to commit uncompilable code even in popular industry-scale open-source software systems.

We fixed all missing dependencies over the commit history of the 38 subject systems studied in our software quality evolution study to find which commits are broken because of a developer's fault. We found that among all analyzed impactful commits, 2% are not compilable. We identified all 385 uncompilable impactful commits and manually inspected them to understand why, when, and how developers introduce compile error. We found that committing too often and too early is the most common reason for committing uncompilable code. We suggested developers to compile the project in a new environment after contributing alone for a period, changing build files, adding new files, and doing maintenance or large refactoring tasks.

47.3.4 *Difference Between Developers in Terms of Impact on Technical Debt*

Two recent extensions [19, 20] of commit-impact analysis study the impact of developers on technical debt in open-source software systems based on their level of involvement and the characteristics of their commits. We investigated whether there is any statistical difference in the amount of change to the technical debt that a commit imposes considering the seniority of the committer and the number of commits she has had by the time of the commit, the interval between the commit and its parent commit, and whether the committer is a core developer of the system based on her commit frequency. Our results showed that while there is no statistically significant difference between core developers and others, in almost all subject systems, some developers increase or decrease the amount of technical debt more than others. We also found that the seniority of the developer has a negative and the interval between the commit and its parent commit has a positive correlation with the increase in the amount of technical debt.

47.3.5 *Visualizing the Evolution*

We have developed interactive web interfaces to illustrate the evolution of different quality attributes and the impact of each developer on software quality. For example, Fig. 47.1 shows the evolution of the size (i.e., NCLOC) and the number of code smells of a subject system's core module. None of two trends increases monotonically over time; however, the size experiences less variation. The ratio of commits that increase the size to the ones that decrease it is 3.00, while this ratio for the number of code smells is 1.45.

The current stable version of the interface allows a user to explore the evolution of a quality metric on a subject system in a form of a scatter plot. A user may hover on each data point to get the information (e.g., author) of the commit. A user may choose to copy the information of a specific data point and use it for further

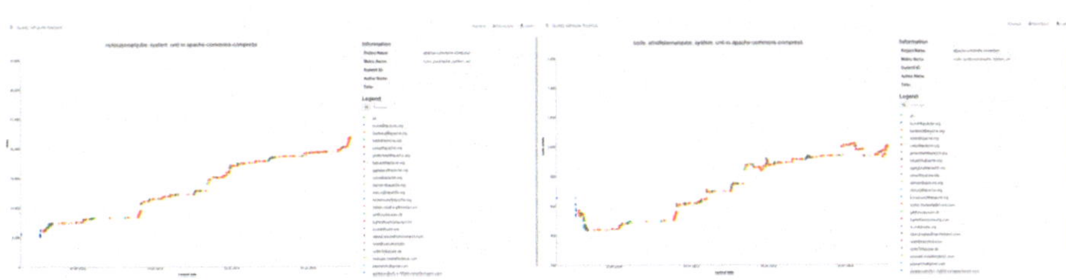

Fig. 47.1 Evolution of size (left) and code smells (right) in a subject system. Each point is a revision. Each color is associated with a developer

investigation (biopsy). There is zooming option on the main chart. We use AngularJS[7]for front-end and Node.js[8] for backend development.

47.3.6 *Industrial Application*

Our integrated tool-based approach allows managers and engineers real-time life cycle assessments of software, systems, and system of systems. We recently delivered advanced tool assessment tutorials to frontline acquisition engineers of a major governmental entity. This led to an in-depth analysis of the quality aspects of an open-source software complex for decisions regarding quality, safety, and security "sniffs" and "taints" to assess an acquisition program of an autonomous system.

47.4 Related Work

The idea of mining the code repository of a software system to gain an insight into how its software quality evolves is not new. However, researchers oftentimes are forced to adjust the scope of their analysis due to scalability challenges involved in large-scale mining studies. This results in either a coarse-grained analysis including only the official releases of the software or a fine-grained analysis (in commit-level granularity) that either skips the source code facts or applies a lightweight static analysis technique, often only on the changed files, over the commit history. There are few studies that discuss automated approaches to collect evolutionary data on software quality from software repositories by commit level. Table 47.2 summarizes

Table 47.2 Comparison of related frameworks

		Program analysis		
Techniques	Year	Type	Compilation	Distributed
SQUAAD	18	Static/dynamic	Yes	Cloud
SmartShark [37, 38]	16/17	Static	–	HPC
HistoryMiner [18]	17	Static	–	–
Candoia [17]	17	Static	–	–
Lisa [9]	17	Static	–	–
QualBoa [11]	17	Static	–	–
Boa [12, 39]	15/17	Static	–	–
MetricMiner [16]	13	Static	–	–

[7]https://angularjs.org/.

[8]https://nodejs.org/.

the most recent tools developed in this area. To the best of our knowledge, SQUAAD is the only tool capable of compiling the source code, distributing the analysis over cloud, and conducting both static and dynamic analysis. Another unique capability of SQUAAD is targeting a module (as opposed to the whole repository) and studying its evolution.

47.5 Conclusions

SQUAAD enables large-scale maintainability analysis over development history of a software system. It targets the whole software (or one of its modules) in a repository, determines its distinct revisions, compiles all revisions, and runs complex static and dynamic analysis on the development history. Its cloud-based infrastructure is accompanied by a data analytics toolset and web interfaces for data visualization. Any organization can apply the analysis conducted by SQUAAD to improve its software and its software engineering, achieve customer satisfaction, and reduce the total cost of ownership. Managers can assess the quality of organization's project types and divisions to understand which quality attributes are being achieved poorly or well. They can also understand which types of processes and project events correlate with which types of quality change and which types of personnel or projects contribute most to quality problems or excellence. Developers can continuously monitor the evolution of software and evaluate their impact on software quality.

Acknowledgment This material is based upon the work supported in part by the US Department of Defense through the Systems Engineering Research Center (SERC) under Contract H98230-08-D 0171. SERC is a federally funded University Affiliated Research Center managed by Stevens Institute of Technology. It was also supported by the National Science Foundation grant CMMI-1408909, Developing a Constructive Logic-Based Theory of Value-Based Systems Engineering.

References

1. Boehm, B. W., Brown, J. R., & Lipow, M. (1976). Quantitative evaluation of software quality. In *Proceedings of the 2nd International Conference on Software Engineering* (pp. 592–605). IEEE Computer Society Press.
2. Mexim, B., & Kessentini, M. (2015). *An introduction to modern software quality assurance. Software quality assurance: In large scale and complex software-intensive systems* (pp. 19–46). Waltham, MA: Morgan Kaufmann.
3. Behnamghader, P., Alfayez, R., Srisopha, K., & Boehm, B. (2017). Towards better understanding of software quality evolution through commit-impact analysis. In *2017 IEEE International Conference on Software Quality, Reliability and Security (QRS)* (pp. 251–262).
4. DAmbros, M., Gall, H., Lanza, M., & Pinzger, M. (2008). Analysing software repositories to understand software evolution. In *Software evolution* (pp. 37–67). Berlin, Germany: Springer.

5. Ganpati, A., Kalia, A., & Singh, H. (2012). A comparative study of maintainability index of open source software. *International Journal of Emerging Technology and Advanced Engineering, 2*(10), 228–230.
6. Le, D. M., Behnamghader, P., Garcia, J., Link, D., Shahbazian, A., & Medvidovic, N. (2015). An empirical study of architectural change in open-source software systems. In *Proceedings of the 12th Working Conference on Mining Software Repositories* (pp. 235–245). IEEE Press.
7. Pinto, G., Torres, W., Fernandes, B., Castor, F., & Barros, R. S. (2015). A large-scale study on the usage of javas concurrent programming constructs. *Journal of Systems and Software, 106*, 59–81.
8. Godfrey, M. W., & Tu, Q. (2000, October). Evolution in open source software: A case study. In *Proceedings of the International Conference on Software Maintenance* (ICSM'00) (p. 131). IEEE Computer Society.
9. Alexandru, C. V., Panichella, S., & Gall, H. C. (2017). Reducing redundancies in multi-revision code analysis. In *2017 IEEE 24th International Conference on Software Analysis, Evolution and Reengineering (SANER)* (pp. 148–159).
10. Bajracharya, S., Ossher, J., & Lopes, C. (2014). Sourcerer: An infrastructure for large-scale collection and analysis of open-source code. *Science of Computer Programming, 79*, 241–259. Experimental Software and Toolkits (EST 4): A special issue of the Workshop on Academic Software Development Tools and Techniques (WASDeTT-3 2010).
11. Diamantopoulos, T., Thomopoulos, K., & Symeonidis, A. (2016). Qualboa: Reusability—Aware recommendations of source code components. In *Proceedings of the 13th International Conference on Mining Software Repositories* (pp. 488–491). ACM.
12. Dyer, R., Nguyen, H. A., Rajan, H., & Nguyen, T. N. (2015). Boa: Ultra-large-scale software repository and source-code mining. *ACM Transactions on Software Engineering and Methodology, 25*(1), 7.
13. Gousios, G., Vasilescu, B., Serebrenik, A., & Zaidman, A. (2014). Lean ghtorrent: Github data on demand. In *Proceedings of the 11th Working Conference on Mining Software Repositories* (pp. 384–387). ACM.
14. Kaur, A., & Chopra, D. (2018). *GCC-git change classifier for extraction and classification of changes in software systems* (pp. 259–267). Singapore, Singapore: Springer.
15. Rozenberg, D., Beschastnikh, I., Kosmale, F., Poser, V., Becker, H., Palyart, M., et al. (2016). Comparing repositories visually with repograms. In *2016 IEEE/ACM 13th Working Conference on Mining Software Repositories (MSR)* (pp. 109–120).
16. Sokol, F. Z., Aniche, M. F., & Gerosa, M. A. (2013). Metricminer: Supporting researchers in mining software repositories. In *2013 IEEE 13th International Working Conference on Source Code Analysis and Manipulation (SCAM)* (pp. 142–146).
17. Tiwari, N. M., Upadhyaya, G., Nguyen, H. A., & Rajan, H. (2017). Candoia: A platform for building and sharing mining software repositories tools as apps. In *2017 IEEE/ACM 14th International Conference on Mining Software Repositories (MSR)* (pp. 53–63).
18. Tufano, M., Palomba, F., Bavota, G., Oliveto, R., Penta, M. D., Lucia, A. D., et al. (2017). When and why your code starts to smell bad (and whether the smells go away). *IEEE Transactions on Software Engineering, 43*(11), 1063–1088.
19. Alfayez, R., Behnamghader, P., Srisopha, K., & Boehm, B. (2017). How does contributors involvement influence open source systems. In *2017 IEEE 28th Annual Software Technology Conference (STC)* (pp. 1–8).
20. Alfayez, R., Behnamghader, P., Srisopha, K., & Boehm, B. (2018). An exploratory study on the influence of developers in Technical Debt. In *2018 International Conference on Technical Debt (TechDebt)* (pp. 1–10).
21. Behnamghader, P., Le, D. M., Garcia, J., Link, D., Shahbazian, A., & Medvidovic, N. (2017). A large-scale study of architectural evolution in open-source software systems. *Empirical Software Engineering, 22*(3), 1146–1193.

22. Bessey, A., Block, K., Chelf, B., Chou, A., Fulton, B., Hallem, S., et al. (2010). A few billion lines of code later: Using static analysis to find bugs in the real world. *Communications of the ACM, 53*(2), 66–75.
23. Kupsch, J. A., & Miller, B. P. (2009). Manual vs. automated vulnerability assessment: A case study. In *First International Workshop on Managing Insider Security Threats (MIST)* (pp. 83–97).
24. Safi, G., Shahbazian, A., Halfond, W. G., & Medvidovic, N. (2015). Detecting event anomalies in event-based systems. In *Proceedings of the 2015 10th Joint Meeting on Foundations of Software Engineering* (pp. 25–37). ACM.
25. Runeson, P., Andersson, C., Thelin, T., Andrews, A., & Berling, T. (2006). What do we know about defect detection methods? [software testing]. *IEEE Software, 23*(3), 82–90.
26. Welker, K. D., Oman, P. W., & Atkinson, G. G. (1997). Development and application of an automated source code maintainability index. *Journal of Software: Evolution and Process, 9*(3), 127–159.
27. Wagner, S., Jürjens, J., Koller, C., & Trischberger, P. (2005). Comparing bug finding tools with reviews and tests. In *IFIP International Conference on Testing of Communicating Systems* (pp. 40–55). Berlin, Germany: Springer.
28. Louridas, P. (2006). Static code analysis. *IEEE Software, 23*(4), 58–61.
29. Le, D. M., Carrillo, C., Capilla, R., & Medvidovic, N. (2016). Relating architectural decay and sustainability of software systems. In *2016 13th Working IEEE/IFIP Conference on Software Architecture (WICSA)* (pp. 178–181).
30. Campbell, A. (2015). *Sonarqube: Open source quality management*. Retrieved from http://tiny.cc/2q4z9x
31. Langhammer, M., Shahbazian, A., Medvidovic, N., & Reussner, R. H. (2016). Automated extraction of rich software models from limited system information. In *2016 13th Working IEEE/IFIP Conference on Software Architecture (WICSA)* (pp. 99–108).
32. Vallée-Rai, R., Co, P., Gagnon, E., Hendren, L., Lam, P., & Sundaresan, V. (1999). Soot—A java bytecode optimization framework. In *Proceedings of the 1999 Conference of the Centre for Advanced Studies on Collaborative Research, CASCON '99* (p. 13). IBM Press.
33. Mahajan, S., Li, B., Behnamghader, P., & Halfond, W. G. (2016). Using visual symptoms for debugging presentation failures in web applications. In *2016 IEEE International Conference on Software Testing, Verification and Validation (ICST)* (pp. 191–201). IEEE.
34. Malaiya, Y. K., Li, M. N., Bieman, J. M., & Karcich, R. (2002). Software reliability growth with test coverage. *IEEE Transactions on Reliability, 51*(4), 420–426.
35. Joshi, P., Naik, M., Park, C. S., & Sen, K. (2009). CalFuzzer: An extensible active testing framework for concurrent programs. In *International Conference on Computer Aided Verification* (pp. 675–681). Berlin, Germany: Springer.
36. Bläsing, T., Batyuk, L., Schmidt, A. D., Camtepe, S. A., & Albayrak, S. (2010). An android application sandbox system for suspicious software detection. In *2010 5th International Conference on Malicious and Unwanted Software, Nancy, Lorraine* (pp. 55–62).
37. Trautsch, F., Herbold, S., Makedonski, P., & Grabowski, J. (2017). Addressing problems with replicability and validity of repository mining studies through a smart data platform. *Empirical Software Engineering, 23*(2), 1036–1083.
38. Trautsch, F., Herbold, S., Makedonski, P., & Grabowski, J. (2016). Addressing problems with external validity of repository mining studies through a smart data platform. In *Proceedings of the 13th International Conference on Mining Software Repositories, MSR '16* (pp. 97–108). New York: ACM.
39. Dyer, R., Nguyen, H. A., Rajan, H., & Nguyen, T. N. (2013). Boa: A language and infrastructure for analyzing ultra-large-scale software repositories. In *Proceedings of the 2013 International Conference on Software Engineering* (pp. 422–431). IEEE Press.

Chapter 48
Shape Inspection Using *N*-View Error Accumulation on Low-Fidelity Point Clouds

Rafael Radkowski

48.1 Introduction

Optical shape inspection refers to a set of methods in which camera images are used to estimate, measure, or reconstruct the surface of a physical object. Application areas for this technology are part inspection in manufacturing, reengineering, or material state tracking. This research focuses on visual deformation detection using 3D scans from commodity range cameras. Processing the 3D scans reveals surface flaws, and a color map rendering visually indicates them so that an inspector can quickly notice the flaws. Typically, laser scanners and monocular *N*-view reconstruction dominate this application field today due to the high-fidelity point clouds these devices and methods yield [1, 2].

Advanced commodity range cameras are slowly gaining ground in inspection and material state tracking. Range cameras such as the Microsoft Kinect or a Dell RealSense D-series camera provide range images of the area in front of the camera. Unfortunately, their depth sensors come with finite sensitivity, which results in a limited depth accuracy. As a consequence, this type of cameras is only usable for a specific task such as state tracking or the detection of significant surface deviation. One of our goals is to determine surface difference pertaining data integration into a digital twin. Typically, inspection data from various sensors is mapped into the shape representation of a digital twin with respect to a specific location on the surface of the product. If the surface deforms due to wear, inspectors should be notified since this can result in data integration problems.

A remaining obstacle is the sensor noise. The noise is considerable high and further limits the fidelity of every scanned surface. Although this issue has several reasons, it is mostly a result of the intended use case: The camera technology was designed to support human-computer interaction (HCI) applications, robot

R. Radkowski (✉)
Iowa State University, Virtual Reality Applications Center, Ames, IA, USA
e-mail: rafael@iastate.edu

S. Adams et al. (eds.), *Systems Engineering in Context*,
https://doi.org/10.1007/978-3-030-00114-8_48

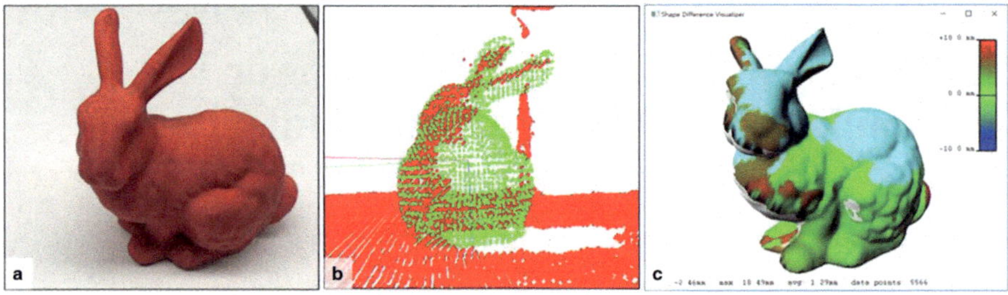

Fig. 48.1 (**a**) A Stanford bunny model with 15 × 9 × 14 cm. (**b**) It is tracked in a point cloud to accumulate point-to-point errors. (**c**) The errors are interpreted as surface deviations and rendered as a color map

navigation, and some others. Optical metrology and related areas were not in focus of the developers. As a result, the camera technology works well for HCI and robotics by supporting a depth accuracy between 2 and 10 mm at a distance of 1 m. Accurate shape analysis, flaw detection, and surface reconstruction, even in this accuracy range, are still challenging and usually require more sophisticated hardware.

This research proposes a method that mitigates depth accuracy problems caused by noise and reflection artefacts by supplementing information from multiple views, object tracking, and a triangulated CAD model (Fig. 48.1). It utilizes object tracking to identify an object of interest and point-to-point registration from multiple views to accumulate the registration errors as deviations into a global deviation model. For error accumulation, the method utilizes a signed distance function (SDF), an approach that was inspired by Newcombe [3] and Curless [4]. The authors suggest a volumetric truncated signed distance function as the basis for surface reconstructions via ray tracing. In this research instead, the approach compares the scanned surface to the surface of a triangulated CAD model using an SDF. The deviation interpretation is also inverted: We assume that the examined surface matches the CAD model surface until the accumulated point-to-point errors indicate differences. The deviations are then incrementally blended into a global model in (u, v)-space, which is a result of a 3D-to-2D parameterization, a parameter model that allows addressing every point on the surface of a 3D model with two parameters u and v ranging from 0 to 1. The resulting deviations appear as a color map on a 3D model.

Related techniques already utilize point cloud-to-point cloud comparison and model-to-point cloud comparison techniques [2, 3]. The suggested method technically belongs to the class of model-to-point cloud comparison, using the single points of a model. However, it favors an incremental approach to estimate a deviation rather than a direct, one-image approach. Additionally, the related research emphasizes sensor noise as the reason for low point cloud fidelity and mitigates its impact using noise models, filters, or statistical surface predictions. For instance, Wells et al. [5] introduce a method that estimates the most likely size, shape, and magnitude of a potential fault. They use an adaptive generalized likelihood ratio technique for that purpose. Awange et al. [6] work with a Groebner basis function to determine the most likely shape of an object in a point cloud. Although they also work with a reference model representing the ideal shape, it is used to cover for missing data rather than supporting reconstruction. Grimm et al. [7] and Shi et al. [8] introduce methods following the same notion. However, they do not focus on detecting and measuring the surface of an object. They focus on shape fitting and model fitting to facilitate robot navigation.

Nguyen et al. [9]. modeled the noise of a Kinect to better predict a surface model. Results show that modeling the noise can undoubtedly yield a smoother surface since noise is a significant reason for rough surface models generated by a Kinect. It is unclear whether the filter facilitates a shape analysis, e.g., to visually detect surface artefacts.

This research contributes an incremental approach for visual surface inspection by accumulating point-to-point errors using an SDF to detect deviations from a nominal CAD model. It addresses problems caused by sensor noise and other reflection artefacts. The remainder of the paper is structured as follows: The next section introduces the method in detail explaining the single steps. Sect. 48.3 describes an experiment that was conducted and presents the results. Sect. 48.4 summarizes this paper and introduces future work.

48.2 Surface Deviation Estimation Method

Figure 48.2 depicts an overview of the process. It is split into four steps: *pose estimation*, *global error calculation* (Sect. 48.2.1), *pose optimization* (Sect. 48.2.2), and *surface deviation estimation* (Sect. 48.2.3). Input is a set of N range images $\mathbf{I}$ and a 3D reference model R, usually a triangulated CAD model. The result is a surface deviation model $\mathbf{D}$ with a deviation d_i for each surface point $\mathbf{p}_i$ of R, represented as parameter model in (u, v)-space.

Figure 48.3 shows a typical setup. For this research, the setup includes a Fotonic P60U range camera with 640 × 480 pixel depth resolution at 30 Hz. The distance between the camera and object varied at approximately 1 m. The object of interest is a printed 3D Stanford bunny model [10].

Note that the method associates surface points $\mathbf{p}$ with graphics pipeline fragments in (u, v)-space and not to the vertices of a triangulated model. Also, the process

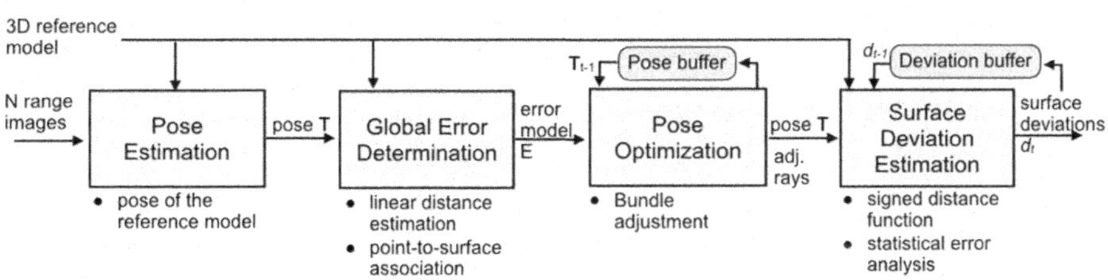

Fig. 48.2 Overview of the surface deviation analysis process

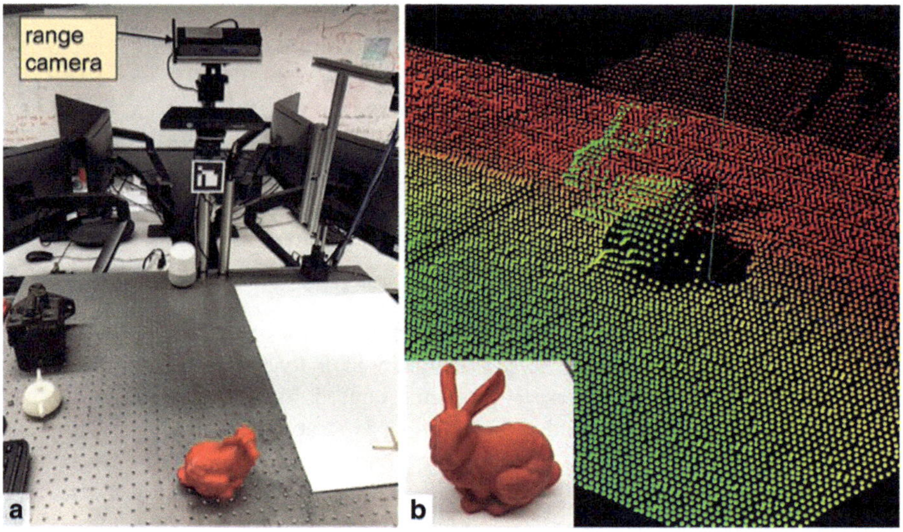

Fig. 48.3 (**a**) A typical setup incorporates a range camera and the object of interest, a Stanford bunny model in this case. (**b**) The point cloud dataset from one range image. Note that the point cloud density was reduced for this figure to enhance visibility

works iteratively, with each new camera frame N_i incrementally contributing to the surface deviation model. To simplify reading the subsequent sections explained the process as a one-step process, with N range images as input.

The objective of the first step is to determine the pose $\mathbf{T}$ of the physical object of interest. Therefore, we use a tracking technology that can actively detect an object and determine its pose. The utilized technology works with point clouds and matches R with its counterpart in a point cloud X yielded from a range image (Fig. 48.1b) [11]. Matching works with surface curvature feature descriptors to describe the characteristics of R and to find the same characteristic pattern in X. As a result, the tracking process can align R with points representing its counterpart in X. For pose refinement, the iterative closest points [12] (ICP) algorithm is applied. The result is a pose $\mathbf{T}_i = [\mathbf{R}|\mathbf{t}]_i$ in homogeneous coordinates with an orientation $\mathbf{R}$ and translation $\mathbf{t}$. Note that this process is not further elaborated. Please find Radkowski et al. [11] for details.

48.2.1 Global Error Estimation

The objective of this step is an error e_i for each surface point $\mathbf{p}_i$ of R mapped to a global error model E which represents the error in (u, v)-parameter space. Therefore, R is first rendered from the current pose $\mathbf{T}_i$ to obtain an ideal range image $\mathbf{I}_R$. This rendering poses a reference for the ideal range. It is pixel-wise compared to the range

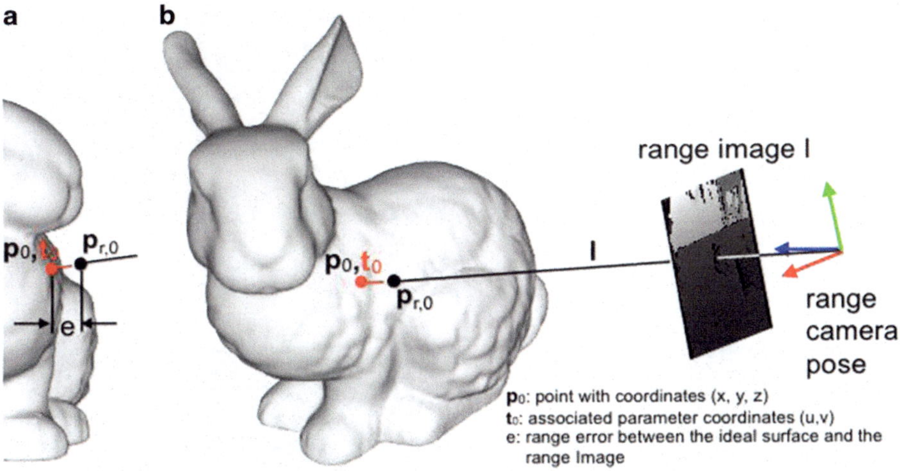

Fig. 48.4 (**a**) The surface error e is the distance between $\mathbf{p}_0$ and $\mathbf{p}_{r,0}$. (**b**) The point $\mathbf{p}_{r,0}$ is the result of the pixel-wise range image projection, and $\mathbf{p}_0$ the related point on the surface of a 3D model. The resulting e is associated with a global parameter coordinate $\mathbf{t}_0 = (u_0, v_0)$

image **I** to calculate e_i. The result is mapped into an error model $\mathbf{E} = (e_0, e_1, \ldots, e_M)$. In detail, the task starts with rendering an ideal range image $\mathbf{I}_R$. The process utilizes the GPU rendering pipeline to raster $\mathbf{I}_R$ from the model R. Technically, it rasterizes a depth image and linearizes the depth data to obtain linear range values. Mathematically, the rendering process is represented as:

$$p_i = T_{\text{Proj}} T_{\text{View}} T \cdot p_{i,\text{model}} \tag{48.1}$$

with the $\mathbf{T}_{\text{Proj}}$, the projection matrix; $\mathbf{T}_{\text{View}}$, the view matrix; and $\mathbf{T}_i$, the pose of the object. The result is a surface point of the model $\mathbf{p}'_i = (x', y', z', w')$ (Fig. 48.4). The projection result is usually represented in normalized device coordinates (NDC) in a range $[-1, 1]$. Also, the z'-value is not linearly distributed over the projection range and has the highest resolution close to the virtual camera. It is necessary for the subsequent comparison to map the z'-value back into linear space:

$$z = \frac{(2fn)}{(f + n - (0.5z' + 0.5) \cdot (f - n))} \tag{48.2}$$

with n and f, the near- and far-clipping plane. The value z is calculated per fragment and represents the range in a linear scale. Although this equation inverts the transformation results obtained by $\mathbf{T}_{\text{proj}}$ and $\mathbf{T}_{\text{View}}$ in Eq. (48.2), the previous transformation is imperative to automatically rasterize R into fragments using the graphics card pipeline. The outcome is $\mathbf{I}_R(u, v)$, which represents the surface points' range.

Fig. 48.5 (**a**) A rendered depth map with linear depth values. (**b**) A depth image captured with a range camera. (**c**) The difference between both rendered as color values

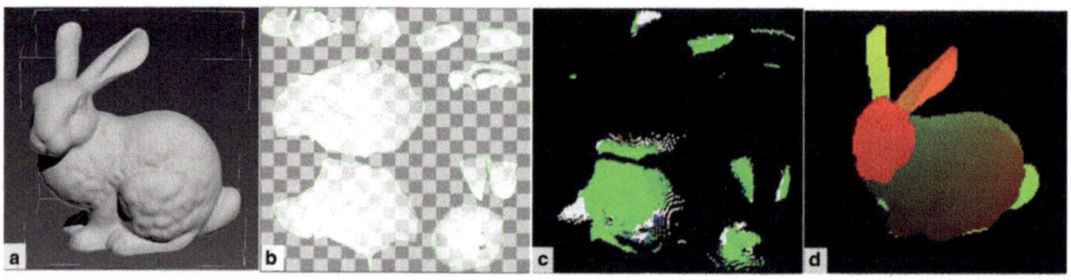

Fig. 48.6 (**a**) The 3D bunny model with seams to (**b**) create a (u, v)-map, which acts as a global reference model to uniquely associate error values **e** to surface points. (**c**) A false color rendering of error values and (**d**) a false color rendering of texture coordinates

One further calculates the error value e_i for each fragment by subtracting the matrix elements in $\mathbf{I}_R(u, v)$ and $\mathbf{I}(u, v)$ pixel-wise (Fig. 48.4) as $e_i = z' - z$, limited to locations (u, v) for which a valid z'-value exists in $\mathbf{I}_R(u, v)$. All other locations are discarded. The result is an intermediate error map $\mathbf{E}_i$.

Figure 48.5a shows the rendered range image $\mathbf{I}_R(u, v)$, Fig. 48.5b a range image fetched from a commodity camera $\mathbf{I}(u, v)$, and Fig. 48.5c the error values e_i rendered as color map. Note that $\mathbf{E}$ still represents each error value in a local image space. Also, we render $\mathbf{I}_R$ with 16-bit depth resolution since the range camera utilized provides 16-bit depth quantization.

Next, the process maps $\mathbf{E}_i$ into a global parameter model. Currently, the local image space prevents one to relate error values from multiple images to one particular point $\mathbf{p}_i$. For this purpose, we map it into the global (u, v)-space or texture coordinates in other words. Texture coordinates pose a 2D parameter space representing each point $\mathbf{p}$ on the model surface with a unique parameter $\mathbf{t}_i = (u_i, v_i)$.

To identify the (u, v)-parameter for a point also takes advantage of the graphics pipeline and its rasterization process. Therefore, a 2D parameter model (texture coordinates) needs to be prepared in advanced using a 3D modeling tool. Figure 48.6a shows the test model. A 2D representation of the final parameters space is shown in Fig. 48.6b. Technically, we store each parameter as texture coordinate $\mathbf{t}_i$ and relate it to a vertex of the 3D model.

During runtime, the graphics pipeline rasterizes the model and, thus, interpolates each (u, v)-parameter for each point $\mathbf{p}_i$ on the surface. This process results in a unique parameter $\mathbf{t}_i$ for each fragment of the surface. Since all texture coordinates belong to the model, we can interpret the (u, v)-space as a global coordinate space $\mathbf{E}_{\text{global}}$. By utilizing this parameter space, each parameter $\mathbf{t}_i$ for each $\mathbf{p}_i$ will be equal, regardless of the viewpoint from which the range camera captures the surface, except for quantization errors. With this parameter space at hand, we map each value in $\mathbf{E}$ into $\mathbf{E}_{\text{global}}$. Figure 48.6d shows a false color rendering of the parameter space. Figure 48.6c shows error values from one range image mapped into the parameter space. Note that both renderings only visualize the result here. They are not used internally because the renderings only support an 8-bit quantization, which is insufficient. Internally, the (u, v)-parameter space (Fig. 48.6d) is rendered in an off-screen process using 32-bit float values and OpenGL's GL_RGBA32F_ARB HDR format.

48.2.2 Pose Optimization

The pose optimization process refines the point-to-point associations by adjusting the virtual rays between the camera and each surface point $\mathbf{p}_i$. We assume that the point-to-point associations are suboptimal due to common tracking errors caused by noise and other disturbances. Tracking errors are usually small but have an impact on the pose $\mathbf{T}$, thus on the surface difference measurement, especially since the approach relies on pixel-wise comparison of image points. To refine them we assume that the connection between the camera and each surface point $\mathbf{p}_i$ forms a ray. Further, all rays from N views that connect to one point $\mathbf{p}_i$ can be co-aligned.

A bundle adjustment (BA) algorithms can align them by refining the bundles [13]. BA minimizes the re-projection error between the image location of observed and predicted points. Figure 48.7 illustrates this for two range images $\mathbf{I}_j$ with $j = (1, 2)$. Assume that several points $\mathbf{p}_{i,j}$ are present from N views j and that all

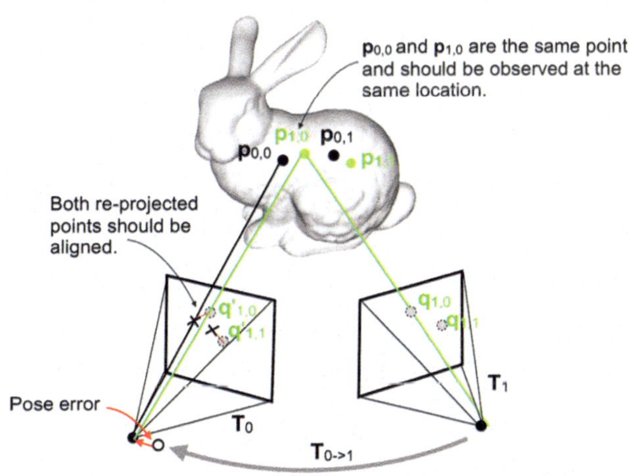

Fig. 48.7 Bundle adjustment is used to reduce the re-projection error that rays from multiple views yield when re-projected

points are associated to the reference model in Cartesian coordinate space. Also, a user moves the physical object between images. Thus, each image shows the object from a different perspective. All poses $\mathbf{T}_j$ between the object and camera are known. Aligning the images by transforming all points from image $\mathbf{I}_1$ to image $\mathbf{I}_0$ should theoretically yield matching re-projections $\mathbf{q}_{0,0}$ and $\mathbf{q}'_{1,0}$. Hence the pose error ΔT does not match accurately. BA adjusts this by refining the rays between the points $\mathbf{p}_{i,j}$ and their re-projected points $\mathbf{q}_{i,j}$ by minimizing the error $\Delta\mathbf{T}$, so that they align in one re-projected point:

$$\min\Delta \mathrm{T} = \min \sum_{i=1}^{N} \sum_{j=1}^{M} \left\| \mathrm{p}_{i,j} - P\left(C_i, \mathrm{p}_{i,j}\right) \right\| \tag{48.3}$$

The BA procedure was slightly adopted for this task. We work with a random sample of M points, pixel-wise selected from all $\mathbf{E}_i$. This approach has three reasons. First, the LM algorithm requires the same number of points and the same points from each image. Since there is no indication whether the range image data is valid for all points, it is a prior unknown how many points in each range image show the physical surface. Thus only a minimum number of M points are selected randomly, and the process is repeated multiple times until the error $\mathbf{\Delta T}$ is at a minimum. Second, a start image is required. The projected points of one start range image need to be considered as correct, and all the other images need to be aligned with this image. Since a "correct image" is an ideal assumption, we select the start image by random and repeat the entire process multiple times. The result that yields the smallest sum of re-projection errors is selected as the final result. Third, point-to-point alignments and the resultant pose typically do not improve with more points. All points need to represent the surface curvatures well. Thus, an overdense point set mostly does not contribute to the surface representation. Additionally, the process is already computational expensive. Using all points will further reduce the performance or render the optimization entirely impractical. Currently, we work with $M = 250$ points and run 25 iterations, numbers which we found empirically.

48.2.3 Surface Deviation Estimation

The objective of this step is to determine a deviation model **D** and to transfer it into a color map which one can project onto the triangulated CAD model. Figure 48.8a–c illustrates the approach for one range image. Assume that the triangulated CAD model gives the ideal surface. The deviation calculation provides a result on a per-fragment basis, and the normal vector $\mathbf{n}_i$ per fragment is also known. Step 2 yields the error e_i for each fragment in direction of a camera ray, and it is stored per surface fragment using $\mathbf{E}_i$ in (u, v)-space. To create an deviation model **D** with a deviation value $d_i(u, v)$ per surface fragment i, all error values are iteratively blend into **D**, as follows:

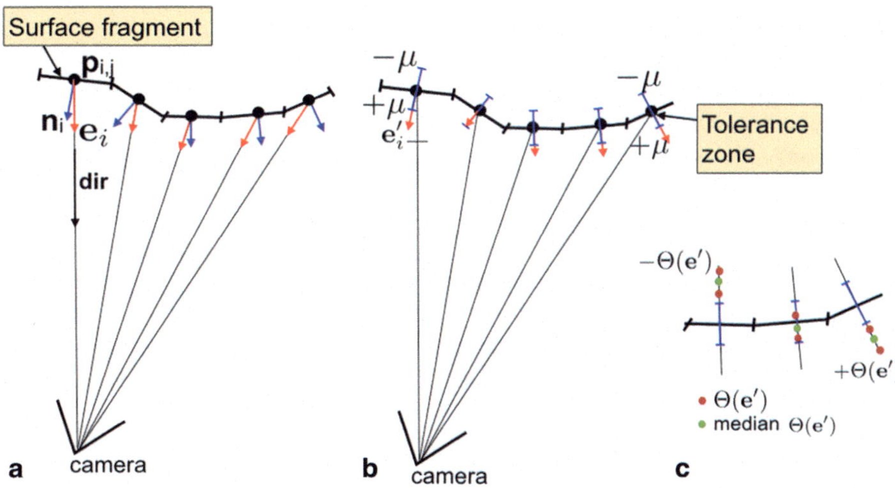

Fig. 48.8 (**a**) Alignment of error values with the direction of a surface normal. (**b**) Signed distance function applied on a per-surface-point grid. (**c**) Accumulation of error values

First, the error values e_i are aligned with the normal vector of the fragment. The values originate from different images, and each image was recorded from a different viewpoint with values e_i representing the error in camera direction (Fig. 48.8b). To align all e_i per point i, they are mapped onto their respective normal vectors $\mathbf{n}_i$. Therefore, the error is transformed into a vector $\mathbf{e}_i = \mathbf{dir}_i\, e_i$ and then projected onto the normal vector: $\mathbf{e}'_i = (\mathbf{e}\ \mathbf{n})/|\mathbf{n}^2|\mathbf{n}$. Additionally, a weight value is calculated for each $\mathbf{e}_i$ with $w = |\mathbf{n}|\ |\mathbf{e}|$. It accounts for the fact that error values are considered as more accurate when the angle between the camera ray and surface normal is small.

Second, the error is mapped into the model using the SDF. The SDF divides the region along the normal vector of each point into tolerance zones of size $\pm\mu$ and computes in which tolerance zone measurement falls. For the deviation analysis, it is only crucial whether the error falls into the first zone around the nominal surface or not. If within the tolerance zone, the error is discarded and set to zero. Outside the zone, the error accounts for a deviation as a multiplier of the tolerance zone. Thus, following this approach, the surface is considered as nominal as long as not a significant number of points are outside the zone $\pm\mu$. At last, the deviation for each point is calculated as:

$$d_j = \underset{(i,j)\in \mathrm{E}_i}{\mathrm{median}} \left(\Theta\left(e'_{i,j}\right)\right) \tag{48.4}$$

Thus, assuming the majority of surface values are correct, d_i should deviate from the nominal surface if the larger number of measurements tends into one direction. Without the SDF, the deviation map accumulates all errors, which would allow one to interpret $\mathbf{D}$ as accumulated variances. A surface artefact would disappear in noise. The final value is stored in a (u, v)-map $\mathbf{D}$. Figure 48.9b shows a color rendering of

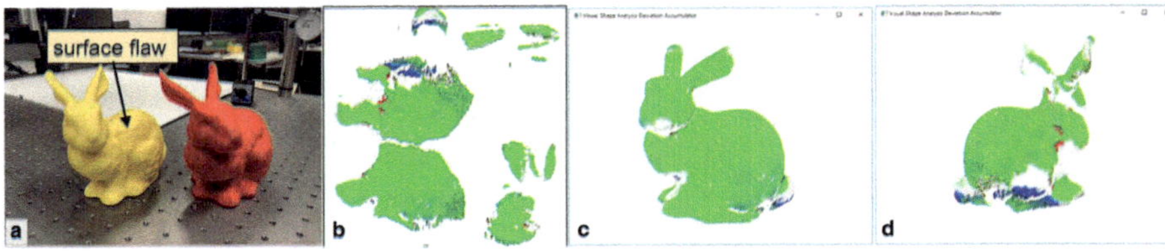

Fig. 48.9 (**a**) The test objects for this research is a 3D-printed Stanford 3D bunny model. (**b**) The *(u, v)*-parameter map with accumulated errors as false color map (**c**) and (**d**) front and backside of the model after eight images where accumulated from different directions

this map, which can also become a texture rendered onto the surface of the 3D model (Fig. 48.9c, d). Note that **D** uses 32-bits floating point values and the color map facilitates rendering only.

48.3 Experiments and Results

We evaluated the outcome of the suggested approach by comparing the deviation maps visually and by comparing the mean depth range. We prepared a test object, a 3D-printed Stanford bunny model with a dent of 2 mm depth. Although the depth of this flaw is within the range of the camera's depth resolution, effects such as noise and point cloud distortions as a result of reflection artefacts veil them in the range image. We assume that the presented method neglects these effects and that the flaw will become observable.

48.3.1 Method

Figure 48.9a shows the test object (yellow) used for this test. Its overall dimensions are 13 × 8 × 15 cm (× width × height). It was printed on a 3D printer using PLA filament. All general dimensions of the printed object were manually measured using a caliper to verify that its dimensions match the CAD model dimensions. We did not find significant differences. The 2 mm dent was added to the CAD model. It is rectangular with 30 × 20 × 2 mm (width × height × depth). The 2 mm depth is used due to the camera specifications. The camera used is a Fotonic P60U with a depth image resolution of 640 × 480@30 fps. The manufacturer promises a depth resolution of 3 mm at a distance of 1 m. However, this value incorporates typical sensor tolerances due to noise and other effects. The 2 mm is at the limit of what the sensor quantization allows.

For the experiment, the bunny model was scanned 20 times from different orientations and staying at different positions in front of the camera. Each position was selected randomly. However, lines on the breadboard ensured that the model remained in an area of 1–1.4 m distance to the range camera. Each scan contains between 8000 and 12,000 valid depth measurements that cover the objects' area. We compared the proposed method to a method that only uses Step 1 and Step 2, denoted as *direct mapping*. It directly maps the error values into (u, v)-space as mean values, without performing a BA or the SDF. The value v was set to 2 mm. A μ value of 0 mm indicates the ideal surface. Since the dent is of 2 mm depth, μ was set to the boundary of the dent.

48.3.2 Results and Discussion

Figure 48.10 shows the results. The upper row shows the direct mapping results and the lower row the results with the suggested method applied. The color mapping shows different hues of green with a color range from −2 mm (blue) to 2 mm (red). Each image column presents a result after the data of a particular number of images were processed. The left column shows the result for only one image. Here, the process skipped BA entirely since it requires a minimum of two images.

From left to right, one can notice that the surface representation becomes smoother. The images at the very left indicate measurement artefacts. Recesses, grooves, and other complex surface structures appear as a slight red hue. A range camera cannot resolve those surface variations from certain angles. The blue areas at the edges of the model are also barely visible to the camera. With more images taken from multiple directions, better data for these regions becomes available. Consequently, the surface tends toward a green hue, which indicates a deviation of 0 mm. Here 0 mm means the error is within the SDF error margin $\pm\mu$. We expected to see

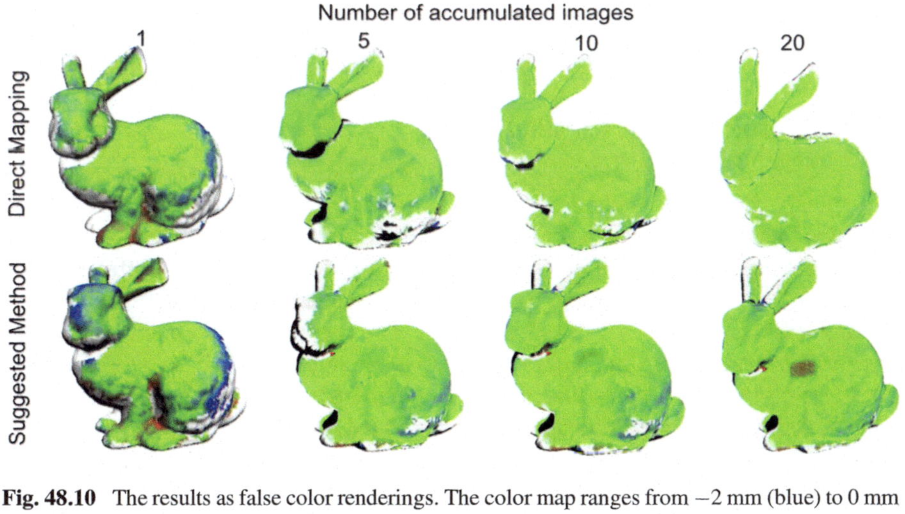

Fig. 48.10 The results as false color renderings. The color map ranges from −2 mm (blue) to 0 mm (green) to +2 mm (red). A white color indicates that no data was available

these results in both cases. Point-to-point associations that the ICP algorithm uses smoothen the results when the process accumulates more datasets. We also expected that false point-to-point associations result in an error more substantial than the camera's depth resolution. As a result, the surface flaw remains hidden with direct mapping applied. On the other hand, the suggested method can unveil the surface artefact. When accumulating more and more image, the method indicates the flaw as a red area. We extracted the mean depth from the flaw at $N = 20$. Therefore, the flaw area was manually marked (rectangle) with ImageJ, and the mean value was extracted. The result gives a flaw depth of +1.821 mm; note that only pixel deviations within the first and third quartiles of the error distribution were counted. Although this result is slightly incorrect, 2 mm was expected; it clearly indicates the deviation from the ideal surface.

A limitation is the small number of objects we worked with. The entire approach was tested with one 3D-printed object—two versions of it—which has a complicated surface shape. The utilized tracking approach can determine a more accurate pose, the more complicated the surface is. Thus, going forward to other shapes is still required to verify this approach.

Currently, the bundle adjustment algorithm is the computational bottleneck of this approach since it takes time for multiple images. It is also sensitive to outliers. Thus, to process 20 range scans correctly is a challenge. The experiment was interrupted several times and needed to be restarted since the Jacobian matrix could not be solved. However, this is an implementation issue and does not pose an obstacle to the approach. The current implemented method is also a naive one, which can be improved in future work.

Thus, the conclusion is that the suggested method facilitates to visually allow surface deviations in comparison to a triangulated CAD model when the surface deviation is of limited dimension with respect to the remaining part. It effectively removes the impact of noise and surface distortions on the range measurement. One caveat, the distance measurement, does not match the expectation.

48.4 Summary and Outlook

In summary, the research showed that the process works as suggested. The lab test indicates that accumulating point-to-point errors facilitates to identify local deviations, which supports the correctness of the approach. However, the limited number of experiments does not allow to generalize the approach further.

The next steps will be of technical nature. The two most important ones are the bundle adjustment algorithm and the *(u, v)*-map parametrization. The bundle adjustment algorithm is a naive one that generates a sparse Jacobian matrix. It takes several minutes to solve this matrix for 20 range images and is sometimes not possible at all. As a next step, the current bundle adjustment algorithm will be replaced with one that works with a dense Jacobian matrix, one that yields results in all cases. Also, we consider a GPU implementation.

Also, the current (u, v)-parameterization approach will be refined. Since the implementation already oversamples the surface and uses floating-point arithmetic, it is unlikely that more matrix elements will mitigate the problem. Since the discretization problems only occur at surfaces with normal vectors that are largely averted from the camera line of sight, e.g., 50–60° and larger, we intend to investigate anisotropic texture filter. Those texture filters yield better texture to fragment mappings, better from a visual point of view. They might also mitigate the problem at hand.

References

1. Jafari, B., Khaloo, A., & Lattanzi, D. (2017). Deformation tracking in 3d point clouds via statistical sampling of direct cloud-to-cloud distances. *Journal of Nondestructive Evaluation, 36*, 65.
2. Jovančević, I., Pham, H.-H., Orteu, J.-J., Gilblas, R., Harvent, J., Maurice, X., et al. (2017). 3d point cloud analysis for detection and characterization of defects on airplane exterior surface. *Journal of Nondestructive Evaluation, 36*, 4–74.
3. Newcombe, R. A., Izadi, S., Hilliges, O., Molyneaux, D., Kim, D., Davison, A. J., et al. (2011). Kinectfusion: Real-time dense surface mapping and tracking. In *2011 10th IEEE International Symposium on Mixed and Augmented Reality* (pp. 127–136).
4. Curless, B., & Levoy, M. (1996). A volumetric method for building complex models from range images. In *Proceedings of the 23rd Annual Conference on Computer Graphics and Interactive Techniques* (pp. 303–312). New York: ACM.
5. Wells, L. J., Shafae, M. S., & Camelio, J. A. (2013). Automated part inspection using 3d point clouds. In *ASME 2013 International Manufacturing Science and Engineering Conference Collocated with the 41st North American Manufacturing Research Conference - Volume 2* (pp. 1–9).
6. Awange, J., Paláncz, B., & Lewis, R. (2014). Maximizing likelihood function for parameter estimation in point clouds via Groebner basis. In H. Hong & C. Yap (Eds.), *Mathematical software – ICMS* (pp. 359–366).
7. Grimm, C. M., Crisco, J. J., & Laidlaw, D. H. (2001). Fitting manifold surfaces to three-dimensional point clouds. *Journal of Biomechanical Engineering, 124*, 1.
8. Shi, Q., Xi, N., & Zhang, C. (2010). Develop a robot-aided area sensing system for 3D shape inspection. *Journal of Manufacturing Science and Engineering, 132*, 1.
9. Nguyen, C., Izadi, S., & Lovell, D. (2012). Modeling kinect sensor noise for improved 3D reconstruction and tracking. In *2012 Second International Conference on 3d Imaging, Modelling, Processing, Visualization and Transmission (3dimpvt)* (pp. 524–530).
10. The Stanford 3d scanning repository. (n.d.). Retrieved January 28, 2018, from http://graphics.stanford.edu/data/3Dscanrep/
11. Radkowski, R., Garrett, T., Ingebrand, J., & Wehr, D. (2016). TrackingExpert - A versatile tracking toolbox for augmented reality. In *2016 International Design Engineering Technical Conferences & Computers and Information in Engineering Conference* (pp. 1–10). New York: ASME.
12. Besl, P., & McKay, N. (1992). A method for registration of 3D-shapes. *IEEE Transactions on Pattern Analysis and Machine Intelligence, 18*, 8.
13. Triggs, B., McLauchlan, P. F., Hartley, R. I., & Fitzgibbon, A. W. (2000). Bundle adjustment - a modern synthesis. In B. Triggs, A. Zisserman, & R. Szeliski (Eds.), *Vision algorithms: Theory and practice* (pp. 298–372). Berlin, Germany: Springer.

Chapter 49
Investigating Approaches to Achieve Modularity Benefits in the Acquisition Ecosystem

Navindran Davendralingam, Cesare Guariniello, Alexandra Dukes, and Daniel DeLaurentis

49.1 Background

The US Department of Defense (DoD) is challenged to affordably address emerging threats, component obsolescence, loss of critical suppliers, and planned technology insertion/upgrades for tightly coupled, highly integrated systems while increasing competition. In response to these challenges, new acquisition strategies call for the incorporation of modular design features in new DoD system designs, through its modular open system approach (MOSA) initiative. MOSA language as articulated in the law, the OSA Contracting Guidebook [1], and the *Defense Acquisition Guidebook* (DAG) [2] includes a description of best practices and benefits, relevant contracting language, and high-level guidance for both the program manager and lead systems engineer on implementing MOSA. Modular design solutions include the use of standardized, modular interfaces to enable "plug-and-play" capabilities while facilitating quick insertion of upgraded or new capabilities on existing platforms. The DoD encourages MOSA adoption for both its major defense acquisition programs (MDAPs) and major automated information systems (MAIS), in pursuit of programmatic benefits associated with modularity and openness. The encouraged adoption of MOSA was in response to perceived benefits of modular and open strategies as evidenced by practical applications in industry. The DoD has identified the following primary benefits of implementing MOSA in an acquisition program:

- *Enhance competition.* Open architecture and modularity are relevant to competitive contracting strategies. The use of open interface standards for military modules permits the independent and competitive development of such modules to more than one party, thereby promoting competitive development of technologies and tendered reduction in cost.

N. Davendralingam (✉) · C. Guariniello · A. Dukes · D. DeLaurentis
Purdue University, West Lafayette, IN, USA
e-mail: davendra@purdue.edu

S. Adams et al. (eds.), *Systems Engineering in Context*,
https://doi.org/10.1007/978-3-030-00114-8_49

- *Facilitate technology refresh.* Adequate process design, standardization, and modularity support easier update of systems without the need to change all components and provide opportunities for technology insertion.
- *Incorporate innovation.* Modularity facilitates life cycle sustainment and operational flexibility, through the ability to plug and play using newer, more capable modules.
- *Enable cost savings/cost avoidance.* Both software and hardware modularity and standardization allow for cost savings and cost avoidance in design, development, and update of systems.
- *Improve interoperability.* Standardization and openness address loosely coupled architectures and facilitate data sharing.

In addition to the listed main benefits, there are ancillary benefits that include (but are not limited to) reduced life cycle costs, shortened acquisition timelines, increased flexibility, incremental approach to capability development, and reduced instances of vendor lock. While the benefits of MOSA address many of the DoD challenges faced in acquisitions programs, there still needs to be reconciliation between an acquisition program's goals and MOSA objectives. *A key issue is the need for the program manager to connect specific program goals to a MOSA-related goals and benefits.* For example, a program manager seeking to incorporate sustainability of a program may wish to pursue a MOSA strategy that focuses on "enhancing competition" as a MOSA benefit that specifically relates to the program goal. When objectives of a program are related to MOSA benefits, the viewpoint of MOSA now changes from being an *end goal* to that of an *enabler* of key acquisition program objectives.

To achieve these MOSA benefits listed above, the DoD established guidelines for the program manager and lead systems engineer of an acquisition program. Core tenets of these guidelines for the program manager are summarized in the following points:

- *Establish an enabling environment*
  - A program manager needs to generate business practices, technology development, test, evaluation, etc. needed for development of open systems.
- *Employ modular design*
  - A program manager needs to employ modular designs. Main characteristics of modular design includes cohesive, encapsulated, self-contained, highly binned.
- *Designate key interfaces*
  - Module interfaces need to be defined to enable designers and system configuration managers, both from a technical and business point of view.

- *Use open standards*
  - Utilization of community accepted standards that are well defined, mature in their use, and available.
- *Certify conformance*
  - Use rigorous assessment mechanisms, interface control management, and proactive conformance testing—this includes comprehensive validations and verification mechanisms.

49.1.1 Current Challenges for MOSA

MOSA guidelines aim to balance the business objectives (open systems) with the technical means to meet its acquisition challenges. However, various issues arise when actual implementation is realized, due to the lack of sufficient guidance and experiential knowledge made available to key stakeholders (e.g., program manager). The complex and often unobvious dependencies between the technical and programmatic components of employing a modular approach can make reaping the benefits of modularity a challenging endeavor for all stakeholders involved. For example, various questions that may come to mind for a program manager may include *who owns the rights to the data at the interfaces? What are good metrics to assess when considering modularity? How do we integrate promising, new modules that do not have a long-term sustainment plan yet? How do we assess technical risks for modular systems?* Furthermore, the diverse context of how the benefits of modularity and openness are perceived, utilized, and pursued across the broader technological ecosystem beyond the defense context makes it difficult for generalized (modular) policies to be implemented. For example, a GAO report detailed the difficulties in adopting modular strategies due to lack of means to modularize existing critical legacy systems, upon which modular systems were to interface.

Currently, MOSA efforts have been under the auspices of open systems architecture (OSA) that largely focuses on the business side of the acquisition process and deals with issues such as data rights and legal matters that are necessary, but not sufficient to achieve the DoD objectives in dealing with challenges faced. MOSA policy in the DoD's 5000.01 guide, specifically expresses (among others) the following consideration: *Acquisition programs shall be managed through the application of a systems engineering approach that optimizes total system performance and minimizes total ownership costs. A modular, open-systems approach shall be employed, where feasible* [2]. Furthermore, a program manager's responsibilities toward open systems are expressed in Section 14 of the DoD 5000.02 (Operation of the Defense Acquisition System) as: *Program managers are responsible for applying open systems approaches in product designs where feasible and cost-effective.* Such policy language provides latitude in the degree to which MOSA is adopted in an acquisition program. However, successful incentivization of MOSA adoption hinges on a keyword from both these policy statements – *feasible*. Through increasing the *number* of available *feasible* solutions to the program manager (and other stakeholders in the acquisition life cycle), successful alignment between MOSA objectives and an acquisition program's objectives can be made. The availability of feasible solutions can only be made clear through key stakeholders' explicit understanding of the impact that MOSA-related technical and programmatic decisions can have throughout the acquisition life cycle and, ultimately, on program outcomes.

49.2 Research Needed

It is this recognition of modularity's diverse contexts, tight coupling between MOSA technical and business components, influence on stakeholder incentive structures, and programmatic implications on various segments of the acquisition life cycle that motivates our research. Our ongoing work, funded by the US DoD Systems Engineering Research Center (SERC), investigates various approaches to achieving the benefits of modularity and openness in the context of an acquisition ecosystem. The knowledge acquisition phase of our research seeks to accomplish two main tasks. The first task is a deep-dive literature review on modularity concepts and relevant case studies that relates modularity to complex system development—this includes (among others) factors that contribute (or even prevent) the active seeking of modular benefits. The second task involves active engagement with SERC collaborators from industry, academia, and government, to further deepen our knowledge base with practical, community-driven perspectives on seeking benefits of modularity. The following phase of research will collate, curate, and distil the acquired knowledge artifacts into two practical and actionable guidance products. The first is a guidance document for program managers and key stakeholders within the defense acquisition community. The second is an active decision support framework that operates as a feedback mechanism that enables acquisition stakeholders to assess the impact that various acquisition actions can have in the pursuit of the benefits of modularity.

In Sect. 49.3 we provide a brief overview of our deep-dive findings—this includes a description of various forms of modularity, advantages and disadvantages, and some examples of ecosystem-level interactions based on published case studies. In Sect. 49.4, we provide an overview of collaborative exchanges and community-driven explorations. In Sect. 49.5, we present features of our guidance document and envisioned decision support framework. Our final section contains a discussion on future work.

49.3 Literature Review: A Deep Dive on Status Quo

Modularity is ubiquitous in complex systems, as evidenced by implementation in architectures ranging from biological constructs [3, 4] to the computer hardware and software systems that we rely on daily [5]. *Openness* on the other hand refers to the use of agreed-upon standards, sources, and architectures as means of interfacing between modules that comprise the complex system [6]. Standards are typically agreed upon through community-driven expert consensus and updated in accordance to perceived benefits for modular interfaces [7]. The division of a complex, monolithic system into a manageable set of smaller constituent systems or "modules" has been an intuitive strategy in dealing with complexity (see Russell [8] for an excellent modern history of modularity). Product modularization is motivated by the need for flexibility in product design changes, in response to evolving customer needs. For example, in 1964, IBM launched its very first modular computer—the System/360—a modular mainframe system that replaced prior, highly integrated designs [8, 9]. Modularization in IBM's computer architecture was in response to users' need to be able to retain existing programs and not have to completely rewrite their programs, should a new mainframe system with greater capabilities be purchased from the manufacturer [9]. In vehicle design, modularity has played a key role in enabling rapid innovation, cost reductions through flexible scheduling and manufacturing, and increased revenue through affording customers' choice of customization [5, 10, 11]. Commercial aircraft design and manufacturing follow the same pattern of modular product development on a global scale, for example, the numerous components of Boeing's aircraft (wing, fuselage, engines, subsystems, etc.) are manufactured separately at multiple global locations, following specified standards for each module. More recently, Boeing and Airbus are recently considering modular strategies to improve passenger experience options for aircraft cabin configurations [12]. These examples, among many others, are exemplars of the widespread use of modularity, in pursuit of its benefits.

Research on modularity has examined ways of categorizing, quantifying, and developing strategies to seek the benefits associated with a modular design. Categorizations, such as those by Huang and Kusniak [13], are motivated by a particular domain applications. In their work, Huang and Kusniak utilize categories of *component swapping*, *component sharing*, and *Bus* modularity for electromechanical application [13]. Miller and Elgård [14], motivated by a literature review on industrial experiences, extend the concept of modularity to include not only physical but functional elements as well. The more recent definitions of Baldwin and Clark [5] provide a broad categorization of modularity in terms of *modularity in design*, *modularity in production*, and *modularity in use* that are applicable to the broader context of the engineering ecosystem. For a comprehensive overview of the spectrum of definitions of modularity used in research, we invite the reader to read Gershenson et al. [15]. A complement to categorization includes the use of various metrics and measures of modularity and openness, to provide an assessment on the degree of modularity being employed, relative to other performance metrics. For example, the US Navy uses an Open Architecture Assessment Tool (OAAT) [16] to assess the degree of openness for selected architectures. In space applications, a work by Tamaskar [17] and Mohsen [18] rationalizes spacecraft modularization through the use of complexity and value-based measures, respectively. Martin and Ishii leverage two developed indices in their method to develop modularized products [19]. Gershenson et al., as part of a compendium literature review series, also provide a deep overview on a spectrum of existing research on measures and design methods used to achieve produce modularity [15]—their review paper culminates with a conclusion that while there are some abstract elements of measures that may be similar, their instantiations differ due to the underlying. Regardless of the diversity in measures and metrics used, the fact remains that the end adoption of modularity and openness is attributed to perceived benefits. In addition to the benefits identified in Section 49.1, modularity and openness also give rise to the following benefits (and potential disadvantages) as well:

(a) *Adaptability and upgrades*—Modular architectures can adapt to future uncertainties through swappable modules that can be upgraded. Furthermore, a system can have reconstituted capabilities when compatible modules are swapped for a different set of capabilities [5, 17, 18].
(b) *Parallel development*—Development of individual modules can be performed independently, due to decoupling of components and separate tasking [20, 21].
(c) *Increased innovation*—Rapid innovative cycles can be brought to bear upon each module where innovation at the modular level can provide incremental add-on innovations to the total architecture (*potential disadvantage*: Interestingly, an analysis of patent data by Fleming [22] reveals that peak innovation breakthroughs have been skewed to integrated designs).

49.3.1 Ecosystem-Level Interactions

Some of the impacts of modularity transcend beyond the immediate product being developed and extend to the broader context of the ecosystem that the modular product interfaces with. We can imagine that each of these modules has not only technical specifications but programmatic implications as how they impact ranging from intellectual property (IP) to the product supply chain and human capital involved. These cross-hierarchical dependencies between product and the broader "ecosystem" have given rise to behaviors, both beneficial and detrimental. In the case of the Daimler-Chrysler merger, the difference in the operations and organizational structure and product structure of each company led to disastrous effects when integrated [23]. The modular architecture of Chrysler's automotive vehicle and organizational setup to support such a product was vastly different than the organizational setup of Daimler that catered to a higher-tier, highly integrated design. Other ecosystem-level effects include the mirroring effect, also known as Conway's law [24], where a product's architecture mirrors the organizational architecture. In the conclusion to Gershenson's paper [15], they revisit key questions brought up by Ulrich and Tung [25] that includes *What is the connection between the organizational structure of the firm and the types of modularity that can be successfully implemented within the firm?*—a question that complements Conway's observation.

Much of the "lesson learned" knowledge related to its implementation and usefulness is dispersed throughout our technological ecosystem. Furthermore, the diversity in types of lessons learned prompts the need for a normative framework to be established. For example, the lesson learned on the use of modularity to cater various customer needs in the automotive industry [11] will be different than expressions of modularity for open versus integrated software development in software engineering, which in turn is different from lessons on incentive structures for contractors to participate in developing open solutions for defense systems [26]. As part of our ongoing efforts to collate, curate, and capture relevant useful knowledge artifacts as it relates to MOSA, our efforts also include engaging the broader community of subject matter experts as well.

49.4 MOSA: Community Perspectives

To reinforce the "lessons learned" from the literature review, and to assure the usefulness and practicality of the products generated in this research, we pursued direct engagement with the community involved in the technical and managerial aspects of projects that can benefit from the MOSA approach. The first step in this phase of community engagement was a workshop that gathered users of MOSA from the government and the military and project managers and subject matter experts from the industry and academia. The workshop, which was held in October 2016, provided experts and stakeholders with a chance to share their experience, discuss the difficulties of the application of the MOSA principles, and provide suggestions for identification of good practices that can enhance the appropriate use of openness and modularity. The first outcome of the workshop was an additional set of practical "lessons learned." Among this, the idea that modularity should not be seen as a measurable output, but as an input necessary to produce functional architecture. This suggests that MOSA should be a means to achieve the goals that project managers care about, summarized in the five benefits of the MOSA approach. Therefore, good modularity corresponds to good architecting. An ecosystem capable of supporting modular, open approach needs to establish a long-term strategy, with drivers and objective for the different stakeholders. The consequences of modularization choices need to be assessed early in the project life cycle, and feedback mechanisms are necessary to help stakeholders understand the consequence of their actions and that of others.

Following the workshop, we initiated a series of collaborative exchanges, where selected stakeholders provided their thoughts on modularity and openness, based on their experience as technical experts of project managers. The outcome of these interactions constituted the backbone of our program manager guidance document, providing more "lessons learned" based on a set of case studies. Besides confirming and expanding on the base of knowledge acquired in the workshop, the collaborative exchanges provided ideas on useful metrics and tools (e.g., design structure matrices) for the assessment of programs. Some common threads among the stakeholder articulations are related to the "pain points" in complex, long projects: program managers frequently face issues due to the timeliness of acquisition, which often fails to follow the speed of innovations. Other considerations necessary to support MOSA practices relate to the different programmatic timescales and to the presence of multiple stakeholders that need to be properly incentivized. The use of COTS and allocation of enough time for prototyp*in*g and testing in early phases of a program can enhance flexibility. These considerations are often overlooked, especially when program managers responsible for early phases of a program are expected to move from the program before its conclusion and might fall into the habit of considering only a limited time horizon. The interactions with stakeholders concluded the phase of data collection in this project and supported the production of MOSA knowledge products.

49.5 SERC RT-185: MOSA Knowledge Products for Defense Acquisitions

49.5.1 Program Manager Guidance Document

The knowledge acquired in our research under a prior funded effort (SERC RT-163), and described in Sections 49.3 and 49.4, resulted in an initial *program manager guidance document* that serves as a living document, into which further informational artifacts in the author's follow on work under the SERC (RT-185) will contribute to. This document is based on the knowledge artifacts gained through deep-dive literature review, case studies, and related considerations collected through interactions with subject matter experts and practitioners. These interactions also allowed for the organization of common threads, including best practices and pain points, which serve as an initial set of guidelines for program managers. Guidance categories in the document included the following:

- *What to measure and why?*—A canvassing of relevant measures used related to modularity, openness, and MOSA in general.
- *Useful strategies exist at different acquisition life cycle phases*—List of strategies, processes, and observations to support appropriate and beneficial adoption of modular designs.
- *Caution! Emergent phenomenon in benefits and risks*—Cases on unanticipated consequences (benefits and detriments) in pursuit of modularity, in part, due to stakeholder dynamics throughout the defense acquisition life cycle.
- *Ouch! Technical and programmatic pain points*—General technical and programmatic pain points in pursuit of modularity.

49.5.2 Decision Support Framework

The knowledge accumulated, both through our currently ongoing deep-dive literature review and engagement with key collaborators, forms a valuable gamut of knowledge artifacts on modularity and openness. However, such information is not actionable when purely treated as unstructured information. Here, accumulated knowledge needs to serve as a general guidance platform, as envisioned in the program manager guidance document, and as a *contextually specific service* that provides key, actionable information relevant to specific queries presented by an acquisitions practitioner. It is this contextual vision that motivates ongoing work toward a MOSA decision support framework, illustrated in Fig. 49.1 below.

Figure 49.1 illustrates our vision of merging of content from collaborative exchanges with relevant subject matter experts and information from our literature-based deep-dive efforts, into a central electronic repository. The subsequent task is to then distill the collected information into the most useful threads of actionable insights to the program manager, given a set of technical and programmatically relevant queries. Depending on the size, scope, and depth of information accumulated, we will employ necessary data filtering and knowledge management strategies to generate actionable sets of information related to the acquisition process. We envision the tool to provide objective feedback such as potential MOSA acquisition strategy options, related antecedent case studies, anticipated risks at various stages of the acquisition process, perceived equity of decisions between stakeholders, and potential impact on desired program outcomes, to the program

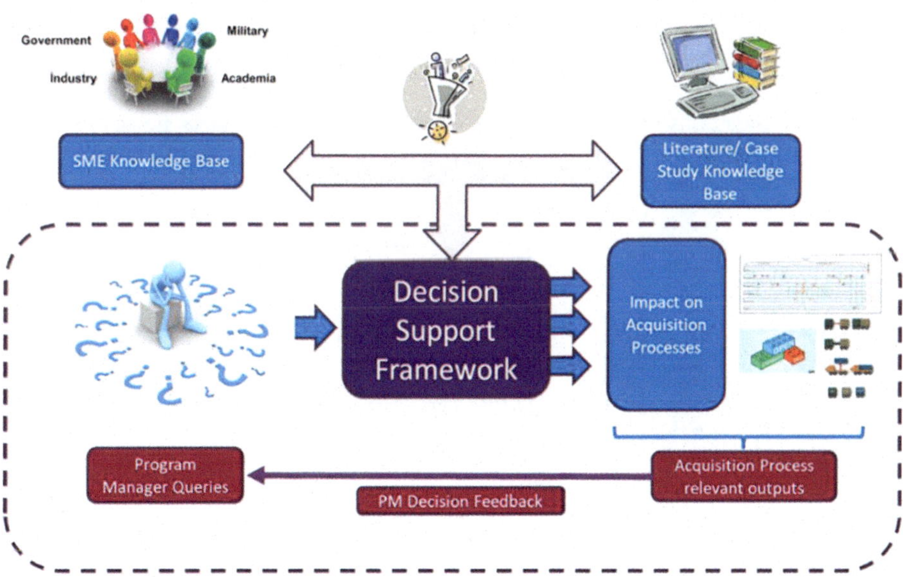

Fig. 49.1 Decision support framework concept

manager and related stakeholders. The feedback loop will enable a subset of the most relevant MOSA strategies to be presented to querying stakeholder during decision-making. Our ongoing research is actively engaged in expanding the contents of our repository through further deep-dive literature review and continuing engagement with additional collaborators.

49.6 Future Work

Our future work will first focus on distilling and curating relevant knowledge artifacts gained through our deep-dive and collaborative exchange endeavors. The effort will include forming a database of case studies and their mappings to relevant sections of the defense acquisition life cycle. Building from the SERC RT-185 work, MOSA-related decision-making scenarios for relevant stakeholders will be identified and prioritized from insights gained from the collaborative exchanges. With knowledge artifacts collected and organized, a programmatic decision support framework will be developed which will allow stakeholders to understand and anticipate technical and programmatic impacts of MOSA strategies toward achieving their intended MOSA benefits. Additionally, this decision support framework will assist stakeholders in exploring the trade-offs between common metrics of program interests, such as cost, schedule, and risk, against various strategies for modularization. The construct of an executable, conceptual model which captures the dynamics and influence mechanisms related to modularity and openness across a range of DoD stakeholders at various stages of the acquisition life cycle will contribute a direct feed mechanism for users that may allow them to see the impact of various MOSA decisions on the acquisition process.

Acknowledgments This material is based upon work supported, in whole or in part, by the US Department of Defense through the Systems Engineering Research Center (SERC) under Contract HQ0034-13-D-0004-0063. SERC is a federally funded university-affiliated research center managed by Stevens Institute of Technology.

References

1. US Department of Defense ODRT. (2013). *Open system architecture contract guidebook for program managers, V 1.1*. Department of Defense.
2. DoD. (2010). *Defense acquisition guidebook*. Washington, DC: Pentagon.
3. Clune, J., Mouret, J., & Lipson, H. (2013). The evolutionary origins of modularity. *Proceedings of the Royal Society B, 280*(1755), 20122863.
4. Newman, M. (2006). Modularity and community structure in networks. *Proceedings of the National Academy of Sciences, 103*(23), 8577–8582.
5. Baldwin, C. Y., & Clark, K. B. (2006). Modularity in the design of complex engineering systems. In *Complex engineered systems [Internet]* (pp. 175–205). Berlin, Germany: Springer.

[cited 2018 Jan 31]. (Understanding complex systems). Retrieved from https://link.springer.com/chapter/10.1007/3-540-32834-3_9
6. Krechmer, K. (2005). The meaning of open standards. In *Proceedings of the 38th Hawaii International Conference on System Sciences.*
7. Ghosh, R. (2005). *An economic basis for open standards*. Maastricht, Netherlands: University Maastricht Economics and Social Research and Training Center of Innovation and Technology.
8. Russell, A. L. (2012). Modularity: An interdisciplinary history of an ordering concept. *Information & Culture, 47*(3), 257–287.
9. Baldwin, C., & Clark, K. (1997). Managing in an age of modularity. *Harvard Business Review., 75*, 84–93.
10. Takeishi, A., & Fujimoto, T. (2001). *Modularization in the auto industry: Interlinked multiple hierarchies of product, production, and supplier systems [Internet]*. Institute of Innovation Research, Hitotsubashi University. [cited 2018 Jan 31]. (IIR Working Paper). Report No.: 1–2. Retrieved from https://ideas.repec.org/p/hit/iirwps/01-02.html
11. Pandremenos, J., Paralikas, J., Salonitis, K., & Chryssolouris, G. (2009). Modularity concepts for the automotive industry: A critical review. *CIRP Journal of Manufacturing Science and Technology, 1*(3), 148–152.
12. Airbus' new jet concept comes with swappable spas and cafés [Internet]. *WIRED*. [cited 2018 Jan 31]. Retrieved from https://www.wired.com/2016/12/airbus-new-jet-concept-features-swappable-spas-cafes/
13. Huang, C.-C., & Kusiak, A. (1998). Modularity in design of products and systems. *IEEE Transactions on Systems, Man, and Cybernetics-Part A: Systems and Humans, 28*(1), 66–77.
14. Miller, T. D., & Elgård, P. (1998). *Defining Modules, Modularity and Modularization-Evolution of the Concept in a Historical Perspective*, Design for Integration in Manufacturing. In Proceedings of the 13th IPS Research Seminar, Fuglsoe.
15. Gershenson, J. K., Prasad, G. J., & Zhang, Y. (2004). Product modularity: Measures and design methods. *Journal of Engineering Design, 15*(1), 33–51.
16. Naval Open Architecture Assessment Team. (2009). *Open architecture assessment tool version 3.0 user's guide.*
17. Tamaskar, S., Neema, K., & DeLaurentis, D. (2014). Framework for measuring complexity of aerospace systems. *Research in Engineering Design, 25*(2), 125–137.
18. Mosleh, M., Dalili, K., & Heydari, B. (2014). Optimal modularity for fractionated spacecraft: The case of system F6. *Procedia Comput Science, 28*, 164–170.
19. Martin, M. V., & Ishii, K. (2002). Design for variety: Developing standardized and modularized product platform architectures. *Research in Engineering Design, 13*(4), 213–235.
20. Baldwin, C. Y., & Clark, K. B. (2006). Modularity in the design of complex engineering systems. In *Complex engineered systems [Internet]* (pp. 175–205). Berlin, Germany: Springer. [cited 2018 Jan 31]. Retrieved from https://link.springer.com/chapter/10.1007/3-540-32834-3_9
21. Drezner, J. A., & Simpson, M. (2017). *Exploring parallel development in the context of agile acquisition [Internet]*. [cited 2018 Feb 1]. Retrieved from https://www.rand.org/pubs/research_reports/RR1808.html
22. Fleming, L., & Sorenson, O. (2001). *The dangers of modularity [Internet]*. Harvard Business Review. [cited 2018 Feb 1]. Retrieved from https://hbr.org/2001/09/the-dangers-of-modularity
23. Ball, J., & Miller, S. (2015). Daimler-Benz, Chrysler Merger fails to live up to expectations [Internet]. *WSJ*. [cited 2018 Feb 1]. Retreived from http://www.wsj.com/articles/SB964560646863325466
24. Conway, M. (1968). How do committees invent? *Datamation, 14*(5), 28–31.
25. Ulrich, K. (1994). Fundamentals of product modularity. In *Management of design [Internet]* (pp. 219–31). Dordrecht, Netherlands: Springer. [cited 2018 Jan 31]. Retrieved from https://link.springer.com/chapter/10.1007/978-94-011-1390-8_12
26. Welby, S. (2014). *Modular open systems architecture in DoD acquisition [Internet]*. Springfield, MA: NDIA. Retrieved from https://www.acq.osd.mil/se/briefs/16943-2014_10_29_NDIA-SEC-Welby-MOSA-vF.pdf

Chapter 50
A Domain Engineering Framework Based on Probabilistic Ontologies for Automated Selection of Features to Reuse

Shou Matsumoto, James D. Lee, Muhammad Imran, Edward Huang, Kathryn B. Laskey, and Paulo C. G. Costa

50.1 Introduction

In this new era of information technology we live in, the process of developing software-intensive systems from scratch is already becoming old-fashioned. Solutions with high quality are regularly available in the World Wide Web as downloadable libraries, frameworks, or open-source software tools. Organizations also frequently maintain their own sets of libraries or repositories of "legacy" systems and reusable software assets which were produced as results of their routine business processes. Consequently, the task of developing a new system is being replaced by the task of adapting and integrating existing solutions to new problems.

In this context, *domain engineering* is a discipline which focuses on reusing domain knowledge in order to produce a family of software-intensive systems [1]. Stated in different words, this discipline comprises techniques for developing a diversity of systems that solve different problems, under the assumption that such problems belong to some specific well-defined domain. The nature of domain engineering involves developing conceptual models to capture vocabulary and meta-information about a particular domain and to elicit common and varying observable characteristics—or *features*—among systems.

On the other hand, *ontologies* (in the context of information technology) are engineering artifacts which consist of formal vocabularies of terms, usually describing specific domain knowledge and accessed by persons or computers sharing a common view or domain application. Various interdisciplinary works addressing the engineering aspects of this field have been held in the recent years by the information systems community [2–6]. The *Web Ontology Language* (OWL) is a standard ontology language which represents classes, properties, and individuals in documents or datasets [7]. Since ontologies are defined as formal specifications of domain

S. Matsumoto (✉) · J. D. Lee · M. Imran · E. Huang · K. B. Laskey · P. C. G. Costa
George Mason University, Fairfax, VA, USA
e-mail: smatsum2@gmu.edu

S. Adams et al. (eds.), *Systems Engineering in Context*,
https://doi.org/10.1007/978-3-030-00114-8_50

knowledge, OWL is a natural candidate for representing conceptual models in domain engineering.

However, traditional ontology languages like OWL have no built-in mechanism for representing or drawing inferences under uncertainty. The lack of mechanisms for treating such uncertainty is often problematic in domain engineering, due to some common characteristics listed below:

- New or emerging problems are unknown to the modeler at the time domain models are built. Consequently, there is always uncertainty due to ignorance or incompleteness of domain knowledge.
- Rules and methods that derive the systems architecture from requirements are a result of complex analysis by engineers, which are often based on heuristics or experts' beliefs. Such beliefs are usually hard to express in terms of deterministic rules of classic logic. This is a relevant source of uncertainty, because it suggests that when problems are recurring, then similar specifications may not always derive the same system architecture.
- Presence of legacy systems is generally associated with lack of up-to-date documentation and personnel with expertise for describing characteristics and behaviors of the system's components for reuse. This also brings uncertainty inherent from ignorance, imprecision, credibility of source, or incompleteness of domain knowledge. This may also be the case of third-party software assets (e.g., downloaded libraries).
- And finally, uncertainty is a ubiquitous property in the real world.

This work examines the applicability of probabilistic ontologies [8] in building domain engineering architectures and models. Probabilistic reasoning can be applied to assist decisions under uncertainty about relationships involving requirements, system artifacts, and other elements in the architecture. The main use case of our model is therefore to be able *to infer which artifacts shall be included in a specific solution, or whether artifacts shall be created or modified, given new emerging problem specifications*. We present an application in the domain of Insider Threat Inference Enterprise Modeling [9].

50.2 Proof of Concept in the Domain of Insider Threat Inference Enterprise Modeling

An *inference enterprise* is an organizational entity or undertaking devoted to making inferences. Engineers in such an enterprise develop *inference enterprise models* to understand the enterprise and to develop predictions of the effects of introducing changes to the enterprise or to predict its performance [9]. This work focuses on inference enterprise models that represent an enterprise that detects, from a large population, the occurrence of some target behavior associated with a potential threat. To do this, the inference enterprise collects data about observables and applies algorithms that process the observables to identify persons of interest who merit further investigation. The main elements of our inference enterprise model are listed below:

- *Population Modeling and Synthesis/Simulation*: step for understanding the population at which the inference enterprise is directed and generates a model for simulating multiple realizations of the population.
- *Fusion and Down-Select:* process of applying rules for fusing relevant observables to identify cases for which an alert is issued. Rules may involve sophisticated machine learning algorithms.
- *Performance Evaluation*: step for estimating performance metrics such as precision, recall, and false-positive rates.

The above elements combined together result in an executable model of an inference enterprise, which is the domain of discourse of our ontology. Our framework therefore combines into a probabilistic ontology the meta-information about such models.

50.2.1 A Tool for Rapid Integration and Analysis: Semantic Testbed for Inference Enterprise Modeling

Semantic Testbed for Inference Enterprise Modeling (STIEM), initially introduced as a multi-modeling framework for executing and analyzing Insider Threat Inference Enterprise Models [9], is a general model-based systems engineering [10] framework and tool for designing, integrating, executing, orchestrating, and analyzing any set of software components with distinct and visible input and output interfaces. STIEM is built on top of Phoenix Integration ModelCenter® [11]. In addition to a wide variety of analysis and simulation tools natively offered by ModelCenter®, STIEM offers a repository of statistical simulation components which are able to reconstruct population data from summary statistics, another repository of machine learning and inference components, a tool and process for quickly adding new components into STIEM's software repository, and open-standard protocols for making distributed and asynchronous execution of components in a workflow.

Figure 50.1 illustrates typical functionalities offered by STIEM. One objective of our domain engineering ontology is to offer a means for automatically selecting a subset of assets from STIEM's software component's repository, so that the selected set of components can be orchestrated via a Multi-model Integration Workflow (upper left corner of Fig. 50.1) for rapid prototyping, distributed computing, and sensitivity analysis.

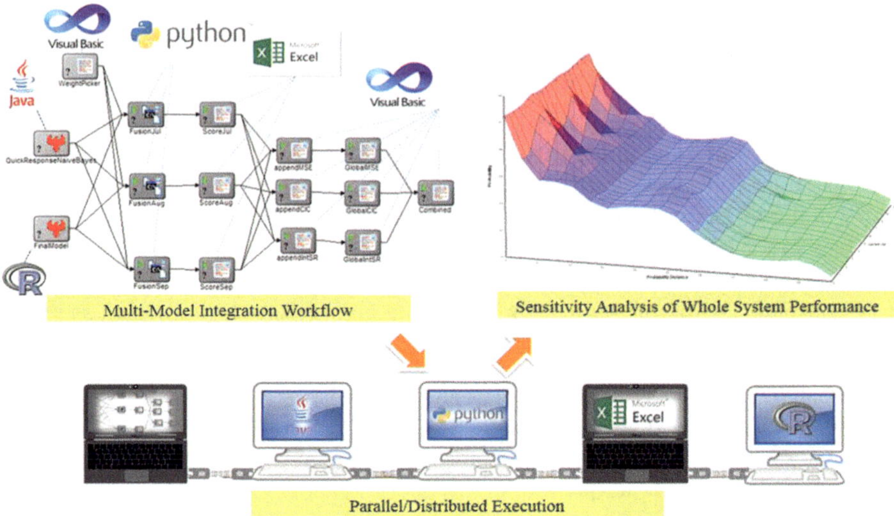

Fig. 50.1 Main functionalities of STIEM

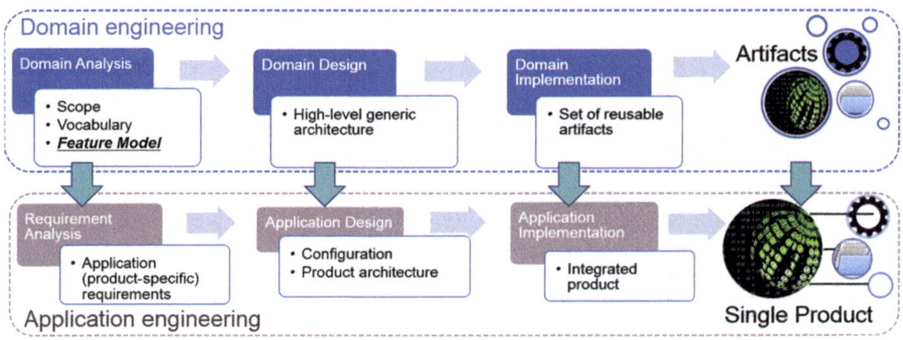

Fig. 50.2 Illustration of relationship between domain and application engineering

50.3 Domain Engineering Principles

The process of domain engineering is often divided into two phases: the "domain engineering" itself (the process of analyzing, architecting, and developing reusable components among the family) and the application engineering (process of producing a single product by integrating and/or customizing reusable components). Proper domain engineering practices enable fast production and customization. Figure 50.2 illustrates how a typical domain engineering process works. The domain engineering phase is usually separated into a domain analysis step, which analyzes the domain scope, produces a domain vocabulary, and specifies a feature model; a domain design step, which generates a high-level generic (abstract) architecture of the family of systems; and a domain implementation, which produces a set of reusable artifacts.

Then, a subset of the artifacts can be reused and integrated during application engineering—which resembles a "normal" systems engineering process—in order to produce a single product/system.

50.3.1 Feature Model

A *feature model* specifies all features in the domain and constraints that form a consistent and operable set of common aspects and variations in a family of systems. Figure 50.3 is a feature model that describes a subset (some features were hidden, for the sake of visibility) of our Insider Threat Inference Enterprise Model domain. Based on the inference enterprise model specification of [9], this family of systems comprises three main features: *Population_synthesis* for reverse-engineering population data from statistics, *Fusion_Down-select* for applying algorithms for classifying users accordingly to desired target behavior, and *Answer_Generation* to estimate the performance of predictions using metrics like precision, recall, and false-positive rates. Additionally, this feature model indicates that *Population_synthesis* requires either *W1008PopSyn* or *MAPopSyn*, which are features representing concrete population simulation algorithms. Similarly, *Fusion_Down-select* and *Answer_Generation* are associated with features related to decision tree (DT), support vector machine (SVM), and hidden Markov model (HMM) algorithms. A miscellaneous feature (*Misc*) which is optional to *Answer_Generation* represents features for converting or translating data types (e.g., *MatrixToCSV* simply translates a two-dimensional matrix to a comma-separated value format).

This feature model was converted to an OWL ontology by following a scheme adapted from [5, 12, 13]. This scheme guarantees that any reasoner of direct semantics of OWL can also verify that a particular subset of features is mutually consistent and complete, accordingly to constraints specified in the feature model.

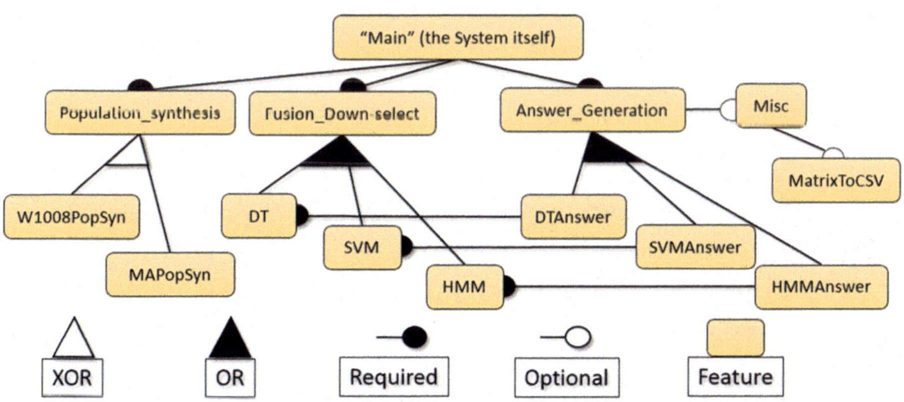

Fig. 50.3 Feature model of a subset of features in Insider Threat Inference Enterprise Modeling domain

50.4 PR-OWL Decision: Ontology Language for Decision Making Under Uncertainty

A *probabilistic ontology* is defined as a formal representation of domain knowledge, expressed in terms of entities, properties, relationships, processes, and events existing in the domain; statistical regularities that characterize such domain; inconclusive, ambiguous, incomplete, unreliable, and dissonant knowledge; and uncertainty about all such forms of knowledge [8]. In such context, the *Probabilistic Web Ontology Language* (PR-OWL) consists of a set of classes and properties (relationships) that collectively form a framework for building and reasoning with probabilistic ontologies, yet keeping syntactical compatibility with OWL and offering meta-level support for specifying probability distributions of OWL properties [14]. *PR-OWL Decision* extends PR-OWL with means for specifying decision variables and utility functions in PR-OWL, and its semantics follow an abstraction inherent from *Multi-Entity Decision Graph* (MEDG) [15].

MEDG is an extension of *Multi-Entity Bayesian Network* (MEBN) [16] and *Influence Diagrams* (ID) [17] for modeling and solving decision problems which require both first-order expressiveness and handling of uncertainty. Knowledge in MEDG is organized as a set of fragments called MEDG fragments (MFrag). Figure 50.4 is an MFrag which is part of the probabilistic ontology produced in this work, and it illustrates typical components of an MFrag.

Context nodes are Boolean expressions representing conditions that must be satisfied to make a distribution in an MFrag valid. In Fig. 50.4, the context nodes are restricting the types of the ordinary variables (nonrandom variables) *rcp* and *feat*, respectively, to *RCP* and *Feature* (which are OWL classes declared in the ontology). *Input nodes* are pointers referencing to some resident node (*hasSuggestedFeature* indicates suggestions—whether requirements fulfill conditions for reusing features—and its probability distribution is declared somewhere else).

Resident nodes are predicates (as in first-order logic) which represent the actual random variables that form the core subject of an MFrag (*hasScoreEntity* represents precision metrics like F_1, and the utility function is proportional to this score). *Decision resident nodes* represent a class of decision variables in the domain

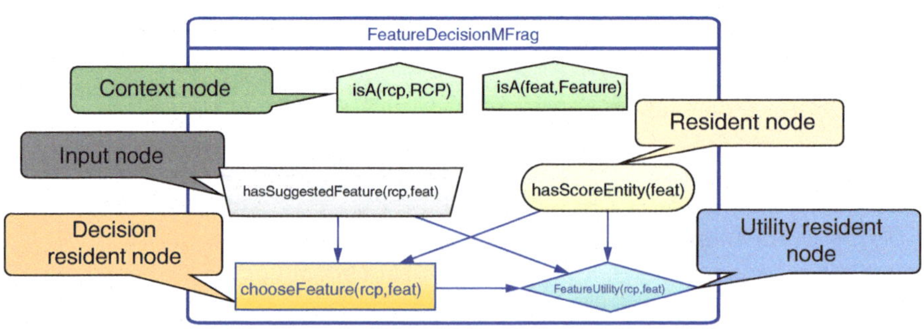

Fig. 50.4 Feature decision MFrag and components in it

(*chooseFeature* represents an actual decision of the inference system to pick a feature to use for a problem, based on probabilities about whether such feature can be suggested, and precision score that components related to such feature have reached). Similarly, *utility resident nodes* represent utility functions to be used for calculating expected utility.

Reasoning or inference in PR-OWL Decision is performed by instantiating MFrags (by instantiating arguments of nodes) in order to generate a *Situation-Specific Influence Diagram* (SSID), which is fundamentally a classic influence diagram which can be solved by any ID algorithm or tool. We mainly use UnBBayes [18] and Netica [19] for solving IDs. We present in the next section an application of this method in the domain of Insider Threat Inference Enterprise Models.

50.5 A Probabilistic Ontology for Automated Selection of Features to Reuse

We developed a PR-OWL Decision ontology which relates the requirements, meta-information about reusable software solutions, feature model, and utility functions, together with respective degrees of uncertainty, in order to be able to use a combination of description logic and Bayesian reasoning for selecting a subset of features that best fits—based on maximum utility criteria—with an emerging problem's specification. The basic ideas are summarized in the following list:

- Domain-level requirements are considered to be properties (i.e., OWL object/data properties) of previously solved problems, and we explicitly represent such properties as random variables—with some probabilities associated—in order to indicate that such properties may recur in future problems.
- Application requirements associated with new emerging problems are represented as findings or evidences to the above random variables. Methods like soft evidence based on Jeffrey's rule [20] shall be used when new requirements do not perfectly match with domain requirements.
- Fragments of expert knowledge (in our context, the experts are generally modelers of inference enterprise models), like probabilistic dependence between domain requirements, how domain requirements relate to a set of features, and which features are suggestible by modelers when some problem recur, are represented as probability distributions in MFrags.
- Experts' preferences, which may relate to metrics about the quality of the solutions, like precision and false-positive rates, were represented as utility functions.
- There was no need for explicitly creating MFrags for consistency rules between features, because such rules were already defined in the feature model OWL ontology (see Sect. 50.3.1), which was entirely imported to the new PR-OWL Decision ontology.

- The decision of whether choosing or not a particular feature to reuse in a new problem is explicitly represented as a decision variable. Therefore, instantiations of such decision variables that maximize the overall expected utility will pinpoint a set of features to be reused. When new problem specifications (i.e., application requirements) are provided to the ontology, reasoning can be applied in order to quickly identify such instantiations.

Once a set of features is selected, we can trace back to the software assets which were derived from such features, in order to determine the set of software assets to be included in the new solution and to perform quick integration and analysis by using tools like STIEM (see Sect. 50.2.1). For simplicity, we consider that a single feature is associated with a single software module/component, so we don't need to represent probabilistic rules associating features to software assets (i.e., once features are selected, respective software assets are immediately determined). This limitation can be relaxed by including additional MFrags for representing such probabilistic rules.

Figure 50.5 shows the MFrags that, when combined with the MFrag of Fig. 50.4, form a consistent MEDG model of our Inference Enterprise Modeling domain. The following list summarizes the probabilistic rules encoded in each MFrag:

- *CopulaMFrag*: it represents a soft/likelihood evidence based on Jeffrey's rule [20] which indicates that if provided statistical data contains marginal distribution

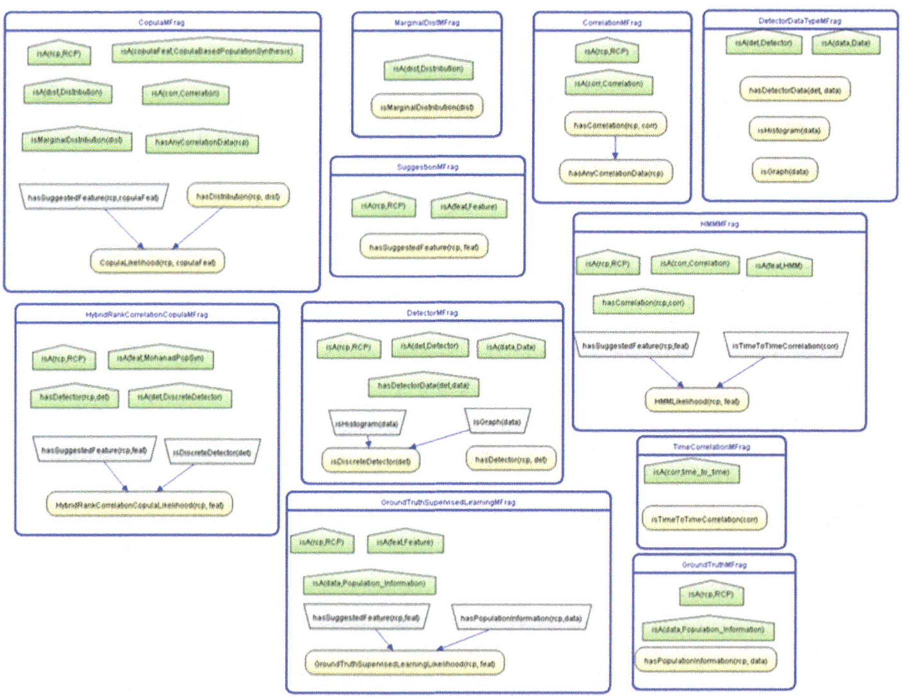

Fig. 50.5 Probabilistic ontology of Insider Threat Inference Enterprise Modeling domain

of variables and some kind of correlation data among them, then the probabilistic reasoning should increase the likelihood of suggesting copula-based [21] solutions.

- *SuggestionMFrag*: it contains the resident node *hasSuggestedFeature*, which indicates whether a new problem is categorized into a class of problems which meet conditions for a feature to be reused. For simplicity, we consider that if *necessary* conditions are met, then it is suggestible for the problem (that's why the name connotes a suggestion). However, this node alone is *not sufficient* for a feature to be actually chosen for reuse. The decision resident node in Fig. 50.4 (*chooseFeature*) is responsible for considering expected utility for actually making the decision/choice of reusing a suggestible feature as part of the solution for a new problem.
- *CorrelationMFrag*: this is used to simulate an existence quantifier, to indicate that if any correlation data exists for a problem, then *hasAnyCorrelationData* is universally true.
- *HybridRankCorrelationCopulaMFrag*: it represents soft/likelihood evidences which indicate that if there is a hybrid data type model (i.e., there is a combination of discrete and continuous data types in input data), then the reasoner should increase the likelihood of *hasSuggestedFeature* for a specific copula sampler which reorders samples based on rank correlations.
- *DetectorMFrag*: it indicates that if available data are histograms or graphs, then the probabilistic reasoner should decrease the likelihood that they are discrete data types (i.e., discrete data types are more likely to come in tables).
- *HMMMFrag*: it indicates that if there's time-to-time correlation data (correlation of variables over time) available, then the probabilistic reasoner should increase the likelihood of *hasSuggestedFeature* for classifiers based on hidden Markov models, which considers variables over time.
- *GroundTruthSupervisedLearningMFrag*: if ground-truth data is available, then the reasoner should increase the likelihood to suggest supervised learning classifiers (e.g., decision tree, support vector machines, etc.).
- *MarginalDistMFrag*, *DetectorDataTypeMFrag*, *TimeCorrelationMFrag*, *and GroundTruthMFrag*: these are auxiliary MFrags that contain resident nodes that represent, respectively, a Boolean variable that indicates whether some probability distribution is a marginal distribution or not, whether the available detector (observable) data are stored as histograms or graphs, whether correlation data is a time-based autocorrelation, and whether population data contains ground-truth information about target behaviors.

Figure 50.6 shows a result from a simple test case. The experiment was performed by inserting a new problem instance to the ontology and by querying the values of utilities and decision variables. We inserted OWL property assertions to indicate that a week-to-week temporal correlation table was available for the new problem, there were some statistics about marginal distribution for some of the observables, and some observables were discrete variables.

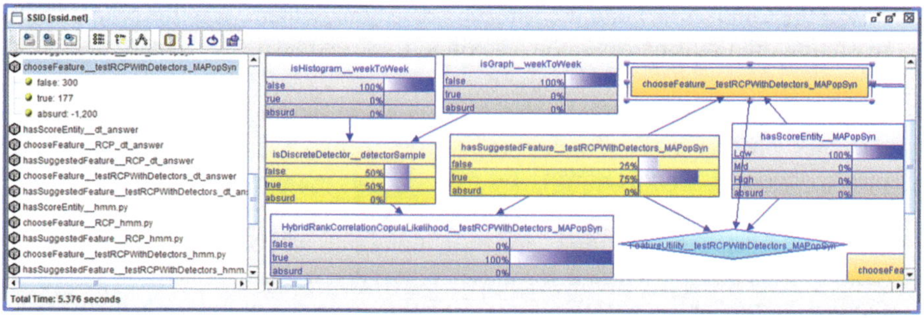

Fig. 50.6 Situation-Specific Influence Diagram (result of inference of PR-OWL Decision), ran in UnBBayes [18]

The reason why this scenario is interesting is because the probabilistic reasoning model may suggest a feature (because it may match reasonably well to the problem), but the decision problem solver (which encodes cost, revenue, or users' preferences in utility functions) may decide not to use the feature. For instance, the results in Fig. 50.6 show that a feature called *MAPopSyn* is suggested with 75% probability (see *hasSuggestedFeature* at the center of the right panel of Fig. 50.6), but this feature will *not* be chosen for reuse, due to low expected utility (*chooseFeature* at the top of the tree at the left of Fig. 50.6 indicates that expected utility of *not* choosing the feature is 300, while the utility for choosing it is 177). This is plausible, because our model associates an overall low preference (utility) to *MAPopSyn*, due to a low precision score obtained when the software associated with this feature was previously used (this is represented in *hasScoreEntity* node) and its difficulty for customization.

50.6 Conclusion and Future Works

This paper described an application of domain engineering techniques, semantic technologies, and probabilistic reasoning models to Insider Threat Inference Enterprise Modeling domain. The software assets considered in this work were mainly algorithms that reconstruct (simulate) population data based on available statistical data or machine algorithms that fuse observables to produce predictions. Since such assets share basic commonalities, they fit seamlessly to principles of domain engineering, which is a discipline for describing and reusing domain knowledge in terms of common and varying features.

Ontologies resulting from this work contain meta-model information which enables probabilistic reasoning on domain engineering models themselves. This was the key factor for automation of application engineering, which is basically a process of automatically classifying new emerging problems (based on the meta-level information about abstract system architecture, domain requirements, and repository of software assets) and selecting a set of software assets to be used in the new solution. Methods presented in this work are applicable to diverse domains that involve reuse of existing software assets or legacy systems, not only those of insider threat and inference enterprise modeling.

Current work assumes that STIEM (presented in Sect. 50.2.1) will be used for quick integration and analysis of the set of software assets selected by the framework, but producing actual workflow models for analysis and simulation in STIEM is still a manual process. As a future work, we plan to semi-automate the production of such STIEM models by adding meta-information about STIEM workflows and input/output mapping as part of our ontology. This should enable our framework to directly produce STIEM analysis models as a result of inference.

Acknowledgments Research reported here was supported under IARPA contract 2016 16031400006. The content is solely the responsibility of the authors and does not necessarily represent the official views of the US Government.

References

1. Pohl, K., Böckle, G., & van der Linden, F. J. (2005). *Software product line engineering: Foundations, principles and techniques*. Berlin, Germany: Springer Science & Business Media.
2. Tudorache, T. (2006). *Employing ontologies for an improved development process in collaborative engineering* (Doctoral dissertation, Technische Universität Berlin, Berlin, Germany). Retrieved from http://dx.doi.org/10.14279/depositonce-1477
3. Mitra, P., Noy, N. F., & Jaiswal, A. R. (2005). Omen: A probabilistic ontology mapping tool. In *The semantic web–ISWC 2005* (Vol. 3729, pp. 537–547).
4. Udrea, O., Yu, D., Hung, E., & Subrahmanian, V. S. (2005). Probabilistic ontologies and relational databases. In *On the move to meaningful internet systems*. Berlin, Germany: Springer-Verlag.
5. Wang, H. H., Li, Y. F., Sun, J., Zhang, H., & Pan, J. (2007). Verifying feature models using OWL. *Journal of Web Semantics, 5*(2), 117–129.
6. Carvalho, R. N., Costa, P. C. G., Laskey, K. B., & Chang, K. C. (2010). PROGNOS: Predictive situational awareness with probabilistic ontologies. In *Proceedings of the 13th International Conference on Information Fusion*. Edinburgh, UK.
7. Carroll, J., Herman, I., & Patel-Schneider, P. F. (2012). *OWL 2 web ontology language* (2nd ed.) [Online]. Retrieved June 20, 2016, from https://www.w3.org/TR/owl2-rdf-based-semantics/
8. Costa, P. C. G. (2005). *Bayesian semantics for the semantic web* (Doctoral dissertation, George Mason University). Retrieved from http://hdl.handle.net/1920/455
9. Huang, E., Zaidi, A., & Laskey, K. (2017). Inference enterprise multimodeling for insider threat detection systems. In A. Madni, B. Boehm, R. Ghanem, D. Erwin, & M. Wheaton (Eds.), *15th Annual Conference on Systems Engineering Research. Disciplinary Convergence in Systems Engineering Research*. Springer.
10. INCOSE. (2007). *Systems Engineering Vision 2020, Document INCOSE-TP-2004-004-02, version 2.03*.
11. ModelCenter® Integrate [Software], Phoenix Integration [Online]. Retrieved February 22, 2018, from http://www.phoenix-int.com/product/modelcenter-integrate/

12. Wang, H., Li, Y. F., Sun, J., & Zhang, H. (2005). Verify feature models using Protégé-OWL. In *Special Interest Tracks and Posters of the 14th International Conference on World Wide Web*. New York.
13. Matsumoto, S., Laskey, K. B., & Costa, P. C. G. (2016). Probabilistic ontologies in domain engineering. In *Presented at the Systems Engineering in DC Conference (SEDC)*. Washington DC.
14. Carvalho, R. N., Laskey, K. B., & Costa, P. C. G. (2010). PR-OWL 2.0-bridging the gap to OWL semantics. In *Proceedings of the 6th International Conference on Uncertainty Reasoning for the Semantic Web*.
15. Matsumoto, S., Laskey, K. B., & Costa, P. C. (2016). PR-OWL decision: Toward reusable ontology language for decision making under uncertainty. In *Semantic technology for intelligence, defense, and security (STIDS)* (pp. 37–45).
16. Laskey, K. B. (2008). MEBN: A language for first-order Bayesian knowledge bases. *Artificial Intelligence, 172*(2), 140–178.
17. Howard, R. A., & Matheson, J. E. (1984/2005). Influence diagrams. In *Readings on the principles and applications of decision analysis II*. Menlo Park, CA: Strategic Decisions Group.
18. Matsumoto, S., Carvalho, R. N., Ladeira, M., Costa, P. C. G., Santos, L. L., Silva, D., et al. (2012). UnBBayes: A java framework for probabilistic models in AI. In K. Cai (Ed.), *Java in academia and research* (2nd ed., pp. 210–246). Kowloon, Hong Kong: iConcept Press.
19. Netica [Software], Norsys Software Corp [Online]. Retrieved February 22, 2018, from http://www.norsys.com/
20. Jeffrey, R. C. (1990). *The logic of decision*. Chicago, IL: University of Chicago Press.
21. Nelsen, R. B. (2007). *An introduction to copulas*. Berlin, Germany: Springer Science & Business Media.

Chapter 51
Rapid Prototyping Insider Threat Inference Enterprise Model Workflows Using Ontology-Template Approach

James D. Lee, Abbas K. Zaidi, and Kathryn B. Laskey

51.1 Introduction

With advances in technology, there are many new and complex ways an insider can be a threat to an organization. As a result, insider threat detection problems require analysts to employ complex solutions that combine vast quantities of data from a variety of sources with subject matter expertise and informal concepts from the insider threat domain while also accounting for uncertainties that stem from noisy data and low likelihood of an insider threat occurring. The term "inference enterprise (IE)" is used to refer to the collection of data, tools, and algorithms that an organization employs to identify suspected insider threats [1]. Below is a representation of an IE that seeks to detect certain behaviors that may indicate a potential threat. A behavior that an IE seeks to identify is called a target behavior. Typically, a collection of algorithms is applied to a combination of behavioral and technical indicators to identify individuals who may be potential threats. These indicators are warning signs, identified by the enterprise, that are thought to be associated with the target behavior in question [2]. The algorithms monitor a set of data, called observables, which come from multiple data sources [1, 2] (see Fig. 51.1), to find patterns that may indicate a potential threat.

For example, an organization concerned with intellectual property theft may identify signs of an employee having intentions of leaving the organization as one of the indicators of the possibility of theft. An observable measure of this indicator would be the user's percentile for a number of visits to job search websites. The enterprise uses this information, along with other indicators and observables, to flag and monitor individuals of concern. Enterprise engineers can measure the effectiveness of the detection process by using performance metrics such as precision, recall,

J. D. Lee (✉) · A. K. Zaidi · K. B. Laskey
George Mason University, Fairfax, VA, USA
e-mail: jlee194@gmu.edu

S. Adams et al. (eds.), *Systems Engineering in Context*,
https://doi.org/10.1007/978-3-030-00114-8_51

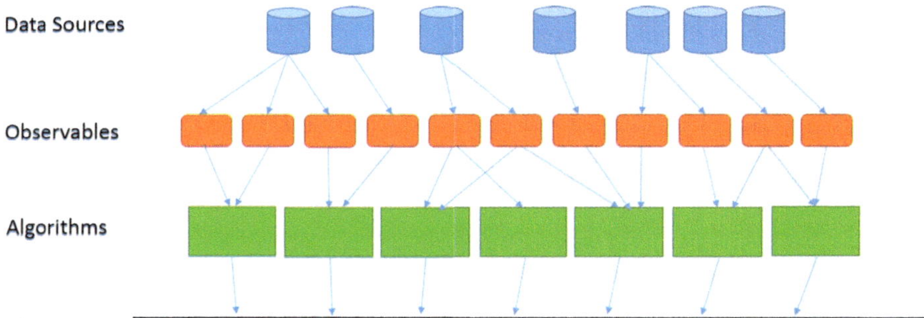

	Indicator 1	Indicator 2	Indicator 3	Indicator 4	Indicator 5	Indicator 6	Indicator 7	Target Behavior
user1	**TRUE**	FALSE	**TRUE**	**TRUE**	FALSE	**TRUE**	**TRUE**	Yes
user2	FALSE	**TRUE**	FALSE	**TRUE**	**TRUE**	FALSE	**TRUE**	Yes
user3	FALSE	FALSE	**TRUE**	FALSE	**TRUE**	**TRUE**	FALSE	No
user4	FALSE	FALSE	FALSE	**TRUE**	FALSE	**TRUE**	**TRUE**	No
user5	FALSE	FALSE	FALSE	FALSE	**TRUE**	FALSE	**TRUE**	No

Fig. 51.1 An inference enterprise

false-positive rate, and F1 score [3], but only after the ground-truth data of target behavior and indicators have been identified sometime in the future.

An inference enterprise model (IEM) provides a more timely assessment of an IE by forecasting its ability to accurately identify threats [1]. An IEM may represent the current operation of an inference enterprise as it is or may represent proposed changes to its algorithms, tools, and processes. That is, an IEM can be used to evaluate the performance and effectiveness of an IE of interest and may also support inference enterprise reengineering. Because an IE draws on a collection of data sources and employs a variety of algorithms, developing an IEM requires a modeling framework that can incorporate multiple data sources and modeling approaches in a semantically verifiable manner. An inference enterprise model (IEM) can be formalized with the help of a workflow language. A workflow can be instantiated with multiple data sources and different parameter values of the detection algorithms. It can then be executed to simulate the given enterprise model [4]. The simulation can be used to evaluate the performance of the inference enterprise on a variety of inference tasks. It can also be used to answer "what-if" questions in support of inference enterprise reengineering.

The rest of the paper is organized as follows: Sect. 51.2 introduces a workflow-building process for inference enterprise models. In Sect. 51.3, we present the architecture of the developed ontology-template-based workflow construction approach. Section 51.4 uses a specific example to demonstrate how this approach can aid a modeler in the development of an insider threat detection model. Finally, the paper concludes with a discussion of the limitations of the proposed approach and suggests future research directions.

51.2 Building an IEM Workflow

An IEM workflow represents the operation of a given inference enterprise and also contains additional elements that are necessary for fulfilling the modeling objectives. The description of the operation of an IE typically includes the target behavior, the associated indicators, the observable data of the indicators, the algorithms used to predict users that may manifest the target behavior, and the performance metrics of interest. Based on this description, an IE modelers develop population synthesis, classification, and performance evaluation modules that make up an IEM workflow.

In order to forecast the performance of an IE, modelers must simulate future population data based on the information provided. Since there is inherent uncertainty about future or unknown data, modelers must consider different assumptions in order to properly estimate what the actual population would look like. Based on these assumptions, modelers build a stochastic population model of the relevant aspects of the organization and use this model to simulate hypothetical populations. Because of privacy or proprietary concerns, data is often redacted or aggregated to exclude any sensitive information about users in the population. Redacted data may include contingency tables calculated from the data, marginal distributions of observables, means and variances of observables, correlations between observables, and correlation across time [1]. Table 51.1 provides a description of techniques that can be used to reconstruct and simulate populations when redacted data is provided. Using the stochastic population model and appropriate population reconstruction technique, we generate multiple simulations of the population with varying assumptions in order to cast an interval wide enough to capture the uncertainty we have about the actual data. This simulation process is referred to as "population synthesis" and is responsible for generating the data necessary for forecasting the performance of an IE.

The classification modules use the generated data to make predictions about the behavior of interest. In some cases, the organization provides details about the classification algorithms that they currently have implemented, and modelers will build modules that mirror the client's implementation while identifying parameters on which to perform sensitivity analysis. In other cases, modelers may be asked to

Table 51.1 Population synthesis methods

Method	Description
Discrete event activity counts	A discrete event simulation used when provided contingency tables of target behavior and observable values
Stochastic optimization	Method to synthesize joint probability distributions when provided contingency tables of target behavior and observable values
Tree-augmented naïve Bayes	A Naïve Bayes augmented tree structure used as the basis for a Monte Carlo simulation. Used when provided distributions of observables conditional on the target behavior
Copula	Used when given correlations and distributions of observables. Methodology to generate samples that match both the provided marginal distributions and the provided correlations between different observables

design new algorithms that would be suitable for the organization's data and objectives. Generally, the classification modules are relatively easier to implement than the population synthesis modules, but the execution of this step can be extremely time-consuming since the model needs to ingest large data files of the simulated population and train and test the classifiers using the specific settings outlined in the IE description. While some classification techniques, such as decision trees, can turn a prediction around relatively quickly, there are other more complex techniques that could take up to days to get a result. Furthermore, when using an IEM for reengineering, it may be necessary to perform extensive tuning and sensitivity analysis to determine the best settings for classifiers. Therefore, modelers can benefit from quicker development and implementation, since some techniques may take much longer to execute and many tuning runs may need to be performed.

For the performance evaluation module, some clients are interested in how the overall system will perform using generic metrics such as precision, recall, false-positive rate, and F1 score. Others may be interested in understanding which observables have a higher impact on the quality of predictions. Calculating performance metrics typically requires only a few lines of code to compare the simulated target behavior and the predicted value for individual users. Analyzing the driving factors behind the performance scores may be very time-consuming.

Once the three components of an IEM workflow are complete, these steps are executed for a number of simulated organizations, and the results are fused to form performance predictions with associated uncertainty intervals. By closely examining the relationship between organizations and performance, the modeler gains knowledge on how different assumptions correlate to results. In addition to varying assumptions about the organization, an IE modeler can also apply different assumptions to other elements in the workflow in order to improve the performance of the inference enterprise by performing sensitivity analysis.

It is evident that the process of designing and building models of an IEM workflow can be very challenging and time-consuming. There are many variations in the ways an inference enterprise can be modeled, and there are many combinations of algorithms and assumptions that a modeler must consider to construct a single workflow. If the modelers can shorten the development and implementation time for workflows, more resources can be allocated to running sensitivity analysis that can inform ways to improve the IEs. In order to enable effective and efficient workflow modeling for IEMs, there is a need not only to facilitate and expedite the workflow-building task, by being able to reuse parts of existing workflows, but also to quickly recognize characteristics of a new problem that may be able to utilize the same underlying assumptions used in existing workflows. This will allow for building more comprehensive and accurate domain models more quickly with minimal need for support from domain experts [5].

As seen from the overall process of building an IEM workflow, it is evident that many modeling decisions depend on the technical aspects of an IE, including the availability of certain data types and technical instructions, such as training and testing period requirements, configurations of the mechanisms to be employed, and types of performance metrics required. The many similarities in the technical traits of each IE suggest that there is an overlap between IEM workflows even when the target behavior is different. This implies that there is an opportunity to reuse components of the previous IEM workflows if the similarities between IE descriptions can be identified.

51.3 Architecture

The methodology described in this section is inspired by an ontology-based meta-modeling approach presented by Mansoor et al. for generalizing existing workflows for future reuse on new problems [5]. The proposed approach introduces the notion of process templates as a collection of generalized fragments of IEM workflows. A process template represents an IEM workflow at a generic level using abstract entities characterizing the executable modules of a workflow. A template can then be instantiated for a new IEM by substituting abstract entities with concrete instances characterizing the new model. A process template provides an effective solution for workflow reuse; it simplifies the workflow construction process by providing information about what to look for while developing a workflow and also enables the modelers to merge workflows from different sources and address new problem instances.

As part of the Scientific Advances to Continuous Insider Threat Evaluation (SCITE) Program supported by IARPA, we have developed several IEMs. It took several months of effort to create and verify each model. Figure 51.2 shows some of the process templates that were created based on the knowledge developed from building IEM workflows. The nodes in these templates represent abstract concepts derived by replacing instances from the workflows. For the purpose of this demonstration, process templates are represented in the form of UML activity diagrams [6].

These process templates (Fig. 51.2) model workflows consisting of three major steps: population synthesis (green box), classification (blue box), and performance evaluation (red box). Each of these templates represents different workflows that consisted of different combinations of modules. Some workflows have multiple population synthesis modules, whereas others have multiple classification modules. The text in the square brackets are placeholders for specific modules: population synthesis ([?popSyn]), classification ([?classification]), and performance evaluation ([?performance]). These placeholders are populated with specific modules when a new problem is encountered.

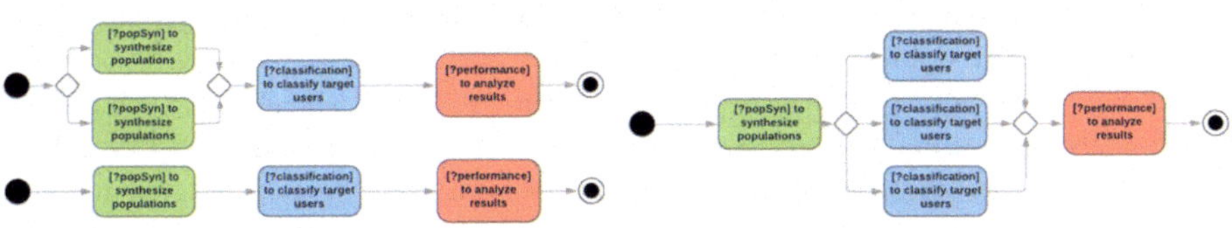

Fig. 51.2 Process templates

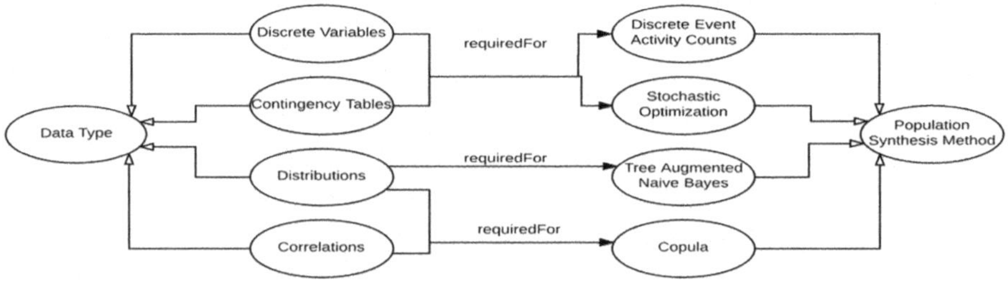

Fig. 51.3 Process ontology: data type and population synthesis method classes

Exploring available knowledge bases for information required for instantiating a process template is also a complex and challenging task. In this work, we are using ontology as the knowledge representation and reuse formalism. The process ontology describes the abstract concepts and relationships characterizing the process of modeling an IE and allows for reasoning about new inference mechanisms for new IE descriptions.

From the concepts defined in the process template, and additional understanding of the general process of constructing workflow for IEMs, a process ontology was created. Figure 51.3 shows part of the ontology that captures the knowledge about different population synthesis methods and its data dependencies. The two parent classes represented here are data type and population synthesis method, and the subclasses are the different data types that modelers have worked with for previous problems and the corresponding population synthesis methods used. The structure of the ontology infers that data type(s) are required for population synthesis method(s). More specifically, for example, distributions are required for tree-augmented naïve Bayes method, while both discrete variables and contingency tables are required for discrete event activity counts and stochastic optimization methods. The process ontology also captures knowledge about previously employed classification and performance evaluation modules. The classification modules consist of a specific method and its settings. For the IEs which were modeled as part of the SCITE program [1], there were four different cases where the specified classification method was a decision tree algorithm, each with different hyper-parameter settings. In the process ontology, each of these four decision tree implementations was cataloged using the hyper-parameter setting information and stored as a distinct module. The performance evaluation modules are either modules that only calculate overall system performance or modules that also evaluate the impact certain observables have on the prediction.

Figure 51.4 shows the architecture of the presented ontology-template-based workflow construction approach.

As shown in Fig. 51.4, workflow construction is a two-phase process consisting of a domain modeling phase and a situation modeling phase. In the domain modeling phase, the process ontology and process template are used to develop a generalized mapping that can be applied to any new IE description compatible with the process ontology. A mapping box defines concepts about the operation of an inference enterprise and relates them to components of the process template that can be used

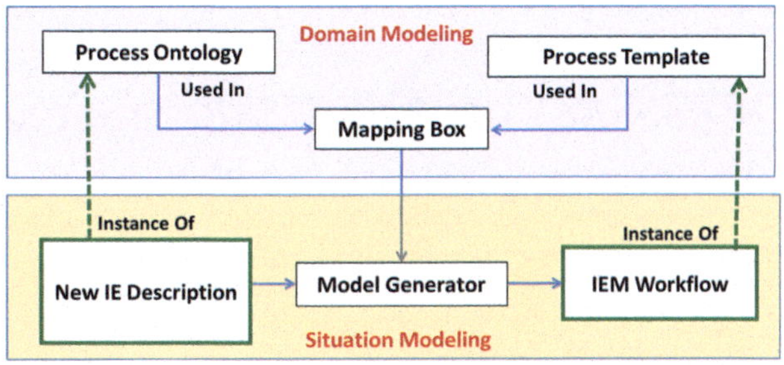

Fig. 51.4 Architecture of ontology-template-based workflow construction

to generate an IEM workflow. Domain modeling is a process done only once (although a domain model can be revised and enhanced as more is learned about the domain). When a mapping box is created, instantiating a workflow from a given instance of the process ontology describing a new inference enterprise becomes a streamlined process. In the situation modeling phase, the model generation step uses the results from the mapping box to instantiate IEM workflows using a workflow language. This ontology-template approach allows an organization to reuse IEM workflows for similar instances of IEs and automatically develops complete or partial solution workflows while saving time and effort [5].

51.4 Application

To illustrate the concepts described in this paper, we have developed a generalization based on the IEM workflows supported by the SCITE program [1] using the domain modeling approach described in the previous section of this paper and used that approach to create a candidate workflow for a new IE description.

The case study provided here uses data from an actual organization with approximately 3000 employees. The inference of interest was to identify employees who will leave the organization in the next calendar month as one of the indicators to flag potential employees who are likely to commit theft of intellectual property. The observables included data on frequent visits and an increase in visits to job search websites, networking websites, webmail sites, apartment search sites, and frequent late arrival, early departure, and their respective times. We were given 7 months worth of the marginal distributions and correlations of the observables for both the group that stays and leaves in the next month. We were asked to train the model on the given months and predict the employees who will leave in the subsequent 3 months using a decision tree classifier with hyper-parameter setting "max_depth" of 6. Finally, we were required to assess the overall prediction performance as well as the impact each observable has on the prediction.

The first step was to create an instance of the new IE in the process ontology and then use the mapping rules to query which previously developed modules could be reused. The information consisted of the types of data provided, including distributions and correlations, requirements to predict using a decision tree classifier, and to

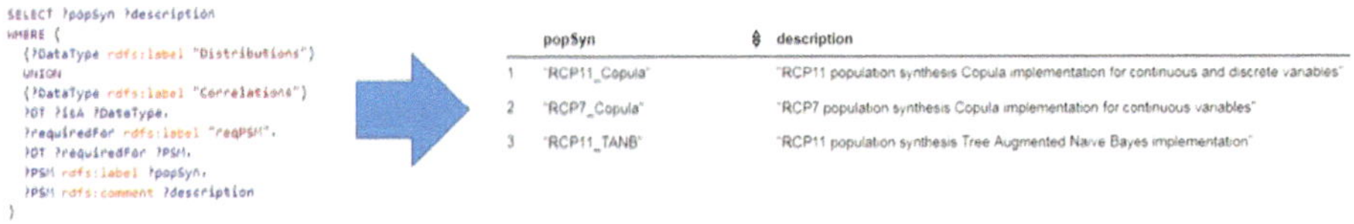

Fig. 51.5 Mapping rule for population synthesis modules

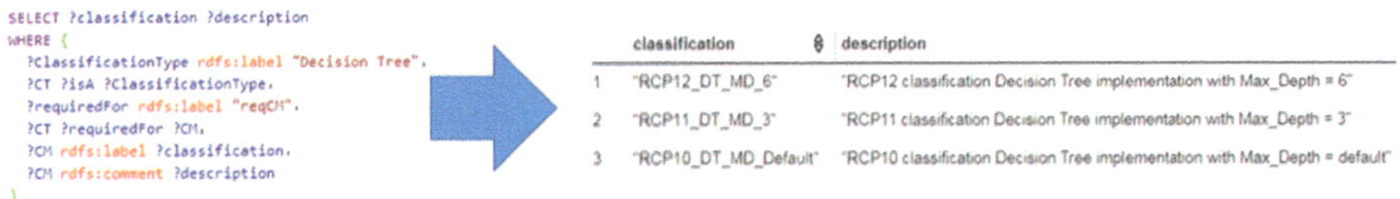

Fig. 51.6 Mapping rule for classification modules

evaluate the overall system performance as well as the impact certain observables have on the overall predictions.

The following (Fig. 51.5) is a mapping rule, expressed in SPARQL [7], that provides the possible population synthesis modules that are compatible with the available data types for the new problem. The rule outputs three population synthesis modules given that the new IE had "distributions" and "correlations" data available. "RCP11_Copula" was a module that used observable distributions and correlations to synthesize populations using a copula implementation for continuous and discrete variables, while "RCP7_Copula" used a copula implementation for continuous variables only. "RCP11_TANB" synthesized populations by using the observable distributions and tree-augmented naïve Bayes method. Since the new IE also had distributions and correlations of both continuous and discrete observable variables, the modeler selected "RCP11_Copula" and "RCP11_TANB" to reuse for the new model workflow.

Similarly, the mapping rules for classification modules were executed to identify the possible modules for the derived workflow. The query for classification modules returned three suggestions (Fig. 51.6), and based on the descriptions of each module, the modeler selected the "RCP12_DT_MD_6" module since the new IE also required the decision tree algorithm's "max_depth" hyper-parameter to be set at 6.

Finally, the query for performance evaluation modules (Fig. 51.7) returned "RCP11_ObsPerf" and "RCP10_ObsPerf" modules that calculated overall system performance as well as observable performance. In this case, there was no distinction between the functionality of the two modules, and therefore the modeler made a choice based on personal preference of the code.

After querying and selecting the appropriate modules for the derived workflow, the process template was determined based on the number of modules selected for each major step. For this example, a process template with two population synthesis modules, one classification module and one performance evaluation module, was identified. The model generating step populated the results from the mapping box in the process template to generate the following derived workflow in Fig. 51.8.

With this approach, the modeler can save significant time. Rather than starting each problem from scratch, the modeler can use the derived workflow as a starting

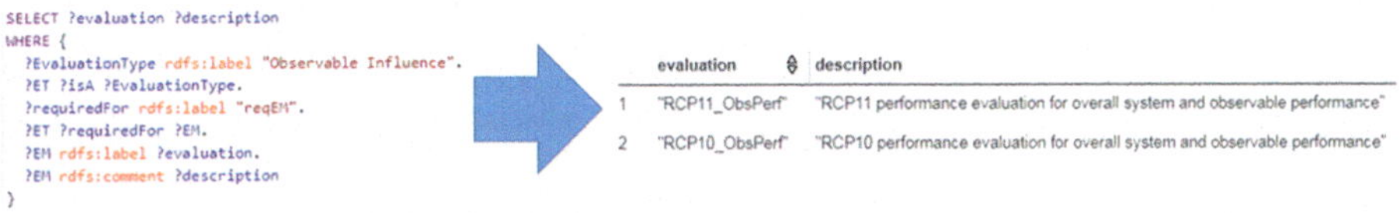

Fig. 51.7 Mapping rule for performance evaluation modules

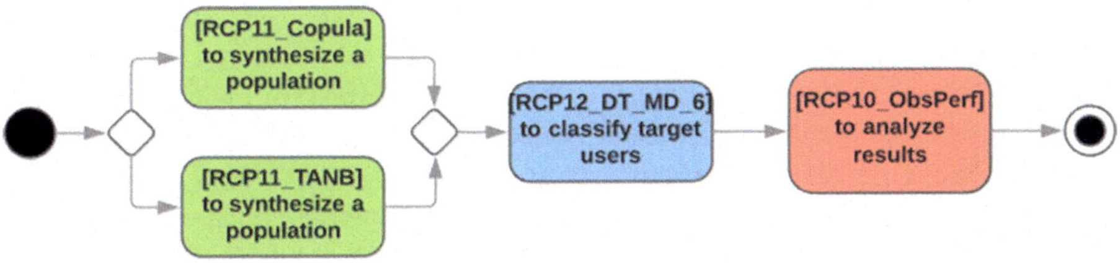

Fig. 51.8 Derived workflow

point. This will allow for more time and effort in examining the workflow to identify if any innovative methodology can be applied or if any improvements can be made. Based on the derived workflow, the modeler will extend previous modules to create new model workflows that are customized for the new IE scenario. The modeler may also make decisions to improve the modules by changing certain values into adjustable variables. By changing the decision tree module's hyper-parameter setting to a variable, the modelers can more effectively reuse the module for different IE scenarios and enable sensitivity analysis for inference enterprise reengineering and improvement. After each iteration of improving and creating new IEM workflows, the extended and newly developed modules are entered into the process ontology and repository of modules to make available for future use.

51.5 Conclusion

An ontology-template-based approach was presented to capture generalized knowledge about specific situations and use that knowledge to help modelers develop workflows quickly. Using a simple example, we explored how a workflow can be instantiated using the process template and process ontology. Our approach has the potential to lead to faster and better analysis of new complex situations that will not only allow the modelers to be more efficient but also enable analysis that can improve the performance of the IEs being modeled.

This paper does not address technical details of how the model generator populates a process template based on the results from the mapping box. We plan to implement this leveraging the functionality of business process language modeling platforms such as MagicDraw [8]. Also, this methodology assumes that a new situation will have overlapping characteristics with previously developed inference enterprise models, which allows the workflow for a new problem to be generated using the current knowledge base. For example, if the modeler was provided with contingency tables for continuous variables, since there is no previous experience on how to solve a problem like this, it would require the modeler to develop new algorithms to synthesize a population based on the provided data. However, once new algorithms are developed, they would then be included in the ontology/knowledge base for future use.

Future work includes integrating the outcome of this research to tools like Semantic Testbed for Inference Enterprise Modeling (STIEM) [4], which can handle multiple models and automate sensitivity analysis. This will allow IE modelers to efficiently model, evaluate, and improve more realistic and complex IEs that have multiple target behaviors and indicators. The work presented in this paper addresses only the workflow-building portion of IE modeling. There are more areas where the IE modeling process can benefit from other semantic knowledge bases. SOFIT ontology [2], developed by Greitzer et al., is a hierarchical representation of indicators and observables related to insider threat and their relationships to certain threat types. By integrating SOFIT ontology with the IE modeling process, we can be informed of all the possible indicators that can be monitored through the provided data and suggest different combinations of the indicators to detect a target behavior, which can both help in improving the performance of an IE.

Acknowledgments Research reported here was supported under IARPA contract 2016 16031400006. The content is solely the responsibility of the authors and does not necessarily represent the official views of the US Government.

References

1. IARPA. (2015). *Broad agency announcement: IARPA-BAA-15-09. Scientific advances to continuous insider threat detection (SCITE) Program.*
2. Greitzer, F. L., Imran, M., Purl, J., Axelrad, E. T., Leong, Y. M., Becker, D. E., et al. (2016). Developing an ontology for individual and organizational sociotechnical indicators of insider threat risk. In *STIDS* (pp. 19–27).
3. Laskey, K., Huang, E., Zaidi, A., Brown, D., Buede, D., Yousefi, B., et al. (2018). *Modeling inference enterprises using multiple interoperating models.* Submitted in INCOSE IS.
4. Huang, E., Zaidi, A. K., & Laskey, K. B. (2017). *Inference enterprise multi-modeling for insider threat detection systems.* Fairfax, VA: George Mason University.
5. Mansoor, F., Zaidi, A. K., Wagenhals, L., & Levis, A. H. (2009). *Meta-modeling the cultural behavior using timed influence nets.* Fairfax, VA: George Mason University.
6. Object Management Group (Ed.). (2005). *Unified modeling language: Superstructure, version 2.0, formal/05–07–04.* Needham, MA: Author.
7. SPARQL Query Language for RDF. Retrieved from https://www.w3.org/TR/rdf-sparql-query/
8. MagicDraw. Retrieved from https://www.nomagic.com/products/magicdraw

Chapter 52
Synthesizing Inference Enterprises from Redacted Data

Bahram Yousefi, Muhammad Imran, and Kathryn B. Laskey

52.1 Introduction

Many enterprises supported by cyber environments (e.g., cloud services) are subject to cyber threats. In fact, each day articles about such incidents can be found in the newspapers. With rising concerns about cyber threats, enterprises are seeking cyber experts to help them mitigate the risk of cyber threats. One particular kind of cyber threat that is proliferating during the recent years is the insider cyber threat. These insiders engage in destructive behaviors such as piracy of industrial secrets through exploiting cyber vulnerabilities. Acts like these are extremely detrimental to industries and agency both financially and reputation-wise. Therefore, organizations are seeking methodologies to advance their ability to screen employees to identify insider threats [1].

In this paper, we use the term "inference enterprise (IE)" to refer to the collection of data, tools, and algorithms that an organization employs to address the insider threat detection problem [2, 3]. While our use of IE is for insider threat identification, the concept of an inference enterprise is also applicable to other problems such as preventing video piracy and cyber loafing at the work place. The components of a generic IE are shown in Fig. 52.1. An inference enterprise makes use of data about various events recorded in logs. The data is compiled, cleaned, and reduced to a data set used for modeling. This process typically involves redacting personally identifiable and/or proprietary information. Depending on the specific requirements of the organization, raw data may be reduced to data summaries such as contingency tables and frequency distributions. The process of collecting and then redacting data is usually performed on the client side. Subsequently, the client provides the redacted data to external subject-matter experts (SME) and modelers to perform some sort of analysis (e.g., threat detection) on the original data. The modeling process may

B. Yousefi (✉) · M. Imran · K. B. Laskey
George Mason University, Fairfax, VA, USA
e-mail: byousefi@gmu.edu

S. Adams et al. (eds.), *Systems Engineering in Context*,
https://doi.org/10.1007/978-3-030-00114-8_52

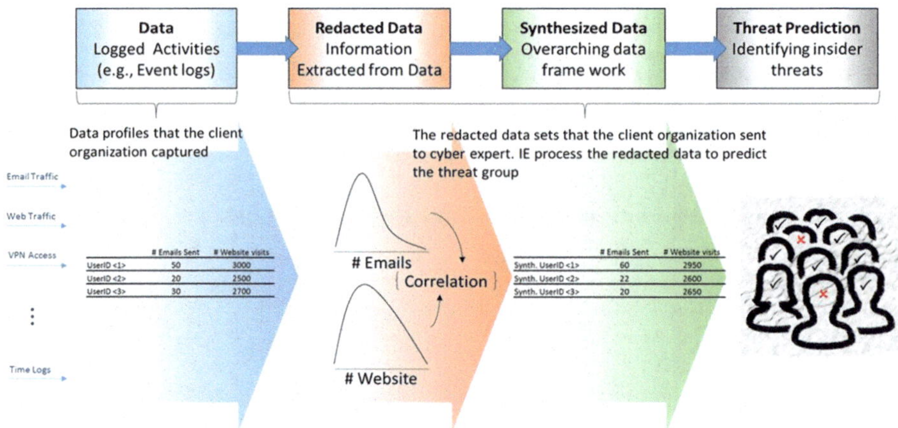

Fig. 52.1 Different components of inference enterprise modeling (IEM)

require reconstructing the original data from the redacted data provided by the client. This process is called population synthesis, because it involves synthesizing a population similar to the actual population with respect to the provided data summaries. Next, the synthesized population is used to perform the analyses. For example, a fusion and down-select algorithm may be used to identify cases that should be flagged as potential insider threats. A challenge in reverse engineering the organizational population is that the redacted data sets may be incomplete and/or inconsistent. Incomplete data sets are missing some necessary data points. Inconsistent sets have no feasible reconstruction that matches all the provided data summaries.

There is no comprehensive previous work available that could be applied to the problem of processing redacted data to synthesize a reconstructed population. However, there are some engineering techniques available that could be leveraged to develop a statistical model of the population and use it to synthesize a population. There are a few methods to estimate the organizational population probability distribution from incomplete data sets, some of which are entropy maximization and likelihood maximization [2].

The first obstacle to population synthesis is either lack of authoritative data or absence of access to authoritative information. Furthermore, it is often the case that owner of such information (e.g., private industries and government agencies) is not willing or able to share the information with external system engineers and architects because of privacy and security concerns. Therefore, the synthesis of an inference enterprise's data to detect pertinent target behaviors is an arduous task. The second roadblock is that the scant data that is provided by the client organization is usually incomplete and even sometimes inconsistent. Subsequently, system engineers must deal with these two problems to enable synthesis of an inference enterprise's data that exhibits the desired behaviors. Third, inference enterprise models may inherently require large-scale synthesis of feature-rich sample cases (e.g., many users each of which has many relevant characteristics over many time periods). These issues make the synthesis of an IE a complicated job for system engineers because there is no scalable methodology available that can handle these issues.

This paper puts forth an approach to population synthesis that is based on nonlinear stochastic optimization. This method can handle the issues mentioned earlier, and it is scalable and thus can be applied to manifold IEs (with consideration of the formatting of the redacted data provided for synthesis). This methodology provides a balance between the native uncertainty about the IE and the scalability of implementation. This is a crucial property because the synthesized enterprise's population must reflect the inherited uncertainty that is caused by various sources such as missing and incomplete data, uncertainty about unobservable aspects of organizational processes, and inherent unpredictability of future events. Sampling marginal histograms while satisfying other properties (provided as summary statistical data) such as means, standard deviations, feature correlations, and temporal correlations is a challenging problem. Moreover, when the provided data summaries are inconsistent, there are no feasible reconstructions, making the problem even more challenging. Subsequently, the developed stochastic optimizer finds optimal samples from the distributions such that they match the provided summary statistics such as mean and standard deviation. This program is implemented in such a way that it could be applied to any redacted data set of the kind used in this research.

The viability of the presented approach is shown through a real-world study of the Insider Threat Evaluation Enterprise. In this research project, employees of an organization are evaluated for unlawful access (misuse) of proprietary and classified information. Only summary statistics of the enterprise were provided because the client company could not share the actual user information to outside organizations. The provided information includes data such as marginal histograms of observables captured in workstations, event logs, and network activity logs. The marginal histograms were discretized into ranges and frequencies. The distribution of observations within each bin was not provided. Therefore, generating random samples that match the given means and standard deviations is a challenging problem because if the bins are immensely wide with few data points in each bin, then the overall distribution of the data points will keep varying upon each iteration of random sampling of data points within bins. This problem is addressed using a nonlinear stochastic optimization program, accompanied by random sampling methods (e.g., a triangular kernel).

The results of the presented IE synthesis method demonstrate scalability of the approach for all the marginal histograms. The random samples are generated using triangular distributions, each of which is fitted to a single bin in the histogram. The distributions are adjusted to meet the desired overall mean and standard deviation of the samples as provided by the client. We will further explain the motivation behind using triangular kernel and stochastic optimization sampling within bins in the methodology section.

One salient issue is that in some cases the optimization program is infeasible. This means a random set of samples cannot be produced such that the overall calculated mean and standard deviation do not precisely match the desired (i.e., given) values.

In order to overcome this limitation, the stochastic optimization program includes a penalty term for deviations from the desired mean and standard deviation values. This approach of using a penalty term has been successful in generating distributions that were reasonably close to the provided values. Specific results for the case study are provided in Sect. 4.

52.2 Related Work

In the following section, we mention some of the other techniques and methodologies used to address somewhat similar problem of simulating data distributions from a given histogram. For example, literature can be found regarding how to construct distributions using given moments or characteristics such as skewness or kurtosis [4]. These techniques are usually based on mathematical models used in the data analysis field. In the case study of this paper, skewness and kurtosis are not provided. The client provided summary statistics such as histograms, means, and variances. Therefore, a decision was made to develop a methodology based on the summary statistics that were supplied.

Jung et al. [5] elaborated on how parametric distributions like Zipf and Uniform can be used to sample from histograms in parallel computing settings. In the case study, the bin widths are in the order of thousands in some cases, assuming Uniform distributions within the bins can produce results that vary widely from sample to sample and for which the expected means and variances do not match the provided summary statistics. As a result, samples drawn from these bins tend to have large deviations from the given means and standard deviations and do not have the correct long-run sampling properties.

Chatterjee and Firat [6] introduced a general procedure to generate data sets with identical summary statistics (i.e., means and standard deviations) but different graphics (i.e., scatterplot) and data patterns. Their procedure uses a genetic algorithm-based approach. In contrast, the data sets that are provided by our client have different means and standard deviations for each feature.

Ricci et al. [7] presented a flexible maximum likelihood estimation method to model heavy-tailed and skewed error distributions in a regression setting. The challenge in the case study presented in this paper is constructing distributions from sparsely binned and skewed histograms by using Monte Carlo simulations. In addition, scalability to multiple types of detector variables is critical in this paper's case study. These two challenges are not addressed by Ricci et al. [7].

52.3 Methodology

The presented methodology uses nonlinear stochastic optimization in conjunction with random sampling from triangular distributions over the binned histograms. The task is to generate marginal samples that match the means and standard deviations provided by the client organization. We developed this approach with two objectives in mind. Firstly, the approach should be scalable to handle sampling of thousands of histograms. Secondly, it should be adaptable to the various structures of histograms, for example, equal size-binned histograms or equal frequency-binned histograms. We have tested this approach by doing multiple experiments and found the approach works well and meets the two previously mentioned objectives.

The rationale behind using this approach is to address a peculiar behavior of the histograms provided to us. The mean statistic of a histogram given to us is almost equal to the sum of all the midpoints of each bin times the number of samples in that bin. An arithmetic representation of this phenomenon for a histogram is shown in Eq. (52.1):

$$\sum_{i=1}^{n} m_i c_i \cong \mu_g \tag{52.1}$$

where m_i is the midpoint and c_i is the count of data points, in the ith bin, and μ_g is the given mean of the histogram.

This suggests that data points drawn from the middle of the each bin will tend to produce a mean statistic very close to the given mean value. However, it is important to introduce some variability in sampling. Therefore instead of repeating the midpoint of each bin times the count for that particular bin, we assume a triangular distribution within each bin of the histogram. From Fig. 52.2, we can also see the bins contain an equal number of data points, but the bins toward the higher end of the data distribution tend to increase in width, which will create a very spread-out triangular probability density function within bins of very large widths. Sampling from the triangular distributions for these bins produces so much variability that the

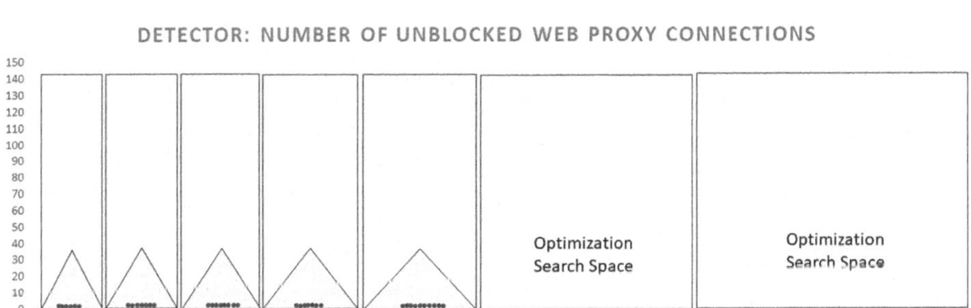

Fig. 52.2 Illustration of the right/top most five bins with the triangular kernel

samples tend not to match the given means and standard deviations. Thus, we use optimization to choose the sample points in these rightmost bins.

One limitation of using nonlinear optimization in this approach is addressing the embedded uncertainty in the sampling process. The samples produced by the method exactly or very closely match the provided summary statistics (i.e., means and standard deviations), whereas the desire is to incorporate realistic sampling uncertainty. Therefore, the stochastic nonlinear program is combined with random sampling of the histograms to better incorporate the native uncertainty involved in the sampling process. Figure 52.2 shows how a triangular distribution is fitted to the first five bins of a notional histogram, in order to produce random samples from each bin. In addition, the remaining bins (i.e., the last two bins in Fig. 52.2) act as a search space for the stochastic optimization program to search and produce data points which satisfy the constraints of matching mean and standard deviation as closely as possible. In a particular histogram, the triangular distribution is configured according to the lower and upper bounds of that bin, and the mode of the fitted triangular distribution is the midpoint of the upper and lower bound of the corresponding bin, and the gray-colored dots represent the scatter of actual data points within bins as shown in Fig. 52.2.

Mostly we encounter heavily skewed histograms such that bins with a very small number of data points have very large ranges of possible values. This phenomenon in histograms makes the task of matching the means and standard deviations difficult. A naïve random sampling process can produce data points from a huge range, leading to large deviations from the given mean and standard deviation. Figure 52.3 shows an example of a long-tailed histogram. In this example, the tail bins are wider and have extremely small data point counts to ranges ratio. Therefore, sampling the last few bins will produce samples with volatile means and standard deviations. To address this problem, nonlinear optimization (see Eq. (52.2)) is used to generate samples from the last few bins to satisfy the constraints of matching the

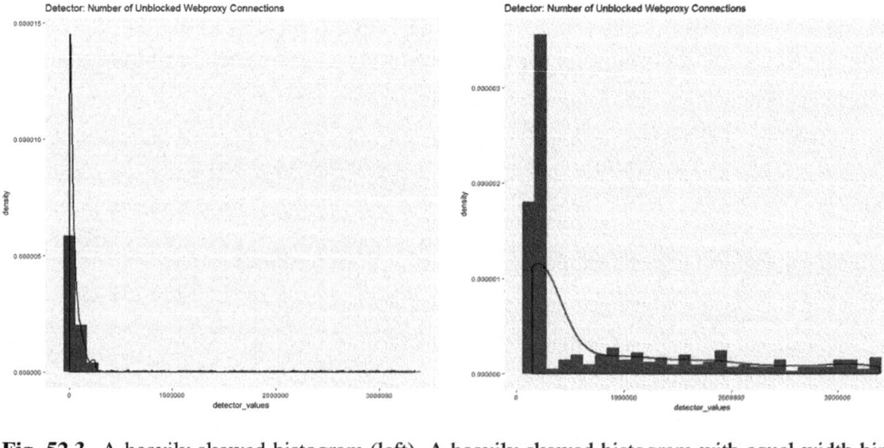

Fig. 52.3 A heavily skewed histogram (left). A heavily skewed histogram with equal width bins (right)

desired mean and standard deviation values. The number of random samples generated from triangular distributions (Fig. 52.2) versus the number of samples generated using the optimization program is crucial for deciding how far off or how closely we want to match the given means and standard deviations. On one hand, when the number of randomly generated samples is larger than the number of optimized samples, the distance between the mean and standard deviation of the synthesized samples and the desired mean and standard deviation may become larger. On the other hand, a larger proportion of optimized samples provides a closer match to the given mean and standard deviation values; however, it may cause over-fitting to the summary statistics, by shrinking the variation in data points. This gives rise to a trade-off between incorporating variation and more closely matching the given statistics. Therefore, the number of data points generated from optimization program is from the rightmost two bins of the histograms because those are the bins which are huge in range and data points drawn from these bins are more suitable to help with meeting the constraints of matching means and standard deviations. Also, histograms with various bin lengths could be used in this approach because the random sampling part will take care of incorporating uncertainty in the data distributions, and the optimization part will ensure matching to the summary statistic. So, we can safely say that this approach is generalizable.

The optimization program is shown in Eq. (52.2). The nonlinear program produces the samples, which are denoted in mean and standard deviation formulas as x_i, $i = 1 \ldots l$, where l is the population size. In the program, k represents the number of bins in each histogram. The value of k is different for each histogram. The parameter n denotes the number of histograms. The objective of the program is to minimize the difference of the calculated mean and standard deviation values. Therefore, the slack variables of the constraints (see Eq. (52.2)) are minimized. In most of the cases, constraints are binding, but there are exceptions that there are no binding constraints, and thus the program may be infeasible. To avoid this limitation, a deviation from the original desired values is permitted.

$$\begin{aligned}
&\text{obj.}: \\
&\min \sum_{n=1}^{k} s_n + a_n + s_n' + a_n' \\
&\text{s.t.}: \\
&\text{mean}_{\text{calculated}} + s_n - a_n = \text{mean}_{\text{given}} \\
&\text{STD}_{\text{calculated}} + s_n' - a_n' = \text{STD}_{\text{given}}
\end{aligned} \tag{52.2}$$

Another method to mitigate the issue of heavily skewed histograms is to construct histograms with equal length bins. Figure 52.3 shows a sample histogram with equal length bins in which the right-hand side bins contain very small counts and even in some cases many have zero counts. Using this kind of bin formatting will help the stochastic optimization by reducing the computational time spent in identifying the feasible data points because the search space is smaller in this particular case. We have devised this approach based on the kind of histograms we currently get. In the past however, we received equal length binned histograms, but it did not provide enough resolution of the data points within the left bins because generally lower bins contained much more data points as compared to the higher bins in the histograms. This structure of histogram is due to heavy skewness in the underlying distributions.

52.4 Case Study

The case study for this paper is adapted from the work performed for the IARPA program, Scientific Advances to Continuous Insider Threat Evaluation (SCITE). For this work, data was redacted to protect potentially sensitive data. Data was provided in the form of summary statistics such as means and standard deviations of all the features which carry the information about employees' cyber activity. These observable features are called detectors. Certain patterns of detectors may be indicative of behaviors of interest, called target behaviors. These target behaviors are the subject of investigation. Table 52.1 shows one of the provided histograms (on the left) and given means and standard deviations of all detectors (on the right). There are about 12,300 histograms provided to us pertaining to 141 detectors' distributions in

Table 52.1 A sample histogram (left) and the corresponding given means and standard deviations (right)

lower	upper	counts
0	0	180
2	1664	148
1664	3582	146
3582	5818	148
5818	7981	148
7981	10894	147
10894	14087	147
14087	17821	148
17821	22188	147
22188	27132	147
27132	33673	147
33673	39973	148
39973	47347	147
47347	56330	147
56330	68003	148
68003	82811	147
82811	103319	147
103319	127445	148
127445	166114	146
166114	259839	148

Detectors	Mean	Std
det001a	70174	145641
det001b	15504	64228
det001c	9167	71502
det002a	2109	12027
det002b	833	9363
det002c	426	6260
det003a	37	38
det003b	9	16
det003c	4	9
det004a	849	674
det004b	227	293
det004c	106	172
det005a	51	163
det005b	52	253
det005c	43	317
det006a	121397	500934
det006b	34266	178882
det006c	19117	196770

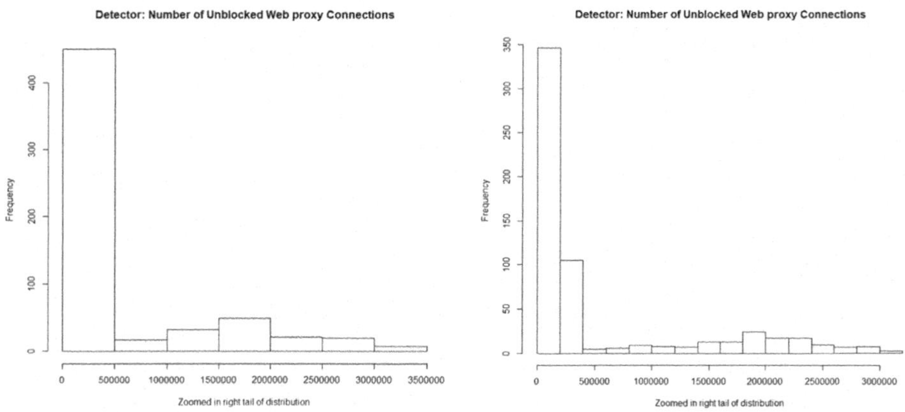

Fig. 52.4 Reconstructed histogram of a certain detector in two different iterations

43 weeks of the data collection period. Each of these histograms has its own bin boundaries. The sampling method is configured to randomly draw more than half of the data points from the triangular distributions. The rest of the data points are produced from the optimization program.

A simple sampling method is used to make a comparison with the stochastic optimization method presented in this paper. The simple method produces samples from triangular distributions fitted to the bins. For example, two random samples were drawn from detector 1 histogram. The first sample had a mean of 131,938 and a standard deviation of 405,978. The second had a mean of 134,837 and a standard deviation of 398,930. This is a difference of 2899 between the two calculated means and 7047 between the standard deviations. The means and standard deviations will vary from iteration to iteration because the simulation process uses a random number generator. We can see the effect of this random variation when we zoom into the right tails of the distribution, as shown in Fig. 52.4. Each panel represents a different iteration of the random generation process. From the visual comparison of the two histograms in Fig. 52.4, we can see there is a significant difference in the data points drawn each time. The mean calculated from these data points is also far from the given mean. The mean calculated from randomly sampled data points is almost twice as large as the given mean for detector 1 which is 70,174, and the standard deviation is approx. 2.7 times the given standard deviation for detector 1 which is 145,640. Based on all the previous discussion in this section and the behavior of the provided data, it suggests that using a more controlled approach where all the bins are sampled using triangular distribution but the data points from last two bins are extracted using optimization may improve the match to a given mean and standard deviation. The reason behind choosing the right most two bins to sample data points using the stochastic optimization search is to capture the data points towards the right tail of the distribution. Capturing the data points in the right tail of the distribution is important because the discriminative model we use down the pipeline of our analysis

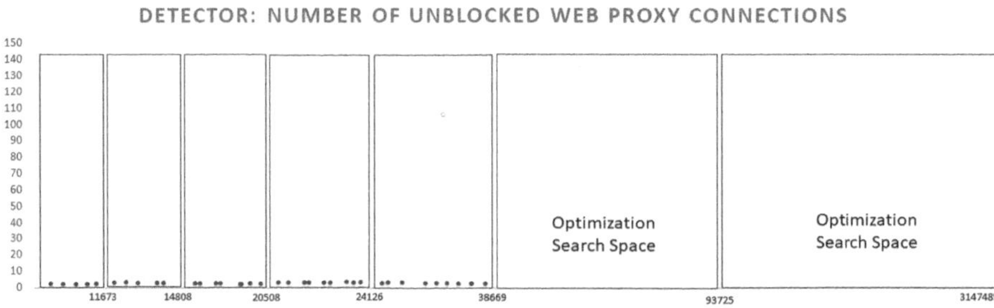

Fig. 52.5 Illustration of the right-/topmost five bins with the uniform kernel

mostly considers outlier data points for classification decision. The last two bins mostly contain the crucial outlier data points, in the provided histograms.

In another example, we compare our methodology with mere random uniform sampling from the bins. Using the same detector 1 histograms, this time we draw uniform random samples from the bins and see that the calculated mean of these data points is 132,834 and the actual given mean of this detector is 70,174. This shows the distribution obtained from uniform random sampling process produces the mean statistics almost twice greater than the actual mean value. Also, the same is the case with standard deviation, i.e., the calculated standard deviation from this simulated distribution is three times greater than the actual given standard deviation value for that particular distribution. A notional representation of the uniformly sampled data points within the bins is shown in Fig. 52.5.

The outcome of this Lagrangian function shows a minimal deviation of at most ratio of 3 (ratio of actual mean or standard deviation over desired mean or standard deviation) in about 10% of the feature samples. The present study contains more than 12,000 histograms (142 features, 2 target groups, and 41 weeks); thus a small deviation is expected in about 1200 sample sets. It is also possible to decrease the degrees of freedom (the number of random samples generated from the pertinent triangular distributions); however this change will reduce the native uncertainty of the generated samples. This is the consequence of generating a larger number of samples using the stochastic optimization program, which finds optimal values only. Despite the fact that there are multiple optimal solutions, the various solutions do not possess the range that a random sample generator produces. Therefore, the trade-off is between matching the desired mean and standard deviations exactly and reproducing the native uncertainty of the samples.

This approach is applied over all the provided histograms to simulate the employees' activities across all the weeks. Each simulated data set represents one simulated organization. Multiple organizations are simulated in this way to provide uncertainty bounds. Table 52.2 displays a single synthetic organization generated using the methodology. This organization contains synthesized detector data for all employees. Each employee exhibits 141 different indicating activities and a ground-truth-based target behavior assignment. In Table 52.2, week column represents the

Table 52.2 A synthesized organization

user_id	week	label	001a	001b	001c	002a	002b	002c	003a	003b	003c	004a	004b	004c
1	9	0	9878	2425	0	81	6	5	81	7	2	1138	0	448
2	9	0	3192	910	645	1001	630	84805	1	15	0	28	189	350
3	9	0	16007	24443	0	2359	166	66	2	0	8	1243	6	0
4	9	0	150996	621436	6382	768	272	0	27	17	2	71	141	0
5	9	0	347	774294	0	2453	1361	0	29	0	2	1970	295	28
6	9	0	92215	24907	10435	86	773	0	8	3	5	1277	179	4
7	9	0	32794	5867	2774	541	414	237	82	0	3	400	85	22
8	9	0	7263	7346	1433	260	109647	0	32	3	0	0	77	0
9	9	0	12355	3096	305	2601	255131	0	81	307	1	476	14	18
10	9	0	54839	1163	0	194	86	0	3	17	1	131	741	242
11	9	0	255666	567	648	356440	619	0	25	0	0	777	13	106
12	9	0	4269	918	1469	5557	90	0	37	1	0	1464	0	12
13	9	0	9920	1573	27	1011	169	16	32	0	1	581	27	42
14	9	0	54834	9518	1231	499	8	29	13	0	0	393	743	48
15	9	0	19416	7545	3999	1419	244	0	24	9	13	69	269	28
16	9	0	8774	265	324	842	12	3	8	2	0	1748	147	0
17	9	0	6270	2721	0	338	109	0	12	12	18	769	0	107
18	9	0	148546	0	0	115559	5	0	19	0	0	347	516	420
19	9	0	31500	7678	0	1013	71	65	15	15	2	1125	14	83
20	9	0	4838	9865	1564	1503	19	0	15	3	8	282	497	9
21	9	0	21112	0	1092	184	41	0	69	3	0	1754	46	12
22	9	0	2502993	5885	1870	143	6	0	2	4	2	848	490	16
23	9	0	75664	0	1039	652	0	0	32	40	0	1478	176	0

week number associated to an employee's enterprise activity at a given time instance. The label column represents a flag of a target behavior, for example, in this case "yes" or "no." The rest of the columns such as 001a, 001b, ... 004c represent the employee's activity data pertinent to various detectors deployed in an inference enterprise or in other words predictive features in an organization-wide data set.

52.5 Conclusions

A novel methodology for synthesizing organizational population from redacted data (i.e., summary statistics) is introduced in this paper. This methodology uses only histograms, means, and standard deviations. When data on correlations is available, this information can be incorporated using a separate methodology not reported here. A comparison is provided between the proposed sampling methodology and the fully random sampling method. It shows that we were able to improve the adherence to provided summary statistics, which ensures close representation of the ground-truth data. A close replication of the original population data leads to further sound and reliable data analysis. This method is scalable in terms of handling large number of histograms and adaptable to various histogram formats (i.e., with different bins' lengths and frequencies).

Acknowledgements Research project used in this paper was supported under IARPA contract 2016 16031400006. The content is solely the responsibility of the authors and does not necessarily represent the official views of the US Government.

References

1. IARPA. (2015). *Scientific advances to continuous insider threat evaluation (SCITE).*
2. Huang, E., Zaidi, A. K., & Laskey, K. B. (2018). Inference enterprise multimodeling for insider threat detection systems. In A. M. Madni, B. Boehm, R. G. Ghanem, D. Erwin, & M. J. Wheaton (Eds.), *Disciplinary convergence in systems engineering research* (pp. 175–186). Cham, Switzerland: Springer.
3. Laskey, K. B., Zaidi, A., Buede, D., Imran, M., Huang, E., Brown, D., et al. (2018). *Modeling inference enterprises using multiple interoperating models.* In Submitted for INCOSE IS2018.
4. Tekel, J., & Cohen, L. (2012). Constructing and estimating probability distributions from moments. In *Automatic target recognition XXII* (Vol. 8391, p. 83910E). International Society for Optics and Photonics.
5. Jung, W., Park, J., & Lee, J. (2014). Versatile and scalable parallel histogram construction. In: *Proceedings of the 23rd International Conference on Parallel Architectures and Compilation.* New York, USA (pp. 127–138).
6. Chatterjee, S., & Firat, A. (2007). Generating data with identical statistics but dissimilar graphics. *The American Statistician, 61*(3), 248–254.
7. Ricci, L., Verardi, V., & Vermandele, C. (2016). *A highly efficient regression estimator for skewed and/or heavy-tailed distributed errors.*

Chapter 53
Cyber-Physical Systems Modeling for Security Using SysML

Bryan T. Carter, Cody H. Fleming, Carl R. Elks, and Georgios Bakirtzis

Abbreviations

CPS	Cyber-physical system
SysML	Systems Modeling Language
STPA-Sec	Systems-Theoretic Process Analysis for Security

53.1 Introduction

Cyber-physical systems (CPS) integrate diverse sets of hardware and software components to provide a particular service or control to a specific physical process. Traditionally, cybersecurity follows a perimeter-defense approach—tactically erecting barriers to adversaries attempting to conduct malicious activity within a system. Securing these systems, therefore, requires a thorough understanding of the system's architecture. However, the physical processes and interactions controlled by CPS necessitate that security measures are also cognizant of the system's expected service, as CPS often perform safety-critical applications where disruptions in the expected service can have catastrophic consequences. In turn, this requires

B. T. Carter
Coordinated Systems Lab, UVA, Charlottesville, VA, USA

C. H. Fleming
Mechanical and Aerospace Engineering, University of Virginia, Charlottesville, VA, USA

C. R. Elks
Dependable Cyber-Physical Systems Lab, VCU, Richmond, VA, USA

G. Bakirtzis (✉)
Coordinated Systems Lab, UVA, Charlottesville, VA, USA

Dependable Cyber-Physical Systems Lab, VCU, Richmond, VA, USA
e-mail: bakirtzisg@ieee.org

S. Adams et al. (eds.), *Systems Engineering in Context*,
https://doi.org/10.1007/978-3-030-00114-8_53

analysts to have access to a well-formed model of not only the system's architecture but also the system's admissible behaviors and requirements such that the model also defines the system's role within its mission. By using model-based cybersecurity analysis, security efforts are able to be moved up in the system's life cycle. This is advantageous for multiple reasons, including lowering costs downstream in the system's life cycle, potentially catching vulnerabilities before they can be exploited, and adopting a proactive, rather than reactive, approach to security. We propose using the Systems Modeling Language (SysML), a graphical, object-oriented language [1], as a tool to represent this proactive, model-driven approach to security analysis.

The need for model-based security analysis is not new to the field of cybersecurity [2]; such methods for assessing the dependability and security of network and computer systems exist. However, research in extending such methods to CPS remains in its infancy. An effective model of a CPS integrates physical process models, the architecture of the system, and the feedback loops between them. However, there is no formulaic approach that can ensure the creation of a sufficient model that encompasses all three parts. In addition, making such a model compatible for security analysis remains an unanswered question. Therefore, we propose a general modeling process that aims to describe the physical processes, architectures, and feedback loops in a manner that facilitates mission-oriented security analysis. This approach builds upon a methodology utilizing stakeholder elicitation exercises and Systems-Theoretic Process Analysis for Security (STPA-Sec) for informing mission context in the model and an architectural taxonomy for defining system components in a security-minded manner.

53.1.1 Contributions of this Research

The central contribution of this paper is a concrete application of a mission-oriented modeling technique using SysML as a modeling tool. The modeling technique described in this paper helps provide the link between mission-level information and the vulnerability analyses described above. Furthermore, the modeling technique answers the need for traceability within the mission-aware philosophy. More specifically, this paper presents an application of the Systems Modeling Language as a graphical representation tool for connecting system architecture information with mission requirements and behavior. However, this modeling methodology is not limited to SysML; its principles can be applied to other graphical modeling schemas.

This paper focuses on applying the modeling methodology to a use case assuming that all the information in the model is known ahead of time. Other works by the authors describe how to generate the mission context information and how the model can be applied to vulnerability analysis [3, 4].

53.2 Background

53.2.1 The Systems Modeling Language

SysML is a general-purpose, graphical modeling language standardized by the Object Management Group (OMG) for model-based systems engineering. More specifically, SysML is based on the concept of UML but implemented to be more abstract and flexible, thus enabling it to be applicable outside of software-centric applications. In our proposed modeling approach, the main purpose of SysML is to facilitate the visualization and communication of a system's design between various stakeholders [5]. SysML achieves this by following a standard set of definitions and specifications intended to minimize ambiguity in design patterns.

SysML is based on a set of diagram types with an associated set of diagram elements that follow a specific syntax with clear semantics. These diagrams represent the structural composition and interconnections between architectural structures, admissible behaviors, requirements, and relationships between these elements within a system. We believe that SysML is an attractive choice for this methodology due to its flexibility for use throughout the system life cycle and its familiarity within the industry. In our modeling approach, using SysML does not limit modeling efforts to a single language, as it is transformable into a graph structure and can be inputted to other analysis techniques through this more generalized schema [3]. But it does provide a medium to reason how requirements are connected to behaviors and, further, how those behaviors are implemented in a modeled system architecture. Additionally, SysML allows us to communicate clearly and in a single visual format the designs as well as their rationale to the various stakeholders. Finally, SysML can take the role of a centralized living document, which starts at the early concept phase by capturing the initial stakeholder requirements, and can assist through the requirements analysis, the architectural design, and the evaluation of the design, to finally model a full system solution to provide a defensible trace of design decisions.

53.2.2 Mission-Aware Cybersecurity and the Need for Traceable Models

One of the core concepts behind mission-aware cybersecurity for CPS is that a system can never be fully secured, especially a CPS composed of numerous heterogeneous, often commercial-off-the-shelf, products. Instead of endlessly trying to secure systems based on emerging threats, tailoring system security to its mission allows the security problem to become more manageable by only focusing efforts on areas that can have the greatest effect on the mission's outcome. This approach aims to adapt a cybersecurity viewpoint that focuses on IT systems, where breaches are equivalent to mission failures, to one that understands that the physical world interactions inherent to CPS create a chance to detect and mitigate attacks such that a mission can still be completed to a reasonable degree of success. This concept is otherwise known as constructing system resiliency.

This mission-aware methodology is based on gathering input directly from mission stakeholders and then using that information to perform a hazard analysis that helps identify the mission-critical portions of a system. This information is combined with architectural information to form a full system model. From there, vulnerability analyses can inform the choices of defense measures based on historically recorded attacks, and experience-based red-team analysis helps inform the choice of resiliency measures within the system. The possible defense and resiliency measures can then be evaluated based on their potential to ensure mission success, cost, and ease of implementation [6].

The information elicitation exercise involves the various mission stakeholders, who give their inputs on mission goals, unacceptable mission outcomes, and critical components to the system. This information is then the input to a Systems-Theoretic Process Analysis for Security (STPA-Sec), which identifies the potentially hazardous states a system can enter and reason about how entering those states can be prevented. More specifically, STPA-Sec is a top-down analysis approach that focuses on the strategic problem of preventing losses rather than the tactical problem of eliminating vulnerabilities [7]. The mission-aware approach uses STPA-Sec to identify the various hazards, control actions, and safety constraints on those control actions associated with the unacceptable losses identified by the mission stakeholders.

In architectural models, security vulnerabilities within a component can be linked to other components via their communication structure. While this is a necessary characteristic of any good model, evaluating a vulnerability in the mission context, however, requires that specific, low-level vulnerabilities within components can be traced to potentially compromised mission requirements. Therefore, we need a systematic process for modeling the mission context, the system architecture, and linking those two together [4].

The proposed approach uses mission stakeholders, and hazard analysis informs the mission context portion of the model. More specifically, the information elicitation exercise captures the mission objectives and its unacceptable outcomes, among other pieces of information, and the STPA-Sec analysis on the CPS within that mission defines its behavior- and system-level requirements. The process of translating these sources of information into model elements is explained later in this paper.

Generating a model of a system's architecture is relatively straightforward; yet different systems engineers can still produce significantly different models of the exact same system. Therefore, there must exist a set of guidelines for representing individual components in a sufficiently detailed manner. For the purposes of this process, the architecture follows a hierarchical format, with the CPS being decomposed into its subsystems, components, and parts. By following a hierarchical structure, we can then attribute specific components or subsystems as being responsible for implementing a specific behavior identified in the STPA-Sec analysis, which correlates to a specific requirement informed by the stakeholder elicitation. Consequently, we create a traceable path from part or component vulnerabilities to mission-level requirements.

53.3 Mission-Aware Modeling Using SysML

A mission-aware SysML model contains information transferred over from prior hazard analyses and requirements elicitations in addition to the system's architecture. These three sources of information are linked together hierarchically to generate a traceable model, such that component- and part-level vulnerabilities can be associated with possible violations of mission-level requirements. Consequently, analysis of low-level vulnerabilities, disruptions, and inadequacies maintains a tangible connection to the mission context, which helps encourage a strategic approach to handling security.

The proposed methodology utilizes only four of the nine diagram types—but all categories of diagrams—within SysML, namely:

- Requirements diagram
- Activity diagram
- Block definition diagram (BDD)
- Internal block diagram (IBD)

These four diagrams map to each of the three sources of information within the mission-aware framework, shown in Table 53.1 below.

We have limited ourselves to these four diagram types simply because they are the minimum set for representing the three components of a mission-oriented model. Other diagram types, such as the parametric diagram, could also prove useful but often assume full knowledge of the CPS design. Since this methodology is designed to be implemented early in a system's life cycle, when the design has not necessarily been finalized, we do not consider the other five diagram types.

We build the SysML model in a treelike structure that mimics the STPA-Sec analysis process. A requirements diagram that contains the mission-level information forms the root of the tree, which branches out into several activity diagrams to represent the behaviors that satisfy the mission-level requirements, which lead to the terminal BDDs and IBDs which contain architectural information. Building the model in this format allows us to represent information in an organized manner in addition to facilitating traceability between model elements. This section presents an application of the modeling methodology using an unmanned air vehicle (UAV) conducting a reconnaissance mission as a use case.

Table 53.1 SysML diagram types mapped to the appropriate piece of mission-aware information

SysML diagram	Captured information
Requirements diagram	Mission-level requirements
Activity diagram	STPA-Sec-derived behavior
BDD and IBD	System architecture

53.3.1 *Mission-Centric Requirements in SysML*

As stated before, the requirements diagram presents the mission-level information obtained from the stakeholder elicitation and STPA-Sec analysis. More specifically, the elements of the STPA-Sec portion of Fig. 53.1 are translated into requirements. The wording of the requirements mirrors the language used in the actual STPA-Sec

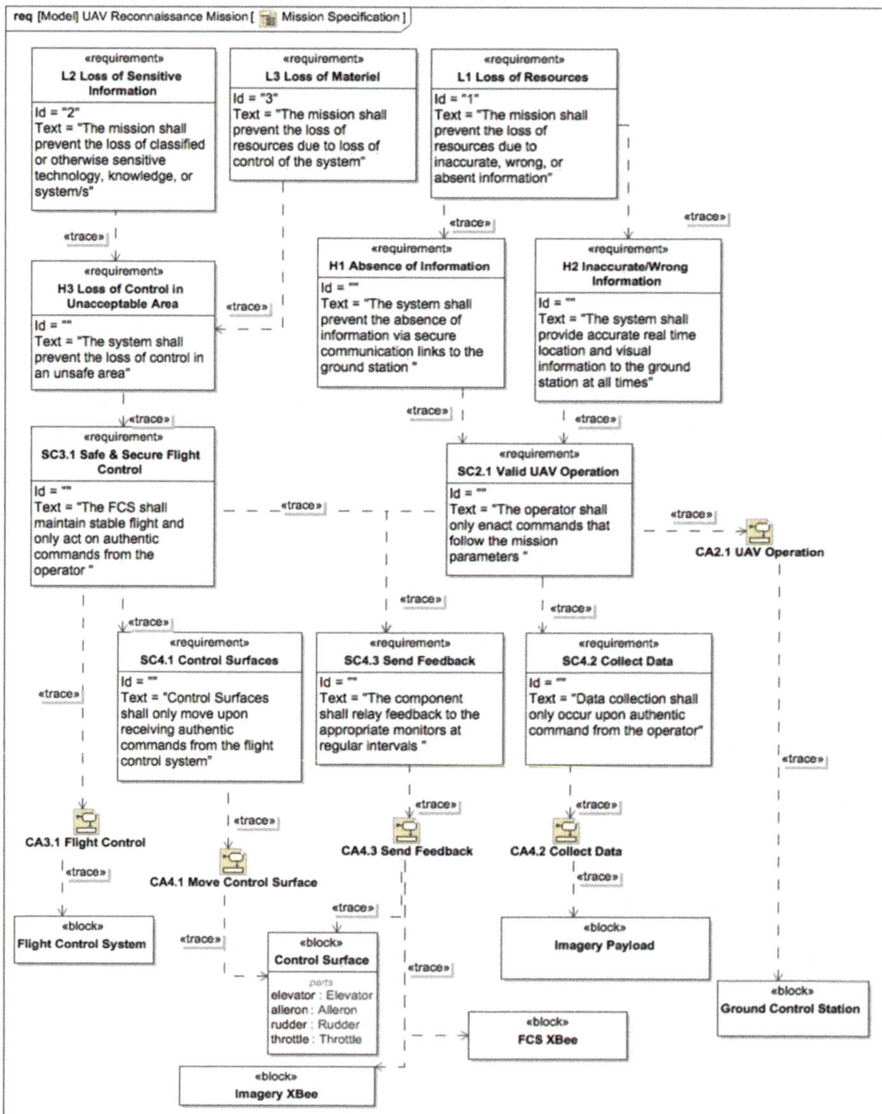

Fig. 53.1 The requirements diagram combined with traces to other model elements

analysis but in a manner more familiar to traditional system requirements prose. For example, an unacceptable loss identified in the STPA-Sec analysis for the UAV mission is a loss of resources due to the lack of reconnaissance information. This loss becomes the requirement, "The mission shall prevent the loss of resources due to inaccurate, wrong, or absent information," in the block labeled *L1 Loss of Resources* (Fig. 53.1).

The basic structure of the requirements diagram follows a standard tree pattern that follows the order of the prior STPA-Sec analysis. The top-level unacceptable losses serve as the root nodes in the requirements diagram tree. The occurrence of one of these unacceptable losses would result in total mission degradation, hence their placement as the roots of the requirements tree. Below the unacceptable losses are the hazardous conditions that could lead to an unacceptable loss. Each hazard could propagate into at least one of the unacceptable losses. Finally, below the hazards are the safety constraints that should be applied to the system's functional behavior. Violations of these could lead to a hazardous condition, which in turn can result in an unacceptable loss. We use the SysML <<trace>> relationship to show the linkages between elements in requirements diagrams. SysML syntax defines the trace relationship as a generic relationship between two elements, which aligns with the STPA-Sec postulation that an unacceptable loss is a possible result of a hazardous condition, not a direct implication.

53.3.2 Functional Behavior in Activity Diagrams

STPA-Sec analysis informs us of the specific control actions that can be taken within a mission; we represent each of these control actions in SysML with its own activity diagram (Fig. 53.2). At the bottom of the requirements diagram tree lie the safety constraints; each safety constraint corresponds to a particular control action and its activity diagram. Within the activity diagram, we use the SysML action, control flow, initial node, and activity final model elements. The initial node element represents the calling of the control action, the action elements represent functional tasks that must be performed to successfully complete the control action, and the activity final element represents the completion of the control action. The control flow element simply indicates the order of tasks within the control action. An example of this is shown for the *move control surface* control action for the UAV in Fig. 53.2.

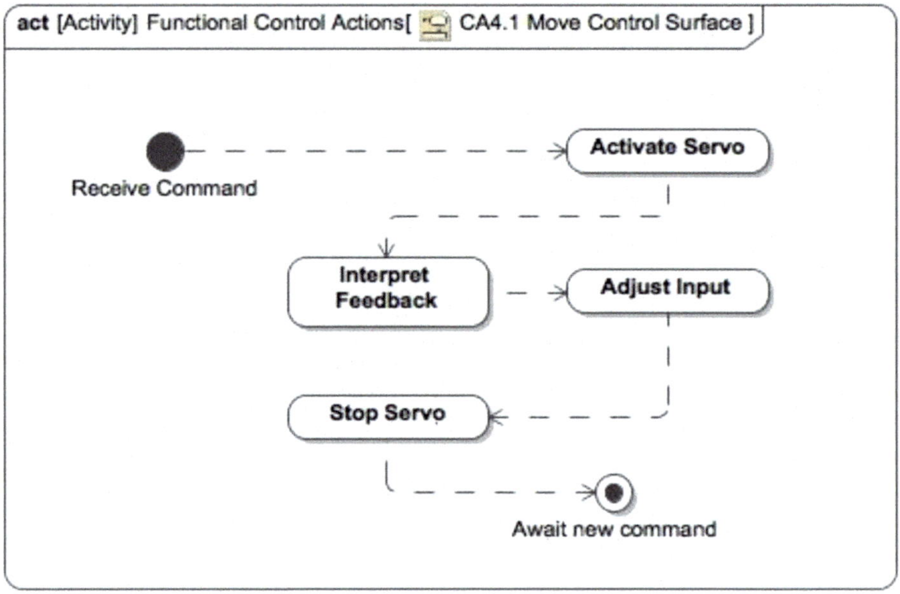

Fig. 53.2 The activity diagram of a control action for a reconnaissance UAV

53.3.3 System Architecture in Block Definition and Internal Block Diagrams

The block definition diagram defines block elements in terms of their features and relationships to other blocks [5]. A block is one of the most versatile SysML elements; it can be used to represent a system, subsystem, component, part, or a number of other things. Our methodology uses block definition diagrams to represent a hierarchical structure of blocks corresponding to the elements that make up a particular system. Using the UAV use case example, the UAV system is composed of four main subsystems, such as the flight control system, which in turn are composed of components, such as a GPS. This structure can be seen in Fig. 53.3.

Furthermore, since this model is intended for security analysis, a component should be characterized by a set of attributes relevant to its security. These component attributes include its operating system, hardware, firmware, software, communication protocols, and entry points [3]. We use SysML part properties to define these attributes for a particular component. To save space, these attributes are only visible for the Adafruit Ultimate GPS (Fig. 53.3).

SysML uses internal block diagrams to specify the internal structure of a defined block [5]. In our methodology, we adapt this slightly to show the communication structure within the system, more specifically, how different parts interact with each other via communication protocols or physical movements. Part property elements represent the different components, and labeled connector elements specify the

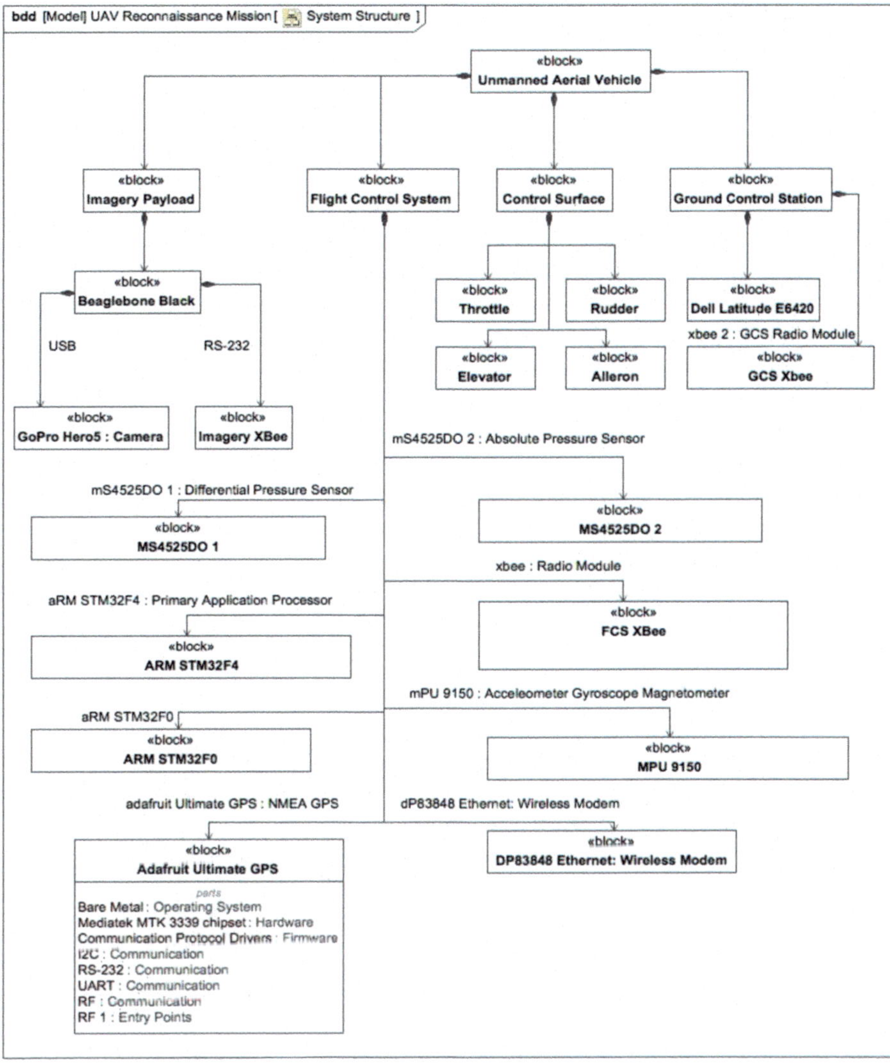

Fig. 53.3 The block definition diagram for a reconnaissance UAV

interaction between two components. For example, the BeagleBone Black imagery processor communicates with the GoPro camera via USB. This relationship can be seen in the BDD in Fig. 53.3. An IBD example is not shown in order to save space.

53.3.4 *Showing System Traceability Graphically*

The SysML requirements diagram is not limited to representing just requirements elements. Our methodology relies on representing traceability; therefore, it would be beneficial to be able to show it visually. The requirements diagram allows model elements like blocks and activity diagrams to be represented alongside system requirements. Hence our secondary usage of the requirements diagram showcases traceability. For the purposes of this paper, this requirements diagram will be referred to as the traceability diagram, seen in Fig. 53.1 above.

This diagram uses the trace relationship between elements for the same reason as in the requirements diagram described earlier. The linkages between components, control action, and requirements are merely just pathways for a vulnerability to impact a mission, not a direct implication of violations. Furthermore, this diagram showcases the key tenet to the mission-aware approach—addressing security in the context of the mission. By presenting the system components with their associated behavior, and mission requirements in a single diagram, we create a powerful tool for visualizing how a simple escalation of privileges within a single computer could propagate through the entire mission.

53.4 Related Works

Modeling of cyber-physical systems has emerged as a valuable area of research—mainly due to the unique set of challenges that it presents. Derler, Lee, and Vincentelli define these challenges as being the result of the "intrinsic heterogeneity, concurrency, and sensitivity to timing," within CPS [8]. They go on to assert that many modeling methodologies do not have sufficiently strong semantics to effectively capture all of the idiosyncrasies of CPS.

While many modeling techniques exist for CPS in general, modeling for security is not as heavily researched. Ten, Liu, and Manimaran describe modeling access points for a SCADA system to guide vulnerability analysis [9]. Brunner, Huber, Sauerwein, and Breu propose a safety and security modeling technique for CPS using UML [10]. Finally, Ouchani and Lenzini also describe a model-based approach to security using SysML that simulates the model and attacks within that framework [11].

53.5 Conclusions

This paper presented an application of a specific modeling technique for facilitating mission-aware security analysis for cyber-physical systems using the Systems Modeling Language. This technique allows for the generation of SysML models that combine system architecture information with its mission context, thus enabling full traceability between component-level vulnerabilities and the potential mission effects of the vulnerability. SysML provides a favorable environment within which to create such a model; however, the modeling methodology is intentionally designed to be translatable to other modeling schemes. This translatability allows the technique to be applied without restricting the modeling language used.

This work forms the groundwork for database-driven vulnerability analysis. More specifically, the architectural information encoded in the SysML model can be used to generate queries into a database of historical attacks and then match potential attack vectors to the system. Since the SysML model traces architecture to mission-level information, the effects of a potential attack vector can be evaluated based on the effect it would have on the overall mission. Such information can be used in a decision support role to inform the choice of defense or resiliency measures within a cyber-physical system.

Acknowledgments This research is based upon the work supported by the Department of Defense through the Systems Engineering Research Center managed by the Stevens Institute of Technology.

References

1. Hause, M. (2006). The SysML modelling language. In *Fifteenth European Systems Engineering Conference*.
2. Nicol, D. M., Sanders, W. H., & Trivedi, K. S. (2004). Model-based evaluation from dependability to security. *IEEE Transactions on Dependable and Secure Computing, 1*(1), 48–65.
3. Bakirtzis, G., Carter, B. T., Elks, C. R., & Fleming, C. H. (2018). A model-based approach to security analysis for cyber-physical systems. In *2018 Annual IEEE International Systems conference (SysCon)* (pp. 1–8). IEEE.
4. Carter, B. T., Bakirtzis, G., Elks, C. R., & Fleming, C. H. (2018). A systems approach for eliciting mission-centric security requirements. In *2018 Annual IEEE International Systems Conference (SysCon)* (pp. 1–8). IEEE.
5. Delligatti, L. (2013). *SysML distilled: A brief guide to the systems modeling language*. Boston: Addison-Wesley.
6. Bakirtzis, G., Carter, B. T., Fleming, C. H., & Elks, C. R. (2017, December). *MISSION AWARE: Evidence-based, mission-centric cybersecurity analysis*. ArXiv-eprints.
7. Young, W., & Leveson, N. (2013). Systems thinking for safety and security. In Proceedings of the 29th Annual Computer Security Applications Conference (ACSAC 2013) (pp. 1–8). ACM.
8. Derler, P., Lee, E. A., & Vincentelli, A. S. (2011). *Addressing modeling challenges in cyber-physical systems*. Technical report no. UCB/EECS-2011-17. Berkeley, CA: Electrical Engineering and Computer Science Department, University of California.
9. Jensen, J. C., Chang, D. H., & Lee, E. A. (2011). A model-based design methodology for cyber-physical systems. In *Proceedings of the International Wireless Communications and Mobile Computing Conference—IWCMC 2011* (pp. 1666–1671).
10. Brunner, M., Huber, M., Sauerwein, C., & Breu, R. (2017). Towards an integrated model for safety and security requirements of cyber-physical systems. In *2017 IEEE International Conference on Software Quality, Reliability and Security Companion (QRS-C)* (pp. 334–340).
11. Ouchani, S., & Lenzini, G. (2014). Attacks generation by detecting attack surfaces. *Procedia Computer Science, 32*, 529–536.

Chapter 54
Bayesian System Averaging: A Grand Theory Unifying Bayesian Forecasting System Method and Bayesian Model Averaging Method

Jie Liu

54.1 Introduction

54.1.1 Reviewing BFS and BMA

When forecasting, people have long learned that human cognition is forever limited and that the only thing for certain in this world is uncertainty. We all know that the forecasting models may approach the "true mechanisms" behind the forecasted events but can never catch them. This is the epistemological base of the contemporary preference for probabilistic forecasts over point forecasts [1]. After realizing the imperfection of our forecasting capabilities, modern forecasters treat such imperfection in two philosophically different manners: (1) When facing a pool of alternative models, some statisticians choose the best one that they *believe* to be the best proxy of the true mechanism, use it to make forecast, and try to *quantify* how far this proxy is away from its master. The distance between the proxy and the master is called model uncertainty, and this statistical approach is called Bayesian forecasting system (BFS) method [2]. (2) When facing a pool of alternative models, some statisticians *believe* that the true mechanism is one of them or a composite of them. The problem is that they don't know which one is the true one or what is the correct way to compose them. The task now becomes to *quantify* the forecaster's confidence for each model to be the true one and *figure out* a "good" way to compose them. Such confidence is called model weight (or pooling weight), and this statistical approach is called forecasting combination (or prediction pools) theory [3], of which the Bayesian model averaging (BMA) is a typical and popular one [4].

Under the theoretical framework of BFS, we denote $W(t)$ as the value of the predict and W and $Y(t)$ as the value of the predictor Y at stage t, for $t = 1, \ldots, T$; and

J. Liu (✉)
Department of Systems & Information Engineering, University of Virginia, Charlottesville, VA, USA
e-mail: jl4eb@virginia.edu

S. Adams et al. (eds.), *Systems Engineering in Context*,
https://doi.org/10.1007/978-3-030-00114-8_54

historical data $D = \{(w(i), y(i)) : i = 0, \ldots, T-1\}$. Our goal is to forecast the value of $W(T)$ at the stage T. We also denote $g(w(T))$ as the prior density function of $W(T)$, $f(y(T)|\, w(T))$ as the likelihood function of $W(T)$ given the value of the predictor $Y(T)$, $\kappa(y(T)) \equiv \int_{-\infty}^{\infty} f(y(T)|w(T))\, g(w(T))\, dw(T)$ as the induced expected density function of the predictor $Y(T)$, and $\lambda_T(y(T))$ as the nonstationary marginal density function of $Y(T)$. The BFS theoretical framework can be summarized in the following two equations:

$$\pi(w(T)|y(T)) = \frac{f(y(T)|w(T))g(w(T))}{\kappa(y(T))}. \tag{54.1}$$

$$\phi_T(w(T)) = \int_{-\infty}^{\infty} \pi(w(T)|y(T))\, \lambda_T(y(T))\, dy(T). \tag{54.2}$$

The $\pi(w(T)|\, y(T))$ in Eq. (54.1) is called "the posterior density function of $W(T)$ given $Y(T)$" or the "short-term forecast." The $\phi_T(w(T))$ in Eq. (54.2) is called "the (nonstationary) predictive density function of $W(T)$" or the "long-term forecast." The difference between short-term and long-term forecasts is whether the specific value or the distribution of the predictor is available to the forecaster.

Under the framework of BMA, *without considering parameter uncertainty*, we denote $W(t)$ as the value of the predictand W at stage t, for $t = 1, \ldots, T$; $\{S_k : k = 1, \ldots, K\}$ as the pool of forecasting systems; $g_k(w(t))$ as the probability density function of $W(t)$ generated by each forecasting system S_k; $P(S_k)$ as the prior probability of S_k being the true mechanism, such that $\sum_{k=1}^{K} P(S_k) = 1$; and historical data $D = \{w(i) : i = 0, \ldots, T-1\}$; $p(w(T)|\, S_k, D)$ as the conditional density function of the predictand of $W(T)$ given forecasting system S_k being the true mechanism and the historical data D. Our goal is to forecast the value of $W(T)$ at the stage T. The BMA theoretical framework *without considering parameter uncertainty* can be summarized in the following equation:

$$\begin{aligned} cg(w(T)|\{S_k\}, D) &= \sum_{k=1}^{K} p(w(T)|S_k, D)\, P(S_k|D) \\ &= \sum_{k=1}^{K} g_k(w(T)) \frac{\prod_{i=0}^{T-1} g_k(w(i))\, P(S_k)}{\sum_{l=1}^{K} \left[\prod_{i=0}^{T-1} g_l(w(i))\right] P(S_l)}. \end{aligned} \tag{54.3}$$

Under the theoretical framework of BMA, *now conside ring parameter uncertainty*, all notations stay the same except for two changes: (a) $g_k(w(t))$ changes into $g_k(w(t)|\, \theta_k)$, the density function of $w(t)$ generated by each forecasting system S_k and given θ_k being the true parameter, and (b) the new notation $p(\theta_k|\, S_k)$ as the prior density function of θ_k being the true parameter under the true system S_k. The BMA theoretical framework *with considering parameter uncertainty* can be summarized in the following equations:

$$\begin{aligned} cg(w(T)|\{S_k\},D) &= \sum_{k=1}^{K} p(w(T)|S_k,D)\, P(S_k|D) \\ &= \frac{1}{p(D)} \sum_{k=1}^{K} \left(\int g_k(w(T)|\theta_k) \prod_{i=0}^{T-1} g_k(w(i)|\theta_k) p(\theta_k|S_k)\, d\theta_k \right) P(S_k), \end{aligned} \tag{54.4}$$

where

$$p(D) = \sum_{k=1}^{K} \left(\int \left[\prod_{i=0}^{T-1} g_k(w(i)|\theta_k) \right] p(\theta_k|S_k)\, d\theta_k \right) P(\, S_k). \tag{54.5}$$

The $cg(w(T)|\,\{S_k\},D)'s$ in Eqs. (54.3) and (54.4) are called "the averaged (coupled) posterior density function of *w*(*T*) given the forecasting system pool $\{S_k\}$ and the historical data *D* (without or with considering parameter uncertainty)."

54.1.2 Comparing BFS and BMA

From the description above, we can tell three important methodological differences between BFS and BMA: (1) Though Bayesian analysis is formally defined as the statistical approach that explicitly considers the prior information ([5], p. 3), the "prior information" in BFS and in BMA are different. In BFS, it refers to the prior density function *g*(*w*(*T*)). But in BMA, the prior density family of the predictand *w* (*T*) is not explicitly considered. Rather, the prior composition of the system pool $\{S_k\}$, as well as the prior probability $P(S_k)$ of S_k being the true mechanism, and the prior density function $p(\theta_k|\,S_k)$ of θ_k being the true parameter under each S_k are explicitly considered. (2) While BFS uses a single forecasting system, BMA incorporates multiple competing forecasting systems. (3) $\pi(w(T)|\,y(T))$, $\phi(w(T))$, and $p(w(T)|\,S_k, D)$ all specify our probabilistic forecast under every single forecasting system but are constructed in different Bayesian ways.

Just as any probability density function, $\pi(w(T)|\,y(T))$, $\phi(w(T))$, or $p(w(T)|\,S_k, D)$ has two unalienable components: (a) the density function family (say, a Gaussian form, a uniform form, etc.) and (b) the parameter θ_k (say, the mean and variance for a Gaussian density, the lower bound and upper bound for a uniform, etc.). To specify a density function, either in a Bayesian manner or not, we have to specify both its density function family and parameter θ_k. While BFS and BMA both utilize the data D to specify the density function family and the parameter θ_k, *the manners in which they address the two tasks are totally different however.*

For the task of specifying density function family, BMA utilizes the data D to calculate $P(S_k|D)$ and *assumes* that each S_k offers a density function family of $p(w(T)|S_k, D)$. But how the density function family is formed is beyond the BMA theory. In contrast, BFS constructs the density function family of $\pi(w(T)|y(T))$ and $\phi(w(T))$ in a typical Bayesian manner described in Eqs. (54.1) and (54.2). In short, while BMA *assumes* the density function family of $p(w(T)|S_k, D)$, BFS *constructs* it in a Bayesian manner. For the task of specifying parameter θ_k under each system (model) S_k, BFS assumes no parameter uncertainty and uses D to estimate a fixed but unknown value of θ_k. After estimating θ_k under each S_k, D has no more use to BFS. In contrast, BMA addresses the issue of parameter uncertainty and treats Θ_k in a typical Bayesian manner that is described in Eqs. (54.3) and (54.4). While BFS *estimates* the fixed parameter θ_k, BMA *models* the variate Θ_k in a Bayesian manner.

Summarizing the above discussions, the $p(w(T)|S_k, D)$ in BMA is a posterior density function in the sense of utilizing the data D to construct the parameter, but not in the sense of using the data D to construct the density function family. The $\pi(w(T)|y(T))$ and $\phi(w(T))$ in BFS are posterior density function and predictive density function in the opposite sense. In the following sections of this paper, we propose and investigate a new theoretical framework and data analysis method called *Bayesian System Averaging (BSA)* that unifies both BFS and BMA.

54.2 Theoretical Framework of BSA

In this section, we first present the theoretical framework of BSA in the context of without considering parameter uncertainty and then expand the results to the context of considering parameter uncertainty.

54.2.1 BSA Without Parameter Uncertainty

Under the theoretical framework of BSA, the coupled posterior density function $c\pi(w(T)|D, y(T), \{S_k\})$ (short-term forecast) and the coupled predictive density function $c\phi_T(w(T)|D, \{S_k\})$ (long-term forecast) are constructed as:

$$c\pi(w(T)|D, y(T), \{S_k\}) = \sum_{k=1}^{K} \pi_k(w(T)|y_k(T)) \, P(S_k|D), \tag{54.6}$$

where

$$P(S_k|D) = \frac{p(D|S_k)\ P(S_k)}{\sum_{l=1}^{K} p(D|S_l)\ P(S_l)}, \tag{54.7}$$

where

$$p(D|S_k) = \prod_{i=0}^{T-1} \pi_k(w(i)|y_k(i)), \tag{54.8}$$

$$\pi_k(w(i)|y_k(T), S_k) = \frac{f_k(y_k(i)|w(i))\ g_k(w(i))}{\kappa_k(y_k(i))}, \text{for } i = 1, \ldots, T; \tag{54.9}$$

and

$$c\phi_T(w(T)|D, \{S_k\}) = \sum_{k=1}^{K} \phi_{kT}(w(T))\ P(S_k|D), \tag{54.10}$$

where

$$P(S_k|D) = \frac{p(D|S_k)\ P(S_k)}{\sum_{l=1}^{K} p(D|S_l)\ P(S_l)}, \tag{54.11}$$

where

$$p(D|S_k) = \prod_{i=0}^{T-1} \phi_{ki}(w(i)), \tag{54.12}$$

$$\phi_{ki}(w(i)) = \int \pi_k(w(i)|y_k(i))\ \lambda_{ki}(y_k(i))\ dy_k(i), \text{for } i = 1, \ldots, T. \tag{54.13}$$

Notice:

The $p(D|S_k)$ in Eq. (54.8) cannot be defined as $p(D|S_k) = \prod_{i=0}^{T-1} f_k(y_k(i)|w(i))g_k(w(i))$ nor can the $p(D|S_k)$ in Eq. (54.12) be defined as $\prod_{i=0}^{T-1} \pi_k(w(i)|y_k(i))\lambda_{ki}(y_k(i))$, because, under the BFS framework, the forecasting system S_k includes not only the prior density function $g_k(w(i))$, the likelihood function $f_k(y_k(i)|w(i))$, and the marginal density function $\lambda_{ki}(y_k(i))$ but also *a specific way to process* $g_k(w(i))$ and $f_k(y_k(i)|w(i))$ that is elaborated in Eq. (54.1), as well as *a specific way to process* $\lambda_{ki}(y_k(i))$ and $\pi_k(w(i)|y_k(i))$ that is elaborated in Eq. (54.2).

54.2.2 *BSA with Parameter Uncertainty*

Now we expand the results in Sect. 54.2.1 to the context of *considering parameter uncertainty* and derive the coupled posterior density function and the coupled predictive density function, respectively, as follows:

$$
\begin{aligned}
c\pi(w(T)|D, y(T), \{S_k\}) &= \sum_{k=1}^{K} \pi_k(w(T)|D, y_k(T), S_k)\, P(S_k|D) \\
&= \frac{1}{p(D)} \sum_{k=1}^{K} \int \pi_k(w(T)|y_k(T), \theta_k) \prod_{i=0}^{T-1} \pi_k(w(i)|y_k(i), \theta_k) \\
&\quad p(\theta_k|S_k)\, d\theta_k \cdot P(S_k),
\end{aligned}
\tag{54.14}
$$

where

$$
\begin{aligned}
p(D) &= \sum_{k=1}^{K} p(D|S_k) P(S_k) \\
&= \sum_{k=1}^{K} \left(\int \left[\prod_{i=0}^{T-1} \pi_k(w(i)|y_k(i), \theta_k) \right] p(\theta_k|S_k)\, d\theta_k \right) P(\ S_k).
\end{aligned}
\tag{54.15}
$$

$$
\begin{aligned}
c\phi_T(w(T)|D, \{S_k\}) &= \sum_{k=1} \phi_{kT}(w(T)|D, (T), S_k)\, P(S_k|D) \\
&= \frac{1}{p(D)} \sum_{k=1}^{K} \int \phi_{kT}(w(T)|\theta_k) \prod_{i=0}^{T-1} \phi_{ki}(w(i)|\theta_k) p(\theta_k|S_k)\, d\theta_k \cdot P(S_k),
\end{aligned}
\tag{54.16}
$$

for

$$
p(D) = \sum_{k=1}^{K} \left(\int \left[\prod_{i=0}^{T-1} \phi_{ki}(w(i)|\theta_k) \right] p(\theta_k|S_k)\, d\theta_k \right) P(S_k).
\tag{54.17}
$$

Proof See Appendix.

54.3 Basic Properties of BSA

In this section, we first choose to present some basic properties of the BSA without considering parameter uncertainty. Two extreme situations are investigated: uninformative forecaster and forecaster with perfect model. Then we expand our investigations to the context of full parameter uncertainty.

54.3.1 Without Parameter Uncertainty

Definition 1 Under the BFS framework, the forecasting system S_k is *uninformative (without parameter uncertainty)* if $f_k(y_k(i)| w(i)) = f_k(y_k(i))$ for all $i = 1,\ldots,T$. Under the BSA framework, the pool of the forecasting systems $\{S_k\}$ is *uninformative (without parameter uncertainty)* if each forecasting system S_k uses a different predictor variate Y_k, but the likelihood functions $f_k(y_k(i)| w(i)) = f_k(y_k(i))$ for all $k = 1,\ldots,K$; $i = 1,\ldots,T$.

Theorem 1 *Without considering parameter uncertainty, BSA framework generalizes BMA framework. The latter is just a special case of the former when the pool of forecasting system $\{S_k\}$ is uninformative.*

Proof See Appendix.

Definition 2 Without considering parameter uncertainty, a forecasting system S_k^M is of *perfect model (is a short-term clairvoyant)* if a deterministic one-to-one relationship d_k exists between the predictand realization $w(i)$ and the predictor realization $y_k(i)$ forever, i.e., $w(i) = d_k(y_k(i))$ for all $i's$.

Theorem 2 *Without considering parameter uncertainty, BSA framework generalizes BMA framework. The latter is just a special case of the former when every forecasting system in the pool $\{S_k\}$ is of perfect model and has the marginal density functions $\lambda_{ki}(y_k(i))$ of the predictor $Y_k(i)$ being stationary* across all *$i's$ for $i = 1,\ldots,T$.*

Proof See Appendix.

54.3.2 Considering Parameter Uncertainty

Definition 3 In the context of parameter uncertainty, under the BFS framework, the forecasting system S_k is uninformative if $f_k\left(y_k(i)|w(i),\theta_k^f\right) = f_k\left(y_k(i)|\theta_k^f\right)$ for all values of $w(i)$ and all values of likelihood parameters θ_k^f. Under the BSA framework, the pool of the forecasting systems $\{S_k\}$ is uninformative if, for each forecasting system S_k, we may have a different predictor variate Y_k, but $f_k\left(y_k(i)|w(i),\theta_k^f\right) = f_k\left(y_k(i)|\theta_k^f\right)$ for all different values of $w(i)$ and θ_k^f and all k.

Theorem 3 *Considering parameter uncertainty, BSA framework generalizes BMA framework. The latter is just a special case of the former when the pool of forecasting system $\{S_k\}$ is uninformative.*

Proof See Appendix.

Definition 4 In the context of parameter uncertainty, under the BFS framework, the forecasting system S_k^M is of perfect model *(is a short-term clairvoyant)* if a deterministic one-to-one relationship d_k exists between the predictand realization $w(i)$ and the predictor realization $y_k(i)$ forever, i.e., $w(i) = d_k(y_k(i))$, for all $i's$ and for all values of likelihood parameters θ_k^f.

Theorem 4 *Considering parameter uncertainty, BSA framework generalizes BMA framework. The latter is just a special case of the former when every forecasting system in the pool* $\{S_k\}$ *is of perfect model and has the marginal density functions* $\lambda_{ki}(y_k(i))$ *of the predictor* $Y_k(i)$ *as well as the prior density function of* θ_k^λ *being stationary* across all $i's$ *for* $i = 1,\ldots,T$.

Proof See Appendix.

54.4 Notes and Conclusion

BSA models have closed-form solutions if the individual BFS models that are coupled have closed-form solutions. By far, we already know that the closed-form solutions are available to several BFS parametric models *without considering parameter uncertainty*. One type is the Gaussian BFS proposed by Krzysztofowicz [2] that has a Gaussian prior density function of the predictand, a Gaussian marginal density function of the predictor, and a normal-linear likelihood functions of the predictand given the predictor. Another type is the BFS model that has a uniform prior density function of the predictand, a uniform marginal density function of the predictor, and various triangular likelihood functions of the predictand given the predictor [6]. Based on these models and using the normal quantile transformation (NQT) technique and the probability integral transformation (PIT) technique, we can derive the much more general meta-Gaussian BFS models and copula-triangular BFS models. These models generate closed-form solutions for the Bayesian posterior density functions and allow any arbitrary continuous distributions for the predictand and predictor, as well as the normal-linear likelihood function or the triangular likelihood functions of the predictand given the predictor in the transformed space after NQT or PIT. The availability of closed-form solutions exempts the calculation of the probabilistic forecasts from numerical integration or the Monte Carlo solution and provides BSA great computational advantage in forecasting practices involving big data analysis and when the computation time and capability are limited.

When parameter uncertainty is considered, the closed-form solutions for BFS models are hardly available (though not completely unavailable). However, the BSA theoretical framework offers a strategy, which is infeasible under the BFS theoretical framework, to *indirectly* address the issue of parameter uncertainty. If we assume the prior density function $p(\theta_k | S_k)$ of the parameter Θ_k given the system S_k to be a discrete function, under such discrete assumption on the parameter Θ_k, a model with a different parameter value could be treated as a distinctive forecasting system and coupled under the BSA framework.

In conclusion, BSA postulates a new theoretical framework and data analysis method for constructing probabilistic forecast given the forecasting system pool $\{S_k\}$ and the historical data D. Comparing to those generated by BFS and BMA methods, the probabilistic forecast generated by BSA method is even more "Bayesian," since the set of the "prior information" BSA explicitly consider is the union of those of BFS and BMA, and the set of the probabilistic forecast components that are revised by BSA in Bayesian manners is even larger than the union of those by BFS and BMA. With and without considering parameter uncertainty, BMA and BFS theories are just special cases of BSA theory with different assumptions: BSA degenerates into BMA in at least two situations—(a) when the pool of forecasting systems, $\{S_k\}$, is uninformative and (b) when every forecasting system is of perfect model and the uncertainty regarding the predictor is stationary. BSA also degenerates into BFS when the pool of forecasting systems $\{S_k\}$ contains only one forecasting system and the corresponding parameters are fixed. BSA is a grand unified theory that generalizes and improves both BMA and BFS, preserving all their major theoretical advantages and disenchanting their major theoretical disadvantages.

Acknowledgment The content of this paper is an extract from the author Jie Liu's Ph.D. dissertation. During the process of developing the BSA theoretical framework, his mentor, Professor Roman Krzysztofowicz, contributed numerous insightful comments that help shaping the BSA theory in its current form, especially pointing out the correct ways to define $p(D|S_k)$, as discussed at the end of Sect. 54.2.1.

Appendix

Proof of Eqs. (54.14)–(54.17)

$$\begin{aligned}
c\pi(w(T)|D, y(T), \{S_k\}) &= \sum_{k=1}^{K} \pi_k(w(T)|D, y_k(T), S_k)\, P(S_k|D) \\
&= \sum_{k=1}^{K} \int \pi_k(w(T)|y_k(T), \theta_k)\, p_k(\theta_k|D)\, d\theta_k \cdot P(S_k|D) \\
&= \sum_{k=1}^{K} \int \pi_k(w(T)|y_k(T), \theta_k)\, \frac{p_k(D|\theta_k)\, p(\theta_k|S_k)}{p(D|S_k)}\, d\theta_k \cdot \frac{p(D|S_k)P(S_k)}{p(D)} \\
&= \frac{1}{p(D)} \sum_{k=1}^{K} \int \pi_k(w(T)|y_k(T), \theta_k) \prod_{i=0}^{T-1} \pi_k(w(i)|y_k(i), \theta_k) p(\theta_k|S_k)\, d\theta_k \cdot P(S_k).
\end{aligned} \tag{54.18}$$

In a similar manner, we can derive,

$$
\begin{aligned}
c\phi_T(w(T)|D,\{S_k\}) &= \sum_{k=1}^{K}\phi_{kT}(w(T)|D,(T),S_k)\,P(S_k|D)\\
&= \frac{1}{p(D)}\sum_{k=1}^{K}\int \phi_{kT}(w(T)|\theta_k)\prod_{i=0}^{T-1}\phi_{ki}(w(i)|\theta_k)p(\theta_k|S_k)\,d\theta_k\cdot P(S_k).
\end{aligned}
\tag{54.19}
$$

Q. E. D.

Proof of Theorem 1

Since $f_k(y_k(i)|\, w(i)) = f_k(y_k(i))$ for all $k = 1,\ldots,K$; $i = 1,\ldots,T$, by Eqs. (54.1) and (54.2), we have

$$
\pi_k(w(i)|y_k(i)) = \frac{f_k(y_k(i)|w(T))\,g_k(w(i))}{\kappa_k(y_k(i))} = g_k(w(i)), \tag{54.20}
$$

$$
\begin{aligned}
\phi_{ki}(w(i)\;|S_k\;) &= \int \pi_k(w(i)|y_k(i))\,\lambda_{ki}(y_k(i))\,dy_k(i)\\
&= \int g_k(w(i))\lambda_{ki}(y_k(i))dy_k(i) = g_k(w(i)).
\end{aligned}
\tag{54.21}
$$

By Eqs. (54.6)–(54.13), we have:

$$
\begin{aligned}
c\pi(w(T)|D,y(T),\{S_k\}) &= c\phi_T(w(T)|D,\{S_k\})\\
&= \sum_{k=1}^{K} g_k(w(T))\frac{\prod_{i=0}^{T-1} g_k(w(i))\,P(S_k)}{\sum_{l=1}^{K}\left[\prod_{i=0}^{T-1} g_l(w(i))\right]P(S_l)}.
\end{aligned}
\tag{54.22}
$$

Comparing Eq. (54.3) with Eq. (54.22), they are essentially the same.
Q. E. D.

Proof of Theorem 2

Under the BSA theoretical framework, if the forecasting systems $\{\boldsymbol{S_M}: \boldsymbol{M} = \mathbf{1},\ldots, \boldsymbol{J}\}$ are all of perfect models, while the systems $\{\boldsymbol{S_N}: \boldsymbol{N} = \boldsymbol{J} + \mathbf{1},\ldots, \boldsymbol{K}\}$ are not, and $\boldsymbol{S_m}$ is one of $\{\boldsymbol{S_M}\}$, while $\boldsymbol{S_n}$ is one of $\{\boldsymbol{S_N}\}$, then for the construction of the coupled predictive density function $\boldsymbol{c\phi(w(T)|D}, \{\boldsymbol{S_k}\})$, we have:

$$\phi_{mi}(w(i)) = \lambda_{mi}\left(d_m^{-1}(w(i))\right)\left|\frac{d}{dw(i)}d_m^{-1}(w(i))\right| = \lambda'_{mi}(w(i)), \tag{54.23}$$

$$P(S_m|D) = \frac{\left[\prod_{i=0}^{T-1}\lambda'_{mi}(w(i))\right]P(S_m)}{\sum_{M=1}^{J}\left[\prod_{i=0}^{T-1}\lambda'_{Mi}(w(i))\right]P(S_M) + \sum_{N=J+1}^{K}\left[\prod_{i=0}^{T-1}\phi_{Ni}(w(i))\right]P(S_N)}, \tag{54.24}$$

$$P(S_n|D) = \frac{\left[\prod_{i=0}^{T-1}\phi_{ni}(w(i))\right]P(S_n)}{\sum_{M=1}^{J}\left[\prod_{i=0}^{T-1}\lambda'_{Mi}(w(i))\right]P(S_M) + \sum_{N=J+1}^{K}\left[\prod_{i=0}^{T-1}\phi_{Ni}(w(i))\right]P(S_N)}, \tag{54.25}$$

$$\begin{aligned} c\phi_T(w(T)|D, \{S_k\}) = &\sum\nolimits_{M=1}^{J}\lambda'_{MT}(w(T))P(S_M|D) \\ &+ \sum\nolimits_{N=J+1}^{K}\phi_{NT}(w(T))P(S_N|D). \end{aligned} \tag{54.26}$$

If all the K forecasting systems in the pool $\{S_k\}$ are with perfect model, we have:

$$c\phi_T(w(T)|D, \{S_k\}) = \sum\nolimits_{m=1}^{K}\lambda'_{mT}(w(T))\frac{\left[\prod_{i=0}^{T-1}\lambda'_{mi}(w(i))\right]P(S_m)}{\sum_{M=1}^{K}\left[\prod_{i=0}^{T-1}\lambda'_{Mi}(w(i))\right]P(S_M)}. \tag{54.27}$$

Comparing Eq. (54.3) with Eq. (54.27), they are essentially the same, *as long as the input uncertainty described by* λ'_{mi} *is stationary*, i.e., $\lambda'_{mi} = \lambda_m$ for all i's.

Q. E. D.

Proof of Theorem 3

Since $f_k\left(y_k(i)|w(i), \theta_k^f\right) = f_k\left(y_k(i)|\theta_k^f\right)$ for all different values of $w(i)$ and θ_k^f and all k's, by Eqs. (54.1) and (54.2), we have:

$$\kappa_k(y_k(i)|\theta_k) = \int f_k\left(y_k(i)|\theta_k^f\right) g_k\left(w(i)|\theta_k^g\right) dw(i) = f_k\left(y_k(i)|\theta_k^f\right), \tag{54.28}$$

$$\pi_k(w(i)|y_k(i),\theta_k) = \frac{f_k\left(y_k(i)|w(i),\theta_k^f\right)\, g_k\left(w(i)|\theta_k^g\right)}{\kappa_k(y_k(i)|\theta_k)}$$
$$= \frac{f_k\left(y_k(i)|\theta_k^f\right)\, g_k\left(w(i)|\theta_k^g\right)}{f_k\left(y_k(i)|\theta_k^f\right)} = g_k\left(w(i)|\theta_k^g\right), \tag{54.29}$$

$$\phi_{ki}(w(i)|\theta_k) = \int g_k\left(w(i)|\theta_k^g\right)\, \lambda_{ki}\left(y_k(i)|\theta_k^\lambda\right)\, dy_k(i) = g_k\left(w(i)|\theta_k^g\right). \tag{54.30}$$

By Eqs. (54.14)–(54.17), we have:

$$c\pi(w(T)|D, y(T), \{S_k\}) = \sum_{k=1}^{K} \pi_k(w(T)|D, y_k(T), S_k)\, P(S_k|D)$$
$$= \frac{1}{p(D)} \sum_{k=1}^{K} \left(\int g_k\left(w(i)|\theta_k^g\right) \prod_{i=0}^{T-1} g_k\left(w(i)|\theta_k^g\right) p\left(\theta_k^g|S_k\right)\, d\theta_k^g \right) \cdot P(S_k), \tag{54.31}$$

for

$$p(D) = \sum_{k=1}^{K} \left(\int \left[\prod_{i=0}^{T-1} g_k\left(w(i)|\theta_k^g\right) \right] p\left(\theta_k^g|S_k\right)\, d\theta_k^g \right) P(S_k). \tag{54.32}$$

$$c\phi_T(w(T)|D, \{S_k\}) = \sum_{k=1}^{K} \phi_{kT}(w(T)|D, (T), S_k)\, P(S_k|D)$$
$$= \frac{1}{p(D)} \sum_{k=1}^{K} \left(\int g_k\left(w(i)|\theta_k^g\right) \prod_{i=0}^{T-1} g_k\left(w(i)|\theta_k^g\right) p\left(\theta_k^g|S_k\right)\, d\theta_k^g \right) \cdot P(S_k), \tag{54.33}$$

for

$$p(D) = \sum_{k=1}^{K} \left(\int \left[\prod_{i=0}^{T-1} g_k\left(w(i)|\theta_k^g\right) \right] p\left(\theta_k^g|S_k\right)\, d\theta_k^g \right) P(S_k). \tag{54.34}$$

Comparing Eqs. (54.4) and (54.5) with Eqs. (54.31) and (54.32) and with Eqs. (54.33) and (54.34), they are essentially the same.

Q. E. D.

Proof of Theorem 4

Since in the pool of forecasting systems, $\{S_k\}$, all forecasting systems are of perfect model, by Eqs. (54.1) and (54.2), we have:

$$\begin{aligned}\kappa_k(y_k(i)|\theta_k) &= \int f_k(y_k(i)|w(i),\theta_k)\, g_k(w(i)|\theta_k)\, dw(i) \\ &= \int \delta(w(i) - d_k(y_k(i)))g_k(w(i)|\theta_k)\, dw(i) \\ &= g_k(d_k(y_k(i))|\theta_k),\end{aligned} \tag{54.35}$$

where δ is the impulse function.

$$\pi_k(w(i)|y_k(i),\theta_k) = \frac{f_k(y_k(i)|w(i),\theta_k)\, g_k(w(i)|\theta_k)}{\kappa_k(y_k(i)|\theta_k)}$$

$$\begin{aligned}&= \frac{f_k(y_k(i)|w(i),\theta_k)\, g_k(d_k(y_k(i)|\theta_k))}{g_k(d_k(y_k(i)|\theta_k))} = \delta(w(i) - d_k(y_k(i))) \\ &= \begin{cases} \infty, & \text{if } w(i) = d_k(y_k(i)), \\ 0 & \text{otherwise.}\end{cases}\end{aligned} \tag{54.36}$$

$$\begin{aligned}\phi_{ki}(w(i)|\theta_k) &= \int \pi_k(w(i)|y_k(i),\theta_k)\, \lambda_{ki}(y_k(i)|\theta_k)dy_k(i) \\ &= \int \delta(w(i) - d_k(y_k(i)))\, \lambda_{ki}(y_k(i)|\theta_k)dy_k(i) \\ &= \int \delta\left(d_k^{-1}(w(i)) - y_k(i)\right)\left|\frac{d}{dw(i)}y_k(i)\right|\, \lambda_{ki}(y_k(i))\, dy_k(i) \\ &= \lambda_{ki}\left(d_k^{-1}(w(i))|\theta_k\right)\left|\frac{d}{dw(i)}d_k^{-1}(w(i))\right| = \lambda_{ki}\left(d_k^{-1}(w(i))|\theta_k^{\lambda}\right)\left|\frac{d}{dw(i)}d_k^{-1}(w(i))\right| \\ &= \lambda_{ki}^{'}\left(w(i)|\theta_k^{\lambda}\right).\end{aligned} \tag{54.37}$$

By Eqs. (54.16) and (54.17), we have:

$$\begin{aligned}c\phi(w(T)|D,\{S_k\}) &= \frac{1}{p(D)}\sum_{k=1}^{K}\left(\int \phi_{kT}(w(T)|\theta_k)\prod_{i=0}^{T-1}\phi_{ki}(w(i)|\theta_k)p(\theta_k|S_k)\, d\theta_k\right)\cdot \\ P(S_k) &= \frac{1}{p(D)}\sum_{k=1}^{K}\left(\int \lambda_{kT}^{'}(w(T)|\theta_k^{\lambda})\prod_{i=0}^{T-1}\lambda_{ki}^{'}(w(i)|\theta_k^{\lambda})p(\theta_k^{\lambda}|S_k)\, d\theta_k^{\lambda}\right)\cdot P(S_k),\end{aligned} \tag{54.38}$$

for

$$p(D) = \sum_{k-1}^{K}\left(\int\left[\prod_{i=0}^{T-1}\lambda_{ki}^{'}(w(i)|\theta_k^{\lambda})\right]p(\theta_k^{\lambda}|S_k)\, d\theta_k^{\lambda}\right)\cdot P(\, S_k). \tag{54.39}$$

Comparing Eqs. (54.4) and (54.5) with Eqs. (54.38) and (54.39), they are essentially the same, *as long as the input uncertainty described by* λ'_{ki} *is stationary*, i.e., $\lambda'_{ki} = \lambda_k$ for all $i's$.

Q. E. D.

References

1. Gneiting, T. (2008). Editorial: probabilistic forecasting. *Journal of the Royal Statistical Society: Series A, 171*, 319–321.
2. Krzysztofowicz, R. (1999). Bayesian forecasting via deterministic model. *Risk Analysis, 19*(4), 739–749.
3. Gneiting, T., & Ranjan, R. (2013). Combining predictive distributions. *Electronic Journal of Statistics, 7*, 1747–1182.
4. Hoeting, J. A., Madigan, D., Raftery, A. E., & Volinsky, C. T. (1999). Bayesian model averaging: A tutorial (with Discussion). *Statistical Science, 14*(4), 382–401.
5. Berger, J. O. (1985). *Statistical decision theory and bayesian analysis* (2nd ed.). New York: Springer.
6. Liu, J. (2018). Bayesian system averaging: A grand theory unifying bayesian forecasting system and bayesian model averaging (Dissertation for the Ph.D. degree in Systems Engineering, Department of Systems and Information Engineering, University of Virginia).

Chapter 55
How Will Humans Cut Through Automated Vehicle Platoons in Mixed Traffic Environments? A Simulation Study of Drivers' Gaze Behaviors Based on the Dynamic Areas of Interest

Xiang Guo, Lian Cui, Brian Park, Wen Ding, Marisa Lockhart, and Inki Kim

55.1 Introduction

55.1.1 Automated Driving and Human Factor Issues

The last few years have witnessed the rapid development of automation in vehicles. However, many steps and advancements in technology will be required before the widespread use of high-performance automated vehicles could possibly occur. Cooperative Adaptive Cruise Control (CACC) is one of these technologies that would be a step toward this goal of creating fully automated vehicles. The idea behind CACC technology is to create connectivity between vehicles using the conjunction of connected and automated vehicle technology in order to synchronize speeds and reduce the time headway (THW) between vehicles. Time headway is measured by taking the time that passes between two vehicles reaching the same location: THW $= t_i - t_{i-1}$, with t_i denoting the time at which the vehicle i passes a certain location, and t_{i-1}, the time at which the vehicle ahead of vehicle i passes the same location [1]. Shorter THW can effectively increase the traffic capacity and lower energy consumption [2]. Fully automated driving is still a long way to go, while several human factor issues remain unresolved.

X. Guo · W. Ding · M. Lockhart · I. Kim (✉)
Department of Systems and Information Engineering, University of Virginia, Charlottesville, VA, USA
e-mail: ik3r@virginia.edu

L. Cui · B. Park
Department of Civil and Environmental Engineering, University of Virginia, Charlottesville, VA, USA

S. Adams et al. (eds.), *Systems Engineering in Context*,
https://doi.org/10.1007/978-3-030-00114-8_55

Inside the Automated Vehicle Problems include overreliance or mistrust on automation, loss of situational awareness, and degradation of driving skills [3], especially when there is a transition between automated driving and manual driving modes. Studies have shown that in highly automated vehicles, drivers tend to do more nondriving-related tasks and take-over driving performance differs from various situations [4].

Outside the Automated Vehicle A mixed traffic situation will emerge when vehicles with automated driving systems interact with traditional vehicles that are not fitted with such automated systems. Some researchers emphasized the central role of human factor discipline in addressing the interaction between automated vehicles, human drivers, and vulnerable road users [5]. A simulation study of analyzing the human drivers' longitudinal control found that the drivers would display a significantly shorter time headway (THW) while driving next to a platoon holding shorter THWs, compared with a platoon whose time headway is set larger [6]. To our knowledge, no study in lateral control in relation to the field has been found.

55.1.2 Eye Tracking as a Behavior Indicator

Apart from driving performance, eye-tracking data is also an important indicator of users' behaviors. Eye tracking for the analysis of gaze behavior is common in much scientific and marketing research, such as human-computer interaction [7]. In more complicated systems, which require users to monitor different regions or handle multiple controllers, eye-tracking data can provide critical information about human behaviors. For example, gaze at the cockpit can indicate the operations of the pilots' in the flight deck [8]. In digital nuclear power plants, researches have shown that eye response measures are sensitive mental workload change [9].

Eye-tracking data is also widely used in human factors in transportation research field. For example, eye-tracking data can be used to evaluate driver distraction [10]. Studies have shown that eye-tracking data can be an effective tool to evaluate driver behavior. In general, people quickly move their eyes across their visible field of view so that they may scan various portions in finer detail. People also divert their attention away from one portion of their visible field to another, indicating changes in mental concentration through an area of interest (AOI).

A proper definition and tracking of AOIs can give quantitative insights into what the observer finds meaningful in the visual scene with respect to the viewer's task performance [11]. By mapping with the AOIs, it is possible to obtain a statistical description of key gaze parameters, which can help explain and interpret users' gaze behavior objectively [12]. One of the more intuitive measures of changes in drivers' gaze at road center is the percentage of road center (PRC). PRC is defined as the percentage of fixations that fall within a predefined road center area during a specific period. PRC has been shown to increase with heightened cognitive demand [13].

Currently the video-based eye-tracking system is the most popular one, which can be divided into two categories: table mounted and head mounted. Table-mounted tracking consists of fixed cameras on the computer screen or on the table. The fixed position allows for easily definable AOIs. However, it is not flexible because the cameras should be attached firmly to the table all the time to avoid the need to recalibrate. Head-mounted eye tracking, which have cameras integrated in a pair of glasses, can overcome the shortcomings mentioned above and provide more flexibility and wider perspective. Compared to static stimuli such as images from fixed-base cameras, the head-mounted eye tracker outputs a relative coordinate system, which makes it hard to define dynamic AOIs (AOIs' size and position will change over time) in long videos.

However, currently available dynamic AOIs come with certain limitation of accuracy in automated detection and recognition. Alternatively, the most previous researcher manually defined AOIs for specific scenarios to assure accuracy. This proved to be highly time-consuming among several researchers and increases the likelihood of annotation inconsistencies. An automatic and accurate way to detect and track AOIs is needed in related research. Apart from manual annotation, there are some semiautomatic or automatic methods that exist. However, to acquire sufficient accuracy, specific markers were used in the experiment in order to be detected [14]. Marker-based method is hard to be applied in general because of the limitation in the numbers and positions of the markers. In addition, the eye-catching markers can be distractive, especially in some visual distraction research [15]. Another automatic method to define AOIs is clustering the raw gaze data without the videos. By clustering all the gaze points into different regions and then mapping them with the definition, it is quicker and can include the gaze points near the AOIs that were not picked up due to calibration error. However, the result of this approach depends highly on the experiment design. This method also has to be double-checked in the videos to assign the right labels, which in return may be more time-consuming [16].

A more practical way is machine learning and computer vision-based method. From some training data with proper definition of AOIs, we can train the detection model and then apply it to all the videos or images. In order to get a satisfying detection result, the training phase is required, which may take some time before it reaches the optimization. The computer vision-based method is more suitable for specific object, such as words in the scene [17]. More importantly, with more training data, machine learning and computer vision-based method is likely to have better performance; there is potential for it to solve the dynamic AOI problems.

55.1.3 Deep Learning Application to AOI Tracking

Deep learning (DL), part of a broader family of machine learning methods, has recently attempted to learn high-level abstractions in data by utilizing hierarchical architectures. It has made great success in computer vision, and it has been predicted

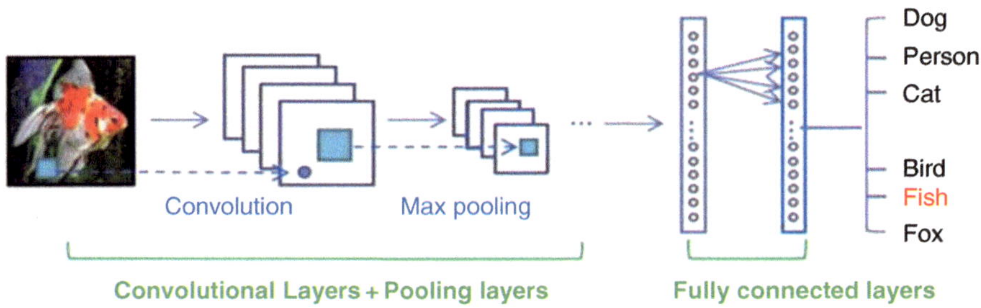

Fig. 55.1 Pipeline of the general CNN architecture [20]

that DL will have many more successes in the near future because it requires very little engineering by hand, so it can easily take advantage of increases in the amount of available computation and data [18]. In recent ImageNet Large-Scale Visual Recognition Challenge (ILSVRC) competitions [19], deep learning methods have been the most widely adopted methods and have achieved top accuracy scores in the rank. The convolutional neural network (CNN) is one of the most notable deep learning approaches where multiple layers are trained in a robust manner. It has been found highly effective and is also the most commonly used in diverse computer vision applications. A CNN consists of three main neural layers, which are convolutional layers, pooling layers, and fully connected layers. Different kinds of layers play different roles. There are two stages for training the network: a forward stage and a backward stage. First, the main goal of the forward stage is to represent the input image with the current parameters (weights and bias) in each layer. Then the prediction output is used to compute the loss cost with the ground truth labels. Second, based on the loss cost, the backward stage computes the gradients of each parameter with chain rules. All the parameters are updated based on the gradients and are prepared for the next forward computation. After sufficient iterations of the forward and backward stages, the network learning can be stopped [20]. The pipeline of the general CNN architecture is shown in Fig. 55.1. Detecting and tracking AOIs can be regarded as a computer vision problem, and DL would be a viable option to solve it.

Other traditional machine algorithms have been applied to human factor issues in transportation. For example, support vector machine (SVM) and random forest (RF) can be effective to estimate driver cognitive workload [21, 22]. Accuracy of these methods is still not high enough, since DL methods outperform other machine learning methods in a public large datasets. Another big advantage of deep learning is that we do not need to manually extract features from the images. The network learns to extract all the effective features during training using its convolution kernels. In fact, deep learning has been widely used in automated driving to perform lane and vehicle detection while running at frame rates required for a real-time system [23]. Disadvantages of deep learning include longer training time and requirement of high computational power. However, with the dramatically increased chip processing abilities (e.g., GPU units), the significantly lowered cost of computing hardware, and the considerable advances in the machine learning algorithms, the disadvantages mentioned are not so critical. Fortunately, many open-source platforms and libraries (TensorFlow, Caffe2, MXNet, etc.) integrated with latest powerful object detection algorithms are available for researchers from other fields. Since DL requires a lot of computational power, determining if a low-cost computer is suitable for AOI tracking still needs to be verified.

55.1.4 Aims of the Study

In this driving simulation study, the major goal was to study human performance when cutting through automated vehicle platoons in mixed traffic environments (both automatic and semiautomatic vehicles) in the perspective of gaze behaviors at dynamic AOIs. To achieve it, the first step was tracking the dynamic AOIs. We explored the effectiveness of DL as an automatic way to AOI tracking in eye-tracking videos, considering the trade-off between speed and accuracy. Only when high accuracy in AOI tracking was achieved could we combine gaze points with AOIs to get a deeper understanding of drivers' gaze patterns.

55.2 Methods

55.2.1 Experimental Design and Participants

A virtual driving simulation environment was created to replicate real-life interactions between human drivers and CACC vehicle platoons. The human drivers' task was to undergo merging lane scenarios of varying headways (0.6 s, 0.8 s, 1.0 s, 1.2 s, and 1.4 s) in order to see how they interact with CACC vehicles. Different time headway was the independent variable. Gaze behavior at AOIs (number of fixation times, average fixation time, percentage of dwelling time in AOIs) was treated as dependent variables. Ten engineering undergraduate students (6 males, 4 females) majoring in Computer Science and Systems, aged from 20 to 22 (mean = 20.9, SD = 0.53) from the University of Virginia, participated in the experiment with the average driving age of 4.6 years (SD = 1.76). The institutional review board approved the research for the Social and Behavioral Sciences (UVA-IRB 2017-0296-00).

55.2.2 Apparatus

Driving Simulator and Eye Tracker The experiment was conducted at the Department of Civil and Environmental Engineering, University of Virginia. The tools

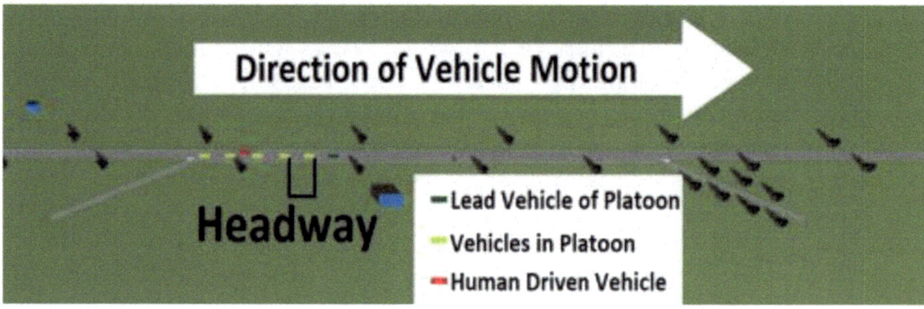

Fig. 55.2 Bird's-eye view of the experiment road

used consisted of a fixed-base driving simulator with an adjustable seat, a Logitech G27 gear, steering wheel accelerator and brake paddle suit, and four computer monitors to display the scenario. The SMI Eye Tracking Glasses 2 was utilized to record eye-tracking behavior. The eye tracker can also output a recording at a 30 Hz frequency with extracted multiple features such as pupil diameter and gaze vector.

Driving Scenario The road was designed as a simple two-lane road to avoid the influence of any other factors that might affect the simulation data results. The human driver's vehicle was initially in the left lane, while a platoon of 8 CACC vehicles was placed in the right lane. The goal for the human driver was to merge through the CACC platoon and exit into an exit ramp. The speed limit was set at 60 km/h to simulate driving on a highway road. There were no cars present on the road aside from the platoon and the human. PreScan was used to design the physical appearance of the driving scenario. Some simple aesthetic features like trees, buildings, and road signs were also added to simulate a real road environment. A bird's-eye view of the road section can be seen in Fig. 55.2.

The CACC vehicles were equipped with sensors (set in PreScan) that enabled them to collect information from each other. Each vehicle received data from the lead vehicle and the one immediately preceding it in the platoon, including its position, speed, and acceleration. This enabled each vehicle to travel safely and react to any changes at the front. When the human driver entered the platoon, the vehicle following would adjust its headway to account for the human driver.

55.2.3 Procedure

Before each subject came into the simulation room, they would be asked some questions about their demographic information. After a general introduction to the experiment, participants were asked to adjust the driving simulator seat to fit with their driving preference. They were then directed to wear the eye-tracking glasses. The first three simulation runs were base cases, in order to familiarize the driver with the simulation. This step can help to avoid mistakes of the operation during their lane change runs. After the three initial simulation runs were complete, the lane change trials began. The eye-tracking glasses were also used during these trials and calibrated before each new trial. The driver was asked to drive as safely and normally as possible, make a lane change through the CACC platoons, and eventually take the exit ramp. Ten to twelve runs were performed for each subject to ensure that each participant experienced multiple trials of each headway. Each trail lasted for about 3 min, and it took about 2 h for an individual to complete all the procedures.

55.3 Results

The data collected included eye-tracking videos and raw gaze point text files. The first step was to detect and track AOI. In this study, we chose two typical and important dynamic AOIs: the vehicle in front and the road center. After successfully tracking AOIs, we analyzed drivers' gaze behavior at these AOIs.

55.3.1 Dynamic AOI Track

The videos recorded by the SMI eye tracker had a video resolution of 960 × 720 (width × height) and a frequency of 30 Hz. Every frame from each video was extracted, and after manually checking the start and end time point, the entirety of the video was analyzed.

The vehicle in front was defined as the closest CACC vehicle in front of the human-driving vehicle. If the human driver overtook a vehicle, the AOI would switch to the vehicle in front of the overtaken vehicle. Namely, the primary platoon vehicle considered was the vehicle always in front of the human-driven vehicle. We applied a small but effective tip to find the road center. The angle of view from the cockpit in the simulator remained unchanged, which meant the center point of the middle monitor could always be regarded as the road center point. If we could track the position of the middle monitor in the video, the road center point could be determined as well, which is an easier way than to directly find the road center. Therefore, the task for finding the road center was changed to tracking the coordinates of the vehicle in front and the middle monitor in the videos.

Deployment of deep learning in TensorFlow, an open-source machine learning framework from Google [24], was deployed on a laptop with Microsoft Windows 10 operating system and NVIDIA GeForce MX150 graphics card with CUDA Toolkit 8.0 driver software. By inputting labelled training data, computers applied statistical learning techniques to automatically identify patterns in data and generate models for prediction. In our case, the two classes we needed were the vehicle in front and the middle monitor. TensorFlow supports data type (of the Pascal VOC

Table 55.1 Tracking performance definition of the vehicle in front

	Positive detection	Negative detection
Event true	Leading vehicle in the image was correctly detected (TP)	Leading vehicle in the image was not detected (FN)
Event false	No leading vehicle, the model detected other objects as vehicle (FP)	No leading vehicle, no detection as well (TN)

dataset), stored in each bounding box annotations' location, class, and image location on file in xml format.

AOI Track Result With all the data available, true positives, true negatives, false positives, and false negatives were recorded. In this case, for example, "event true/false" means whether there is a leading vehicle in the image; "positive/negative detection" distinguishes whether there is a detected box for the leading vehicle (see Table 55.1 for details of each example). Based on this definition, we have Accuracy = (TP + TN)/(TP + TN + FP + FN), Precision = TP/(TP + FP), and Recall = TP/(TP + FN). In our case, TN is always 0 because the middle monitor and leading vehicle are in the video at all times. A very high performance was achieved for tracking the middle monitor; the average accuracy for ten participants' videos was 99.85% (SD = 0.41%). Precision and recall were 99.98% (SD = 0.09%) and 99.87% (SD = 0.39%), respectively. When tracking the vehicle in front, the accuracy was 71.95% (SD = 14.57%), precision 99.73% (SD = 0.57%), and recall 72.15% (SD = 14.75%). It has to be noticed that the standard deviation is large, which indicates that there are wide individual differences.

55.3.2 *Gaze Behavior Analysis*

Gaze behavior analysis in this section was based on the AOI tracking results. Different time headway was treated as the sole independent variable. Gaze behavior (*number of fixation times, average fixation time, the percentage of dwelling time in AOIs*) at both AOIs (road center and the leading vehicle) during each experiment was analyzed as dependent variables. For road center gaze behavior, the perceived distance is important for the driver to cut through, while the road center can serve as a frame of reference for visual attention in a primary driving task.

Road Center AOI While PRC is conceptually easy to understand, the definition of road center varies considerably across studies. Previous studies compared the radius of the road center circle 8°, 12°, and 16°, and suggested shorter radiuses such as 8° are not optimal [25]. We chose 16° as the radius of the road center. No significance was found in PRC (F (4, 78) = 1.69, $p = 0.161$) or the number of fixation times (F (4, 78) = 0.47, $p = 0.759$). Results showed that different levels of THW can result in different average fixation times (F (4, 78) = 2.60, $p < 0.05$).

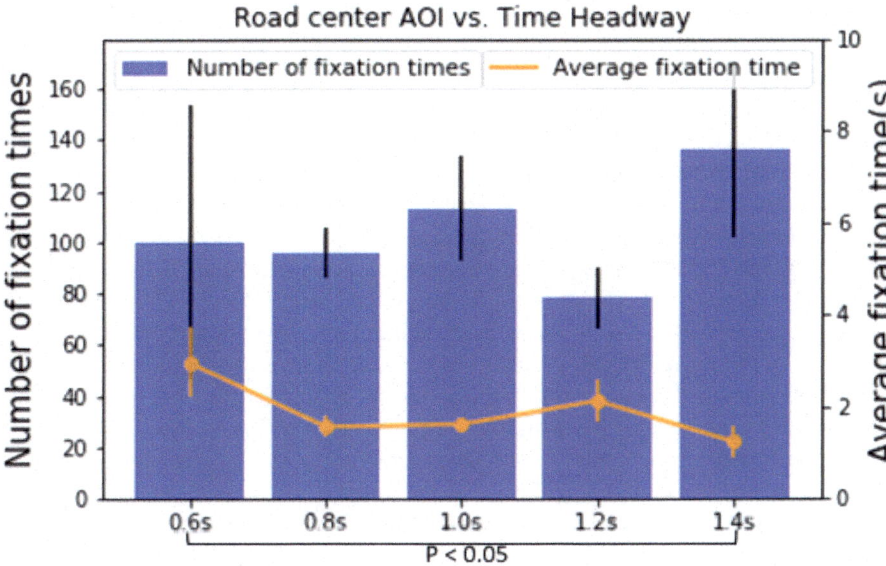

Fig. 55.3 Gaze behavior at road center AOI

Below is a pair-wise test comparing the very low THW 0.6 s and the very high THW 1.4 s ($p < 0.05$) (see Fig. 55.3).

Vehicle in Front AOI Considering the accuracy (71.95% ± 14.57%) of AOI of vehicle in front, seven trials' data with accuracy lower than 50% were not taken into analysis. Results showed that there were significant differences in the percentage of dwelling time in AOI between different THW ($F\,(4, 67) = 3.12$, $p < 0.05$). Though no further significance was found in the pair-wise test, however, with the increase in THW, percentage of dwelling time at the leading vehicle decreased (except for THW = 1.2 s). No significance was found for number of fixation times ($F\,(4, 75) = 0.93$, $p = 0.450$) and average fixation time ($F\,(4, 76) = 1.86$, $p = 0.127$).

55.4 Discussion and Summary

The detection and tracking results of dynamic AOI verified that deep learning could be a powerful way to detect any dynamic AOI in eye-tracking videos. For a relatively easier object such as the middle monitor in this study, deep learning can output near-perfect results. Having only used 1000 images from a single participant out of the 500,000+ collected, it is fair to say our results may not be used to represent all of the participants. Given the nature of deep learning, providing more representative training data could help improve the accuracy even further. Also, other slower but more accurate models from TensorFlow such as mask R-CNN can be applied to get better results. To collect more precise gaze behavior during analysis, a method for better AOI detection is required. Although to achieve 100% accuracy is not practical, this method of deep learning can provide rough results automatically, which could be beneficial to pilot analysis on dynamic AOI and reduce manual labor.

The analysis of gaze behavior did not show many significant results partially due to the limited number of participants, but some trends are still visible. With shorter THW, drivers frequently looked at the leading vehicle, as well as spent more time in each fixation; this means that drivers tend to allocate more visual attention to the CACC platoons if the THW is smaller. To human drivers, this may cause additional visual distraction. Another interesting finding is that with lower THW, the average dwelling time in road center increased and the number of gaze in and out of this AOI decreased, resulting in the PRC remaining at a certain level. It could be a compensation behavior for drivers. According to previous research, this phenomenon may be caused by raised cognitive workload [25]. Whether this would be beneficial to driving performance depends on the level of workload. Driving data was also collected from the driving simulator. In the future, we plan to include vehicle position to have a greater depth of understanding of drivers' behavior when driving through automated vehicle platoons.

Acknowledgments The authors would like to thank Katie Asmussen, Gabriella Greiner, Christian Halsey, Angely Jomuad, and Christian Vila from the Department of Civil and Environment Engineering at the University of Virginia for their help in data collection.

References

1. Vogel, K. (2003). A comparison of headway and time to collision as safety indicators. *Accident Analysis & Prevention, 35*(3), 427–433.
2. Cui, L., & Park, B. B. (2017). *Assessing eco-driving behaviors using driving simulator*. Transportation Research Record (No. 17-06132).
3. Parasuraman, R., Sheridan, T. B., & Wickens, C. D. (2000). A model for types and levels of human interaction with automation. *IEEE Transactions on Systems, Man, and Cybernetics-Part A: Systems and Humans, 30*(3), 286–297.
4. Merat, N., Jamson, A. H., Lai, F. C., & Carsten, O. (2012). Highly automated driving, secondary task performance, and driver state. *Human Factors, 54*(5), 762–771.
5. Kyriakidis, M., de Winter, J. C., Stanton, N., Bellet, T., van Arem, B., Brookhuis, K., et al. (2017). A human factors perspective on automated driving. *Theoretical Issues in Ergonomics Science*, 1–27. https://doi.org/10.1080/1463922X.2017.1293187
6. Gouy, M., Wiedemann, K., Stevens, A., Brunett, G., & Reed, N. (2014). Driving next to automated vehicle platoons: How do short time headways influence non-platoon drivers' longitudinal control? *Transportation Research Part F: Traffic Psychology and Behaviour, 27*, 264–273.
7. Poole, A., & Ball, L. J. (2006). Eye tracking in HCI and usability research. *Encyclopedia of Human Computer Interaction, 1*, 211–219.
8. Weibel, N., Fouse, A., Emmenegger, C., Kimmich, S., & Hutchins, E. (2012). Let's look at the cockpit: Exploring mobile eye-tracking for observational research on the flight deck. In *Proceedings of the Symposium on Eye Tracking Research and Applications* (pp. 107–114). ACM.

9. Gao, Q., Wang, Y., Song, F., Li, Z., & Dong, X. (2013). Mental workload measurement for emergency operating procedures in digital nuclear power plants. *Ergonomics, 56*(7), 1070–1085.
10. Hurtado, S., & Chiasson, S. (2016). An eye-tracking evaluation of driver distraction and unfamiliar road signs. In *Proceedings of the 8th International Conference on Automotive User Interfaces and Interactive Vehicular Applications* (pp. 153–160). ACM.
11. Duchowski, A. T. (2007). *Eye tracking methodology. Theory and practice* (p. 328). Berlin, Germany: Springer.
12. Xu, Q., Guo, T. Y., Shao, F., & Jiang, X. J. (2017). Division of area of fixation interest for real vehicle driving tests. *Mathematical Problems in Engineering, 2017*, 10.
13. Engström, J., Johansson, E., & Östlund, J. (2005). Effects of visual and cognitive load in real and simulated motorway driving. *Transportation Research Part F: Traffic Psychology and Behaviour, 8*(2), 97–120.
14. Pfeiffer, T., Renner, P., & Pfeiffer-Leßmann, N. (2016). EyeSee3D 2.0: Model-based real-time analysis of mobile eye-tracking in static and dynamic three-dimensional scenes. In *Proceedings of the Ninth Biennial ACM Symposium on Eye Tracking Research & Applications* (pp. 189–196). ACM.
15. Kurzhals, K., Hlawatsch, M., Seeger, C., & Weiskopf, D. (2017). Visual analytics for mobile eye tracking. *IEEE Transactions on Visualization and Computer Graphics, 23*(1), 301–310.
16. Tsang, H. Y., Tory, M., & Swindells, C. (2010). ESeeTrack: Visualizing sequential fixation patterns. *IEEE Transactions on Visualization and Computer Graphics, 16*(6), 953–962.
17. Kobayashi, T., Toyamaya, T., Shafait, F., Iwamura, M., Kise, K., & Dengel, A. (2012). Recognizing words in scenes with a head-mounted eye-tracker. In *2012 10th IAPR International Workshop on Document Analysis Systems (DAS)* (pp. 333–338). IEEE.
18. LeCun, Y., Bengio, Y., & Hinton, G. (2015). Deep learning. *Nature, 521*(7553), 436.
19. Russakovsky, O., Deng, J., Su, H., Krause, J., Satheesh, S., Ma, S., et al. (2015). Imagenet large scale visual recognition challenge. *International Journal of Computer Vision, 115*(3), 211–252.
20. Guo, Y., Liu, Y., Oerlemans, A., Lao, S., Wu, S., & Lew, M. S. (2016). Deep learning for visual understanding: A review. *Neurocomputing, 187*, 27–48.
21. Liao, Y., Li, S. E., Wang, W., Wang, Y., Li, G., & Cheng, B. (2016). Detection of driver cognitive distraction: A comparison study of stop-controlled intersection and speed-limited highway. *IEEE Transactions on Intelligent Transportation Systems, 17*(6), 1628–1637.
22. Koma, H., Harada, T., Yoshizawa, A., & Iwasaki, H. (2016). Considering eye movement type when applying random forest to detect cognitive distraction. In *2016 IEEE 15th International Conference on Cognitive Informatics & Cognitive Computing (ICCI* CC)* (pp. 377–382). IEEE.
23. Huval, B., Wang, T., Tandon, S., Kiske, J., Song, W., Pazhayampallil, J., et al. (2015). *An empirical evaluation of deep learning on highway driving*. arXiv preprint arXiv:1504.01716.
24. Abadi, M., Barham, P., Chen, J., Chen, Z., Davis, A., Dean, J., et al. (2016). TensorFlow: A system for large-scale machine learning. In *OSDI* (Vol. 16, pp. 265–283).
25. Wang, Y., Reimer, B., Dobres, J., & Mehler, B. (2014). The sensitivity of different methodologies for characterizing drivers' gaze concentration under increased cognitive demand. *Transportation Research Part F: Traffic Psychology and Behaviour, 26*, 227–237.

Index

S. Adams et al. (eds.), *Systems Engineering in Context*,
https://doi.org/10.1007/978-3-030-00114-8

D

E

T

Printed by Printforce, the Netherlands